T0192687

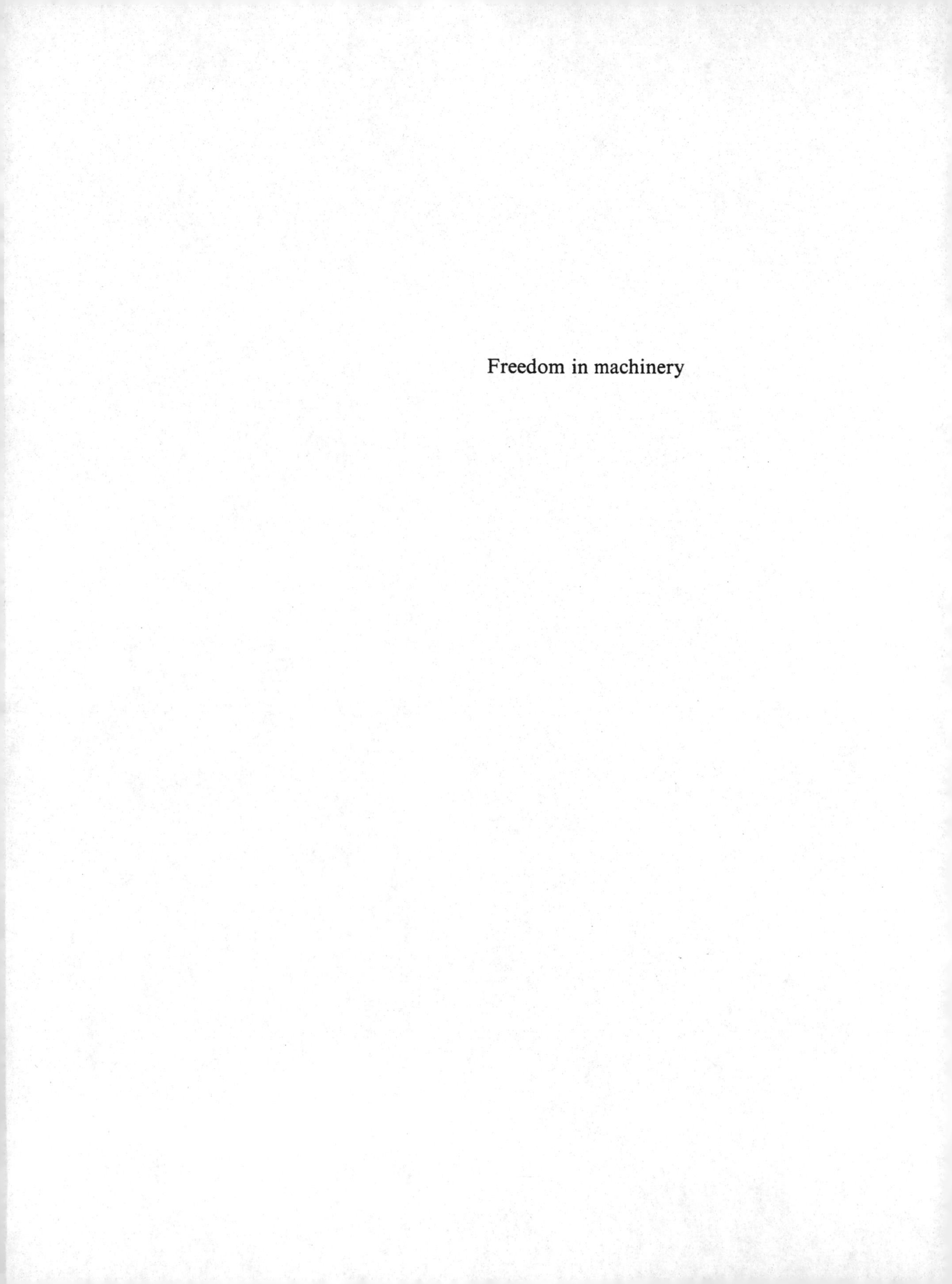

Freedom in machinery

# *FREEDOM IN MACHINERY*

## *Volume 1 (1984) and Volume 2 (1990) Combined*

JACK PHILLIPS
*Associate Professor in the Theory of Machines*
*University of Sydney*

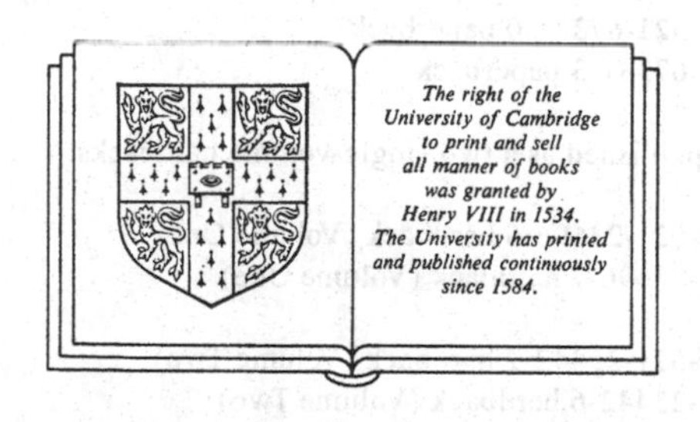
The right of the
University of Cambridge
to print and sell
all manner of books
was granted by
Henry VIII in 1534.
The University has printed
and published continuously
since 1584.

**Cambridge University Press**
*Cambridge*
*London New York New Rochelle*
*Melbourne Sydney*

CAMBRIDGE UNIVERSITY PRESS
Cambridge, New York, Melbourne, Madrid, Cape Town, Singapore, São Paulo

Cambridge University Press
The Edinburgh Building, Cambridge CB2 2RU, UK

Published in the United States of America by Cambridge University Press, New York

www.cambridge.org
Information on this title: www.cambridge.org/9780521673310



This digitally printed first paperback version 2007

*A catalogue record for this publication is available from the British Library*

Set ISBNS (combined two-volume set) are as follows:

ISBN-13 978-0-521-67331-0 paperback
ISBN-10 0-521-67331-3 paperback

Also originally published as a two single-volume hardbacks:

ISBN-13 978-0-521-23696-6 hardback (Volume One)
ISBN-10 0-521-23696-7 hardback (Volume One)

ISBN-13 978-0-521-25442-7 hardback (Volume Two)
ISBN-10 0-521-25442-6 hardback (Volume Two)

Freedom in machinery: Volume 1
*Introducing screw theory*

This book deals with questions of freedom and constraint in machinery. It asks, for example, whether the smooth working of a machine will depend entirely upon the accuracy of its construction. As it answers such questions, it explores the geometrical interstices of the so-called screw systems at the joints in mechanism. It combines, in three dimensions directly, the twin sciences of the kinetics and the statics of rigid contacting bodies, striking deeply into the foundations of both. It introduces the idea that the kinetostatics of spatical mechanism is a valuable discipline, thus setting down for further development the beginnings of screw theory. A special feature of this book is the profusion throughout of its spectacular line drawings, which excite and assist the imagination. This is important in the general area of modern machinery and robots where an ability to think and to design such systems until the geometrical nature of the practical joints and the spatical movements of the various parts of modern machinery is properly understood, these areas make this book an important and timely addition to the literature. This book, with its main accent upon geometry, makes its thrust at those levels of our understanding that lie below the algebra and the mere programming of modern computers. The author's direct personal style makes even the more difficult of his geometrical and other ideas accessiblc to thc wide range of mathematicians, research workers, and mechanical engineers who will read this book. Designers and the builders of robots will study the material carefully.

# Contents

VOLUME ONE

# *Preface*

In April 1976 I began to see that this book was already begun. I found myself at that time drawing and writing a scatter of sketches and notes that gradually grew. During the years, however, I have never felt that finality was achieved. Even now, as I write no more and delete no more, the work is incomplete. It is no more than the limited remarks that I can make in the complicated circumstances of the wide field of TMM, the Theory of Mechanism and Machines.

I wish to acknowledge the primary influence of the International Federation for the Theory of Mechanisms and Machines. Without IFToMM, without its wide membership among the territories of the world, and without the magnificent opportunity this body has given me to consult with my colleagues in western and eastern Europe, the USSR, India, Asia, and in the Americas, this book would never have been attempted. Let me mention here with gratitude the encouragement and the general assistance I have had from the many individual scientific workers of IFToMM, and from their institutions. It should be needless to say that we (these colleagues and I) collectively encourage the life of IFToMM and trust that that life will continue to flourish.

At home I must mention my special colleague and sometimes collaborator Professor Kenneth Hunt of Monash University at Clayton in Australia. Without reserve – or so it has seemed to me – he and I have exchanged our views and openly expressed our occasional disagreements for a period now extending over twenty years. Our meetings have been often and our correspondence voluminous, and I thank him here for his major contribution to my ideas and work.

Next I come to the University of Sydney where my colleague Associate Professor Arthur Sherwood and a small but important group of doctoral and other students have made their contributions to the ideas expressed in this book. In chronological order, and extending across the years from 1963 to the present time, I wish to mention the following: Michael Skreiner, Brian Mahony, Kenneth Waldron, Friedjhof Sticher, Bruce Hockey, Kim White, Andrei Lozzi, Bill Fisher, John Gal, and Yu Hon-Cheung. I have tried to mention in discussion or by means of footnote those ideas that have come from them. In Sydney also I wish to thank Dr J. Eddie Baker of the University of New South Wales; his early encouragement and his recent direct assistance with the material of chapter 20 is much appreciated.

On the technical side I acknowledge the valuable contributions of Hasso Nibbe. During a period of more than ten years, he and I have worked amicably and constructively together at the University of Sydney to produce, not only most of the pieces of apparatus that appear in the photographs dividing the chapters that follow (and a great deal more that is not recorded here), but also the many drawings for the figures of the book. With a rare combination of skill, patience, and a sympathetic treatment of sometimes rough material, he has carefully traced the figures in black ink ready for publication. The excellence of the photographs is due to the special skills and enthusiasm of Trevor Shearing.

I wish to thank the editorial and other staff at Cambridge University Press for their patience with the manuscript, their ready responsiveness with me at great distance, and their fine professional work.

There has been, throughout, the continuous joyful support of my good friend Elayne Russell. Without that the present author might well have become by now a much less unruffled man. I thank her for what she has done for my equanimity and I offer her now this my dedication.

Jack Phillips, Sydney, March 1984.

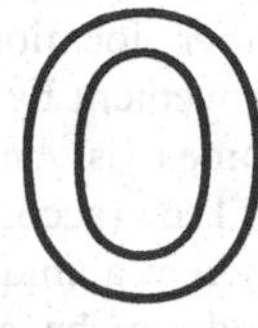

# General introduction

### Some observations

01. This book deals with mechanism. Mechanism is presented here as the geometric essence of machinery. The study of mechanism is important because the geometry of mechanical motion is often the crux of a real machine's design. The kind of motion I call mechanical motion is that which occurs between the rigid, contacting bodies or the material links of mechanism. This occurs in conjunction with two phenomena that happen together at the places of contact between the bodies. Firstly, transmission and other forces operate between the bodies as the motion occurs; and secondly, the shapes of the surfaces there guide the bodies as they move. A contact is a constraint upon a body's motion. Each movable body in machinery suffers its constraints accordingly. If we wish to penetrate the deeper issues at work in real machinery, if we wish to control (or at least not to be bamboozled by) the overwhelming complexity of the constraints upon the mechanical movements occurring constantly within machinery and thereby all about us, we must grapple with these constraints as directly as we can.

02. The digital computer demands on the part of its machine-designing users a ruthless competence in the algebraic processes needed for the manipulation of mechanical information and its numerical analysis. It is accordingly fashionable just now in the field of the theory of machines not so much to denigrate as simply to ignore the main bases in actual mechanical motion from which these algebraic processes grow. The main bases are essentially pictorial, geometrical. They arise from natural philosophy. Students in the mechanical sciences are becoming increasingly unable to contemplate a piece of ordinary reality in machinery accordingly, and to extract from that reality the geometric essence of it. It is of course true that without algebra there can be no programme, no numerical data, and no numerical result; but without an underlying geometry of the reality there can be no applicable algebra. Without a diagram we cannot write an equation. But without geometry we cannot even begin to draw.

03. I am not referring here to the ordinary capacity of a skilled draughtsman (or of a machine) to make an intelligible drawing of some component for production. I refer to the much more powerful, and almost lost ability to see the geometric crux of some mechanical matter in the working imagination, and to nail it down. It is not sufficient to counter here by arguing that the basic geometry, like the algebra, has its own appointed and equal place in the scheme of things. One might even argue that the algebra is the geometry, and that from the mathematical point of view there is no valid distinction. But suppose I ask: what is safety pin? Without first describing safety pin – by a picture or by words or both, for which we are obliged to use quite ordinary and thereby somewhat disreputable Euclidean concepts such as helical bend, curved surface, convex point, tapered slot – no amount of algebraic equationry about the pin and its mode of operation will be in any way intelligible. Between machinery and ourselves there is a kind of membrane, wrapped around the machine itself, and this will not be punctured by the mere wielding on our part of an irrelevant algebraic symbolism. A spade, we say, is a spade. To describe spade, indeed to invent spade, one must actually approach the thing and push through with a firm geometric imagination.

04. Even spade however derives its identity not from itself alone but from its mechanical contact with what it digs. A spade and its soil, a cam and its follower, a worm and its wheel, these are the pairs of contacting links which begin to characterize for us the mechanical movements of which I speak. Since Reuleaux (1875) we have known how to regard the motion of any part of a machine as occurring against a frame attached to any other part of the same machine. The idea of frame itself, of course, came very much earlier from Descartes. Having established such bases, relative motion and the relative movements between contacting and noncontacting parts of machinery are the next matters presenting themselves for consideration when we begin to study the mechanics of machinery.

05. On beginning, however, we find straight away that the instantaneous motions and the various finite movements are very complicated. It is true that in certain simplified studies in the largely illusory field of planar mechanism, where a symmetry of the moving parts about the reference plane and an absence of clearances that might lead to spatial backlash at the joints are both prerequisites, the authors can throw away even the most ordinary of the complicated motions – even the well-known motion of a nut upon its bolt. It is not only that these authors can do this. They must do it. Any screwing motion is a spatial motion, so the least appearance of a screwing of any kind must be removed from consideration in any planar study. To gain an effective overview of machinery we must step both up and out of the plane. We must look directly at all of the kinds of spatial motion that can occur, and do occur, in the fields (for example) of cultivation, harvesting, winnowing, knitting, processing, printing, logging, mining, and leatherworking machinery. In ordinary engineering practice, even at the domestic level (can opener, egg beater, cork puller, telephone), there is no escape from the fact that most of the motions and movements are wholly spatial. Pure rotation about some axis, or pure translation parallel to some line, or the simple rectilinear combinations of these that can be argued by planar thinking, occur only rarely within and at the business end of serious machinery.

06. In certain specialised fields, for example engines and accurate machine tools, immense efforts are made to achieve straight-line and circular motions of points within the moving parts of the machinery. Great attention is paid to parallelisms, concentricities, and right angularities in this class of machinery. The machinery is thereby expensive. It has, however, the special advantage of being explicable in terms of planar kinematics at a fairly elementary level, and its parts can be drawn more easily in the office and produced more easily in production. But when, in order to improve the economies of its construction, we try to understand the principles of its constraint, we walk straight into a trouble that is bewildering. The difficulty is that the geometry of planar motion is a drastically simplified form of the corresponding geometry in space. When we seek to write a coherent and comprehensive theory about the nature of the freedoms and constraints in machinery whose actions and motions are held to be only two-dimensional, we write a degenerate science. The science contains many exceptions and hardly any rules; and we find that many call it, not the science of the kinetostatics of planar mechanism or some such respectable thing, but, more simply and quite derogatorily, linkages and all that.

**On the history of screws**

07. Now in the matter of movement, Chasles explained with his celebrated theorem of 1830 that, in order to take a body from one location in space to any other location, it is always possible to effect the movement by a single finite twisting of the body about some axis. And in the matter of motion, Mozzi showed in 1763 (according to Hunt (1978)) that, at any instant, there is a unique axis in a moving body about which the body can be seen at the instant to be twisting. The rate of the twisting can be seen by us today as a dual-vector quantity because it can be nominated by the combination of (*a*) an angular velocity of the whole body about the mentioned axis, and (*b*) a linear velocity of all of the points in the body parallel to the same axis. What the various workers and their coworkers have shown, in effect, is that there is no instantaneous motion of a rigid body which cannot be seen most simply as a twisting motion about some particular axis. We call this axis the instantaneous screw axis, or, for short, the screw, about which the twisting of the body at the instant is occurring.

08. It is interesting and important to find that there were in parallel with these developments the work of Poinsot in 1803 (1848) and those of Möbius (1837) and Plücker (1869). These works, as was mentioned by Ball (1875), contributed greatly to the formulation of clear ideas about the summation of a number of forces upon a body. These workers began to speak about a wrench, another dual-vector quantity comprising (*a*) a resultant force upon the whole body acting along a certain axis, and (*b*) a couple acting parallel with the force and applicable at every point within the body. What these workers and coworkers have shown, in effect, is that there is no system of forces upon a body that cannot be seen as a wrench acting along some particular axis. This axis we call the central axis of the wrench, or, for short, the screw, along or about which the wrench upon the body is instantaneously acting.

09. In general, of course, when a wrench acts upon a screw upon a body, the body is twisting about some other screw; or, correspondingly, when a body is twisting about some screw, a wrench might well be occurring about some other screw. We can say now, after Newton (1687, 1726) and Euler (1765), that wrenches cause linear and angular accelerations of a body and thus changes in its rates of twisting. Equally we can argue that changes in the rates of twisting of a body demand the mobilisation from somewhere of the necessary wrenches; and, in the presence of gravity, the mechanics of the matter is fairly complicated. We can see however that, while the screws about which a body is free to twist at an instant are definitively determined by the possible, infinitesimal, crosswise movements which might occur at the constraints upon the body, the screws

upon which a wrench at the instant might be acting are similarly determined by the normal reaction forces and the friction forces which might be mobilised (*a*) at the same constraints, and (*b*) at those other places where external or 'applied' forces may be acting. In circumstances of zero friction, a wrench upon a certain screw within a body may occur in the absence of any change in the rate of twisting about some other screw about which the body is free at the instant to twist. Equal and opposite wrenches upon the same screw, and equal and opposite rates of twisting about some other screw in the absence of friction, may similarly occur together in a joint between two links. The two screws are then said to be reciprocal, and the mechanics of the reciprocity of the two is not only not complicated, but also directly relevant to important aspects of machine design.

10. According to Dimentberg (1965), Study in Germany (1901) was the first mathematician – after Clifford [1] who wrote about quaternions in 1873 – to exploit the evident analogy between (*a*) the screw and the small twist of instantaneous kinematics and (*b*) the screw and the wrench of statics. Study wrote a useful algebra of complex numbers based, not upon the more commonly used $x+iy$, where $i^2=-1$, but upon the more exotic $x+\varepsilon y$, were $\varepsilon^2=0$. Complex numbers such as $x+\varepsilon y$ are called by certain modern writers, for example Keler [2], dual numbers. It can be shown that, if two dual vectors represented by the symbols $\boldsymbol{\omega}+\varepsilon\boldsymbol{\tau}$ (a rate of twisting) and $\mathbf{F}+\varepsilon\mathbf{C}$ (a wrench) reside upon a pair of screws that are reciprocal, their scalar product, which can be defined in such a way that it corresponds exactly with the ordinary scalar product of a pair of single vectors, will be zero. Roth [3], Keler [4], and others, have recently attacked the problem of adding algebraically the successive finite screw displacements that might be suffered by a rigid body, and they have used for this the connected ideas of the dual angle between a pair of skew lines and the dual angular displacement of a rigid body. The logic of many of these and other such manipulations was clearly explained by Rooney [5] in 1978, and organized explanations appear in the work of Bottema and Roth (1979). I discuss such matters against the background of real mechanical devices and I mention the mechanics of friction-prone machinery in chapters 10, 12, 17, and 19. In chapter 19 I try to make a connected summary of the many mathematical methods.

11. Study also employed ideas in common with Kotelnikov, of Kazan [6], who wrote in 1895; Study established practical uses in mechanics for Kotelnikov's principle of transference. Associated with this principle is a device called by Keler and others the spherical indicatrix. A set of screws in space is collected together for analysis; maintaining both direction and pitch, the axes of the screws are collected to intersect at a common point at the centre of a sphere; relationships between the respective traces of the collected screws upon the surface of the sphere become aspects of the spherical geometry there; and these relationships reflect in a compact way the original locations and orientations of the real screws. Such devices as these, in conjunction with the complex numbers and vectors involving $\varepsilon$, can be used to advantage in simplifying analyses in mechanics and thus in mechanism. One of the things I would like to do in this book is not so much to operate these devices, but to buttress them with the kind of physical understanding that can come directly from natural philosophy.

12. Ball, in Ireland (1875, 1900), began to establish a firm physical base for the mathematics of screws when he wrote his various historic treatises upon the small oscillations of a rigid body, propounding at the same time his fundamental theory about the reciprocity of screws. He also introduced the centrally important idea that, for each of the six states of freedom of a body – the degrees of freedom of a free body can range across the integers from unity to six inclusive – there is associated a unique system of screws or a screw system. Dimentberg in the USSR (1965), and Hunt in Australia (1978), each made further contributions to the mathematics surrounding these phenomena, especially where the phenomena bear upon the matters of machine design.

13. During the sixties and the early seventies various workers in the USA made a number of advances. Probably for the first time and with the aid of computers and computer methods, Yang [7] and Uicker [8] and others have shown how analyses can be made of the forces and couples required at the frictionless joints in massive machinery. Their work has relied of course on their ability or otherwise to make a kinematic analysis of the particular mechanism involved, and to this extent they have relied very heavily upon the matrix methods for the transfer of coordinates that were introduced by Denavit [9] somewhat earlier. In West Germany at the same time, studies were undertaken by Keler [10] of the wrenches that must be at work at the moving joints of mechanism in the presence of friction. Keler has used a very direct kind of dual-vector algebra derived directly from Study (1901) for this purpose. Each of these workers has found his own way of avoiding the problems involved in making analyses of rigid bodies moving under the influence of applied forces and complicated constraints, and together they have made substantial progress towards solving what might be called the machinal problem. It remains true however that, while Lagrange (1788) showed how to ignore the forces and couples at the constraints of a machine-supported moving member, and while at least

some of those workers have used Lagrangian methods, we need to confront these forces and couples as directly as we can. We must discover, not only what these forces and couples are, but also how they relate to the cyclic and other movements in machinery. These data and that relationship are together the crux of some of the main matters at issue here.

**On the question of ourselves and our machinery**

14. It is interesting to note that, with the possible exception of Reuleaux, none of the mentioned writers state as a reason for writing the fact that machine design is a strangely reactive business. There is a struggle between the creative ideas of machine designers and their recalcitrant, real machinery on the one hand, and there is another, reactive struggle between the nature of the newly created machinery and the resistant aspects of our aesthetic judgement on the other. I would like to think I am addressing this complicated, somewhat painful, and continuous interaction between ourselves and our machinery in at least a preliminary way. It seems to me that, unless we attend to this aspect of our technical studies and activities, we shall understand ourselves and our culture, and the extent to which that culture determines us and our social behaviour, somewhat less fully than we might. The same thing happens of course with music: we make it, we judge it, we hate some of it; it exists; we apply it, we sustain it; we keep it or we destroy or forget it. And yet, because we must, we live by it. It is in the nature of man that he makes music: it is in the nature of man that he makes machinery too.

**Kinetostatics and dynamics**

15. I implied somewhat earlier that whereas the transmission forces in mechanism operate through – that is, from one side to the other of – the tangent planes at the points of contact between the members, thereby forming the basis for a *statics of massless mechanism*, the local linear velocities at the points of contact occur within those tangent planes, and thereby form the basis for a *kinematics of the rigid members*. With this general dichotomy between what I call the action and the motion clear, I wish to speak in summary now about the contents of this book. I mean by the term *kinetostatics*, incidentally, the study of the angular and the linear velocities and thus the kinematics of mechanism on the one hand, this being taken in close conjunction with the study of the forces and couples and thus the statics of massless mechanism on the other. These two opposing yet complementary areas of study are directly related through Ball's conceptual, geometric device, the reciprocity of screws. It must be said here that all matters of *dynamics* – and such matters naturally appear in massive machinery – can be superimposed upon the foundations in statics and kinematics about to be discussed. Matters of power, energy, work, acceleration, and momentum are clearly unavoidable even in this restricted study, and all five of these find their places here wherever they are relevant.

**Guide and synopsis**

16. Next I wish to say that each of the twenty three chapters which follow has been written in such a way that I can hope it can stand alone. Due to what might be called the overall circularity of the presentation however, every chapter depends to some extent on almost every other. For this reason I have continuously employed a device for immediate reference. I might suddenly say, in chapter 10 for example, (§12.28): this will refer the reader to chapter 12 paragraph 28, where some related remark may be relevant or otherwise helpful. Correspondingly the diagrams of one chapter may well be quoted and further discussed in another: each figure (or diagram) has a separate pair of numbers, see figure 15.10 for example. The first number of the pair identifies the chapter in which the figure appears, and the second is its serial number there. There is a bibliography of books quoted at the end of the book and references to this are made by the year of publication, set in brackets: thus (1936) for example. Whenever journal articles are quoted they are given a serial number within the chapter, set in square brackets, thus [8] for example, and listed at the end not of the book but of the chapter. The chapters may appear at first glance to be in no clear logical order. There is however an underlying logic to the chapters, and I hope that the following remarks about this may be useful.

17. In the first two chapters I offer an elementary account of machinery and its mechanism as I see it. This approach might be held by some to be ingenuous: no real attempt is made to separate machinery itself from its designers, namely us, and inherent in the writing is my expressed belief that that is not unreasonable. In chapter 5 there is a discussion based on elementary principles of the instantaneous screw axis for the motion of a rigid body, and in chapter 9 this matter is taken up in similar style to discuss the $\infty^3$ of right lines that must exist in a moving body (§3.01). Naturally there are certain seminal chapters suitable for reading first by different kinds of reader. Chapter 10 for example might be read with interest by the worker in pure mechanics. It deals with the analogies connecting the sister fields of statics and kinematics, introduces an early study of screws and of the action and motion duals, and begins to discuss the matter of reciprocity. Chapters 6 and 7, which deal with the quasi-general systems of motion screws at a generalised joint and with the nature of possible joints in real machinery, might appeal as a pair of chapters to the designer of new mechanism. The

ubiquitous cylindroid and its implications for machinery are examined in chapter 15 and throughout the middle chapters. The all-pervasive and most important matter of reciprocity is introduced in a serious way in chapters 14 and 23. In chapter 20 the reader will find a collection of overconstrained yet mobile linkages (whose mobilities are negative), and a glimpse is given there of the geometry of the 4, 5, and 6-link loops from which these interesting and often useful linkages spring. In chapters 17 and 18 the interrelated matters of singular events and the backlash at joints with clearances are discussed and, although this latter is done in some detail, it is by no means exhausted there. There are in chapter 19 some organised remarks about the methods in mathematics that are either currently used or recommended for future manipulation of mechanism and thus of machine design. One important part of the machinal problem, that of predicting the dynamic forces which must be mobilized at the joints in the mechanism to produce the accelerations required of the moving masses in the actual machinery, is also discussed in chapter 19. Robots and manipulators are mentioned at various places throughout the book; and the matter of gears in chapter 22.

18. Looked at overall, the book begins with an introduction to questions of freedom and constraint, ends with a somewhat abstract discussion of the crucial matter of reciprocity across the screw systems, and presents in the middle an appeal to the natural philosophy of *freedom in machinery*, both for the forces and couples freely to do their work at the joints, and for the required spatial movements freely to occur among the members. Throughout the work an attempt is made to ensure that ordinary rigour is not lost – and this despite a widespread use of visual devices and the deliberate absence of much in the way of written algebra. I hope to provide insight into the geometrical bases underlying all mechanical motion and action. I hope to make the kind of contribution to our collective wisdom that might result in better machinery. I think that machinery ought to be made harmonious within itself, better controlled in its actions and motions, and thereby better accepted by us as an integral, interesting, and satisfying part of our material culture.

### Notes and references


[1] Clifford, W.K., Preliminary sketch of biquaternions, *Proceedings of the London Math. Society*, Vol. **4** (64, 65), p. 381–95, 1873. See also Hamilton (1899), *Elements of quaternions*. Refer §0.10.

[2] Keler, M., Die Verwendung dual-komplexer Grössen in Geometrie, Kinematik und Mechanik, *Feinwerktechnik*, Band **74,** Heft 1, p. 31–41, 1970. Refer §0.10.

[3] Roth, B., On the screw axes and other special lines associated with spatial displacements of a rigid body, *Journal of Engineering for Industry*, Vol. **89,** (*Trans ASME* Series B Vol. **89**), p. 102–10, February 1967. Roth also mentions Halphen, M., Sur la théorie du déplacement, *Nouvelles Annales de Math.*, Vol. **3,** p. 296–9, 1882. See also Bottema and Roth (1979), *Theoretical kinematics*. Refer §0.10.

[4] Keller, M., Dual-vector half-tangents for the representation of the finite motion of rigid bodies, *Environment and Planning B*, Vol. **6,** p. 403–12, Pion, 1979. Refer §0.10.

[5] A set of three review papers by Rooney, directly relevant here and a valuable guide to the literature, may be found as follows. Rooney, J., A survey of representations of spatial rotation about a fixed point, *Environment and Planning B*, Vol. **4,** p. 185–210, Pion, 1977; Rooney, J., A comparison of representations of general screw displacement, *Ibid.*, Vol. **5,** p. 45–88, 1978; Rooney, J., On the three types of complex number and planar transformations, *Ibid.*, Vol. **5,** p. 89–99, 1978. Refer §0.10.

[6] Kotelnikov of Kazan. Very few writers appear to have actually read the original works of this early Russian writer. I quote directly from Dimentberg (1968) as follows: Kotelnikov, A.P., *Vintouoye schisleniye i nekotoryye prilozheniya yego k geometrii i mechanike*, (Screw calculus and some of its applications to geometry and mechanics), Kazan, 1895. Kotelnikov, A.P., Teoriya vektorov i kompleksnyye chisla, (Theory of vectors and complex numbers), Collection entitled, *Nekotoryye primeneniya idey Lobachevskogo v mechanike i fizike*, (Certain applications of Lobachevski's ideas in mechanics and physics), Gosekhizdat, 1950. Refer §0.11.

[7] Yang, A.T., *Application of quaternion algebra and dual numbers to the analysis of spatial mechanisms*, doctoral dissertation, Columbia University, New York, 1963. Yang A.T. and Freudenstein, F., Application of dual-number quaternion algebra to the analysis of spatial mechanisms, *Journal of Applied Mechanics*, Vol. **31,** (*Trans ASME* Series E Vol. **86**), p. 300–8, June 1964. Yang, A.T., Displacement analysis of spatial 5-link mechanisms using (3×3) matrices with dual-number elements, *Journal of Engineering for Industry*, Vol. **91,** (*Trans ASME* Series B Vol. **91**), p. 152–7, February 1969. Refer §0.13.

[8] Uicker, J.J., Denavit, J., and Hartenberg, R.S., An iterative method for the displacement analysis of spatial mechanisms, *Journal of Applied Mechanics*, Vol. **31,** (*Trans ASME* Series E Vol. **86**), p. 309–14, June 1964. Denavit, J., Hartenberg, R.S., Razi, R., and Uicker, J.J., Velocity, acceleration, and static force analyses of spatial linkages, *Journal of Applied Mechanics*, Vol. **32,** (*Trans ASME* Series E Vol. **87**), p. 903–10, December 1965. Uicker, J.J., Dynamic behaviour of spatial linkages, part 1, exact equations of motion, *Journal of Engineering for Industry*, Vol. **91,** (*Trans ASME* Series B Vol. **91**), p. 251–65, February 1969. Refer §0.13.

[9] Denavit, J. and Hartenberg, R.S., A kinematic notation for lower-pair mechanisms based on matrices, *Journal of Applied Mechanics*, Vol. **22,** (ASME), p. 215–21, June 1955. See also Hartenberg and Denavit (1964), *Kinematic synthesis of linkages*. Refer §0.13.

[10] Keler, M., Kinematik und Statik einschliesslich Reibung in Mechanismen nullter Ordnung mit Schraubengrössen und dual-komplexen Vektoren, *Feinwerktechnik + Micronic*, Jahrgang 1973, Heft **1,** p. 7–17, 1973. Refer §0.13.

1. Seven links connected serially into a loop by means of seven revolutes make up a mechanism of fundamental importance. It may be assembled in any one of its configurations and its mobility is in general unity.

# 1 Mechanism and the mobility of mechanism

### Some opening remarks

01. Some successful mechanical devices function smoothly however poorly they are made, while others do this only by virtue of the accurate construction and fitting of their moving parts. In this chapter I try to quantify such phenomena. The term *mechanism* is introduced, and I allow some ambiguity about the exact meaning of it. I speak about links and joints, and discuss the elementary principles of freedom and constraint in mechanism, mentioning some of the difficulties met. We find that, while acceptable analyses of existing mechanism can often be made quite easily, we cannot without insight and imagination make effective syntheses of new mechanism. A range of relevant abstractions is made, some quantifications are proposed, and criteria are established for the *mobility* of mechanism. This then becomes the main subject for study. It needs to be explained however that mobility is not related with matters of mass, inertia, acceleration, or the natural laws of Newton.

02. All matters of dynamics in the mechanics of mechanism, and these have just been mentioned parenthetically, are immensely important in the successful design of working machinery. They are however ignored in the pure geometrical considerations which follow soon. I am dealing exclusively in chapter 1 with the *kinematics* of machinery. Yet, as soon as one thinks of a moving, massless mechanism transmitting power, with work being done as the forces are transmitted, one begins to see that the kinematics of the relative motions on the one hand, and the *statics* of the transmitted forces on the other, are always related in machinery (§ 13.04). We correspondingly find a special term appearing in the literature, *the kinetostatics of mechanism* [1], introduced by Federhofer (1932). We can, if we wish, see this latter as referring to the study of a composite set of mechanical phenomena, combining both the kinematical and the statical. The combination of the phenomena does determine the accelerations and thus the accelerating forces and couples called into play by the existence of the accelerations. These latter are always present and often changing in the massive parts of real machinery.

03. Two things need to be said about the dynamics of machinery. First it should be said that, with the relationships established between the motions and the masses of the moving parts, we can with the principle of d'Alembert to hand always reduce any matter of dynamics to one in statics (§ 12.61). Second it should be said that, with the celebrated two sorts of problem in dynamics distinguished by Wittenbauer (1923), we can see in machinery that it is the motions, with their attendant complicated accelerations, which oblige the various accelerating forces and couples to be mobilised according to the natural laws of Newton and Euler (§ 19.17). The motions are geometrically determined by the shapes of the various parts in contact in the moving machinery (and by the masses), and the forces can be seen to follow as a result of the relative motions. It is not, in machinery, as it is in ballistics for example, that the motions are determined by the forces. It is in any event a fact that, in the creative mental processes of machine conception and in machine design, it is the motions, not the forces, which predominate among the schemes of the active thinkers (§ 1.60). Motion demands to be studied first; I propose to accede to that demand.

04. Mobility, as I have said, is not related to questions of dynamics. It is, however, related to statics, and thus to motion. It is, indeed, a kinetostatical phenomenon. We could enquire here about the meanings of such apparently well known words as engine, machine, machinery, mechanism, linkage; for there is the practical question now of how these and other such words should be used in the present context. I shall merely assert at this early stage however, (*a*) that, pervading both the man-made and the natural machinery which surrounds us everywhere, there is the geometrical essence of that machinery, namely *mechanism*; and (*b*) that the principles of the working of mechanism in general, and of the separable pieces of mechanism in particular, can be studied in isolation. Some meanings for the others of the more important words will be offered later.

**Links and joints**

05. Begin by contemplating the mechanism shown in figure 1.01. Although it displays a crooked or bent appearance, all of its parts are hard and rigid. It 'works' in the sense that when the operating handle at $H$ is moved the spherical ball at $B$ is able freely to plunge into (or out of) the fitted hole in the curved tube. When motion occurs, a hinging action takes place at $C$, while a combination of sliding and turning takes place at $A$. The tube is an integral part of the base frame of the mechanism; its inner surface embraces the sliding ball around most of its circular periphery, but the edges of the slot which have been cut in the tube to make way for the movement do not touch the ball-stem. The mechanism is 'constrained' in the sense that for every chosen location of the handle there is a corresponding predictable location not only for the ball in its hole but also for the whole of the rigid member 3 of which the ball is a part. For precision in the following discussion, let $B_3$ be a point fixed in the centre of the ball (a part of the member 3 of the mechanism), and let $H_2$ be another point fixed as shown in the centre line of the member 2. Note that the subscripts indicate the members within which the mentioned points are fixed. There is, incidentally, another point $B_1$ shown at the centre of the ball. Its particular significance will be clarified soon.

06. Although in figure 1.01 the point $H_2$ is shown to be outside the closed physical boundary of the real material of the member 2, it is 'in' the member 2 in the sense that we allow the boundary of a member, or of a *body*, in the context of mechanism, to be outwardly extendable into space as far as we wish. Within the imagined extended boundary at the surface of the extended body the imagined material of the body is held to be rigid (chapter 5); nominated points within the body are imagined to be embedded there and thus to move as the body moves. Such extended bodies in mechanism freely interpenetrate one another; they can severally and at the same time occupy the same zones in space, and these imagined interpenetrating bodies do not hinder the separate motions of one another. Note for example that the point $B_1$, while *at* the centre of the ball of the body 3, is *in* the body 1; as the point $B_3$ moves upwards or downwards along the curved centre line of the tube, the point $B_1$ remains at rest.

Figure 1.01 (§ 1.05) The paths of some points in a mechanism whose links are crooked and rigid (§ 7.43). The mobility $M$ of the mechanism is unity (§ 1.22).

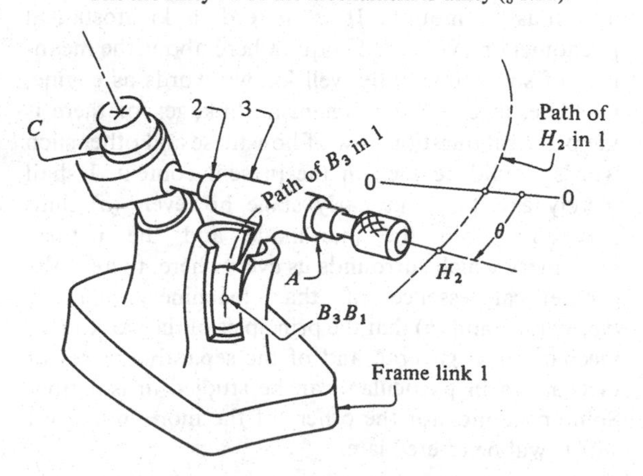


07. It might be mentioned here that when we are working the other way, that is, from some envisioned new mechanism whose dimensions are not yet certain towards some proposed machine which is not yet built, and are thus involved in the process of synthesis as distinct from the simpler process of analysis being followed here, the success or failure of the proposed machine may well depend upon which parts of the extended bodies of the alterable mechanism are in fact allowed to become real. The question indeed becomes: where can we put the necessary metal in the machine so that, (*a*) the members will be strong enough, and (*b*) the members will not collide with one another?

08. If we knew for example that the apparatus in figure 1.01 was not simply some explanatory diagram but some actual piece of machinery, we would presume that somebody's thought had been given to the question of where (and of why) the slots should be cut in the sides of the tube which can be seen. One slot was clearly cut to make way for the ball-stem of the plunging member 3, we would say, while the other was cut perhaps for some other reason. In fact it was cut to render visible in figure 1.01 the spherical shape of the surface of the ball, but it may have been cut to make way for some protruding button or boss yet to be attached at the visible surface there.

09. There remains in any event the mechanical fact that the relative motions of the imagined interpenetrating bodies in mechanism are determined by the actual shapes of the physical boundaries of the real members of the mechanism where they come into moving and changing contact with one another. There are important assumptions implicit here about the rigidity and irregularity of members, about clearances, and about the practical questions of friction, lubrication, and jamming in actual machinery. These assumptions are dealt with more fully elsewhere, but it can be said in summary now: (*a*) that all motion in constrained mechanism is determined by the real members or *links* of the mechanism, in contact with one another at the *joints* of the mechanism; and (*b*) that, in mechanism, there is no thing existing which cannot be seen as a link or a joint.

10. The important word *link* here introduced, is nearly though not exactly synonymous with those other

widely used words, part, component-part, and member. My own view is that, whereas the special word 'body' concentrates attention upon the imagined, rigid, extended material of the member, the special word 'link' concentrates attention more upon the dimensions which geometrically describe and relate with one another the fixed shapes of the mating surfaces of the member. From the word link is derived the technical word *linkage*. Incidentally, and contrary to some others, I will almost always mean, when I use that word, a dimensioned line-diagram of a mechanism, or of an equivalent mechanism. Members are drawn as mere skeletal lines in a linkage and are sometimes replaced with whole new sets of equivalent links; the members have their necessary fixed dimensions in a linkage, but they have no 'bulk' or body at all.

11. The other important word *joint* has here been chosen to mean that neighbourhood of contact, that line or area or volume swept out by that point or line or area, or collection of points, or all or some of these, wherein two links can and do come into changing and moving contact with one another, contact which may be either direct or indirect. I am being careful not to offer a clear or a short definition here. The concept joint is a complicated one, and which aspect of it might become most useful in an argument depends upon the nature of the argument (chapter 7 and, for a particular reference, § 10.61). In this chapter and elsewhere I do employ however the special term *simple joint*. I mean by this the same thing roughly as other writers mean when they employ that other, equally well used term, *pair or kinematic pair*. There can be no joint without at least two links (the pair of them), and no more than two links can be involved in any one simple joint. If the joint between the two *main or reference links* of a joint has involved the insertion of some other, so called *intermediate links*, which will mean in fact that the joint in question has become a collection of simple joints connected together by some links (§ 1.38), I call the overall joint a *complex joint* (§ 1.31, § 2.37), and that is a different matter. Each link of a simple joint provides at its relevant extremity one *element* at the joint (§ 6.02). The pair of joint elements there, namely that pair of mating surfaces and its changing and moving *contact patch* at the joint (chapter 7 and, in particular, § 7.37), constitute the main aspects of what I mean when I speak about a simple joint.

12. In figure 1.01 for example (ignoring the shapes of the cut away slots in the tube) the spherical ball at the lower extremity of the coupler link 3 makes contact along a circular line around the inside of the circular hole that is at an extremity at the right-hand end of the frame link 1. It is of course axiomatic in joints that the shape of the changing contact patch at the joint is identical for each of the mating surfaces at any instant (§ 7.37 *et seq.*), but it might be noticed in this particular case that, as the mechanism moves, the contact patch (which is and remains a circle) moves not only across one of the mating surfaces but across both of them (§ 7.17, § 7.41). Let us call this joint the joint at *B*.

13. We can nominate in figure 1.01 the already mentioned point $B_1$ at *B*. This point is by definition fixed in the frame link 1. In the shown configuration of the mechanism the point $B_1$ at *B* is instantaneously coincident with the point $B_3$ at *B* (§ 12.10) but, should the link 3 move, the point $B_3$ would move with it, leaving the point $B_1$ to remain fixed, relative to 1, on the centre line of the hole. If $B_3$ were moved away from $B_1$, the simple appellation *B* without subscript could not be used for the points; it could only be used in a vague way to indicate the general neighbourhood of the joint.

14. Having seen in this way that there is a joint at *B*, which has some characteristics worthy of special study (§ 7.43), let us see that at *C* and *A* there are also joints of the mechanism. At *C* there is a thing like an ordinary hinge, which is accordingly called a hinge; though often in the literature also a turning pair or a *revolute*. At *A* there is a joint like the joints of a simple telescope, and this is called in the literature a *cylindric joint*, or a *cylinder pair*, or some other combination of these and similar terms. The instantaneous shapes of the contact patches at *A* and *B* and *C* in figure 1.01 are all shown in their spatial relationships with one another in the figure 1.02. I shall be returning later (§ 1.49) for a fresh consideration of these and of the 'spots' that can be seen to infest the contact patches there.

**Loops and the paths of points**

15. There are only three joints in the mechanism in figure 1.01, and there are three links. Note that topologically speaking the links are connected by the joints into a *loop*. This word 'loop' is also an important

Figure 1.02 (§ 1.14) The joints being perfect, this figure shows the areas and lines of contact at the joints of the mechanism in figure 1.01. The shown spots are mentioned later, at § 1.49 (§ 6.03).

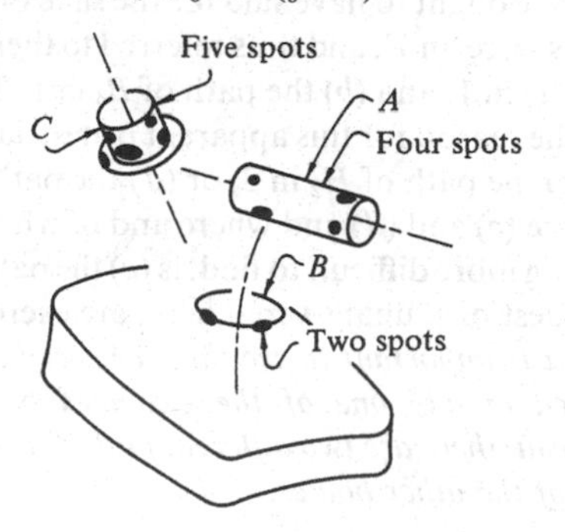

word in the literature. The particular arrangement of links in figure 1.01 may be called a 3-link loop. It is sometimes called a 3-bar chain, and other such terms and phrases are used to describe the general arrangement. A whole selected series of 3-link loops is shown in the forthcoming figures 1.03, from (*a*) to (*j*). There are also some in figures 1.10, 2.01 and 22.01. The mechanical characteristics of these loops will be studied not now but later, at § 1.25, § 1.47, § 1.50, and even later.

16. The path of the point $H_2$ in figure 1.01 is circular and roughly horizontal, while the path of $B_3$ (along the curved centre line of the hole) is almost straight and roughly vertical. It is clear from the picture not only (*a*) that for each position of $H_2$ along its path, there is a corresponding position of $B_3$, but also (*b*) that an actual series of corresponding positions could be discovered by 'construction', that is, by constructing on paper the geometrical essence of the successive movements, or, if necessary, by constructing in three dimensions the mechanism itself, whereupon we would have at hand a convenient 'mechanical computer' for the solution of this most difficult problem; or (*c*) that we could write some algebra. In this latter event we would write, if we could, what is called a closure equation for the linkage (§ 19.47, § 20.09); and we would intend, subsequently, to 'solve' this equation by writing a digital computer programme to get some actual numbers printed out. In any event it is this predictability of the position of $B_3$, and thereby of the location of the ball in its hole, which is a feature of what is meant by the term 'constrained' in the present context. It behoves us now to look more carefully at what exactly is meant by this.

17. Please note in passing here that whereas I will usually speak on the one hand of the *position* of a point, I will usually speak on the other of the *location* of a body. This choice of words helps to indicate the fact that, while a point needs only three parameters to be fixed in Euclidean space, the whole of a rigid body needs six (§ 4.01, § 4.04).

18. We spoke only very loosely at § 1.16 about the paths of the points $H_2$ and $B_3$. It was not stated, only implied, that the mentioned paths were in the frame-body 1. We ought to have said for the sake of clarity that the paths were 'in 1', and thus referred to them as, (*a*) the path of $H_2$ in 1, and (*b*) the path of $B_3$ in 1. To see more clearly the reason for this apparent pedestrianism, think about (*c*) the path of $H_2$ in 3, or (*d*) the path of $H_1$ in 2. Where are (*c*) and (*d*) and where and of what shape, for something more difficult to find, is (*e*) the path of $B_1$ in 3? These questions, unanswered here, are exercises for the reader: *it is important to see that for every chosen point embedded in any one of the extended bodies in this mechanism, there are two different paths for it, one traced in each of the other bodies.*

**Constraint and the idea of mobility**

19. If there exists a recognised technical term *constrained* in the scientific literature to do with mechanism, the term is so ill-defined and loosely used that each reader must try to decide what is meant by it. Keeping in mind that the word 'line' in the present context – namely that of the paths of points – will usually mean a curved or a 'twisted' line or a 'curve' in space, namely a line which in general cannot be contained within a plane, I personally would like to say, of this debated matter of constraint, the following: *if all of the paths, in all of the other links of the mechanism, of all of the points of any selected one of the links are only curves, that is, if none of the paths in any of the links are surfaces or volumes, the mechanism is either just constrained or overconstrained.*

20. Refer to figure 19.10. There, a certain body 2 may be seen to be tracing out a collection of the line-paths of its points in another body 1. Such a collection of paths I call the body's *track* in 1, and in the circumstances of figure 19.10 each line-path of the track is wholly determined by the system of links and joints connecting the bodies 2 and 1. Whenever the track of a body consists of predetermined, curved lines like this, and is *repeatable* in this way, the bodies 2 and 1 must both be members of some one mechanism. Otherwise the mentioned track would contain the paths of points which, like a tangled ball of string, could be drawn anywhere. The body 2 in figure 19.10 does not have the freedom to move in any other way; every point within it is obliged to move along its predetermined path, which is a curve; and when *every* such moving body in a mechanism is similarly restricted in its capacity for relative motion with respect to every other body of the mechanism, the mechanism as a whole, as proposed above, might be said to be constrained or overconstrained. I must explain however that many mechanisms by the definition of § 1.19 would be *just constrained*, while others by the same definition would be *overconstrained*. I imply by this latter that the overconstrained mechanisms suffer, internally and at their joints, 'redundant constraints', which are, from the kinematical point of view, unnecessary for their proper and sufficient working. There is in the field of ordinary engineering practice an abundance of such mechanisms (chapter 2). A system of joints within a mechanism may similarly be said to be 'overclosed'; other such adjectives and phrases abound in the context of freedom and constraint. One hears for example: overconstrained (used already above); overstiff; necessary; sufficient; free-running; restrained; obstructed; jammed; and a host of others. And the question must be asked: are there, in this context of careful speech, precise meanings for any of these various words?

21. Precise meanings can be found if we break off now to introduce a new term, *mobility*. The actual word 'mobility', carefully chosen from the English dictionary to describe the fundamental idea of mobility, was used by myself and my colleagues from about 1963 onwards, one of its first appearances occurring in a work of Waldron [2]. It was always sharply distinguished by us from another word we used, *connectivity*, the different meaning of which will be dealt with later (§ 1.31). Mobility is a relevant, accurately-definable property of a mechanism which can be quantified. It is quoted as a whole, dimensionless number, and it may be of any value, either positive or negative. *The general mobility M of a mechanism is that whole number of independent parameters required in the absence of special conditions to locate (or to 'fix') in Euclidean space all of the other links of the mechanism with respect to any chosen link.*

22. The full meaning and the wider ramifications of this compacted statement are not easy to digest immediately; the matter needs to be elucidated. To start off quickly, however, it might be useful to see that the mechanism in figure 1.01 has a mobility of unity. Let us choose for convenience the frame link 1 as the 'chosen' link, and regard it fixed: set up, in other words, a fixed frame of reference in the frame link 1. Next insert a value of the variable parameter $\theta$ by moving the centre line of the operating handle at $H$ through some nominated angle $\theta$ from some arbitrarily chosen origin location *o–o*, which is fixed in space with respect to the frame of reference. As soon as this one parameter has thus been nominated, all of the 'other' links of the mechanism have freely arrived at their fixed locations in space with respect to the chosen link. The nomination of only one parameter has been enough; there are no other variable parameters requiring to be given a value; all of the links are located (namely, fixed) and the mechanism can be seen to have a mobility of only one namely $+1$, or unity (§ 1.34, § 4.02).

23. Please note, in § 1.22 above, that the links of the mechanism were held to arrive *freely* at their respective fixed locations. A way to see what was meant by this is to notice that the mechanism in figure 1.01 functions smoothly despite the crookednesses which are apparent. The axis of the hinge at $C$, that of the cylindric joint at $A$, and the centre line of the curved tube at $B$, are neither horizontal nor vertical, nor rectilinearly arranged nor intersecting; the actual shape of the curve at the centre line of the tube, moreover, is not even nominated. The point being made is that, given that the joints themselves are satisfactorily manufactured, these various axes and centre lines of the mechanism could, within limits, be altered and moved about in manufacture without disturbing the smooth working of the real machine. This is the crux of an important distinction which must be made between the 'general' mobility $M$ of a mobile mechanism, which is what we have been discussing here, and the 'apparent' mobility (say $M_a$) of some overconstrained mechanism which, despite internal redundancies, is also mobile (§ 2.05).

24. In the mechanism in figure 1.01 the mobility $M$ is unity, and motion within the mechanism can occur despite any alterations made or errors due to faulty manufacture. In the mechanism shown in figure 2.01(*a*), however, motion within the mechanism is quite impossible unless the set of cylindrical sliders at $F$ are arranged to be exactly concentric with the hinge at $D$. If this special geometrical requirement were to be provided in figure 2.01(*a*), motion would be possible and the apparent mobility $M_a$ of the mechanism would be unity. We shall see later that the general mobility $M$ of the mechanism in figure 2.01(*a*), which is a number characteristic of the mechanism and independent of whether the special requirement for motion is there or not, is considerably less than zero. If the special requirement for motion were there, $M$ (which is less than zero) would be less than $M_a$ (which would be unity), and the mechanism, in the sense of § 1.20, would be overconstrained (§ 2.05).

**The freedom of joints**

25. The mobility $M$ of a mechanism clearly depends upon the number and arrangement of its links and upon the kinds of joints which are employed. Refer to the figures 1.03 where a series of 3-link loops of mobility unity is drawn. In a more thorough study of the question of mobility, which is due to begin quite

Figure 1.03(*a*) (*b*) (*c*) (§1.25) Various 3-link loops of mobility unity. Note the *f*-numbers at the joints, and please observe the constrained, mobility-unity motion.

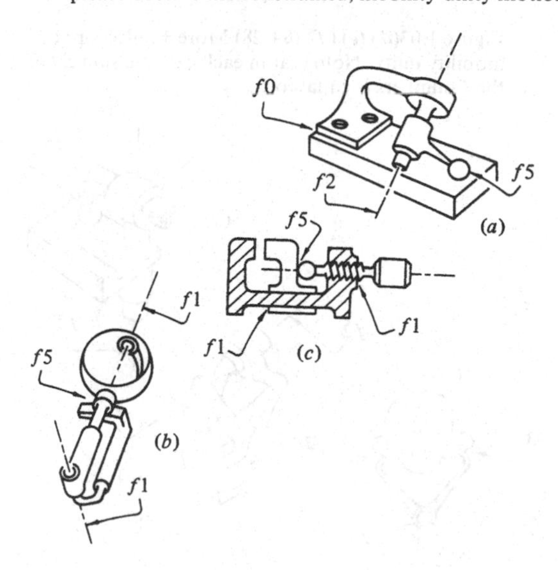

soon – at § 1.32 *et seq.* – we can return to the actual mechanisms there, but please look now exclusively at the joints. Wherever there is in the figures 1.03 a joint which is marked with the symbol $f1$, there is either (*a*) an ordinary screw joint or a *helical joint*, of some finite pitch (mm/rad); or (*b*) a hinge or revolute, which is simply a screw joint of zero pitch; or (*c*) a spline or a sliding joint, which is otherwise known as a *prismatic joint*. And a prismatic joint, as one may easily see, is simply a screw, or a helical joint of infinite pitch. All three of these joints have the same property in common, namely their capacity to have their joint elements – and, thereby, their two members or links – fixed relatively to one another by the nomination of only one parameter. One needs simply to agree that, if one member of the hinge or of the screw joint were turned or twisted through some certain angle from some origin location fixed in the other member, or if one member of the prismatic joint were similarly slid for some certain distance, the turned or twisted or slid member of the joint would thereupon become completely fixed with respect to the other member. There is in each of these three cases the freedom within the joint to make this one, and only one, manoeuvre.

26. Corresponding remarks could be made about the minimum number of parameters required to fix each of the joints in the various figures which variously bear the symbols, $f2$, $f3$, *et seq.*, but this would be repetitive in the circumstances. The reader should either go directly to chapter 6 where this whole question is separately and exhaustively treated, or simply note the following. In chapters 6 and 7 a discussion is offered about the nature of joints in the kinematics of mechanism, and it is there explained how every joint of whatever kind possesses a property called its freedom

Figure 1.03(*d*) (*e*) (*f*) (§ 1.28) More 3-link loops of mobility unity. Note that in each case the sum $\Sigma f$ of the $f$-numbers is equal to 7.

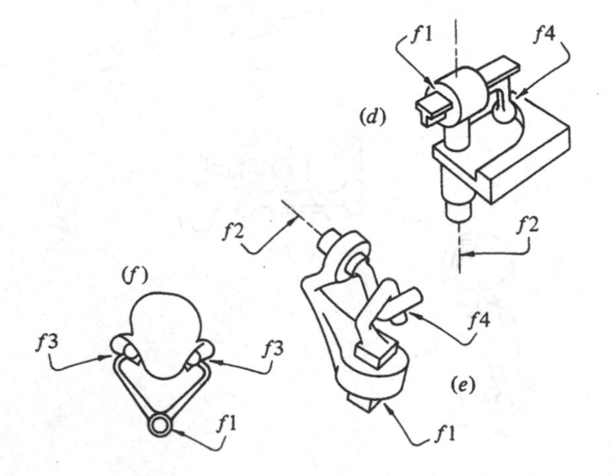


(§ 6.16). *The freedom f of a joint is that whole number of independent parameters required to locate in Euclidean space either one of the links of the joint with respect to the other link of the joint.*

27. It should be mentioned before going on that this symbol $f$ for the freedom of a joint will be used in different ways. Suppose that the freedom $f$ of some joint is 3. We may (*a*) say that, in this joint, $f=3$; or (*b*) say that this is an $f3$ joint; or (*c*) point to the joint in a diagram using the naked symbol $f3$, as we have done in the figures 1.03 for example; or (*d*) write with subscripts that the values $f_1$, $f_2$, $f_3$, etc. are the $f$-values of a set of consecutively numbered joints. It is not until § 1.33 of the present chapter, however, that this latter device is first employed.

28. Please notice also that the discussion has switched to joints, but this is necessary because an understanding of the mobility of mechanism depends upon an understanding of the freedom of joints. To understand this, one must be quite clear about the meanings contained in the important statement made at the end of § 1.26. To use the definition of freedom mentioned there to find the freedom $f$ of some joint, the joint in question must be regarded independently of the action of any mechanism within which it may be seen to be working. It must be excised and extracted, as it were, for independent examination. One link of the excised joint must be imagined to be fixed, in the jaws of a vice for example, while the other link, the 'free' one, must be imagined to be progressively moved stage by stage from some general starting location into some possible general nominated location (§ 1.29), each separate movement of the motion corresponding with the insertion of one or another of the required independent parameters. A parameter, in this context, is any one independent movement to which we can put a single number: an angular movement, for example, between the starting location of some straight line in a body and its finishing location, the two line-locations not necessarily intersecting, or the distance travelled by some point in a body along a nominated or a mechanically determined line-path, the path not necessarily being straight (§ 4.02). Owing to the learned idea among us – which is there, chiefly, because we cannot deny it – that only six parameters are ever needed to lock into a fixed location any rigid body initially free in Euclidean space, *the freedom f of a joint, we say, can never be less than zero nor greater than six.*

29. In the above paragraph I spoke about a nominated location of the free link which was *possible*. In this connection it should be explained that, in most joints in ordinary engineering practice, there are surfaces, irrelevant to the question of freedom, which can contact one another in a well-conditioned way (§ 7.36),

and which could be called, in the context, the *stop surfaces* at the joint. These surfaces are there, either at or near or far away from the contact patch, either to prevent the joint physically from coming apart, or because they are unavoidable when looked at from a topological point of view, or to impose some required mechanical limit upon the otherwise 'free geometrical' travel of the joint. In an ordinary telescope, for example, all of the cylindric joints have a limited axial travel; in a ball and socket joint, because the ball must have a stem, and because the socket must have a sufficient surface, the stem of the ball will strike the body of the socket around a certain periphery of the necessary hole, and the particular shape of this hole will limit accordingly the otherwise free spherical motion of the ball. Quite independently of these stop surfaces, however, which are simply matters of practicality, it can be argued that, *if the freedom of the joint is f, the number of independent locations of the free link of the joint within the fixed link of the joint is infinity raised to the power f, namely* $\infty^f$. For a further reading on this and related matters, please go to § 4.04.

30. If the joint in question is rigid, that is if the two links of the joint are in fact one link, the freedom, or the *f*-value of the joint, is zero; and we note here that $\infty^0$ can take the value unity. If, on the other hand, the two links make no contact at the joint, that is, if the two links of the joint are not in fact joined at all, the *f*-value of the joint is clearly six. In between there is the whole range of the five integers, going step by step from one to five inclusive, and this is the range which is of practical significance for us. Each real joint in mechanism will have an *f*-value of one or another of these integers: there is no such thing as a 'fractional' joint. It is true that a given joint may, by virtue of suddenly making or breaking some extra point or line or area of contact within itself, suddenly change from one *f*-value to another, but such complicated joints (which can change catastrophically their kinematic characteristics) are beyond the scope of this discussion. Refer however to my remarks at § 2.53 and to the worked exercise 3 about the mousetrap at § 2.54.

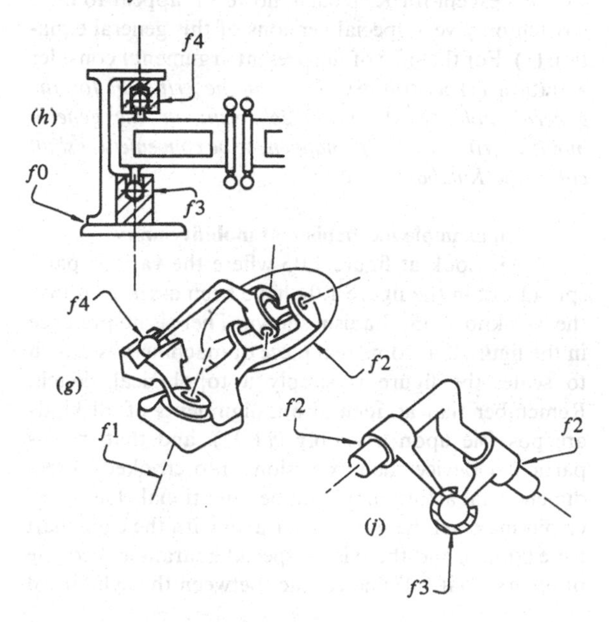


Figure 1.03(*g*) (*h*) (*j*) (§ 1.30) These are the last of the figures 1.03. Note the *f*0 joint at (*h*), and the *f*2 complex joint at (*g*). All three loops are of mobility unity.

31. Let us touch by way of example now upon the matter of *connectivity*; it was mentioned last at § 1.21. Take the mechanism shown in figure 19.10 and notice that the two links 1 and 2, although connected, are out of physical contact with one another. Next imagine this mechanism to be entirely enclosed within a flexible black bag, except for some suitable rigid extensions to the two links 1 and 2 which protrude through small holes beyond the bag. The black bag and its contents can now be seen as an unknown 'joint' between the members 1 and 2, which could be tested for its freedom *f* by the method outlined at § 1.28. Because of the nature of the track of 2 in 1 (§ 1.20), the freedom *f* of the bag and its contents will be found upon a testing to be unity. We say accordingly that the connectivity $C_{12}$ of the two links 1 and 2 is 1. Connectivity is thus seen to be the same thing as the freedom *f* of the 'joint in the bag', and it is defined in the same way (§ 1.26). But think now about the connectivity $C_{15}$: the two links 1 and 5, like 1 and 2, are, despite their physical contact with one another, which is irrelevant, connected together by the whole mechanism; their connectivity $C_{15}$ is not 1 as before, but 2. If the particular ball and socket joint that is seen to be working at the actual region of physical contact (at *F*) between the two links 1 and 5 of the mechanism were to be excised now and extracted for independent examination according to the method of § 1.28, with or without a black bag, its freedom – or its connectivity, if you like – would be found to be, of course, not 2 but 3. The somewhat complicated question of connectivity is mentioned again at § 2.39 and discussed in greater detail in chapters 20 and 23 (§ 20.17, § 23.43).

## A criterion for mobility

32. Without going further now into the question of joints, we are in a position to connect algebraically the general mobility *M* of a mechanism with the number of its links and the *f*-values of its joints. The argument goes as follows. Imagine some mechanism consisting of *n* links and *g* joints. In the simple 3-link loops so far considered and illustrated in the figures 1.03, $n=3$, and

$g=3$; but in general, of course, a mechanism may have more than one loop, and, given that 'open' and 'composite' loops can occur, any other number of joints. Refer for example to the mechanism taken apart and shown in figure 1.04, where $n=6$, and $g=8$. Suppose next that the imagined mechanism was similarly taken apart, and that its $n$ links, numbered consecutively from 1 to $n$, were separated, and that the other $(n-1)$ of them were simply 'waiting' in space to be reassembled with the chosen, namely the fixed link 1. Each of the waiting links would have six degrees of freedom with respect to the fixed link in such a circumstance, so the total degrees of freedom of the $(n-1)$ links with respect to the fixed link, namely the mobility of the mechanism in this undone condition, would be $6\ (n-1)$, namely $6n-6$ (§ 1.21). In the particular case of the mechanism of figure 1.04, the mobility of that mechanism, in its undone condition, is clearly 6 less than 36, namely 30.

33. Suppose next that we begin to assemble the imagined mechanism by effecting (that is by joining), one by one, the $g$ joints. Let the joints be numbered consecutively from 1 to $g$, and let their corresponding $f$-values be $f_1\, f_2 \ldots f_i \ldots f_g$. As soon as the first joint (whose freedom $f$ is $f_1$) is effected, the total degrees of freedom of the movable links, namely the mobility of the imaginary undone mechanism, is reduced by $(6-f_1)$; and the mobility becomes $(6n-6)-(6-f_1)$. In the mechanism of figure 1.04, for example, we could begin by sliding the valve 2 into its guide 1; as soon as that first joint (whose $f$-value $f_1=2$) is effected, the general mobility in figure 1.04 would be reduced by $(6-2)$ namely 4, becoming $(30-4)$, namely 26.

34. Returning to the general argument and to the imagined mechanism, we can now see that, if the proposed assembly of the $n$ links were taken to completion, and quite independently of the order in which the $g$ joints were effected, the mobility of the assembled mechanism would be

$$M = 6n-6-\sum_1^g (6-f_i)$$

$$= 6n-6g-6+\sum f$$

$$(1) \qquad = 6(n-g-1)+\sum f$$

which is an important result. This equation (1) is variously stated and variously known in the literature. Heidebroek [3] makes direct reference to the matter mentioning his colleague Kutzbach in a collection of papers edited by him after a local VDI conference in 1933, and Kutzbach himself [4] betrays the beginnings of his ideas in 1929. Grübler (1921) offered a similar though simpler criterion in his second book; it was written however for planar mechanism (§ 2.36). Assur (of Leningrad) is said by some to have written earlier, and Manalescu [5], writing in 1968, mentions even Chebyshev as having been the 'initiator' of today's presentation of the formula. Manolescu mentions a host of contributors, including Kutzbach [6], and reflects the general confusion. Refer also to Volmer *et al.* (1969), where the above equation (1) appears as Müller's equation (2.5) in chapter 2. Müller writes a short history, with a number of references beginning with Chebyshev in 1869, about the development of the various early criteria, and mentions among others Artobolevski and Dobrolovski as having been active contributors during the period 1934–6. Most of these workers except for Kutzbach, however, appear to have written only very special versions of this general equation (1). For the sake of the present argument I consider equation (1) as follows: *I call it the criterion for the general mobility M of a mechanism, or the general mobility criterion, or, if it happens to be convenient, I shall call it the Kutzbach criterion.*

Figure 1.04 (§ 1.32) Six links waiting for assembly into the mechanism shown in figure 1.05.

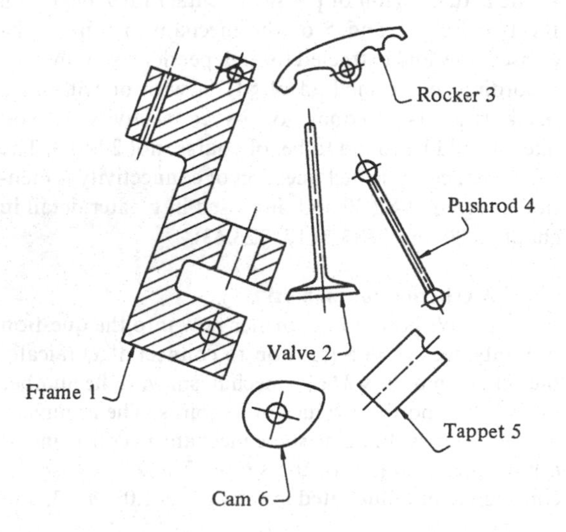


### An example mechanism of mobility unity

35. Look at figure 1.05 where the various parts spread out in the figure 1.04 have been assembled into the well known mechanism shown. There is no pretence in the figure that some real piece of machinery is drawn to scale; the figure is simply a topological sketch. Remember that accidental misalignments of all kinds are possible upon assembly (§ 1.23), and that in this particular device such excursions into crooked, three-dimensional action may even be intentional. The rocker-pin may well be put non-parallel with the cam-shaft for example, and there is no special guarantee (because of errors) that full line contact between the cylindrical

cam-face and its flat follower will be achieved. Pay attention to the *f*-numbers indicated at the joints, and, knowing that these have been ascribed not according to some infallible law but simply by me exercizing my judgement, consider whether you agree with them. Substitute my data into the mobility criterion (1), and find that the general mobility of the assembled mechanism is, according to me, $M=6(6-8-1)+22=+4$. Look at the definition for mobility at § 1.21, think about fixing all of the other links in figure 1.05 with respect to some chosen link (choose for convenience the frame link 1), and notice that all of the four, the +4, parameters in this particular case can be chosen to be purely angular parameters. To fix with respect to frame all of the movable links in figure 1.06, one may, for example, insert the following four 'inputs' at the mechanism: (*i*) rotate the cam-shaft from some origin location through some nominated angle, thus causing (but not yet definitively in every case) each of the movable links to move from its origin location towards some other final 'fixed' location; (*ii*) angularly locate the tappet which is next otherwise free-rotating; (*iii*) similarly locate the push rod; and (*iv*) similarly locate the valve. These four may freely be seen to be the four 'mobilities' of this particular mechanism whose mobility *M*, as we have seen, is +4. Notice also that, because the mobility is positive, and, keeping in mind the general assumptions upon which the general criterion of equation (1) was derived (§ 1.36), the mechanism of figure 1.05 is free to execute its action in such a way that it can behave as a truly spatial mechanism. Although figure 1.05 is drawn as though there is to be a central plane of symmetry in the plane of the paper, it should be clear (*a*) that no such special geometrical conditions were assumed in the above analysis; and (*b*) that such conditions are not necessary for an overall smooth working of the device.

Figure 1.05 (§ 1.35) A well known mechanism of six links, eight joints, and of mobility 4.

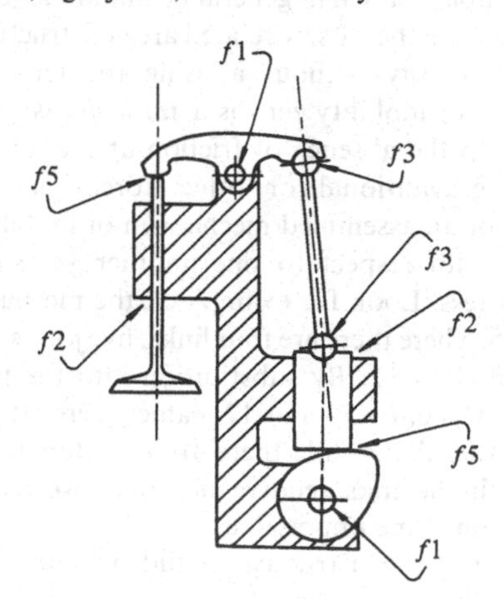


36. Parenthetically now, and with respect to the mentioned question of what the fundamental assumptions are upon which the general mobility criterion was first devised (§ 1.32, § 1.35), there is somewhat more to be answered here than meets the eye. One obvious assumption is that all joints are geometrically free to be effected; that is, that independently of constraint or overconstraint, it is physically possible to effect each joint. It is assumed in figure 1.04 for example that the respective central axes of the cam-shaft and the tappet-follower are close enough to be intersecting so that the tappet will never drop off the cam. This, of course, is a simple and a necessary assumption. Another assumption is that the effecting of any one joint will not lock up an overconstraint in one loop while permitting an underconstraint in another. Unfortunately this latter is not well dealt with among the assumptions, and we can get into some difficulty by blindly applying the general mobility criterion of § 1.34 to some multi-loop mechanisms. Refer for example to the 'mixed' mechanism displayed in figure 20.01 and discussed nearby in chapter 20. See also the worked exercise 2 at § 2.52. Such matters as these are beyond the scope of this chapter 1 and I can do no more, just now, than to offer an assurance here that the example mechanism of figure 1.05 is quite free of such difficulties (§ 20.05).

37. It might be argued in figure 1.05 that, if it were not for the three irrelevant freedoms to rotate of the three rod-like links which have been mentioned (all three of which are axi-symmetric about their own centre lines and may therefore be, as rotators, disregarded), the mechanical connection between the angular input at the cam and the linear output at the valve could be a truly constrained one (§ 1.19). It is only in the kinematical sense, incidentally, that these freedoms to rotate are disregarded here; in any real piece of machinery, such freedoms to rotate of the mentioned members are of great importance for the even distribution of wear. This leads us to see nevertheless that, if each one of these three links were either feathered in its cylindrical guide or otherwise suitably prevented from freely rotating, the general mobility *M* of the mechanism could be exactly unity. Look at figure 1.06. Three rigid cylindrical pegs, laterally protruding from the three mentioned members, are caused to slide in cut slots in the frame which are not only not straight but also generally shaped in cross section in such a way that each peg (which must of course be given an infinitesimal clearance at its slot) makes contact with the frame at one spot only – at one or the other side of its slot. Having thus with these pegs and slots reduced the mobility *M* to unity, we can now see that, despite the pegs and the slots, there are still no

geometrical features of the mechanism that need to be aligned, or set coplanar with, or set at right angles to, or set in any other special way with respect to frame or to one another, to ensure a smooth working of the relatively moving members. The mobility is unity: *and it is characteristic of a mechanism of mobility unity that, however poorly it may be constructed, it can always be assembled in any one of its single infinity of configurations and then be caused to execute its movements smoothly, there being no distortion of any of its links and no advantage taken of the clearances at its joints.*

**Diagrammatic methods for the kinematic chain**

38. During the assembly in figure 1.05 of the scattered parts of figure 1.04, figure 1.07 could have been used as a kind of recipe to show how the various parts should go together, the quoted *f*-numbers in the circles corresponding at the appropriate joints. The figure is self-explanatory in the circumstances and such a diagram represents the *kinematic chain.* Notice that there are, in this particular chain, not only one but a number of loops. Notice also that there are not just links with two joint elements but also links with three and four joint elements. Some writers, for example Rosenauer and Willis (1953), refer to links as being binary, ternary, quaternary, etc.; but I prefer, especially when the link has more than four joint elements, either directly to quote the number of joint elements upon the link, or to speak about the *jointivity of the link.* There is at § 2.41, somewhat tentatively suggested, another, opposite, term, the *linkivity of a joint*, and the meaning enclosed within that term could be held to coincide with ideas about the kinematic chain put forward by Davies and Crossley [7] in 1968. These authors proposed, in effect, using circles for links and lines for joints. Although this opposite method departs in my view too far from the appearance of physical reality to be ordinarily useful in the practical business of machine design, it is logical and certainly instructive when contemplated carefully (§ 2.40). *It is important to see that a kinematic chain is just as much a system of joints linked by links as it is a system of links joined by joints.*

Figure 1.06 (§ 1.37) The valve mechanism of the previous figure modified in such a way that its mobility $M$ is reduced to unity.

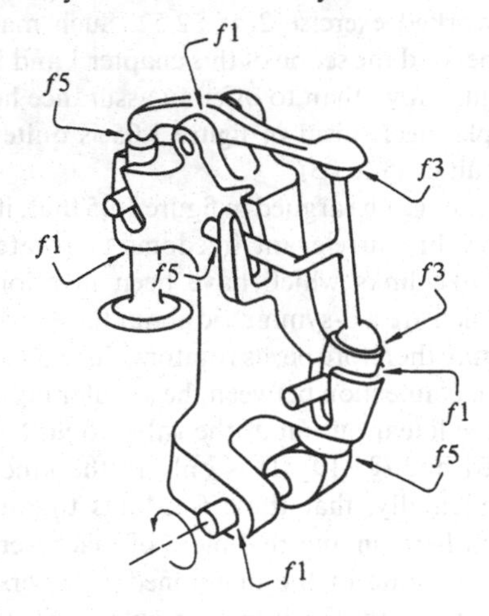


Figure 1.07 (§ 1.38) Kinematic chain for the assembled mechanism shown in figure 1.05.

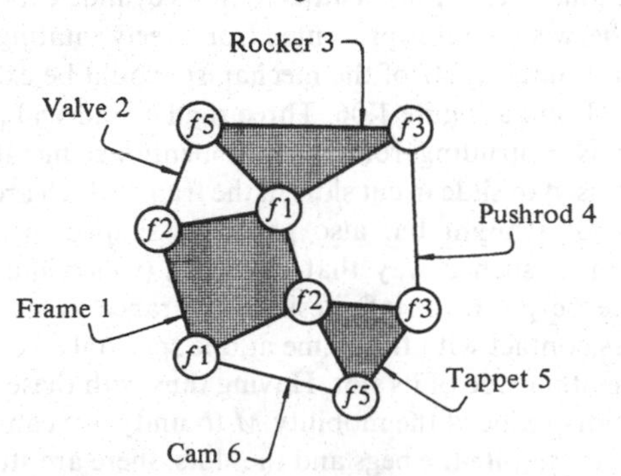


39. At this point the reader should, as an exercise: (*a*) construct a kinematic chain for the mechanism shown in figure 1.06; (*b*) check by substitution into the mobility criterion that the general mobility $M$ there is, as has been claimed, unity; and (*c*) find out and think about what would happen if any selected one of the joints in figure 1.06 were to have its *f*-number reduced by say one, to the next lowest *f*-number.

**Zero mobility and modes of assembly**

40. It should be clear by now that, if $M$ is not unity but zero in a mechanism, the mechanism will in general be assemblable (in one or another of its possible configurations) without distortion of its links and without reliance upon the clearances at its joints. But once assembled, it will in general be immobile. Borrowing a term from the closely related area of structures, we can certainly say without abusing the term that a mechanism of mobility zero is a *statically determinate structure*. In the absence of friction at the joints, and with no gravitational or other forces acting, the members of an assembled mechanism of mobility zero are fixed with respect to one another yet suffer no internal stress. Look for example at the mechanism in figure 1.08. There there are four links, five joints, and, by inspection, $\Sigma f = 12$. By substitution into the mobility criterion, $M = 6(4-5-1)+12$, namely zero. It is a fact in figure 1.08 that, while there are no internal stresses locked up in the links, relative motion between the links is at the same time impossible.

41. It will be instructive to study in figure 1.08 one

of the ways of assembly of this particular mechanism. Referring to figure 1.08, imagine that the four links have been taken apart and that they are waiting in space, ready for reassembly. Now take the link 1, lay it down in figure 1.08, thus choosing it to be the frame link. Effect the $f3$ joint at $A$ and then the $f4$ joint at $B$, thus connecting the link 2 to frame in such a way that it can freely pivot about what clearly becomes a true 'equivalent hinge' along the line $A$–$B$. The point $E_2$, which is at the centre of the spherical socket in 2, is free, at this stage, to execute a circle in space concentric with, and normal to, the line $A$–$B$.

42. Now effect the hinge at $C$, thus permitting the link 3 to carry the straight line of its hinge axis at $D$–$D$ through a complete revolution – please remember here my remarks at § 1.06. Thus see how $D$–$D$ can describe a hyperboloid of revolution about the central axis $C$–$C$. Next effect the hinge at $D$ by inserting there the relevant portion of 4, and see that the point $E_4$, which is at the centre of the spherical ball in 4, is now free to execute (*a*) a circle relative to 3 concentric with and normal to the hinge axis $D$–$D$, and thus (*b*) the surface of a skew torus described in fixed space in such a location that its axis of symmetry is $C$–$C$. Refer to figure 1.09.

43. It should next be clear that, wherever the circle described by $E_2$ in § 1.41 can cut the surface of the torus described by $E_4$ in § 1.42, there is a possible configuration for the final job of putting the ball at $E_4$ into the socket at $E_2$. Whether or not the mentioned circle can cut the mentioned torus, and the question of how many times it cuts it if it does, are questions which clearly depend upon the shapes and dimensions of the various links. It is known, looking again at figure 1.09, that a circle of this kind can cut a torus of this kind in a maximum of four points, and, depending upon whether the circle misses the torus or is tangential to it (§ 2.46), intersects it once and thereby twice, or both, or intersects the torus four times, there are either zero or any number up to four different ways, or *modes*, in which this mechanism of mobility zero (in figure 1.08) can be assembled. For a source of these visual, geometrical ideas about the question of modes of assembly in a mechanism, the reader is referred to Torfason and Crossley [8] who worked together in 1969.

44. It is important to recognise here that various modes of assembly are also available in mechanisms of mobility unity. To demonstrate this we shall take for a simple example the crank–rocker mechanism of figure 1.11(*a*). Fix not only the frame of this mechanism of mobility unity but also its input link, by fixing the variable angle $\theta$ at some particular value. Next ask the following question: in how many different ways, copying the given kinematic chain, can, (*a*) the two fixed links which have now become the frame, and (*b*) the remaining $(n-2)$ links of the mechanism, be put together into what must clearly be a rigid structure of mobility zero? The answer is some number which could be as high as 2; for we are looking here, in this simple example, at the question of a sphere (centre $B$ and radius $B$–$C$) being cut perhaps at two points ($P$ and $Q$ which are not shown) by a circle (traced by the free socket at $C$ about its axis at $D$–$D$). Having seen this, we must look next at the question of whether or not, after release again of the input link, the mechanism can transport the assembled joint at $C$ from $P$ to $Q$ without transgressing one of the 'limits of travel' of its links (§ 2.42, § 18.04). It may or may not be able to do this according to whether the

Figure 1.08 (§ 1.40) A mechanism of mobility zero. This statically determinate structure can be assembled, as can other mechanisms, in a number of different modes. The different modes are illustrated in figure 1.09.

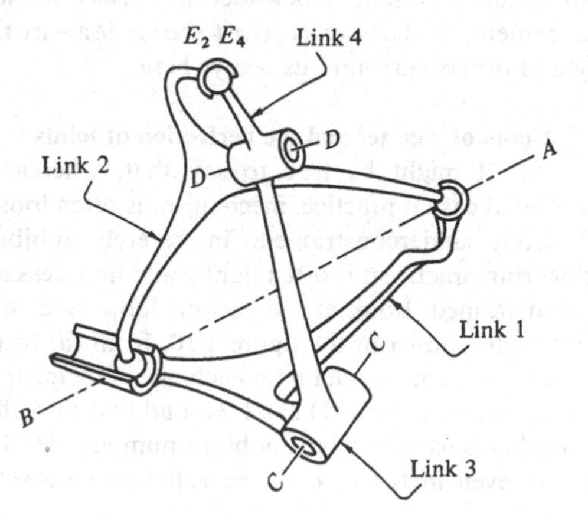


Figure 1.09 (§ 1.43) This sketch is illustrating the different modes that are possible in figure 1.08. Whereas the skew torus is coaxial with the joint centre line $CC$ at $C$, the circle is coaxial with the line segment $A$–$B$ that joins the centres $A$ and $B$ of the relevant balls.

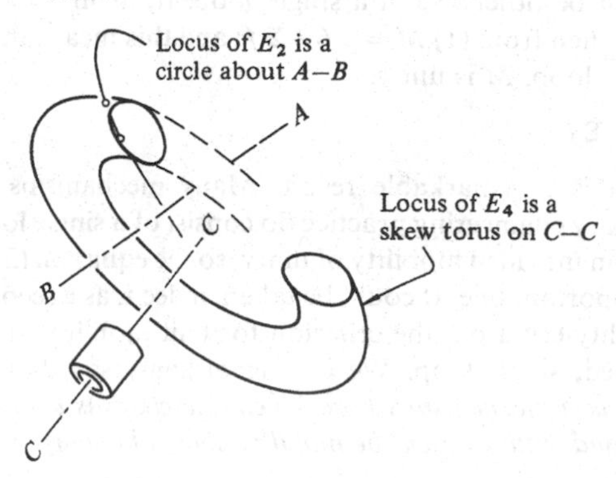

mentioned circle traced by the free socket at $C$ traverses the solid, circular torus traced by the free ball at $C$ along only one continuous segment of itself, or along two.

45. We can thus see, by way of this example: (*a*) that various modes are possible in mechanisms of mobility unity; (*b*) that for each mode there will be an independent set of a single infinity of configurations (§ 1.37); (*c*) that for each given value of an input angle $\theta$ there may in any event be a number of different values for an output angle $\delta$; (*d*) that this whole matter can become very complicated if we look too deeply into it; and (*e*) that the circumstance of there being a multi-valued $\delta$ for each given $\theta$ is associated with the algebraic degree of the closure equation for the linkage and its real and imaginary roots (§ 1.16, § 19.50). One can also see that, except in the event of there being a 'change point' somewhere in the motion of the mechanism (§ 2.24, § 18.01), it is not physically possible for the mechanism to escape by itself from operating one of its modes into operating another; it must at some joint be uncoupled, moved to another mode, then coupled-up again. At the risk of repeating material from other chapters it should be mentioned here that, while finite multiples of infinity are simply infinity, and thus that the total number of configurations of a multi-moded mechanism of mobility unity is simply $\infty^1$, the total number of configurations of a mechanism of mobility zero is $\infty^0$, namely unity, *or some other finite number*. This latter accords exactly with the material of § 1.43. We shall see indeed that, while finite multiples of infinity are not important in the context of mechanism, powers of infinity such as those already mentioned, and $\infty^2$ and $\infty^3$ etc., are. The matter is looked at in chapter 4, and dealt with in many other places.

**Mechanisms consisting of a single loop**

46. It might be convenient now to look at that large class of mechanisms which consist of a single loop. If the links of a mechanism are connected into a single loop, the links are all binary links and the number of joints required is the same as the number of links: it cannot be otherwise in a single loop. If, as in a loop, $n = g$, then from (1) $M = -6 + \Sigma f$; and this means that, if, in a loop, $M$ is unity,

$$(2) \quad \Sigma f = 7,$$

which is a remarkable result. Many mechanisms in ordinary engineering practice do consist of a single loop with an intended mobility of unity, so the equation (2) is an important one. It could be taken, indeed, as a second mobility criterion, the criterion for unit mobility in an isolated, single loop. We can accordingly say that, *if there is a mechanism whose kinematic chain is a single loop and within which the mobility should be unity, then the sum $\Sigma f$ of the freedoms at the joints should be exactly seven.*

47. Let us look now at the 3-link loops of alleged mobility unity set out in the figures 1.03. Most of them are self-explanatory. Some of them are recognisable as pieces of existing machinery, while others are not. The mechanism of figure 1.03(*g*), for example, clearly represents the working parts of a gramophone, while an elaboration or duplication of the kinematic chain in figure 1.03(*j*) may or may not be the key to some ingenious new coupling. As well as the 'serious' joints whose *f*-numbers range from one to five inclusive, examples of the 'trivial' joints, of *f*-zero and *f* 6, are included also, both for the sake of completeness and for fun. Notice also that in one place a convenient group of simple joints has been collected together into one: the *f* 2 'universal' joint of the gramophone is an example of the complex joint to be discussed at § 2.37. The reader might usefully wonder at this stage how many physically different kinds of joint can be found in current practice (or in yet to be invented practice) for each different *f*-number, and in how many different ways three chosen *f*-numbers can be added exactly to seven. Also in the figures 1.03 one should wonder whether the gramophone mechanism is not indeed a 4-link loop, and whether we ought to have spoken first, in this discussion, even more trivially about the many 2-link loops which exist in engineering practice. Another matter of variety to be seen in the figures 1.03 is the question of which link of a nominated kinematic chain might be chosen as the fixed link: which *inversion* of the mechanism, in other words, might be interesting, or relevant, or existing? The idea of inversion, incidentally, which is mentioned here and should be self-explanatory in the context, is of fairly recent origin. Reuleaux introduced the concept first in his celebrated work (1875), and standard texts have dealt with this important matter ever since. With these and added questions, it can be seen that the few collected pictures of the figures 1.03 represent a bewildering variety of such arrangements, and it is instructive to be made aware that they and others surround us everywhere.

**Spots of contact and the perfection of joints**

48. It might be true to say that, whereas in uninhibited casual practice, mechanism is often loosely and safely underconstrained, in severely inhibited engineering practice it is often tightly and unnecessarily overconstrained. Look at the various loops and other mechanisms shown in the figure 1.10, from (*a*) to (*g*). Construct a kinematic chain for each one of them, apply the Kutzbach criterion (1) of § 1.34, and find that all of the mechanisms have high mobility numbers $M$. This reflects – even in the case of the roller and shovel in

figure 1.10(*g*), where each link of the mechanism makes a joint with each other link of the mechanism – the truth of the opening part of the above remark. These casual mechanisms are, however, almost without exception, 'gravity closed' devices. This means that gravity along with friction applies at most of the joints in the figures 1.10, thereby suppressing the unwanted though possible motions which might otherwise occur. Notice also in the figures 1.10 that there are questions arising about the *stability* of the mechanisms or of the linkages. This latter, somewhat complicated matter also depends upon force; indeed it depends upon the changing configuration of the whole system of forces acting as the mechanism moves; it is not considered here. There are also displayed in the figures 1.03 and 1.10 questions surrounding the *conditioning of contacts* in mechanism, but this matter also, depending as it does partly upon force and partly upon metrical possibilities for motion, must be left for consideration elsewhere (chapter 6, § 7.36, chapters 17, 18).

49. While viewing the figures 1.10, it will be convenient to contemplate the important idea discussed at length in chapter 6 and mentioned again at § 8.07 that, if there were no gravity and no friction, if, nevertheless, the contacting members in mechanism adhered somehow to one another, and if we assumed (as we should) that no member is manufactured accurately nor any joint element other than irregularly, then we could think in terms of *spots of contact* at the joints between the members. It must be said that Hertzian areas of contact between elastic bodies of high Young's modulus are being conjured up in the imagination here but the idea of 'area' is being minimised. The term 'spot' is in fact a euphemism in the present context, for I am not yet prepared to use, at this particular place, the only other available word, point. Refer to chapter 6. Look again at the spots in figure 1.02, which are drawn within the contact patches there, and refer again to the figures 1.10. Refer also to § 7.06.

50. The shaft of the shovel in figure 1.10, the tip of the umbrella and the convex underside of the spoon all make contact with their mating members at one spot only (note the joints *f* 5); the handle of the umbrella, the handle of the spoon and the top end of the meat hook all make contact at two spots only (note the joints *f* 4); the underside of the umbrella bowl and the bulbous tip of the compass point both make contact at three spots (note the joints *f* 3); the log makes contact with its one-piece trestle at four spots (note the joint *f* 2). While the cup contacts its saucer at five spots, three of them distributed across the undulating bottom of the roughly cylindrical depression and the other two distributed

Figure 1.10(*a*) (*b*) (*c*) (§ 1.48) Various casual mechanisms. All of these have high mobility numbers *M*. Note that most of the joints are force closed (§ 2.21), and that gravity is apparent. Refer also § 2.37, § 14.38.

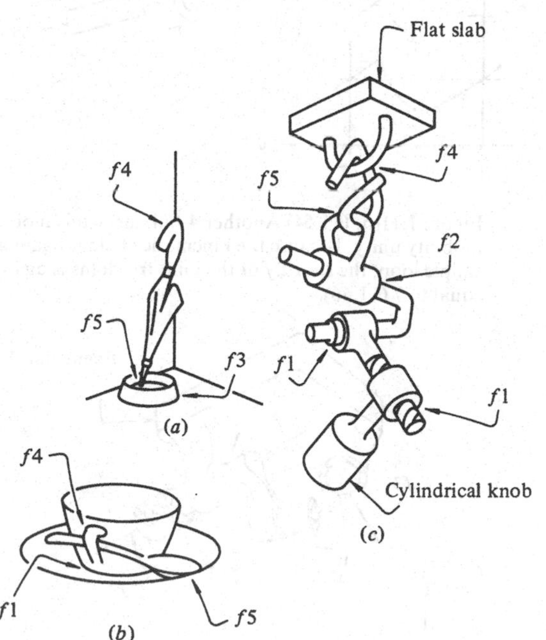


Figure 1.10(*d*) (*e*) (*f*) (*g*) (§ 1.50) The log and the trestle at (*f*) approximates an *f* 2 cylindric joint. The garden roller and its shovel at (*g*) is a mechanism with a completed kinematic chain; note the six joints marked *j* at (*g*) (§ 2.23). Refer also § 1.58.

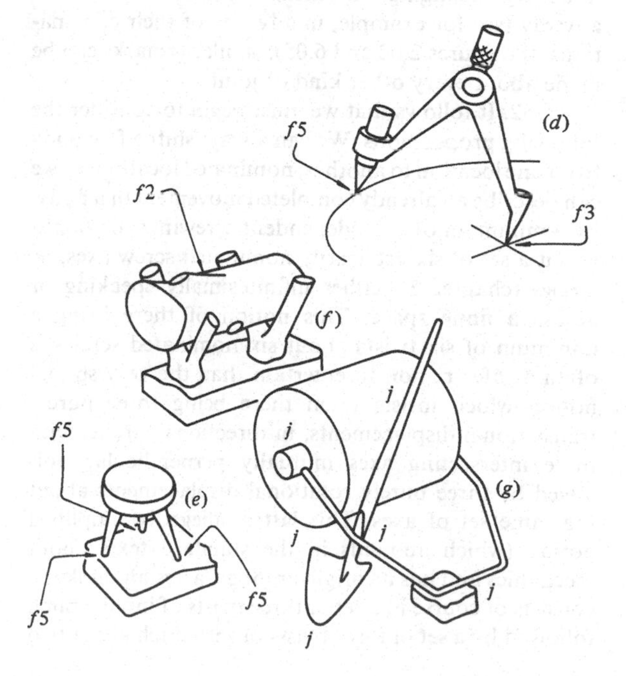

around the non-circular circumference of it, and here we note a joint which approximates to an ordinary hinge ($f$1). It is safe to argue (chapter 6) that each new spot of contact between any pair of irregular bodies destroys one of the six relative degrees of freedom which otherwise occur in nature between the bodies. This presumably means, of course, that no two bodies can make contact at more than six points – unless of course the bodies bend. But here we are dealing with rigid bodies; we are not yet ready to speak about the flexible or elastic bodies (§ 2.47).

51. Notwithstanding my own remarks at § 1.35 where I spoke about the four 'purely angular' parameters, it is necessary to explain why it is that, in general in the present context, I never automatically refer to the separate freedoms, for example of a cylindric joint, as being 'one of translation along and one of rotation around' the central axis of the joint. The remarks which follow in this and the next paragraph are in some ways premature but, unless certain misconceptions are to be permitted, they must be made now. The rough log on its rough trestle in figure 1.10($f$) has the freedom at an instant to twist about a whole infinity of different screw axes (§ 5.50) upon a cylindroid (§ 15.13), no one of which is necessarily along the central axis – wherever that is exactly – of the log, and only maybe two of which are of zero pitch (§ 15.49). The log will seldom simply slide along or turn about a central axis; it will, most often, move in a much more general manner. Given either that irregularities of this kind always occur in allegedly 'regular' joints, or that joints are most often continuously changing in their intended movements anyway (see for example, in advance of their explanations, the figures 2.02 and 6.05), similar remarks can be made about every other kind of joint.

52. It follows that we must begin to consider the following propositions. We can always shift a free body from one location to another, nominated location, or we can describe an already completed movement of a body, by a minimum of six independent screwings, or *twists*, about a set of six arbitrarily nominated screw axes, or *screws* (chapter 5), either infinitesimally speaking or across a finite space. This notion of there being a minimum of six twists about six nominated screws is often a safer notion to entertain than the very special notion which insists upon there being three purely translational displacements, in directions parallel with three intersecting axes mutually perpendicular, followed by three purely rotational displacements about the same set of axes. This latter, allegedly simplified notion (which abounds in the standard texts about mechanics and has its origin among Cartesian thinkers) consists of course in a set of three twists of infinite pitch followed by a set of three twists of zero pitch about two superimposed sets of three mutually perpendicular axes; and in a general circumstance such a special choice of the nominated six screws can lead very easily to serious mistakes in logical thinking. For further readings about this important matter, go to § 2.15 where the question of passive freedoms is discussed, to § 2.39 where the complicated motions of complex joints are mentioned (see the figures 2.02), and to § 2.58 where it is explained how there are two different ways in which we appear to be obliged, in mechanism, to envisage the set of the six screws. Also, of course, the above remarks are germane to the whole of this book.

53. The § 1.50 was written openly to create the impression that there is associated with each freedom number $f$ at a joint an appropriate number of spots of contact $c$, in such a way that the relation

(3) $f+c = 6$

pertains at any one joint. The relation (3) has not been proven here, only mooted. But if it is true, and it is

Figure 1.11($a$) (§ 1.54) A 4-link mechanism of mobility unity. It features revolute joints at $A$ and $D$ and a pegged ball at $C$. There is a discussion about its modes of assembly at § 1.45.

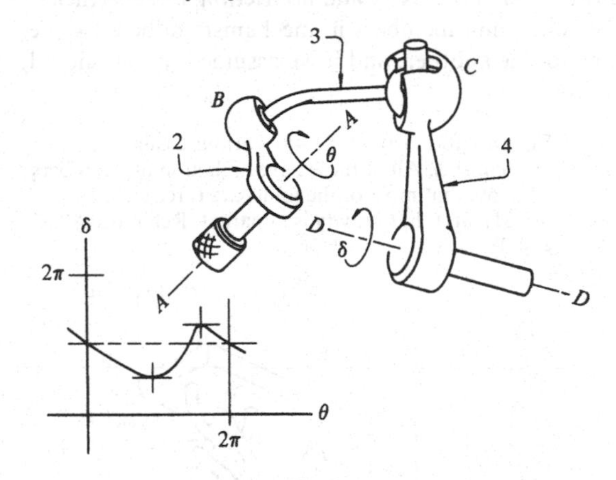


Figure 1.11($b$) (§ 1.54) Another 4-link mechanism of mobility unity. Because the kinematic chain is again a simple loop, the sum $\Sigma f$ of the joint freedoms is again equal to 7 (§ 1.46).

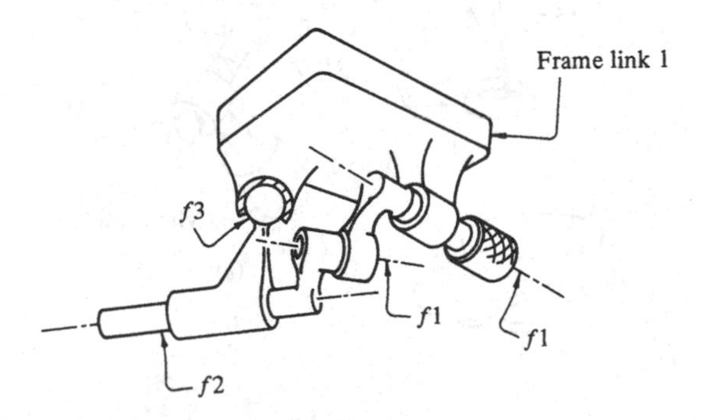

(chapter 6), given some necessary assumptions, it is also true that, as a joint approaches 'perfection' in its construction, some of the spots begin to coalesce. In the case of a well made ball and socket joint, for example, where both parts are accurately spherical and the clearances adequate, the three spots will coalesce into only one triple-spot. In a well-made ordinary cylindric joint (refer to figure 1.02) where both parts are accurately cylindrical and the diametral clearance is adequate, the four spots will coalesce into only two double-spots, one double-spot at each end of the joint with one of them displaced some angular amount from the other circumferentially. Refer also the figure 17.09 (*a*) and to the discussion in § 17.27 dealing with the case of an ordinary hinge, with five spots. Similar remarks can be made about all joints. What happens at the joint *B* for example in figure 1.02, when the ball is sufficiently spherical and the hole sufficiently circular? Before we leave this paragraph, please become aware of the ambiguities here: if the bodies are rigid the spots will be points and they may never coalesce (§ 6.06); yet if the bodies are flexible it could be argued that the spots might grow to cover the whole available area at the contact patch, in which case the 'spots' as such would disappear (§ 6.01). But such ambiguities are not uncommon in the field of natural philosophy: there is no way, except by closing one's eyes to certain aspects of the matter (and we shall need to do precisely that), of avoiding them.

Figure 1.11(*c*) (§ 1.54) A 5-link loop of mobility unity. Think about this: uncoupled at *C* the centre of the ball $C_3$ may traverse the surface of a skew torus while the centre of the socket $C_4$ may traverse the surface of another skew torus. There are closed and twisted curves existing where the tori intersect, and these formations of a possible path for *C* in 1 occur at a maximum of four places. Notice again that $\Sigma f$ equals 7, and refer again to § 1.45.

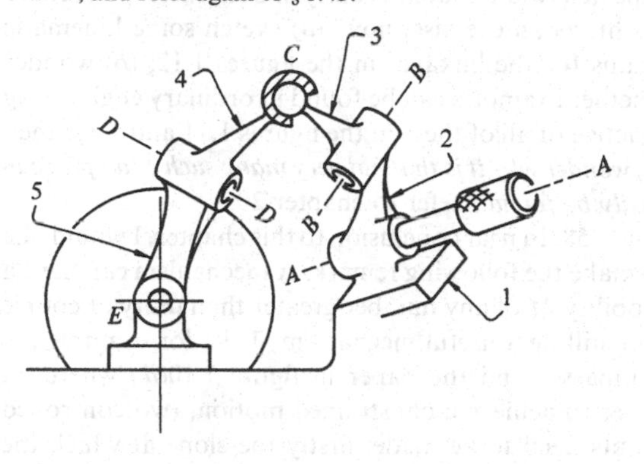


Figure 1.11(*d*) (§ 1.54) A 6-link loop of mobility unity. This mechanism exhibits certain geometrical specialities, and these produce the occurrence of a passive freedom at *B* (§ 2.20, § 2.32). See the similar picture in figure 2.01(*e*). See also § 2.22 for a mention of bifurcation.

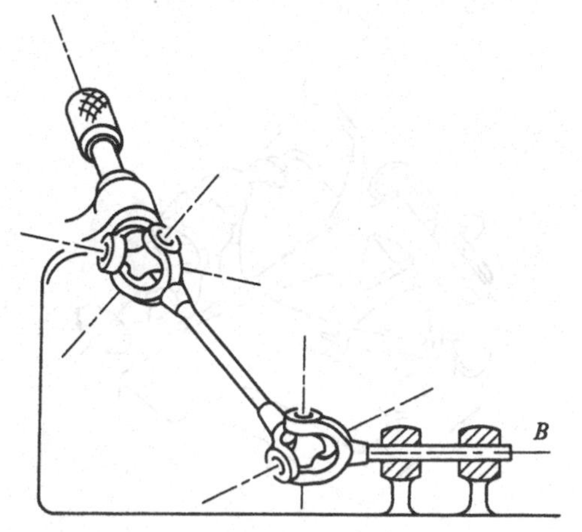


### Some wider considerations and a summary

54. Having thus spoken in a preliminary way about the question of spots of contact in the absence of flexure, and leaving until a later discussion the proliferation of spots in the event of overconstraint (§ 2.26), it is time now to show some single loop and multi-loop mechanisms of mobility unity where the numbers of links involved are greater than three. Look first at the mechanisms of the figures 1.11, ranging from (*a*) to (*d*), where a selection of 4-, 5-, and 6-link, single loops are shown. See next from equation (2) at § 1.46 that, for mobility unity in an isolated loop where *f* 6 joints are forbidden, the maximum number of links is seven, and thus see that the mechanism of photograph 1 (the frontispiece to this chapter) is an important member of this collection. Photograph 1 shows a working model of the so called 7*R*-loop and, whereas in this model all of the seven joints are revolutes namely screw joints of zero pitch, the mobility would still be unity if all of the screw joints were of any or of changing pitch.

55. In figure 1.07 there are two examples of what might be called screw joints of changing pitch; they are the slotted and pegged cylindric joints which can be seen there, both marked with the appropriate symbol *f* 1. In the 'fully' general *f* 1 joint, incidentally, not only will the pitch of the screwing change in magnitude as the joint engages in its motion, but also the axis of the screwing will change in location (§ 2.64, § 6.58, § 7.04). If, in any event, we take any seven ordinary screw joints – namely any seven sets of a nut and a matching bolt – and arrange them into a single loop, and if next we link them together with a suitable seven, randomly chosen binary links, we can produce what could be called the 7*H*-loop, *H* standing for the word helical (§ 1.25, § 2.56, § 7.12).

Although it is not reported in his final work, which was posthumously put together by his assistant workers (1956), Beyer built such a loop in Munich in about 1952. The direct application in useful practice of the general 7*H*-loop is probably nil, but its general significance from the algebraic point of view and its importance in robotics is clearly immense. It should be obvious that, once a closure equation for the 7*H* (or even for the 7*R*) can be written, the workings of the simplier, single-loop linkages of mobility unity could be clarified by being categorized: certain of the so-called fixed parameters (that is of the dimensions of the links) of the 7*H* could be put to zero, and the closure equation for one or other of the simpler, derived linkages would immediately emerge. Refer to [9] and to Duffy (1980).

56. Each of the mechanisms in the figures 1.11 could be discussed at length and all kinds of useful comments could be made, especially about the mechanical generality of the figures and the special cases (geometrically speaking) of the 7*H*-loop, from which many other loops and multi-loop mechanisms of mobility unity can be derived. It will be seen for example that three intersecting revolutes arranged in series are equivalent to a single ball and socket joint. One thing however to note parenthetically is that within some of these mechanisms the joints are *bifurcated*. Take for example, the bifurcated hinges of the universals and the bifurcated cylindric joint at the output of the 6-link loop in figure 1.11(*d*), and notice that when looked at from a purist point of view, such splittings and separations of the parts of a joint could be held to confuse the issue of constraint (§ 2.22). Look also at figure 2.01(*h*).

57. Notice next, as a matter to be distinguished from bifurcation, the circumstance at the general area *E* of the toggle mechanism in figure 1.12(*a*). The figures 1.12 are of multi-loop mechanisms of mobility unity. The thing at *E* is not a bifurcated hinge, nor is it a complex joint between two members of the mechanism; *it is two separate hinges*. Before one can draw a kinematic chain for this particular, multi-loop mechanism one must decide which of the three members at *E* carries fixed to it the one single pin which acts for both of the hinges there. We might also remark upon the arrangement in figure 1.12(*b*). This is a representation of a working model built by myself in Sydney of a 'fully completed', multi-loop, 4-link mechanism where each link makes contact at a joint with each other link of the mechanism (§ 2.23). The four internal, or subsidiary, 3-link loops which can be identified within the pictured linkage are, of course, 'interwoven' with one another, and cannot be regarded as being alone, as are for example the isolated 3-link loops of the figures 1.03. The Kutzbach criterion, necessarily applied to the whole of the mechanisms at the figures 1.12, will reveal the expected result, that *M* is unity in both cases. The reader might, as an exercise, now: (*a*) sketch some kinematic chains for the linkages in the figures 1.12; (*b*) wonder whether examples can be found in ordinary engineering practice of all of these in the figures 1.11 and 1.12; then (*c*) *wonder why it is that not very many such examples can easily be found.* Refer to chapter 20.

58. In near conclusion to this chapter, I would like to make the following remark. A mechanism can have a mobility *M* of any number greater than unity of course and still be a useful mechanism. Take for example the compasses and the paper in figure 1.10(*d*) where, in order to achieve a constrained motion, two controlled inputs need to be made: firstly the slope at which the

Figure 1.12(*a*) (§ 1.57) A multi-loop mechanism of mobility unity. Within the 5-link, interior loop *ABCDE*, there is a double hinge at *E*.

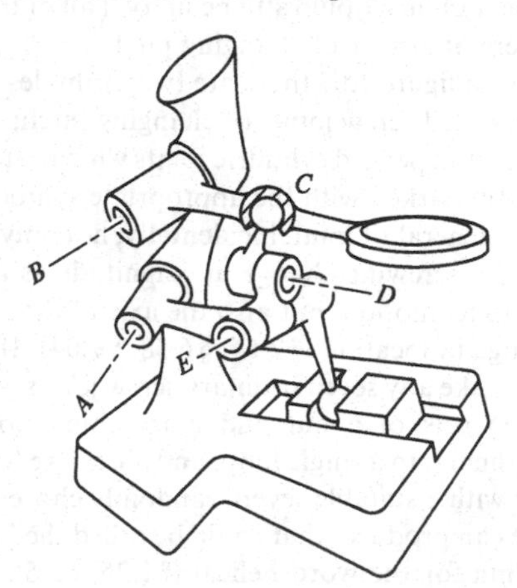


Figure 1.12(*b*) (§ 1.57) Another multi-loop mechanism of mobility unity. It has four links and a completed kinematic chain (§ 2.23). Think about the path traced by the ball at *B*.

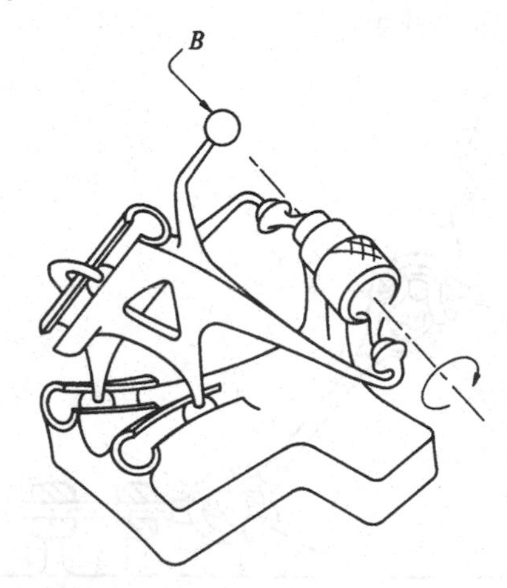

already adjusted instrument is to be held by hand; and secondly the insertion of the circle-drawing action. There are many such devices in ordinary and casual engineering practice and in the wide realm of nature and in the biomechanism (§ 2.60).

59. Whenever the mobility $M$ of a mechanism of rigid links is unity, however, the mechanism can: *(a) always be assembled in any one of its single infinity of configurations without destroying either its links or the necessary clearances at its joints; (b) function with a single input in such a way that the number of spots of contact at the joints can remain exactly as predicted by equation (3); and (c) sustain inaccuracies of overall dimension, namely crookednesses, irregularities in the mating surfaces at its joints, and, within limits, accidental damage, without disturbing either its capacity (a) or its capacity (b).*

**The influence of Reuleaux**

60. Much of what I have said already – since § 1.48 especially, and please read § 1.52 – has relied upon my underlying thesis here that the members at joints in mechanism are touching one another at spots of contact. Joints have been seen by me accordingly as being, of necessity, *force closed.* Now Reuleaux in his collected works (1875), I quote from the Kennedy translation (1876), appears to have believed on the contrary – and many have followed him (they still do) in this – that 'pair closure' at joints, which he so strongly advocated and very clearly prognosticated for the whole of machinery practice, would in some miraculous way do away with the clear need we have today to consider the forces and wrenches and thus the statics in machinery (§ 10.59, § 13.05). Pair closure, or *form closure* as I have called it in § 2.21, was seen by Reuleaux in terms of a perfected, geometrical accuracy (§ 6.02). He appears not to have given much consideration to the need for clearances at joints or the inevitability of wear. He speaks introductorily in that book and again in his important chapter 6, about the 'primitive' nature of force closure, yet he mentions that the primitives, like children, saw *motion* first while being unaware of forces – especially the immense forces and the powers of nature and thus the forces in machinery too. And I agree with that.

61. But Reuleaux thereby gets into a dilemma. He says that the primitives (like children) saw motion first, and I agree. But then he says that, because they saw this motion first, they engaged in the primitive business of force closure with its 'clicking and clacking', by means, presumably, and in my terms (chapter 10), of the *action* at the joints. Later we, unlike the primitives he said, discovered the immense importance of the forces in machinery and thus began to see our way towards perfection, through pair closure and a purer understanding – through the pure science of kinematics! In saying that the ultimate replacement of all force closures by form closures will achieve the perfection we not only seek but shall inevitably find, by saying that the replacement of force closure by form closure is the hallmark of all progress towards perfection, he is really saying, it seems to me, that total constraint is our objective and that, moreover, overconstraint will do no harm. I must remark (*a*) that, despite what I have to say at § 2.27, Reuleaux was unaware of both freedom and constraint as we know of those phenomena today; (*b*) that he thus became beguiled by the uncommon idea that a suitably developed science of kinematics might well do without a consideration of the statics; and (*c*) that, and this is my view, *if the statics and the kinematics of a mechanism are both to have a better chance of being, as we say, determinate, and if all other things are equal, sufficient constraint is better than overconstraint; this means in effect that, if a mechanism is to be employed for some particular purpose, its general mobility M should be accurately known, be understood, and be, above all, appropriate.*

**Notes and references**

[1] Bogolubov (1964) speaks well of Federhofer (1932). Federhofer, author of *Graphische Kinematik und Kinetostatik*, appears not only to be an early originator of the term kinetostatics but also to be the first exponent of that science as it applies among the links and joints of planar mechanism. See also Whittaker (1927), *Analytical Dynamics*, and read the closing words of his § 26; Whittaker remarks there, and I paraphrase him here, that it is the reactions at the constraints in problems of mechanics, namely those reactions so carefully avoided by the equations of motion of Lagrange, that lie at the root of that branch of mechanics known as kinetostatics. Another more recent presentation of the term kinetostatics may be found in chapter 7 (headed Getriebedynamik) of Volmer *et al.*, *Getriebetechnik* (1968), where the term kinetostics is used, as is usual now in German, to mean the statical analysis for force in mechanism in the absence of mass and the presence of motion. Erdmann, A.G. and Sandor, G.N., in Kineto-elastodynamics – a review of the state of the art and trends, *Mechanism and Machine Theory*, Vol. **7**, p. 19–33, 1972, do not mention the term kinetostatics; and nor should they; the main subject matter of this well celebrated paper is not directly related to that that I have at issue here. Refer § 1.02.

[2] Waldron, K.J., The constraint analysis of mechanisms, *Journal of Mechanisms*, Vol. **1**, p. 101–14, Pergamon, 1966. Refer § 1.21.

[3] Heidebroek, E., Gegenwartsfragen in Bau von Maschinenteilen. This is the report of a local VDI conference held in Stuttgart during April of 1933. *VDI Zeitschrift*, Band **77**, Nr 43, p. 1165–71, October 1933. See the section entitled 'Einselfragen aus dem Gebiet der Maschinenteile'. Refer § 1.34.

[4] Kutzbach, K., Mechanische Leitungsverzweigung; ihre Gesetze und Anwendungen, *Maschinbau der Betrieb*, Band **8,** Heft 21, p. 710–6, November 1929. Refer § 1.34.

[5] Manolescu, N.I., For a united point of view in the study of the structural analysis of kinematic chains and mechanisms, *Journal of Mechanisms*, Vol. **3**, p. 149–69, Pergamon, 1968. Refer § 1.34.

[6] In his paper at [5] above Manolescu quotes the same paper by Kutzbach as I do at my note [4]. Refer § 1.34.

[7] Trevor H. Davies and F. Erskine Crossley, Structural analysis of plane linkages by Franke's condensed notation, *Journal of Mechanisms*, Vol. **1**, p. 171–83, Pergamon, 1966. The authors quote Franke, *Vom Aufbau der Getriebe*, VDI, 1958. Refer § 1.38.

[8] Torfason, L.E. and Crossley, F.R.E., The intersection of solids shown by electronic analog for mechanism simulation, *Journal of Engineering for Industry*, Vol. **93,** (*Trans ASME* Series B Vol. **93**), p. 17–26, February 1971, and, by the same authors in the same issue of the same journal, Stereoscopic drawings made by analog computer of three-dimensional surfaces generated by spatial mechanism, *Ibid.*, Vol. **93,** p. 239–50, February 1971. Refer § 1.43.

[9] Duffy, J. and Crane, C., A displacement analysis of the general spatial 7-link, 7R mechanism, *Mechanism and Machine Theory*, Vol. 15, p. 153–69, Pergamon, 1980. The gist of the same paper may be found in Duffy's book (1980), *Analysis of mechanisms and robot manipulators.* Refer § 1.55.

2. The spherical crank rocker mechanism that consists of four rigid links and four revolutes with their axes intersecting at a point does not function satisfactorily unless the accuracy of its manufacture is adequate.

# 2 Overconstraint and the nature of mechanical motion

**Transition comment**

01. The broad matters discussed in chapter 1, namely those of freedom, mobility and the need for sufficient constraint in mechanism, are of great importance in the wide realm of practical machine design. They are derived however from a somewhat narrow set of assumptions regarding (*a*) rigidity in the absence of elasticity, (*b*) crookedness in the absence of accuracy, and (*c*) direct contact at joints in the absence of lubricated clearances. In chapter 1 there is thus developed what might be seen to be a somewhat simple theory of constraint. It is a simple theory; it is directly applicable; but only in a very rough way is it able to predict, for much of our ordinary machinery, the actual mechanical behaviour of the moving parts.

02. We often see in ordinary working machinery the unhappy effects of friction due to eccentricities of loading (§ 17.24), the consequent continuous likelihood of jamming, chatter, and wear, and the necessity for lavish lubrication at badly affected joints. Such phenomena will be evident whenever the kinematic design for a smooth transmission of the wrenches has been poor (§ 17.26), or, alternatively, whenever the jamming phenomena themselves have been an intended activity of the machine (§ 10.61). Although such matters are important and need to be studied, they are in a curious way irrelevant here. They are not the matters being implied at § 2.01. I tried to imply at § 2.01 that nearly always in a drawing for some ordinary machine there will be some special aspects of the geometrical layout or arrangement of the parts that are difficult or troublesome to obtain by ordinary machining methods in the manufacture of the machine. Such special geometrical aspects often interfere, moreover, with an easy application of the general mobility criterion (§ 1.34). Sometimes, indeed, they appear to render it quite inapplicable; clearly we need to look at this.

03. If a machine intended for mobility unity accidentally becomes immobile, owing perhaps to some overlooked mistake in its design, it will in general be quite incapable of movement until the offending mistake has been rectified. Please go back to § 1.39, and see there for example that, if in figure 1.06 the ball and socket at the lower end of the pegged push-rod could not be allowed for some reason to be as it is a spherical joint of $f3$, but was obliged for some other reason to be a cylindric joint of only $f2$, the whole machine would be immobilized and something would need to be done to restore its capacity for motion.

04. If however, and this is contrary to the gist of the above example, some special geometrical arrangements were carefully devised and built into the layout of the links of some machine, the machine might (*a*) grossly violate the criterion for general mobility $M$, yet (*b*) by virtue of flexibility, clearances, and the mentioned special layout, be mobile throughout the full cycle of its motion. Such special arrangements are made, of course, at the expense of the care which must be taken to avoid in the actual machine any accidental damage that might distort or destroy the special arrangements (§ 1.59); but such arrangements are very commonly employed in ordinary engineering practice. Please glance for examples at all of the figures 2.01 from (*a*) to (*m*), then look at the figure 2.01(*a*). Notice in that case that, unless the centre $D_1$ of the pivot at $D$ is made to coincide with the centre $D_2$ of the curved and interslotted members at $F$, the mechanism will not be mobile (§ 1.23).

**Overconstraint and the idea of apparent mobility**

05. Suppose for another example that the cam in figure 1.06 were the kind of cam with a box-like follower where there was, allegedly, 'positive line contact' maintained at both the top and the bottom of the box as is shown in the figure 2.01(*b*). The $f$-number at the joint could not be other than 4, and overall the mobility $M$ would be zero but, provided the centre line of the cam shaft and the centre line of the follower were accurately square with one another, the mechanism would 'work' all right, with an apparent mobility $M_a$ of unity.

06. There are, as I have said (§ 2.04), in ordinary engineering practice many such contrived arrangements whose mechanisms function with an $M$ of zero or less,

while exhibiting, for the full cycles of their motions, an apparent mobility $M_a$ of unity or more. The 'extra' mobilities, which account for the sometimes surprising difference between the $M_a$ and the $M$, I shall call $M_s$. *Provided the $M_a$ is not greater than unity, such extra, or special mobilities $M_s$ are always obtained in conjunction with overconstraint, and the kinds of mechanism exhibiting such extra mobilities are, by definition, overconstrained.*

07. We can measure and at the same time define the apparent mobility $M_a$ of any linkage or mechanism by physically checking, against whichever link is chosen to be the fixed link, the actual availability of independent freedoms among the remaining movable links (§ 1.21). Imagine that we are to examine some overconstrained linkage, and that this work of checking the availability of the freedoms is to be done with the real, manufactured machine to hand. Remember that the required geometrical conditions will not have been provided accurately, because this is never possible (§ 1.49, § 6.05), and we shall suppose also that the necessary compensating clearances at the joints are not sufficient either. As the machine is moved by hand through the cycle of its motion, we the manual operators would detect, independently of the uneven effects due to gravity, both 'tight' and 'easy' patches in its motion; these would be due to the 'redundant constraints'. Redundant constraints were mentioned first at § 1.20. Distinct from the arbitrarily located, and currently irrelevant, stop surfaces at the joints (§ 1.29), there are the functionally important, *guide surfaces* (§ 7.36), and these latter will be rubbing against one another tangentially while under crosswise compression due to elastic distortion of the 'rigid' links. Refer again to § 1.23 where this matter was, in a somewhat more primitive way, mooted originally. I can now offer a definition, and I wish to state the following: *if the observed mobility $M_a$ of a mechanism is, (a) unity, and (b) greater than the general mobility $M$ as calculated by substitution into the general mobility criterion, the mechanism is in one respect or another overconstrained.*

Figure 2.01(*a*) (§ 2.04) A grossly overconstrained, 3-link mechanism requiring the geometrical speciality (among others) that the axes at $D_1$ and $D_2$ should coincide at $D$. Motion can otherwise not occur. There are discussions at § 1.23 and § 2.25.

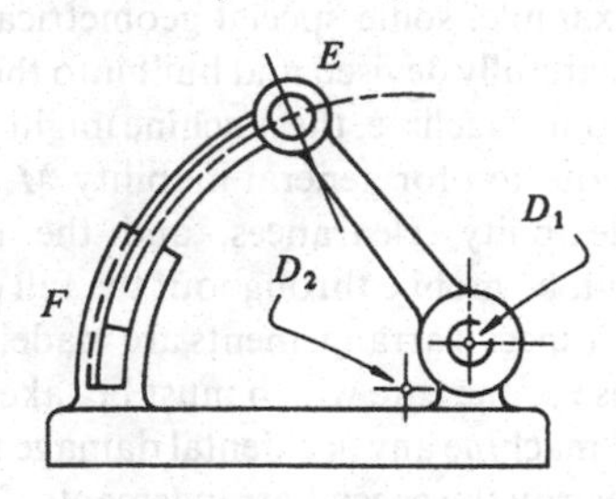


Figure 2.01(*b*) (§ 2.05) This overconstrained cam and follower mechanism depends for its smooth operation upon (*a*) the perpendicularity of the cam and the follower shafts, and (*b*) the accurate cylindricallity of the cam upon the cam shaft. Line contact is presumed to occur between the outside working edge of the cam and the inside working edge of the box follower at both top and bottom of the cam.

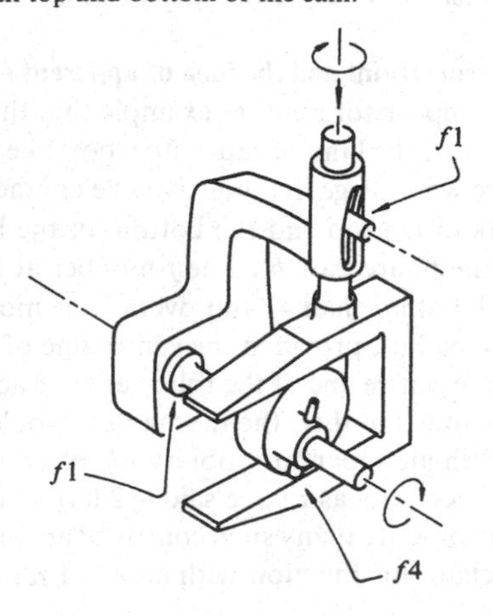


### Full cycle and transitory, apparent mobilities

08. Many mechanisms are frankly designed to exhibit throughout the whole of their cycles the special geometrical conditions which give rise to the special extra mobilities mentioned at § 2.06 and exemplified at § 2.07. Other mechanisms, on the other hand, often by accident but also sometimes by design, are able only *suddenly* to exhibit, at certain isolated configurations in their cycles, their special conditions and consequent extra mobilities $M_s$. These latter are those which are said to suffer their extra mobilities $M_s$, and thus their apparent mobilities $M_a$, only transitorily. We need, in the present context, two separate adjectives to describe and to distinguish the two different kinds of apparent mobility $M_a$. We need the term *full cycle* for the former, which has already been mentioned and used, and we can use the term *transitory* for the latter.

09. The matter of transitory extra mobilities is not very easy to understand immediately, especially when for example $M_a$ becomes 'suddenly' unity while $M$ is zero (see the worked exercise 1 continued, at § 2.46); or when for example in a planar 4$R$-loop of certain link proportions all of the revolute axes suddenly become coplanar. Its general implications will not be apparent either. There was a hint of these matters however at § 1.44, and an ongoing reference there to § 2.46. Refer also to my remarks at § 1.45 about the modes of a linkage, and about the possibility of a 'change point'

occurring at a certain configuration of a linkage, at which configuration it might be free to choose between one of its modes of operation and another. The whole matter is looked at more carefully at § 2.44 *et seq.*, and it is taken up again to become an important part of the subject matter of chapter 18.

10. It must be said parenthetically here that my definition for overconstraint at § 2.07 is certainly not an all embracing one, in the sense that it might have been written to embrace all cases of full cycle or suddenly occurring special mobilities $M_s$. I have carefully avoided inclusion of all cases where a mechanism, whose $M_a$ may be as high as 2 or even higher, may have its $M$ already unity or even higher. This has been done for two reasons, (*a*) to keep faith with speakers of the English language, who would surely assert that the idea of high-mobility motion on the part of a mechanism is hardly consistent with the idea of overconstraint, and (*b*) to keep faith with my own remarks at § 1.19.

11. The most important thing to see however is that a special geometrical condition, or a *geometric speciality*, as I shall call it, may occur throughout the whole cycle of a motion as we have seen, or only transitorily (§ 2.08). All geometric specialities, whether consciously introduced by the designer or occurring accidentally, give rise to special extra mobilities $M_s$: *these always increase the apparent mobility* $M_a$ *of mechanism, never decrease it.*

**On the idea of geometric speciality**

12. At this stage, of course, I have not yet explained what a geometric speciality is. I have dissembled quite circularly at § 2.11, describing it as whatever it is that causes the somewhat unexpected special mobilities $M_s$ which occur sometimes in mechanism. Please refer in advance to the question concerning the pair of universal joints put at the middle of § 2.20. See there for example that the right angularity of the intersecting centre lines at the cruciform pieces in the joints is *not* a geometric speciality. It might be true to say that the whole material of this book is dealing with the following, difficult question: *what particular circumstance constitutes a geometric speciality in a mechanism, and what other circumstance, while having the appearance of constituting such a speciality, does not?*

13. I have said in effect that the presence or absence of a geometric speciality in a piece of non-moving machinery is difficult to detect by a mere observation of the static geometry; and it cannot be done by the easy application of some simple algebraic criterion either. Fundamentally it is a matter of the projective dependence or independence of lines (chapter 10), but there is no place here for a discussion of that. We can, however, write here the simple equation

(1) $M_a = M + M_s$

to clarify the terminology. $M_a$ is always equal to or greater than $M$, because the added $M_s$ is always some *extra* number of mobilities which (magically occurring, in defiance of the general criterion, either for full cycle or only transitorily) is always, when not zero, positive. On those occasions when $M_a$ is greater than $M$ for a full cycle of the motion, the mechanism, often overconstrained, can always be seen in the long run to have been designed somehow conservatively, beyond the bald requirements of the general mobility criterion of § 1.34.

14. It will in any event be clear that, whereas some outright mistake in design which reduces the mobility $M$ of an intended mechanism below its intended value is always detectable by the general mobility criterion, and is, thereby, eradicable, a geometric speciality in a mechanism, which gets there either by design or by accident, and which permits unexpected apparent mobilities in contravention of the criterion, is not detected by it. This general statement sums up the gist of what has only been implied before: *the general mobility criterion, although of great importance for strengthening the wisdom of the analyst and the designer by being properly and usefully applicable in all of the general circumstances, is not a one-shot formula for solution of all of the real problems of practical machine design.*

**The locking in of passive freedoms**

15. What happens in the design of mechanisms is that accidents occur, either by accident or design: and I mean by this that it can happen – it does, in fact, often happen – that certain freedoms in some of the joints become, as we say loosely, *passive*. If a passive freedom exists that is, if a joint is not obliged by the motion of its parent mechanism to exploit one of its nominated capacities for action (§ 1.28, but also § 2.31 *et seq.*), it, the passive freedom, got there (and here is another circular statement) by virtue of some geometric speciality.

16. This kind of occurrence is common in design because designers so often forsake the full scope of three dimensions to make their plans on a drawing board in only two. Planar and other allegedly simple mechanisms, such as spherical mechanisms, are the result (§ 2.29, § 2.30); and these, if they obey the general criterion, are awash with passive freedoms. The passive freedoms exhibited in these very ordinary cases are mostly of the very ordinary kind. They can be exemplified by the $f2$ cylindric joint which simply rotates as though it were an ordinary $f1$ hinge. Passive freedoms of this most ordinary kind also regularly occur, however, in the genuine 'spatial' mechanisms, and the matter is quite complicated. Refer to the chapters 18 and 20. But refer also to Hunt (1978), where that author

refers, at his § 2.4, to *inactive* freedoms at a pair, and to the possibility at the instant of *locking* the freedoms also. There is a difference of terminology here. Whereas Hunt is speaking there about his stationary and uncertainty configurations, I am speaking here about the matter of overconstraint. Please refer also to my § 2.32 to § 2.35 inclusive, where the somewhat confused meaning of the loose term passive freedom is more thoroughly discussed.

17. I think that in design a designer sometimes looks at a passive freedom which has become apparent, and then, by provision of extra guide surfaces at the joint, obliges it to 'behave itself'; I think that the designer sometimes thinks that, given the flexibilities among the links and the clearances at the joints, unless he 'locks it in' in this way, a passive freedom will escape his designer's control (§ 1.60). I wish to say in any event that the occurrence of overconstraint in machinery is always a consequence of (*a*) design, and (*b*) the locking in of passive freedoms. *Such lockings in are always done by the designer: he provides at the joints displaying the passive freedoms extra, and perhaps unnecessary, guide surfaces there; and these are added by him in such a way that the f-numbers at the particular joints are effectively and thus irrevocably reduced.*

### Some examples of overconstraint in working machinery

18. Some examples of overconstrained mechanisms whose apparent mobilities $M_a$ are unity or more are shown in the figures 2.01. Almost without exception their general mobilities $M$ are zero or less. Take for example the 4-link loop at 2.01(*c*). If all of its hinges are seen to be $f1$ hinges, its general mobility $M$ is clearly $-2$. If, however, all three axes of the hinges are seen to be normal to the plane of the wall and thereby parallel, and if the edges of the prismatic hexagon at $D$ are seen to be parallel with the plane of the wall, then the apparent mobility $M_a$ of the loop will clearly be unity. Now in this case it is likely (or so I intend to argue) that some designer saw the hexagonal shaft safely reciprocating in an entirely satisfactory circular hole, feared that it may rotate, and, by arranging to cut a very expensive hexagonal hole for it, *locked in its freedom to rotate, which was passive anyway.*

19. By inspection one can also see in 2.01(*c*) that the hinge at $C$ could afford to be an $f2$ cylindric; but one should also notice that, if the hinges at $B$ and $C$ were *both* permitted to become cylindrics (and this would, by the general criterion, be proper for unit mobility), the mechanism would exhibit a dangerous new, full cycle, extra mobility $M_s$; and this would be the single new freedom for the whole hexagonal rod and its mated hexagonal rocker to drift together and away from the wall horizontally. So what, precisely, are the geometric specialities here? What are the necessary and/or sufficient conditions for mobility in the circumstances? Take a look at the overconstrained yet mobile $4R$-loop in figure 2.01(*d*), and see that we cannot answer such questions here. Refer to chapter 20.

20. Take the next mechanism in figure 2.01(*e*), which is a modification of the one in figure 1.11(*d*). Here the cylindric joint at output has been buttressed against a possible axial movement: the output shaft has been 'guided into its proper working' by provision of the annular thrust faces at either end of the joint at $B$: the designer has seen the danger, and has, with caution, *locked in the freedom to reciprocate, which was passive anyway.* The $f$-number of the joint at $B$ has been reduced, from $f2$ to $f1$, and the mechanism has thus become overconstrained. In this particular case, in

Figure 2.01(*c*) (§ 2.18) A planar, 4-link loop with the designation *RRPR* (§ 7.12). The mechanism is grossly overconstrained, and one piece of clear evidence for this is the hexagonal shaft in its hexagonal hole. Why not a round shaft, or a round hole, or both?

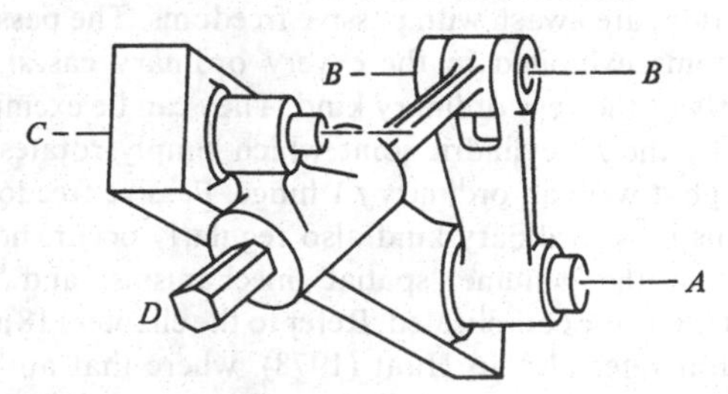


Figure 2.01(*d*) (§ 2.19) This is the celebrated mechanism of Bennett. Refer to chapter 20. Suitably proportioned it is an overconstrained loop of four revolutes that is, geometrically speaking (§ 1.06), mobile throughout a full cycle of motion. Provided the material links are constructed in such a way that they do not clash with one another (§ 1.07, § 1.29), full revolutions can occur between the participating members at each of the four joints.

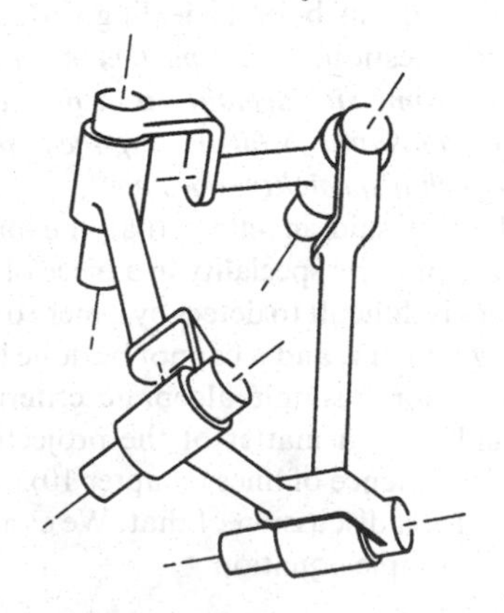

figure 2.01(*e*), we might ask whether the precise nature of the geometric speciality which permits the full cycle, unexpected extra mobility $M_s$ of unity is the condition (*a*) that the hinge axes intersect one another in sets of three at the centres of the Hooke's joints, or (*b*) that the hinge axes of the cruciform pieces intersect one another in sets of two at right angles, or (*c*) that both of these circumstances obtain. Are they in any event the necessary or sufficient conditions, or both? Take a look at the overconstrained yet mobile 6*R*-loop in figure 20.07, and see that these and other such questions cannot be answered here. Refer to chapter 20.

**Form closure at joints in machinery**

21. I should mention here another matter which is marginally related. Due to the accelerating and transmission forces associated with most movement in machinery (§ 1.02), and for other more prosaic reasons, many joints in working machinery need to be, not simply *force closed*, by means of gravity, springs, or transmission forces (§ 1.48, § 1.60), but *form closed*. This means that one member of the joint needs to be imprisoned somehow within the other member to prevent it either from falling out or being driven out by the accelerating or other forces. Compare for example the ball and socket joints in figure 1.05, where force closure due to the valve spring (which is not shown) is adequate, and the ball and socket joints in figure 1.08, where form closure is necessary.

22. Generally speaking, the careful form closure of joints in the process of design does not interfere with the mobility of mechanism, or deny the relevance of the equation (3) at § 1.53; but in the bifurcation of the hinges in the 6-link loop in figure 1.11(*d*), for example, which is (in a way) a kind of form closure against the likelihood of uncontrollable flexure of the links, it could be argued that the two separate parts of the bifurcated hinges can never be quite collinear, in which case overconstraint (in the presence of flexing) has already become apparent. We are moving here, however, into the macroscopic realm of the geometry of mating surfaces at the contact patches within the joints. It might be wise not to do that here but to go direct to the chapters 6 and 7 for a much better discussion of such matters.

**The question of completed chains**

23. We can extract for comment the gear box shown in figure 2.01(*f*). Remarks about its not unreasonable level of overconstraint are made elsewhere, but here we should note as a matter of special interest its kinematic chain. Except for the joint freedoms, the kinematic chains for, (*a*) this box in figure 2.01(*f*), (*b*) the shovel and the garden roller in figure 1.10(*g*), and (*c*) the nameless mechanism in figure 1.11(*f*), are all the same. They are all of four links, and, in each case, each link makes a joint with every other link of the mechanism: they are all *completed* chains. In such mechanisms with five or more links, incidentally, the kinematic chain cannot be drawn in a 'flat' manner using two dimensions only. A study of the mobility of these completed and other, *almost completed* chains is an interesting, though possibly useless exercise [1]. Why are such chains not used more extensively in engineering practice? And anyway, what has the question of being fully completed got to do with the question of overconstraint? The answer is, of course, that the more joints for a given number of links, the higher must be their separate freedoms, unless we are not concerned, as designers, about the rapid growth of overconstraint.

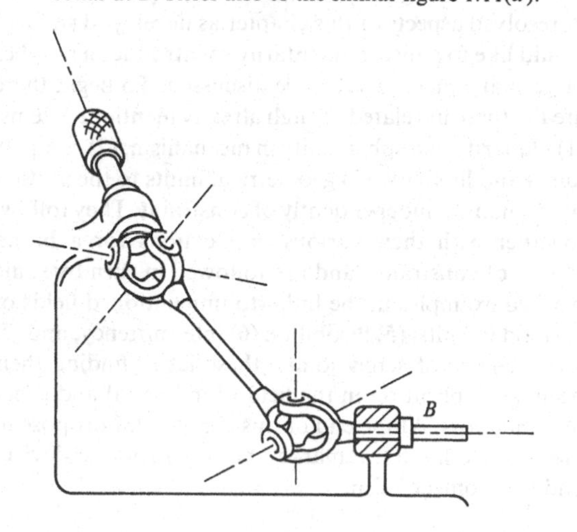


Figure 2.01(*e*) (§ 2.20) Hooke joints connected in this manner produce no axial oscillation at the output shaft at *B*. There is a passive freedom locked in by the collar at *B*. Refer also to the similar figure 1.11(*d*).

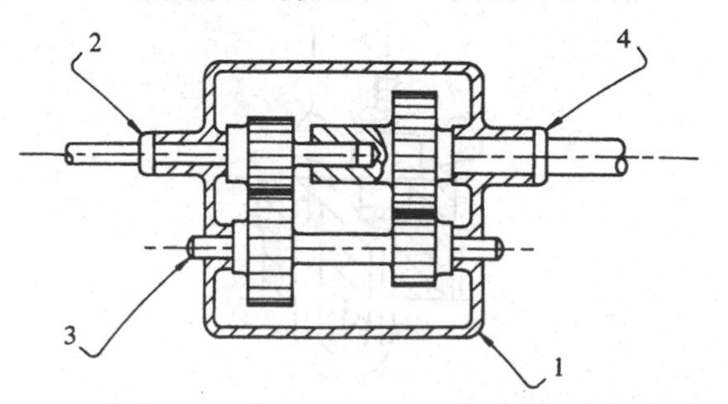


Figure 2.01(*f*) (§ 2.23) An overconstrained, completed, 4-link mechanism of apparent mobility unity (§ 1.23). There are six joints, each link enjoying a joint with each of the other links. Refer also to figures 1.10(*f*) and 1.11(*e*).

**Gross overconstraint and consequent unwanted spots**

24. Look at the *J*-bolt assembly in figure 2.01(*g*). Although by inspection its apparent mobility $M_a$ is unity, gross overconstraint in this example will render the mechanism (as drawn) quite useless in ordinary practice. Either the shaft of the bolt, or the boss of the wing nut, should be clear of the hole to avoid the necessity for concentricity of these parts. The J-end of the bolt should engage not in a fitted hole as drawn but in a suitable slot radially arranged and with a suitable clearance, this latter to avoid the necessity for accurate parallelism and equidistance between the two sets of centre lines. Is this thing a simple 3-link loop, or must it be seen as some multi-looped arrangement? Construct in any event a kinematic chain for the mechanism, and detect by inspection then the full extent of its absurdity.

25. The device in figure 2.01(*a*) has already been mentioned at § 1.23, and similar remarks apply. Unless there are some very good reasons for the present state of the mechanism, the designer must be said to have been severely inhibited in his search for its overall freedom; by virtue of its gross overconstraint, the device would be costly to manufacture. Suppose there were a cylindric joint at *E*. In view of the fact that both of its freedoms would be passive, is there any point in it? Or should there be, indeed, yet another passive joint introduced somewhere between the joint at *D* and the joint at *E*? Notice incidentally that the pin at *E* can be set at any angle; or, given the likelihood of other errors, can it? What can be done with this thing? In order to have a sensible answer for that, we must ask first of all the most important question: what is the mechanism for?

26. As we contemplate, in general, the overconstrained mechanisms shown in the figures 2.01 with an $M_a$ of unity (and in figure 20.02, for another, related example, where the $M_a$ is zero), and keeping in mind the remarks about spots of contact at § 1.51 and § 1.52, we can say here without going any further that, whereas the accurate construction of machinable joints in mechanism where *M* is equal to or greater than zero tends to coalesce the spots of contact into double and triple spots within the contact patches, the provision of overconstraint (along with the necessary flexibility of members) tends to multiply the spots at the joints beyond the minimum number otherwise appropriate.

Figure 2.01(*g*) (§ 2.24) Only a poor designer would envisage a hook like this. Critical remarks about the lack of freedoms here are made at § 2.24.

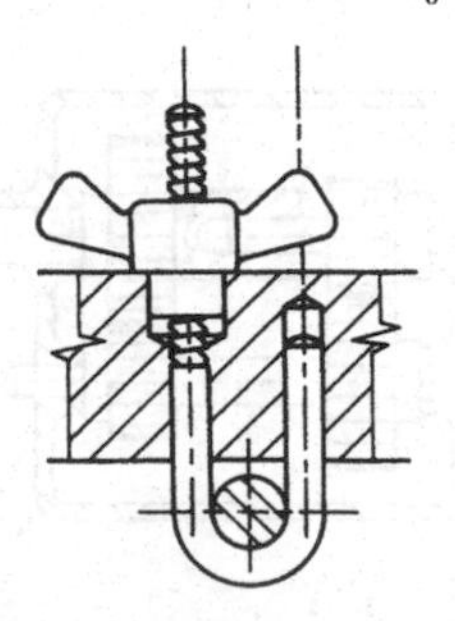

27. It will of course be evident that this whole matter of spots is a very complicated one; how for example is the flexing of links to be balanced against the clearances and backlash at joints? It is also evident that overconstraint is a matter born of synthesis (namely design), not a matter of nature discovered by analysis. Although important advances have been made across the years, and fashions have changed along with social and other circumstances, it must nevertheless be said that our applicable theoretical knowledge in the field of overconstrained mechanism is fairly limited. Our grip upon the necessary art of making an apt decision is limited too. It is accordingly proper to say here that, given all of the real complexities of practical machine design, it behoves us to employ with wisdom and caution our knowledge of the field. In so far as it may be possible we should try to devise, and to apply, sensible overall principles of freedom and constraint which are applicable equally (*a*) for analysing existing mechanism in whatever the states of its constraint, and (*b*) for synthesizing new mechanism. It should also be said of design in the field of freedom and constraint that a balanced compromise between the ordinariness of what is clearly necessary and the excitement of what may indeed be possible is the way towards the combination of elegance and efficiency that characterizes (in my view) effective machinery.

**More transition comment**

28. Having summarized in this way the gist of the unresolved aspects of this chapter as developed so far, I would like to pause momentarily to introduce a number of germane matters yet to be discussed. To begin there are the three unrelated though already mentioned items: (1) planarity and sphericality in mechanism; (2) complex joints and linkivity; (3) geometrical limits to the motion of mechanism independently of constraint. They follow, together with their various implications for a broad theory of constraint, and are followed in turn by some worked examples in the hitherto unmentioned fields of (4) modal limits; (5) flexibility; (6) intermittency; and (7) series chains of screw joints, these latter finding their peculiar application in the field of industrial and other manipulators. At the end of this chapter 2, I propose to make some limited remarks about (8) robotic devices and (9) biomechanism.

### Spherical and planar mechanism

29. At § 2.15 some remarks were made about the sudden occurrence of passive freedoms at the joints of mechanism when, due to an altered or accidental construction of the members, the layout, and thus the overall motion of the mechanism became in some ways special. One way to look at the overall motion in mechanism is to study the pattern of the paths of the moving points (§ 1.18, § 1.19, § 19.45). If the apparent mobility $M_a$ of a mechanism is unity, the paths traced in the other links by each point in a chosen link are in general twisted curves; otherwise the paths are curved surfaces or bounded volumes. Now if the paths of all of the points in any one chosen link of a mechanism, traced in all of the other links of the mechanism, can be seen always to be arranged in such a way that each one of the paths resides on the surface of one of a set of concentric spheres, the relative motion between the links is said to be *spherical*, and the mechanism is said to be a *spherical mechanism*. This implies, incidentally, that in a spherical mechanism there is always a unique point $Q$ at the centre of the spheres where the velocities relative to each one of the links of all of the points $Q_1$, $Q_2$, $Q_3$, etc. are zero (§ 1.13). Further implications are (*a*) that all of the instantaneous screw axes in a spherical mechanism are of zero pitch, that is, that the links simply rotate relative to one another; and (*b*) that the thus produced axes of pure rotation continuously intersect one another at the central point $Q$ (§ 8.11). In figure 2.01 the items at (*h*) and (*j*) exhibit sphericality. Such pronounced geometric specialities as are there apparent render the study of spherical mechanism a highly specialised one. Spherical mechanism is dealt with in other places, and in great detail, by many other writers.

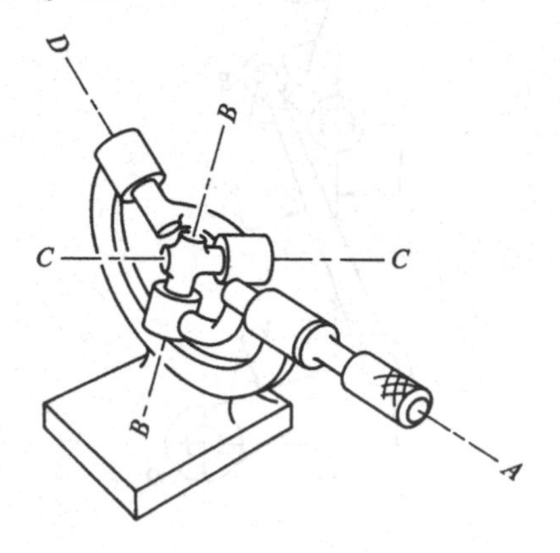


Figure 2.01(*h*) (§ 2.29) This is a spherical 4*R*-loop because all four of the axes of the revolutes intersect at a common point. The loop is overconstrained because $\Sigma f$ is less than 7 (§ 1.46); $\Sigma f$ is only 4. This figure is also mentioned for various reasons at § 2.22, § 2.37, and § 7.16.

30. When the point of intersection $Q$ of the intersecting axes in spherical mechanism is taken to infinity however, we arrive at the doubly special field of planarity, planar motion, and plane, or *planar mechanism*; and here, as might be expected, ambiguous situations and paradoxical behaviours abound due to the thoroughly degenerate geometry. It is not surprising to find that whole books, often quite loosely argued, are written in this difficult, special field. Discussion for example about the ordinary action of a nut upon a bolt is forbidden in the field of planar mechanism for that is a spatial matter, and such limitations often render the whole subject bewildering.

31. I am proposing now, for the sake of discussion in the following paragraphs, to take the spatial 4-link loop of mobility unity in figure 1.11(*a*), and to alter the construction of its members in such a way that the mechanism becomes planar. Please notice in figure 1.11(*a*) that $\Sigma f = 7$, and that, due to the prevailing 'spatiality' of the movement, all of the available free-

Figure 2.01(*j*) (§ 2.29) Another 4*R*-loop exhibiting sphericallity. The mechanism is a crank rocker, but notice that the energy input is occurring at the joint at *B*, which is remote from frame. Read about actuated joints at § 10.61.

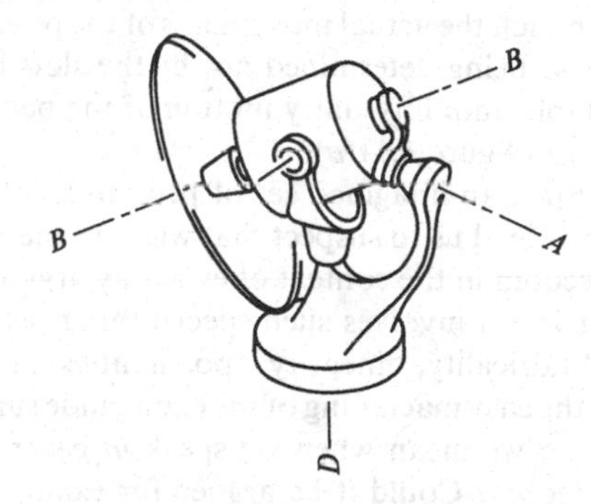


Figure 2.01(*k*) (§ 2.29) An overconstrained, 3-link, spherical mechanism whose apparent mobility is unity. Its geometrical speciality is the continuous line contact (and this is accurately achievable) between the sets of contacting teeth. Refer also § 22.01.

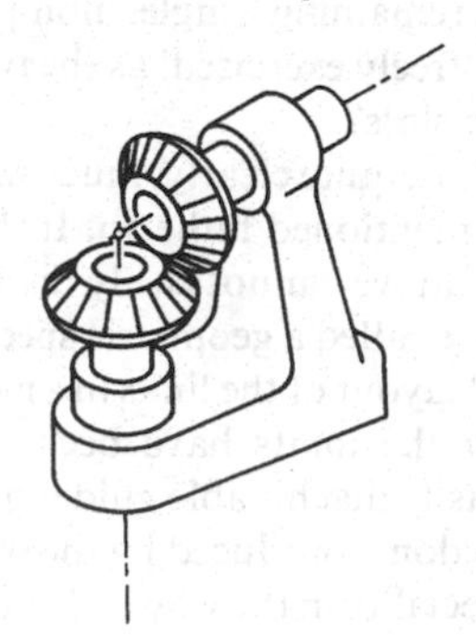

doms at the various joints are being fully exercised. In view of my own remarks at § 1.51, however, what exactly do I mean by this?

**More about passive freedoms**

32. What, for that matter, do any of us mean when we say for example in figure 1.11(*a*) that all of the three freedoms at the ball and socket at *B* are being exercised? In the context of such machine-manufactured joints as these, which are, geometrically speaking, 'highly special' joints, we mean that we have arbitrarily and somewhat thoughtlessly chosen three intersecting axes fixed in the socket and passing through the centre of the ball and can see that, to get from one location in the socket to the next, the ball must suffer a twist (of zero pitch) about each of these three axes (§ 19.28). But how should we consider the matter if the joint at *B* were not such a tidy spherical thing, but some force closed joint like the joint between the stool and the undulating floor in figure 1.10(*e*), which is also an *f*3-joint? Or, what is even more confusing, how would we consider the same question, that of the geometrical separability of the freedoms at the joint, if the joint at *B* in figure 1.11(*a*) were a specially constructed joint like the joint in figure 7.11? We must imagine for this particular question that the two curved slots in the socket at *B* have been cut exactly to match the actual movements of the pegs in the sockets these being determined not by the slots but by the repeatable, mobility-unity motion of the particular mechanism in figure 1.11(*a*).

33. Such an imagined set of pegs and slots at *B* would surely lead us to suspect that what we mean by a passive freedom in the context of everyday argument is something which involves such special things as circularity, cylindricality, linearity – possibilities in other words for the easy machining of the extra guide surfaces. But what do we mean when we speak *in general* of a passive freedom? Could it be argued for example that the curved guide surfaces specially cut in the socket at *B*, as explained above, are 'extra guide surfaces' provided by the designer for the 'locking in' (§ 2.17) of all of the passive freedoms there, namely two of them, leaving aside the one remaining single, non-passive freedom which is being 'freely exercised' as the two pegs traverse their respective slots?

34. There is, quite clearly, much more to all of this than has been mentioned hitherto. It should indeed be clear by now that we cannot easily distinguish between (*a*) what we have called a geometric speciality in the way of some special layout of the links in a mechanism where some or all of the joints have been modified by the provision of easily machinable guide surfaces to lock in the passive freedoms produced by the speciality; and (*b*) a geometric speciality in the way of some joint with extra guide surfaces specially constructed to suit exactly the particular geometrical movements of its parent mechanism. Both of these kinds of special circumstance accord with the definitive statements made at § 2.07 and § 2.17.

35. With all of this complicated matter and at least some of its inherent dangers clearly in mind, we can now say with a clear conscience that, if the two hinge axes at *A* and *D* in figure 1.11(*a*) were made parallel, and if the centre line of the slot at *C* and the centres of the balls at *B* and *C* were all made coplanar and in a plane normal to the axes at *A* and *D*, then the balls at *B* and *C* would have all of their freedoms relative to their respective sockets, except for their respective single freedoms to rotate about axes parallel with the hinges at *A* and *D*, rendered passive. There would be two freedoms rendered passive at *B*, and one at *C*. Having arrived at such a special circumstance, we the designers might replace the balls and sockets with ordinary hinges set parallel with the hinges at *A* and *D*, and thus effectively lock in the passive freedoms there. Refer to figure 2.01(*m*), which shows an example of the planar, 4-link, 4*R*-loop.

36. Having done such a thing as this we might next argue that, because we are now working (maybe thankfully) in two dimensions only, where the total degrees of freedom of a free lamina is only 3 not 6, a suitably simplified criterion applies which says, using the multiplier 3 not 6, that the mobility *M* can be written, $M = 3(n-g-1) + \Sigma f$. A person is quite free to do this, and to hold that such a criterion remains available and useful for all mechanisms which are undeniably planar or spherical, and thus to conclude among other things that the general mobility of the

Figure 2.01(*m*) (§ 2.35) The planar, 4-link, 4*R*-loop. This figure is mentioned again at § 17.02, where, in respect of the joints in spatial mechanism, the related questions of clearance and backlash are introduced.

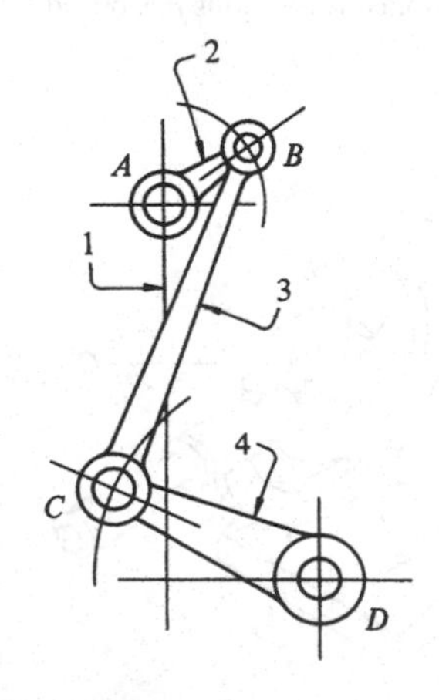

planar 4-link 4*R*-loop is unity. It will be needless to say however that any general theory of mechanism based upon such flimsy foundation will be doomed to being (at the very least) shot through with ambiguity, contradictions, and paradox. Grübler (1917) is said to have been the first to recommend this mobility criterion for planar mechanism, and he did so in these or similar terms.

**On the question of complex joints**

37. I move now, at last, to the question of the complex joint. I wish to note first of all that the adjective 'complex' as used in the present context is in no way related to the same word which is used as a noun, which appears for example at chapter 9. We can simply say, by way of definition that *if there are one or more other members intervening between the two main, or reference members of a joint, the joint is a complex one.* A very simple example has already been shown in figure 1.03($g$) and mentioned at § 1.47. It is the well known universal, the Hooke, or Cardan joint, where the cruciform piece 'intervenes' between the two main members of the joint which are the shafted clevises; look also at the stylized figure 2.01($h$), where the topological arrangement of the three members of this joint can more clearly be seen. If the intervening members of a complex joint are joined together in this way, that is, by a series of simple joints 'in series', and this is more blatantly suggested next by figure 1.10($c$), the freedom of the complex joint will, in general, be the sum of the intervening joint freedoms, *or six, whichever is least.* When we look at a simple joint of freedom $f$ already excised from its mechanism (§ 1.28), we see that it is by itself a mechanism, with $n=2$, and $g=1$; and the mobility of the excised joint is, from (1) at § 1.34, $M=\Sigma f$, which is simply $f$. We can similarly calculate for a complex joint of the 'series' kind, an already mentioned example of which appears in figure 1.10($c$), that its mobility is $\Sigma f$, which is, in the mentioned case, 13. However the *freedom* between the cylindrical knob at the bottom of the open chain in figure 1.10($c$) and the flat slab at the top of it, which is the same thing as the freedom of the complex joint between those two main members of the joint, is not 13 but only 6.

38. Apart from the naturally accepted fact that the freedom of a joint can never exceed six (§ 1.28), there are other, more subtle reasons for the fact that the freedom of a complex joint may often be less than the mobility of the joint when excised and seen as a mechanism. Take the cases of the two complex joints excised and hanging at ($a$) and ($b$) in the figures 2.02. They are both of the series type, and their mobilities $M$ are 6 and 4 respectively. *Their freedoms f, however, are only 5 and 3 respectively.* In both of the cases there is an axis with a pitch (the pitch is zero) about which the member 2 may screw with respect to the member 1, and that screw axis is common to the two 'screw systems' enjoyed respectively by the two series-connected, simple joints. Here we are blundering into a matter beyond the scope of the present chapter however. It is discussed by Waldron [2], and well-enough explained, the writer hopes, in the chapters 7 and 23 (§ 7.78).

39. The idea of connectivity was first mentioned at § 1.31; and the stylized figures in ($c$) and ($d$) of the figures 2.02 show next how it is, in general, that any mechanism can be seen as a complex joint between any pair of its members. It will be seen accordingly that *in any mechanism, the connectivity between a pair of its members is the same thing as the freedom of the complex joint between them.* In the two examples shown in ($c$) and ($d$), the mechanism is the same 4-link *RSSR*-loop, whose mobility $M$ is 2; but the freedoms of the complex joints in ($c$) and ($d$), namely the connectivities between the reference links at these two joints, are different; they are, respectively, 1 and 2.

40. We appear to have concluded here the following: *any joint can be seen as a mechanism and any mechanism can be seen as a joint.* We have distinguished by careful definition however between the mobility $M$ of a mechanism and the freedom $f$ of a joint, and those definitions hold (§ 1.21, § 1.26). Without confusion, therefore, the above identity can easily be accepted. In the context, however, where the idea of mobility is

Figure 2.02 (§ 2.38) Each of these assemblies is a complex joint between the roof (the member 1) and the block (the member 2) of the joint. Although drawn in a stylised manner and thus appearing flat, the drawings all refer to spatial arrangements of the component parts and thus of the simple joints (§ 2.39, § 2.40).

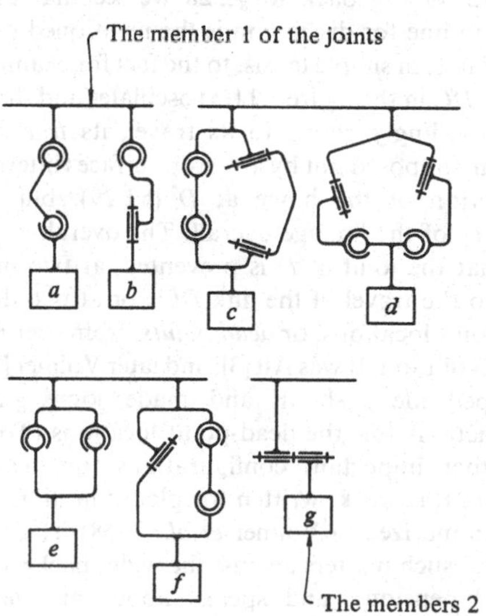

confused by overconstraint, and the idea of freedom is confused by what we might call 'complexity', it will nevertheless be useful to notice that, whereas a noticed apparent mobility $M_a$ in a mechanism (due to the occurrence of a geometric speciality) is always greater than the general mobility $M$, *the freedom f of a complex joint (due to a variety of complicated reasons) may be greater than, equal to, or less than the general mobility M of the mechanism of the joint*. Take a look in figure 2.02 first at the item (*e*), where $M$ is zero but $M_a = 1$ and $f = 1$, and then at item (*f*), where $M = 1$ and $f$ is zero.

41. Going back to §1.38 where the relatively unimportant idea of linkivity was first proposed, it can now be seen that the following definition might be useful: *the linkivity of a joint is that number of those members considered by the analyser or designer to be the members of the joint which is under consideration.* With this definition, the linkivity of all simple joints becomes 2, that of the Hooke joint 3, and that of the joint in figure 2.02(*d*) for example, 4. The concept of linkivity is clearly a reasonable one, but is it useful? Could it be used, for example, to clarify the case of the very commonly seen joint in figure 2.02(*g*), or do we simply say, referring to the figures 1.03 again, that the number of links $n$ in this 3-link loop (of mobility 2) is 3, and get on with the job in some other way of discovering why it is true what we know to be true, namely that the freedom $f$ of the complex joint in figure 2.02(*g*) is only unity? By such unanswered questions as these I openly imply (*a*) that there are quite difficult questions to be encountered in quite simple mechanism, and (*b*) that those questions are not all answered adequately by the elementary ideas available here in this introductory chapter 2 (§7.78).

### Dead-point and change-point configurations

42. Going back to §2.28 we see that the next matter in line for discussion is the mentioned item (3). This relates, in simple terms, to the fact for example that the link $DC$ in the figure 1.11(*a*) oscillates and that there are, accordingly, limits to its travel, its *travel limits*. These are imposed not by any stop surfaces interrupting the motion of the hinge at $D$ (§1.29), but by the geometry of the linkage overall. The overall motion is such that the joint at $D$ is prevented, at two different limits to the travel of the link $DC$, or at two different dead-point locations, or *dead-points*, from ever making a full revolution. It was Alt [3], and later Volmer [4], who developed ideas about, and made some graphical constructions for, the dead-point locations (Totlagen) and other important configurations for the planar 4$R$-loop; this work, written for planar motion only, is well summarized in Volmer *et al.* (1968). Hunt (1978) discusses such matters against the wider background of three dimensions, and speaks about his *stationary configurations*. This form of words is, in many respects but not entirely in my view, sufficient to cover all of the corresponding circumstances in that area (§18.01).

43. Now if $\theta$ were an input angle at $A$ in figure 1.11(*a*) and $\delta$ the output angle at $D$, an equation of the form $\delta = f(\theta)$ would be called a *closure equation* for the linkage (§19.47). With this in mind, please think about (*a*) how the dead points could be discovered algebraically, first by differentiating with respect to $\theta$ the closure equation; (*b*) the possibilities for static force analyses (§12.54), and, with a couple acting not at $A$ but at $D$, questions of stability and instability of the linkage at its dead-point configurations (§18.02); (*c*) why this mechanism can nevertheless be used successfully for the treadle drive to a spinning wheel or a sewing machine; (*d*) whether or not the backlash in the linkage due to clearances at the joints would be greatest at the dead points or at those configurations where the transmission angles at $B$ and $C$ are the least (§17.19). Such considerations and a host of others are interesting and important, but unrelated to mobility. The mere fact that a mechanism cannot be moved under certain circumstances – try for example to drive this mechanism, with no initial motion, from $D$ at a dead point – does not mean that its mobility might not be unity or even greater.

44. It should be clear on the other hand that what happens at a *change point* in the motion of mechanism *is* related to the question of mobility. Hunt (1978) speaks about *uncertainty configurations* here, but both of these terms refer to those configurations which can occur in certain linkages where, at the instant, the linkage can change from one of its modes of operation into another (§1.45). At a change point (or uncertainty configuration), the configuration is most often such that extra mobilities $M_s$ become apparent transitorily; these mobilities, which are special in the sense that they are always produced by the sudden occurrence of some geometric speciality (§2.11), need to be handled very carefully. We shall consider now a pair of simple but confusing examples. The first of the pair relates to a highly special case of the matter alluded to at §2.43, while the second relates to the matter mentioned here and deals with a mechanism of mobility zero. Both examples are dealt with in a somewhat pragmatic way, with no cohesive philosophy in evidence; but please refer, in due course, to chapter 18.

### Worked exercise on figures 2.03 and 2.04

45. Worked exercise 1: Comment upon the following facts: (*a*) in figure 2.03 there is a 4-link loop in which, although $\Sigma f = 7$ and $M$ is unity, motion is impossible; (*b*) in figure 2.04 there is a 4-link loop in which, although $\Sigma f = 6$ and $M$ is zero, motion can occur; (*c*) there are special conditions prevailing in both

cases which allow these anomalies to occur, and they are: (*d*) in figure 2.03 the coupler *BC* is only as long as the shortest distance between the two crank circles at *B* and *C*, coaxial respectively with their revolute axes at *A* and *D*; (*e*) in figure 2.04 there is a straight line which can be drawn from the centre of the ball at *B* to cut all three of the revolute axes at *A* and *C* and *D*. We can comment as follows. With respect to the mechanism in figure 2.03, refer first to figure 1.11(*a*) where, for an input angle $\theta$ at *A*, and an output angle $\delta$ at *D*, a plot of $\delta$ against $\theta$ for this 'crank rocker' mechanism – the shaft at *A* can make a complete revolution while the shaft at *D* cannot – will appear typically as is shown at the inset. If on the other hand the linkage is so proportioned that the motion is of the 'double rocker' kind where only oscillations can occur at *A* and *D*, the corresponding plot will have the appearance of the loop which is shown at the inset in figure 2.03; if further the linkage is so proportioned that the limits to the oscillatory motions (namely the dead points) of the cranks at *A* and *D* are brought so close together that the ranges of variation for $\delta$ and $\theta$ both disappear, the loop in the plot becomes a point; this is the general condition prevailing which is mentioned at (*d*) above; refer to chapter 18.

46. Worked exercise 1, continued: With respect to the mechanism in figure 2.04, refer first to the figures 1.08 and 1.09, and note that, given that the line *A–B* is a hinge, these two figures deal with the same 4-link, *RRRS*-loop as the one which is shown in figure 2.04, which is, in general, a statically determinate structure. We shall show however that, in figure 2.04, the circle described by the centre $B_2$ of the ball (§ 1.42) is tangential to the skew torus described by the centre $B_3$ of the socket (§ 1.43), and thus that, transitorily (§ 2.08), $M_a$ is unity and motion is possible. Disconnect the spherical joint at *B* and consider separately the possibilities for instantaneous motion first of the point $B_3$, then of the point $B_2$. Keeping the hinge at *D* inoperative, note that an infinitessimal displacement at $B_3$ must be perpendicular to the broken line *B–x*. Releasing now the hinge at *D*, but keeping next the hinge at *C* inoperative, note that a similar displacement at $B_3$ must again be perpendicular to *B–x*; these two displacements at $B_3$ are of course in different directions, but together they define the tangent plane to the torus there. Now notice that, by virtue of the same line *B–x*, an infinitessimal displacement along its circular path of the point $B_2$ must also be perpendicular to *B–x*; the circle is thus tangential to the torus. Observe that the 'extent' of the motion in the two directions away from the assembly configuration of the mechanism and the 'sogginess of the limits' to the motion in the different directions will depend, not only upon the clearances at the joints and the flexibilities of the links, but also (and more importantly) upon the difference between the changing curvatures of the circle and of the surface of the torus at the point of tangency. It should be clear in conclusion that, whereas the special condition in figure 2.04 mentioned at (*e*) above does constitute a geometric speciality, the special condition in figure 2.03 mentioned at (*d*) above does not (§ 10.57).

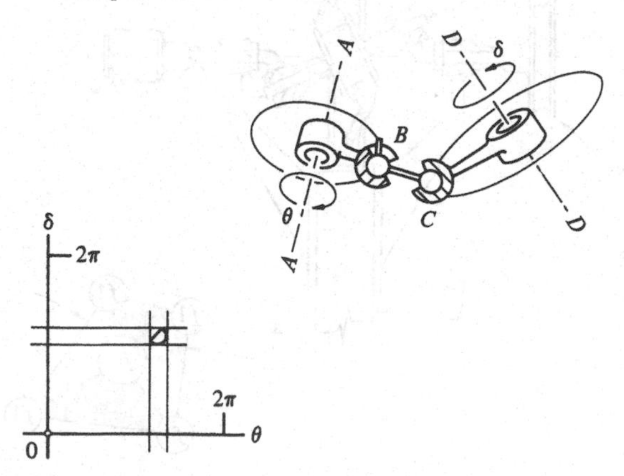


Figure 2.03 (§ 2.45) In the 4-link mechanism shown, $\Sigma f$ equals 7, and the mobility *M* is unity. Special proportions of the links restrict the limits of travel, however, and the mechanism cannot move. Refer to chapter 18.

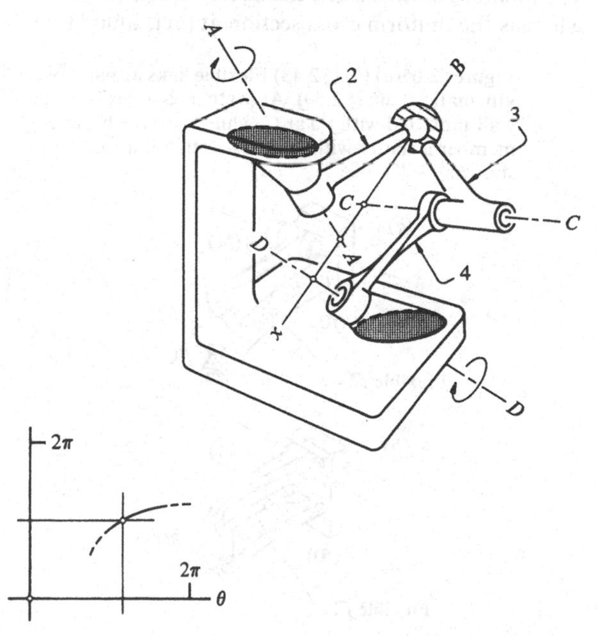


Figure 2.04 (§ 2.46) In the 4-link mechanism shown, $\Sigma f$ equals 6, and the mobility *M* is zero. There is a geometric speciality inherent in the configuration at assembly, however, and a transient apparent mobility $M_a$ of unity obtains (§ 2.45, § 3.43, § 10.57).

**On the question of flexible links**

47. The difficulty in dealing with flexibility derives at least partly from assumptions made about the absence of it. We can however separate (*a*) 'ordinary' links, whose elastic deformations under ordinary working loads are of the same order of magnitude as the clearances at the joints; and (*b*) 'flexible' links, which are so proportioned that they can suffer, by intention, gross elastic deformations of one or another kind. On considering such a flexible link, one can always imagine two rigid bodies of suitable shape and convenient dimensions embedded at two separate neighbourhoods within the link, and then nominate the *f*-number that (due to the link's capacity to suffer the gross deformations) can be seen to exist between the neighbourhoods. This means in effect that any zone of a flexible link can be characterized by the *f*-number of what might be called the *distributed joint* between a pair of neighbourhoods nominated within the zone. If the two neighbourhoods are chosen to include respectively the rigid joint elements at two existing joints upon the link (§ 1.11) we could next, in effect, break that zone of the link into two pieces and join them together again with a replacement equivalent joint with that *f*-number characteristic of the zone. In this way we can begin to speak intelligently, not only about flexible links whose jointivity may be only 2 or maybe more (§ 1.38), but also about the mobility of mechanism containing flexible links.

48. Take for example the two flexible links shown stippled which appear at the figures 2.05(*a*) and 2.05(*b*). The jointivity of each link is 2 (they are both binary links), and in each case the link is one of a 3-link loop. The links are both of some elastic isotropic material but, whereas the uniform cross section at (*a*) is annular and the link is tubular, the uniform cross section at (*b*) is shaped like a dumb-bell, and the link is flat. At (*a*) there are two freedoms to suffer a gross deformation; they are a pair of unrelated bendings about, say, the horizontal and the vertical neutral axes respectively of the cross section of the link. At (*b*) there are also two such freedoms; in this case however they are, first, a freedom to bend about the horizontal neutral axis of the dumb-bell section and second, a freedom to twist in torsion about a central longitudinal axis of the link. It will be observed, of course, that at (*a*) the link is stiff against torsion, and that at (*b*) the link is stiff against bending about a vertical axis, so that these particular capacities for relative motion between a pair of neighbourhoods within the links are negligible. In each case, accordingly, both in figure 2.05(*a*) and figure 2.05(*b*), the appropriate *f*-number for the distributed joint, which is characteristic for the whole of the flexible link, is *f* 2.

Figures 2.05(*a*) (*b*) (§ 2.48) Flexible links as parts of a kinematic chain (§ 2.50). At (*a*) there is a flexible tube wall-mounted with *f* 0 at *C*, while at (*b*) the flexible member is flatter with its cross-section shaped as shown.

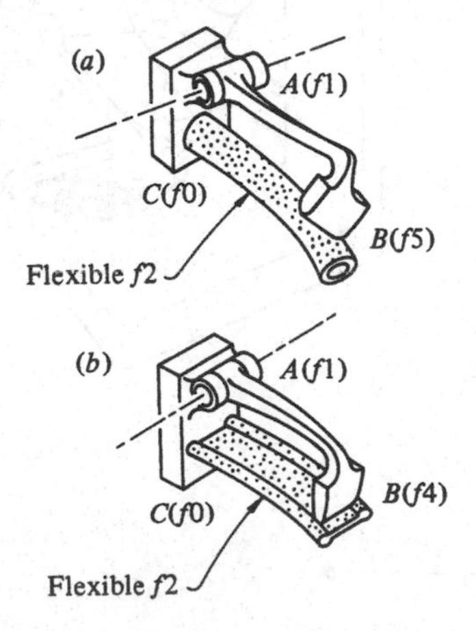


**Worked exercise on figures 2.05 and 2.06**

49. Worked exercise 2: Comment in turn upon the mechanisms shown in the figures 2.05(*a*), 2.05(*b*), and 2.06. The first two of these are well enough described in the previous paragraph, and a description for figure 2.06 follows. The stippled area in figure 2.06 is the diametral cross section of an annulus-shaped flexible piece 5, which is firmly attached at its outer circumferential periphery to the inside of a circular hole in the fixed wall 1 and at its inside hole to the outside of a central boss

Figure 2.06 (§ 2.51) This mechanism permits the insertion of a rotary drive through a flexible gas-tight seal in a wall. It relates to the worked exercise 2 beginning at § 2.49.

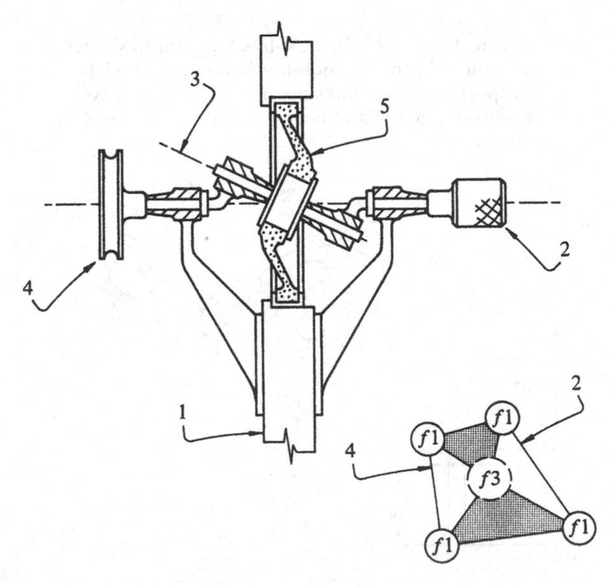

upon the nutating shaft 3. A cranked shaft 2, mounted on fixed bearings and running concentric with the circular hole in the wall, delivers its rotary motion via two long collinear hinges on 3 to the similarly cranked shaft 4, which is mounted collinear with 2 but at the other side of the wall. The whole arrangement is point-symmetric about the centre point $Q$ of the central boss on 3, and the continuing motion is such that the centre line of the straight shaft 3 is caused to generate a right circular cone whose vertex is at $Q$. The purpose of the mechanism is to offer a gas-tight seal at the wall which is non-rubbing and which does not interrupt the transmission of a rotary motion through the wall.

50. Worked exercise 2, continued: Given the remarks already made at § 2.47, the mechanism in figure 2.05(*a*) is a clear example of a differential loop where $\Sigma f=8$, and where the mobility $M$ is accordingly $+2$. At the $f5$ joint at $B$ there is, for every angular input at $A$, a single infinity of positions for the outer end of the tube. There is, however, for every input at $A$, a 'central' position along the sloping, straight, knife-edged knob at $B$ where the overall bending of the flexible link is least, and where (in the absence of friction) the flexible link will be in stable equilibrium. Moving to the loop in figure 2.05(*b*), however, we see that, if either or both ends of the horizontal, straight, knife-edged knob at $B$ are designed to make contact with the ridges at the sides of the flexible strip, there would be an overconstraint in attendance; and this would be, probably, not only unnecessary, but also the likely cause of mechanical trouble in the circumstances. If the joint at $B$ were to have been a ball against a plate ($f5$) centrally arranged, the bending freedom would be exercised but the torsion freedom, free to be exercised, would, due to symmetry, remain passive. The equilibrium with respect to torsion, however, would be unstable, so the provision of a knife edge at $B$ (two spots of contact along the knife edge instead of only one at a central point) would safely lock in this passive freedom. The non-movement of the imagined $f1$ hinge for torsion, longitudinally arranged within the link, would be obliged thereby to remain non-moving, namely passive, and this, in the likely practical cricumstances, would not be unreasonable. It should next be clear however that any intended contact between one or other end of the knife-edged knob and the ridges of the flexible link – contact could occur at one end only unless the material of the links is crushable – would cause an overconstraint without which the mechanism would be a better one. In figure 2.05(*b*) as drawn, $\Sigma f=7$, $M$ is unity, but, by virtue of an exact symmetry which would be achievable only with great difficulty in ordinary practice, we could have actual contact with the ridges (one at a time of course), overconstraint, and an apparent full cycle mobility $M_a$ as $M$ was before, namely unity. The mechanism is planar.

51. Worked exercise 2, continued: We can comment on the mechanism of the gas tight seal, shown in figure 2.06 and described in § 2.49, as follows. Leaving aside the flexible diaphragm 5, we see that the remaining rigid parts of the mechanism make up a $4R$-loop which is not only special by being spherical, but doubly special in that the four hinge axes are collinear in pairs in such a way that they are, as well, coplanar. Although $\Sigma f$ is only 4 for the rigid parts, and $M$ is accordingly only $-2$, the apparent mobility $M_a$ is $+2$. This becomes clear when we see that, independently of the smooth transmission of the rotary motion from 2 to 4, the shaft at 3 and its boss is able freely to rotate in its bearings and thus to rotate about its own nutating axis. Let us say that the diaphragm is stiff in torsion against a rotation of its inner periphery or ring against its outer periphery or ring and thus that its characterizing $f$-number is at most only 5. It follows that, given a steady rotary input at 2, insertion of the diaphragm into the mechanism ensures a steady rotation (relative to its bearings) of the shafted boss at 3. This in turn ensures that (relative to the wall) the angular velocity of the boss is always in a direction along the intersection of two known planes, the plane through $Q$ of the wall, and the plane through $Q$ containing the centre lines of the four revolute axes (§ 12.12).

52. Worked exercise 2, continued: If next, due, say, to some spokes, the diaphragm in figure 2.06 is held to be stiff in compression against a radial displacement of its inner ring against its outer ring, its characterizing $f$-number will be seen to be at most only 3. These three, of the inner ring with respect to the outer ring, can be seen to be, (*a*) two independent freedoms to rotate about a pair of intersecting diameters; and (*b*) one independent freedom to plunge bodily along the undistorted centre line. It can now be seen that, upon insertion of the diaphragm, even more overconstraint is introduced into the already overconstrained mechanism: the centre point $Q_3$ of the boss will not be exactly coincident with the centre point $Q_1$ of the hole, and the inner and the outer rings of the undistorted diaphragm will not be exactly concentric either. Even so we might argue next that the two freedoms mentioned at (*a*) above will duly be exercised, while the single freedom mentioned at (*b*) will be passive. This mechanism is a very cunningly compacted confusion of freedom from overconstraint on the one hand, and locked-in passive freedoms on the other, and there is not much advantage to be gained in applying the general criterion to it. It is, in the sense of § 1.36, a mixed mechanism of more than one loop which cannot be handled easily by the mere blanket application of our primitive formulae. Note the kinematic

chain in figure 2.06, and note that, given this, $M = -5$. $M_a$, of course, is unity. An exercise for the reader: what kind of joint with *f*3 (made of rigid parts and either simple or complex) can be imagined to replace exactly the mechanical action of the diaphragm as here explained?

**Intermittency in mechanism**

53. It was mentioned at §1.30 that, although by definition there is no such thing as a joint with a fractional *f*-number, there are in ordinary practice a host of working joints whose *f*-numbers change, from time to time and in a catastrophic way, from one integer value to another. It should be clear by now that whenever the functioning of mechanism requires the sudden making or breaking of contact between a pair of links the mooted phenomenon occurs, and we have what might be called *intermittency*. The topology of the intermittent mechanism must be seen to require at least two different kinematic chains for its clarification, or the single kinematic chain must be seen to include a number of special joints whose *f*-numbers change intermittently. One might distinguish the *repeating* intermittency, at escapements and ratchets and so on in clocks and winches and so on, and the *non-repeating* intermittency which occurs so often in nature and in casual mechanisms such as those of the figures 1.10. Printing, textile, and engine machinery abounds in repeating intermittency. Occasional as distinct from regular intermittency is also widely employed in those most ancient of man-made machines, locks and traps. A long series of synthetic leaps, each following a sudden insight or a flash of imagination, surely lies behind the simplicity and elegance of the best of these devices.

**Worked exercise on the figures 2.07**

54. Worked exercise 3: Describe the stages of operation of the domestic mousetrap, paying attention at each stage to the changing nature of the joints and the kinematic chain. Refer to the series of figures 2.07. In the set configuration (*a*) there is a force closed *f*5 joint at *B*, and two force closed *f*4 joints at *C* and *D*. See the *f*4 joint in figure 1.10(*c*). The mobility at configuration (*a*), from §1.34, is +3; and these three mobilities can be argued to be residing at the 'input' link 3. But how can we describe the three parameters? One sliding and two rollings, or two slidings and one rolling, or what? There is in fact no simple way, by means of simple length and angle parameters, to speak separately for example of translations of 3 along, and rotations of 3 around, one or another of a triad of Cartesian axes fixed in the frame 1 (§1.51, §4.02). Soon after the bait is depressed, one of the originally made, two points of contact at *D* is broken, the joint at *D* becomes *f*5, and the pierced lug 4 becomes suddenly free to slide towards the tip of the member 3. Refer to the configuration (*b*). In this configuration the mobility of the mechanism, again from §1.34, is +4, and, whereas the location of the 'input' link, the link 4, is easy to nominate by means of a pure angle parameter at the pivot *E*, the consequent freedom of the link 3 with respect to frame, its connectivity (§2.39), which is again *f*3, is again impossible to describe by means of pure length or angle parameters only. Upon release of the member 3 at *D*, the joint at *D* becomes *f*6, and the new linkage *ABC* begins to operate, see (*c*). The mobility of the mechanism at (*c*) is again +4. At (*c*) the location of the 'input' link, link 2, is easy to nominate by means of a pure angle at the pivot *A*, but again the consequent connectivity of 3 with respect to frame (which is again *f*3, assuming that both points of contact remain extant at *C*) is impossible to nominate by means of pure length or angle parameters only. Next, see (*d*), a 2-link open chain goes into operation until, see (*e*), closure upon the mouse 5 takes place to complete a new and final kinematic chain. Being of a biological nature (§2.64), the joints at *F* and *G* are open to wide interpretation. It is however not unreasonable to argue that the final linkage, *AFG*, is a 3-link loop of mobility zero.

Figure 2.07 (§2.54) Intermittent events in the cycle of operation of a domestic mousetrap. The kinematic chain suffers a series of catastrophic changes. These are displayed at the inset, right.

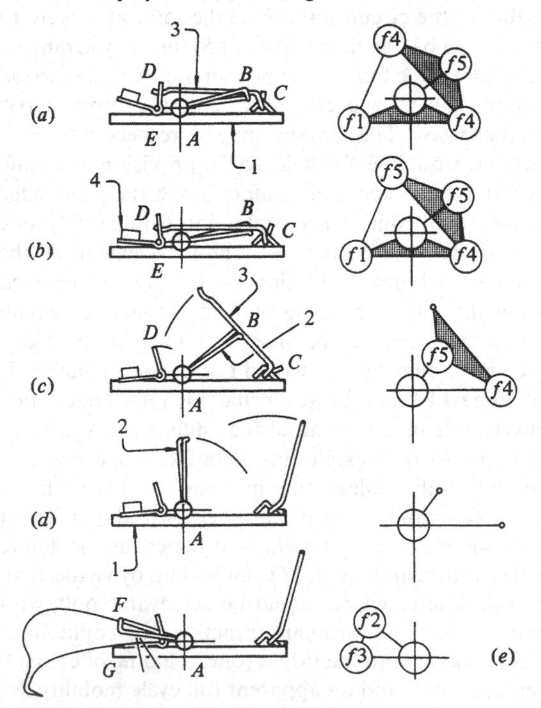

### The inevitable idea of a twist about a screw

55. The mousetrap exercise at § 2.54, although designed primarily to exemplify the various phenomena of intermittency, also throws into clearer focus the impossibility of describing the instantaneous capacity for motion of many of the members in practical mechanism by exclusive use of the simple, uncluttered parameters, namely pure linear, and/or pure angular, finite or infinitesimal displacement. The same thing can clearly be seen at the simple cases (*c*) and (*d*) among the complex joints at the figures 2.02. Although at (*c*) in figure 2.02 the connectivity of the joint is only 1, at (*d*) in figure 2.02 it is only 2, and at (*a*) and (*b*) and (*c*) in figure 2.07 the connectivity under consideration is in each case 3, in none of these cases is it possible, in general, to effect a pure translation or a pure rotation of the moving member with respect to the fixed member. In all of the six cases quoted the two kinds of 'pure' movement must be combined at the instant inextricably with one another, like they are, for example, when we consider the capacity for motion of an ordinary movable nut engaged upon the thread of a matching fixed bolt. It will be shown in chapter 6 that, whereas in the first case there is only one way for the member to 'twist about its only available screw' ($\infty^0$ being, in this case, unity), and whereas in the second there is an $\infty^1$ ways of doing that action, there is for the member 3 of the mousetrap, under the conditions specified, an $\infty^2$ of ways to twist about the 'available screws' at the instant under consideration. There is clearly much to be said about the theory of screws which has not been mentioned hitherto, but please be referred again to § 1.25, and wonder whether the 'purest' kind of motion is neither translation nor rotation, but a 'screwing' kind of motion, and please be aware that, merely upon the elementary bases of these chapters 1 and 2, the following worked exercise about a closed loop of ordinary screw joints is quite explicable.

### Worked exercise on figure 2.08

56. Worked exercise 4: The apparatus in figure 2.08 consists of an ordinary bolt 7 rigidly attached at the extremity of the seventh link 7 of an open chain of seven links connected by six screw joints whose locations and pitches are variously arranged. The frame link 1 of the open chain has rigidly attached at its extremity a split nut 1 which is matched in size and pitch with the mentioned bolt. The questions are (*a*) is it possible for the bolt to be brought by hand into mating contact with the waiting nut; (*b*) if that action can be achieved, what is the mobility of the closed loop resulting; (*c*) given this, why is it that, after fixing six parameters by choosing in effect six suitable angular insertions at the six screw joints and thus 'fixing' the bolt in Euclidean space, the bolt can still move by twisting upon the screw (namely by screwing upon the thread) of its nut? Some suitable answers are offered in the following paragraph.

57. Worked exercise 4, continued: The offered answers are (*a*) yes it is possible because, by virtue of the six *f* 1 joints in series (§ 2.37), the bolt has six degrees of freedom with respect to the nut, and can thus be put (either head to rear or head to front, and here we might begin to think about the design of industrial manipulators) in the open nut; (*b*) by the criterion for mobility of § 1.46 the mobility $M$ of the resulting loop is unity, and this means that the loop can freely traverse its continuous single infinity of configurations in whichever mode it happens to have become assembled (§ 1.45); there are, of course, other modes of assembly but we cannot say how many for we do not know how many real roots there are of the closure equation for this 7*R*-loop (§ 2.43, § 1.53); we have not written the closure equation, and certainly have not solved it (§ 19.48); we don't even have a clear idea of this very difficult problem geometrically; (*c*) the bolt is able to twist upon its screw in the nut because, among the $\infty^6$ of different locations it was free to occupy before it was put in the nut (§ 4.03), there is a single infinity of different locations it can occupy in the nut; for each of these locations in the nut there is one or more different configurations of the loop (§ 1.45) and it will not be surprising to find that the linkage can move smoothly in its mode from one configuration to the next; various discontinuities can occur in the motion (chapter 18), but such matters are too difficult for consideration here. All this means, in effect, that merely to put the bolt somewhere in the nut requires the fixing of only five parameters. We could, in other words, first render inoperative any one of the

Figure 2.08 (§ 2.56) The question here is whether or not the seventh bolt can be manoeuvred into a mating contact with the open, seventh nut. The relevance of this apparatus to the many problems of robotics is intended to be obvious. Please refer to § 1.46, § 1.53, § 2.43, § 2.59, and § 4.03.

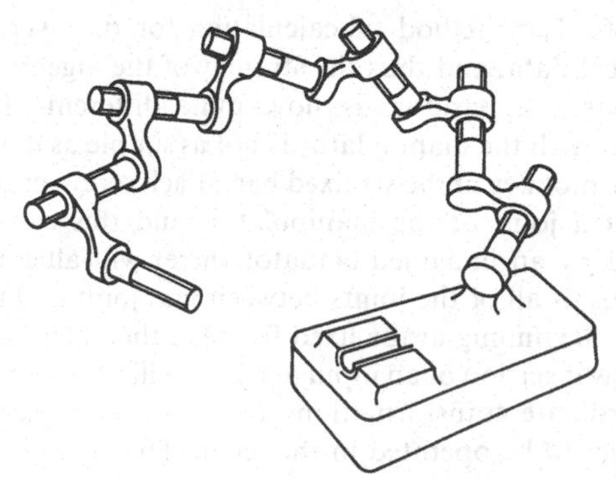

screw joints of the chain, then still be able to put the bolt somewhere in the nut.

### The different arrangements of six screws

58. At § 1.52 I introduced the idea of there being six independent screws available for the displacement, by six successive screw-movements, of a body from one location to another. Let there be – and for the sake of drama here, I speak fancifully – a monkey in a cage. Let there be six fixed bars in the cage set generally, at various angles in space, and let the bars be threaded, each with some arbitrarily chosen pitch. Imagine that, when the monkey swings one-handedly on a bar, he swings like a rigid body in a screw-movement whose pitch is the pitch of the bar. I wish to make three assertions: (*a*) that, by using the bars in turn, and by choosing judiciously the magnitudes of the six successive screw-movements, the monkey can displace himself from any original location to any chosen new location in the cage; (*b*) that the order in which he chooses the bars will determine the screw movements; but (*c*) that, if we wished to make some numerical calculations about the additions of the screw-movements, the algebra would not be unduly difficult (§ 19.28). All six of the threaded bars would remain fixed with respect to one another and to the cage; the main frame of reference would never need to alter.

59. Such a fixed arrangement of six screws fixed in space is not a familiar arrangement in ordinary engineering practice however. In some devised, monkey-displacing machine, for example, the six screws for the six screw-movements would be, most likely, in the form of six screw joints arranged in series. Some of them would be of zero pitch perhaps and some of them may intersect and thus be combined into cylindric or ball or other joints; but the joints would be in series, for otherwise the construction of any imagined apparatus would be exceedingly and quite unnecessarily complicated. Refer to figure 2.08: replace the last free bolt of the open chain by a mechanical hand, the hand to hold the monkey. Now see the apparatus there as a generalized manipulator with six degrees of freedom at the gripper.

60. The method of calculation for discovering numerical data, and thus the strategy of the algebra at the devised apparatus, is now quite different. The problem with the manipulator is not as simple as it was with the monkey at the six fixed bars. Each screw-movement at a joint of the manipulator (and this can be inserted by an imagined actuator there) will alter the locations of all of the joints between that joint and the gripper. Beginning at the fixed frame of the apparatus, each new insertion of an input at a joint will require a set of coordinate transformations for each of the screw joints yet to be operated in the chain. The chain is an open chain and, although the order of the inserted inputs is not important once they are determined, the complications of the algebra needed to determine the inputs required for a given displacement from one location to another at the gripper will be evident.

### Manipulators and robots

61. Some of the algebraic methods which can be used in the design of and for the control of manipulators and robots are mentioned and discussed at chapter 19. Here I would like to say the following. As soon as a manipulator arm, attached at its frame, has grasped with its gripper a rigid body whose location is known with respect to the said frame, there is a kinematic chain existing which is a closed loop. If the body about to be grasped by the gripper is an axi-symmetric one, such as a rivet, and if the grasping is to occur at some required location upon the body, say at the shank and directly under the head of the rivet, the closed loop consisting of the arrived manipulator arm and the grasped rivet can be seen as a loop of mobility unity. The point I wish to make is that, in the same way as the apparatus in figure 2.08 can be closed and then transported through a single infinity of different configurations, any manipulator arm with six degrees of freedom at its gripper can grasp its given rivet in a single infinity of different ways.

62. The joint between the gripper and the rivet can be seen in other words as a revolute. This means (*a*) that, because the actuated arm can arrange its many parts and approach the waiting rivet in many different ways, more than one mode of the mentioned loop can occur (§ 1.44, § 1.45). It means also (*b*) that, if the complicated matters at (*a*) are understood, there is a whole field of algebraic endeavour available for manipulator and robot design. The basis of the algebra is the closure equations that either can or cannot be written for spatial loops with simple and complex joints, and upon the multiple roots of those equations. Refer to § 19.51, and to Duffy (1980).

63. I have used the term *actuated joint*, and I have implied in the too brief remarks above that the philosophy of manipulator and manipulator-joint design is thoroughly settled, well understood by engineers. It is not; there are many problems and deep uncertainties. It can be said however that the elementary material of these two chapters 1 and 2, with its remarks about the nature of mechanism, mobility, kinds of joints, simplicity and complexity, guiding and stopping surfaces, flexible and other links, and intermittency, is germane to the broad area of robotics. There is of course no reason to believe that robots (which are machines) should resemble us or the animals, both of which are also machines; but the occurrence of anthropomorphism in our thinking and the consequent discussion about its

appropriateness in design is almost inescapable. I wish to remark now about the actuated joints of man and the animals.

**Biomechanism**

64. At § 1.55 I spoke about the 'fully general', *f* 1 joint, wherein the axis of the screwing changes its location and its pitch as soon as the joint engages in a finite movement; see also § 6.58. Such joints and others much more complicated between the contacting and the non-contacting bones of the vertebrates and, to a lesser extent, between the rigid, moving parts of the crustaceans, abound among the animals (§ 6.07, § 2.37, § 10.62). At all of these joints, both simple and complex, there is a lubricant between the guiding surfaces and at the spots of contact (§ 2.07, § 1.49); at all of them there are limits set to the geometry of the motion – the ligaments do this (§ 1.29); and at all of them there are muscle actuators connected to the bones or members at the joint by means of tendons. These latter have the capacity to control the movements at the joints, to control, in effect, the *f*-numbers there, and to produce a continuous mechanical activity quite beyond the wit of us to copy.

65. As in all joints in all mechanism, however, there is at an instant at a biomechanical joint an axis with a pitch about which the relative motion is occurring; that axis may be called the *motion screw at the instant*; refer to chapters 5 and 6, and to § 10.60. Similarly there is at an instant at a biomechanical joint (as there is at all joints) another axis with a pitch along which a pair of equal and opposite wrenches are acting; these wrenches are the resultants of the elementary forces existing at the points of contact at the joint, the forces occurring by virtue of the muscle-produced tensions in the tendons and by virtue of the external, transmitted loads. The common axis of the two wrenches at the joint (along with its pitch which is the pitch of the wrench) may be called the *action screw at the instant*; refer to chapters 3 and 10, and to § 10.61. It is only seldom in mechanism of any kind, and never, incidentally, in planar mechanism (§ 17.09), that the motion screw and the action screw coincide. The mechanics of the relationship between these two screws at a joint – they are in general skew with one another and of different pitches – forms an important part of the theory of mechanism and machines. The two screws must, accordingly, play an important role in all of our first attempts to understand the complicated mechanics of biomechanical joints. Refer to the chapters 14 and 17, and, for a beginning, to § 3.01.

**Notes and references**

[1] Phillips, J.R., On the completeness of kinematic chains of mobility unity, *Environment and Planning B*, Vol. **6**, p. 441–6, Pion, 1979. Refer § 2.23.

[2] Waldron, K.J., The constraint analysis of mechanisms, *Journal of Mechanisms*, Vol. 1, p. 101–14, Pergamon, 1966. Refer § 2.38.

[3] Alt, H., Das Konstruieren von Gelenkvierecken unter Benutzung einer Kurventafel, *VDI Zeitscrift*, Vol. **85**, p. 69–82, 1941. Refer § 2.42.

[4] Volmer, J., Die Konstruktion von Gelenkvierecken mit Hilfe von Kurventefeln, *Maschinenhautechnik*, Band **7**, Heft 7, p. 399–403, 1958. Refer § 2.42.

3. The right line modelled here has its usual rectangular display of velocity (or moment) vectors and it bears its distinctive geometrical relationship with the motion (or the action) screw.

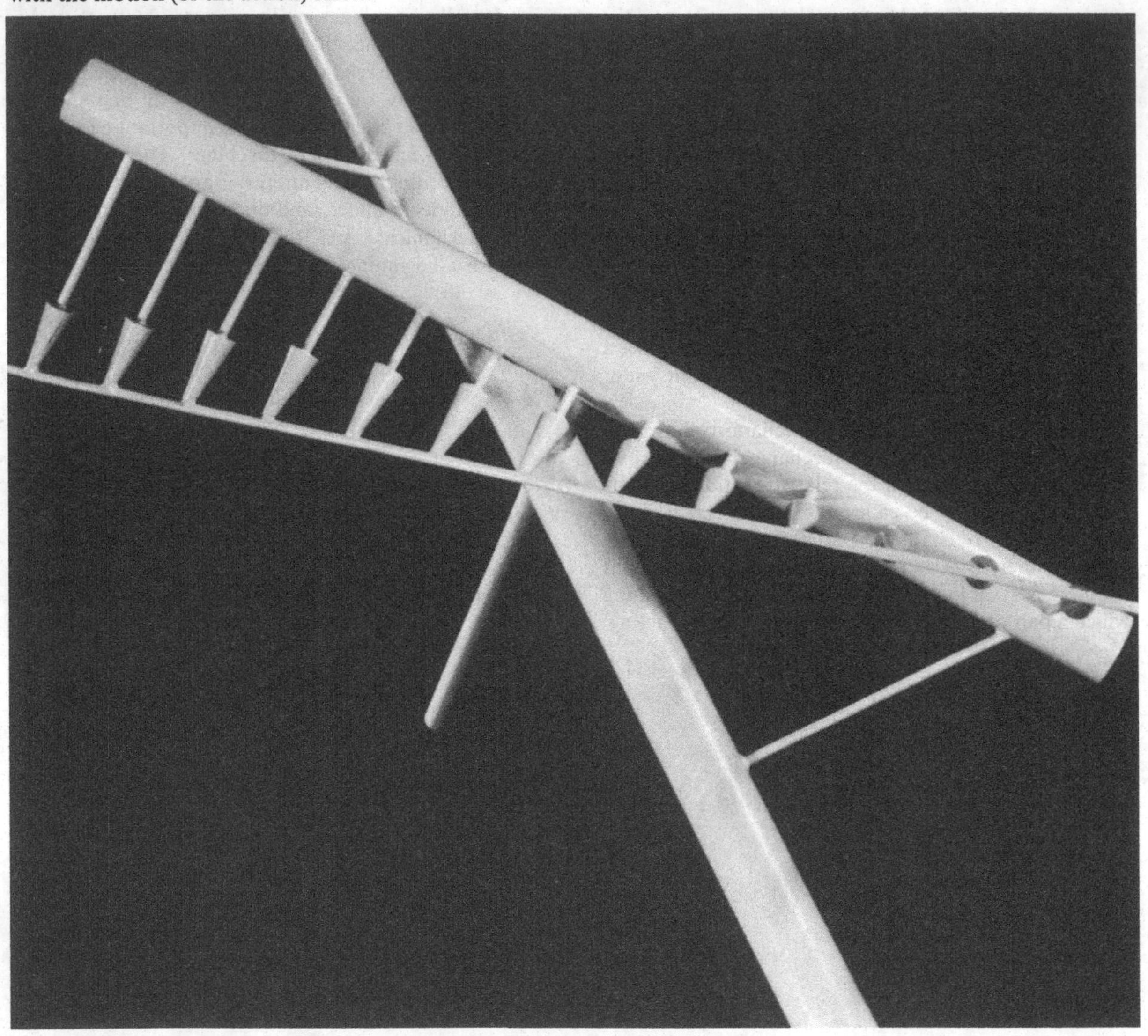

# 3 Some of the various lines in a moving body

## Introduction

01. I define the meanings of and occasionally use in this chapter two contrived adjectives: motional and actional. I use the two adjectives to distinguish two different kinds of straight line. Both of these I call, not because they are both straight but because they share another important characteristic, *right lines*. I speak about (*a*) motional right lines, and (*b*) actional right lines. The definition of the first kind of line comes from within the kinematics of its circumstance, while that of the second springs from the statics. The two kinds of line are similar and closely related with one another. At the beginning, to avoid confusion, I shall speak about them separately; I begin by discussing (*a*) the motional right line. I continue to do that until we reach § 3.43. For the sake of brevity throughout the passages leading up to § 3.43, however, I call the motional right line by its shorter name, which is common to both of the lines, right line. *A right line in a moving body is any straight line which joins two points in the body whose linear velocities are perpendicular to the line.* Along any such right line all points have linear velocities perpendicular to the line (§ 5.26, § 21.08); the tips of all of the attached velocity vectors along the line are in a straight line also (§ 5.25); there is an $\infty^3$ of right lines at an instant (§ 9.29); and, for a note about the terminology, please refer to § 10.50. I choose the simple, movable mechanism whose description follows as a vehicle for argument. The mechanism has four links and its mobility is unity (§ 1.34). Refer to figure 3.01.

## Right lines and n-lines in a body with 1 °F

02. Attached to the frame member 1, in figure 3.01, by means of two hinges at $A$ and $B$, there are two ball-ended members 2 and 3. The balls are engaged in sockets in the movable member 4 at $D$ and $C$. The member 4 is in direct contact with the frame at $E$ where a single, pointed prong on 4 is free to slide across the flat surface there of 1. At $A$ and $B$ the joints are $f1$; at $C$ and $D$ they are $f3$; and at $E$ the joint is $f5$; the mobility of the mechanism is, by § 1.34, $M=6(4-5-1)+13$, namely unity. It follows that the member 4 and both of the other moving members suffer constrained motion as the mechanism moves, that no stresses are locked into the members by virtue of the assembly (§ 1.55), and that the curved paths in the frame 1 of all of the points in the member 4 are predetermined. We shall pay attention now to the motion of the member 4 relative to the member 1 and look for some right lines. An implied assumption is that 1 is the frame of reference for all linear velocities.

03. It will be clear upon inspection that, given a rotating input at $A$ as shown, the instantaneous velocity $\mathbf{v}_{1D4}$ (namely the linear velocity relative to 1 of the point $D_4$) at $D$ will be occurring as shown, namely in a direction normal to the plane defined by, (*a*) the axis of the hinge at $A$, and (*b*) the point $D_2$ at the centre of the ball; and these two latter are functions, not of the motion of the mechanism, but of the configuration of the mechanism at the instant. The points $D_2$ and $D_4$ are not only instantaneously but also continuously coincident incidentally (§ 1.06), and, for a general explanation of the subscripts used for velocity, the reader is referred to § 12.08 *et seq.* The instantaneous velocity $\mathbf{v}_{1E4}$ is not yet known without analysis, but we do know that it will be occurring in one or another of the directions indicated by the planar pencil of possible velocity vectors drawn upon the flat surface of the member 1 at the point of contact there, $E$.

04. Erect at $E$ a normal $n$–$E$–$n$ to the planar pencil there, and let this normal pass through a point $F_4$ which is a fixed point in the member 4 (§ 1.06). Before considering the velocity $\mathbf{v}_{1F4}$ at $F$, please refer to the passages § 5.01 and § 5.03, and see (in figure 3.01) that, because the component of the velocity at $E$ along the line $EF$ is zero, the component of the velocity at $F$ along the same line is also zero. It follows that the velocity $\mathbf{v}_{1F4}$ at $F$ is perpendicular to the line $EF$. The line $EF$, or $n$–$E$–$n$, is a right line in the body 4; but, for reasons to be explained immediately, it can also be called an *n-line* in the body 4, and a common normal line or a *contact normal* (§ 3.06).

05. Independently of the degrees of freedom of

the body, *I wish to define an n-line in a body as a line which is a right line, not merely by virtue of some particular instantaneous motion of the body, but by virtue of some constraint upon the body*. What this will mean is that, whereas all *n*-lines in a body will immediately become right lines as soon as the body begins to move, not all right lines in a moving body will have been *n*-lines before the body began to move (§3.24). The line *n–E–n* in figure 3.01 is an *n*-line by this definition, for it is a right line by virtue of a constraint, namely the point of contact with the body 1 at *E*. Whereas the number $\infty^3$ of right lines in a moving body is independent of the degrees of freedom available to the body, the number of *n*-lines in a body does depend upon that (§3.44).

06. There are often, however, as we shall see, many more *n*-lines in a body than lines such as *n–E–n*. The *n*-line *n–E–n* is special: it is normal, not only to the surface of the member 1 at *E*, but also to the surface of the blunted point of the member 4 at *E*. It is called, accordingly, *a common normal line or a contact normal*. The gist of the matter at issue is illustrated in figure 6.03. There, two bodies 1 and 2 are in contact at *A*, the common tangent plane at *A* is α, and the common normal line or the contact normal (or the *n*-line at *A*) is *n–A–n*. Please peruse also §8.32 through §8.36. This passage refers to figures 8.05 to 8.08 inclusive. A few remarks are made there which are relevant here.

Figure 3.01 (§3.02) A 4-link mechanism of mobility unity. The link, the member, or the body 4 is being examined for its instantaneous motion with respect to the body 1. Please note the velocity vectors drawn at the centres *C* and *D* of the balls, and at the other two points *G* and *H* in the body 4; at each of these points a circle is mounted whose plane is normal to the relevant velocity vector and whose centre is at the relevant point; the lines *n–n*, which reside as diameters of the circles, are right lines in the body 4.

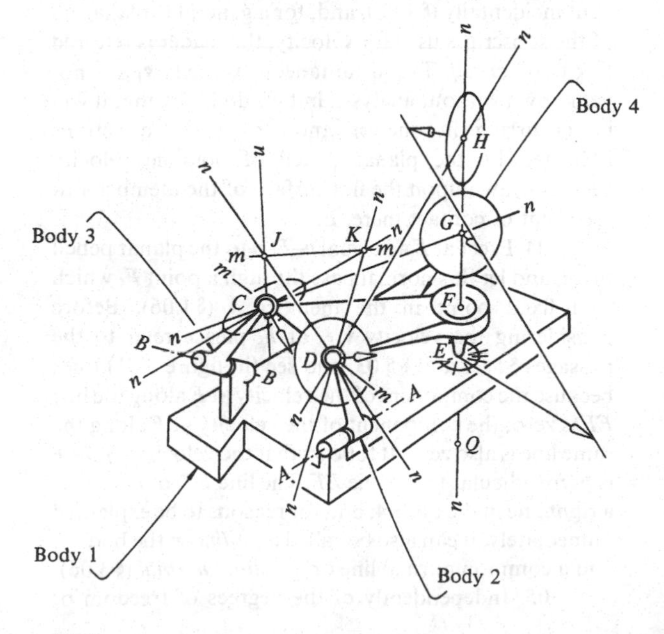


07. Coming back to figure 3.01, and by using again the argument of §5.03, we can next see that, quite independently of points of contact directly between the bodies, there are whole planar pencils in the body 4 of right lines which are also *n*-lines. Two of them are centred respectively at the two points $C_4$ and $D_4$. The planes of the planar pencils there are normal respective to the known directions of the only possible velocity vectors there, $\mathbf{v}_{1C4}$ and $\mathbf{v}_{1D4}$. For each line in each of these two planar pencils it can be seen that, because there is one velocity perpendicular to it, namely the one whose direction is known, all velocities along it will be perpendicular to it (§5.03). The right lines of the planar pencils at *C* and *D* are also *n*-lines in the body for the reason that we do not need to think about the particular *motion* being undertaken by the body in order to see this fact. The stated fact is a function primarily of the constraints; it is only secondarily a function of the only possible motion which can occur (§3.05).

08. The only right lines I have shown so far in the body 4 in figure 3.01 are, (*a*) the single one at *E*; and (*b*) the two planar pencils of them at *C* and *D*. Although there are many more of them, all of which are also *n*-lines in the body 4, and although they have not all been pointed out as yet, it can be seen straight away that, *whereas for any right line in a body established by virtue of some constraint (namely for any n-line) there is a continuous series of planar pencils of possible velocity vectors normal to it and distributed at all points along its length, there is for any velocity vector in a body established by virtue of some constraint a single planar pencil of right lines which is the set of n-lines normal to it at its basepoint.*

**An important theorem**

09. We should be able to prove the theorem that, if two intersecting right lines exist at an instant, all lines of the planar pencil defined by the two existing right lines will be right lines at that instant also. This theorem flows in fact from other, more mature considerations outlined in chapter 9 and elsewhere (§9.10, §10.15); but it can, and will be proven here by using the simplest of elementary ideas.

10. If two right lines do intersect, the common velocity vector at the point of intersection must, for obvious reasons, be normal to the plane of the two right lines; otherwise it could not be perpendicular to both of them (§5.01). In figure 3.02 I have drawn in the plane of the paper two intersecting right lines *n–AB–n* and *n–AC–n*; they intersect at *A*. The common velocity

vector at $A$ is pointing downwards, normal to the plane of the paper, and appears therefore as a point in figure 3.02. I have erected at $B$ and $C$ (which are any two other points upon the right lines) two velocity vectors $\mathbf{v}_B$ and $\mathbf{v}_C$ which are, (*a*) consistent with their relating to points upon a right line; and (*b*) consistent with one another. Note that their components in the plane of the paper are perpendicular to the lines *n–AB–n* and *n–AC–n* respectively, and that their components along the rigid line BC in the moving body are equal (§ 5.03).

11. I have chosen next in figure 3.02 any other point $D$ in the plane of the paper, and have shown by construction in the figure how to find the component in the plane of the paper of the velocity $\mathbf{v}_D$ at that point. What might be called the axiomatic principle of rigidity (§ 5.03) has been applied along the lines *BD* and *CD* in the plane of the paper, and, from the two components thus derived, the vector at $D$ has been constructed. Component velocities are shown in the figure by means of the thickened segments of line; and please be reminded that, whereas at $B$ and $C$ the components were derived from the whole velocities $\mathbf{v}_B$ and $\mathbf{v}_C$, at $D$ the whole velocity $\mathbf{v}_D$ was derived from its components.

12. Now it can be shown in figure 3.02 by ordinary Euclidean geometry, and independently of what the components normal to the plane of the paper might be, (*a*) that the discovered component of $\mathbf{v}_D$ in the plane of the paper is proportional to its distance *A–D* from $A$, (*b*) that the same constant of proportionality applies along all such lines as *AD* radiating from $A$, and (*c*) *that the angle between the vector* $\mathbf{v}_D$ *and the line AD is a right angle*. It follows that the line *AD*, which is any line through $A$ in the plane of the paper, is a right line.

13. A simple extension of these analyses will also show, (*a*) that any line in the plane of the paper which does not pass through $A$, such as the line *BC* in the figure, cannot be a right line; and (*b*) that any line through $A$ which is not in the plane of the paper cannot be a right line either (§ 9.10, § 9.11). We can say, therefore, and with more precision now, that, *if two right lines are known to intersect at a point A, then (a) those lines and only those of the planar pencil at A defined by the two are right lines through A, and (b) no other lines in the plane of the pencil can be right lines.*

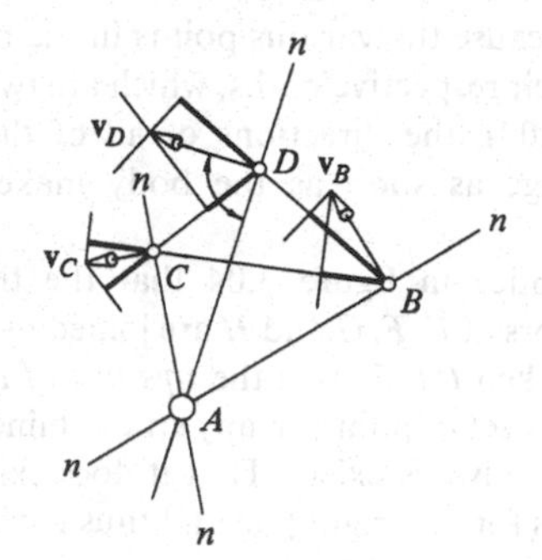


Figure 3.02 (§ 3.10) Illustrating a proof of the theorem that right lines exist in a moving body in such a way that, normal to every velocity vector at its basepoint, there is a planar pencil of right lines intersecting at the basepoint.

**The whole multitude of right lines in a moving body**

14. Returning again to figure 3.01, we can see that, because the right line *n–EF–n* is parallel with neither of the planar pencils of right lines at $C$ and $D$, it must intersect them both. The two points of intersection are indicated $G$ and $H$ in the figure, and it follows from the theorem just proved (§ 3.13) that there are planar pencils of right lines in the body 4 at the two points $G$ and $H$. Refer to figure 3.03.

15. Let us also note in figure 3.01 that the discovered planar pencils at $C$ and $D$ intersect one another along a line which is marked *m–JK–m*. At $J$ (which is any point along the line *m–JK–m*) there are two known intersecting right lines, namely *n–JC–n* and *n–JD–n*, so there is a planar pencil of right lines at $J$ (§ 3.13). But $K$ is also any point along the line *m–JK–m* where a similar argument applies, and it follows that, at all points along the line *m–JK–m*, there are planar pencils of right lines.

16. Next we can notice that any pair of planar

Figure 3.03 (§ 3.14) The distribution of the $\infty^3$ of *n*-lines in the body 4 in figure 3.01. Whereas the lines marked *m–m* are not members of the planar pencils at the nominated points and are not *n*-lines, all of those remaining, the unmarked ones, are. There is only 1 °F extant at the body 4: this means that every one of the $\infty^3$ of right lines which will exist as soon as the body moves can be seen as *n*-lines in the body before the body moves.

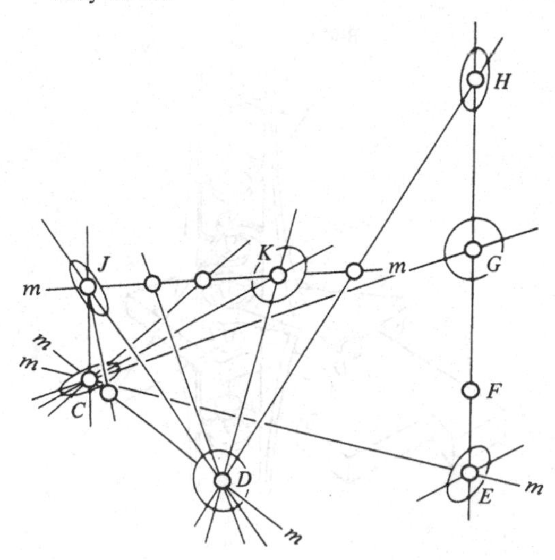

pencils (such as those at $J$ and $K$) along the line $m$–$JK$–$m$ intersect one another along the line $CD$ which joins the centres of the balls; $CD$ is marked in the figure $m$–$CD$–$m$ because it is plain that, like $m$–$KJ$–$m$, this is a line of points at each one of which there is a planar pencil of right lines.

17. Notice next in figure 3.01 that, whereas all lines $n$–$n$, which are right lines (and $n$-lines) in the body 4, contain, at each point along their lengths, one generator of the planar pencil of right lines existing at that point, the lines $m$–$m$, which are not right lines in the body 4, do not display this special property.

18. Notice also that, by continuous extension of the above argument, we could go from point to point throughout the whole of Euclidean space and find that, *at every point in the body 4, there is a planar pencil of right lines*. It follows that, if any line $m$–$m$ is drawn through the body 4, there will be, at all points along it, a planar pencil of right lines existing; in certain special cases of these lines $m$–$m$, however, the planar pencils at all points along it will include the line itself; and in these special cases the line will be a right line $n$–$n$. It will be shown in chapter 9 that, independently of the degrees of freedom being enjoyed by a moving body, the number of right lines in it will be $\infty^3$, and that the whole ensemble of these $\infty^3$ of lines always falls into a certain pattern of lines called a linear complex. In our case here, where the degrees of freedom of the moving body 4 is only 1 °F, where the nature of the constraints has predetermined

Figure 3.04 (§ 3.19) An imagined set of prongs on the body 4 and platforms on the body 1 mounted along the contact normal $n$–$EF$–$n$. Any contact normal in a pair of contacting bodies (and, by implication, this one) is by definition an $n$-line in both of the bodies. Please read also § 7.20.

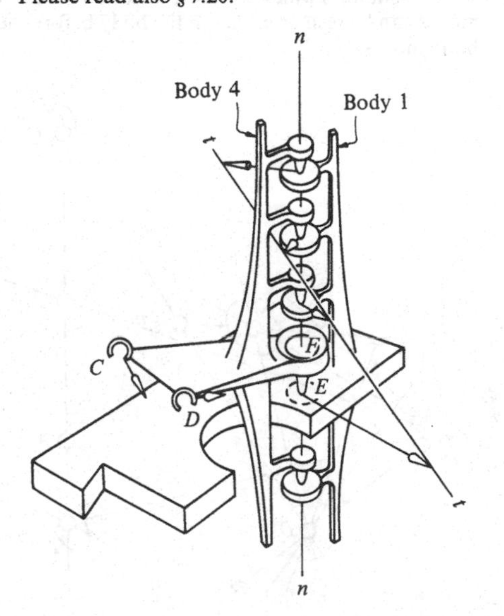


the shape of the curved paths of every point in the body 4 (§ 3.02), *all of the right lines in the body 4 are also n-lines, which means that we have a whole linear complex of n-lines existing in the body*. See § 10.09, § 10.10, and please refer to figure 3.03.

**The nature of the relative motion at an n-line**

19. Notice now in figure 3.03 that the directions of the velocities at $G_4$ and $H_4$ can be determined. They are (*a*) normal to their respective planar pencils of right lines at the two points $G$ and $H$, (*b*) perpendicular to their common right line $n$–$EFGH$–$n$, and (c) *not parallel with one another*. Refer to figure 3.04.

20. We show in figure 3.04 that, if two bodies 1 and 4 are in contact at a point $E$, and if $n$–$E$–$n$ is the common normal there, then, without disturbing the instantaneous motion, or the capacity for instantaneous motion (§ 3.34), we can imagine that the two bodies are actually built as shown, with a continuous series of points of contact along the line $n$–$E$–$n$. The imagined points of contact must be imagined in such a way that all of the tips of the sharp prongs on 4 are in a straight line, and that all of the tangent planes at the points of contact are parallel with one another. To ensure this latter we imagine that all of the 'platforms' on 1 in figure 3.01 are flat, and that they are all normal to the line $n$–$E$–$n$.

21. As we have seen at § 3.19, however, the velocity vectors at the points of contact need not be parallel with one another; they need only reside in parallel planes. This possibility (which occurs in the general case) is illustrated in figure 3.04. Please look again at the contact normal $n$–$EGH$–$n$ in figure 3.01, see it as a line fixed in the moving body 4, and try to imagine its instantaneous motion; whereas the point $G_4$ is moving instantaneously towards the left background of the picture, the point $H_4$ is moving instantaneously (and contemporaneously) towards the right foreground of the picture; the angle of the line $n$–$EGH$–$n$ to the vertical is changing and the line as a whole is moving forwards and away from its reference location. *Do be aware however that, as soon as it makes a finite movement, the line ceases to be a right line and an n-line*. Unless there is some special case (and this case here is not a special case), and because the various points in the body 4 are traversing their respective paths, which are twisted lines in space (§ 3.02), the directions of all of the velocity vectors change as soon as the body makes a finite movement.

22. Notice in figure 3.04 that the tips of the velocity vectors at $E$, $F$, $G$ and $H$ are joined by a straight line; it is marked $t$–$t$. This is the *tips line of the vectors* which, for any set of points in any straight line in a rigid moving body, always exists. That it does exist for any line (and thus for any right line and thus for any $n$-line)

in a rigid moving body is proven at chapter 5; but here in chapter 3 the reader might try to imagine, not the truth of this particular fact, but the truth of the matter mooted at §3.21 and explained at §3.23.

23. Try to imagine, at an instant, every one of the right lines in the moving body 4 in figure 3.03 to be equipped with a column of prongs as is shown at figure 3.04, and try to imagine that same set of right lines in the fixed body 1 to be similarly equipped with columns of matching platforms, so built to ensure that all of the tangent planes are correct at the instant. Imagine also that the prongs, while free to slide without friction across the surfaces of the platforms, are attracted in some magnetic way to remain in contact as they slide. If you can imagine all of this you will see that the two bodies 1 and 4, thus equipped, would have no need of the old connecting links at 2 and 3 or of the old point of contact at $E$ to be able to reproduce their original, instantaneous, relative motion with one degree of freedom.

24. It might be mentioned again that, in the event of 1 °F only, all of the right lines in a body are $n$-lines also. This follows from the fact that, when a body has only 1 °F or is overconstrained yet mobile, all of the paths of all of the points within it are predetermined by the nature of the constraints (§1.20, §2.04, §3.02). When a body has more than 1 °F, however, most of its points will have a choice of paths at the instant; this means that, before the body moves, there is only a limited number of its $\infty^3$ of right lines which can be predicted; these latter are the $n$-lines existing in the body.

25. But if you guessed at §3.23 that the provision of all of those rows of prongs and platforms was redundant in the circumstances, you would have been guessing correctly. How many of them, namely how many $n$-lines (each with only one pair of prongs and platforms), would be enough? The answer (with certain reservations) is only five; but the reasoning behind this answer is not being offered at this stage. Take a hint however at §10.12.

26. In chapter 5 it is shown that such an instantaneous motion as that of the whole body 4 with respect to the whole body 1 – or, indeed, any relative motion between a pair of rigid bodies – can be seen as a screwing, or a twisting motion, about an axis with a pitch. Here in chapter 3, without knowing where that axis is, we might simply imagine that the body 4 is connected to the body 1 by an actual nut and an actual bolt arranged to engage one another along that mentioned axis with the pitches matching (chapter 5). The idea is materialised and shown in figure 3.05. Such a connection, which can be shown to provide exactly the same capacity for instantaneous motion of the body 4 as do the two links 2 and 3 and the single point of contact at $E$, can be called, in the circumstances, a reciprocal connection, as in Hunt (1978), or a reciprocal joint, between the bodies. The significance of this remark will be made apparent later (chapters 3, 10, 14, 23).

### Mechanical substitutions at existing n-lines

27. Consider now what would happen at $E$ in figure 3.01 if ball and socket joints were introduced at $E$ and $Q$ along the $n$-line $n$–$E$–$n$ in such a way that the rigid body $EQ$ became a new link 5. Refer to figures 3.06($a$) and 3.06($b$). Relative to 1, the point $E_5$ and thus the point $E_4$ would be obliged to travel somewhere upon the sphere at centre $Q$ and of radius $Q$–$E$. The equivalent tangent plane at $E$ would be exactly the same as it is in

Figure 3.05 (§3.26) At the instant under consideration the body 4 can be seen to be twisting about an axis with a pitch, and that axis has been materialised here as an imaginary nut and a bolt. Note that the linear velocities shown are consistent with one another and with the pitch of the screw axis. Refer to §5.02.

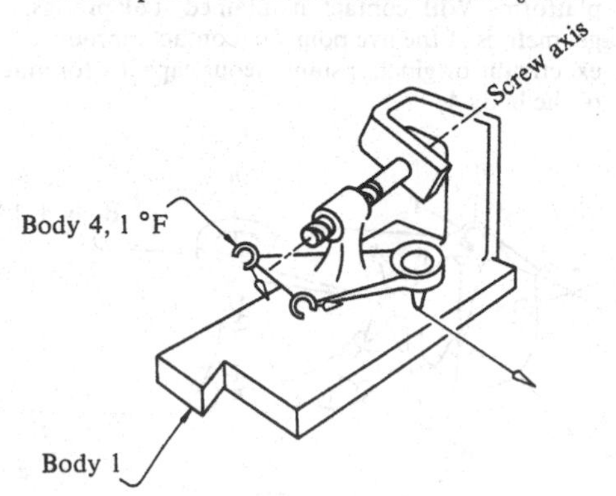


Figure 3.06 (§3.27) These illustrate that, at an instant, a double ball-ended rod mounted anywhere along a contact normal line can act as a mechanical substitution for the actual point of contact between a pair of bodies. They are also showing the reverse proposition, that a suitably chosen point of contact between a pair of bodies can substitute for an actual, double ball-ended connection.

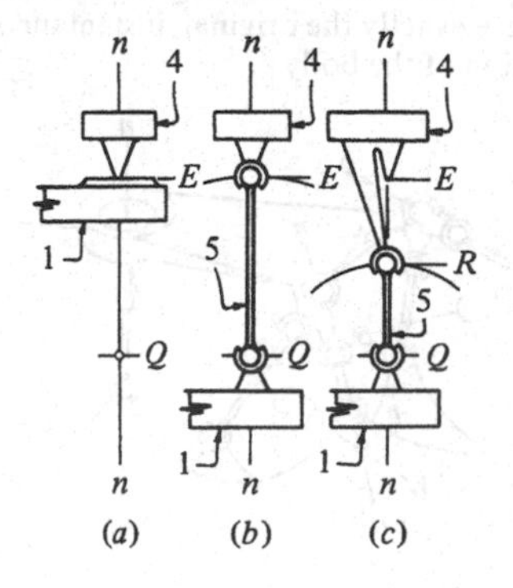

figure 3.01, normal to the $n$-line $n–E–n$, and the instantaneous capacity for motion of the body 4 would be unchanged. Notice that, unless the length $Q–E$ of the new link 5 is zero, the actual position of $Q$ along the $n$-line $n–E–n$ (and thus the radius $Q–E$ of the sphere at $Q$) is, for infinitesimal displacements and instantaneous velocities, irrelevant.

28. But refer to figure 3.06($c$) and notice also that, if we put the upper ball of the member 5 not at $E$ but at some other point say $R$ along the vertical $n$-line $n–E–n$ and provided a mating socket there, the capacity for instantaneous motion of the body 4 would still remain unchanged. By virtue of the horizontal instantaneous velocity of $R_4$ upon the surface of its sphere at centre $Q$ and of radius $Q–R$, there is an $n$-line $n–QRE–n$ as shown, and the velocity at $E_4$ is accordingly perpendicular to it, namely horizontal as before. This argument has shown that, *if an n-line exists in a body, it is possible in the imagination to connect that body to frame by means of a ball-ended rod such as the rod QR in the n-line n–E–n, without disturbing the instantaneous capacity for relative motion of the bodies.*

29. At § 3.23 I asked the reader to imagine a multitude of columns of prongs and platforms, spread out along all of the $n$-lines which were evident to him in figures 3.01 and 3.03, and to see that such a conglomeration of apparatus would not interfere with the instantaneous relative motion of the bodies 1 and 4. And now I ask a similar favour. Please imagine, along all of the $n$-lines in the body 4 which may be evident, that ball-ended rods like the rod $QR$ in figure 3.06($c$) are inserted, and become aware, then, that the action of none of these inserted rods will interfere with the already existing capacity for instantaneous motion. In fact you might be so kind as to imagine all of the ball-ended rods and all of the prongs and platforms to be in operation simultaneously. Refer to the figures 3.07 however, and begin to suspect from them that, with certain reservations which have yet to be mentioned (§ 3.32, § 10.12), *either only five ball-ended rods, or only five sets of individual prongs and platforms, will be sufficient to support the body 4 in such a way that its capacity for instantaneous motion will remain unchanged.*

30. A check with the aid of the general mobility criterion (§ 1.34) upon the two connectivities $C_{14}$ in the figures 3.07 will reveal that they are both unity. In making this check please do not forget the earlier remarks at § 1.37 about the occurrence of irrelevant rotations; notice, in other words, the five freedoms to rotate of the five ball-ended rods. If however one of the rods in figure 3.07($a$) and one of the prongs in figure 3.07($b$) were removed, the connectivities $C_{14}$ would be increased to 2, and there would be not one, but two, degrees of freedom between the bodies, 2 °F. If one more of the rods and one more of the prongs were next removed, we would arrive at 3 °F in both figures, and so on; this would continue all the way to a total absence of rods and prongs, and 6 °F between the bodies. All of this could be monitored accurately by the general mobility criterion.

31. But if there were six (or even seven or more) ball-ended rods in figure 3.07($a$), and six (or even seven or more) pairs of prongs and platforms in figure 3.07($b$), so arranged that they all corresponded with $n$-lines existing in the body 4 of the original mechanism in figure 3.01, *the general mobility criterion would give us faulty information.* We found from § 3.23 and from § 3.29 that a suitably selected multiplicity of rod or prong connections would not interfere with an apparent (though transitory) connectivity between the bodies 1 and 4 of unity in the figures 3.07, yet we might have believed from earlier experience that even one extra rod or prong in the figures 3.07 would immobilise the motion (chapter 1), and we know that the general mobility criterion would predict just that. There appear to be some special conditions here. Have we found, in the sense of § 2.12,

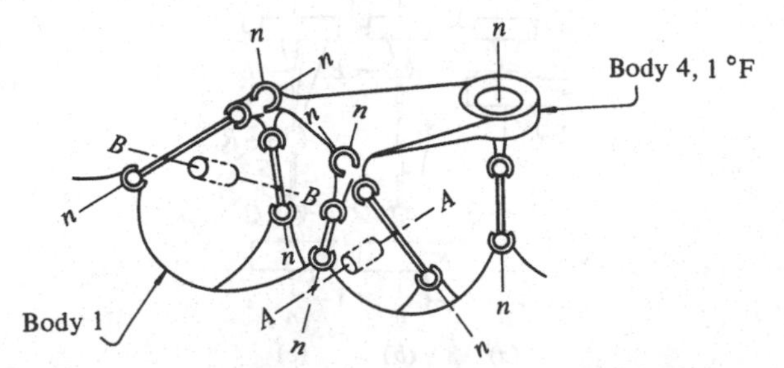


Figure 3.07($a$) (§ 3.29) The body 4 supported upon a set of only five double ball-ended rods. Together the rods are a mechanical substitution at the instant for the links of the original mechanism; they reproduce exactly the original, instantaneous capacity for motion of the body 4.

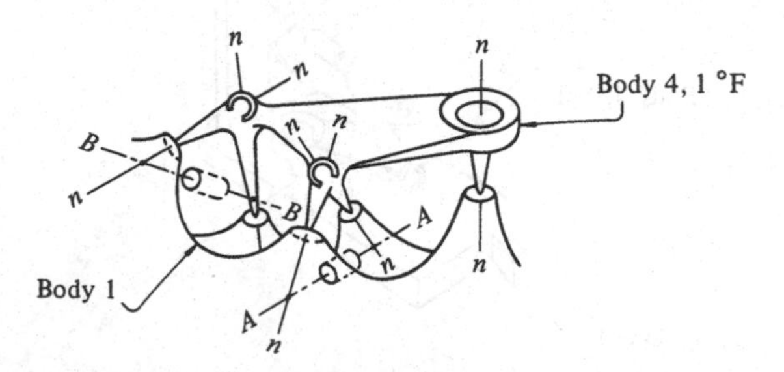


Figure 3.07($b$) (§ 3.29) The body 4 supported upon a set of only five individual prongs and platforms. With contact maintained at all prongs, the geometries at the five points of contact reproduce exactly the original, instantaneous capacity for motion of the body 4.

some new kind of geometric speciality in the present circumstance?

32. It is well known for example that six ball-ended rods, suitably arranged, will immobilise a six-component wrench-measuring apparatus in its mount, so there does appear to be some special circumstance here. The matter at issue is, in fact, as follows: *it is the fundamental geometrical matter of the projective dependence, or the projective independence, of lines.* A few elementary aspects of this need to be mentioned now. If in the figures 3.07 for example, more than two *n*-lines from the same planar pencil had been chosen to be among the chosen five, the set of five *n*-lines would not have been projectively independent of one another and there would have been an apparent, transitory connectivity $C_{14}$ not of unity but of 2 extant. If in the figures 3.07 for another example, an extra rod or prong were added which belonged to the linear complex of *n*-lines which was indicated but only partly revealed in figure 3.03 it would not be projectively independent of the other five and thus, transitorily, it would not reduce to zero the connectivity $C_{14}$ of unity which is extant. Refer to the chapters 7, 10, 11, and 23 for further discussion of this most important yet somewhat complicated matter. *The matter is connected with the question of geometric speciality, inextricably.* See § 2.12.

**Revolute axes, *r*-lines, and the actional right lines**

33. Whenever we find that a body's degrees of freedom with respect to another body, its *f* °F, has been limited to some number *f* less than six, the limitation has been achieved somehow by means of *constraints*. What this inevitably means is that the body has either been put into contact directly with the other body at various places or points (chapter 6), or it has been put into contact with it through other, intervening bodies, or both; and this latter is precisely what was done with the body 4 in respect of the body 1 in figure 3.01. The science of the matter of connectivity, incidentally, deals with this same problem of discovering and of quoting the relative degrees of freedom between a pair of connected bodies (§ 1.31, § 2.39, § 3.34, § 20.07).

34. I have already spoken about a body's ICRM, its instantaneous capacity for relative motion with respect to some other body (§ 3.28). Whenever I use this term ICRM, I shall always be implying something more than the body's mere *f* °F with respect to the other body or its connectivity *C* with that body, both of which latter are the same thing namely the same mere integer *f*. I will always imply by ICRM, not simply that some number *f* between zero and six requires to be quoted, but that some clear and comprehensive *graphical description* of the possibilities for relative motion at the instant is required to be seen. What I want to stress is that, when a body has more than one degree of freedom with respect to another body, there is certainly more than two ways in which it is free to move with respect to that body; and what I am seeking to explain is that all of the multiplied infinities of possibilities for motion at the instant can be presented in the form of a pictorial layout in space of all of those possibilities. That *layout* is what I have in mind when I speak about a capacity for relative motion at an instant. Now it so happens that such a layout can always be offered. We can always show (either graphically or mathematically) the whole set of the so-called 'available screws for the motion' at the instant; this set can be called, and it is so called, *the screw system for the motion at the joint between the bodies* (§ 2.40, § 6.07, § 10.68, § 23.02). What we shall see however is that this whole system of screws can always be represented as it were, or in fact provided by, a minimum number of imagined screw joints series-connected with one another between the body in question and its reference body. Please read also § 4.04.

35. Whenever a body's instantaneous capacity for relative motion is so severely restricted that only one degree of freedom remains (and this is always the case for any one of a pair of bodies in a constrained or overconstrained mechanism where the apparent mobility is unity), the above mentioned layout can always be seen as follows: *it is the sole possibility of being able to twist about an already determined, single, available screw.* Such a circumstance is illustrated in figure 3.05, where the two relevant bodies are 1 and 4, where the relevant degrees of freedom is 1 °F where the connectivity $C_{14}$ is accordingly unity, and where the already determined, single, imagined, available screw at the instant has been materialised and is clearly shown. No information was offered at § 3.26 about the exact location or the pitch of this imagined screw, and there is none being offered here; it is simply being stated that, with unique location and pitch, such a single, imagined screw does exist; but please refer in time to the chapters 5 and 6.

36. Whenever a body's instantaneous capacity for relative motion is less restricted than that, its screw system can be better described by using the following form of words: *it is the multiple possibility of being able to twist about a very large number of screws, and these in turn can be represented by a minimum number f of screw joints, all series connected to frame as is shown in figure 3.08.* Provided the screws which are pictured in figure 3.08 are projectively independent of one another (chapters 10, 23), the minimum number *f* of them is a measure of the degrees of freedom. If there are *f* screw joints (and *f* can never be greater than 6), there are *f* degrees of freedom. The number *f* can now be seen as the number that describes (but only in a numerical way) the instantaneous capacity for relative motion; the reader

will see in retrospect that, at § 3.26 and § 3.35 where $f$ was unity, we were dealing with only a special case of the somewhat wider circumstance being mentioned here.

37. Under certain specifiable conditions, however, which unfortunately cannot be explained just yet (§ 23.44), the set of $f$ series-connected screws to frame can be chosen in such a way that all of the screws are hinges, namely screws of zero pitch; these are the sets of prospective revolutes, or of *r*-lines, of which I wish to speak. The word revolute, incidentally, is synonymous with the word hinge; so also is a revolute simply the same thing as a turning pair. Refer to § 7.12 and the table 7.01 for a clarification of this. One thing which can be said with certainty now is that, *if the constraints between the bodies take the form of being, merely, direct points of contact between the bodies, and if the number of these does not exceed three, then the ICRM can always be exactly duplicated with an equivalent set of series-connected screws of zero pitch, namely a set of series-connected revolutes.* Refer in advance to figure 3.13(*a*). *If the number of direct points of contact between the bodies does exceed three, then the ICRM can only sometimes be exactly duplicated by a set of series-connected revolutes.*

38. It can moreover be said in the circumstances that, if the minimum number of series-connected revolutes is $f$, the number of direct points of contact from which the number $f$ of revolutes was derived will have been $(6-f)$. Please note in figures 3.13(*a*) and 3.13(*b*) that a moving body 2 is hung up there upon a fixed body 1 via three series-connected revolutes; it is a fact in those two figures that the three revolutes were derived from the $(6-3)$, namely 3, direct points of contact which can be seen. Please take a look also at the two similar figures in chapter 14 where, in figure 14.09 there are two revolutes between the bodies, and in figure 14.10 there are four. The numbers of direct points of contact from which the revolutes were derived in those two cases are, $(6-2)$ namely 4, and $(6-4)$ namely 2, respectively.

Figure 3.08 (§ 3.36) The body 2, series connected to its frame 1 by the three screw joints which are generally disposed and have their several pitches, has 3 °F relative to that frame. The body has the capacity to twist about any one of the $\infty^2$ of the different screws of the relevant screw system. Please read § 6.49.

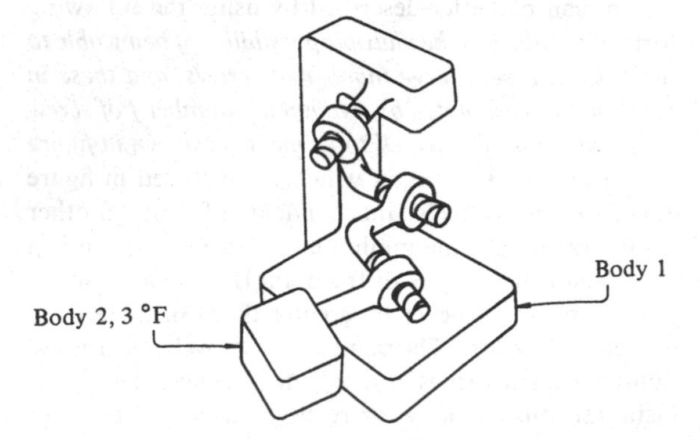


39. The revolute axes in the mentioned figures are marked with the nomenclature *r–r*; *the lines r–r, which can be called r-lines in the bodies, have the special property that each of them are cut by all of the n-lines in the body.* An *r*-line, evidently, is a particular kind of line in a body, wholly determined by the nature of the constraints, along any one of which an actual revolute joint between the body and the frame can be inserted at the instant, while the overall possibility for motion at the instant is not entirely destroyed. The matter must be put in this negative way because, if the body which is somehow connected to frame has, say, $f$ °F with respect to frame (§ 3.33), the insertion of only one revolute along some chosen *r*-line existing in the body will not be enough to reproduce the $f$ °F (§ 3.38). Such an isolated insertion will, indeed, destroy all of the body's $f$ °F except for one. Provided, (*a*) that sufficient *r*-lines exist, and (*b*) that enough of them, projectively independent, are materialised in this way into a series-connected set of revolute joints at the instant, the set can exactly reproduce the original ICRM of the body and its frame. This applies, incidentally, independently of the way in which the body is originally connected to its frame (§ 3.33, § 3.37 and the figures 3.07); *but there is no guarantee that such a substitution of revolutes can always be achieved, because there is no guarantee that the necessary r-lines will always exist in the body.*

40. The previous paragraph has served to show that the appearance or otherwise of *r*-lines in a body is something of a mystery; *but it can be said correspondingly that the appearance or otherwise of n-lines in a body is something of a mystery too.* Consider this: a body may be connected to its frame by means of some collection of links and joints which exclude any direct points of contact between the body and its frame (§ 3.33). Under such a set of circumstances we might ask the obvious question: can the ICRM of the body (with its $f$ °F) be exactly reproduced by the substitution of $(6-f)$ direct points of contact between the body and its frame? The answer to this question is: if the freedom of the body is less than three, yes; but if the freedom of the body is three or greater, only sometimes. This means that, *if the freedom of a body is three or greater, there is no guarantee that such a substitution of direct points of contact can always be achieved, because there is no guarantee that the necessary n-lines will always exist in the body.*

41. Please refer again to the § 3.37; *and come to the important conclusion that n-lines and r-lines in a body are curiously alike.* While there is clearly much more to this matter of their similarity than meets the eye at the moment, please look next at the two 2-link mechanisms

in the figures 3.09. It should be seen in the mentioned figures that, while the five inserted revolutes in figure 3.09(*b*) are in series with one another between the body 2 and the frame 1, the five inserted points of contact in figure 3.09(*a*) are in parallel with one another between the bodies 1 and 2. Let us however, and more importantly, see in the figures that, *whereas each r-line cuts all of the n-lines which are extant, each n-line cuts all of the r-lines which are extant*. It will be shown in due course that this statement, which summarises an aspect of the *reciprocity* of these two kinds of line, applies throughout the whole of the field of mechanism; but a convincing proof of it must wait for later argument.

42. It is important also to notice with respect to this matter of reciprocity that, whereas in figure 3.09(*a*), where 1 °F is extant, there is only one *r*-line and a vast multiplicity of *n*-lines, there is in figure 3.09(*b*), where 5 °F are extant, the opposite. Notice moreover that, if a steady force **F** were applied from the body 1 upon the body 2 along any or all of the *n*-lines in either one of the figures, this force or those forces would result in no relative rotation of the bodies about any of the revolutes. Notice also that, if a relative angular velocity $\omega$ were extant at the instant at any one or all of the revolutes, it would result in no work being done by the forces **F** at the frictionless prongs and platforms. Instantaneously, the tips of the prongs would be traversing paths parallel with the surfaces of the platforms there; they would neither be departing from the surfaces nor would they be digging in. One should also be aware that, while a force along a line is by definition (§ 10.40) *a wrench of zero pitch* along that line, a rate of pure rotation, an angular velocity alone, about a line is by similar definition (§ 10.41) *a rate of twisting of zero pitch* along that line. One begins to see accordingly that there is a sameness about *n*-lines on the one hand and *r*-lines on the other. Until these lines can both be seen as *screws* of zero pitch, however, the matter of their reciprocity being hinted at here will not be fully clarified (chapter 14). The following paragraph will not be fully clear to many readers either, but for compelling reasons, I wish to write it now. Please either ignore the following paragraph if it is unintelligible, or do a little preliminary reading of chapter 10.

43. I announced by definition at § 3.01 that a right line in a body is any straight line which joins two points in the body whose linear velocities at the instant are perpendicular to the line; I proposed at the same place the extended term, *motional right line* for that line. We noted later that, while all *n*-lines are by definition right lines, namely motional right lines, all motional right lines are not necessarily *n*-lines (§ 3.05). I now announce by an extension of the definition that a right line in a body can also be a line which joins two points in the body whose *moment vectors* at the instant are perpendicular to the line, and I propose the extended term, *actional right line* for such a line. I moreover assert that, whereas all *r*-lines in a body are right lines, namely actional right lines, all actional right lines are not necessarily *r*-lines. In this way I am forecasting that, in

Figure 3.09(*a*) (§ 3.41) Five inserted prongs and platforms mounted in parallel. Frictionless, they are collectively reciprocal to the frictionless revolute. Note that no combination of the forces **F** that might be generated at the prongs will result in a turning at the revolute, and that no angular velocity $\omega$ at the revolute will cause the forces **F** to do any work at the points of contact.

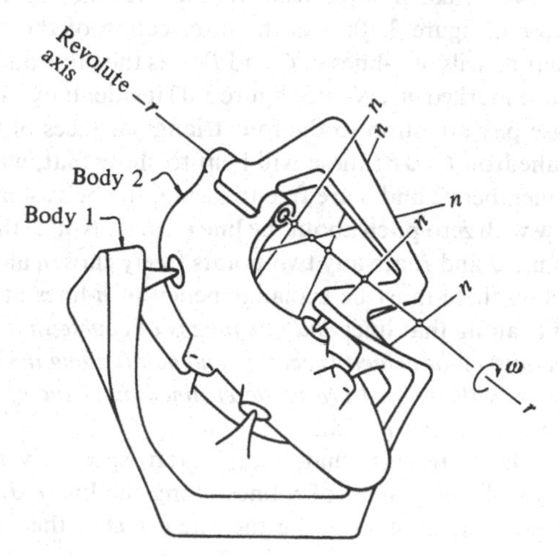


Figure 3.09(*b*) (§ 3.41) The five series-connected revolutes between the body 2 and frame 1 provide a joint which is reciprocal to the given single prong and its platform. No combination of angular velocities $\omega$ at the revolutes (at the instant) will cause a coming away or a digging in at the prong, and any force **F** at the prong will cause no rotation at the revolutes. Please read also § 14.47.

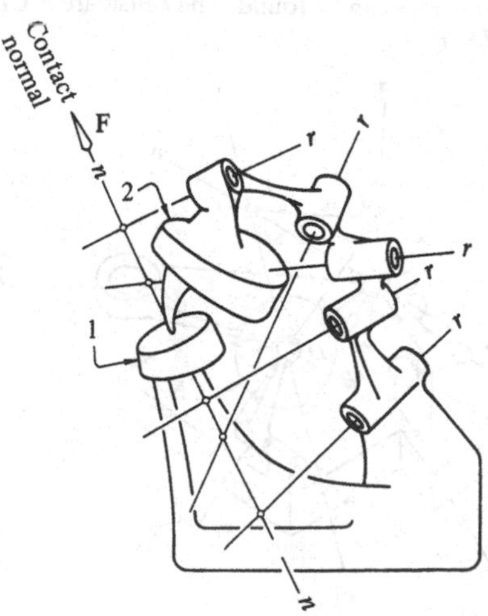

all essential particulars, *r*-lines will be found to be like *n*-lines. Please go to chapter 10, and refer to § 14.37 through § 14.44, for a further clarification of this important idea. Please note here that, whereas some *n*-lines may be actually 'clothed in reality' and thus become *real contact normals*, some *r*-lines may similarly become, in design, *real revolute axes*. Look at the line *B–x* in figure 2.04 for example: it is an *n*-line in the bodies 1 and 3, and it could become a contact normal there; if it did, the revolute axes at the hinges shown could 'destroy' any force along the line *B–x* (§ 14.18).

**The various lines in a body with 2°F**

44. For a long time in this chapter, the mechanism of figure 3.01 has been used as a vehicle for the argument. The degrees of freedom of the body 4 was, and has remained, at unity. An immense number of *n*-lines were identified in the body 4 in figure 3.01, and some of them are shown in figure 3.03. We found moreover that all of the right lines in the body 4 in figure 3.01 were *n*-lines also. With regard to *r*-lines, however, and remembering § 3.35, one can see that, unless the single screw of figure 3.05 turned out by chance to be a screw of zero pitch, *there are no r-lines which might have been found by us in the body 4 at the figure 3.01, for none exist.*

45. If however we removed in figure 3.01 the prong at *E* and thus reduced the mechanism to that which is shown in figure 3.10, the degrees of freedom of the body 4 would be increased to 2°F. Under these new circumstances, in figure 3.10 (where there is no gravity), let us look at the now more limited distribution of *n*-lines in the body 4 and find, perhaps, some *r*-lines there. Let us be aware as before that the implied frame of reference for the movable body 4 is the fixed link 1 of the mechanism, and that this means as usual that any *n*-lines or *r*-lines in the body 4 are also *n*-lines or *r*-lines in the body 1.

Figure 3.10 (§ 3.45) This is the mechanism of figure 3.01 without the prong at *E*. The body 4 now has 2°F relative to frame, and the circumstances are such that two *r*-lines can be found. The *r*-lines are *r–CD–r* and *r–JK–r*.

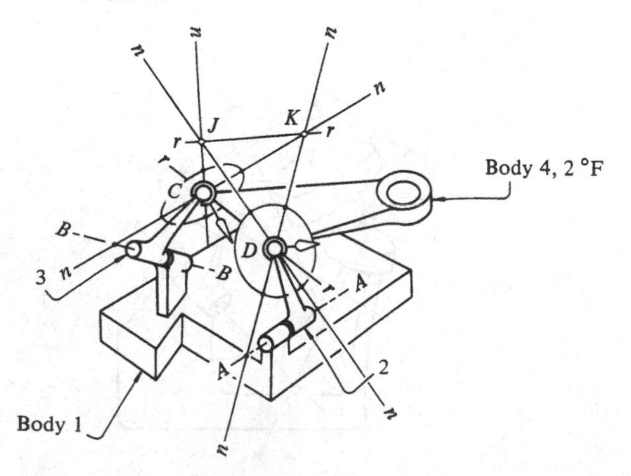


46. The only sets of *n*-lines that are immediately visible in figure 3.10 are the two planar pencils of them which were earlier identified at *C* and *D* (§ 3.06). As before, the same argument applies for these two planar pencils because the directions of the likely velocities at *C* and *D* remain as they were. *It is however now and newly evident that, if the two links 2 and 3 were both to remain at rest, the body 4 would be free to rotate (namely to screw with zero pitch) about the line CD.* It is for this reason that the line *CD* is marked *r–CD–r* in figure 3.10. Thus we note that at least one *r*-line has been found in the body 4 at this particular configuration of this particular mechanism in figure 3.10.

47. We have already noted that the directions of the velocities of the points $C_4$ and $D_4$ in figure 3.10 are irrevocably determined by the nature of the mechanism, and now, by § 5.25 (which is illustrated in figure 5.01), we can see that the directions of the velocities of all of the points in 4 which are along the straight line *r–CD–r* in 4 are also irrevocably determined by the nature of the mechanism. It would appear, however, upon first sight that, for every other point in the body 4, there is a choice for the direction of its velocity. By combining a freely chosen, instantaneous angular velocity about *r–CD–r* with a freely chosen, instantaneous linear velocity at *C* say, each point in the body might be seen to have a single infinity of choices for its linear velocity; indeed it would appear to have a planar pencil of choices. *So we might expect to find through every other point in 4, not a single infinity, but only a single one, of n-lines, normal to the planar pencil there of the possible directions of its velocity.* Please refer to § 3.07.

48. Take a look however at the line marked *r–JK–r* in figure 3.10: it is the intersection of the two planar pencils of *n*-lines at *C* and *D*; it is the same line as the line marked *m–JK–m* in figure 3.03 incidentally. And please pay attention to the four triangular faces of the tetrahedron *CDJK*; these will help to show that, when the members 2 and 3 are free to swing, the body 4 may screw with zero pitch about the line *r–JK–r*. Notice that, because *J* and *K* are any two points freely chosen along *r–JK–r*, there must be a planar pencil of *n*-lines at all points along that line: *and this means of course that the directions of the velocity vectors at all points along the line r–JK–r in the body 4 are predetermined by virtue of the nature of the mechanism.*

49. Notice also that, because, retrospectively, any pair of planar pencils of *n*-lines along the line *r–JK–r* intersect one another along the line *r–CD–r*, there are

planar pencils of $n$-lines at all points along that line also (§ 3.12). This we already know (§ 3.49) *for we have already found that the directions of the velocity vectors at all points along the line r–CD–r are predetermined by virtue of the nature of the mechanism.*

50. Let us look now at a modification of the mechanism to see how it is in figure 3.10 that, *although the directions of the velocities at the vast majority of points in the body 4 are unpredictable given the nature of the mechanism, the directions of the velocities at all points along the two lines r–CD–r and r–JK–r in the body 4 are predetermined given the nature of the mechanism.* Refer to figure 3.11.

51. Let us identify and mount in the imagination a new, intervening body, the rigid tetrahedron *CDJK*, and let us call it the body 5. Let us imagine that, along the edge *r–JK–r* of the body 5, a hinge is inserted directly between the frame 1 and the body 5. Let us also imagine that, along the edge *r–CD–r* of the same tetrahedron, another hinge is inserted to join it to the body 4, thus removing from consideration the bodies 2 and 3. Please pay attention to the two lines *r–JK–r* and *r–CD–r*: *please see them as being fixed in the body 4, and be aware that we are about to study the instantaneous motion in the body 1 of these two straight lines of points which are fixed in the body 4.* Refer to § 1.18.

52. The body 4 is connected to frame via two, series-connected revolutes in figure 3.11, and this coincides with the circumstance of there being 2°F at the body 4: the number $f$ is 2, and the screws of minimum number $f$ have both been chosen to be screws of zero pitch (§ 3.34). Let there be two angular velocities $\omega_{15}$ and $\omega_{54}$ of any chosen values occurring at the hinges simultaneously (§ 12.11), and let us see that this will

Figure 3.11 (§ 3.50) The body 4 of the mechanism in figure 3.10 suspended upon two series-connected revolutes to frame. These have been materialised at the two $r$-lines discovered at § 3.48. The revolutes produce thereby exactly the same instantaneous capacity for relative motion of the body 4 as did the members 2 and 3. Refer also § 15.46.

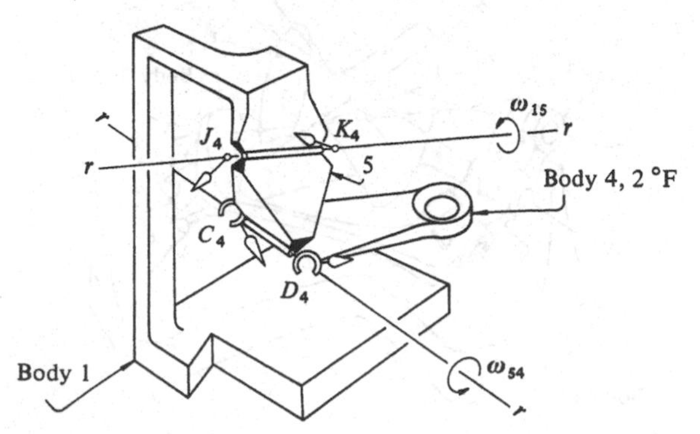


correspond with a randomly chosen, instantaneous motion for the body 4. To study the effect of these angular velocities at the instant, let us apply first an infinitesimal angular displacement of some magnitude about the line *r–CD–r* while the revolute at *r–JK–r* is immobilised, then next an infinitesimal angular displacement of some other magnitude about the line *r–JK–r* while the revolute at *r–CD–r* is immobilised. Such a staged manoeuvre will show very clearly that, while the ratio of the magnitudes of the two displacements will determine the directions of the instantaneous linear velocity vectors at all points in the body 4, *it will have no effect upon the directions of the linear velocity vectors at those particular points which are distributed along the two mentioned r-lines in the body 4.* These latter coincide respectively with the two, series-connected revolutes which are shown.

53. It now becomes evident in figure 3.10 (*a*) that there are two $r$-lines associated with the instantaneous capacity for motion of the body 4; (*b*) that neither of these lines is more important than the other; and (*c*) that, while there are very many $n$-lines in this body in these circumstances of 2 °F, the whole set of them consists simply of all those lines which can be drawn to cut both of the $r$-lines. I explain elsewhere how such an ensemble of lines constitutes an $\infty^2$ of lines, that it is called a *hyperbolic linear congruence*, and how the two $r$-lines in figure 3.10 are the two directrices of the congruence there (§ 10.09). The reader should be warned here that, although there is always a linear congruence of $n$-lines in a body with 2 °F, the directrices of the congruence are not always real lines. If the directrices are imaginary lines the congruence is called an *elliptic linear congruence*, and we would see under those circumstances, not only that the linear congruence would look different, but also that there would be no $r$-lines (no directrices) to be found in the body with 2 °F (§ 3.39, § 3.54, § 10.27, § 11.24). Please be reminded that, as soon as the body 4 in figure 3.10 begins to move, the number of right lines existing at an instant becomes the usual number, namely $\infty^3$ (§ 3.05).

54. A practical exercise. Approximated ball and socket joints to join round wooden rods can be made with a pencil sharpener, drilled holes, flexible string, and adhesive tape. See the inset in figure 3.12. Connect a pair of bodies – let us call them 1 and 2 – by means of four, non-intersecting, ball-ended rods, and examine the 2 °F action. Having chosen a reference location for the body 2 with respect to the body 1, can you see two $r$-lines which cut the centre lines (namely the $n$-lines) of all four of your rods? Or are there no such $r$-lines in your apparatus? Or is there, by some extraordinary chance, only one of them? For further reading, go to chapter 10. Please be reminded here that my remarks at § 3.43,

which explained how, while any $n$-line can be seen as a motional right line, any $r$-line can be seen as an actional right line, are applicable generally and thus in the present case as well. Consider the system of four forces which may be obliged to act along the centre lines of the four ball-ended rods and whose separate magnitudes are open to choice, and consider whether any one of these forces and thus any combination of them can have a moment about any one of the $r$-lines, which latter are the actional right lines here. For reference reading, go to chapter 10. Refer also § 11.29.

**The central, symmetrical case of 3 °F**

55. The object of the illustrated apparatus in figure 3.13(*a*) is to set up a body 2 with 3 °F with respect to a body 1 in such a way that, within it, three $n$-lines do exist (§ 3.40). The simplest way for ensuring this is to bring the chosen body 2 into contact with the body 1 at three direct points of contact. In the figure, this has been done; the three points of contact are at *A, B* and *C*; the three resulting $n$-lines are marked *n–A–n, n–B–n*, etc.; and please note in the figure that for convenience there are prongs upon the body 2 and flats upon the body 1 (§ 3.05, § 8.09).

56. Following now the gist of the argument introduced at § 3.38 and more or less concluded at § 3.41, we can see that any imagined hinge which connects the bodies 1 and 2, and which is located along an $r$-line cutting all three of the $n$-lines, will be a hinge which could operate at the instant without disturbing the maintenance of contact at any one of the points of contact. The main question I wish to ask in the present circumstance is the following: given that we may wish to reproduce exactly the limited ICRM between the bodies 1 and 2 which is mechanically determined by the three given points of contact between the bodies, by providing instead a number of hinges *r–r* in series with one another as shown, what is the minimum number $f$ of hinges required (§ 3.37)?

57. I would like to prove, in other words, for this particular case of exactly three points of contact between the bodies, the hitherto unsubstantiated statement made at the beginning of § 3.38. Indeed I would like to deal with what might now be suspected to be the *central, symmetrical case* (chapter 6), where exactly three points of contact between the bodies is matched by an equal number, namely three, of equivalent hinges. This will lead to the presentation here of a most important theorem.

58. Before beginning to offer the promised proof, I would like to ask the same question of the case in a different way. Please refer to § 1.31 where a complex joint in a black bag is discussed in relation to the matter of connectivity. Please also imagine that the three prongs in figure 3.13(*a*) are magnetically attracted at the three flats, so that, despite any crosswise sliding which may occur, contact there will not be broken (§ 6.15, § 8.12). Putting the same question now in other terms, I wish to ask: supposing that the three-point-contact joint between the bodies 1 and 2 were in a black bag in figure 3.13(*a*), and that it was to be replaced in its entirety by a number of hinges *r–r* mounted in series between the bodies 1 and 2 in the same black bag; *how many hinges would be necessary and sufficient to provide exactly the*

Figure 3.12 (§ 3.54) Such an apparatus as this can easily be built. If you grasp the body 4 and wrench it in such a way (and there is an $\infty^3$ of such ways) that no angular velocity occurs about either of the two $r$-lines shown, you are applying a wrench which could have been mobilised via the original links 2 and 3 in figure 3.10. For wrench, refer to § 10.02.

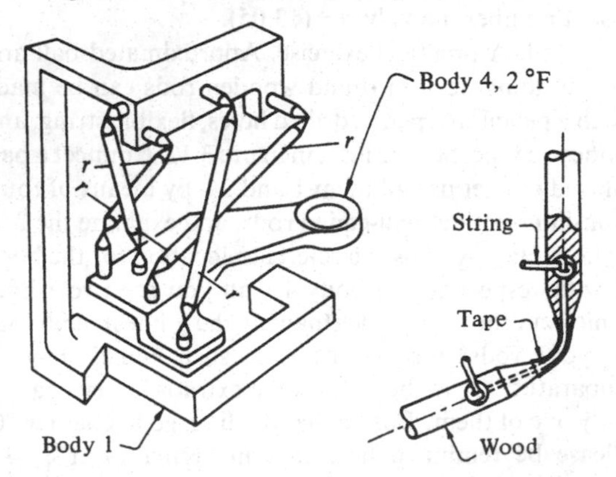


Figure 3.13(*a*) (§ 3.55) A set of three revolutes, series connected to frame, being equivalent at the instant to three points of contact at prongs. Maintaining contact at the prongs, and in its shown configuration, the apparatus is transitorily mobile (§ 2.08). Refer also § 7.52.

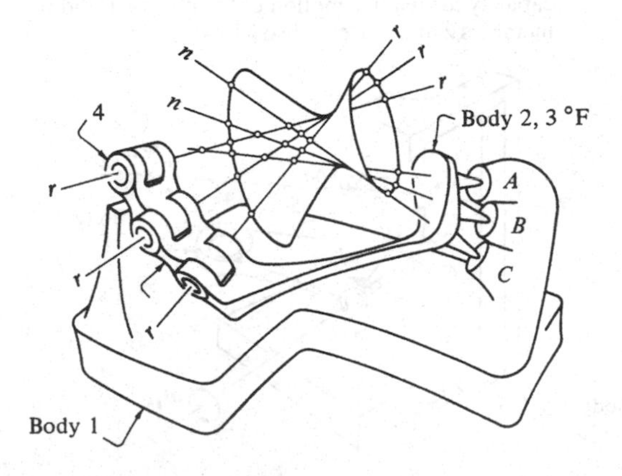

*same 'feel' at the instant, an investigator, blind because of the bag, not being able to tell the difference merely by operating the joint?*

59. One way to offer a proof that the answer to this question is three is to argue as follows. Taking the point $A_2$ at $A$, and arguing always in terms of infinitesimal displacements, we can see that provision of a single hinge *r–r* will allow $A_2$ to be moved only along the arc of a circle about *r–r*, and thus not to be provided with the freedom it must have in reality to be moved to any new position in its immediate neighborhood. Provision of a second hinge *r–r* (in series with the first) would clearly allow that freedom however. Having thus provided with two hinges the freedom which $A_2$ requires for an arbitrary repositioning, we can look next at the point $B_2$. We see here that, $A_2$ having been freely repositioned using the first two hinges, the total freedom for any arbitrary repositioning of $B_2$ can be provided with only one extra hinge, for $B_2$ upon its flat need now only be free to move along an arc (§ 5.01). Next we can see by the same token that, $A_2$ and $B_2$ having been freely repositioned in this way, $C_2$ will have already taken up its only available position to suit. *Three hinges in figure 3.13(a) are, accordingly, necessary and sufficient for the relocation of the body 2 into any new, neighboring location which may arbitrarily be chosen for it.*

60. Another way to check that the answer to the question of § 3.58 is three, is to argue the matter in terms of mobility and connectivity. Refer to § 1.34 for a statement of the general criterion for mobility. The connectivity of the joint between the bodies 1 and 2 with three prongs only is the same as the mobility of the mechanism. The connectivity, $C_{12} = M = 6(2-3-1)+15$, namely 3. Three hinges, series connected between the bodies and in the absence of prongs, would, we have said, provide the same mobility of the mechanism, and, once again, the connectivity between the bodies 1 and 2 would be the same as the mobility (§ 2.37). We can check this by writing for three hinges only that the connectivity, $C_{12} = M = 6(4-3-1)+3$, namely 3, as before. What cannot be seen, however, by the use of this particular argument is: *that the three prongs and the three hinges can exist, at the instant, simultaneously.*

61. Refer now to figure 3.13(*b*) and see that this is no more than a different drawing of figure 3.13(*a*). It is designed to show more clearly however the central importance of the *hyperboloid* involved (chapter 10), and the reader might care to study the various aspects of this figure carefully. There are some relevant properties of the hyperboloid which might be briefly quoted here. They are: (*a*) that any three lines will, in general, define a hyperboloid; this can be seen if one cares to draw all of the lines which can be drawn to cut a given set of three lines; (*b*) the 'throat' of the hyperboloid is, in general, elliptical; (*c*) any generator of either one of the two families of generators upon the hyperboloid will cut every generator of the other family. Each family of generators upon a hyperboloid is called by the name *regulus*; there are, thereby, two reguli of lines which can be drawn upon the surface of any hyperboloid, and such reguli, as we shall see, figure largely and continuously in the ICRM of mechanically connected bodies (chapters 6, 10, and § 11.04).

**Another important theorem**

62. Having inserted actual hinges along any three of the *r*-lines which cut respectively all three of the *n*-lines bespoken by the prongs in the figures 3.13, and having proven that only three such hinges are sufficient for reproducing exactly the ICRM of the two contacting bodies, we can see next: (*a*) that additional hinges, mounted along other *r*-lines also drawn to cut the three bespoken *n*-lines, could be physically inserted into the imagined apparatus without destroying (or enhancing, either) the ICRM; (*b*) that a whole single infinity of such *r*-lines, and thus of possible additional hinges, exists; and (*c*) that these *r*-lines are the generators of a regulus upon the surface of the hyperboloid defined by the three *n*-lines. *We can see accordingly that we could entertain in the imagination the physical existence of the whole regulus of hinges simultaneously.*

Figure 3.13(*b*) (§ 3.61) A redrawing of figure 3.13(*a*). Please note that the actual hinges and prongs may be located anywhere along their *r*-lines and *n*-lines respectively without altering the instantaneous geometry. Please read also § 7.52.

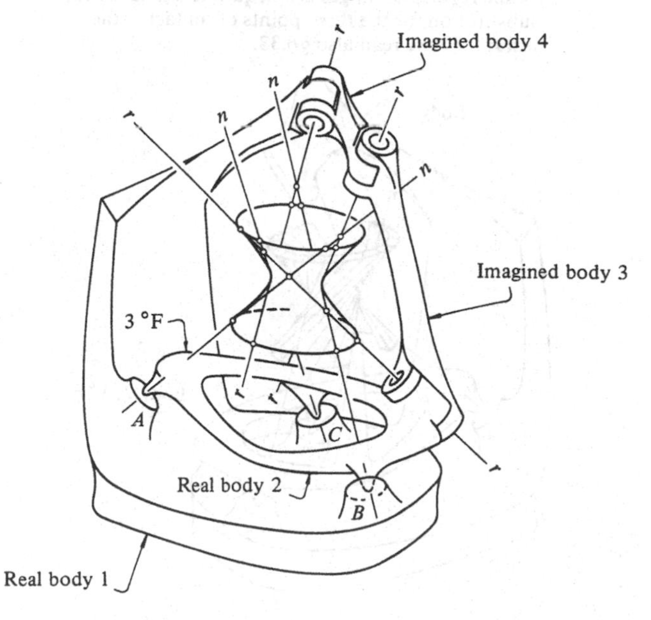

63. Refer to figure 3.14 which is, as will be seen, directly derived from figure 3.13(*b*). Please imagine that, like a hinged and twisted watchband or some such thing, the ribbon of the infinity of hinges between the bodies is available for the motion; and be aware that any chosen three of those hinges are sufficient to duplicate the ICRM provided independently by the prongs. Please notice (and refer again to § 3.43) that some steady force applicator, such as a hydraulic jack or a suitably shaped compression spring, could be inserted between the platform and the prong at any one or all three of the pairs of prongs and platforms without causing a rotation about any one of the hinges of the ribbon. Please consider in the statics here the balance of the forces only; the stability or otherwise of the balance is irrelevant.

64. It should next be clear that, given this infinity of hinges (or of *r*-lines) in figure 3.14, there is a corresponding infinity of *n*-lines, which are the generators of the other regulus upon the same hyperboloid. What this means is that an important theorem has been proven: *given any three lines in a body (projectively independent, chapter 10) which are bespoken by virtue of some existing constraints to be n-lines in the body, the whole regulus of lines to which the three bespoken lines belong are also n-lines in the body.*

65. This is a widely applicable theorem. It applies both independently of the nature of the constraints and independently of the degrees of freedom being enjoyed by the body. Unless the freedom is equal to or less than 3 °F however, there cannot be as many as three projectively independent *n*-lines bespoken in the body; this fact is, of course, a natural restriction upon the applicability of the theorem.

Figure 3.14 (§ 3.63) A regulus of imaginery hinge axes in series. It physically connects the bodies 1 and 2. The whole regulus of hinges is an equivalent mechanical substitution for the three points of contact at the instant. Please read also § 6.33.

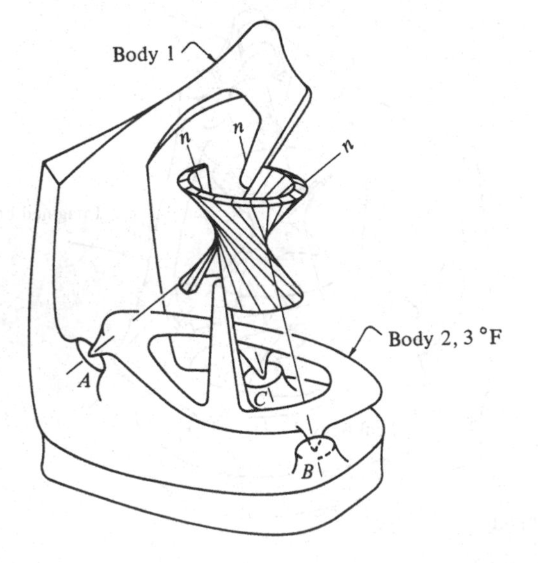


66. The theorem of § 3.64 will find its first application in chapters 6 and 7, where the question (among others) of the total list of the possible kinds of simple joint between a pair of members is pursued. The theorem will be embroidered, and the matter elaborated in chapter 9, where the pattern of all of the right lines in a moving body is discussed (§ 9.31).

67. Please notice now (and refer again to § 3.63) that some steady force applicator could be inserted at each one of the whole single infinity of equivalent pairs of prongs and platforms in figure 3.14, each one of which might be imagined to be put anywhere along its parent *n*-line. *And we could, if we wish, imagine the single infinity of prongs and platforms to be distributed neatly around the elliptical throat, or along the line of striction which is different (§11.09), of the hyperboloid.* Refer to chapter 6.

**Some concluding remarks**

68. It may appear that I have approached the various phenomena discussed in this chapter 3 in a somewhat disorganised manner. Indeed it could be argued that the approach I have taken is, from the pure mathematical point of view, a sadly misguided one. Summarising, however, the following can be said: (*a*) it has been shown that a special kind of line in a body which is determined by the constraints, namely an *n*-line, is a regularly occurring and probably useful kind of line in mechanism (§ 3.06); (*b*) it has been suggested that an equivalent hinge line, or an *r*-line in a body, is a line which is strangely akin to an *n*-line, and explained that any *r*-line in a body must cut all the *n*-lines and vice versa (§ 3.41); (*c*) two important theorems about the overall layout of *n*-lines in a body have been presented (§ 3.13, § 3.64); (*d*) a selection of workable mechanisms have been put together and dissected; and (*e*) *the beginnings of a geometric pattern have emerged; it has been found, in some special cases only, that, associated with 1 °F, 2 °F, and 3 °F of a body in mechanism, there are ensembles of n-lines existing which are, respectively, linear complex, linear congruence, regulus.*

69. We shall see in later chapters that these ensembles of lines figure largely and continuously in the theory of freedom and constraint. They exist and interact in our geometrical analyses, and their ubiquity among the geometrical essences of machinery cannot be denied (§ 1.04). There is a rigid hierarchy of number among them, which has yet to be explored, but let me mention here that, whereas a linear complex has an $\infty^3$ of lines, a linear congruence and a regulus have $\infty^2$ and

$\infty^1$ of lines respectively. The smooth simplicity of their interrelationships and the wide generality of their importance cannot be easily seen however, and in this chapter 3 only a few limited glimpses of all of this have been revealed (§10.09).

70. One trouble is, in developing a convincing relationship between the general theory of screws – Ball (1900), Dimentberg (1965), Hunt (1978) – and the ordinary practice of machine design is that, in ordinary machinery, isolated direct spots of contact between the members, actual physical screws of zero pitch, and various simple combinations of these, predominate. The profusion of points of contact, hinge joints, pairs of balls and sockets, prismatic joints and the like lead us to approach the general theory of screws with a somewhat limited outlook. Although it is quite possible, for example, for a body with 2°F to have no *r*-lines at all in it (§3.53, §15.49), and for a body with 5°F not to have even one *n*-line in it (§3.37, §23.38), such occurrences are not commonly met or recognised in ordinary machinery. To provide convincing examples of such extraordinary things is not being contemplated at this stage. The writer would much rather run the risk of being thought to be limited and misguided than to run the worse risk of being seen to be exotic or impractical. Please refer however to chapters 10, 14, 15, and 23.

4. Given that the grains are small enough and that the world is big enough, it may usefully be argued that the number of grains of sand in the world is not infinity but infinity cubed.

# 4 Enumerative geometry and the powers of infinity

**Some first thoughts**

01. If we agree (and we do) that no more than three parameters are needed to fix a single point in our Euclidean 3-space, we can argue that there are, all counted, an $\infty^3$ of points. Next we can say (and we do) that, because the number of different radii available for the drawing of a sphere about a point is $\infty^1$, there are, all counted, an $\infty^4$ of spheres. Statements like that are typical of enumerative geometry, a method of mathematics much to be used in this book.

02. A parameter, in this metrical context, is a thing to which we can unambiguously ascribe a single number to correspond exactly with its size. Parameters turn out to be, either *(a)* a distance along some line either straight or curved which is measured from some origin point in the line, or *(b)* a magnitude of some angle across some surface either planar or ruled which is measured from some origin line in the surface. There appear to be no other kinds of parameter.

03. When we say that there exists for example an $\infty^n$ of some geometrical entity we mean, quite simply, that a minimum number $n$ of parameters is needed to specify a nominated one of the entities from among its family. The family in question is then called an $n$-parameter family. The spheres in space, accordingly, are a 4-parameter family: any one of the spheres can be nominated by the use of only four parameters. Within a Cartesian frame, for example, a set of three, $x$, $y$, $z$, length-parameters could position the centre, while another length-parameter $r$ could tell the radius of the sphere. Alternatively, within a cylindrical system of coordinates a set of three $x$, $r$, $\theta$, of parameters could position the centre of the sphere while another, say $a$, could nominate its radius. The spheres in any event remain the same spheres; they exist independently of the parameters which may be chosen to choose among them.

04. When we say correspondingly that there are for example an $\infty^f$ of locations available for a bounded, irregular, rigid body whose degrees of freedom with respect to some frame is $f$, we are speaking again of $f$ parameters being needed to locate the body with respect to that frame (§ 1.28). Reflecting upon the matter we find that a minimum of three points fixed in the body (§ 5.30), or a single point at the intersection of two lines fixed in the body with segments of those lines arranged like a *tick* and fixed at the point, are each of them examples of necessary and sufficient devices for representing the body geometrically. For other mentions of this tick, refer to § 10.64, § 19.31, and § 23.05. Any rigid body with less than six degrees of freedom with respect to some frame will be restricted in its *capacity for motion* (§ 3.34); there will be, in all such cases, the evident existence of some mechanical connection between the body and that frame (§ 23.05). If that mechanical connection (or those mechanical constraints) effectively predetermine say $c$ of the six parameters necessary to locate the body, the $f$ of the parameters remaining will always be the number six minus $c$ (§ 1.52). Mechanical constraints are invariably of the nature that their zone of effectiveness in space is limited, however (§ 1.29), and this fact will always restrict the body in question to moving within its *zone of movement*. These latter remarks and this bland term, zone of movement – they hide a multitude of difficulties (§ 10.64) – apply to all bodies mechanically connected in any way to frame, even if the body's degrees of freedom with respect to frame happens to be six (§ 2.37): please refer to figure 1.10(*c*) and see there, if you can, the boundaries of the limited zone of movement of the suspended cylindrical knob. Take account of the shapes of the connected pieces, let none of the joints requiring force closure lose contact, and be aware as you think that I am thinking here about the dexterity of a robot. This term, zone of movement, is used by me to describe, not the layout of the ISAs which are available to the body and thus its capacity for motion (§ 3.34), *but the bounded envelope in space outside of which the bounded body is not free to protrude itself, but within which the body has, independently of the restricted size or the shape of the envelope, its whole $\infty^f$ of possible locations.*

05. The number $\infty^n$ when $n$ is unity or more, in other words, is unaffected by any division by a finite

number. One third of all of the spheres in space, for example, is not one third of $\infty^4$ which is an absurdity, but $\infty^4$ simply; the possible locations in space for the cylindrical knob of the apparatus in figure 1.10(*c*), despite its limited zone of movement, is the whole of $\infty^6$ simply. The whole matter of enumerative geometry itself, however, and here I quote a major source of reference for this material, Gilbert and Cohn Vossen (1936), is not so simple as that. There is almost always the trouble that parameters for an entity may be offered redundantly.

**The redundancy of parameters**

06. Suppose we speak about the circles upon the spheres in space. Because each sphere is cut by an $\infty^3$ of planes (§ 4.11), there is an $\infty^3$ of circles upon each sphere. But does this mean that there is an $\infty^4$ times $\infty^3$, namely an $\infty^7$, of circles? Refer to figure 4.01 and see that for each circle upon a sphere there is an $\infty^1$ of spheres. It follows that, by the above false reckoning, each circle in space has been counted an $\infty^1$ of times too often. There are, accordingly, not an $\infty^7$, but only an $\infty^6$ of circles which can be drawn in space.

07. This fact can be confirmed by seeing that, to nominate a given circle in Euclidean space, the following six, non-redundant parameters are enough: a set of three length-parameters, say $x$, $y$, $z$, to position the centre of the circle in a Cartesian frame; one length-parameter say $r$ for the radius; then two more, both angle-parameters to orient the plane of the circle unambiguously.

08. To see another example of the same artifice, take the circle defined by three points. There being $\infty^3$ of ways of choosing each point, there are an $\infty^9$ or triplets of such points. But each circle has an $\infty^3$ of such triplets of points upon its circumference, so the total number of circles which can be drawn is $\infty^9$ divided by $\infty^3$, namely $\infty^6$, as before.

Figure 4.01 (§ 4.06) A single infinity of spheres may be drawn to contain the same circle. In the figure here a single circle appears as a straight line segment *C–C*. The number of circles that may be drawn in space is not $\infty^7$, as one could be led to expect, but only $\infty^6$. Refer to § 4.07 and § 4.08.

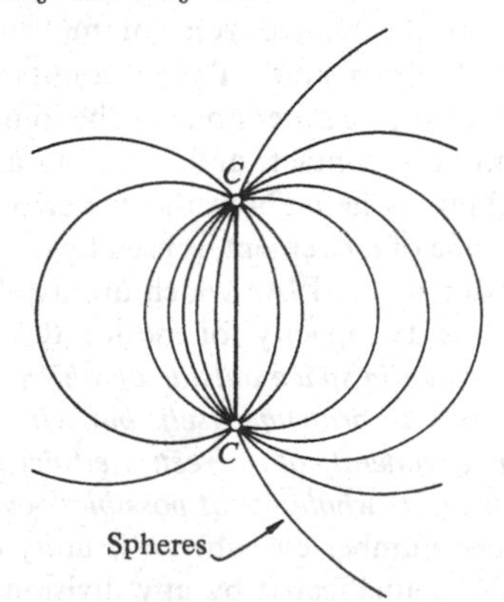


**On the question of points and lines in space**

09. Although not exclusively so, much of the enumerative argument in this book, derived as it is from the physics of forces and angular velocities, is concerned with various ensembles of lines (§ 10.09), and thereby to some extent with planes and other ruled surfaces. Refer to § 3.64, § 10.26, § 15.12. I wish to include here some examples from that domain.

10. Let me deal first with the number of straight lines which can be drawn in our ordinary, Euclidean space. Take in the space a rectangular, Cartesian frame with axes $x, y, z$ and see that, because the points in space are a 3-parameter family (§ 4.03), there is an $\infty^3$ of points. It should next be clear that, in the same way as there is an $\infty^2$ of planes which can be drawn to contain a given point in space, there is an $\infty^2$ of lines (these together can be called a *star* of lines) which can be drawn to pass through a given point. It might be argued to follow that, because there is an $\infty^3$ of stars, the total number of lines in space is $\infty^3$ times $\infty^2$ namely $\infty^5$; but that would be wrong, for a redundancy is thereby missed. The redundancy resides in the fact that, along each of the lines in space, there is an $\infty^1$ of points; each line has thus been counted an $\infty^1$ of times too often. It follows that the correct number for the number of straight lines in space is not $\infty^5$, as falsely told above, but only $\infty^4$. This result should be a well remembered one; it is used by me as a basis for argument very often (§ 19.25).

11. About planes we can say that, in the same way as there is through a given point in space an $\infty^2$ of planes (§ 4.10), there is an $\infty^2$ of orientations in space for a plane through a given point. But for each orientation there is an $\infty^1$ of planes which are parallel. There are, accordingly, an $\infty^3$ of planes which can be drawn in space. It is instructive here to wonder about the argument that, because there is an $\infty^3$ of planes in the whole of space, there will be an $\infty^3$ of planes which can be drawn to cut any given sphere (§ 4.06).

**The regulus and the linear complex of lines**

12. This and the remaining paragraphs of the present chapter might be omitted upon a first reading. A regulus of lines is defined by a set of three skew lines in space (§ 10.11); this particular ensemble of lines is a very important one in the general context of freedom and constraint (§ 3.62, § 10.22). The regulus is a ruled surface comprising a single infinity of lines ($\infty^1$) and the shape of the surface is that of the hyperboloid. The hyperboloid is otherwise known as the only ruled quadric surface of the second order, its algebraic equation in Cartesian coor-

dinates being of the second degree (§ 11.13). As is well known, each hyperboloid carries two reguli upon its surface, a right handed and a left handed one. It will be interesting to know how many different hyperboloids can be drawn in space. Whatever that number is, it will be the same number as the number of reguli which can be drawn; all finite multiples of $\infty^n$ are simply a repeat of the same number, $\infty^n$.

13. Paraphrasing the previously quoted Gilbert and Cohn Vossen (1936), I say the following: a regulus is defined by three skew straight lines; space contains $\infty^{(4\times3)}$ namely $\infty^{12}$ triplets of straight lines; but since every generator of a regulus is a member of a one-parameter family, $\infty^3$ triplets of straight lines define the same regulus; hence there are in space $\infty^9$ of reguli. There are accordingly an $\infty^9$ of hyperboloids which can be drawn in space, and upon each of these there are two reguli.

14. While not overlooking the intermediate, linear congruence (§ 10.09), another important ensemble of lines in space is the linear complex (§ 10.09). It is an axi-symmetric group of lines comprising only a very few of the $\infty^4$ of lines which are available in space. The lines are defined (in conjunction with a given central axis and certain geometrical conditions) by a certain restrictive equation which excludes all other lines, namely $r = p \tan \theta$, where $p$ is the pitch of the complex and $\theta$ is the helix angle (§ 9.15), and there is only an $\infty^3$ of lines. The linear complex can also be defined by the simple equation $P = kL$, where $k = -p$, when Plücker's line-coordinates are used (§ 11.44). Please ask yourself the question: if a complex comprises an $\infty^3$ of lines, how many lines are not of the complex? An answer to this question is discussed at § 9.44.

15. Another interesting question for us is the question of how many reguli can be drawn in space using only the generators of a given complex. To put the same question into other words: how many different reguli exist in a given, single complex? By an argument similar to that at § 4.13 it is easy to reason as follows: a complex contains an $\infty^{(3\times3)}$ namely an $\infty^9$ of triplets of straight lines; but an $\infty^3$ of triplets of straight lines define the same regulus; hence there is in a given complex an $\infty^6$ of reguli. It is not so easy to show however where and of what shape the hyperboloids of the reguli are, and that more difficult matter is dealt with elsewhere (§ 9.31). It can be explained in a similar way that there is an $\infty^3$ of reguli in a linear congruence, the congruence containing only an $\infty^2$ of the lines of space (§ 11.27).

5. A geometrical construction for finding the ISA or the motion screw of a moving body at an instant may be modelled by means of a wire framework. Here the interrelated linear velocities at three known points in the body are the only data.

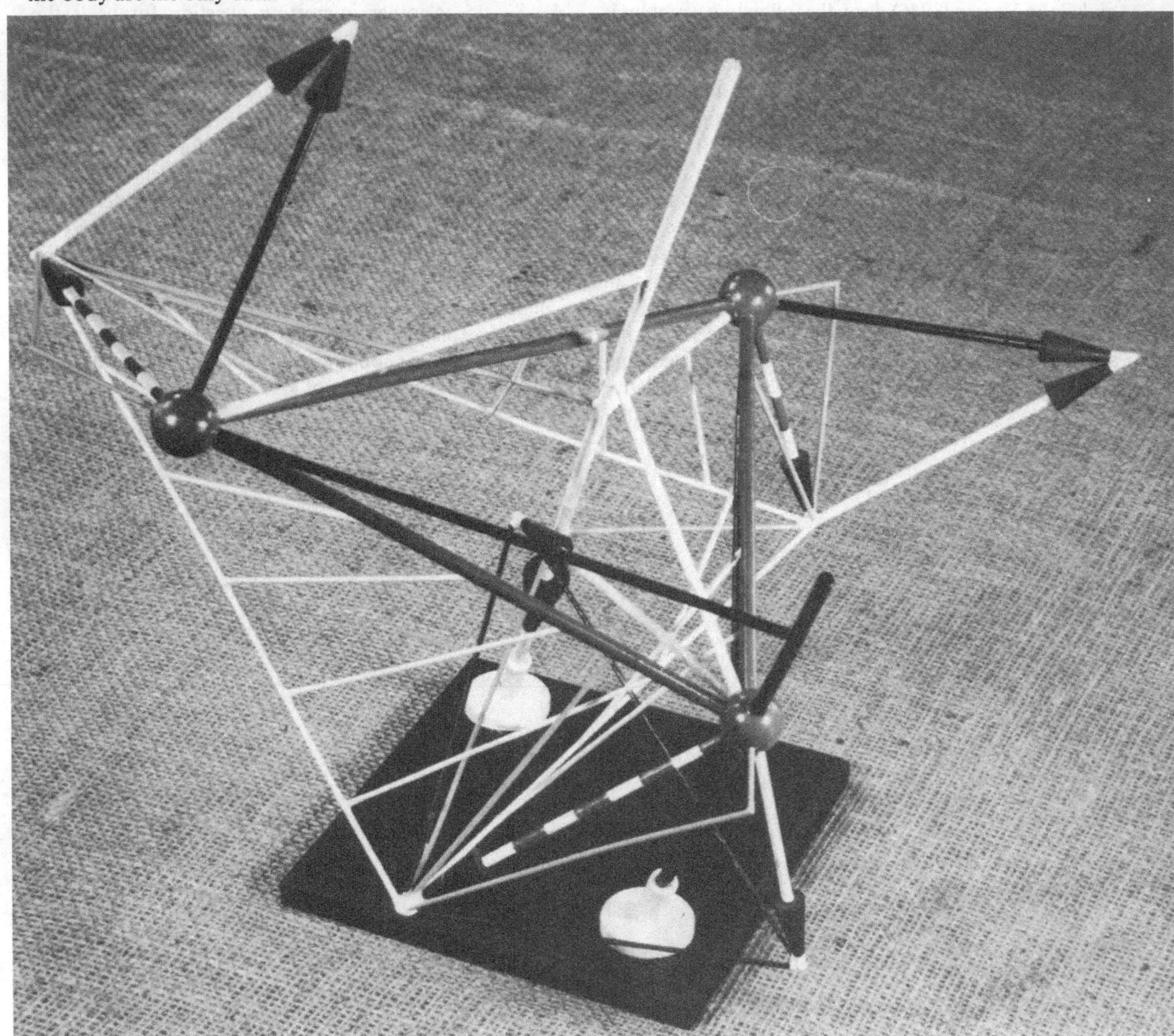

# 5 Rigidity and the instantaneous screw axis

### Some implications of a well-known idea

01. A widespread intuitive idea (and thus a useful definition of) the rigidity of a rigid body is as follows: having fixed any two points $A$ and $B$ in a moving body we argue that, if the body is rigid, the distance from $A$ to $B$ remains constant throughout a movement.

02. A number of important conclusions can be drawn from this fundamental statement. This chapter is devoted to drawing at least one of them, namely that, at any given instant in the motion of a rigid body, an axis with a pitch exists which is unique at the instant. The axis may be called the instantaneous screw axis for the body, or, more simply and often (§ 6.07), the screw in space about which the body is twisting (§ 5.50). Each of its aspects – orientation, location, and pitch – must be known before connections can be made between the so-called angular velocity of the body and the linear velocities of points such as $A$ and $B$ within it.

03. Because the points $A$ and $B$ in their line $AB$ cannot move closer together or further apart, the component along the line of the velocity at $B$ must be the same, at any instant, as the component along the line of the velocity at $A$. Applying this argument to other points along the line, we see that, *for all points along a straight line in a rigid body, the velocity vectors at the points have equal components along the line*. This is a general result. It holds for any line which may be drawn in a rigid body.

### The beginnings of a logical argument

04. Let us call the velocities at $A$ and $B$, to follow convention, $\mathbf{v}_{1A}$ and $\mathbf{v}_{1B}$. They are thus referred to the same frame of reference 1, which is some body 1 other than the moving body 2; although both $A$ and $B$ are fixed in 2, the subscripts 2 for $A$ and $B$ have been omitted here (§ 12.08, § 12.10).

05. Having noted in our imaginations the directions and magnitudes for such linear velocities as these, we usually and conveniently draw the relevant vector arrows with their basepoints attached at the points to which the vectors respectively refer, and accordingly draw the arrowheads or *tips* of the vectors at those ends of the vector arrows that are remote from the points. This usual convention is followed in figure 5.01: the line $AB$ is shown with the vectors $\mathbf{v}_{1A}$ and $\mathbf{v}_{1B}$ attached at $A$ and $B$. Perpendiculars dropped from the tips of those two vectors onto the line $AB$ are marked in the figure with the symbol $d$, while the two corresponding right angles are marked in the conventional manner 90°; the equal components along the line of the velocities at $A$ and $B$ appear as thickened segments of the line.

06. We now consider the velocities relative to 1 of other points along the line $AB$. It will be shown in the following argument that, if the velocity vector at each other point on the line $AB$ is drawn, the tips of all of the vectors will be in a straight line; this line, marked $t$–$t$ in figure 5.01, joins the tips of the vectors at $A$ and $B$. Please refer also to § 3.22.

07. We have already said and note again in recapitulation that unless the segment of line $A$–$B$ is shrinking or stretching (which it is not because the body is rigid), the component of $\mathbf{v}_{1B}$ along the line must be the same as that of $\mathbf{v}_{1A}$, and, because this argument applies for all such pairs of points along the line, we know that the velocity components along the line of all of the points in the line are equal. It follows from this that, if, as we shall prove, the tips of the vectors are distributed along $t$–$t$, and if a series of equally spaced points along $AB$ is taken, the tips of the corresponding vectors will be equally spaced along $t$–$t$.

### Retreat for simplicity to a point with zero velocity

08. To show now that the tips of all of the vectors along $AB$ are in the same straight line $t$–$t$, we may begin by subtracting at all points the vector $\mathbf{v}_{1A}$. This manoeuvre, later to be reversed, is exactly equivalent to putting the observer upon a new frame of reference 3 which, relative to the old frame 1, is moving bodily in such a way that all points within it have the same velocity $\mathbf{v}_{1A}$. The new velocity at $A$ namely $\mathbf{v}_{3A}$ is thus zero, while that at $B$ is $(\mathbf{v}_{1B}-\mathbf{v}_{1A})$ namely $\mathbf{v}_{3B}$ as shown. The vector $\mathbf{v}_{3B}$ at $B$ must be perpendicular to the line $AB$ because its component along the line must be the same

as the component of $\mathbf{v}_{3A}$ along the line, namely zero; this argument applies for all such points as $B$ along the line. It follows that, relative to the frame of reference 3, all of the vectors along the line $AB$ are perpendicular to the line.

09. What we need to show now in figure 5.01 is that all of these vectors which are perpendicular to the line $AB$ are also parallel with one another and proportional in magnitude directly to their distances from $A$. This could be said to follow from the well known relationship in pivoted plane motion that the linear velocity $\mathbf{v}_{3B}$ of a point $B$ in a body 2 is proportional to the directed distance $\mathbf{r}_B$ of the point from the pivot, such that $\mathbf{v}_{3B}=\boldsymbol{\omega}_{32}\times\mathbf{r}_B$, the constant of proportionality being the angular velocity $\boldsymbol{\omega}_{32}$ of the pivoted body. But to argue thus in the present context would be to beg the question; we do not even know that there is a pivot; we only know that there is a single point $A$ in the body 2 whose velocity is zero. This might make it possible, for example, that the body 2 may be swivelling in some mysterious 'spherical' manner about the point at $A$.

10. In figure 5.02 we can look more clearly at the body 2 with its line $AB$ moving relatively to the new frame 3. The circle $\alpha_1$ at centre $A$ and radius $A$–$B$ has been drawn in the plane which contains the vector $\mathbf{v}_{3B}$ at $B$; and the line $A$–$a$ has been erected at $A$ normal to that plane. The vector $\mathbf{v}_{3A}$ at $A$ is zero and the vector $\mathbf{v}_{3B}$ being perpendicular to $AB$ at $B$ is tangential to the circle there. This plane containing $AB$ and the vector $\mathbf{v}_{3B}$ and the circle called $\alpha_1$ will be called whenever convenient the plane $\alpha_1$.

Figure 5.01 (§ 5.05) The straight line array of the velocity vectors at points long any straight line $AB$ in a moving body. The tips of the vectors are along another straight line; that line is marked $t$–$t$ here (§ 5.25).

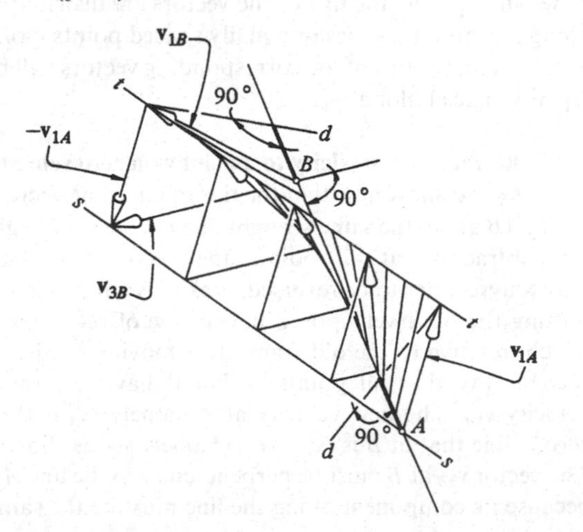


11. Another circle $\alpha_2$, also at centre $A$ and of radius $A$–$B$, has been drawn in the plane $BAa$; this circle also gives its name to the plane it occupies. If we now take any point $D$ in the plane $\alpha_2$ we see that, by virtue of rigidity along $BD$, its velocity must be perpendicular to $BD$ and that, by similar reasoning along $AD$, its velocity must be perpendicular to $AD$. Accordingly, all points $D$ in the plane $\alpha_2$ have velocity vectors normal to that plane, and it follows from this that all of the vectors at points along the line $AB$ are parallel.

12. We wish to show, however, not only that the velocities at these points are parallel with one another but also that their magnitudes are directly proportional to their distances from $A$. In showing this in the following paragraphs we shall also show incidentally (referring to figure 5.02) that, if there exists one point $A$ in a rigid body whose velocity is zero, there will also exist a single infinity (chapter 4) of other points in the body whose velocities are zero. This single infinity of points is concentrated exclusively in, and distributed evenly along, some straight line such as $A$–$x$ which is shown in figure 5.02; $A$–$x$ contains the point $A$, resides in the plane $\alpha_2$, and is inclined at some angle $\xi$ to $A$–$a$. The angle $\xi$ depends upon the velocities at the instant of points in the body other than those at $A$ and $B$.

13. Consider the point $E$ in the body 2; it has been chosen for convenience and without loss of generality to be on the circle $\alpha_1$ in such a position that the angle $BAE$ is a right angle. The triangle $BAE$ is coplanar in $\alpha_1$ with the vector $\mathbf{v}_{3B}$ at $B$, and it is easy to see by virtue of rigidity along $BE$ that the tip of the vector $\mathbf{v}_{3E}$ at $E$ must reside somewhere in the plane marked $\alpha_3$. But by virtue of rigidity along $AE$ the tip of the vector $\mathbf{v}_{3E}$ at $E$ must also reside in the plane marked $\alpha_4$. Accordingly its tip must be chosen to be somewhere along the line marked $e$–$e$ in the figure, $e$–$e$ being at the intersection of $\alpha_3$ and $\alpha_4$. We could choose the velocity $\mathbf{v}_{3E}$ at $E$ to be in $\alpha_1$

Figure 5.02 (§ 5.10) The line segment $A$–$B$ in the body 2 moving relatively to a new frame 3, in such a way that the velocity $\mathbf{v}_{3A2}$ is now zero (§ 12.09).

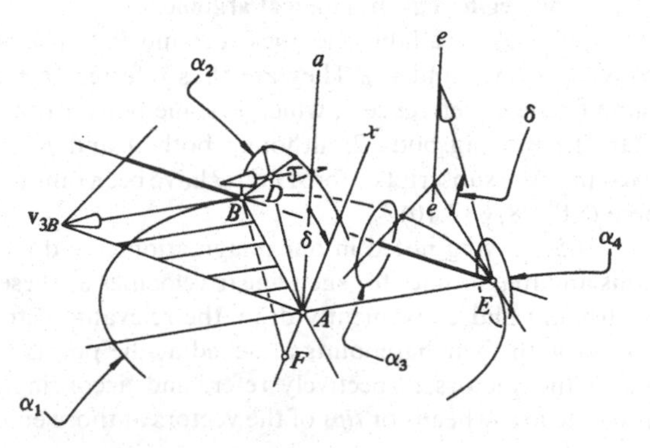

namely to be in the same plane with $\mathbf{v}_{1B}$, in which case it would need to have the same magnitude as $\mathbf{v}_{3B}$; but we shall choose it to be as shown, inclined at some angle $\xi$ to the mentioned plane; we note accordingly that $|\mathbf{v}_{3E}| = |\mathbf{v}_{3B}| \sec \xi$.

**The locus of other points of velocity zero**

14. Erect now a line $A$–$x$ in the plane $\alpha_2$ inclined to the line $A$–$a$ at the same angle $\xi$ as shown. Drop a perpendicular from $B$ to meet $A$–$x$ in $F$, and consider next the velocity $\mathbf{v}_{3F}$ at $F$.

15. It was shown at §5.11 that the velocities at all points such as $D$ in the plane $\alpha_2$ are normal to $\alpha_2$, and we note that the point $F$ is such a point. But by virtue of rigidity along the line $FE$, which is inclined at some angle to the plane $\alpha_2$, $\mathbf{v}_{3F}$ can have no component in that direction. The velocity $\mathbf{v}_{3F}$ at $F$ is accordingly zero. Having exploited our limited choice for the velocity at $E$ by choosing the angle $\xi$, and thus determined the velocities at all other points in the body incidentally, it begins to appear now that the line $A$–$x$ is important.

16. We note next that $A$–$x$ is normal to the unique pair of parallel planes that can be drawn to contain the vectors at $B$ and $E$, and that the vectors at $B$ and $E$ are not only perpendicular to $A$–$x$, but also, because $|FB| = |AE| \cos \xi$, proportional in magnitude to their distances from that line.

17. We wish to show next that the velocity relative to 3 of any point in the body 2 is perpendicular to $A$–$x$ and proportional in magnitude to its distance from that line. This can be done conveniently by first redrawing figure 5.02 in such a way that the two circles of radii $F$–$B$ and $A$–$E$, centred at $F$ and $A$ respectively in the line $A$–$x$, are drawn in the mentioned parallel planes normal to the line $A$–$x$. See figure 5.03.

Figure 5.03 (§5.17) The axis $x$–$A$–$x$ is here identified as that straight line in the body 2 that contains all those points in the body 2 having a linear velocity relative to 3 of zero (§5.24).

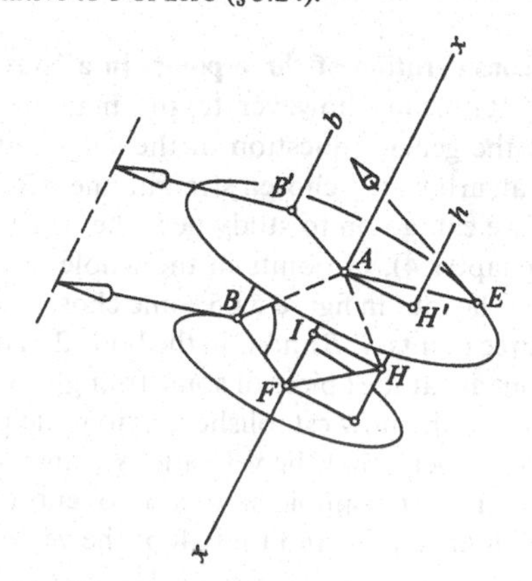


18. Throw up in figure 5.03 the line $B$–$b$ parallel to $A$–$x$ and consider the point $B'$ in the plane of the circle at $A$. We see, by virtue of rigidity along $BB'$ and $FB'$, that the vector at $B'$ must be parallel with that at $B$. But by virtue of the known proportions already mentioned (§5.16), and by rigidity along $EB'$, it is easy to show that the vector at $B'$ is also equal to that at $B$. We have thus shown that through any point such as $B$ in the body, where the vector has a certain magnitude and is perpendicular to $A$–$x$, a whole line of such points can be found parallel with $A$–$x$.

19. It remains now to show in figure 5.03 that, if we choose any other point say $H$ in the body, the velocity at $H$ is perpendicular to $A$–$x$ and proportional to its distance from $A$–$x$. That $\mathbf{v}_{3H}$ is perpendicular to $A$–$x$ is easy to show by using $AH$ and $FH$; by virtue of the velocities of zero at $A$ and $F$, the vector $\mathbf{v}_{3H}$ at $H$ is perpendicular to both of these lines; it is thus perpendicular to $A$–$x$. Throw up next the line $H$–$h$ parallel with $A$–$x$ to cut the plane of the circle at $A$ in $H'$. Look at figure 5.04, which is a projection along $A$–$x$ of figure 5.03, and consider the velocity $\mathbf{v}_{3H'}$ of the point $H'$.

20. In figure 5.04 the plane $B'EH'$ is in the plane of the paper and, by laying out components first along $B'$ and next along $B'H'$ and $EH'$, the orientation and the magnitude of the velocity vector at $H'$ can be constructed graphically as shown. Not only does the angle marked $\zeta$ in figure 5.04 turn out to be a right angle (as we have already shown that it must be), but also we find by ordinary plane geometry in figure 5.04 that the magnitude of the vector at $H'$ (and therefore of the vector at $H$) is directly proportional to its distance $I$–$H$ from the line $A$–$x$. Not surprisingly the constant of proportionality is $|\mathbf{v}_{3B}|/F$–$B$, a constant across the whole of the body 2. It could be measured in rad/sec.

21. It follows of course that any point which is actually in the line $A$–$x$ in the body 2, being at zero distance from the line $A$–$x$, has the velocity zero relative to 3. There are no other such points in the body, so the line $A$–$x$ can now be seen as a special line.

Figure 5.04 (§5.19) This construction in the plane of the paper shows how great the velocity at $H'$ must be, given the velocities at $B'$ and $E$. It can be shown that the angle $\zeta$ must be a right angle (§5.20).

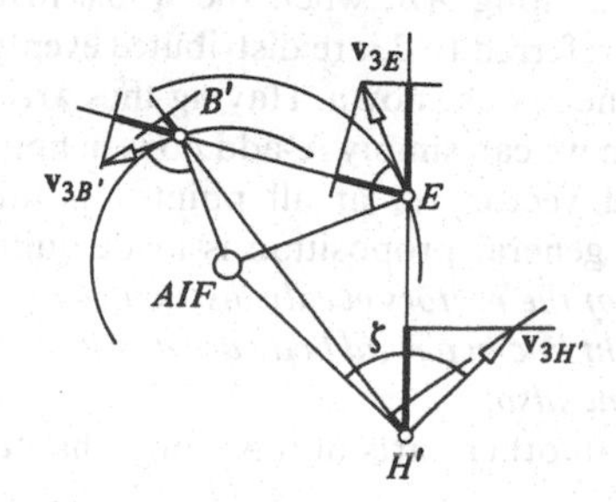

**Return for clarity to general spatial motion**

22. It is a special line however only at the instant and only in respect of the carefully chosen frame of reference 3. In general it will disappear as soon as any finite movement of the body begins. Referring again to figure 5.01 we note that the velocity $\mathbf{v}_{1A}$ at $A$ is not necessarily constant in time. It is clearly a fact that, unless the frame of reference 3 is chosen to be a continuously changing frame to suit at all times the changing velocity at $A$, the line $A$–$x$ will not only not continue to pass through $A$ but also it will in general cease to exist at all. There remains for us the awkward fact that, unless a special frame of reference is carefully chosen for each instant during the general motion of a rigid body such as 2, there is in general no point in the body with a velocity of zero, and therefore no straight line or 'axis' of such points about which the body might be said to be 'rotating'. To say it again: the existence of such a straight line or axis of points of zero velocity in a rigid body depends entirely upon the special circumstance (either existing or contrived) of there being at least one point in the body with zero velocity.

23. Yet at this critical stage of the ongoing argument the line $A$–$x$ is called by some writers the 'axis of rotation' of the body. There is presumed to be no important problem in the field of engineering endeavour where such an axis cannot be found; it is even held by some that there always exists in a rigid body some point possessed of zero velocity; the quantity 'angular velocity about an axis of rotation' is defined; the vector equation $\mathbf{v} = \boldsymbol{\omega} \times \mathbf{r}$ is written; questions about the dynamics of certain pivoted rotating bodies are successfully pursued; and the whole other matter of the generality of spatial motion is ignored. But in the study of mechanism, as we shall see, such primitive analyses are not enough.

24. We have shown at this juncture nevertheless that, *if there is at least one point in a rigid body with a velocity of zero, there is a single infinity of such points along a straight line which is unique in the body, and that the velocities of all other points in the body are then perpendicular to and proportional to their distances from that line.*

25. We have accordingly and incidentally shown what was wanted in figure 5.01, namely that the tips of the vectors along $AB$, when the whole instantaneous motion is referred to 3, are distributed evenly along the straight line $s$–$s$ as shown. Having thus arrived at that conclusion we can simply re-add now, in figure 5.01, the subtracted vector $\mathbf{v}_{1A}$ at all points. In this way the following general proposition is made quite clear: *the tips of all of the vectors evenly distributed at points along any straight line in a rigid body are distributed evenly in a straight line also.*

26. In other parts of this work (chapters 3, 9, 21) the geometry of such lines as $AB$ and their arrays of velocity vectors is studied in greater detail. Refer for example to figure 21.02 where the important angle $\delta$ for an array is shown and defined. Notice also that if the components of velocity along the line are zero, that is if the angle $\delta$ in figure 21.02 is a right angle, the line becomes an interesting and useful line composed of points whose vectors are all perpendicular to the line (chapter 3). Such special lines are called right lines (or sometimes also $n$-lines or even contact normal lines), and their numbers in a moving body are limited; their usefulness, and their distribution at an instant throughout the whole space of a moving body (§ 1.06), are dealt with in chapters 3 and 9.

**The various geometric arrangements of vectors**

27. It might be useful to break in here with the following remark. Unlike for example a number of force vectors, each one of which may be imagined to be separately applied anywhere along any chosen line of action and with any magnitude on a body, a number of velocity vectors relating to some collection of points fixed in a rigid body are obliged by its nature to be arranged according to some pattern. We are investigating here what that pattern must be and in so doing become aware of the plain fact that different kinds of vector , although alike in many respects, appear to occur and behave in nature in different ways. It is probably useless to ask, however, why it is that the different kinds of vector, based as they are upon our various fundamental notions such as force, velocity, etc., appear to differ from one another: they differ because we choose them to differ. In each field of mechanics – in statics, kinematics, kinetics – we begin (or began) by defining concepts and by claiming to 'know' something. All we are claiming to know in the current context is the meanings of the words rigidity and velocity; with respect to velocity there is of course one thing we know with certainty, that velocity – linear velocity, that is – is not the same as, or even akin to, force. Please refer to § 10.03.

**A consideration of three points in a body**

28. Returning however to the main argument, now that the general question of the single infinity of points in an arbitrarily chosen straight line $AB$ has been clarified, we can go on to study next the triple infinity, the $\infty^3$ (chapter 4), of points in the whole of the rigid body 2. We lay out in figure 5.05 some chosen velocities for any three points $A$, $B$ and $C$ in the body 2 that are not in a line but are at the apices of some triangle; we appeal in so doing to the now established, axiomatic principle of rigidity, namely, that the velocities $\mathbf{v}_{1B}$ and $\mathbf{v}_{1C}$ of the points $B$ and $C$ for example have components along the line $BC$ that are equal, and that all of the vectors along

*BC* have their tips distributed in a straight line array (§ 5.03, § 5.25). We thus ensure that the chosen set of three velocities (and thus those for all of the points in the plane of the triangle *ABC*) is a mutually consistent set. In figure 5.05 the relevant pairs of components of the three chosen velocities at *A*, *B*, and *C* are shown by the thickened segments of line.

29. It should be mentioned here that the pair of component velocities shown at *B* for example in figure 5.05 do not recombine upon addition to produce the total velocity at *B*. These components are not components in that sense; they might be better described as resolved parts of, or projections of, the vector at *B*, but to follow common practice and for the sake of brevity the acceptable word 'component' is being used in this chapter throughout.

30. Before going on from figure 5.05 some important matters of detail need to be noticed. Notice that, if the velocity at *A* is chosen first, our choice for the velocity at *B* is limited (§ 5.03); the velocity vector at *B* must have its tip somewhere in the plane $\alpha_1$, which is drawn normal to the line *AB* at the tip of that vector's already determined component along *AB*. Notice also that, if the vectors at *A* and *B* are both already chosen, our choice for the vector at *C* is doubly limited; the vector at *C* is obliged to have its tip somewhere along the line of intersection of the two correspondingly relevant planes $\alpha_2$ and $\alpha_3$ drawn normal to the lines *BC* and *CA* respectively. It is significant next to notice that, if a velocity were now to be chosen at some fourth point say *D* in the body (at the apex *D* of some tetrahedron *ABCD*), there would be no freedom of choice for us at all: *the velocities at all points D in a rigid body are determined as soon as a mutually consistent set of three velocities is chosen at A and B and C.*

**Relevance of the velocity polygon**

31. Despite an earlier remark to the contrary (§ 5.27), I would like to introduce the notion now that a linear velocity vector (which is a 'bound' vector in the sense that it is bound to its relevant basepoint in the moving body) may nevertheless be *plucked* at an instant like a flower and be carried without loss of meaning (by maintaining magnitude and direction as it goes) to some other basepoint of our choice. There the carried vector may be *planted* at the instant along with some other such vectors to form the basis of what we already know (chapter 12) to be a *linear velocity polygon* for the body or for the mechanism. Any new basepoint chosen under these circumstances may be called *the origin of the velocity polygon.*

32. Go now to figure 5.06, which is a redrawing to the same scale of the main velocity vectors at *A*, *B* and *C* worked out in figure 5.05. Pluck these vectors $\mathbf{v}_{1A}$, $\mathbf{v}_{1B}$ and $\mathbf{v}_{1C}$ and rename them so that for each vector the number in its subscript appears at its basepoint while the letter in its subscript appears at its tip. The vector $\mathbf{v}_{1A}$ for example becomes the directed distance **1a**, $\mathbf{v}_{1B}$ becomes the directed distance **1b**, and so on. Plant the three vectors now with their basepoints together at some new

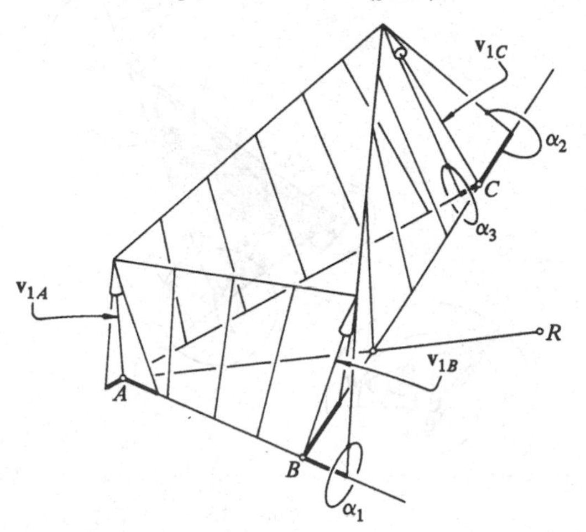


Figure 5.05 (§ 5.28) Velocity vectors at three points *A*, *B*, and *C* in the moving body 2. Velocity vectors attached at all points along the sides of the triangle *ABC* have their tips along the sides of another triangle, which triangle is in general neither congruent, similar, nor parallel with *ABC* (§ 5.30).

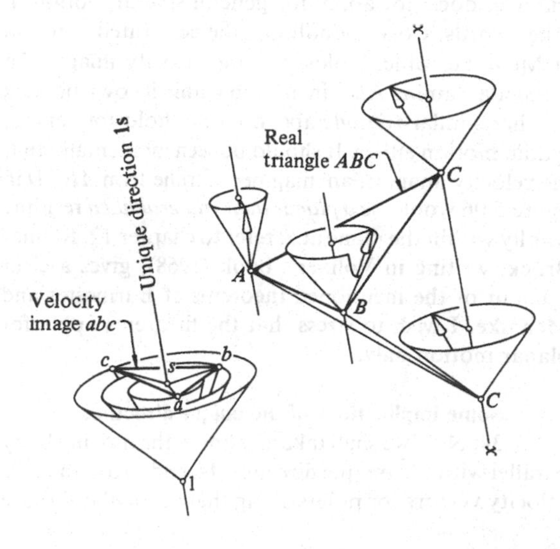


Figure 5.06 (§ 5.32) The velocity vectors at *A*, *B*, and *C* plucked and replanted at an origin 1. The triangle *abc* is the velocity image of the real triangle *ABC*, and the velocity vector $\mathbf{v}_{1s}$ is the least velocity vector. For further reading about the velocity image, please refer to § 12.14 *et seq.*

basepoint called 1 and note the plane $\alpha$ which is drawn to contain the tips *a, b* and *c* of the vectors in their new locations. Drop next a directed normal **1s** from the basepoint 1 to the plane $\alpha$ (so that *s* is in $\alpha$), and note that this **1s** now nominates a new, unique direction along which the components of the three velocity vectors are equal [1]. The triangle *abc* in the polygon is the *velocity image* of the body *ABC*, or of the body 2 (§ 12.14); the basepoint 1 at the origin of the polygon is, in fact, the velocity image of the fixed body 1 (§ 12.28); and the directed distance **1s**, as we shall see, is indicating the direction of the angular velocity $\omega_{12}$. *This latter quantity has been, hitherto, not only unknown but also not even defined.*

**The inapplicability of Burmester and Mehmke**

33. By drawing lines from point to point in the plane *ABC*, say through the points *ABCAR* . . . etc. in figure 5.05, it is easy to extend the earlier argument that velocity vectors along a line in a moving body have their tips in a line to make the wider statement now that the velocity vectors across a plane in a moving body have their tips in a plane. In general, of course, the plane containing the tips of the vectors will not be parallel with the plane of the points; it is only in the special case where the triangle *ABC* in the body has been chosen by chance to be normal to the unique direction **1s** that the two mentioned planes will be parallel. It follows from what has been said that, if the body 2 is represented by all four apices of some tetrahedron *ABCD* in figure 5.06, the tips of the vectors whose basepoints are at these four apices will form, in general, *another tetrahedron which will not be similar to ABCD*. The celebrated theorem of Burmester, which holds that, if a triangular lamina *ABC* is in motion within its own plane, the tips of the vectors at the apices of the triangle form another similar triangle, does not apply for general spatial motion in other words. Correspondingly, the celebrated theorem of Mehmke, which holds that the velocity image of a triangular lamina ABC in motion within its own plane is another similar triangle abc, does not hold for general spatial motion either. It should be seen incidentally that the velocity image of an imagined tetrahedron *ABCD* in figure 5.06 would be *a planar quadrilateral abcd* residing wholly within the plane at $\alpha$; refer to chapter 12. Reimar Brock, writing in Volmer's book (1968), gives a clear account of the mentioned theorems of Burmester and Mehmke. I wish to stress that the theorems apply for planar motion only.

**Some implications of the unique direction**

34. Next we shall take any line in the moving body parallel with the unique direction **1s**, and prove that the velocity vectors for points along that particular line are not only equally inclined to the line but are parallel with one another. We can see straight away from figure 5.01 that, if they have the same component along the line (which they must), they cannot be equal with one another without being parallel.

35. Having drawn the normal **1s** against the plane $\alpha$ in figure 5.06, throw up parallel with **1s** in the body 2 the line *x–C–x* through *C*, and nominate a second point *C* along this line. We now wish to prove that the velocities at all points such as *C* along the line *x–C–x* are equal and parallel.

36. See figure 5.07 as a projection of the whole picture in figure 5.06 onto a plane normal to the unique direction **1s**. The conically shaped pieces which were drawn for convenience in figure 5.06 all become circles upon projection, and these circles can be seen in figure 5.07. Please note in figure 5.07 that the original pairs of equal components along the lines *AB, BC*, etc. have projected into different although still equal pairs of components along the projected lines *A′B′, B′C′*, etc.

37. Given the velocities at *A* and *B* in figure 5.06 it is now clear that the projected velocity at *C′* in figure 5.07 (for all points *C* along *x–C–x* in figure 5.06) must always be of the same magnitude and in the same direction. It follows that all velocities along the line *x–C–x* must be coplanar, equal, and parallel.

38. It can next be seen quite clearly that, if all of the velocity vectors at all of the points in the moving

Figure 5.07 (§ 5.36) An orthogonal projection of figure 5.06 taken in the unique direction **1s** of the angular velocity. The stippled triangles abc and *A′B′C′* are similar. The point *S′* (which corresponds with *s*) in the figure is also unique; this can be located by constructing any two of the right angles which are shown.

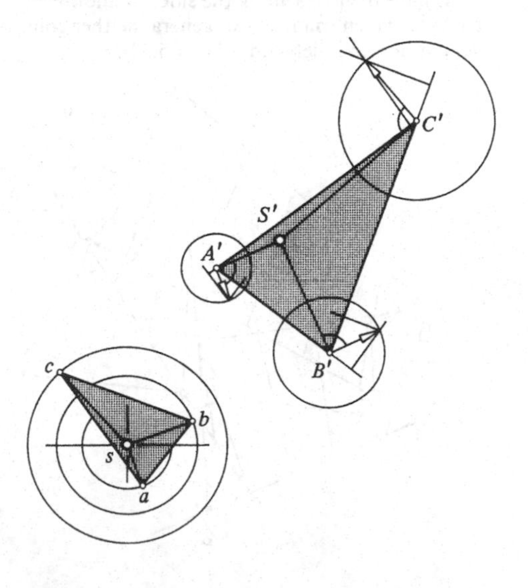

body 2 are plucked and replanted together at some common basepoint 1, the tips of all of the replanted vectors will reside in the same plane; the plane of course is the plane $\alpha$ of figure 5.06. In general, the directed line **1s** in figure 5.06 represents all those velocity vectors of least magnitude that can exist in the moving body. Note that in general the vector **1s** is of non-zero length; *this means of course that in general there are no points in a moving body whose velocities are zero.*

### Location of the axis that is unique

39. An instructive thing to do in figure 5.06 is to subtract from all vectors the special vector **1s**. By thus specially changing the frame of reference we are reducing the otherwise spatial motion to planar motion and might under those circumstances just as well draw a plan of the resulting vectors upon the reference plane. In which case, of course, we simply arrive again at figure 5.07, within which we should now note the point marked $S'$.

40. Given the special relationships between the various components in figure 5.07 and given that the perpendiculars to the projected vectors at $A'$ and $B'$ have intersected at the point $S'$, it now follows by ordinary Euclidean geometry in figure 5.07 that, if a perpendicular is similarly dropped from $C'$, that perpendicular will pass through $S'$. The point $S'$ is thus unique in the reference plane because, having arbitrarily chosen $A$ and $B$ in the body, $C$ was any other point.

41. Looking next at the lines $A'S'$, $B'S'$ and $C'S'$ on the projection plane in figure 5.07 and noting that the projected vectors at $A'$, $B'$ and $C'$ are all respectively at right angles to these lines, we see by the axiomatic principle of rigidity that they are, all three of them, lines in the projection plane whose sets of component vectors in the directions of their lengths are zero.

Figure 5.08 (§ 5.42) Longitudinal view of the instantaneous screw axis. All points $S$ along it are projecting into the point $S'$.

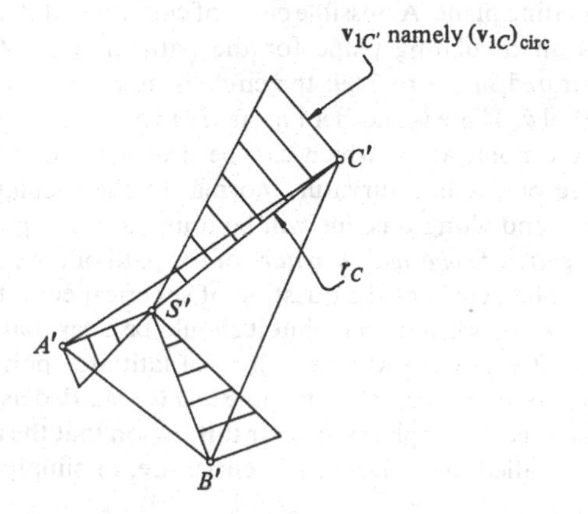


42. It follows inescapably now that the projected velocity at $S'$ is zero, that $S'$ in this respect is unique in the projection plane, and that the projected velocities for all other points in the projection plane are arranged as shown in the planar figure 5.08.

43. Looking again at figure 5.06 in conjunction with its projected counterpart in figure 5.07 we now note that, if there are lines such as $C$–$c$ in the body along each one of which all of the vectors are equal and parallel (this has now been proven), then there is accordingly a line through $S'$ in the direction **1s** along which all of the velocity vectors, at a single infinity of points $S$, *are equal, parallel, collinear with the line, and minimum in magnitude for the body.*

### The ISA and a definition for angular velocity

44. This special line through $S'$, which is unique at the instant in the moving body, along with its *pitch*, which will be mentioned shortly, is called the *instantaneous screw axis for the relative motion of the bodies 1 and 2.* This line with its pitch is of fundamental importance in the kinematics of mechanism. Refer again to §3.23 and to figure 3.05 where the two bodies under consideration are the members 1 and 4 of the mechanism shown in figure 3.01, and where the $ISA_{14}$ is materialised at the instant into an actual nut and a bolt; please note also the directions and the magnitudes of the indicated velocities there. Before returning here to the mentioned matter of pitch, however, some further remarks will need to be made.

45. Nowhere in this chapter have we accepted as axiomatic that the circumferential component of $\mathbf{v}_{1A}$ (whose true length appears for example in figures 5.07 and 5.08) may be written $(\mathbf{v}_{1A})_{circ} = \boldsymbol{\omega}_{12} \times \mathbf{r}_A$, where $\mathbf{r}_A$ is the directed distance from some origin-point on the ISA to the representative point $A$ in the moving body 2. We have studiously avoided, in fact, using any of the fundamental properties (such as this one) of the ISA. *We can in fact legitimately argue now that angular velocity has just been defined in figure 5.07. It has been defined in terms of the linear velocities of some selected points in the moving body and the axiomatic principle of rigidity.*

46. Angular velocity is thus seen to be the constant of proportionality which connects such velocities as $(\mathbf{v}_{1C})_{circ}$ with its radial distance $\mathbf{r}_C$ from the ISA namely $S'$–$C'$ in figure 5.07. It follows from this (and from conventions with respect to the directions taken by 'angular' vectors) that the angular velocity $\omega_{12}$ of the body is in the direction of the $ISA_{12}$ and vice versa. In figure 5.09 we summarise pictorially the now-established situation.

47. Knowing now what is meant by the angular velocity $\omega_{12}$ of the body 2, and using the special symbol

$\tau_{12}$ to designate that unique, least linear velocity of all those points $S$ in 2 along the $ISA_{12}$, the total linear velocity at $A$ may now be written: $\mathbf{v}_{1A} = \omega_{12} \times \mathbf{r}_A + \tau_{12}$.

48. The angular velocity $\omega_{12}$ of the body 2 is clearly a property of the whole body; its vector is a free vector whose magnitude and direction must be quoted but whose position need not. It has been accordingly shown in figure 5.09 to be acting 'along' either or both of two different though parallel lines; it is shown to be acting along the ISA (which is tidy), but it is also shown to be acting 'at' a generally chosen point $Q$ in the body which is not in the ISA. This latter way of attaching the vector $\omega_{12}$ at some convenient basepoint $Q$ in the body is an entirely legitimate way of showing the angular velocity of the whole of the body. We shall return to this matter at §5.57.

49. The quantity $\tau_{12}$, which might be called the velocity of the 'axial sliding', is also a property of the whole body. The component in the direction of the ISA

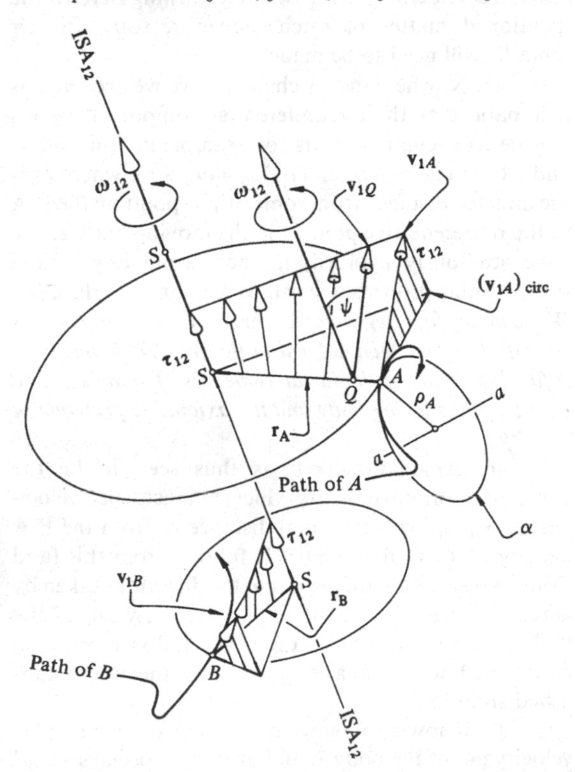


Figure 5.09 (§5.48) Panoramic view of the $ISA_{12}$. The points $A$ and $B$ are fixed in the moving body 2; the velocities $\mathbf{v}_{1A}$ and $\mathbf{v}_{1B}$ are tangent to the paths in 1 of $A$ and $B$ respectively. Note the distinction between the radius vector $\mathbf{r}_A$ and the radius of circular curvature $\rho_A$ of the path of $A$ (§5.52). For explanations of the dual vectors (the motion dual) that can be seen at $Q$, please refer to §5.57, §10.47, and §10.50 *et seq.*

of the linear velocity of every point in the body has this value $\tau_{12}$ (§10.45). Its dimensions are $LT^{-1}$.

50. We can now define the pitch $p$ of an instantaneous screwing by quoting the relationship: $\tau_{12} = p\,\omega_{12}$. The pitch p of an ISA accordingly has the dimensions L; pitch is a scalar not a vector and is usually quoted mm/rad (§10.47, §10.48). It follows from this definition of pitch incidentally that, while a motion of pure rotation can be seen as a screwing of zero pitch about a discoverable ISA like the line $A$–$x$ in figure 5.03, a motion of pure translation (in which every point in the moving body has the same linear velocity with respect to frame) can be seen, either as a screwing of zero pitch about an ISA which is perpendicular to the direction of motion and infinitely far away, or a screwing of infinite pitch about some axis which is parallel to the direction of motion and is somewhere 'close at hand'.

51. Such ambiguities as this latter are naturally characteristic of analyses made of planar mechanism (§2.30). They must be seen for exactly what they are, degeneracies of the general geometry which can make no sense in isolation. They are also, however, important special aspects of the generality, and they must be studied carefully (§23.57).

### Curvature of the paths of points

52. Before commenting upon the directed distance $\mathbf{r}_A$ in figure 5.09, and in order to distinguish this $\mathbf{r}_A$ from another distance with which it is often confused, we need to introduce a special term, the *polar axis of curvature of the path of a point*, and the following remarks are designed to make clear what this axis is. Whenever a point such as $A$ is travelling a twisted path in space there always exists a unique plane containing (*a*) the point itself; (*b*) the tangent to the path of the point, along which line the velocity vector may be properly drawn; and (*c*) the *normal* to the path of the point, along which line will be found the *centre of circular curvature* of the path; the plane accordingly contains the circle of curvature and is called the osculating plane. A possible circle of curvature at $A$, and thus an osculating plane for the path of $A$ at $A$, is illustrated in figure 5.09; the circle is marked with the symbol $\alpha$. There is another line, marked $a$–$a$ in the figure for the point at $A$, which can be drawn through the centre of circular curvature normal to the osculating plane, and along this line will be found another point, the *centre of spherical curvature* of the path of the point at $A$. The details of the question of spherical curvature need not concern us here, but it should be clear that the circle of curvature will be a circle of latitude upon the sphere of curvature when the line $a$–$a$ is regarded as the polar axis of the sphere. It is for this reason that the axis $a$–$a$ is called the polar axis of curvature, or simply the

*axis of curvature*, of the path of $A$ at $A$. It is erroneously held by some to be a kind of 'axis of rotation' for the point at $A$.

53. In the special case of screwing of zero pitch about an ISA that is fixed, in which case the ISA becomes in effect a pivot and the motion becomes a limited form of planar motion, the axes of curvature for the paths at all points in the body become collinear with one another at the ISA. It is the treatment in isolation of this most particular case which causes the confusion often existing between such distances as $r_A$ and $\rho_A$, both of which are illustrated in figure 5.09. Whereas the radius $r_A$ is used as we have seen to connect the angular velocity $\omega_{12}$ with the linear velocity at $A$, the radius $\rho_A$ of the circle of curvature of the path at $A$ is associated with an entirely different matter the normal or the centripetal component of the linear acceleration at $A$, which is of course irrelevant in the present context.

54. In order to reveal more clearly the question at issue here let us consider a slightly more complicated motion, the motion of a continuous screwing at some finite pitch about an ISA which is fixed. In this particular case any chosen point in the moving body – it could be some point in a nut moving on a bolt which is fixed – will be travelling a regular helical path. If the radius of the helix is $r$ and the pitch of the screwing $p$, the radius of circular curvature of the path at all points along its length will be found to be, $\rho = r + p^2/r$. By virtue of symmetry, incidentally, the centres of circular and spherical curvature coincide. We see that the axes of curvature of the paths are at all points skew with the ISA, being inclined at the helix angle $\theta$ pertaining. Note the values of $\rho$ for the special values zero, unity, infinity, etc. of both $p$ and $r$.

55. The suggested possible paths for the two points $A$ and $B$ in figure 5.09, however, have been drawn to reveal the most general possibility. *Such twisted paths in space are not only entirely possible for points in the links of mechanism; they are indeed typical of the paths of such points; and in these most general of motions it is surely quite clear to see that the directed distance $r_A$ is measured, not from the axis of curvature of the path of A at A, but from the ISA$_{12}$.*

56. It is also important to notice that the curvature with respect to 1 of the path of a point such as $A$ in 2 need not be concave towards the ISA; it may be either concave or convex; and the reader is reminded here of the geometry of the inflexion circle for the relative motion of two laminae in the plane. One could refer in English for this latter to Rosenauer and Willis (1952). One is reminded also, referring for convenience to figure 17.02(*a*), that, while the coupler link 3 of a planar crank rocker mechanism rotates about its pole $P_{13}$ with respect to the frame link 1 (that is, screws with zero pitch about its ISA$_{13}$, which latter is, as we know, parallel with the pins of the mechanism), the crank pin $B$ of the mechanism (which is, as we can see, a point in the planar coupler link) travels in a circular path with respect to frame, and the centre of curvature of this path is at the crank shaft $A$. We are looking in figure 5.09 at the spatial analogue of such geometry and there are, of course, many difficulties. While the distance $A$–$B$ in figure 5.09 should be held in the mind's eye to remain a constant, for otherwise the body 2 could not be rigid (§ 5.01), the separate paths for the two points $A$ and $B$ in the body 1 should be seen to have been drawn by me quite arbitrarily.

### The classical mathematical view

57. On the question of the point $Q$ in figure 5.09, it should be seen that the combination of the two vectors $\mathbf{v}_{1Q}$ and $\omega_{12}$ at $Q$ are sufficient to nominate completely the instantaneous velocity-motion of the whole body. In the classical works of mathematical mechanics (a good example is the work of Synge and Griffith (1949), where the question of a nominated spatial movement of a body from one location to another is considered), an arbitrarily chosen point such as $Q$ is taken through its nominated linear displacement. An important theorem due to Euler is called upon to explain that, once the point $Q$ has completed its displacement (having taken the whole body by pure translation with it), an axis through the achieved new position of $Q$ can always be found about which the body may next be rotated to achieve the nominated new location of the whole body. It is then explained that the extent and the direction of this final rotation about an axis through $Q$ is independent of the choice of $Q$, and it is upon this basis (having moved from the finite to the infinitesimal) that the concept angular velocity for the body is defined. In section 10.5 of the above mentioned work this matter is very well explained by the authors. Indeed they go even further at the end of their section 10.5 to deal in effect with the ISA; they call it there the 'axis of the screw displacement'.

### Closure

58. There is much more to be said about the ISA, the motion screw: the reversibility of its nomenclature with respect to the two bodies necessarily involved – the terms ISA$_{12}$ and ISA$_{21}$ both refer to the same screw; the two ruled surfaces traced by the ISA in the two bodies respectively as the bodies move – these are the axodes; the instantaneous relationships of the several ISAs or motion screws in a moving or movable mechanism – here we find for example the first of the two theorems of three axes (chapter 13); the system of the $\infty^2$ of motional right lines surrounding an ISA – this

system turns out to be a linear complex (chapter 9); the seven general systems of the motion screws and of the reciprocal, action screws in the various circumstances of freedom and constraint (chapters 6, 23). All of these and related other matters are pursued somewhere in the present work. The reader is invited to peruse chapter 8 as a next introductory piece. What can be said in conclusion here however is: *whenever a body moves or is about to move in a constrained manner relative to its frame-body (its frame of reference), the ISA exists and that ISA is, in general, unique.*

### Notes and references

[1] See the Appendix 1, Some notes on the linear complex, in Sticher, F.C.O., *Contributions to geometric interpretations and methods in the study of spatial linkages*, doctoral dissertation, University of Sydney, 1972. Refer § 5.32.

6. A rigid five-legged table with all of its feet in contact with the floor has, at any instant, one degree of freedom only. Under the same conditions a four-legged table has two degrees of freedom and a whole cylindroid of motion screws available.

# Irregularity and the freedoms within a joint

## On regularity, and the irregularity at surfaces

01. The idea that regularity is a possibility derives directly from the foundations of Euclidean geometry. We allow by it such concepts as the straightness of a straight line, the flatness of a plane, the sphericality of a sphere. We can extend the notion to allow in the imagination such other things as, the exact ellipticality of a plane section through an ellipsoid, or the exact equi-angularity at the apices of some regular polyhedron. Extending the meaning of regularity into the realm of joints, we can envisage if we wish, and please refer to figure 6.01, that the convex conical surface at the joint element on link 1 and the flat plane surface at the joint element on link 2 can make, with one another, *continuous line contact*. We can similarly imagine that the convex surface of the accurately constructed spherical ball on 2 can make, with the concave surface of the matchingly constructed, equi-radial, spherical socket sunk in 3, *continuous surface contact*. These and other such imagined, continuous contacts between a pair of joint elements in mechanism – they are not single point contacts or even multiple point contacts but contacts involving $\infty^1$ or $\infty^2$ of points – can exist in our imaginations only by virtue of the wholly idealised, but well established concepts due to Euclid.

02. But we know that no mould can ever exactly fit its pattern, that no casting can ever exactly fit its mould, that no welded construction ever matches its production drawing, and that no subsequent machining is ever done on any job accurately. There is the philosophical difficulty that, however clearly it may be delineated and thus made ready for measurement, no dimension of a body can ever be exactly known. The general fuzziness which accordingly characterises our descriptions of the shapes of matching bodies is openly implied by our always-quoted pair of tolerance limits within which each actual dimension is intended or believed to lie; this question of tolerance limits is not so much a question of the microstructure of the material, or of elasticity, or of shrinkage, but of number. The question most often is: to what accuracy, in minuscule parts of a millimeter or of a degree, are we prepared to go in the offering of a guessed linear or angular measurement? Whereas it is claimed by certain enthusiastic naturalists that nature never drew a straight line, it should be seen by them, and by us too, that engineers never drew one either. If two allegedly identical, regular Euclidean shapes cannot be measured accurately, or even accurately compared, it is surely safer, and certainly more convenient for the sake of the present study, not only not to presume their identicallity and thus their regularity, but also actively to deny it.

03. Notwithstanding that, and presuming still the presence of rigidity (§ 5.01), I shall find it necessary in chapter 7 to argue that a supposedly irregular surface, such as that for example of a carved stone, is matched by the *approximately regular* surface of some consciously constructed copy of it. I shall be saying for example that the rigid, homogeneous material of some convex bolt at some member is enclosed within its surface and displaying approximate regularity because the bolt has been designed and built to fit and to move within some concave nut at some other member. The idea of approximate regularity at a joint permits, as we shall see, the idea that, as one area or spot or point of contact is caused to be broken, another one can smoothly be allowed to take its place (§ 1.49, § 7.35, § 7.36). It will moreover be seen in chapter 7 that a thing called *contact patch* (§ 1.14), a geometrical abstraction which is mentioned here and dealt with later (§ 7.37), is a thing which can exist and can be stood alone, independently at the joint elements. Looked at from the point of view of either one of the joint elements at a joint at an instant, the contact patch always has the same shape. Let us however not forget, (*a*) that the contact patch may or may not remain invariant as a joint engages in its motion, and (*b*) that it may or may not remain in a fixed relationship with one or the other or both of the joint elements (§ 1.12). This matter will be seen to loom much larger later, at § 7.16 *et seq*.

04. We do find in the study of joints that the elusive idea of regularity is bound up with the inevitable

idea of *fitting*. In a joint one can always identify the two surfaces at the two mating elements of the joint. This circumstance obtrudes whenever we try to formulate a relevant definition for regularity at the joint. We become involved, not solely with some single surface existing alone, but with a relationship between a pair of surfaces. The relationship is always between, either (*a*) some known or proposed surface and some ideal pattern which can be described in Euclidean terms (§ 6.01), or (*b*) some known or proposed surface and some other which can only be described numerically (§ 6.03).

05. The notion is expressed in other places that it is impossible for two irregular, rigid bodies to make contact with one another at more than six points (§ 1.50, § 3.28, § 8.01). The notion implies of course that, if a pair of rigid bodies makes contact at seven points or more, the bodies cannot be irregular. A unifying approach to the matter is to argue that all rigid bodies are necessarily irregular. This idea, *the inevitability of irregularity*, is a reasonable one and, after this discussion, I propose to accept it, to enlarge upon its axiomatic aspects and to employ it in this chapter 6, not merely to negate the possibility of there ever being as many as seven points of contact between a pair of bodies, but to build a useful theory about the nature of practical joints in real machinery.

06. It is thus foreshadowed – for I have spoken here about the impossibility of regularity at the surfaces of real bodies (§ 6.05) and in chapter 1 about the kinematic significance of the real surfaces of bodies in contact (§ 1.09) – that, in the absence of flexibility, the three matters of rigidity, irregularity and machinal accuracy must go hand in hand (§ 1.57). This bears upon the various remarks in chapter 1 about the occurrence of 'spots'. These spots are held in chapter 1 to be distributed about the contact patches at the joints in mechanism wherever the mechanical construction of the joint makes line or surface contact likely. Please refer to these passages, § 1.11 to § 1.14 inclusive and § 1.48 *et seq.*, in conjunction with the somewhat tighter argument due to follow soon. I am about to speak, from § 6.13 onwards, about contact between rigid, irregular bodies exclusively in terms of points of contact. I shall now, accordingly and actively, make the related assumption: *rigidity and irregularity are always extant and it follows that, at joints in mechanism, line or surface contact is never possible.*

Figure 6.01 (§ 6.01) Continuous line contact at one joint and continuous surface contact at another. To achieve such continuous contacts in the imagination we need to envisage Euclidean regularity at the mating surfaces; accurately made spheres, cones, planes, etc. must be seen to exist.

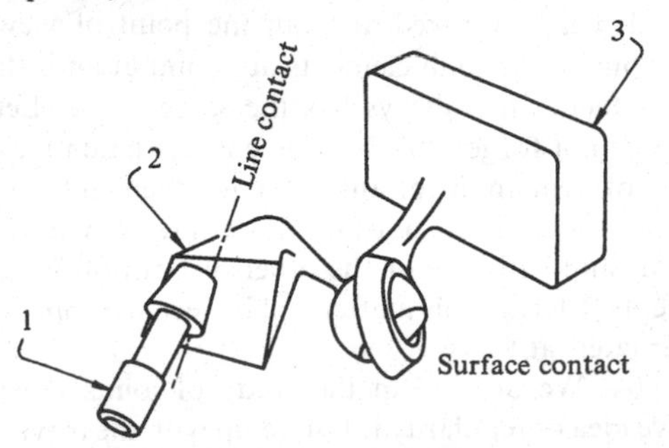


**Various other preliminary remarks**

07. In this chapter I make a general kinematic investigation of the simple joints. I take two irregular bodies in contact at one, two, three, etc. direct points of contact. The discussion introduces and embraces various systems of available screws for the possible relative motions of the contacting bodies at the instant under consideration (§ 5.02). Such systems of screws are always to be found wherever rigid bodies are connected with one another in any way mechanically. The screws themselves I begin to call, a little later, motion screws (§ 10.47), and these are, quite simply, the same as the ISAs of chapter 5 (§ 5.44). My reason for the gradual introduction here of the term *motion screw* is that such a screw needs to be distinguished in the mechanics of mechanism from another kind of screw I shall later call an *action screw* (§ 2.65, § 3.43, § 10.48).

08. Unlike, however, the general systems of screws discussed in chapter 23, which are general not only in many other ways but also in the way that they can be related to the capacity for motion of a rigid body in whatever manner it may be connected to its reference body (via intermediate bodies or chains of such bodies or otherwise), the systems discovered here are only of a quasi-general nature. The plain fact here is that direct points of contact always exist between the bodies, namely between the moving body and the reference body at the joint, and this ensures that the systems are such that there exists in the moving body at the instant points whose paths in the reference body are restricted to a surface at the instant (§ 23.45).

09. That this quasi-generality is a special feature of the systems of motion screws discovered here needs to be kept in mind. The matter is mentioned regularly throughout the chapter; the matter simplifies rather than complicates the systems and its significance, which is well reasoned and carefully presented by Hunt in section 12.11 of his recent work (1978), can either be clarified there or slowly as we go.

10. In the forthcoming material I first consider zero points of contact between the bodies. Following that, the strategy of the argument is that the number of points of contact between the bodies is progressively

increased. For brevity, and for a suitable continuity all the way from zero to six (and, to a certain extent, beyond), prior knowledge is assumed. Prior knowledge from chapter 5, for example, involving crucial facts about the ISA, is openly employed, and some from chapters 9 and 11 is assumed as well. It would be best if the reader had digested the crux of chapter 5 before proceeding here and he should, from chapter 3, both know and be able to distinguish between the terms contact normal and *n*-line. The cylindroid, too, is an important geometrical concept employed for clarification in the present chapter, but suitable references for information about this are made at the proper places. I wish to recommend here, however, as I did at the end of chapter 5, a first reading of chapter 8. As already said, it is not absolutely necessary to study the various aspects of the overall matter in any particular order. The reader may, though at his risk, read at will.

11. Looking forward further, one should be aware that in chapter 7 these quasi-general systems of chapter 6 are permitted to collapse into a set of carefully chosen and very special degeneracies. These reveal there, not only the well known simple joints of engineering practice (and some others), but also their logical origin in screw theory. This leads in turn to a rational way for categorising the practical and other joints in mechanism in terms of the systems of motion screws attached at their contact patches, and thus, incidentally, to an overall recognition of the fact in the kinematics of mechanism that for a full understanding of the phenomena a study of the general and the special screw systems is almost unavoidable (§ 7.15).

12. It is important for the sake of continuity in this chapter 6 to remove the more difficult material to appendices. There are three of these. They are to be found, not at the end of the book, but at the end of the chapter. Headed 6A, 6B and 6C, they may be omitted upon a first reading but they are important, for without them a sufficiently rounded explanation of the quasi-general systems of motion screws will not have been provided.

### Consider first two irregular bodies free in space

13. The general relative motion of two bodies is dissected in some detail at chapter 5. It is shown there to be quite natural (and in the context I would like to say canonical, chapter 10) to look at the relative 'velocity' motion in terms of the ISA. If a known relative motion is already occurring, which will often mean incidentally that the mobility of some guiding mechanism is either unity (§ 12.28, § 19.45) or less than unity (§ 1.56, § 2.06), we can look at the motion in terms of the *ISA which is extant*. If the relative motion is as yet unknown because some number more than one of degrees of freedom is involved, we can study the various possibilities for that motion in terms of the *ISAs which are available*. Often, in these latter circumstances, the two bodies are known or presumed to be connected in some way (§ 13.11, § 19.45). If they are not connected at all, however, the following relevant questions can still be asked.

14. *In how many different ways, at an instant, with a body 2 free in space, is the body 2 free to screw (or can the body 2 be caused to screw, or to be twisted) with respect to the body 1?* Where, moreover, are the available ISAs for the motion, or what, in other words, is the capacity for motion of the body 2 (§ 3.34, § 4.04)? We shall answer these questions before we go on to allow, successively, an increasing number of points of contact to occur between the bodies.

15. One way to answer the questions is to argue as follows. Refer to figure 6.02. Choose any point $Q$ in space and see, preliminarily, that, passing through $Q$, there is an $\infty^2$ of lines (§ 4.10). Each one of these is a possible location, or a possible 'seat', for an $ISA_{12}$. See also that, along each one of these lines, the pitch of the possible screwing may be taken at any value, from negative through zero to positive infinity. This means that, along any one of the lines, the pitch $p$ (mm/rad) may be taken at any one of an $\infty^1$ of values. There is, accordingly, an infinity to the power $(2+1)$, namely an $\infty^3$, of available ISAs passing through this representative point $Q$; and there is, thereby, an $\infty^3$ of possible ISAs passing through every point in space. Next, because there is a total of $\infty^3$ of points (§ 4.01), and making sure that each of the lines in space is counted

Figure 6.02 (§ 6.15) Two rigid, irregular bodies free in space. Their capacity for relative motion can be measured by the fact that there is at an instant an $\infty^5$ of ISAs or motion screws available. These screws constitute all of the screws in space, and together they may be called the 6-system of screws.

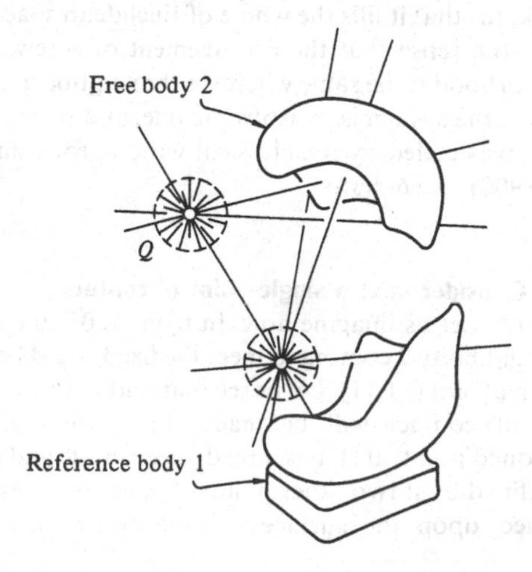

only once (§4.10), we see that infinity to the power $(3+3-1)$, namely $\infty^5$, is the total number of ISAs in space about which the body 2 is free to screw, or to be twisted, with respect to the body 1. This line of argument is consistent with the otherwise evident fact that, along every one of the $\infty^4$ of lines in space (§4.10), there is a single infinity, namely $\infty^1$, of ISAs available. Thus the answer to the first question can be seen again to be: infinity to the power $(4+1)$, namely $\infty^5$.

16. We are moving here (although as yet unspokenly) into the realms of the general screw systems (chapter 23) and in these realms it is important not to be dumbfounded by the immensity of the numbers we find. We are looking, here, for the bare essentials only of the geometries pervading, and, to defend against the very large numbers mentioned, we shall pay attention firstly and mainly to the relatively small, integral powers of infinity which abound. These powers (or indices or exponents) are connected throughout the theory of screws by a simple and useful logic. The power 5 in the very large number $\infty^5$, for example, can be taken and used to find a direct measure of what we shall call the *freedom* of the body 2 with respect to the body 1. This freedom is the same as that freedom discussed at § 1.26 and § 3.33 incidentally. We shall say in general and without exception that the freedom $f$ of the body 2 with respect to the body 1 can be stated as follows: *it is that number f which is one more than the power of infinity that is telling, in the particular circumstances of the constraint, the extent of the multiple infinitudes of ISAs available for the relative motion.*

17. I want the reader to see, in the present particular case, namely the case of the two non-connected bodies, that, *with zero points of contact extant, there are six degrees of freedom, there being an $\infty^5$ of ISAs available for the motion.* About the particular system of the ISAs, or, more simply, about the particular system of screws existing here, I would like to remark, (*a*) that it fills the whole of Euclidean space, (*b*) that in the sense that the arrangement of screws at a neighborhood is the same wherever the neighborhood is in space, the system is an isotropic one, and (*c*) that the system was called by the classical writers, for example Ball (1900), *the 6-system.*

**Consider next a single point of contact**

18. Let us imagine now, in figure 6.03, the same free, rigid body 2 coming to meet the fixed, rigid body 1 to form a joint (§ 1.11). Let us see that, suddenly, a single point of contact will be made. In figure 6.03 the mentioned point, at $A$, has already been made and there are defined by it two other points $A_1$ and $A_2$. These are attached upon the surfaces of the bodies 1 and 2 respectively, and, at $A$, they are instantaneously coincident (§ 1.05, § 12.07).

19. Now, before we allow any other points of contact to occur, let us suspend events until we ask again the relevant questions: *in how many different ways, at this particular instant under consideration when $A_1$ and $A_2$ are together at A, can the body 2 be screwing relative to 1 in such a way that the contact at A remains?* Or, to put the same question in similar, though different, words: *given that the point of contact at A, once made (by whatever a movement), cannot be unmade, in how many different ways can the body next move at that instant by sliding at A, relative to the body 1?* We should ask, moreover, about the ICRM, about the layout in space of the system of the available ISAs (§ 3.34, § 4.04).

20. Both italicised questions have the same answer and we shall quite naturally, and as before, be attempting to answer the questions in terms of the $ISA_{12}$. We shall see that, whereas the number of possible ISAs for the instantaneous motion in this early case of a single point of contact is still almost unbelievably large, the number of ISAs available will diminish step by step as we increase the number of points of contact between the bodies. Finally we shall see, when six points of contact have been made (§ 6.61), that relative motion between the bodies has become impossible.

21. It should be made clear that the discussion is

Figure 6.03 (§ 6.18) Two bodies in contact at a single point. The moving body has 5°F with respect to the reference body, there is an $\infty^4$ of motion screws available, and we are investigating here the quasi-general 5-system of screws.

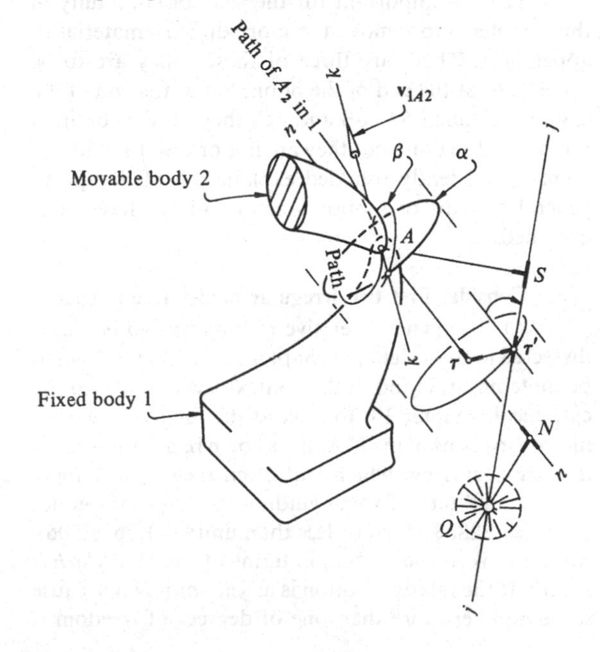

concerning instantaneous motion. The fact, in figure 6.03 that $A_2$ will move in some manner towards, then slide away from $A_1$, and that other, continuously new, contiguous points of contact $A$ will be in the process of formation as the sliding, irregular surfaces move through time, does not concern us here. Except in so far as we must be able to see that such continuous touching and sliding at undulating surfaces can, and does occur in mechanism, and that it needs to be studied somewhere as a separate issue (§8.35, §19.60), details of this aspect of the matter are not relevant here. To say it again, we are looking at *motion at an instant*. Another way of saying this is to say that we are looking at displacement and small increments of displacement only. Often, however, we shall 'divide through' by the corresponding small increments of time and thus speak, either qualitatively or quantitatively, about *velocity at the instant*. We shall mean by that particular phrase that we are not interested, at the moment, in such things as the acceleration of the point $A_2$, whose various components and derivatives determine the shape, as distinct from the direction, of the path at $A$ of $A_2$ in 1 (§1.18).

22. Refer again to figure 6.03. An example of a possible path for $A_2$ in 1 is shown by the twisted line. This path must, for obvious reasons, exist wholly outside the physical boundary of 1, but the path is shown in the figure to be touching 1 at $A$. The shown example path thus gives the beginnings of an answer to the first of the two questions put at §6.19. It should be clear however that the main matter at issue is not the general appearance of the path as it passes through the depicted point of contact $A$, or the question of whether or not the bodies remain for long in contact after the instant under consideration, but the mere direction of the path at $A$. The shown velocity vector at $A$, marked $\mathbf{v}_{1A2}$ in figure 6.03, is naturally drawn to emanate from $A$ and to be tangential to the shown example path. But, provided it resides within the plane marked $\alpha$, which is that unique plane tangential to both of the contacting surfaces at $A$ and accordingly called the *tangent plane at the point of contact*, the mentioned vector may take any one of the $\infty^1$ of directions which are available there (§3.04).

23. It has been said above that all of the possible lines that might contain the mentioned velocity vector at $A$, all of which must pass through $A$ and be upon the tangent plane at $A$, combine to form a planar pencil of lines which wholly resides within that plane. It follows, accordingly, that the two questions put at §6.19 can now be combined into one, and the combined question can be re-put as follows: *about how many different ISAs can the body 2 be caused to screw, at the particular instant depicted by figure 6.03, while we ensure that the velocity vector at A lies somewhere within the tangent plane at A?* The answer to this question, like the answer to the earlier one (the one italicised at §6.14), is a very large number. We shall see however that the number is very much smaller than previously. Refer to figure 6.03.

24. Choose as before any point $Q$ in space and draw any straight line $j$–$Q$–$j$ to pass through $Q$. Drop from $A$ a perpendicular $AS$ to meet this line in $S$. At $A$ erect a plane $\beta$ normal to the line segment $S$–$A$ and note the straight line $k$–$A$–$k$ in which this plane cuts the plane $\alpha$. Now imagine $j$–$Q$–$j$ (which is any line through $Q$) to indicate the location of a possible $\text{ISA}_{12}$ and see that, among the single infinity of velocity vectors which could be drawn at $A$ in the plane marked $\alpha$, the one directed along the line $A$–$k$ is the only one consistent with the possibility that the line $j$–$Q$–$j$ might contain an $\text{ISA}_{12}$. Note next that, (*a*) the length of the radius vector $S$–$A$ (see also an identical radius vector $S$–$A$ in figure 5.09), and (*b*) the magnitude of the angle between $A$–$k$ and $S$–$j$, will together determine the pitch $p$ of the screwing. We can thus see that, although any one of the $\infty^2$ of lines through $Q$ – they collectively form a star there (§4.10) – can be a seat for an $\text{ISA}_{12}$, the pitch is determined as soon as the seat is chosen. This means that the number of available ISAs which can be drawn through $Q$ is only $\infty^2$.

25. There is however an $\infty^3$ of points such as $Q$ (§4.01) and so, making sure that each of the lines in space is counted only once (§4.10), the total number of ISAs which can be chosen is: infinity to the power $(2+3-1)$, namely $\infty^4$. We have shown in other words that, *with one point of contact extant, there are five degrees of freedom, there being an $\infty^4$ of ISAs available for the motion*. This algebraic result should now be compared with the corresponding earlier result at §6.17. The actual, geometrical arrangement in space of these $\infty^4$ of ISAs or motion screws will need to be known and clearly envisaged by us, however; this aspect of the matter is discussed in appendix 6A, beginning at §6A.01. It is discovered there (in a very broad summary) that, *without change of character as we go from one end to the other of the infinitely long, contact normal n–n, the system of screws is axi-symmetric about that line*. The system of screws is a 5-system and, from the shape of the system, it is also clear that, so far as the capacity for instantaneous motion of the body 2 is concerned, the actual position of the point of contact $A$ along the line $n$–$n$ is irrelevant. *The capacity for instantaneous motion of the body 2, which is summarised by the system of screws extant (the quasi-general 5-system), is wholly and solely determined by the location in space of the single, contact normal n–n.*

**Two points of contact between the bodies**

26. Having thus discussed the implications of a

single point of contact between the bodies, let us look now at the question of what relative motion may be possible after a second point of contact $B$ has been made. We can rest assured, incidentally, that the newly arriving point $B_2$ can be brought into contact with its waiting point $B_1$ in such a way that the vector $\mathbf{v}_{1B2}$ at $B$ can lie somewhere within the tangent plane at $B$. It is a fact, in other words, that a 'flying' $B_2$ can always be led to 'land' at $B$ smoothly. A proof of this fact will be found among the following paragraphs; as soon as we have found even one available ISA for the motion, we have found a confirmation of the fact. Refer to figure 6.04. Take as before a representative point $Q$ and ask the question: *how many possible ISAs (or available screws) can be drawn through $Q$ to satisfy the new conditions prevailing, namely that there are now two contact normals in the body 2?* Please remain looking at figure 6.04.

27. It will be clear first of all that there is only one line which can be drawn through $Q$ to cut both of the contact normals, and that this, which is marked $p=0$ in the figure, is the one and only screw of zero pitch which can be drawn to pass through $Q$. There is, similarly, only one screw of infinite pitch which can be drawn through $Q$, and this, which is marked $p=\pm\infty$ because it may be of either sign, is parallel with the common perpendicular $z$–$z$ between the two $n$-lines.

Figure 6.04 (§ 6.26) Two bodies in contact at two points. The movable body has 4°F with respect to the reference body, there is an $\infty^3$ of motion screws available, and we are investigating here the quasi-general 4-system of screws. For the ellipse that is shown please refer to § 15.34.

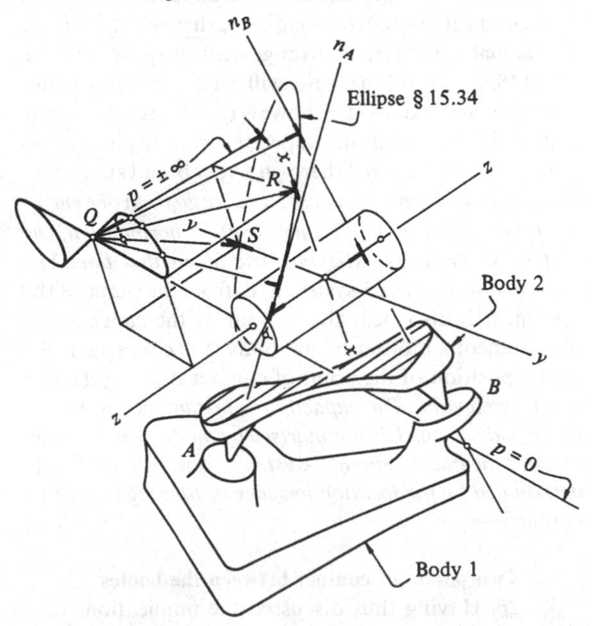


28. Now it is a fact that, whatever value of the pitch we choose, we find in every case that only one screw of that pitch can be drawn through $Q$. Much of the detail of why, or even how, this is so is difficult to see however. Ball, in his book (1900), deals with the matter and Hunt, in his (1978), clarifies the question by writing the necessary algebra for the general 4-system and describing the layout of the screws of that system verbally and accordingly. I speak about it at § 15.33 *et seq.*, where a study of certain aspects of the cylindroid is undertaken. Refer also to § 23.50. Relying upon these and other works I offer a fairly bulky study of the quasi-general 4-system of motion screws in my appendix 6B. The study derives, of course, directly from the circumstances of the two points of contact which is currently being considered, and it begins with a further look at figure 6.04. I give a summary of the main conclusions of that study in the next paragraph, § 6.29, so that it can be read conveniently, and quite quickly, soon. In any event, and, to some extent, independently of all that, the following remark can now be made: if it is true that only one screw can be drawn through $Q$ for each value of the pitch, then there is through $Q$ only an $\infty^1$ of screws available for the motion. We can conclude accordingly that, because there is an $\infty^3$ of points such as $Q$, and, making sure as before that each of the available screws in space is counted only once (§ 4.10), the total number of ISAs available is: infinity to the power $(1+3-1)$, namely $\infty^3$. I have thus shown here, although by reference partly, that, *with two points of contact extant, there are four degrees of freedom, there being an $\infty^3$ of ISAs available for the motion.*

29. For the earlier mentioned 6-system, and for the 5-system, remarks were made at § 6.17 and § 6.25 about the natures of the symmetries obtaining. Extracting the necesary material from appendix 6B, we can now make similar, though somewhat more complicated remarks about the symmetries of the 4-system. Referring to figures 6.04, 6B.01 etc., we should see first of all that, looking at the conical and hyperboloidal surfaces, ignoring the values of the pitches upon the generators of the cones and ignoring the right handed or left handed sweeps of the reguli upon the hyperboloids, *the common perpendicular between the two contact normals is not some mere unimportant line, but a main axis of symmetry of the whole system.* Next we should see that there is at O, midway between the common normals, a central point of the system, and that, at this $O$ and normal to $z$–O–$z$, there is a plane which divides the whole system into two identical halves. The system is line-symmetric about the axis $z$–O–$z$, but not axi-symmetric about it; given that quadrants are designated by the intersecting planes $z$–O–$x$ and $z$–O–$y$, the contents of each quadrant is line-reflected through the axis $z$–$z$ into its diagonally

opposite quadrant. *There is, accordingly, what can be called a rotary symmetry of the second order about the axis z–O–z*. The whole system is point-symmetric about the central point O, however, and this means in effect that, *although the two halves of the system on either side of the central plane x–O–y are mirror images of one another, these halves are displaced from one another by 90° of rotation about the central axis z–O–z*. It might be noted in passing that the symmetries here outlined are exactly those exhibited by that very much simpler figure, the well-known cylindroid (chapter 15); the significance of this remark will become apparent later (§ 6.52, § 7.94).

**In recapitulation**

30. Before going on to consider a third point of contact between the bodies, let us summarise the main information established so far. We have constructed, in effect, table 6.01. The last row of figures at the table has not been discovered by us. It was simply stated on each occasion that the relevant screw system had a name; I referred obliquely to its general importance against a wider background (chapter 23) and we made in each case a brief exploration of the geometry. It must be reiterated here that the vast generality and the wide applicability of these and the other screw systems in the mechanics of mechanism (the kinematics, the statics, the kinetics, and their interrelationships) is carefully being understated here. We are working within a strictly limited area of the wide field of the mechanics of mechanism, namely that of the *contact geometry at joints*; as was said before (§ 6.08), this limits our wider view somewhat.

**Three points of contact between the bodies**

31. I would like to investigate now the somewhat more mysterious three points of contact between the bodies. It will in some respects be seen as the most mysterious because, although the number of points of contact is becoming larger as we go, the cases for four and five points of contact become, henceforth, progressively easier, not more difficult, to understand. By virtue of the overall symmetry in this particular case (three points of contact will lead us to study the quasi-general 3-system, six minus three being three), there is a kind of knotty centrality about the matters at issue which not only demands to be understood for its own sake, but stands also, as is seen elsewhere, significant in a central way for the study of joints and mechanism (chapter 7, § 10.66, § 14.36).

Table 6.01 (§ 6.30) A summary of the conclusions reached so far. The gaps in the table are to be filled as we read from § 6.31.

| | | | | |
|---|---|---|---|---|
| Number of contact points between the bodies | 0 | 1 | 2 | 3 |
| Number of ISAs to be found through the point $Q$ | $\infty^3$ | $\infty^2$ | $\infty^1$ | ? |
| Number of locations available for ISAs of any pitch | $\infty^4$ | $\infty^2$ | $\infty^0$ | ? |
| Number of ISAs (or screws) available for the motion | $\infty^5$ | $\infty^4$ | $\infty^3$ | ? |
| The accepted name of the screw system involved | 6 | 5 | 4 | ? |

32. Let us set up, to begin, in figure 6.05(*a*), three points of contact between the bodies 1 and 2, and erect the three resulting contact normals there. Because it will be clear by now that the exact location of a point of contact and its tangent plane along an already established contact normal is irrelevant so far as the capacity for instantaneous motion of the moving body is concerned (chapter 3 and figure 3.04), the actual points of contact and their tangent planes are next ignored in figure 6.05(*a*).

33. Before we ask this time the usual question about how many available screws might be drawn through a representative point $Q$, however, let us look, with the eyes of some experience now, at the three contact normals which are extant and see (*a*) that that set of lines which can be drawn to cut all three of the contact normals, namely that set of lines about each one

Figure 6.05(*a*) (§ 6.32) Two bodies in contact at three points. The movable body has 3°F with respect to the reference body, there is an $\infty^2$ of motion screws available, and we are investigating here the quasi-general 3-system of screws.

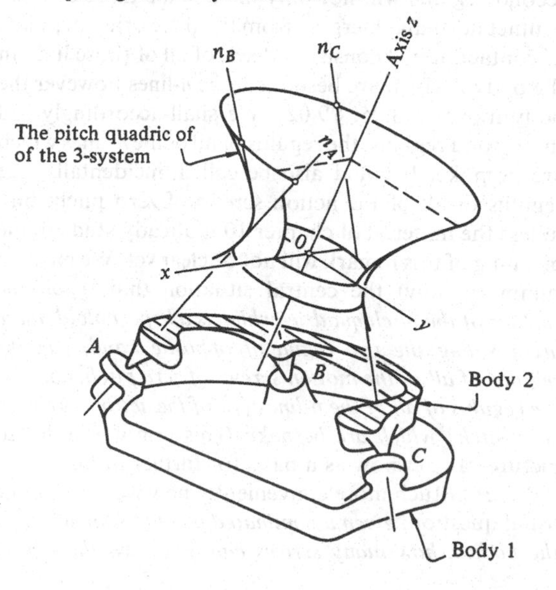

of which the body 2 can screw with zero pitch, is a regulus upon that same hyperboloid which is, in fact, defined in this manner by the contact normals themselves (§ 3.39); and (*b*) that the regulus thus constructed might be called, for the sake of present argument, *the regulus of all of the motion screws of zero pitch.* Please be aware that much of the argument since § 6.31 is outlined in other words at § 3.39 *et seq.*, where the lines called *r*-lines there are the motion screws of zero pitch being mentioned here. There is no contradiction among the arguments, and figure 3.13(*a*) is relevant.

34. Refer to figure 6.05(*a*). Be aware that the mentioned hyperboloid upon which the regulus resides has a central point *O*, and that its regular set of Cartesian axes at that point can be determined. Appealing to known geometrical properties of the hyperboloid (§ 11.08), we can (*a*) construct the three common perpendiculars between the three pairs of contact normals; (*b*) erect at the mid-points of these the three planes which are normal to them respectively; (*c*) note the point where the three planes intersect one another; and thus (*d*) locate the central point *O*. Although the following is not so easy to do (§ 11.61, § 23.37, and Ball § 175), we can next erect the axis *z*–O–*z* along the central, longitudinal axis of the hyperboloid, and erect its companion axes *x*–O–*x* and *y*–O–*y* coincident with the minor and the major axes respectively at the elliptical throat. Please notice that all of this is well enough shown in figure 6.05(*a*).

35. Let us call the particular hyperboloid just located, for a reason best to be found at Ball § 173, *the pitch quadric*. We shall next erect upon the surface of this pitch quadric that regulus which is opposite to the regulus of all of the motion screws of zero pitch. This second regulus will not only include the three original contact normals emerging from the three original points of contact, it will consist, indeed, of all of those lines in the body 2 which are bespoke to be *n*-lines however the body moves (§ 3.05, § 9.02). We shall accordingly call this second regulus, the regulus of all of the *n*-lines which are bespoke. It could also be called, incidentally, the regulus of all of the action screws of zero pitch, but, unless the material of chapter 10 is already studied, the meaning of this remark will not be clear yet. We have (in summary now) the central situation that, *upon the surface of the pitch quadric, which is a hyperboloid, there are residing the two reguli of opposite hand, (a) the regulus of all of the motion screws of zero pitch, and (b) the regulus of all of the n-lines (or of the action screws of zero pitch) which are bespoke*. This firmly established picture can act now as a basis for further argument.

36. Much more conveniently now we can ask the usual question: *given a nominated point Q somewhere in the bodies, how many screws can be drawn through Q which are available screws for the instantaneous motion?* Looking at figure 6.05(*a*) we can see straight away that, if we look in the first instance, as we did before, for available screws through *Q* which are of zero pitch, we will, in general, find none. It is true of course that, if the point *Q* were specially chosen to be upon the surface of the pitch quadric just located, we would find one, but even in this special case only one will be found; it would be that single, relevant member of the regulus of all of the motion screws of zero pitch. *We have, in summary, said here that, whereas in general there are to be found through Q no screws of zero pitch for the instantaneous motion, there are in any event no more than one of them which can be found.*

37. But next we must ask the remaining questions: are there available screws of other pitches which can be found to pass through *Q*, and, if there are, how many of them are there, and what are the values of their pitches? Refer to table 6.01. Relying upon the unproven but likely continuity of this unfinished table, we might expect to find, with respect to screws through *Q*, that there are an $\infty^0$ of screws of various pitches which can be drawn through *Q*. But what is this $\infty^0$? What is the precise finite number?

**One way into a first study of the 3-system**

38. Choose any pair of motion screws of zero pitch from the pitch quadric, namely any pair from the whole regulus of them, and construct the common perpendicular. The common perpendicular will not, in general, pass through the central point *O* of the pitch quadric (§ 11.08); nor will it pass, however, more than a certain finite distance away from it (§ 11.11). Next erect upon that common perpendicular that cylindroid of screws defined by the two already selected motion screws of zero pitch (§ 14.40). The nodal line of the cylindroid and the common perpendicular will be collinear (§ 15.05), the length 2*B* of the nodal line will never be less than the length from foot to foot of the common perpendicular (§ 15.12), and the central points of the cylindroid and of the common perpendicular will coincide (§ 15.14). Because every screw of the cylindroid is linearly dependent upon the two selected motion screws of zero pitch (§ 10.14), each one of them will be a member of the quasi-general 3-system of motion screws we are exploring. Refer to the figure 6.05(*b*), where the cylindroid of which I speak is designated $C_1$.

39. It should be remembered that, while every conceivable aspect of the said 3-system is wholly determined merely by the regulus of motion screws of zero pitch already established by us upon the pitch quadric, we have achieved, as yet, only a very weak picture of the layout of this system in the Cartesian space. We have, for example, no idea as yet, of whether

or not various groups of screws of equal, non-zero pitch exist, or of how they are arranged if they do exist. In the following paragraphs I intend to point one way towards a clarification of this; I hope, in so doing, that the reader will bear with my particular way of doing it, for there are, doubtless, many other ways. Please note that, by introducing the cylindroid $C_1$ of §6.38, I have introduced already one single infinity of pairs of screws of different, finite, non-zero pitches (§15.49), and that every one of these will need to be incorporated somehow into an overall pattern which is intelligible.

40. I now choose a point $Q$ anywhere along the nodal line of $C_1$, and note the two screws there. I call them $Q$-(*ii*) and $Q$-(*iii*). They reside upon the two generators of $C_1$ which intersect at $Q$ and, while $Q$-(*iii*) is destined never to cut the surface of the pitch quadric, or so it would appear in figure 6.05(*b*), $Q$-(*ii*) is destined to do so twice. Please be aware that, at my appendix 6C, these and other such matters are discussed more exhaustively. Now it is a fact that, if through a general point $Q$ one common perpendicular between the lines of a regulus can be drawn, only one other can be drawn

Figure 6.05(*b*) (§6.38) The ellipse is the throat ellipse of the pitch quadric. At a general point $Q$ anywhere within a certain central zone of a 3-system, the nodal lines of three different cylindroids of screws intersect. Two of such a set of three cylindroids, marked $C_1$ and $C_2$ in the figure, are shown to be intersecting in the screw $Q$-(*iii*) which is common to both. The two other pairs of cylindroids, $C_2$ and $C_3$, and $C_3$ and $C_1$, intersect in the screws $Q$-(*i*) and $Q$-(*ii*) respectively, and the three shown screws are the three screws through $Q$. Please note that the panoramic view of the Cartesian triad $x$, $y$, $z$ is the same here as the one in figure 6.05(*a*).

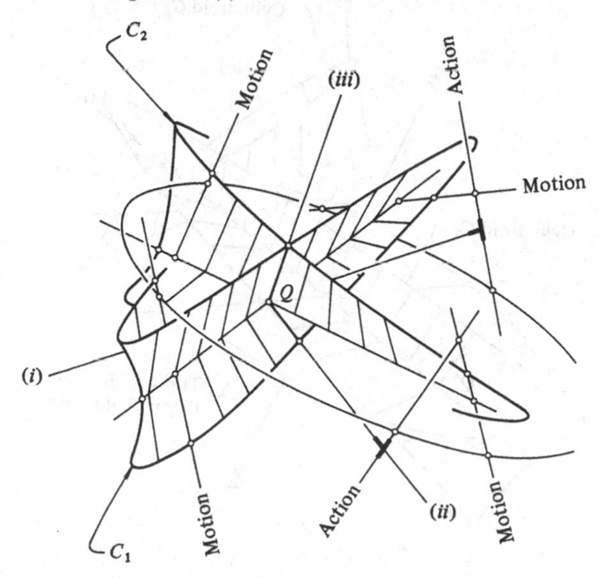


(§11.11). This other common perpendicular, through $Q$ and between a pair of motion screws of zero pitch, is drawn in figure 6.05(*b*); it carries a similarly constructed cylindroid of motion screws which is marked $C_2$. Next it is a fact that the cylindroids $C_1$ and $C_2$ will intersect along a common generator, namely $Q$-(*iii*), thus confirming that the screw $Q$-(*iii*) is an available screw of the 3-system. At $Q$, however, there is also the generator $Q$-(*i*) of the cylindroid $C_2$, and this screw, like $Q$-(*ii*) of $C_1$, is destined to cut the surface of the pitch quadric twice. We thus have, intersecting at $Q$, *three screws of non-zero pitch, and they are not, in general, either of equal pitches or mutually perpendicular*.

41. Any pair of intersecting screws defines a cylindroid (§15.18); so through $Q$ now, and normal to the plane defined by any two of the three screws at $Q$, there is the nodal line of a cylindroid. There are, accordingly, *three* cylindroids of motion screws with their nodal lines intersecting at $Q$. Only two of them are shown in figure 6.05(*b*) however; the third, $C_3$, is not drawn there. The three intersecting nodal lines at $Q$, like the three screws there, are not in general intersecting perpendicularly; but, because each of them is normal, in a cyclic manner, to the plane defined by the 'other' two – $C_1$ is normal to the plane defined by $Q$-(*ii*) and $Q$-(*iii*), etc. – the angular relationships between the triplet of motion screws at $Q$ and the triplet of nodal lines at $Q$ are the same.

42. The sum of the pitches of intersecting pairs of generators is a constant for any one cylindroid (§15.49), so the sum of the pitches of all three screws through $Q$ is a constant for all positions of $Q$ along the nodal line of $C_1$. Although, quite clearly, there is a limited, symmetrical, central zone within which a point $Q$ can be chosen upon a pair of common perpendiculars in the manner described and beyond which it cannot, this sum of the pitches turns out to be a constant for all general points $Q$ (§6C.06). It might also be noted that in the limit as $Q$ is taken closer and closer to the origin of coordinates $O$ the three screws arrive at a mutual orthogonality, each one of them coalescing with the nodal line defined by the other two. Accordingly there is both a screw and a nodal line along each of the three principal axes of the Cartesian system. The three screws become there the *three principal screws* of the 3-system and the three nodal lines become the nodal lines of the *three principal cylindroids* (Ball §174, Hunt §12.6). It remains to remark that, if the point $Q$ is taken inside that central zone (which zone may protrude, incidentally, outside the elliptical throat of the pitch quadric), the screws $Q$-(*i*) and $Q$-(*ii*) both intersect the pitch quadric twice. The intersections occur at places which may be described as follows: *$Q$-(i) and $Q$-(ii) both cut two action screws upon the pitch quadric, namely two screws of the*

*opposite regulus to that of the motion screws, perpendicularly.*

43. This means that, to locate at least two of the available three of the motion screws through $Q$, namely $Q$-(*i*) and $Q$-(*ii*), we could draw the two (the only two) common perpendiculars through $Q$ which could be drawn between pairs of action screws of zero pitch. The entire family of the latter reside, as we know, as members of the opposite, original regulus upon the pitch quadric (§ 6.35). Next we could mount at $Q$, normal to the plane of the two discovered screws, the direction of the nodal line of the cylindroid $C_3$. We could moreover calculate the mid point and the length $2B$ of this $C_3$ from the pitches of and the angle between the two discovered screws $Q$-(*i*) and $Q$-(*ii*), because those two screws, as already explained, define the cylindroid $C_3$. Through each point $Q$ where common perpendiculars between generators of the pitch quadric can be drawn (and there are an $\infty^3$ of such points, all inside the mentioned, central zone), there pass the nodal lines of three cyclindroids; but along each nodal line a single infinity of points such as $Q$ exist. So the total number of cylindroids is, infinity to the power $(3-1)$, namely an $\infty^2$. Each cylindroid carries an $\infty^1$ of motion screws (chapter 15); but, as we have seen, each screws is a generator of an $\infty^1$ of different cylindroids, so the total number of screws passing within the mentioned, central zone is, infinity to the power $(2+1-1)$, namely an $\infty^2$. We shall see in due course that these together constitute all of the screws of the quasi-general and, indeed, of the general, 3-system of motion screws.

44. In summary I have said so far, albeit somewhat sketchily, that, through every general point $Q$ within a certain, central zone in the Cartesian space, there can be drawn three but no more than three available screws, that these are of different, finite pitches, but that, incidentally, and independent of position within the central zone, the sum of the three pitches at $Q$ remains a constant.

45. I wish to explain now that, while some one screw such as the screw $Q$-(*iii*) will always exist wherever the point $Q$ may happen to be, the other two, the two such as $Q$-(*i*) and $Q$-(*ii*), may be either, (*a*) intersecting at $Q$ and of different pitches as above; (*b*) coalesced at $Q$ and of equal pitches; or (*c*) *both unreal.* If $Q$ is taken anywhere outside a certain, central, symmetrical zone – this zone is, indeed, the same zone as the zone I have mentioned at § 6.42 – we get into a quite different situation. Outside the mentioned zone, not three but only one real screw exists at each point $Q$. To begin to see a few of the main aspects of this exclusive (or, if you wish, inclusive) zone, its significance, and some of the special intersections of three real screws which occur within it, please refer to figure 6.05(*c*). Please refer also, but in due course, to appendix 6C; it begins at § 6C.01. In figure 6.05(*c*) a certain, special cylindroid has been drawn. Its nodal line is not along a common perpendicular between any pair of motion screws of zero pitch, but along that particular common perpendicular between a pair of such screws which coincides with the $y$-axis. This $C_2$ is, indeed, one of the three principal cylindroids of the system (§ 6.42). Please become aware now that, if $a$, $b$, and $c$ are the half axes of the pitch quadric measured in the directions $x$, $y$ and $z$ respectively, the three available screws along those axes which are called the principal screws have pitches $p_\alpha$, $p_\beta$ and $p_\gamma$ respectively, and that these can be determined, in terms of $a$, $b$ and $c$, easily (§ 6.42, § 6C.09, § 11.13). If $a$ is shorter than $b$, $p_\alpha$ is greater than $p_\beta$. Given the right-handedness of the system of $n$-lines in figure 6.05(*a*), $p_\gamma$ is negative and thus, in the present investigation, the least of all three. Now the maximum and minimum pitches of our principal cylindroid $C_2$ in figure 6.05(*c*) are $p_\alpha$ and $p_\beta$ (§ 15.49). It follows that, somewhere along its nodal line, on either side of its centre and equidistant from it, there are two points where a generator of pitch $p_\beta$ is extant. These two points – they are both inside the above-mentioned symmetrical central zone – are special. They are singular points of the 3-system. I wish to call these points, after Hunt, $Q'$ and $Q''$.

Figure 6.05(*c*) (§ 6.45) The principal cylindroid on $y$–O–$y$ being intersected at the special point $Q'$ by another of the three cylindroids there. This particular intersection reveals the special regulus of screws that contains all screws of the principal pitch $p_\beta$; see figure 6.05(*d*). Please note here and at the next figure that the panoramic view of the Cartesian triad $x$, $y$, $z$ has been taken from a new direction.

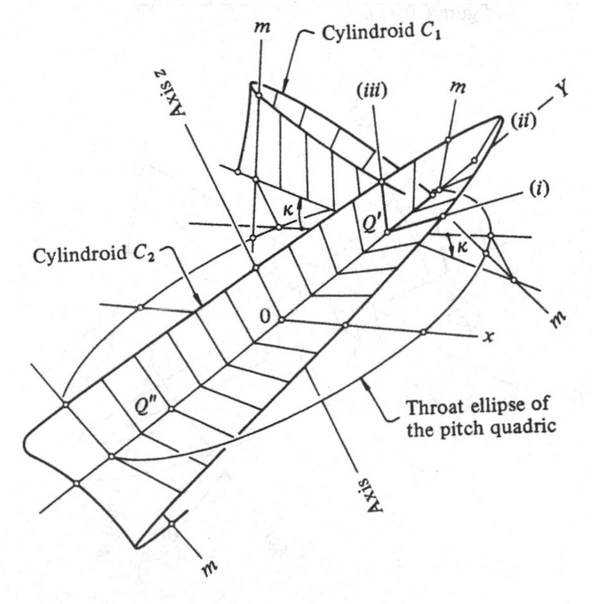

46. Let us take a moving point $Q$ now, begin with it at O, and let it move from O towards $Q'$ along the line O–$y$. At each new position of $Q$ imagine a second cylindroid $C_1$; let $C_1$ be mounted upon the other common perpendicular through $Q$ (drawn, as before, between its particular pair of motion screws of zero pitch) and notice that the symmetry of the regulus is such that the nodal line of this new $C_1$ remains perpendicular to that of $C_2$. It tilts, however, as it goes and, were it to go all the way from O to the extreme generator of $C_2$, which generator is beyond the vertex of the throat ellipse, as shown in figure 6.05($c$), it would tilt through the total angle of 45° from O–$x$ (§15.49). There at that extreme generator, at that end-point of the nodal line of $C_2$, the two screws $Q$-($iii$) and $Q$-($i$) would become of equal pitch, namely $(p_\alpha+p_\beta)/2$, and coalesce. We would be, at that point, at a point upon the boundary of the said symmetrical central zone. It should be mentioned here that, as $Q$ passes the vertex of the throat ellipse of the pitch quadric and thus moves beyond the confines of the pitch quadric, it still remains within the cylindroid $C_2$. It goes to a region of $C_2$, however, where the cylindroid $C_1$ has ceased to have generators of zero pitch. The cylindroid $C_1$ does not disappear as $Q$ passes the said vertex; it remains defined by its two principal screws whose signs however are now the same, $Q$-($iii$), whose pitch continues to vary as $Q$ moves towards the end-point, and $Q$-($ii$), whose pitch $p_\beta$ remains the same. It should also be mentioned that, at all positions of $Q$ along O-$y$ there are three screws through $Q$ as explained in general at §6.41. The three screws are $Q$-($iii$), along that single generator which is common to $C_1$ and $C_2$; $Q$-($ii$), along the other generator of $C_1$ at $Q$, which, as will be seen, is, at all points $Q$, collinear with O–$y$ and thus with the principal screw where the pitch is obliged to be $p_\beta$; and $Q$-($i$), along the other generator of $C_2$ at $Q$. When $Q$ arrives at $Q'$, however, something quite spectacular happens: $Q$-($i$) suddenly obtains the same value of pitch, namely $p_\beta$, as has the screw $Q$-($ii$), and, centred at $Q$, containing the $y$-axis, and inclined at the angle $\kappa$ to the $xy$-plane, there springs into being a whole planar pencil of screws of the same pitch $p_\beta$. See in advance figure 6.05($d$). Subsequently and until the end-point of $C_2$, where, as we have said, the two screws $Q$-($iii$) and $Q$-($i$) become of equal pitch and coalesce, we revert to the usual three screws through $Q$. The singular point $Q'$ is not a point upon the boundary of, but a point within, the said symmetrical central zone.

Figure 6.05($d$) (§6.47) The degenerate or special regulus of screws of pitch $p_\beta$. It divides the two parts of the 3-system from one another; see figure 6C.05. There is one set of a single infinity of screws coplanar at $Q'$ and another, identical set of screws coplanar at $Q''$. There is thus a single infinity of screws of the regulus all together, and the regulus might, on that account, better be called a special regulus (§11.14). The ellipse that can be seen is the throat ellipse of the pitch quadric; it resides in the $xy$-plane.

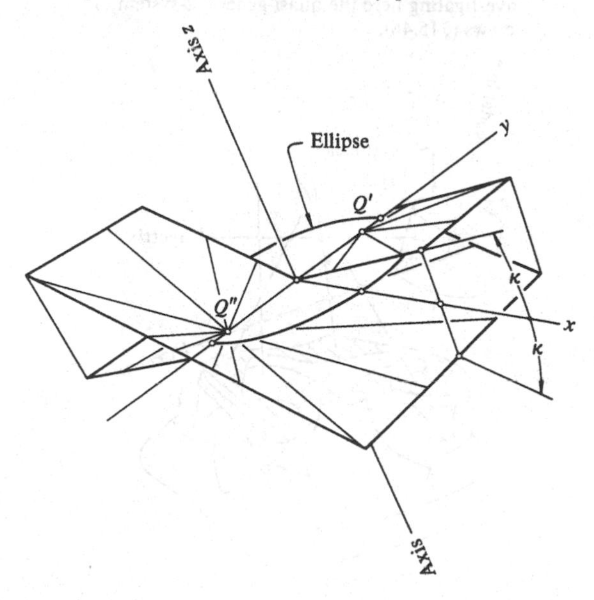


47. It is clear from symmetry that, if $Q$ were moved from O along the $y$-axis in the other, negative direction, a similar spectacular event would occur at $Q''$. Another planar pencil of screws of pitch $p_\beta$ would suddenly appear. This pencil would be inclined at the same angle $\kappa$, but it would be symmetrically opposite against the $xy$-plane as shown in figure 6.05($d$). This discovered pair of planar pencils of screws of pitch $p_\beta$ can next be seen, not illogically, as a regulus; its throat ellipse is the line segment $Q'$–$Q''$, an ellipse of zero width (§6C.08, §11.15). We know already that all of the screws of *zero* pitch are upon a regulus, a regulus upon the pitch quadric (§6.35). It can be shown moreover, and I do so at §6C.11, that every set of screws of the *same* pitch resides upon a regulus. I shall not attempt to describe just yet all details of the geometry of these reguli and their complicated interrelationships, but the following overall statements might now be made: *(a) within the 3-system there is an* $\infty^2$ *of screws; (b) the 3-system consists of an* $\infty^2$ *of cylindroids; (c) screws of every pitch within the range from* $p_\alpha$ *to* $p_\gamma$ *inclusive exist, but there are no screws existing whose pitches lie outside that range; (d) for each existing pitch there is a single infinity of screws of that pitch; (e) every screw of whatever pitch passes through the mentioned, symmetrical, central zone; and (f) each single infinity of screws of the same pitch resides upon a regulus whose centre is at the centre of the pitch quadric.*

48. The said reguli, however, are not only concentric, as stated above; they are coaxially arranged in two

sets. Those carrying screws whose pitches are less than $p_\beta$ (these include the regulus of the screws of zero pitch) are coaxially arranged about the axis $z$–O–$z$. Those carrying screws whose pitches are greater than $p_\beta$ are coaxially arranged about the axis $x$–O–$x$. The already mentioned, degenerate regulus residing upon the pair of planes, which is shown at 6.05($d$), upon which reside the screws of pitch $p_\beta$, acts as a kind of divider or borderline between the two sets of reguli. Please refer in advance to figures 6C.05 and 6C.06, where the two coaxial sets of the corresponding hyperboloids can be seen. Upon these the mentioned reguli reside. The hyperboloids, even within the coaxial sets, intersect one another in a complicated manner and the 3-system is not an easy thing to digest in one gulp. *The envelopes of the two sets of the throat ellipses, however, which are two closed symmetrical planar curves, wrap up like string around a parcel the central symmetrical zone within which there are three screws passing through a general point Q. At the surface of the said zone there are only two, and outside the zone there is only one, screw – one motion screw or an ISA – which can be drawn through a general point Q.*

**Return to the main argument**

49. We see accordingly that, because there is an $\infty^3$ of points such as $Q$, and, making sure as before that each of the available screws in space is counted only once (§ 4.10), the total number of available ISAs is infinity to the power $(0+3-1)$, namely $\infty^2$. *With three points of contact extant, there are three degrees of freedom, there being an* $\infty^2$ *of ISAs available for the motion.*

50. I wish to remark now upon the shape of this quasi-general 3-system, with its mentioned $\infty^2$ of screws. The matter is discussed both here and at the appendix 6C. We see that all of the motion screws can, among other ways, be grouped together upon a set of reguli whose hyperboloids are concentric. Each hyperboloid carries upon its surface a regulus of screws of the same pitch. In our case here the pitches of the screws have ranged from some minimum finite value negative $p_\gamma$ to some maximum finite value positive $p_\alpha$, there having been no available screws of pitches with values beyond this range. There are, accordingly, and in general, no screws at all of infinite pitch; unless, in particular, either $p_\alpha$ or $p_\beta$ or $p_\gamma$, or a selected pair of them, or all of them, are themselves infinite (§ 23.60 *et seq.*). This should well be clear from figure 6.05($a$), for it is evident upon inspection there that, unless the tangent planes at the prongs and platforms are intersecting in the same line or are parallel or are otherwise specially arranged (§ 7.67), $p_\alpha$ and $p_\beta$ are both finite. It might be observed in general that, if we the observers were far away from the central point of a 3-system, at such a distance that the central zone of § 6.48 appeared to be small, the whole system of lines of the system would resemble a star (§ 4.10). There are most certainly no lines (or locations or seats) in the system containing screws of all pitches. There are, in other words, no lines in the movable body 2 which can be taken as an ISA of any pitch. Does the number of these latter lines (and this is an exercise for the reader) work out to be $\infty^{-2}$; and does that number, which accords with the scheme of table 6.01 at § 6.30, make any sense in the circumstances?

51. In any event we can say with respect to the symmetries that, *(a) there is a central point of symmetry O, about which the whole system is point-symmetric; (b) there is no axis of rotary symmetry and no axis of axi-symmetry of the type encountered in the 4-system; and (c) that, whereas a first group of concentric hyperboloids is coaxial about one axis x–O–x of the Cartesian system, a second group is coaxial about another axis z–O–z of the same system, there being no hyperboloids coaxial about the third axis y–O–y.* Refer for comparison to § 6.29.

**Four points of contact between the bodies**

52. At this stage of the ongoing argument we should ($a$) be reminded that the matter at issue here is the freedoms at the simple joints of mechanism; ($b$) set up for consideration now four arbitrarily chosen points of contact between two bodies and erect the four contact normals there; (refer to figure 6.06); ($c$) check that the contact normals are linearly independent of one another

Figure 6.06 (§ 6.52) Two bodies in contact at four points. The movable body has 2°F with respect to the reference body, there is an $\infty^1$ of motion screws available (upon a cylindroid), and we are investigating here the quasi-general 2-system of screws (§ 15.46).

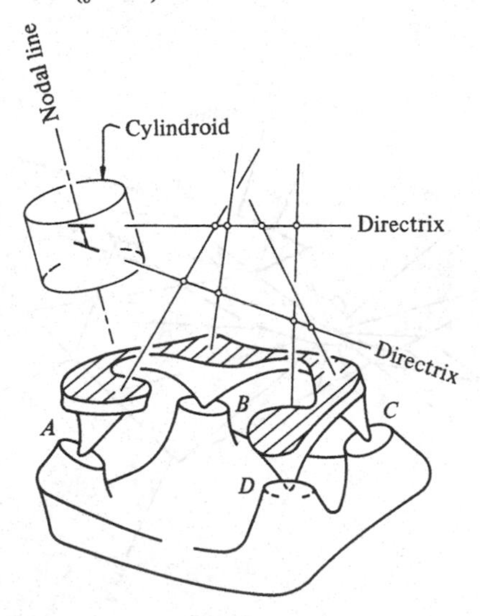

by removing from consideration any one of them which accidentally belongs to the regulus defined by the other three (§ 7.53); (*d*) notice that to begin the argument here by quoting a generally chosen point $Q$ in the bodies will be useless; then (*e*) refer in due course to the materials of the chapters 14 and 15 (§ 6.53).

53. While chapters 14 deals in a preliminary way with the idea of reciprocity, chapter 15 deals with the origins in mechanics and the geometry of the cylindroid. Reference to § 14.37 *et seq.* will reveal a discussion which could lead to another discussion about the reciprocity between four *n*-lines, which are given, and two *r*-lines, which latter may or may not exist according to the geometrical circumstances. Reference to § 15.46 *et seq.* will reveal other material which explains (*a*) how a cylindroid can be constructed if we are given four *n*-lines (namely four action screws of zero pitch) that are reciprocal to it and thus to its motion screws, and (*b*) what can be done if the cylindroid itself contains no *r*-lines namely no motion screws of zero pitch. Perusal of passages such as these in the chapters mentioned, in conjunction with an understanding of the cylindroid itself and its geometrical properties, will reveal the following: *the quasi-general 2-system of motion screws which is associated with the four points of contact being set up here is none other than the cylindroid.*

54. Refer to figure 6.06; recall the well-known theorem which states that, given any four straight lines in space, no more than two straight lines may be drawn to cut all four of them (§ 11.20); and note in the figure the two lines marked *directrix*. They are indeed the directrices of the linear congruence of *n*-lines associated with the four contact normals which are nominated there and they are, as well, the seats for the two motion screws of zero pitch which are available for the relative motion of the two bodies at the instant (§ 3.50, § 14.37). Refer to figure 15.03 to see how it is that the nomination of any two representative screws of equal pitch will determine all of the other screws of a cylindroid (§ 10.03, § 10.14) and thus see how the cylindroid of motion screws which is merely indicted in figure 6.06 can be constructed.

55. If the two directrices in figure 6.06 cannot be drawn, in which case they are imaginary directrices and the linear congruence is not hyperbolic but elliptical (§ 10.09, § 11.24 and elsewhere), we can construct a pair of motion screws of equal pitch of some other, non-zero value (§ 15.48), and thus construct the cylindroid of available motion screws according to the method suggested in figure 15.03 and recapitulated at § 15.13.

56. As we can see from § 15.15 *et seq.*, the cylindroid of motion screws may be, in this case of the four points of contact between two bodies, of the most general kind. It may or may not be possessed of generators of zero pitch and its nodal line and its length $2B$ will bear no easy-to-see relationship with the layout of the four contact normals. That there may be no motion screws of zero pitch means that it will only sometimes be possible to substitute for the four prongs and platforms at the points of contact in figure 6.06 two hinges, series connected to the frame. Please read again in this connection the conditional statements carefully made at § 3.37 and § 23.44, and be aware that, if there were an appendix D to be written to deal with the more difficult aspects here, it would closely resemble the mentioned parts of the chapters 14, 15 and 23.

57. We can say in any event however that, *with four points of contact extant, there are two degrees of freedom, there being an* $\infty^1$ *of ISAs available for the motion.* Like the 4-system whose symmetries were summarised at § 6.29, the 2-system of motion screws in the same context has an identical set of symmetries: *the 2-system (which is the cylindroid) has a central point of symmetry, an axis of rotary symmetry of the second order along its nodal line, and a central plane of symmetry against which its two mirror-image parts face one another and are twisted 90° apart.*

**Five points of contact between the bodies**

58. It was mentioned at § 3.49 and it can be deduced from other known facts that a body with two degrees of freedom can only move in such a way that, *at every point within it, there is, in general, a planar pencil of possible velocity vectors normal to the single, bespoken n-line which passes through that point.* When the thus formed linear congruence of bespoken *n*-lines in the body is a hyperbolic linear congruence however (there is in general an even chance that that will be so, § 11.21), there are two straight lines of points in the body which are obliged to travel in certain predetermined directions § 3.51, § 23.19). These straight lines of points are the directrices of the hyperbolic congruence and they are of course not only that single infinity of points in the body which are special in the body, but together they form the two *r*-lines which may or may not exist in the body (§ 3.51, and figure 3.13). To say it again: *it is in general true that every point in a body with 2°F is free to move, at the instant, in a single infinity of different directions, and that those directions are always distributed upon the planar pencil of lines which is normal, at the point, to the single n-line which passes through that point.*

59. However, if any chosen one of the general points in a body 2 with 2°F is obliged for some reason to move in a certain direction (this could be achieved, for example, by attaching a ball at the chosen point in 2 and causing the ball to slide in a suitably directed tube in 1), *then every other point in the body 2 would have the direction of its velocity at the instant determined also.* This follows from the principle of rigidity (§ 5.03). It

follows accordingly that, if a fifth point of contact $E$ between a pair of bodies 1 and 2 is added to an already existing four (in such a way that the tangent plane at $E$, or the platform there, is not coplanar with the already existing planar pencil of possible velocity vectors there), then, *not only will only one direction be available for the velocity of the prong $E_2$ at $E$, namely the line of intersection of the existing planar pencil and the newly inserted platform there, but only one direction will be available at the instant for the velocities at all of the other points in the body 2 as well.*

60. Now this is precisely the crux of the circumstance treated at the chapter 5; there it was shown that, if a body 2 is obliged to move in such a way that every point within it has a direction specified for its instantaneous linear velocity at the instant, *and this, as we have seen above, is achieved with five points of contact between the pair of bodies*, then there is only one possible screw available for the instantaneous motion. This single possible screw, which changes its location and its pitch, of course, as finite movements occur (§ 2.64), might be called the quasi-general 1-system of screws in the circumstances; it is also, in this particular case, the quite general 1-system of chapter 23, for it may be of any pitch. In any event we can say in general that, *with five points of contact extant, there is one degree of freedom, there being an $\infty^0$ (namely only a single one) of ISAs available for the motion.*

**Six points of contact between the bodies**

61. It is now quite clear to see that, if a sixth point of contact is permitted to occur between the bodies, relative motion becomes no longer possible. The matter is that, if the sixth point of contact is say $F_1$ the predetermined path of $F_2$ in 1 will cut the surface of 1 at $F$: in general the path of $F_2$ in 1 will not, and it cannot be made to be, tangential to the surface there. Please refer again to the matter of flying and landing at § 6.22, and refer to figure 6.07. If the mentioned path does happen to fall exactly parallel with the tangent plane at $F$ (there is a single chance in a single infinity of chances that this will occur by accident), then we can argue if we wish, and we do in chapter 7, that either the body 1 or the body 2 is not perfectly irregular.

Figure 6.07 (§ 6.61) A rigid body 2 having five points of contact with its reference body 1 slides with one degree of freedom until a sixth point of contact (at $F$) suddenly occurs. With $F_1$ and $F_2$ together at $F$, relative motion between the bodies is no longer possible (§ 6.63).

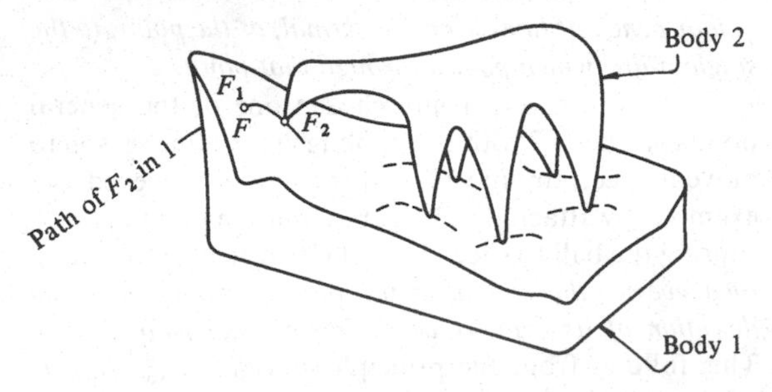


62. It is important to mention with regard to figure 6.07 that the paths of the five feet drawn upon the surface of the body 1 are not the paths of points fixed in the body 2. The protuberances I have drawn for the 'legs' of the body 2 are all rounded (§ 8.35) so the broken lines are the paths, not of the tips of five sharp prongs, but of five crosswise-sliding points of contact. Each one of these latter move with respect to both bodies as the bodies move and the whole matter could become quite complicated if we wished to study, not as we have here the mere collision of $F_2$ with $F_1$ at $F$, but the so-called gross motion of some body 2 as it might move through space and time with these contraints and this single degree of freedom (§ 19.60).

**On the question of the immobilisation of a body**

63. Following my argument offered above about the final establishment of a sixth point of contact by a well conditioned collision at the sixth point (§ 7.26), we can see that, keeping five points of contact extant, we can always withdraw from the sixth. Refer for body closure to § 2.21 and for stop surfaces to § 1.29. Thus a seventh point for stopping, always achieved in such a way that the new contact normal there is projectively dependent upon the given six, will be enough to hold the body 2 against the body 1 provided withdrawal at any one of the six will result in collision at the seventh. Under these special circumstances of there being more than six points we are in a wholly diferent arena. Given rigidity, no more than six points can be 'active' at the same time, and the problem I feel sure is quite complicated. Nevertheless something at least qualitative might soon be said about it with the help of screw theory. One might refer to the works of Somov (who wrote in 1897 and 1899); these are quoted and interpreted by Lakshminarayana [1]. Refer also to the more recent works by Dizioglu and Cordes of Braunschweig; these authors claim to have shown by screw theory that seven points for gripping are always enough to immobilize a generally shaped body [2]. Ohwovoriole [3] has recently worked in the area of robotic grippers, and he claims also to have made some progress with the problem of the seven points. Are seven points, however, sufficient to immobilize a spherical body? And how does a robot grip a cylindrical piece? Dizioglu [4] discusses special questions such as these, and the whole matter remains, in many ways, open. A question I would like to ask is this: are the four fingers, the thumb and the two lower

protuberances surrounding the palm of one's hand sufficient to grasp an object and to apply to that object, in the absence or the presence of friction, *any wrench?*

### Notes and references


[1] Lakshminarayana, K., Mechanics of form closure, ASME paper no. 78–DET–32, prepared for presentation at the Design Engineering Technical Conference, ASME, Minneapolis, Minn., September 1978. See abstract, *Mechanical Engineering (ASME)*, Vol. **100**, p. 90–1, December 1978. Refer § 6.63.

[2] Cordes, P., Ein Beitrag zu Beweglicheitsuntersuchungen in Elementenparen bei Vorgabe diskreter Stüzpunkte, *Berichte Nr 321*, VDI, 1979. Cordes quotes in this paper an unpublished work of Lakshminarayana (above) who appears to have worked for a time with Professor Dizioglu (below) at TH Braunschweig. Refer § 6.63.

[3] Ohwovoriole, Morgan S., *An Extension of Screw Theory and its Application to the Automation of Industrial Assemblies*, doctoral dissertation, Stanford University 1980. Republished as Memo AIM–338 of the Stanford Artificial Intelligence Laboratory, Department of Computer Science Report No STAN–CS–80–809. Refer § 6.63.

[4] Dizioglu, B., Kinematische und statische Grundlagen des Spannens und Positionierens im Maschinenbau, insbesondere bei Werkzeugmaschinen, *Berichte nr 281*, VDI, 1977. Refer § 6.63.


# *Appendix 6A*

# *Notes on the quasi-general 5-system*

01. The first reference to this appendix is made at § 6.25, and its conclusions are needed first at § 7.48. It concerns the distribution in space of the $\infty^4$ of ISAs or motion screws which have been mentioned, and the distributions of the pitches among them. Refer to figure 6.03. The line *n–n*, which is normal to the tangent plane at *A*, is a contact normal and thus an *n*-line in the body 2 (§ 3.06). The common perpendicular $\Gamma\Gamma'$ between it and the example $ISA_{12}$ or motion screw along *j–j* locates the centre point $\Gamma$, for that particular ISA, of this central contact normal line (§ 16.09, § 21.03). See also figure 9.04, where the symbols $\Gamma$ and $\Gamma'$ bear exactly the same significance as they do in figure 6.03. It should of course be mentioned that, whereas in chapter 9 I am speaking about the $\infty^3$ of right lines in a body surrounding a given ISA or motion screw, I am speaking here about an opposite question: *where are all of the possible $\infty^4$ of ISAs or motion screws in space surrounding a single given n-line which, by virtue of the particular form of constraint upon the body, namely a single point of contact, is bespoke in the body at the instant under consideration?*

02. It will next be seen, paying attention to sign, that the pitch of the screwing along *j–j* is directly determined by the length of the line segment $\Gamma$–$\Gamma'$ and the magnitude of the angle $\gamma$. In fact, if $\Gamma$–$\Gamma'$ were written *r*, the pitch of the screwing along *j–j* could be written, $p = r \tan \gamma$. Please note that the angle $\gamma$ in figure 6.03 and in the present context is the complement of the angle $\theta$; $\theta$ is the helix angle, used extensively at chapter 9 and elsewhere (§ 9.15, § 11.47). It follows that, for each value of the pitch, there is a whole linear complex of lines about the central axis *n–n*; these lines give the locations of the possible ISAs of that pitch and the pitch of the complex here is the same as the pitch of the possible ISAs which make it up. For general readings about the linear line complex please go to the whole of chapter 9, to § 10.07 *et seq.* and to § 11.30 *et seq.* All of this means that, at a give radius *r* from the line *n–n*, tangential to the surface of a right-circular cylinder there and binormal to the relevant helices upon that cylinder, the ISAs of a given pitch are arranged as is

shown by the members of the reguli in figure 9.07. There is in figure 9.07 an $\infty^1$ of reguli, but there is in this 5-system here an $\infty^1$ of complexes, one complex for each value of the pitch. The complexes are coaxial about *n–n*; each has an $\infty^3$ of ISAs; and all of this information is consistent with the statement made at § 6.25 that every line in space, including the line *n–n* itself, can be an ISA (§ 9.44). The pitch of the available ISA along *n–n* is zero, incidentally. This follows from the circumstance of there being the point of contact between the bodies there; it is this particular feature of the 5-system here which renders it quasi-general (§ 23.41). The linear complex of screws of zero pitch is a special linear complex (§ 9.41), which means that all available screws of zero pitch actually cut the line *n–n*; we find correspondingly that all of the screws of infinite pitch are found everywhere in the infinite number of planes which are normal to the line *n–n* (§ 9.43).

03. It has been shown at § 6A.02 that all of the possible ISAs (or available screws) of a given pitch are upon a linear complex of that pitch about *n–n*. *It follows that all of the available screws of a given pitch, which pass through a generally chosen point Q, are coplanar* (§ 9.47, § 11.35). We can see independently, (*a*) that all lines through *Q* which are drawn to cut the contact normal *n–n* will form at *Q* a planar pencil of available screws of zero pitch (§ 9.41); and (*b*) that all lines through *Q* which are drawn to be perpendicular with *n–n* will form at *Q* a planar pencil of available screws of infinite pitch, positive and/or negative (§ 9.43). Referring however to figure 6A.01 which is another version of figure 6.03 concentrating upon the perpendicular *Q–N* dropped from *Q* onto the central axis *n–n* of the system, we see that all planar pencils between the above mentioned two at *Q* intersect one another along the line *QN* and contain respectively the available screws of all of the pitches of intermediate value.

Figure 6A.01 (§ 6A.03) Two selected sets of screws of the quasi-general 5-system (§ 23.13). Screws of all other pitches belong in this figure as well, of course, but none of them are shown.

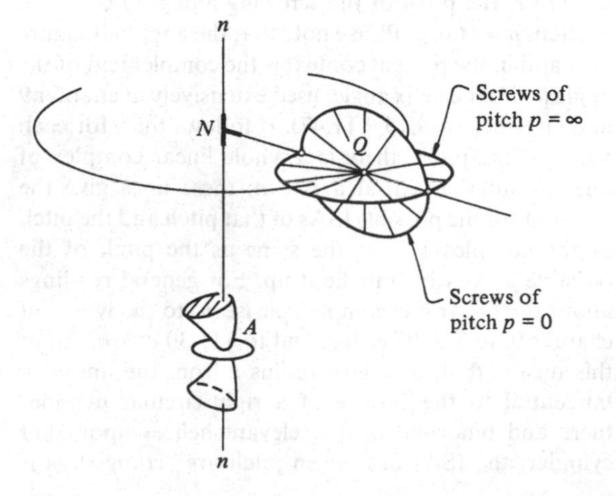


04. This overall, sketched-in picture of the distribution of the pitches at *Q* is consistent with the other evident fact that the line *QN* itself, like all such lines which cut *n–n* at right angles, can be a possible ISA, or an available screw, *of any pitch*. This latter fact is consistent in its turn with the fact that all of the lines cutting *n–n* at right angles belong to all of the single infinity of complexes which are mentioned by implication at the beginning of § 6A.02. We are speaking about that special set of lines in the body 2, about any one of which the body 2 may purely rotate, or parallel with any one of which it may purely slide, and about any one of which it may, accordingly, *screw with any pitch at all*. Whereas there were at § 6.15, in the earlier case of zero points of contact, an $\infty^4$ of such lines, there are here, with a single point of contact, only an $\infty^2$ of them. The material at § 9.15 may be a useful, general reference here, but be reminded again that, whereas we are dealing here, in this case of the single point of contact, with the various linear complexes of possible ISAs of a given pitch, all coaxially surrounding the single given *n*-line, we are dealing in chapter 9 with the single linear complex of *n*-lines coaxially surrounding a given ISA.

# Appendix 6B

# Notes on the quasi-general 4-system

01. The first reference to this appendix is made at § 6.28, and its conclusions are needed first at § 7.27. It concerns the distribution in space of the $\infty^3$ of ISAs or motion screws which have been mentioned, and the distributions of the pitches among them. Refer to figures 6.04 and 6B.01. Choose the point $O$ along the common perpendicular line $z$–$z$ midway between the contact normals $n_A$ and $n_B$, thus establishing an origin O upon an axis $z$–O–$z$. Next erect at O two more axes $x$–O–$x$ and $y$–O–$y$ in such a way (*a*) that the three axes are mutually perpendicular, and (*b*) that, while $x$–O–$x$ bisects that particular angle between the contact normals which makes the successive, finite, screw displacements from $n_A$ through $x$–O–$x$ to $n_B$ left handed (§ 15.05), $y$–O–$y$ bisects the other. We have thus set up a triad of Cartesian axes about each axis of which the pair

Figure 6B.01 (§ 6B.01) Various screws of the 4-system of motion screws drawn to known lines, axes, etc. from a generally chosen point $Q$. The cone of screws at $Q$ is intersecting the $xy$-plane in the ellipse $RLMNS$. Some generators of the reciprocal cylindroid of action screws are also shown (§ 6B.14, § 6B.15).

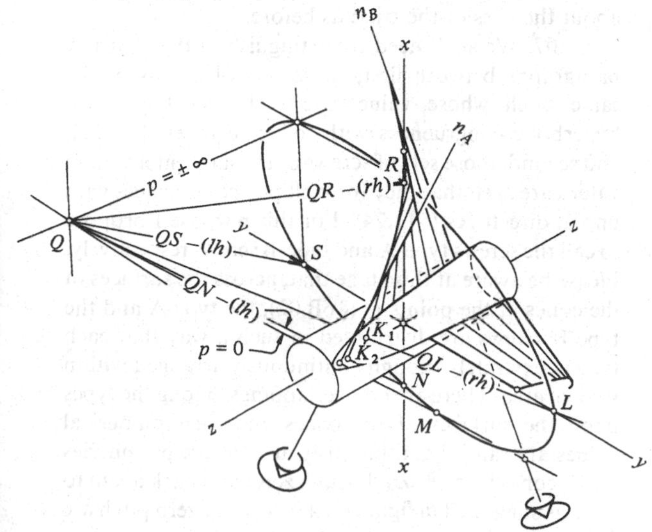


of contact normals $n_A$ and $n_B$ are line-symmetrically arranged. Choose as before, at figure 6B.01 a general point $Q$ in space.

02. Now drop perpendiculars from $Q$ onto $x$–O–$x$ and onto $y$–O–$y$, and observe, by paying attention to the symmetry (or by making at the bench a simple wire model to discover the symmetry), that both of these perpendiculars are locations for available screws through $Q$; they are marked $QR$ and $QS$ in figure 6.04. Their pitches are in general different and, in our particular case, they are of opposite hand; they are marked *QR-(rh)* and *QS-(lh)* in figure 6B.01. Drop also from $Q$ the two perpendiculars onto the two contact normals $n_A$ and $n_B$. These perpendiculars are also locations for available screws through $Q$; the reader should, for the sake of exercise here, (*a*) see for himself why this latter is so, and (*b*) examine the handedness of the screws (§ 15.40). Observe next the cone which I have sketched in figure 6.04 to contain upon its surface the six found and mentioned screws through $Q$. Both branches of this sketched cone have been truncated by planes parallel with the $xy$-plane, incidentally, and the truncations appear elliptical; but even right sections through the cone would be indeed elliptical because it is, in general, an elliptical, second order cone. With this picture of the cone in mind, please become aware next that it was shown by Ball (1900) that, *there being one generator for each pitch, the generators of such a cone at $Q$ contain respectively each of the single screws of each of the possible pitches which can be drawn through $Q$*. The pitches range across the whole spectrum from negative infinity through zero to positive infinity, and the whole corresponding single infinity of screws covers the whole surface of the cone; it makes up the complete set of available screws which can be drawn through $Q$ (§ 15.33).

03. There is, of course, such a cone of screws at all points $Q$ in space and the total set of the $\infty^3$ of screws which is under consideration resides upon an arrangement of lines which (by virtue of the cone at all points $Q$) is called a *quadratic complex* (§ 11.57). Notice that there is always one generator of the cone, with $p = \pm\infty$, which is parallel with $z$–O–$z$. Notice also that (and here is another exercise for the reader), when $Q$ is upon one or other of the $n$-lines, the cone degenerates into two planes. Notice most importantly that, when $Q$ is anywhere upon the obviously important, central axis $z$–O–$z$ of the complex, the cone degenerates into $z$–O–$z$ itself. This means that the body 2 can screw with whatever pitch it pleases (including zero and infinity) about this line $z$–O–$z$. The line $z$–O–$z$ is a line in the body which is unique at the instant and it is, to say it again, *the only line about which the body 2 may rotate, or along which the body 2 may slide, or both, without disturbing the*

*maintenance of contact at the two points A and B*. There is an $\infty^0$ (namely only one in this case) of such lines. Compare this remark with the remarks made (*a*) at the penultimate sentence of § 6.15 where the relevant number mentioned was $\infty^4$, and (*b*) near the middle of § 6A.04 where it was in that case $\infty^2$; be aware that, as this discussion goes on, the body 2 is becoming more and more restricted in its capacity for instantaneous motion.

04. After setting up the case of the single point of contact at § 6.18, we investigated soon afterwards and without too much difficulty, not only all of the available screws through $Q$ (§ 6.24), but also the distribution in space of the available screws of a given pitch (§ 6A.02). Here, however, with two points of contact we have blundered into a system of screws which is a different, more complicated system, and one more difficult to penetrate visually than was the relatively simple 5-system we looked at before. This is a case of the 4-system (§ 23.30) and to locate by mere visualization the layout of the sets of the available screws of a given pitch (say $p=p_g$) is not so easy here. In figure 6B.02 however, each of the two $n$-lines of figure 6.04 are shown equipped with a representative example, (*a*) of the $\infty^2$ of coaxial helices

Figure 6B.02 (§ 6B.04) A screw through a new $Q$ of pitch $p=p_g$, the various reguli, the two directrices $w_1$ and $w_2$, and the elliptical cone at $Q$, are all sketched into the figure only approximately (§ 6B.06).

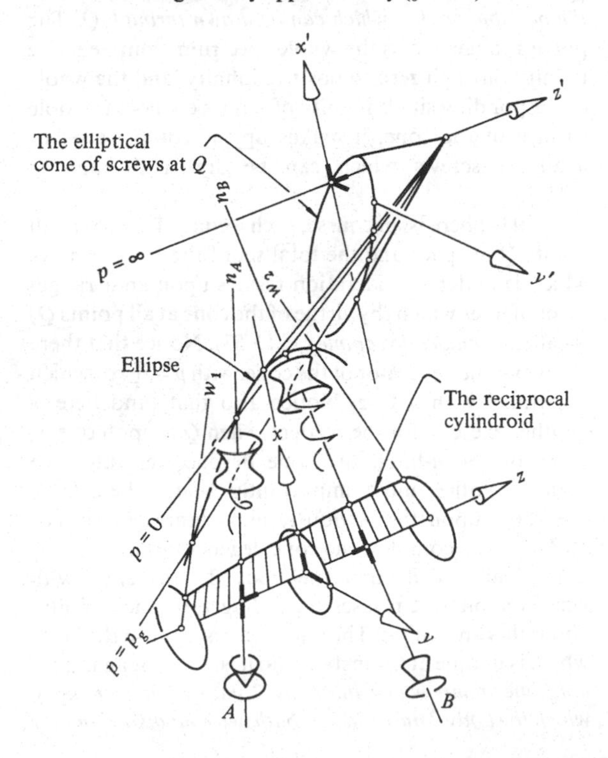


of pitch $1/p_g$, tangent to one of which each screw of the given pitch $p_g$ must lie, and (*b*) the $\infty^2$ of reguli which show the possible locations, so far as that $n$-line alone is concerned, for screws of the mentioned pitch $p_g$. Now any line, such as the one marked $p=p_g$ in the figure, which can be drawn to be a member, both of a regulus on $n_A$, and of a regulus on $n_B$, is clearly an available screw of the given pitch; our problem is to see how each complete set of such screws is arranged.

05. Let us see first of all that for each given value of the pitch there is an $\infty^2$ of available screws, there being upon each regulus on $n_A$ only one line which can be collinear with a line of another regulus on $n_B$. In the special case of the given pitch being zero, the sets of helices surrounding the $n$-lines collapse into sets of concentric circles coaxially arranged and the sets of reguli collapse into sets of intersecting lines which become special linear complexes (§ 9.41). It should be clear accordingly that, *for zero pitch*, the set of the $\infty^2$ of available screws is a linear congruence – a hyperbolic linear congruence with real directrices (§ 11.22) – whose directrices are the two $n$-lines. This latter could otherwise, and quite simply, have been shown by arguing as we did in § 6.27 that all screws of zero pitch must cut both of the $n$-lines.

06. Now it will not be shown here but it can be shown that not only for the special value zero of the pitch but for all other given values of the pitch as well, (*a*) the available screws of that pitch reside upon the lines of a linear congruence, whose directrices are either real if the pitch belongs within a certain range of values, and unreal if it does not; (*b*) the two directrices of the relevant congruence, if real, cut the line $z$–O–$z$ perpendicularly; (in this connection please note the directrices $w_1$ and $w_2$ drawn for the lines $p=p_g$ in figure 6B.02); and (*c*) the arrangement of the directrices is line-symmetric about the axes of the triad as before.

07. We shall need to distinguish in the next few paragraphs between first, those sets of screws of the same pitch whose values are such that they form hyperbolic congruences with real directrices (§ 11.22), and second, those sets of screws of the same pitch whose values are such that they form elliptic congruences with unreal directrices (§ 11.24). For this purpose I propose to call the screws type-A and type-B screws respectively. Please be aware at this stage that, across the surfaces of the cones at the points $Q$ (§ 6B.02), the type-A and the type-B screws are distributed in such a way that each type is separately though continuously arranged within its own area. There are no overlappings among the types upon the surfaces of the cones and their numerical values are ranged continuously around the peripheries of the cones from $\pm\infty$, through zero, and back again to $\pm\infty$ as suggested in figure 6.04. Screws of zero pitch are

clearly type-A screws in our case and screws of infinite pitch are clearly type-B screws because they are all parallel with one another. What we have not discussed, just yet, is the question of where the two dividing lines are, on the cones, between the type-A and the type-B screws. Please forgive my double use of the symbols A and B just here; there is no connection between the *A* and *B* of the two points of contact and the A and B of the two types of screws.

08. Look at the truncated, right-circular cylinder sketched in figure 6.04, which is centred upon *O* and coaxial with *z*–O–*z*. It is mooting (refer to figure 15.06) a cylindroid, located there and in that way. Now pay attention to this: *if we drew the two mentioned directrices for each one of the single infinity of different values of the pitch which make up the type-A screws, this single infinity of directrices would generate the mooted cylindroid.* The fact remains unproven here, but please read Ball (1900) and Hunt (1978). both of whom discuss the matter exhaustively. The matter is, in my view, one of the more startling aspects of the whole pattern of instantaneous motion in spatial kinematics. It is of course related to the main subject matter of chapter 14, namely the reciprocity of screws, and it will be mentioned again in the present chapter when I come to speak about the clearly related question, not of two, but of four points of contact at § 6.44. The reader may refer, also, to my remarks from § 23.30 to § 23.36 inclusive. Just here, however, let me make the following, elementary observations: the nodal line of the cylindroid is *z*–O–*z*, its centre is at O, the axes *x*–O–*x* and *y*–O–*y* and the two contact normals $n_A$ and $n_B$ are all generators of it, and its shape, as we know, is wholly determined by the location in space of the two contact normals (§ 15.13).

09. If it can now be seen in retrospect (*a*) that the two points of contact and the tangent planes at A and B determine the two *n*-lines; (*b*) that the two *n*-lines with their unique common perpendicular *z*–O–*z* determine the mentioned cylindroid which is symmetrical about them; (*c*) that the single infinity of symmetrically arranged pairs of conjugate generators of the cylindroid are the pairs of directrices of the single infinity of hyperbolic congruences of type-A screws whose pitches belong within the mentioned (though as yet undetermined) range of values, there being one congruence for each pitch; and (*d*) that every one of these congruences is line-symmetrically arranged about all three of the carefully chosen Cartesian axes, then the following can be seen also. Refer to figure 6B.03.

10. The lines $w_1$ and $w_2$ in figure 6B.03 are the pair of example directrices for that set of type-A screws of the given pitch $p_g$ extracted directly from figure 6B.02. Choose two points $G_1$ which are equidistant from O upon the axis *y*–O–*y* and erect through them the two lines *g*–$G_1$–*g* to cut the two directrices as shown. These lines, which both contain available screws of the given pitch, are diametrically opposite generators of the same regulus of the hyperboloid $\alpha_1$ which is shown, $w_1$ and $w_2$ belonging to the other regulus of the same hyperboloid. All members of the regulus $\alpha_1$ of which the lines *g*–$G_1$–*g* are members cut both $w_1$ and $w_2$; they are, accordingly, all members of the relevant congruence and thereby all of the same pitch. For each choice of pair of points *G* there is another such regulus of screws of that same pitch $p_g$; so there are, concentric at O and coaxial with *x*–O–*x*, a single infinity of such hyperboloids. See the regulus $\alpha_2$. We could equally well have chosen pairs of points *H* equidistant from O along *x*–O–*x* and these pairs of points would in a similar way have yielded another nest of hyperboloids concentric at O and coaxial along *y*–O–*y*; two of them are merely indicated in figure 6B.03 by their throat ellipses and by right sections marked $\beta_1$ and $\beta_2$. It follows that, concentric at O and coaxial in two nests along *x*–O–*x* and *y*–O–*y*, there is an $\infty^1$ of reguli of screws of the same, type-A pitch $p_g$. Please notice in passing here that, if the pairs of points *G* or *H* are at zero distance from O along their respective axes, they determine the same special regulus; the regulus resides upon that hyperboloid at the 'interface' between

Figure 6B.03 (§ 6B.09) This is a composite sketch showing (*a*) a hyperbolic congruence of type-A screws of the same pitch consisting of the intersecting reguli $\alpha$ and $\beta$, and (*b*) an elliptical congruence of type-B screws of some other pitch consisting of the non-intersecting reguli $\gamma$.

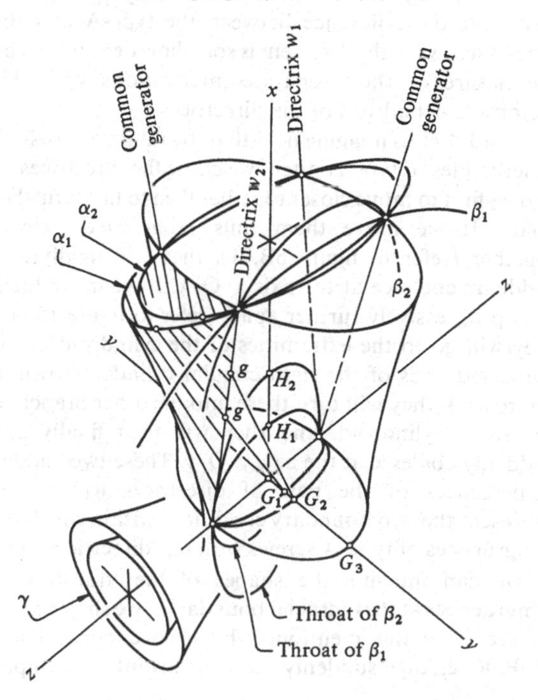

the nests. This latter is simply those two planes which intersect along *z*–O–*z* and contain, respectively, the two directrices $w_1$ and $w_2$ of the congruence. If the pairs of points *G* and *H* are at infinite distances they determine another special regulus; this regulus resides upon a pair of parallel planes parallel with the *xy*-plane and this pair again contains the two directrices. Please refer to § 7.68 (which discusses not this but related matters), to the figures 7.12, and § 11.15.

11. In summary now it should be seen from figure 6B.03 (*a*) that all of the hyperboloids $\alpha$ and $\beta$, including the two special, interfacial ones just mentioned, intersect along the two lines $w_1$ and $w_2$ which are the directrices of the relevant congruence; (*b*) that, although these directrices are not themselves members of the relevant reguli, all of the lines of the reguli (which contain all of the available screws of this particular congruence of screws) cut them both; and, accordingly, (*c*) that such a complete set of reguli upon their hyperboloids $\alpha$ and $\beta$ is another way of presenting all of the screws of a given type-A pitch.

12. There are however a single infinity of such sets of such reguli, one set for each such type-A pitch, and whereas the reguli of one given pitch intersect one another along the directrices of their congruence as shown in figure 6B.03 the reguli of other type-A pitches intersect one another along other pairs of directrices. Do not forget that the pairs of directrices of which I speak together constitute all of the generators of the cylindroid, the so called, reciprocal cylindroid which was mentioned first at § 6B.08. It will surely be suspected now that the difference between the type-A and the type-B screws in the 4-system is somehow determined by the nature of their separate intersections with this reciprocal cylindroid of the directrices.

13. Let us imagine now that, by ranging across all other values of type-A pitch, we cause the directrices $w_1$ and $w_2$ first to move closer together then to move further apart. If we move them thus progressively closer together (refer to figure 6B.01), they will finally and suddenly coalesce at the axis *x*–O–*x*. If we move them thus progressively further apart (refer to figure 15.06), they will go to the extremities of the cylindroid at the truncated ends of the right-circular cylinder shown in figure 6.04, they will turn there into the other branch of the same cylindroid, and they will then finally and suddenly coalesce at the axis *y*–O–*y*. These two sudden coalescences of the pair of directrices $w_1$ and $w_2$ represent the two boundary conditions within which the congruences of type-A screws with real directrices exist. If we can imagine the shapes of the 'transitional' congruences at these crucial boundaries, we shall be able to see how the mentioned hyperbolic congruences (§ 6B.06) change suddenly but logically into the elliptic congruences of type-B screws where the directrices have become unreal (§ 6B.07, § 6B.17, § 6B.20).

14. In the limit, as the pair of directrices approaches coalescence (at *x*–O–*x* or *y*–O–*y*), the congruence approaches that transition condition where it consists of the following: *all of those lines through all points along the relevant axis which are tangential to the surface of the cylindroid there.* It follows that the transitional congruences are identical in shape with one another, are line-symmetric about the axis *z*–O–*z* and are displaced thereabout exactly 90° apart. For algebraic reasons beyond the scope of this discussion, these transitional congruences are called, in the mathematical literature, parabolic linear congruences (§ 11.23). The pitches of the screws upon the lines of these two transitional, parabolic congruences are not the same however. The pitches are equal in magnitude but opposite in sign to the maximum and minimum values of pitch upon the reciprocal cylindroid (§ 15.49). These latter occur at the two central generators of the cylindroid which intersect one another at the centre perpendicularly (§ 15.49). Because we have in our case two generators of zero pitch upon the cylindroid, this being due to the two actual points of contact between the bodies which give rise to the contact normals there and thus to the mentioned generators (§ 10.14), one of the transitional values of the pitch will be negative while the other will be positive. The value zero of the pitch will be, as already mentioned, among the type-A values. The pitches $p_\alpha$ and $p_\beta$ of the screws of the two transitional congruences are the pitches of the two principal screws of the 4-system. For an explanation of the mentioned matters of sign, please go to § 23.71.

15. Look again at figure 6B.01 and see the elliptical trace upon the *xy*-plane of the cone at *Q*. Notice that, although the two screws through *Q* which are marked *QR*-(*rh*) and *QS*-(*lh*) cut the axes *x*–O–*x* and *y*–O–*y* respectively (because they were, indeed, earlier drawn to cut them perpendicularly), they cannot be the transitional screws upon the cone which divide the type-A from the type-B screws. The reason is that, although they cut the mentioned axes and thus the pairs of double directrices lying along those axes (§ 6B.13), they are not tangential to the surface of the cylindroid. Indeed they very clearly pierce the cylindroid at *R* and *S*. The two transitional screws we are seeking are elsewhere upon the cone: they are at the locations *QL*-(*rh*) and *QN*-(*lh*). Both to be found at the lower part of the picture in figure 6B.01, they are cutting their different double directrices at *L* and *N* respectively, and both of them are tangential to the surface of the cylindroid at those respective points.

16. This is not so easy to see but, before going on to clarify a little more the exact layout of these

transitional screws $QL$ and $QN$ upon the cone at $Q$ and in the cylindroid, I wish to remark parenthetically that the screw $QM$, which is of zero pitch because it was earlier drawn to cut both of the contact normals, is, as was mentioned previously, among the type-A screws, so the following can be said: *the range of type-A screws extends across that sector of the surface of the cone at Q which extends from L to N through M.*

17. It is time now also to mention that, in accordance with a theorem due to Plücker explained at § 15.36, each one of the single infinity of available screws through $Q$ must cut one of the generators of the cylindroid perpendicularly. Although it is not very easy to visualize without a model of the cylindroid (and there is such a model displayed in photograph 15, which is not very difficult to build), $QL$ is cutting its generator perpendicularly at $K_1$ while $QN$ is cutting its perpendicularly at $K_2$. This permits us to see that (*a*) while all of the type-A screws from $L$ to $N$ through $M$ on the cone cut the cylindroid three times, once at the generator to which it is perpendicular and twice at the two directrices of its relevant hyperbolic congruence, (*b*) all of the type-B screws from $L$ to $N$ through $R$ and $S$ on the cone cut the cylindroid only once, at the generator to which it is perpendicular. This fact that the type-B screws upon the cone are more nearly parallel with the central axis $z$–O–$z$ than are the type-A screws is a fact which helps us to see more clearly how the as yet unmentioned elliptic congruences of type-B screws fit into the general scheme. A picture and a sketchy explanation of these is offered in the following few paragraphs.

18. Refer to figure 6B.04. This is an orthogonal view of a few aspects of a quasi-general 4-system of screws being looked at along the axis $z$–O–$z$. It is a picture obtained by graphical construction of three representative points $Q$ which are equidistant in the same direction from the $xy$-plane. The $xy$-plane itself resides in the plane of the paper and through each point $Q$ a number of screws of the system has been drawn. The screws are shown to extend from the particular $Q$ itself to the ellipse of intersection at the $xy$-plane. One such ellipse – it could be drawn for the point $Q_2$ approximately – has already been used for illustration in the trimetric view in figure 6B.01. The proportions are such that, while the distance $Q_1$–$Q_2$ (parallel to $y$–O–$y$) is 50 mm, all three points $Q$ are 50 mm away from the $xy$-plane and towards the foreground of the picture. The contact normal $n_A$ is 15 mm on this side while the contact normal $n_B$ is 15 mm on the other side of the $xy$-plane, so the distance between $n_A$ and $n_B$ (along the axis $z$–O–$z$) is 30 mm.

Figure 6B.04 (§ 6B.18) Ellipses of intersection between the cones of screws at three points $Q$ and the $xy$-plane. The three points $Q$ are equidistant from, and on the same side of, the $xy$-plane.

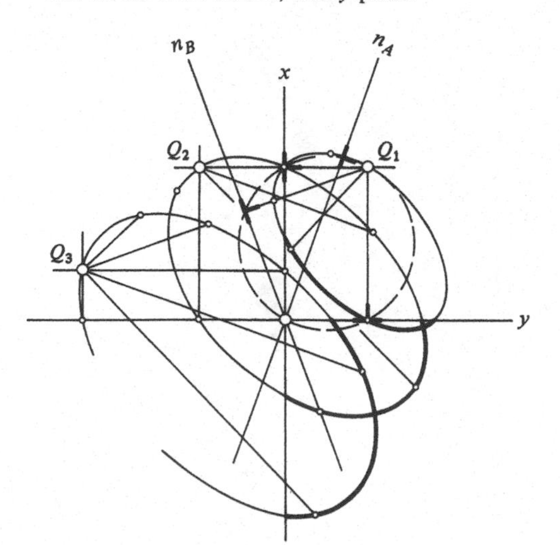


19. Some hints about the method of construction follow. Drop perpendiculars from any chosen point $Q$ onto $n_A$, $n_B$, O–$x$, and O–$y$. The chosen $Q$ and the four feet of the four perpendiculars are upon a circle. This is the same circle with the same $Q$ that appears in figures 15.11, 15.12, and 15.13. Please peruse the whole of chapter 15, but study the passages from § 15.26 to § 15.45 inclusive. These latter will clarify the construction in figure 6B.04. They will explain, moreover, why I can say here, as I do, that the thickened portions of the ellipses in figure 6B.04 coincide with the ranges of the type-A screws upon the cones.

20. If the thickened portions of the peripheries of the cones in figure 6B.04 indicate the presence of type-A screws, the thinner portions indicate the type-B screws. I have said that the vertices $Q$ of the cones in figure 6B.04 are in a plane parallel with the plane of the paper, so we can see next that the type-B screws of these cones are, generally speaking, more nearly parallel with the axis $z$–O–$z$ than the type-A screws are. This will help to make it clear that the central axes of the single infinity of elliptical congruences (upon whose sets of generators the type-B screws of the respective pitches lie) all lie collinear along the central axis $z$–O–$z$ of the system. Two reguli $\gamma_1$ and $\gamma_2$ of one such elliptical congruence of type-B screws are indicated in figure 6B.03. The two are indicated by means of right sections only (taken at the throat and at the foreground) of the two relevant hyperboloids. These hyperboloids $\gamma$ do not intersect one another as the hyperboloids $\alpha$ and $\beta$ do, for, as has been said, the relevant congruences, those for the type-B screws, have no real directrices (§ 11.24).

21. Three cones of screws can be seen in figure 6B.04. Their ellipses of intersection with the $xy$-plane all have their major axes inclined at 45° as shown. But

please note that the vertices $Q$ in figure 6B.04 are all in the same 'quadrant'. They are all 'above' the $yz$-plane; they are all on 'this side' of the $xy$-plane; they are all, indeed, on this side of the plane containing $n_A$ which is parallel with the $xy$-plane. It will be found however upon inspection that, if the points $Q$ were to be reflected in the line $z$–O–$z$, the sizes, the shapes and the orientations of the corresponding ellipses would all remain the same. Indeed it will be found that, for all points $Q$ in any one plane parallel with the $xy$-plane, the ellipses of intersection of the cones with the $xy$-plane will all be of the same shape and orientation as is shown in the case of the points $Q$ in figure 6B.04.

22. If a set of points $Q$ in any one plane parallel to the $xy$-plane were to be reflected in the $xy$-plane however, the ellipses of intersection would all have orientations opposite to that which is shown in figure 6B.04. This can best be shown by figure 6B.05: there are two points, $Q$ and $Q'$, reflections of one another in the $xy$-plane and the corresponding ellipses of intersection are also shown. We see that the major axes of the ellipses are mutually perpendicular. The main matter being canvassed here is that the whole 4-system of motion screws does not consist of two equal halves mirror-imaged in the $xy$-plane. Its two equal halves do exist on either side of the $xy$-plane, they are identical with and facing one another, but they are displaced (by rotation about $z$–O–$z$) exactly 90° apart.

23. Next it should be seen that, as the points $Q$ become further and further away from the $xy$-plane (both on this side and on the other), the cones become more and more circular and the proportion of type-A screws upon the cones diminish. In the limit, as the distances from the $xy$-plane become infinite, the screws upon the cones become parallel and the pitches of all of the screws become infinite (§ 15.44). Please note at § 15.44 however that the ellipse which is mentioned there is not the ellipse of intersection with the $xy$-plane; it is the ellipse of intersection with the cylindroid and is shown in figure 6.04. Also it should be noted that, if the points $Q$ are taken closer and closer towards the $xy$-plane, the cones become, although not all of the same shape, more and more flattened. When the points $Q$ are taken inside the limits $\pm B$ (that is inside the extreme generators of the reciprocal cylindroid) the ellipses of intersection at the $xy$-plane become hyperbolas and, in the limit, when the points $Q$ arrive at the $xy$-plane, the ellipses of intersection become mere pairs of intersecting lines.

24. In this appendix we have been examining in a visual way the general appearance of the quasi-general 4-system of motion screws that springs into being as soon as two points of contact between a pair of rigid bodies are established. For those who wish to see more deeply into this most complicated of the screw systems, I refer them to Hunt's Cartesian equation for the hyperboloids of the general 4-system which is quoted at my § 23.50, and to his Plücker equations for the same system which appear at my § 23.53. The relevant original reference is Hunt (1978), chapter 12. Finally I wish to say that a summary of this appendix appears as a short statement about the symmetries of the quasi-general 4-system written at § 6.29. The reader may care to peruse, also, my remarks about the general question of four degrees of freedom; the remarks extend from § 23.30 to § 23.35 inclusive.

Figure 6B.05 (§ 6B.22) Ellipses of intersection between the cones of screws at the two points $Q_1$ and $Q_1'$ and the $xy$-plane. The two points are reflections of one another in the $xy$-plane.

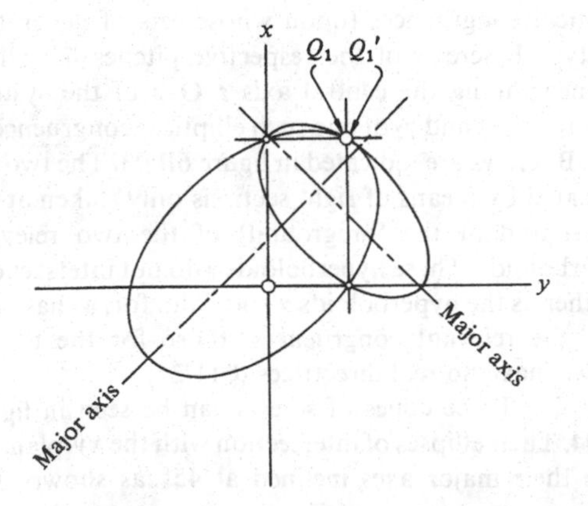

# Appendix 6C

# Notes on the quasi-general 3-system

01. One way to begin to deal with the matters relegated to this appendix is to work with a set of numerical values. I can do this with the help of Ball's Cartesian equation for the hyperboloids of the reguli of the 3-system (1900). This equation is quoted and proved by Hunt (1978) in his chapter 12. It is explained and requoted by me at my §23.49. I shall choose the three principal screws of a 3-system of motion screws by nominating suitable dimensions for the mechanical apparatus which appears in figure 6C.01. The three screw joints in series in figure 6C.01, mutually perpendicular and intersecting at the single point O, have their pitches as follows: $p_\alpha$ along the axis for $x$, $+6$ mm/rad; $p_\beta$ along the axis for $y$, $+3$ mm/rad; and $p_\gamma$ along the axis for $z$, $-4$ mm/rad. The movable body 2 has 3 °F with respect to the fixed body 1 and we shall investigate here the 3-system of motion screws which exists at the drawn configuration of the apparatus. Please note that, by choosing $p_\alpha$ to be positive, $p_\beta$ to be less than $p_\alpha$ and $p_\gamma$ to be negative, I have, (*a*) accorded with convention, (*b*) ensured that the pitch quadric will be real (§23.24). I have ensured, in other words, that this numerical example will be compatible with the quasi-generality of the 3-system defined by the apparatus in figure 6.05(*a*).

Figure 6C.01 (§6C.01) The movable body 2 has 3 °F with respect to the fixed body 1, and we are investigating here the 3-system of motion screws that is available at the instant. The pitches at the screw joints, $p_\alpha$, $p_\beta$, and $p_\gamma$, are the pitches of the three principal screws of the system.

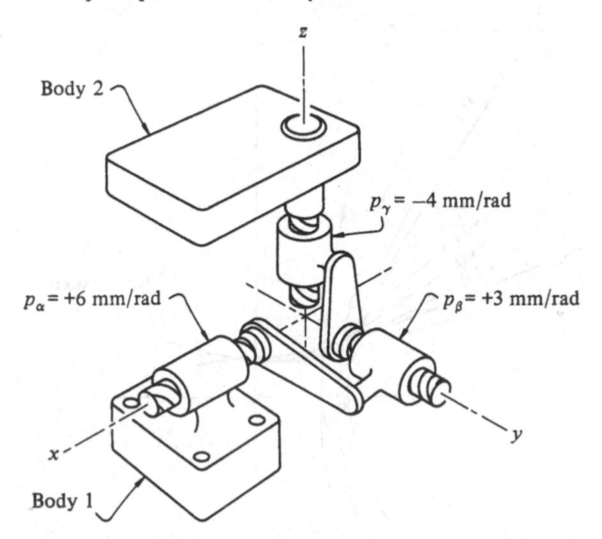


02. Ball's equation, (2) at §23.49, clearly represents a series of concentric quadric surfaces, there being one surface for each value of the pitch $p$. By virtue of the signs of the coefficients for all those values of $p$ which produce real surfaces, however, the surfaces are all hyperboloids. By inserting the chosen numerical values of §6C.01, and by substituting $y=z=0$, $z=x=0$ and $x=y=0$, we can get the following expressions for $x$, $y$ and $z$ at the vertices of the throat ellipses for the reguli of screws of given pitches $p$:

$$(1)\quad \begin{aligned} x^2 &= (3-p)(p+4) \\ y^2 &= (6-p)(p+4) \\ z^2 &= (6-p)(p-3). \end{aligned}$$

For values of pitch between $-4$ and $+3$, values for $z$ are unreal; for values of pitch between $+3$ and $+6$, values for $x$ are unreal. For all values of pitch between $-4$ and $+6$, real values for $y$ exist. For all values of pitch outside the range from $-4$ to $+6$, values for $x$, $y$ and $z$ are all unreal. A direct plot of $x$, $y$ and $z$ against $p$ will reveal the three half circles shown in figure 6C.02. If next we plot with these data selected throat ellipses existing in the principal planes, we can arrive at figure 6C.03. We find that, while there are ellipses in the $xy$- and the $yz$-planes, none exist in the $zx$-plane. There are in other words none of the concentric hyperboloids whose axes are coaxial with the axis for $y$. This circumstance of there being

Figure 6C.02 (§6C.02) The three half-circles plotting $x$, $y$, and $z$ of the equations (1) against $p$. The lengths $a$ and $b$ are the minor and the major half-axes respectively of the throat ellipse of the pitch quadric (§11.13).

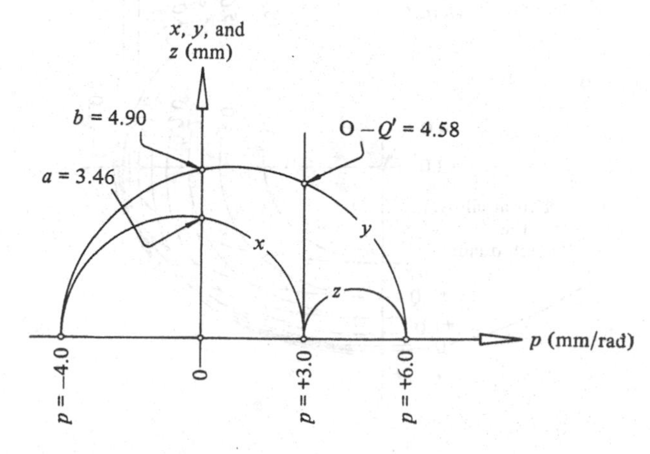

none of the concentric hyperboloids coaxial with one of the principal axes is a characteristic of the 3-system. It is a fact however that, if the principal screws are quoted in the conventional manner (§ 23.24), the naked axis will always be the axis for $y$.

03. To get an idea of the shapes of the hyperboloids, we can take a section through the whole system at the $zx$-plane. We can do this by putting $y=0$ into Ball's equation. That substitution will reveal the hyperbolas of intersection of the hyperboloids with the $zx$-plane and, by a usual method, we can find that the slopes of their asymptotes are given by:

$$\text{(2)} \quad \text{slope} = [(p+4)/(6-p)]^{\frac{1}{2}}.$$

With these data it is easy to plot figure 6C.04. It shows the traces upon the $zx$-plane of the complete range of the hyperboloids for pitches in steps of unity from $p_\alpha$ to $p_\gamma$. We note that for $p=p_\alpha=+6$, a representative hyperboloid begins as a line collinear with the axis for $x$ but, as $p$ decreases towards $p=p_\beta=+3$, the hyperboloid swells out (about that axis) until, at $p=p_\beta=+3$, it degenerates into its pair of intersecting planes (§ 6.47). As $p$ next continues to decrease a new hyperboloid takes shape about the axis for $z$. This then proceeds to shrink until, at $p=p_\gamma=-4$, it collapses into a line, collinear this time with the axis for $z$. A section taken the other way, across the $yz$-plane, will reveal a similar pattern: except for the pitches $p=+6$ to $+4$ where the slopes are unreal, and after the slope for the pitch $p=p_\beta=+3$ where the slope is infinity, the slopes of the asymptotes ($dy/dz$) diminish steadily to zero as the pitch decreases to $p=p_\gamma=-4$.

04. It is clear that, within a certain, central zone of the system, the intersections among the hyperboloids are complicated and not easy to understand. Outside that zone, however, it is possible to see that the hyperboloids appear in relation to one another as is shown very approximately in the sketch in figure 6C.05. Each successive hyperboloid is wholly 'outside' its predecessor (or wholly 'inside' as the case may be), and no intersections are apparent. The picture in figure 6C.05 is consistent with my statements at § 6.45 where I say among other things that, outside a certain, central zone, only one real screw can be found to pass through a generally chosen point $Q$. We show in figure 6C.05 in other words that, in the space outside the central zone, space is filled with an $\alpha^2$ of screws of various pitches which never intersect. In the following paragraphs I try

Figure 6C.03 (§ 6C.02) Some plotted throat ellipses for the hyperboloids of the selected 3-system. Note that the ellipses for the pitches +0.5 mm/rad and −0.5 mm/rad (these being taken for the sake of example) both cut the pitch quadric four times. The singular point $Q'$ and one of the two vertices $V$ of the throat ellipse of the pitch quadric can be seen in each of the two orthogonal views.

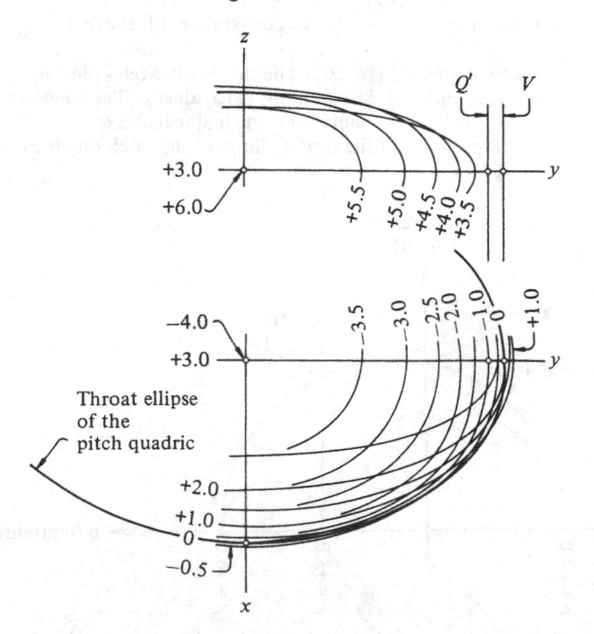


Figure 6C.04 (§ 6C.03) A plane section taken at the $xz$-plane through some of the hyperboloids of the selected 3-system. The curves that appear (and these are all hyperbolas) have been plotted by substitution into Ball's equation (2) at § 23.49. Each hyperbola is marked according to its relevant value of $p$. Note that the particular values −0.5 and +4.5 reveal the two hyperboloids that are tangential with Hunt's sextic at the isolated points respectively, $x=\pm 3.5$ on the $x$-axis and $z=\pm 1.5$ on the $z$-axis. Please note that the same particular values of $p$ are seen to be special in figures 6C.02 and 6C.03 as well.

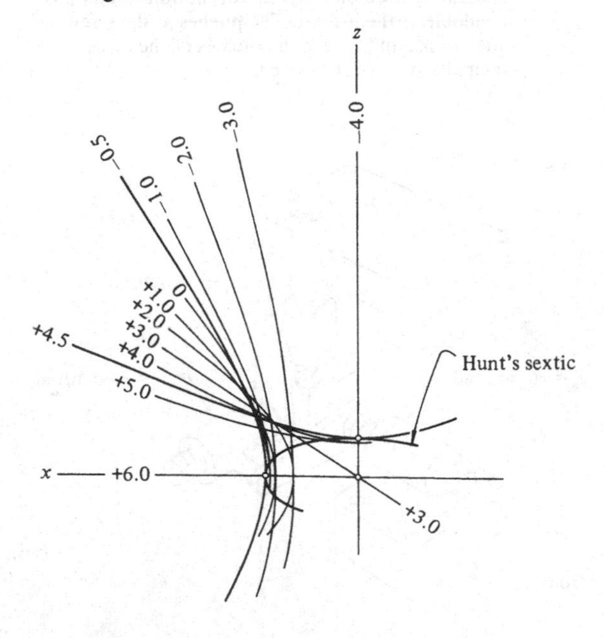

to deal on the other hand with the space inside the central zone. Intersections between the hyperboloids do occur within that space and the matter needs to be explained.

05. Both Ball and Hunt state – see Hunt's example 12B.8 (1978) – that any two hyperboloids of a 3-system will intersect in a sphere. Hunt gives the radius of the sphere in terms of (*a*) the three pitches of the three principal screws of the particular 3-system, and (*b*) the two pitches of the screws upon the intersecting hyperboloids. The equation to a sphere of intersection is easy to obtain from Ball's equation (2) at my §23.49 and, as will be seen, the radius of the sphere may or may not be real according to the pair of hyperboloids which are chosen. Let me note in addition however that any pair of hyperboloids of a 3-system will have their axes either (*a*) collinear, or (*b*) intersecting perpendicularly (§6C.03). Refer to figure 6C.06. Suppose that the symmetrical, twisted curve marked $\zeta$ in figure 6C.06 is the curve of intersection of the two hyperboloids 1 and 2 whose axes are collinear along the axis for $x$. The curve $\zeta$ has two equal symmetrical branches, each in itself a closed curve, but because the other is underneath the shown hyperboloids only the one can be seen. Let us take a

Figure 6C.05 (§6C.04) Some truncated hyperboloids of the selected 3-system roughly sketched in relation to one another. Note that the systems of generators are left handed systems and that, on the particular regulus that is simply a pair of planes, the generators intersect in two sets at $Q'$ and $Q''$. Because it would be too confusing (and, may I say, too difficult to draw), the exact detail of the intersections of the hyperboloids has been avoided here. Refer for an early impression of those intersections to figure 6C.06.

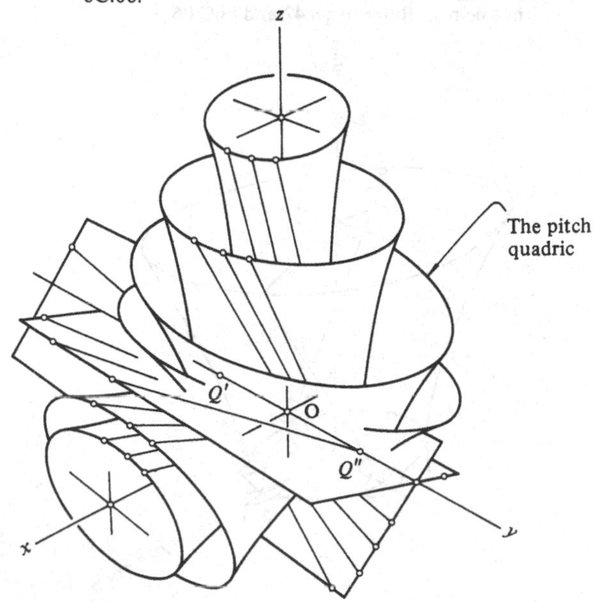


general point $Q$ on this curve and note that, among the hyperboloids surrounding the axis for $z$, there will be at least one (namely the hyperboloid 3) which contains this point. There will indeed be no more than this single hyperboloid 3, because Ball's equation (2) at §23.49 is, as both Ball and Hunt point out, a cubic in $p$. The hyperboloid 3 will cut the hyperboloids 1 and 2 in two other, symmetrical, twisted curves. These two will both contain the same point $Q$. It can next be seen by symmetry that all three curves ($\zeta$ and the other two) must, because they reside on the same sphere of radius O–$Q$, be identical. The same thing can be shown by considering a second point $Q$ on the same curve $\zeta$. *It follows that through any general point $Q$ where hyperboloids intersect there are three hyperboloids intersecting; the three intersect moreover in the same curve of intersection on the same sphere of radius O–$Q$.*

06. Please note in figure 6C.06 the three screws which have been drawn to pass through $Q$. They are all motion screws. Each screw is a generator of its relevant regulus upon its own hyperboloid and the pitches of the three screws are, of course, different. Hunt claims (and I do too) that the sum of the pitches of the three screws at any generally chosen point $Q$ in a 3-system is always the same (§6.42). Whether the screws be real or not, he says, the sum is the same as the sum of the pitches of the three principal screws. Refer to his example 12B.5 (1978) where he says, and I agree, that this can be shown from Ball's equation. Hunt also reports (1970) that the lines of an entire 3-system, the seats of all of the screws, constitute a (3,2) congruence (§11.18). It might next be

Figure 6C.06 (§6C.05) Three hyperboloids of a 3-system intersecting in the same twisted curve $\zeta$. The curve $\zeta$, in two branches, resides upon the sphere of radius O–$Q$. The seats $Q$-(*i*), $Q$-(*ii*), etc. of the three screws through $Q$ are the seats for the motion screws through $Q$.

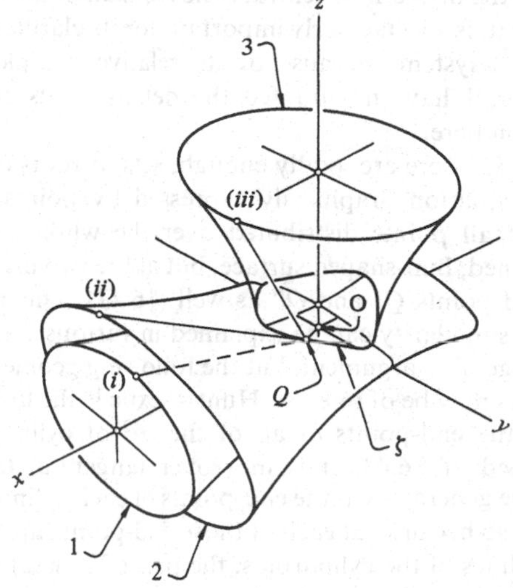

explained here that, through $Q$ in figure 6C.06, three action screws could also be drawn. They are not shown but if they were they would be members of the opposite reguli upon the three hyperboloids respectively, and their pitches would be equal in magnitude but of opposite sign to the three motion screws respectively (§ 14.29). With them and the motion screws in mind it can now be seen that the screws of any 3-system of motion crews and the screws of their reciprocal system of action screws reside upon the same set of hyperboloidal surfaces. The action system and the motion system in any case of 3 °F, in other words, are always of the same shape (§ 7.71). All of what I have just reported here coincides with the series of observations made by me at § 6.42.

07. We have already said that the three screws through $Q$ are not always all real (§ 6.43); they can only all be real, as we have seen, when $Q$ is within the mentioned, and as yet mysterious, 'central zone' (§ 6.42, § 6.45). In order to locate the boundary of that zone Hunt has done the following. Refer to Hunt (1978), example 12B.4. He has rewritten Ball's equation (2) at § 23.49 in such a way that it becomes a conventionally expressed cubic in $p$. Then, by considering the condition for two of its three roots to coincide, he derives the equation for a trilaterally symmetrical, closed, sextic surface which I would like to call now, Hunt's sextic. Hunt's sextic describes a bun-shaped surface which exactly envelopes the two series of elliptical throats which are pictured (by way of numerical example only) in figure 6C.03. A plane section through the sextic, taken at the $zx$-plane, is shown by the closed curve in figure 6C.04, the curve is not an ellipse. Through all points $Q$ that reside exactly upon the said surface, there are, in effect, only two screws passing; two of the three screws are there identical and collinear. Hunt's sextic, which divides the mysterious central zone from all of the space outside it, is spectacularly important for its clarification of the 3-system. Because of its relative complexity, however, I have not quoted the details of its actual equation here.

08. There are, oddly enough, sets of roots of the sextic equation graphically expressed by points, not only at all points distributed over the whole of the mentioned, bun-shaped surface, but at the two interior, isolated points $Q'$ and $Q''$ as well (§ 6.47). This unexpected singularity can be explained in various ways by pure algebraic argument, but the following geometrical remarks may be of interest. Hunt's sextic is the locus of all of the end-points of all of the $\infty^2$ of cylindroids described at § 6.43. It is moreover tangential to the extreme generators at the end-points of each cylindroid. This is so because, at each of those end-points upon the nodal lines of the cylindroids, the pair of generators of the cylindroid at the points in question is coalesced (§ 6.45). At both $Q'$ and $Q''$, however, there are singular cylindroids of the system whose lengths $2B$ are zero; they are the planar pencils of screws of pitch $p_\beta$ described already at § 6.47 (§ 15.49). For a parallel reading about the above mentioned singular points, please go to Ball § 180; there and in a footnote there, Ball not only mentions the two points but also gives an algebraic expression for the distance O–$Q'$.

09. I would like to remark upon the three half-axes of the pitch quadric (§ 6.45). Please refer in advance to § 7.56, § 7.57, and § 11.13. See also figures 7.09 and 11.02. Go to figure 6.05($a$) and imagine the three prongs and platforms there to be, not as they are shown, but at new points $A$, $B$, and $C$ as shown in figure 6C.07. The points of contact in figure 6C.07 are at the vertices of the throat ellipse of the pitch quadric and the slopes of the contact normals there can be seen to be, regardless of sign, $c/b$, $c/a$, and $c/b$ respectively (§ 11.13). Keeping contact at the three prongs, now give the body 2 in figure 6C.07 an infinitesimal twist about the axis for $x$; please find upon doing this that the pitch of the twisting is, $p_\alpha = bc/a$, right handed. Use a similar argument for the pitches of the twistings about the axes for $y$ and $z$ respectively and find all together that

(3) $$p_\alpha = +bc/a$$
$$p_\beta = +ca/b$$
$$p_\gamma = -ab/c.$$

Figure 6C.07 (§ 6C.09) Relationships of the pitch quadric. The sum of the pitches of the three principal screws intersecting at O is $(a^2b^2+b^2c^2-c^2a^2)/abc$. This sum is, as we know, a constant for the three pitches of the three screws intersecting at a general point $Q$. Two of the three screws at $Q$ may or may not be real. Refer to § 6.42 and § 6C.06.

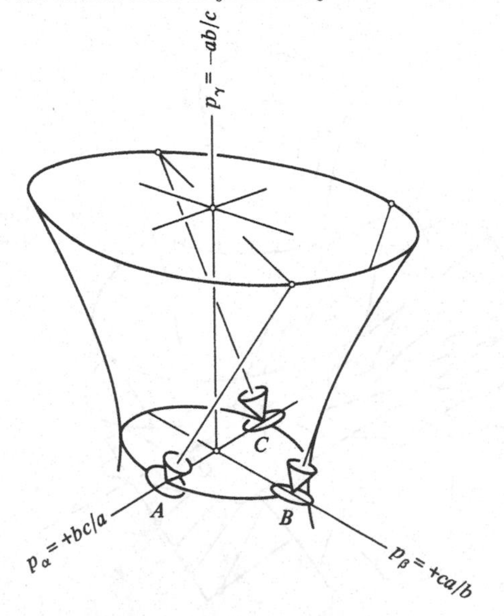

Solving these equations (3) for the half-axes $a$, $b$ and $c$ of the pitch quadric, we can next find that

$$(4) \quad \begin{aligned} a^2 &= -p_\beta p_\gamma \\ b^2 &= -p_\gamma p_\alpha \\ c^2 &= +p_\alpha p_\beta. \end{aligned}$$

One might go for a parallel reading to Ball (1900) § 173. Please compare these equations (4) with the equations (8) at § 11.54. At § 11.51 *et seq.* I am discussing the Plücker equations to a regulus. Please note, while ignoring the fact that the sign of each $k$ at (8) at § 11.54 is opposite to that of its corresponding $p$ here, that the sets of equations are the same. I am canvassing an argument, by saying this, that the symbolism of the so called Plücker or the line-coordinate algebra coincides very closely with the fundamental bases of screw theory (§ 19.21).

10. I would like to discuss the question of how a whole 3-system of screws might be constructed from any three given screws. Refer to figure 3.08. Please read § 3.37, which touches upon the question of quasi-generality; it refers, indeed, to the question of the reality or otherwise of the pitch quadric. Please read also Ball (1900) § 175. Ball proves there the theorem I have quoted at § 11.08, and he describes in detail (there and elsewhere) what I have merely summarised here. The method runs as follows: take the three pairs of screws from the three given screws and construct the three cylindroids determined by those pairs (§ 15.18); note that the point of intersection of the three central planes of symmetry of the three cylindroids is the central point of the whole 3-system (§ 11.08). Continue to construct cylindroids in this way upon all new pairs of available screws until all $\infty^2$ of the cylindroids have been constructed; thus the whole 3-system of screws may be found. For a parallel reading about this way in which any general 3-system can be seen as an $\infty^2$ of cylindroids, go to Hunt (1978) § 12.6. Hunt proves there incidentally and by a cunning method that all screws of the same pitch within a 3-system reside upon the same regulus and that the hyperboloids of the single infinity of reguli are concentric. I offer a similar argument about the same fact at § 6C.11. For further reading about the concentricity of the hyperboloids, and for other demonstrations of the fact (in the context of statics), please go to § 10.36 and to § 10.66 *et seq.* At chapter 10 I deal with the kinetostatic analogies which lie at the root of reciprocity and some of the mechanical aspects, namely some of the implications for mechanics, of screw theory.

11. At § 6.47 I promised to give a proof here that all screws of the same pitch in a 3-system reside upon the same regulus and that the hyperboloids of the various reguli are concentric. Please read § 6C.10 and go to figure 6C.07. Remain aware that we are dealing here in this chapter 6 with the quasi-general systems only, namely those systems containing screws of zero pitch. Although the following argument applies for such a quasi-general system only, it does indeed produce a result which applies for general systems also (§ 23.24). Erect the cylindroid of screws upon the pair of contact normals (action screws of zero pitch) at $A$ and $C$, and note the principal action screw upon the axis for $z$. Its pitch will be $(-p_\gamma)$, namely positive. Next take the contact normal at $B$ in conjunction with the principal action screw just found and reconstruct (by symmetry) the action screw of zero pitch at $D$. We thus see two of the principal cylindroids of the action system with their nodal lines upon the axes for $x$ and $y$ respectively and intersecting perpendicularly at O. Take now the four action screws of some small, nominated (say positive) pitch in the close neighborhoods of $A$, $B$, $C$ and $D$ respectively, and notice that, because of the peculiar nature of the geometry of the cylindroid (chapter 15), these screws may cut their respective axes for $x$ and $y$ either inside or outside the elliptical throat of the pitch quadric. Whatever they do, however, they will do in symmetrical pairs; see figure 6C.03. It is now necessary to see ($a$) that all four action screws of this new, small, positive pitch reside upon the same regulus – and, by symmetry, this is easy to see; ($b$) that this new regulus is concentric and, in general, coaxial, with the original regulus upon the pitch quadric – this is also easy to see; and ($c$) that all screws of the system of that new pitch reside upon that new regulus. This latter could not be otherwise for, if some of those screws of the same, small, positive pitch were not upon the said regulus, there would be cylindroids of screws in the system whose central planes of symmetry did not contain the central point of the system O (§ 11.08).

12. In this appendix we have been extending our examination of the quasi-general 3-system of motion screws which springs into being as soon as three points of contact between a pair of rigid bodies are established. It is true of course that everything I have said here and at § 6.38 *et seq.* is encapsulated into Ball's equation (2) at § 23.49. It is however necessary in my view not only to know the algebra but also to have a visual impression of the geometric reality which is described by the algebra for, unless such a visual impression is achieved, we shall, collectively, be that much less effective as synthesisers of new mechanism. There is of course, and finally, the completely general (as distinct from the quasi-general) 3-system. For some remarks about that and for some related ideas about the important matter of reciprocity involved therein, the reader may go to § 23.23 *et seq.*

7. New joints may be synthesised by means of the method of button pairs. Joints (whether simple or complex) may be categorised according to the natures of the two, mutually reciprocal, systems of screws that they exhibit.

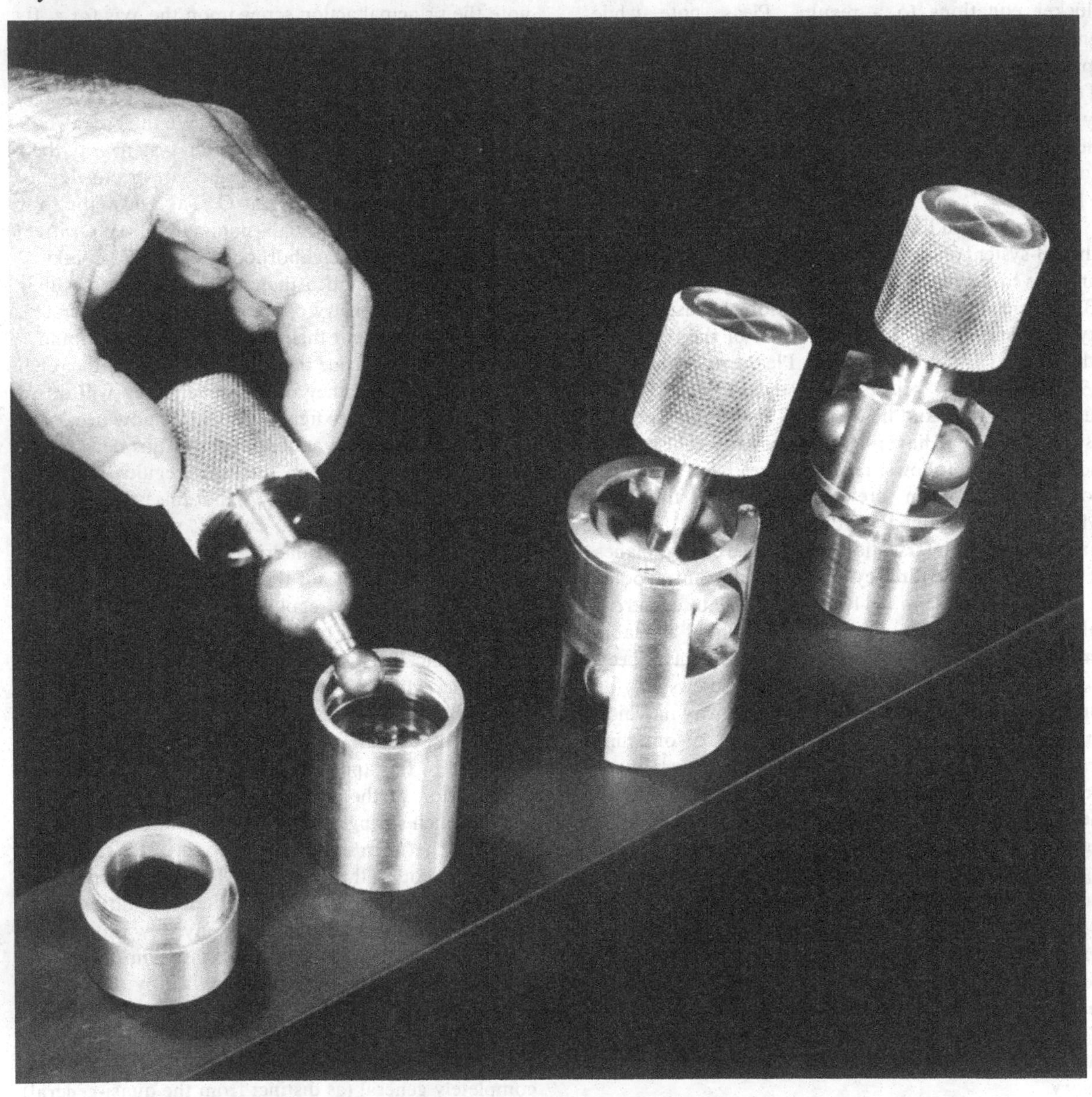

# 7 The possibilities in reality for practical joints

### Some remarks connecting with chapter 6

01. Associated with each set of points of contact and its associated tangent planes and contact normals at a simple joint in mechanism, is a system of motion screws; this is explained in chapter 6. The so-called system of screws or the screw system is simply the set of all of the ISAs which are available for relative motion at the joint. It is taken for granted in chapter 6 that the simple joint is considered alone (§ 1.28), and that contact is not being lost at any of the points of contact at the joint (§ 6.15, § 8.12).

02. Provided each one of the contact normals at a joint is projectively independent of the other contact normals at the joint (§ 3.32), the system of motion screws at the joint takes for its numerical name the number six minus the number of points of contact. This matter is thoroughly canvassed in chapter 6; with two points of contact there is discovered and explored the quasi-general 4-system for example, with three, the quasi-general 3-system, and so on.

03. It is important to see that in general the systems of screws of which I speak in chapter 6 – they are the systems of motion screws – contain no screws which lie along the contact normals at the points of contact at the joints. The contact normals at any one joint belong as screws of zero pitch to other systems of screws, which are, as we shall see, reciprocal to the mentioned systems. The general matter of the reciprocity of the screw systems is more fully discussed in other places, notably chapters 10, 14 and 23; but one might notice in passing here that any pure force applied at a point of contact at a joint, which will in the absence of friction be pure and be acting along the common normal there (§ 10.59), can cause no twisting about any one of the motion screws of the screw system at the joint.

04. In chapter 6 there are descriptions given of the screw systems which exist instantaneously at the joints. There is however no consideration given to the question of finite relative movements. No account is taken of how, or even of whether or not, the systems of screws may change as the various simple joints engage upon their motions. No attention is paid either to the possibilities for (or the practicalities of) the special cases of the joints in question. Both of these are new matters which need to be considered now.

05. I shall be appealing here in this chapter to some aspects of the projective dependence and independence of lines and screws (chapter 10). I intend in this way and by other means to explore the important question of whether or not a joint and indeed, which if any joints, might be possible or practical in real machinery. The chapter gets off to a slow start, unfortunately, but the reason for this will become apparent, I think, whenever a reader goes to the end of it to try to digest there unaided the whole gist of it. Before beginning, I wish to examine what is meant in the present context by the two mentioned words, possible and practical. What is meant, as I intend to show, is a matter mainly of convenience; likely meanings of the words are related to the maturity or otherwise of our general theory and to the need or otherwise for accurate repeatability at the joint; and both of these are related to our available technology.

### The possibility and practicality of joints

06. Whatever 'works' to our satisfaction is clearly possible and thereby useful; this broad principle is applicable on many occasions, both in the careful design of casual machinery and in the more careless business of casual machine design. Look at the $f5$ joint at $B$ in the mousetrap in figure 2.07 for example; look at the $f1$ joint between the cup and the saucer in figure 1.10(*b*); and compare these with the various joints in figures 1.10(*a*) and 1.10(*e*). Despite such comparisons which need to be made in this general area of roughness versus accuracy (§ 6.02) and of higher and lower joints or pairs (§ 7.15), it is a fact in all of these mentioned cases that the actual shapes of the surfaces at the elements of the joints are not very important in the circumstances. It is clear that the designer of such joints in such machinery need not care very much about the exact geometric layout of the screw systems at his joints, or about their possible

changes as the various relative displacements at the joints are caused or happen to occur.

07. Indeed it is taken for granted in much of such machinery that spring, gravity and other force closures at such rough joints as these will effect and maintain the necessary points of contact there, while friction will prevent or retard all of the unwanted though irrelevant motions which may occur. This general circumstance pervades (*a*) the whole of nature – which latter is held by some to be 'natural' by virtue of its not having been designed; (*b*) much of our carefully designed and manufactured machinery which is, accordingly, or presumably, unnatural; and (*c*) most of our manipulative contact with nature via that machinery (§ 10.61).

08. Often however, accurately dimensioned movements are required to occur. These are worked out against an accurately specified pattern in time and they must be repeatable or repetitive. In these other, quite different circumstances a designer must be able accurately to predict the behaviour at the joints of his mechanism, and he must be able safely to rely upon each joint's ability in the real machine to repeat its action regularly.

09. Thus we arrive at a stage where one might say the following: whether one thinks or works in terms of pure geometry, graphics of various kinds, or algebraically, the easiest kinds of joints to deal with in dimensional design are the following: *they are those kinds wherein the screw system existing at the joint remains invariant as the joint engages in its motion.*

10. Typically though, when seeking numerical answers to questions about dimensional movement in a chosen type of mechanism undergoing a dimensional synthesis, we step from joint to joint across the links by means of algebraic matrices of various kinds (§ 19.52). The purpose of this activity is to relate the changing location of each next link of the chain with respect to each previous link and thus, of course, to the fixed location of the frame link. Ultimately, an algebraic closure of the linkage might be achieved (by a multiplication of all of the matrices together around the completed loop or loops of the assembled mechanism) and a so called closure equation might result. An excellent elementary account of how this might be done in certain cases by one of the earlier methods is given in chapter 12 and in some related appendices of the book by Hartenberg and Denavit (1964); there it can be seen quite clearly that although screw systems as such are never overtly mentioned in that text, my remarks at § 7.09 obtain.

11. It will be seen from that account moreover that, if the system of motion screws at the joint were to remain not only invariant but also fixed in location with respect to one or both of the links of the joint, the joint would be even easier to handle by the known algebraic methods. A perusal in advance of § 7.91 will give an early idea of what I mean by 'fixed in location' in the present context. Joints with invariant screw systems, where the system of screws might be fixed with respect to only one of the links of the joint are, however, impossible (§ 7.97). The only joints of this kind which can exist are those where the invariant system is movable with respect to both of the links, or fixed with respect to both of them.

12. So next – and here I wish to make a naive remark – we can arrive at the following, even tighter definition of a practical joint which can be handled easily: *there exists a particular, special kind of joint in mechanism wherein the system of motion screws remains not only invariant but also fixed in location with respect to both of the links (or members) at the joint.* This latter class of joint comprises, I believe, the six well known lower or 'surface contact' joints or pairs which are widely described by the symbols *S*, *E*, *C*, *H*, *R*, *P*. They are discussed by Waldron [1], who believes he may have shown by motor algebra (§ 19.07, § 19.14) that these six lower joints are the only ones of their kind available. They are also discussed by me at the end of the present chapter, from § 7.91 onwards, and I believe I may have shown, by other methods, the same result. I say in any event now that the six lower simple joints are these: *they are the spherical, ebene, cylindric, helical, revolute, and prismatic joints.* Refer to table 7.01.

13. On this overall question of the possibility and practicality of joints in mechanism we have arrived now at the following main conclusions: (*a*) that the higher simple joints or pairs such as those already mentioned at § 7.06 and the *f* 2 joint between the log and the trestle in figure 1.10(*f*), can often perform their intended functions satisfactorily, and that they are, accordingly,

Table 7.01 (§ 7.12) The six lower joints or pairs arranged according to their *f*-numbers which range from 1 to 3. Letter-symbols for the six joints are listed here and some of the graphical devices for indicating the joints stylistically are also shown (§ 7.17, § 7.65).

| | | | |
|---|---|---|---|
| *f*3 | Spherical *S* | Ebene *E* | |
| *f*2 | Cylindrical *C* | | |
| *f*1 | Helical *H* | Revolute *R* | Prismatic *P* |

possible because they are useful and vice versa; and (*b*) that the lower simple joints or pairs, which are listed at §7.12 and displayed in table 7.01, are practical because they are predictable and repetitive in their motions, because they can be manufactured easily by existing methods in technology, and because they are the joints most easily handled by known methods in the algebraic analysis and synthesis of mechanism.

### An extended classification of joints

14. It should be clear however that between the two mentioned well established categories of joint, namely the higher and the lower, there might be inserted one other. That other is suggested at §7.09 and delineated more clearly at §7.11. I would like to propose for it the middle term of the following list: *higher, intermediate, lower*.

15. I propose in other words: (*a*) to restrict the use of the term *higher joint or pair* to that class of joint wherein the system of motion screws is both changeable and movable with respect to both members at the joint; (*b*) to introduce the term *intermediate joint or pair* to refer to that class of joint wherein the system of motion screws is invariant but movable with respect to both members at the joint (§7.11); and (*c*) to retain the term *lower joint or pair* with its present meaning. With (*c*) I am keeping faith with Waldron [2], but I argue that at a lower joint, the system of motion screws will be invariant and be fixed with respect to both of the members at the joint (§7.11, §7.17). Refer to table 7.02 where various examples of these three are offered. Please also peruse §7.97; the material there might be seen as an appendix to §7.11.

Table 7.02 (§7.15) This table displays and describes some examples of the three kinds of simple joint: the higher, the intermediate, and the lower. Please note that, although the stylised devices depicting the various joints appear to be flat, the joints themselves exist and operate in all three dimensions.

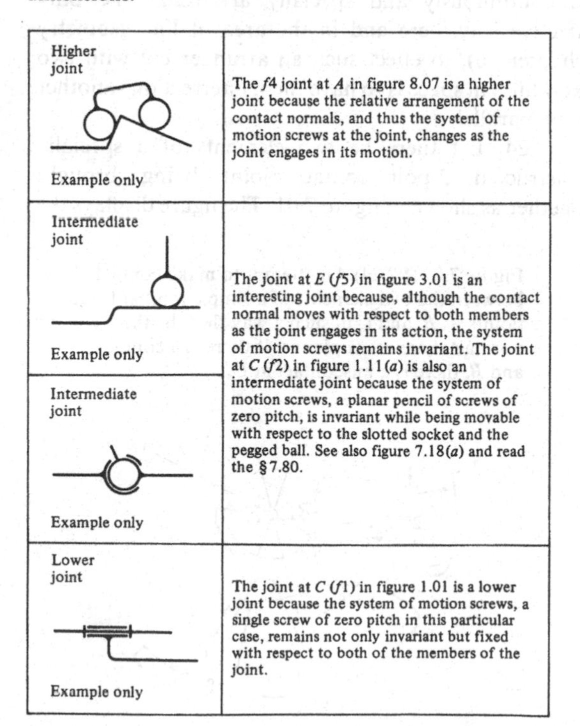

| | |
|---|---|
| Higher joint (Example only) | The $f4$ joint at $A$ in figure 8.07 is a higher joint because the relative arrangement of the contact normals, and thus the system of motion screws at the joint, changes as the joint engages in its motion. |
| Intermediate joint (Example only) | The joint at $E$ ($f5$) in figure 3.01 is an interesting joint because, although the contact normal moves with respect to both members as the joint engages in its action, the system of motion screws remains invariant. The joint at $C$ ($f2$) in figure 1.11(*a*) is also an intermediate joint because the system of motion screws, a planar pencil of screws of zero pitch, is invariant while being movable with respect to the slotted socket and the pegged ball. See also figure 7.18(*a*) and read the §7.80. |
| Intermediate joint (Example only) | |
| Lower joint (Example only) | The joint at $C$ ($f1$) in figure 1.01 is a lower joint because the system of motion screws, a single screw of zero pitch in this particular case, remains not only invariant but fixed with respect to both of the members of the joint. |

### Remarks in passing

16. Whereas (*a*) the three categories listed above are applicable not only to the simple joints but by easy extension to the complex joints as well, noting for example that the joint via $B$ and $C$ at the Hooke joint between the shafts $A$ and $D$ in figure 2.01(*h*) is an intermediate, complex joint whether the centre lines at $B$ and $C$ intersect or not (§2.37, §7.80); and whereas (*b*) upper case letters have been ascribed in the technical literature only and inclusively to the six lower pairs, these separate matters are in no way related to (*c*) the degrees of freedom, the $°F$, or the $f$-number at the joints. In table 7.01 the six lower pairs are sorted out according to their $f$-numbers; these data might be kept in mind as the following argument develops.

17. We might also clarify here the origins of the meanings ascribed by consensus to the two words, *lower* and *higher* in the present context. Whereas the word lower appears earlier to have been applied exclusively to joints whose $f$-number is only 1, see for example Grübler (1917), Steeds (1930) and Rosenauer and Willis (1952), it has since been applied by consensus to those simple joints enjoying *surface contact* at the interface between the elements of the joint, see for example Hain (1961), Hartenberg and Denavit (1964) and Waldron [1]. The word higher on the other hand has always, until now, been used for all of those pairs which were not lower pairs.

18. Following this mention of lower pairs and the question of surface contact, I would like to mention now, in advance of its further discussion at §7.37, the idea of *contact patch*. I ask the reader at this early stage to view this as a surface of points, or a line or lines of points, or a single point, or any two or all of these, *which enjoys a geometrical identity of its own*. Please see a contact patch as independent of the elements at the joint; but be aware at the same time (*a*) that it is upon the respective surfaces of the mating elements that it has its being (§1.12, §1.52, §6.03), and (*b*) that its shape, whatever it is, will somehow be related to the system of motion screws existing at the joint.

**Prongs, platforms, facets, and button pairs**

19. I have already said that I hope to explore in this chapter the possibilities for practical joints in real machinery (§7.05). To do this I shall need to engage in a new, synthetical kind of argument; it is different from, and more difficult and more uncertain than, the analytical argument engaged in hitherto. Also, and despite my remarks at §6.06 and other places about the absolute inevitability of rigidity and irregularity at joints, I shall need to speak in this chapter openly and straightforwardly about line and surface contact as well as point contact between the elements of practical joints in real machinery (§6.03). I shall need to entertain some apparently self-contradictory ideas, such as the approximate regularity of surfaces mentioned already at §6.03, and to employ some new terms about to be explained. Please go back to §1.53.

20. At §3.20, §6.32 and at other places I mention or imply the existence of *prongs and platforms* in and among the bodies. My intention with these two terms in chapter 3 and elsewhere is to convey more clearly the idea of what an *n*-line in mechanism is and of how the relative motion between a pair of bodies might exhibit itself along an *n*-line (§3.06, §3.19 *et seq.*). An *n*-line exists, of course, always in two bodies, never in one alone, so the prongs are on one of the bodies while the platforms are on the other. Please refer to figure 3.04 where the general air of unreality of these things, the prongs and platforms, is made quite evident. Rows of prongs and platforms such as these are used most often as figments of the imagination (§3.23); they have no obvious counterpart in actual physical reality. The occasional *real prong against its real facet*, however, will exist at the actual material surfaces of two contacting bodies. There is such a pair for example in figure 3.04, where the real body 4 touches the real body 1 at *E*.

21. In figures such as 3.13(*b*) and 3.14, accordingly, the shown flat areas might better be called *facets* upon the surfaces of the bigger bodies there. Facets are finite, flat areas, existing in fact or imagined for convenience to be cut into the otherwise undulating surface of a body which does exist in reality. The term *facet* in other words refers to a special kind of platform; it is, in a sense, a 'real' platform. At a facet, the common normal at the prong's tip is normal to the surface of the body as it was before (§3.20); if in reality there is friction in evidence at the real surface of some body, then the imagined facet at a point of contact may be tilted in the imagination accordingly (§10.59).

22. In certain circumstances in the business of synthesis, however, a prong and its companion facet can be imagined to exist, if we wish, anywhere along its established *n*-line (§6.32). In this sense the prong and the facet together can be seen as a permanently connected pair which can be slid as an entity anywhere along the *n*-line in question. This entity when reduced to infinitesimal proportions I propose to call a *button pair*. I mean by this new term the following: a button pair comprises two infinitesimal pieces of convex surface in contact at a point, with an infinitesimal piece of tangent plane at the point of contact and with a short piece of the contact normal having the capacity to remain collinear with the *n*-line concerned as the button pair is free to be slid along it. Upon arrival at its decided destination, in the process of synthesis, the button pair may operate (within its infinitesimal area) exactly as a prong upon a facet. Circularly arranged with respect to a prong and a facet at *A*, some button pairs are shown in figure 7.02.

**On trying to set up an intermediate f4 joint**

23. Refer to figure 6.04 and note that, in an *f*4 joint in general, there exist two contact normals, both of which are *n*-lines at the respective points of contact, and that these two *n*-lines are in general skew. Refer to §1.26 for a definition of the *f*-number at a joint and to §6.30 for a useful summary. Be aware also that, whenever such a general *f*4 joint engages in a finite movement, the initial geometrical relationship between the two *n*-lines will vary. This means that the 4-system of motion screws which is based upon the two *n*-lines will vary also, and that, accordingly, the joint will be a higher joint (§6.26 *et seq.*, §7.15). There appears, upon inspection, to be no escape from this unless the two *n*-lines can in some way be continuously and specially arranged. The only effective way, here and in the area of line geometry (chapter 10), to effect such an arrangement with two skew lines in space is to have them intersect one another or be parallel.

24. Let there be two elements of a specially constructed 2-point-contact joint being brought together as shown in figure 7.01. The figure displays the

Figure 7.01 (§7.24) An attempt to make contact between the two elements of a simple joint at three points *A*, *B*, and *C* in such a way that the three contact normals there are coplanar. If there is a contact at *A* and *B*, there is a miscontact at *C*.

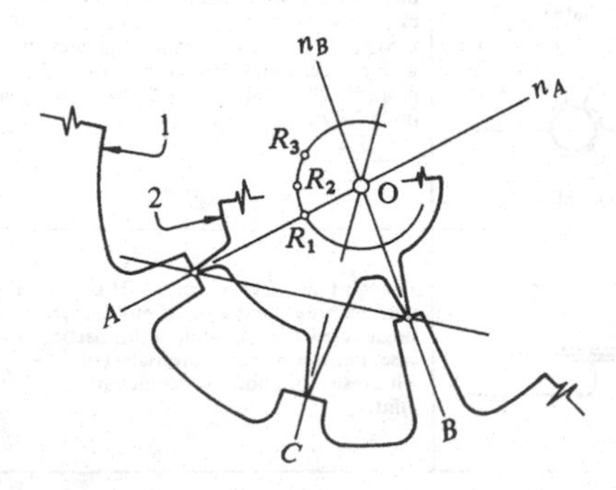

special requirements at the joint. In figure 7.01 a plane cross-section is taken through the two facets upon the element 1 of the joint and the two corresponding prongs of the element 2 are in that plane. The mentioned prongs are at $A$ and $B$. The figure thus shows how the two n-lines at the intended points of contact intersect. There is moreover at the joint in figure 7.01 a third, incipient point of contact at the facet $C$, my intention being there to provide some extra guidance at the joint (§2.21). Remember next that both of the elements at this special joint are rigid (§5.01) and that they are, despite a careful effort on the part of the constructor to make them fit with one another (§6.04), irregular. If, due to irregularity, contact is preliminarily made at $A$ and $B$, there will be an error, a shortfall as shown at $C$. Alternatively, the prong at $C$ may pierce the facet at $C$ before the contacts can be made at $A$ and $B$ (§1.06). In whatever way the inevitable miscontact at $C$ may occur and then to be taken up, the following argument applies.

25. Next we discover that, in whatever way we try to seat the prong at $C$ by effecting a small 'final twist of adjustment' about some available axis at the joint, we destroy the intended coplanarity of the *n*-lines, or their intended intersection, or both. Another way of saying this is to say the following: *the effecting of a sudden, well-conditioned contact at C by means of a small twist of the element 2 about some ISA available at the joint is impossible.*

26. A proof of the above important fact is about to be offered, and implied in the proof is an explanation of the term 'well conditioned'. The proof is somewhat extended because I wish to make as I go some relevant observations about the system of motion screws which is extant at this particular joint. The system is, as we shall see, not a 3-system as might be expected from the three points of contact which have been nominated, but simply a special 4-system. This fact needs to be explained and the system described. The details may be tiresome for some, however, and the whole of the following section may be omitted on a first reading. Let me mention here however that *if a contact is to be a well conditioned one, the path of the approaching incipient contact point at the tip of the prong is near to being normal to the waiting surface, while in the case of a poorly conditioned one the path of the tip of the prong is near to being parallel, namely tangential to the waiting surface.*

**On the projective (or linear) dependence of lines**

27. In looking in figure 7.01 for motion screws about which a small twist of final adjustment might be tried in order to seat the errant prong at $C$ (while keeping, of course, the contacts already made at $A$ and $B$), let us look first for screws of zero pitch. Following the method of §6B.05, screws of zero pitch will be seen to inhabit (*a*) all $\infty^2$ of the star of lines at the intersection $O$ of the two contact normals at $A$ and $B$, and (*b*) all $\infty^2$ of the plane of lines which can be drawn in the plane of the paper. These lines together make up the $\infty^2$ of generators of a special linear congruence whose directrices (the two contact normals at $A$ and $B$) intersect (§11.29). However any small twist (which would be a pure rotation in this case) about any one of these $\infty^2$ of motion screws of zero pitch, even a twist about the 'best' of them, namely the one which exists along the line $AB$, *will result in a small displacement at the prong at C which will be parallel with the surface of the facet there and be thereby useless for effecting a sudden contact there.*

28. Looking next at the motion screws of infinite pitch in figure 7.01, we see that the $\infty^2$ of these exist normal to the plane of the paper and piercing every point upon it (§11.28). We see moreover that these screws are no better than those at §7.27 as vehicles for a final twist of adjustment (always a pure translation in this case) to try to seat the prong at $C$. Any resulting small displacement of the prong at $C$ *will again be parallel with the surface of the facet there, and be thereby useless for effecting a sudden contact there.*

29. Looking next at the possibility that some motion screw or screws of finite pitch might solve the problem in figure 7.01, let us nominate some finite pitch say $p_i$, and ask whether a screw of such a pitch might pass through a nominated point say $R_1$ on $OA$ and at the circle which is in the plane of the paper. The axis of such a screw must (*a*) be perpendicular to the line $AR_1$ (see figure 5.08 and its relevant text at §5.42), and (*b*) be inclined to the plane of the paper at such an angle $\theta_i$ that a small displacement vector at $B$ can reside within the surface of the facet there. There is a unique solution $\theta_i$ for each $p_i$ at each point $R_1$ along $OA$ in circumstances such as these (see figure 6.03 and its relevant text at §6.24), so a single motion screw of each pitch $p_i$ can always be found to pass through each point such as $R_1$ in the line $OA$, there being an $\infty^3$ of motion screws all together (§6.30).

30. Refer next to figure 3.02 and its related text from §3.10 to §3.12 inclusive. By applying exactly the same argument to the rigid triangle $ABC$ and the centre $O$ in figure 7.01 we can see that any small twist of final adjustment about the unique screw of pitch $p_i$, which passes through any nominated point $R$ anywhere in the plane of the paper, *will result in a small displacement at the prong at C parallel with the surface of the facet there, and will thereby be useless for effecting a sudden contact there.*

31. It should now be evident that there is a difficulty here: *the n-line at C is projectively dependent upon the two n-lines at A and B.* That is why no amount of fiddling with a small twist or a combination of small

twists will seat the prong at $C$. Indeed if a few more of the lines through $O$ that are coplanar with the intersecting $n$-lines at $A$ and $B$ were to be provided next with prongs and facets, the difficulty would still remain. Look at figure 7.02 and consider whether any small twist of final adjustment would succeed in seating *any* of the extra prongs upon their companion facets there.

32. It is not very difficult next to see that the whole planar pencil of projectively interdependent $n$-lines at $O$, each line with its prong and facet, and, if we wish to imagine it, each with its sliding button pair, is a special 2-system of screws defined by the two $n$-lines at $A$ and $B$. The planar pencil is indeed a cylindroid of zero length made up of screws of zero pitch (§15.50) and this special cylindroid is the system of screws reciprocal to the special 4-system of available ISAs extant at this special joint. I am however in process of describing, not this reciprocal 2-system of action screws I have mentioned here just incidentally, but the 4-system of motion screws available at the joint (§6.07, §14.25).

33. Notice next that, for each point $R$ on the circle shown in figure 7.01 (they are marked $R_1$, $R_2$, etc.), there will be a screw of pitch $p_i$ as before, each screw at its angle $\theta_i$ to the plane of the paper and perpendicular to its radial line $OR$. Seeing that the actual magnitude of the radial distances to the facets, O–$A$, O–$B$, O–$C$, etc., are irrelevant in the present argument, given in other words that all of the possible $n$-lines of §7.28 have sliding button pairs to be fixed wherever we wish along these lines, the above statement follows from radial symmetry about the centre O, if from nothing else. This means that, for each pitch $p_i$ from negative infinity through zero to positive infinity there is an infinity of concentric,

Figure 7.02 (§7.31) On some of the single infinity of generators of the planar pencil of $n$-lines intersecting at the centre here there are erected fixed prongs and platforms, and on others sliding button pairs. The whole thing is a simple, $f$4 joint; refer to figure 7.03.

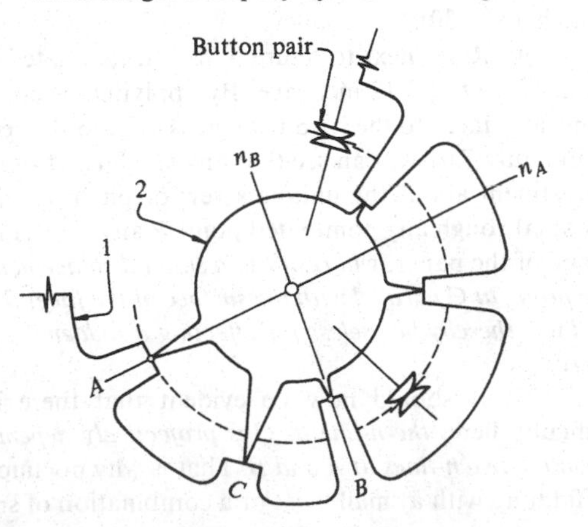


circular hyperboloids whose reguli make up elliptic congruences of screws whose pitch is $p_i$; the single axis of concentricity for these $\infty^2$ of hyperboloids and $\infty^1$ of elliptic congruences is the line through O normal to the plane of the paper; the common centre point of all of them is the central point of the system O; and the plane of the paper is the plane of symmetry dividing this whole 4-system of screws into its mirror-image halves as explained at appendix 6B. There is also a short summary of that explanation at §6.29.

34. The 4-system just described is a *first special 4-system*. See §12.9.1 on p. 366 of Hunt (1978), where his $h'_\beta = h'_\alpha =$ zero (§15.50). See also my §23.72 and be aware that, whereas Hunt uses the symbol $h$ for pitch (sometimes primed, thus $h'$), I use the symbol $p$. In both of our works, however, the subscripts $\alpha$, $\beta$ and $\gamma$ are used to distinguish the one, two or three necessary and sufficient, principal screws of a system (§6B.14, §6.45, §23.47 *et seq.*, and §23.71). Any $h_\alpha$ or $h'_\alpha$ of his corresponds to a $p_\alpha$ of mine. Because it is a fact in this special 4-system that the mentioned set of concentric and coaxial hyperboloids are all circular, there is a pure axi-symmetry of the system about its central axis. The rotary symmetry of the second order which was mentioned at §6.29 has disappeared. It so happens also that, because the pitches of the two principal screws of the system, $p_\alpha$ and $p_\beta$, are not only equal and finite but both zero, the $\infty^2$ of screws in the plane of the paper in figure 7.01 are all screws of zero pitch. The system of motion screws at this particular joint is, in other words, a somewhat special case of the first special 4-system. Refer to my remarks at §6.08 about the quasi-general nature of all of the systems of motion screws existing at the simple joints.

### A reversal of the argument: the idea of contact patch

35. Please read §7.31 again and be prepared now to hear it in a different way. If the opposing sets of prongs and facets in figure 7.02 are manufactured to match one another, that is, if an attempt is made to see that they will *fit* (§6.03, §6.04), the resulting approximate regularity of the mating surfaces at the elements of the joint will display the following advantage in the action of real machinery: whichever two of the prongs come into contact at any one fitting of the pieces, a subsequent twist about any one of the available screws at the instant will not be obstructed by a sudden collision of any one of the prongs with its corresponding facet: *the construction will have ensured that all of the incipient contacts between all of the prongs and facets are poorly conditioned ones.*

36. When any machine constructor sets out to make a practical, smoothly running, $f$1 screw joint, for

example, with its inevitable backlash and its five points of contact (§ 1.53), his objective with taps and dies or with the screw-cutting attachments will be to arrange that, whenever a sixth point of contact accidentally occurs within the joint, the velocity at that sixth point will never be in a direction other than close to the tangent plane at that sixth point. He will ensure in other words that any accidental hindrances will not be stopping hindrances (§ 1.29) but guiding ones (§ 6.03). See figure 14.01($a$). *This general objective, namely the attempted achievement of poorly conditioned contacts across all guide surfaces within all joints, with well conditioned contacts being destined to occur at stop surfaces on the other hand, will be seen to pervade the whole of machinery practice.*

37. It now becomes evident that an entity entitled contact patch deserves to be considered. Please entertain the following preliminary ideas. The contact patch is *(a) a surface or some other thing associated only with guiding at the joint; (b) a surface or some other thing attached to neither of the elements at the joint*. Independently of whether it consists of a continuous surface of $\infty^2$ of button pairs or a continuous line of $\infty^1$ of button pairs or of isolated single button pairs or any combination of these, the contact patch is *(c) a geometric entity which can be described in Euclidean terms; (d) a shape which can be characterised merely by the set of contact normals associated with its set of button pairs; and (e) a thing which will determine, we suspect, the system of motion screws or the screw system for motion extant at the joint*. Do not forget however that, while the set of contact normals will be part of the system of action screws at a joint, the system of motion screws will not in general include these contact normals.

### A practical intermediate $f4$ joint

38. Suppose now we wished to invent an $f4$ joint of the kind wherein the $f4$-action described at § 7.34 could continuously occur. Suppose in other words we wished to invent an $f4$ joint of the mentioned kind in such a way that, even after the occurrence of a finite relative movement, the 4-system of available screws at the instant would remain exactly as before.

39. We have already shown in effect at § 7.31 that, if the $n$-lines and their sliding button pairs at the joint in figure 7.02 were multiplied to fill the whole circumference at the joint, and if all of the button pairs were arranged at the same one radius from O, we would arrive without loss of generality at the new joint illustrated in figure 7.03. In figure 7.03 an external, circular, knife-edged disc makes *continuous line contact* with a matching internal, circular, knife-edged hole of infinitesimal length.

40. It appears to be obvious that the circular shape of this line of contact is the only shape which will permit an uninterrupted twisting of zero pitch about the unique axis through O which is normal to the plane of the circle. On further examination, it is the only continuous shape which will permit a small twist about the screw of arbitrarily chosen pitch $p_i$ taken through the arbitrarily chosen point $R$ in the plane of the circle of contact (§ 7.29). For all of these latter possibilities, however, an actual finite twist would displace the disc from the circular hole and thus destroy the joint.

41. For *continuous* twisting of this latter kind, beyond the realm of the infinitesimal, it is clearly necessary to sweep the periphery of the circular hole in a direction perpendicular to itself. This can be done in the most general way by rotating the whole periphery (the circle) about an axis which may move, but which must remain within the plane of the circle. In this way we could sweep out a tubular hole which could also be seen as the envelope traced out as a sphere of the circle's radius moved in such a way that its centre followed some twisted curve which passes through O and is normal to the plane of the circle at $O$. Refer to figure 7.04($a$) and then to figures 1.01 and 1.02.

Figure 7.03 (§ 7.39) A circular knife edged disc making $f4$ contact with a circular knife edged hole.

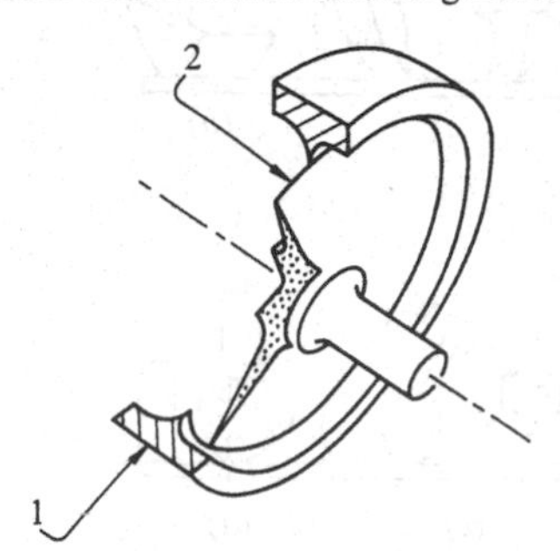


Figure 7.04($a$) (§ 7.41) The intermediate $f4$ ball-in-tube. This joint is derived from the impractical, $f4$ joint in figure 7.03.

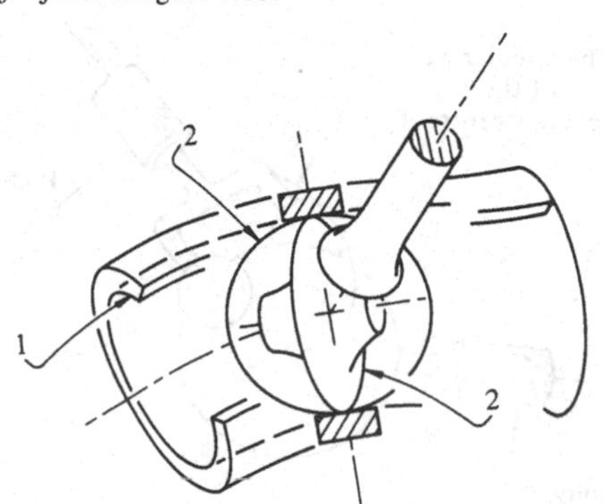

42. In figure 7.04($a$) we see, either the circular knife-edged disc at 2 free to move with 4 °F within the tubular hole at 1, or we see the whole spherical surface at 2 which is free to move in exactly the same way. If we wished to ensure the continuity of the circular line of contact between the members, *if we wished to ensure in other words the maintenance of an unchanging, continuous, circular contact patch of the kind which is becoming evident*, we would need indeed to sweep out on 2 the spherical surface shown. This act appears to be not only the natural thing to do in practical circumstances, but also the only thing. Minor interruptions could be allowed to occur upon the major surfaces however (§ 1.05).

43. We have in this way arrived at the intermediate $f4$ joint which is illustrated at $B$ in figure 1.01. Please read my remarks at § 1.12, where the nature of the contact patch at $B$ is mentioned, and where it is said that the circle of contact slides across both elements at the joint as the joint engages in its motion. It is clear in this case that the interrupted circle of button pairs at the contact patch determines the special 4-system of motion screws which is extant, and that that system of screws is fixed with respect to the contact patch, namely the sliding circle of the button pairs. Accordingly the joint is, in any one of the events in figure 7.04($a$), an intermediate $f4$ joint.

**Other intermediate, simple $f4$ joints**

44. Two n-lines can have other lines projectively dependent upon them only if they intersect (§ 10.15). A practical example of this has led to the joint in figure 7.04($a$). If the lines intersect at infinity however, there arises the special case which leads to figure 7.05. At ($a$) in figure 7.05 the knife-edged piece is straight and the flat piece is necessarily flat. It has not been necessary – or even possible without destroying one of the available four degrees of freedom – to sweep out the sphere of figure 7.04($a$) in this particular case. The flat shape of the tubular hole and the flat shape of the swept out sphere have become identical in the limiting circumstances and, unless we wish to produce a different joint by accident, the well-known $f3$, $E$-pair (which has appeared already in table 7.01), we should forbear to complete the sphere (§ 7.65).

45. We could, however, rotate the circle of infinite radius not about its own diameter, but about another line parallel with its straight circumference and at a fixed or varying finite distance from it. We could in that way produce any one of the other joints depicted in figure 7.05. At ($c$) and ($d$) the swept out prismatic pieces are of constant breadth; this provides the possibility for body closure at those joints as is suggested there. Please note however that, *if the pieces are of constant breadth, the geometry is such that the alternative straight line of contact is not only parallel with the original one but also resides in the unique plane determined by the n-lines at the original row (and at the alternative row) of the button pairs.*

46. In every case in the figures 7.05 the 4-system extant is a *second special 4-system*. See Hunt (1978) § 12.9.2, where his $h'_\alpha$ (my $p_\alpha$) is not simply finite but zero, and where his $h'_\beta$ (my $p_\beta$) remains infinite. Please refer also to my § 23.73. The hyperbolic congruence of screws of zero pitch has degenerated into all of those $\infty^2$ of lines which can be drawn in the plane marked $\alpha$ in figure 7.05($d$). The elliptical congruences of screws of finite pitch have all degenerated into those lines in the plane marked $\alpha$ which are parallel with the straight line of the contact patch. Screws of infinite pitch are everywhere parallel with the plane marked $\beta$ of the flat element of the joint. The central axis of the general 4-system has disappeared as such; it might be conceived to exist now

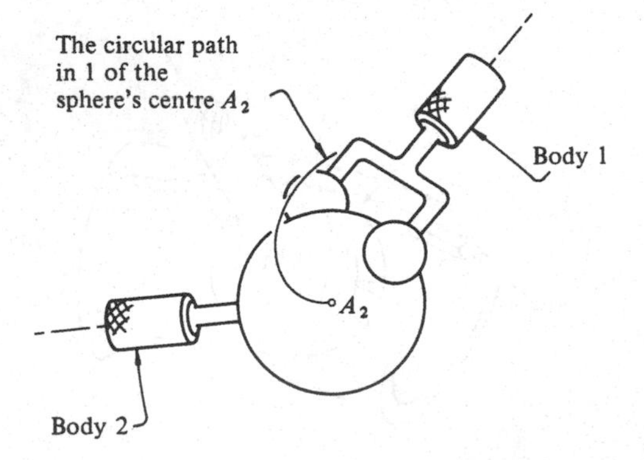


Figure 7.04($b$) (§ 7.41) Another intermediate $f4$ ball-in-tube. The details of this joint, equivalent to that in figure 7.04($a$), are mentioned at § 7.47.

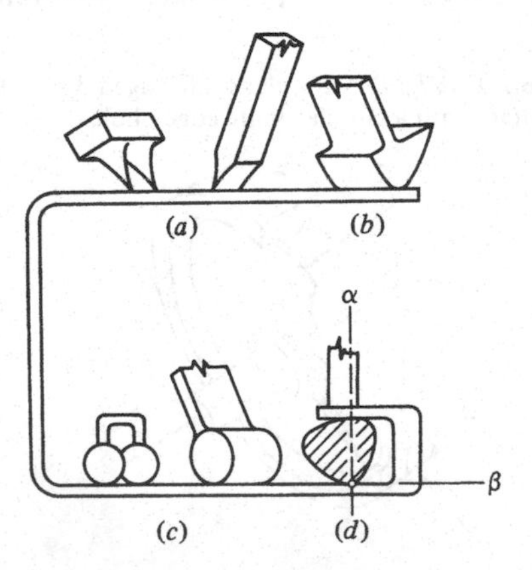


Figure 7.05 (§ 7.44) Other intermediate, simple, $f4$ joints.

along every line in $\alpha$ parallel with the line of the contact patch. The rotary symmetry of the second order about the central axis of the system has also disappeared of course, leaving only mirror-image symmetry about the plane $\alpha$. This degenerate 4-system remains invariant at all of the joints in figure 7.05, but it can be seen in every one of the cases there to move with respect to both elements of the joint. *The joints in figure 7.05 are all other intermediate, simple, f4 joints accordingly.*

47. But do we know that the joints in figure 7.05 are *all* of the other intermediate simple *f* 4 joints? That question could be a matter for study. Let the common, flat joint-element in the figures 7.05 become spherical, convex or concave towards the other, mating elements. The two prongs at (*a*) and the two balls of the dumb-bell at (*c*) will ride upon the spherical surface in a proper, 2-point manner, and the joints there will both remain intermediate *f* 4 joints. But refer to figure 7.04(*b*) and ask the question: is not this simply a special case of figure 7.04(*a*)? Hold the dumb-bell fixed, and study the locus of the centre of the sphere.

**The simple, form closed, and complex *f* 5 joint**

48. In the case of five degrees of freedom with only one point of contact there is only one contact normal and no *n*-lines except that one can be drawn which are projectively dependent upon it. The system of motion screws associated with this simple *n*-line is the quasi-general 5-system whose characteristics are summarised at § 6.25. In Hunt's terminology it is a *general 5-system*, where the pitch $p_\alpha$ of the single principal screw happens to be zero (§ 23.78). The contact patch in other words consists of a single isolated button pair and there appears to be no way in which a simple *f* 5 joint can be constructed where the screw system might, either (*a*) change, or (*b*) not move with respect to both elements of the joint. The joint with a single point of contact, accordingly, is the one and only intermediate simple *f* 5 joint.

Figure 7.06 (§ 7.49) The intermediate, simple, *f* 5 joint in (*a*) its force closed, and (*b*) its form closed, forms. There is in each case only one point of contact (and that is shown), and a single contact normal (§ 10.74).

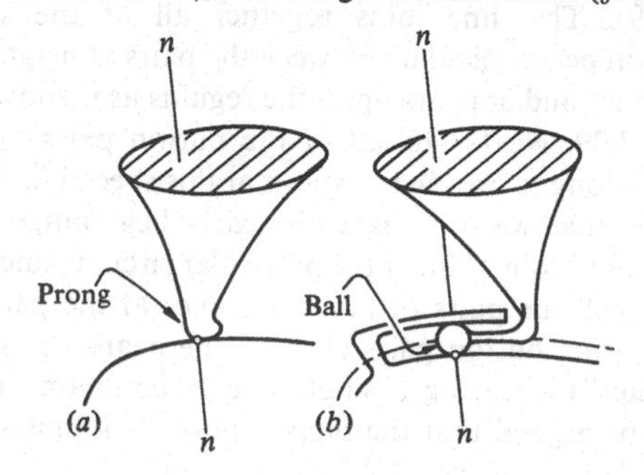


49. In the absence of force closure or magical, magnetic contact (§ 6.19, § 8.07), an *f* 5 joint might need to be seen as being form or body closed (§ 2.21). This can be envisioned in the following manner. Refer to figure 7.06. At the left a single point of contact is being maintained between a prong on a body 2 and the convex surface of some general shape of a body 1. At the right a cavern-like slot has been cut by a spherical cutter moving with its centre upon the original shape of the convex surface at the left, while a ball 2 of the size of the cutter is free to move inside the body 1 as shown. Only one active point of contact between the ball 2 and the body 1 is shown, it being clear that the alternative point of contact, which is momentarily inactive, will give, if activated, the same contact normal and thus the same quasi-general 5-system of motion screws as does the active one. *The simple f5 joint appears, therefore, necessarily and in any event, to be an intermediate, simple joint.*

50. Take a look at the joint in figure 7.07. If we mounted upon the line *x*–*x* of the centres of the balls the centreless but axi-symmetrical, quasi-general 5-system of § 6.25, we would set up the system of motion screws for the complex joint between the bodies 1 and 2. We shall return to this matter later (§ 7.78) but for the meantime please see this joint as an example of an intermediate, *complex*, *f* 5 joint.

**On the projective dependence of lines at an *f* 3 joint**

51. It might be fair to say that, having looked at the *f* 5 and the *f* 4, we might look in the same way now at

Figure 7.07 (§ 7.50) A well known joint between the bodies 1 and 2. It is a complex, intermediate, *f* 5 joint. Please refer to figure 7.10 to find that the stars of lines at the balls are stars both of *r*-lines and of *n*-lines. Please read also § 7.78.

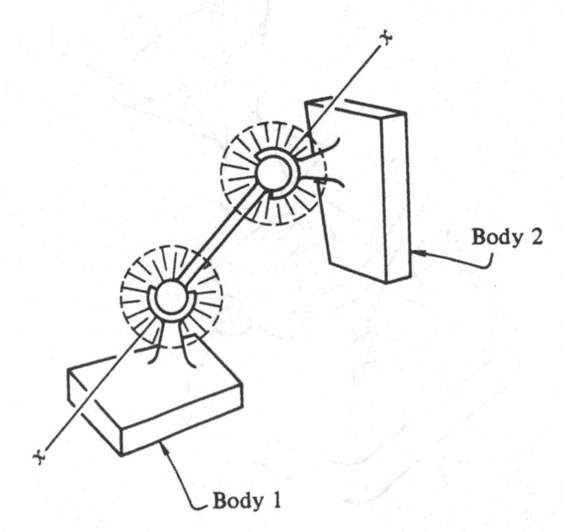

the practical $f3$ joints. For a consistent approach this is necessary; but the issues are more complicated at this middle stage of the argument and the reader might be wise to consult the relevant parts of chapter 23 and that part of chapter 6 which deals with the quasi-general 3-system before going on any further here (§6.31). The following material can almost stand up alone however. It deals directly by example with another way of examining in a physical manner the projective dependence of a fourth line upon a given three (§10.22), so perhaps the question of where and in which direction to read is open.

52. Take any three facets on a fixed body 1 and three prongs on another body 2 in such a way that none of the three $n$-lines either intersect or are parallel. Refer to the figures 3.13, look at the three lines $r$–$r$, all of which cut all of the $n$-lines, and notice the following fact. A small twist of zero pitch, of 2 relative to 1, about any one of the $r$-lines will result in a set of interrelated movements at the three prongs all of which are (*a*) small and (*b*) poorly conditioned with respect to the facets there. None of the prongs will either lift off or dig in at their facets; the overall infinitesimal motion of the body 2 will not be hindered at the facets.

53. See next in figure 7.08 that, provided we arrange it in such a place and with its facet so oriented that the $n$-line there cuts all of the mentioned $r$-lines, a fourth real prong upon a fourth real facet can be inserted into the imagined apparatus of §7.52 without disturbing the established condition that a small rotation about any one of the $r$-lines will not be hindered at the facets. In figure 7.08 this new fourth prong is the one at $D$. Due to rigidity and irregularity, it, unlike the others, has not been able to seat itself (§7.24). It should be clear that no number of small separate twists of zero pitch about any one of the mentioned $r$-lines, taken either separately or added together, will achieve a sudden, well conditioned contact at the fourth prong $D$ – unless, of course, a contact is lost at $A$ or $B$ or $C$.

**Figure 7.08 (§7.53) In this specially constructed apparatus any fourth contact normal is designed to be a member of the regulus defined by the other three. It is accordingly not possible (unless we dispense with rigidity) to achieve simultaneous contact at all four prongs (§23.23). Can a rigid four-legged table not rock on a dead flat floor?**

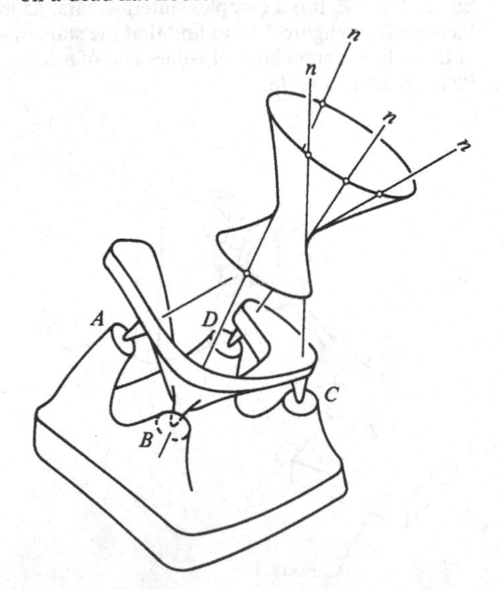


54. Go next to chapter 6, note the pitch quadric of the quasi-general 3-system illustrated in figure 6.09, and note the regulus of $n$-lines and the opposite regulus of $r$-lines upon its surface. Please see there (§6.29), or with the aid of those parts of chapter 15 where the addition of small twists is dealt with early (§15.04), or with the aid of both, (*a*) that every one of the $\infty^2$ of motion screws of the quasi-general 3-system can be 'operated' (namely, twisted about) if we wish, by judiciously choosing and suitably adding together separate amounts of small twist about any three of the $r$-lines of the whole regulus upon the pitch quadric; and accordingly, (*b*) that, *given that the contacts at A and B and C in figure 7.07 continue to be maintained, the prong at D can never be seated upon its facet, whatever we do.*

55. This demonstration has shown, in effect, what it means for a fourth line to be projectively dependent upon a given three lines which themselves are projectively independent of one another. The fact is that *if a fourth line belongs as a member of the regulus defined by the other three it is projectively dependent upon them.* This sums up precisely the 'trouble' we have been having with this particular 4-point-contact $f3$ joint. Refer again however to my remarks about the contact patch at §7.35 *et seq.*, and please see there that, in the special area of the continuity of motion at the practical joints in real machinery, this whole matter is not so much a trouble but a boon. Please see also §10.24 and §10.25.

### Towards a practical $f3$ joint

56. Complete the regulus of $n$-lines in figure 7.08 and construct upon its surface the line of striction (§11.09). This line joins together all of the short common perpendiculars between the pairs of neighboring $n$-lines and appears upon the regulus as is shown in figure 7.09. Next slide all of the button pairs of the $n$-lines along to terminal stations at this special line and observe that we now have the early beginnings of a practical $f3$ joint. Under the particular circumstances of figure 7.09 the pairs of tangent planes at the pairs of neighboring button pairs all have the nearest possible approach to bearing a single line in common and it could be argued that the biggest possible infinitesimal

twists between the bodies 2 and 1 can be obtained with this arrangement of the button pairs. It is true that the button pairs could have been assembled along some other curved line surrounding the surface of the hyperboloid, but more of this alternative arrangement later (§ 7.61).

57. Using material of § 11.13 and § 6C.09, it can be found very easily in figure 7.09 that all three of the separate small twists of the body 2 about the shown axes *x*, *y* and *z* correspond exactly with the three principal screws of the particular 3-system of motion screws which is extant. But with this most general arrangement of the regulus of interdependent *n*-lines it is clearly not possible ever to achieve anything other than an infinitesimal twist. All finite twists in figure 7.09, of whatever separate or additive composition, will result in an immediate dislocation of the joint.

58. As a first step towards achieving a joint which is capable of continuous motion here, let us choose – there are a number of different possibilities (§ 7.67) – that the set of three facets and prongs in figure 7.09 are set out in such a way that the hyperboloid becomes a circular-throated, not an elliptical-throated one. Let us arrange, in other words, a hyperboloid of revolution. In this way, the line of striction will coincide exactly with the circular throat. Once again however, and even under these restricted circumstances, the now circular line of the button pairs is necessarily 'jagged'; the tangent planes at the neighboring button pairs along the curved line of the circle still bear no line in common and a curious 'frictionful' rotation about the axis *z* is the only possibility for finite motion.

59. In order to put the button pairs into that kind of smooth harmony with one another which will permit a finite twisting about the axis *z*, we could arrange that all of the *n*-lines of the now circular regulus cut the circle at its throat perpendicularly. When we do that, which is possible by choosing the three original prongs and facets (and another one) in such a way that their respective *n*-lines *intersect*, either the circle at the throat reduces to a point, or the throat of the hyperboloid (which becomes a point) becomes displaced away from the plane of the circle; see figure 7.10.

60. Choosing either possibility, we can see next that, among this new regulus of lines (which is simply a circular cone) there is an infinity of pairs of separated *n*-lines intersecting at the same point. It follows that the whole $\infty^2$ of lines intersecting at the vertex of the cone are projectively (or linearly, chapter 10) dependent upon any selected three of them, both in the general mathematical sense of chapter 10 and in the sense of my remarks at § 7.39. The regulus has become a star at the vertex (§ 4.10) and all of the *n*-lines there have become *r*-lines also (§ 6.35).

61. Quite clearly, after a few more steps which require no mention here, *we can stumble upon the f3, S-pair of table 7.01*. In this way we might see the first of the lower joints emerging from screw theory. Although this one may appear to have emerged almost by accident, actually it was by design; the nature of this might bear a minor comment here. The process of the design of this joint, namely the process of its discovery by considering a series of more and more special arrangements of the three original contact normals, has been like climbing a tree to discover one of its extremities. This way of discovering all the possible practical joints is not rigorous and certainly not as predictable as the opposite process of discovering the trunk of a tree by

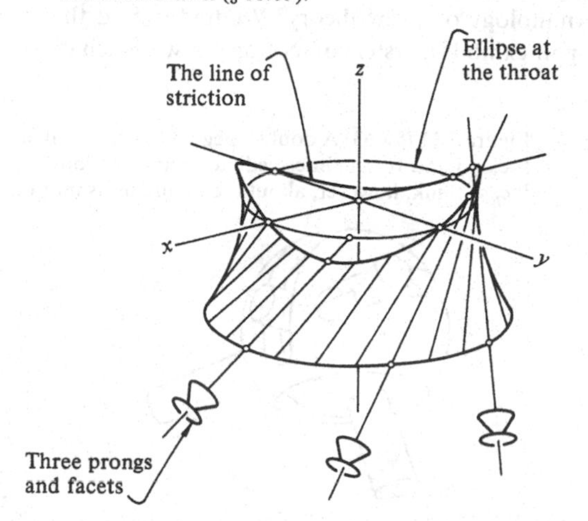


Figure 7.09 (§ 7.57) Three contact normals define a regulus. On the regulus shown here the line of striction has been drawn (§ 11.09).

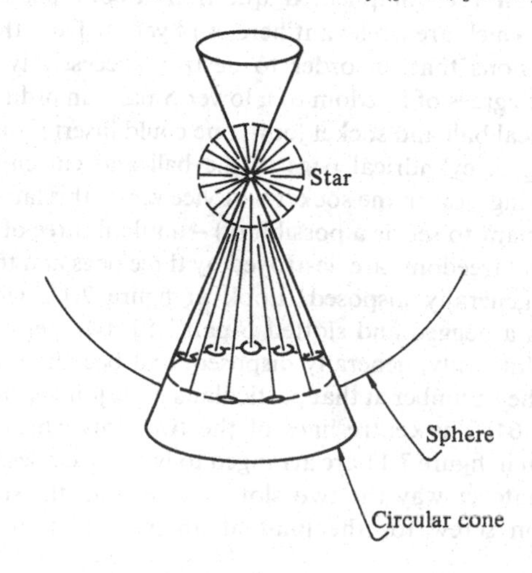


Figure 7.10 (§ 7.59) A circular cone of *n*-lines becomes a star, not only of *n*-lines but of *r*-lines too, and the *f* 3, *S*-pair emerges from screw theory (§ 7.62, § 23.62).

climbing down. But we have found a practical, simple $f3$ joint and its system of motion screws is a certain *second special 3-system*. Refer to Hunt (1978), § 12.7.2, and to my § 23.62: in order to achieve a second special 3-system we set the three pitches of the three principal screws equal to one another; here, however, all three pitches are set to zero.

**A comment in retrospect**

62. Throughout the above remarks there has been a tendency to deal, not with the particular system of motion screws which might have been the one exhibited by the particular sought-after joint, but with a particular arrangement of the one, two, or three contact normals which seemed, in the circumstances of the synthesis, to have been special. I have tended to deal, in other words, not directly with the system of motion screws which might have been possible at the joint, but with the bare bones of the system of action screws (chapter 10). The system of action screws does not exclude (§ 7.03) but includes the contact normals; the whole system of action screws becomes extant, indeed, as soon as the contact normals are set up (§ 7.69, § 10.66 *et seq.*). On each of the *n*-lines projectively dependent upon the chosen contact normals – these are the action screws of zero pitch – one may imagine a button pair; the whole system of these *n*-lines, with all of the button pairs appropriately arranged, has indeed determined the shapes of the contact patches we have already found. A similar strategy is employed from the forthcoming § 7.65 onwards. In the interim however, let us break to look at a relevant but digressionary matter.

**The pegged and slotted, spherical, simple joints**

63. Please recall the discussions about the lockings in of passive freedoms distributed throughout the early parts of chapter 2 and refer to § 2.32. One can leave aside all the complicated questions about passivity itself, which are irrelevant here, and yet see from those discussions that, in order to destroy successively the three degrees of freedom of a lower *S*-pair, an ordinary spherical ball and socket joint, one could insert radially arranged, cylindrical pegs in the ball and cut curved matching slots in the socket – or vice versa (this latter is important to see as a possibility) – until all three of the original freedoms are destroyed by three pegs and three slots generally disposed. Look at figure 7.11, which shows a pegged and slotted *S*-pair with two pegs and two slots only, generally disposed, and become aware that the *f*-number at that particular simple joint is unity.

64. The centre lines of the two slots which are shown in figure 7.11 are arranged to be great circles but, in whatever way the two slots may be cut, the single motion screw for the joint at an instant – it passes through the centre of the ball and it is of zero pitch – is movable (*a*) with respect to the ball and its pegs namely the member 1, and (*b*) with respect to the socket and its slots, namely the member 2. The joint in figure 7.11 and all such joints with two slots are, in general, intermediate simple $f1$ joints accordingly. The joints are easily constructed and thus can be seen to have their practical and useful applications in machinery. With one peg missing the joints would become $f2$ joints with one peg only, and such joints are intermediate joints as well; please refer to § 7.80 and see figure 7.18(*a*). In figure 7.18(*b*) we have, in effect, an intermediate simple $f2$ joint where a single slot in the spherical socket is a non-great-circle slot. It is possible to arrange a lower joint as follows. If we were to allow the two great-circle slots in figure 7.11 to coalesce and thus allow the two pegs to run in the same great-circle slot, the single available motion screw would become fixed with respect to both of the members at the joint and we would have arrived at the *R*-pair; it is the *R*-pair already listed in table 7.01.

**Other practical, simple, $f3$ joints**

65. We have seen that, if we seek a surface to surface contact at a practical $f3$ joint, there is no way in which a set of three generally disposed contact normals can become the basis for it (§ 7.57); and we appear to have shown that the only way to achieve such a joint is to arrange that the three contact normals intersect one another at a single point (§ 7.62). When we set the three contact normals to intersect at infinity we arrive at the special case, the *E*-pair, and that is listed in table 7.01. From the geometrical point of view one begins to wonder, next, about our criteria for judgement when we list the *S*-pair and the *E*-pair as being separate. Are our criteria based upon Euclidean geometry and its logic; is the question a mere matter of machine tools and our methods of machining; can the matter be clarified by the terminology of screw theory? We find indeed that the *E*-pair exhibits a system of motion screws which is not a

Figure 7.11 (§ 7.63) A double, peg and slotted, ball and socket joint. It is an intermediate, simple, $f1$ joint. Please think, however, about the nature of its motion.

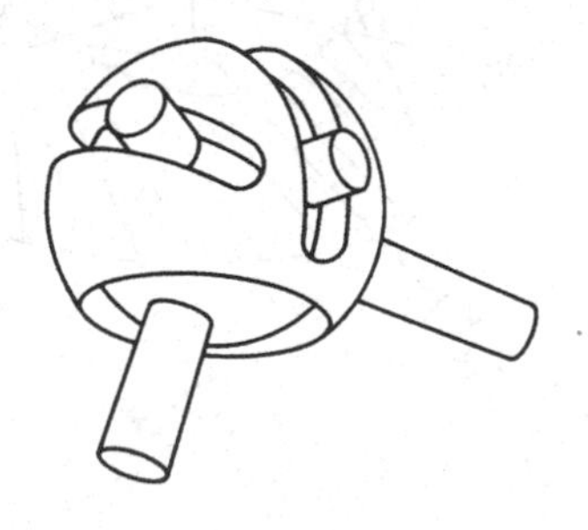

second (§ 7.61), but a *fifth special 3-system* and I propose that this clear difference between the two joints be accepted now as a resolution. Please refer to § 23.65.

66. We have spoken about the two lower $f3$ joints; but what do we know about the intermediate and the higher $f3$ joints? It appears to be true – I speak geometrically now – that each line of any set of three lines will be projectively independent of the other two unless the three lines are coplanar and they intersect, the common point of intersection occurring at some local point or at infinity. We discovered some peculiar things about a coplanar, intersecting set of three lines at § 7.23 *et seq.*; that same matter of the projective or the linear dependence of three lines is discussed in further detail at § 10.15, § 13.03, and elsewhere. *What we have not done, however, is to ask about the ways in which we can be special with the layout of three contact normals in space without destroying the mutual independence of the three; we might try to do this now.* The object of the exercise is to try to discover whether or not there may be available not only those two lower $f3$ joints (the *S*-pair and the *E*-pair) but also, perhaps, some intermediate $f3$ joints, which might be available and useful for practical machinery too. Please read as a preliminary § 10.11 and § 11.04 about the generation of a regulus by means of any three projectively independent lines, and look at figure 6.05. Recall also that for only two projectively independent lines we asked the corresponding questions at § 7.23.

67. Removing from consideration for the meanwhile the occurrence of right angularities because they are in general irrelevant in the area of line geometry, it would appear that at least some of the ways in which we can be special with three projectively or linearly independent lines may be listed as follows: (*a*) the three lines, while noncoplanar, might intersect at a single point – note the *S*-pair at § 7.61; (*b*) the three lines, while noncoplanar, might be parallel with one another – note the *E*-pair at § 7.65; (*c*) *there might be one intersection*, involving a pair of the lines, or *there might be two intersections*, one of the lines intersecting both of the others; (*d*) *there might be a single parallelism*, involving, either a pair of the lines, or one of the lines and the plane of intersection of the other two; (*e*) *there might be two parallelisms*, first between a pair of the lines, and next between the remaining line and the plane of the pair; (*f*) *there might be three intersections*, the three lines being necessarily coplanar in that case; (*g*) *the three lines might reside in parallel planes*; or (*h*) *various symmetries might occur*, the lines might for example be so arranged that they are generators of a regulus which is circular.

68. Keeping in mind that we are laying out a set of three contact normals here and that, once laid out, there will spring into being among them a single infinity (a regulus) of *n*-lines projectively or linearly dependent upon the set (§ 7.69), look now at the figures 7.12. From (*a*) to (*h*) they are a series of sketches of degenerate or otherwise special reguli corresponding to the items (*a*) to (*h*) at § 7.67. They are all reguli of *n*-lines, namely reguli of action screws of zero pitch. We have produced the following: (*a*) a star; (*b*) a parallel field; (*c*) a regulus upon two intersecting planes with two polar points; (*d*) a regulus upon two intersecting planes with only one polar point; (*e*) a regulus upon a pair of parallel planes; (*f*) a planar field; (*g*) a regulus upon a parabolic hyperboloid; (*h*) a regulus upon a hyperboloid of revolution. Beginning at § 11.14, the reader will find a more ordered account of the special and degenerate reguli.

69. Any set of three projectively independent screws defines a 3-system, in the manner suggested for example at § 3.36 and in figure 3.08 (§ 7.54). Three projectively independent screws of zero pitch, all proposed contact normals and all, thereby, action screws of zero pitch, have been used to define the above reguli. It

Figure 7.12 (§ 7.68) The various special ways of arranging a set of three proposed, contact normals. Each way produces a degenerate or a special regulus of *n*-lines (§ 9.38, § 11.14).

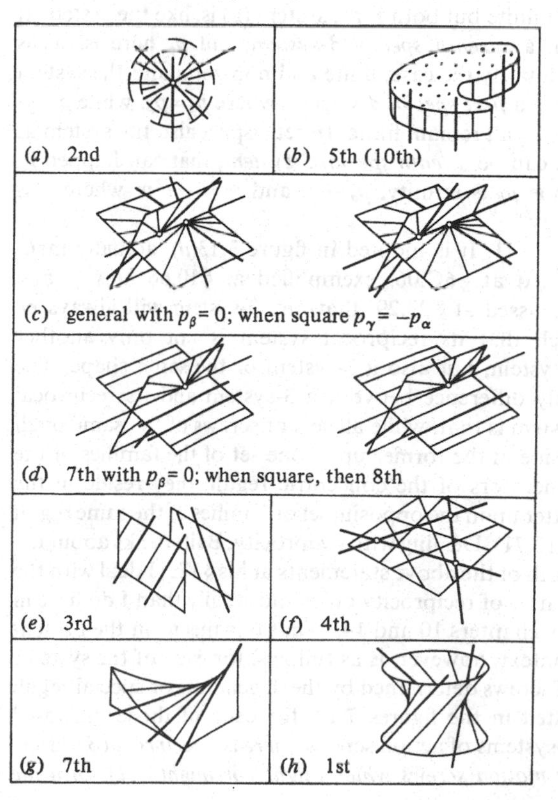

can be said, therefore, that each of the above reguli defines its own 3-system (§ 6.39). The reguli are, indeed, when seen as hyperboloidal surfaces, the pitch quadrics of their respective action systems (§ 6.35). The 3-systems thus defined are mostly special in one or another way because the reguli are mostly special; they are all systems, do not forget, not of motion screws as discussed for example at § 6.39, but of action screws.

70. I shall now attempt, according to the scheme of classification proposed and adopted by Hunt (1978) and summarised by me at § 23.61 *et seq.*, to identify the 3-systems: system (*a*) is a *second special 3-system*, where all of the pitches $p$ of the principal screws, $p_\alpha$, $p_\beta$ and $p_\gamma$, are not simply equal and finite but equal and zero; system (*b*) is a *fifth special 3-system*, where $p_\alpha$ and $p_\beta$ are both infinity, while $p_\gamma$ is not merely finite but zero; system (*c*) is a *general 3-system*, where $p_\beta$ happens to be zero and where, if the planes are at right angles to one another, $p_\gamma = -p_\alpha$; system (*d*) *is a seventh special 3-system*, this occurring when $p_\alpha \rightarrow \infty$ and $p_\gamma = -k^2 p_\alpha$, while $p_\beta$ remains finite, but in our case here $p_\beta$ is not merely finite but zero and, if the two planes are at right angles to one another, $k$ is unity and the system is an *eighth special 3-system*; system (*e*) is a *third special 3-system*, where $p_\alpha$ is infinity, while $p_\beta$ and $p_\gamma$ (unequal) remain finite; system (*f*) is a *fourth special 3-system*, where $p_\alpha$ is infinity, while $p_\beta$ and $p_\gamma$ are not merely equal and finite but both zero; system (*g*) is, like the system at (*d*), a *seventh special 3-system*, but $p_\beta$ here is, as is ordinarily the case, finite and non-zero; and the system (*h*) is a *first special 3-system*, where $p_\alpha = p_\beta$, while $p_\alpha$, $p_\beta$ and $p_\gamma$ all remain finite. In retrospect also the system at (*b*) can be a *tenth special 3-system*; that can happen if, while $p_\beta$ is infinity, $p_\alpha \rightarrow \infty$ and $p_\gamma = -k^2 p_\alpha$ where $k$ is unity.

71. It is mooted in figure 3.13(*a*), actually mentioned at § 6C.06, exemplified at § 10.66 *et seq.* and discussed at § 23.29, that any 3-system will always be such that its reciprocal system is not only another 3-system, but also a 3-system of the same shape. The only difference between a 3-system and its reciprocal system is that, while all sets of screws of the same pitch reside in the former upon one set of the families of the generators of the concentric reguli, they reside in the latter upon the opposite set of families of the same reguli (§ 10.71). Dealing with reciprocity, Ball speaks about the truth of the above statements at his § 172; I deal with the matter of reciprocity much more fully than I do here in my chapters 10 and 14. What this means in the present context, however, is as follows: for each of the systems of screws determined by the degenerate or special reguli listed in the figures 7.12, for each of those proposed 3-systems of action screws, *there is a reciprocal 3-system of motion screws which might (or might not) form the available system of motion screws at some simple, useful f3 joint yet to be discovered.*

**Some of the simple, f3 joints that can exist**

72. In table 7.03 I show what I see to be a number of practical possibilities for the simple $f3$ joints. The nearer the sketched arrangements are to the left hand side of the table the more interesting they may be presumed to be; for, as I have said, their analyses there are easier (§ 7.13). Please note the column mentioning the particular special 3-system that springs from the regulus of action screws of zero pitch which is extant, and find that the only special 3-systems missing from the table are the sixth and the ninth. Notice also, however, that these are the only two which contain no screws of zero pitch. I would like to comment on a few of the joints which appear at the table, and the commentaries follow.

73. The only intermediate, simple joint which I have found for the table is the one I have called the 2-ball

Table 7.03 (§ 7.72) Summarising the argument since § 7.65, and showing what I can see so far of the $f3$ simple joints. The table is not complete. Nor is it properly organised, for many questions remain unanswered here.

| Which special 3-system of Hunt | | Lower simple | Intermediate simple | Higher simple |
|---|---|---|---|---|
| (*a*) Star | 2 |  Spherical ball | | |
| (*b*) Parallel field | 5<br>10 |  Ebene flat | | |
| (*c*) Two planes with two polar points | G | | |  Cylindriconic |
| (*d*) Two planes with one polar point | 7<br>8 | | 2-ball cylindric  | 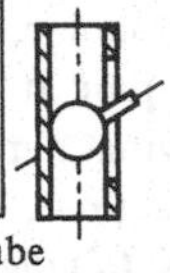 Slotted tube |
| (*e*) Two parallel planes | 3<br>7 | | | 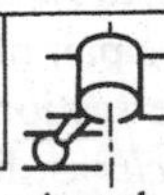 Plunging universal |
| (*f*) Planar field | 4 | | | |
| (*g*) Parabolic hyperboloid | 7 | | | 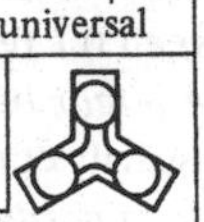 3-ball polycentric |
| (*h*) Circular hyperboloid | 1 | Nil | Nil | 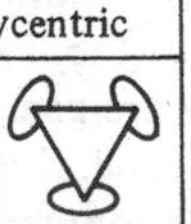 Tripodic (?) joint |

cylindric joint appearing at (*d*). Illustrated more clearly in figure 7.13(*a*), it is a joint easy to construct, easy to analyse, and possibly of use. Its fixed, special regulus of *n*-lines for the action system is shown at the left in figure 7.13(*b*). The said regulus – it resides upon a pair of intersecting planes – is such that the two planes intersect perpendicularly, and there is only one polar point. This is the pitch quadric for an eighth special 3-system where $p_\beta$ is not simply finite but zero. The line *n–n* is that unique line of this eighth special 3-system of action screws which carries screws of all pitches; it is that line along which the joint may sustain a wrench of any pitch without moving. Everywhere parallel with *n–n* are action screws of infinite pitch and along any one of these the joint may sustain a pure couple without moving. The line *r–r* at the right in figure 7.13(*b*) is the corresponding unique line in the reciprocal system of motion screws; about *r–r* the joint may suffer a rate of twisting of any pitch. Everywhere parallel with *r–r* are motion screws of infinite pitch; in the direction of any one of these the member 2 of the joint may suffer a pure translation relative to member 1. It is not necessary for the intermediate status of this joint that the points of contact at *A* and *B* be equidistant from the central axis of 1; it is only necessary that the bored holes be cylindrical and coaxial. Please note the similar but higher joint at (*c*) in table 7.03 whose non-cylindrical surfaces have rendered it a higher joint. There are of course in the two 3-systems mentioned screws other than those which are mentioned here. There is an $\infty^1$ of screws each of finite non-zero pitch and, for each pitch, there is a regulus of screws upon a parabolic hyperboloid. The single infinity of parabolic hyperboloids share a common vertex at the centre of the ball. Please refer to § 23.68.

Figure 7.13(*a*) (§ 7.73) The intermediate, 2-ball cylindric, *f*3 joint. The bigger ball accurately fits the cylindrical hole and the smaller ball accurately fits the annular hole. The contact patch, which remains of fixed shape, consists of (*a*) the great circle at *A* where the bigger ball makes line contact with 1, and (*b*) the two isolated points where the smaller ball makes contact not only with the outer wall of the annular hole at *B* but also diametrically opposite at the inner wall (§ 7.37). Given not only this accuracy of construction but also rigidity and the necessary small clearances, one can conceive of the three required, active points of contact as being, for instance, one alone at *B* and two others coalesced at *A* (§ 1.53, § 6.03, § 6.63).

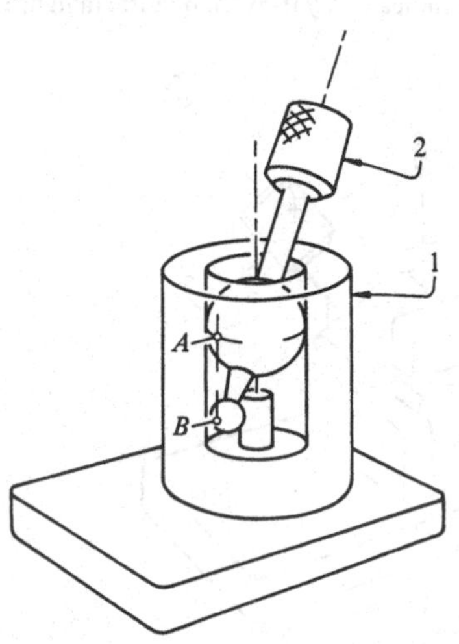


74. The higher *f*3 joint at (*f*) in table 7.03 and illustrated more clearly in figure 7.14(*a*) is interesting. The joint is a relatively well-known joint in ordinary engineering practice, being used sometimes, at slow speed, as an approximate constant-velocity coupling which is capable of 'plunging'. The mechanics of this joint is complicated. It is not well understood, but please note that when the two members are arranged co-axially the system of action screws exhibited is a fourth special 3-system (§ 23.64). When displaced however, the three contact normals exist no longer coplanar but merely in parallel planes, the regulus of *n*-lines becoming of such a shape that it resides upon a parabolic hyperboloid (§ 11.13, § 11.15) and the system defined becomes, at least in general, a seventh special 3-system (§ 23.67, § 23.68). In figure 7.14(*b*) is sketched first three contact normals at *A*, *B* and *C* displaced, all of which are horizontal in the figure, and second, some of the available *r*-lines, each of which cuts all of the *n*-lines and all of which reside, as we can see, upon the same parabolic hyperboloid. Please refer to § 11.05 for some information about the construction in figure 7.14(*b*), being aware at the same time that the matter here is not

Figure 7.13(*b*) (§ 7.73) The pitch quadrics of the 3-systems of screws for action and motion respectively at the joint that is illustrated in figure 7.13(*a*). The systems are both eighth special 3-systems with $p_\beta$ = zero, and the lines marked *n–n* and *r–r* respectively are those unique axes of the systems that contain screws of all pitches.

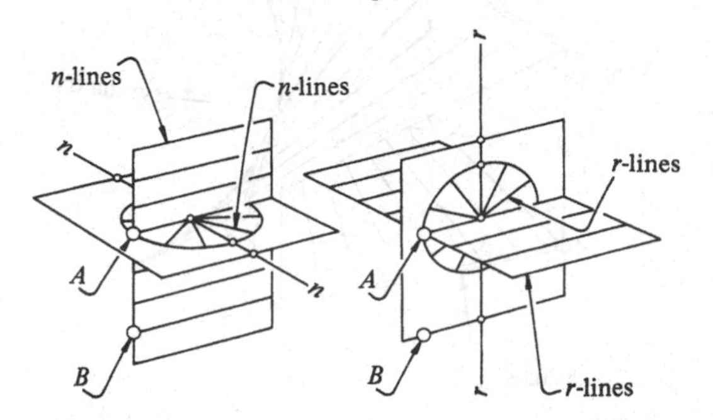

Figure 7.14(*a*) (§7.74) The 3-ball polycentric, a higher, simple *f* 3 joint. It derives – first flat and then tilted – from the fourth and then the seventh special 3-systems of Hunt (§23.61).

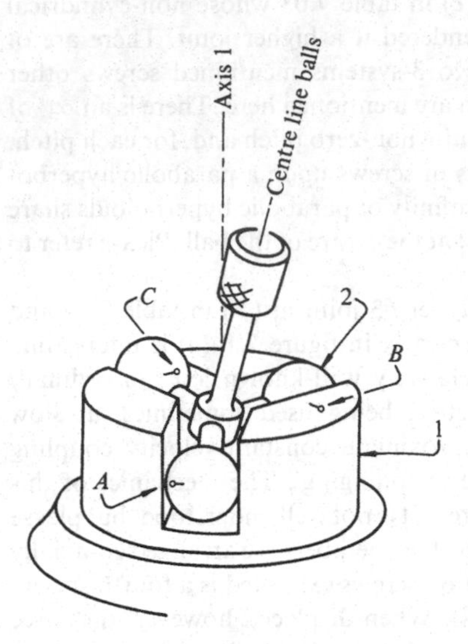


Figure 7.14(*b*) (§7.74) Based upon the three *n*-lines drawn at the three points of contact *A*, *B*, and *C* at the joint in figure 7.14(*a*), we see here an approximate picture of the pitch quadric for the seventh special 3-system of motion screws that inhabits that joint. The regulus resides upon a parabolic hyperboloid whose axis of symmetry is somewhere approximately as shown.

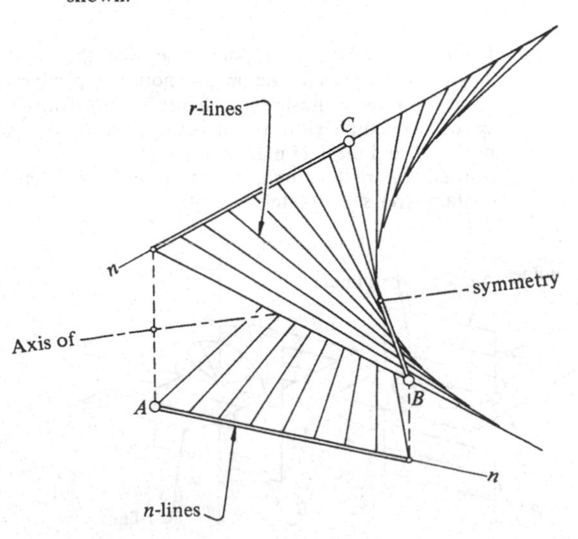


symmetrical and that the resulting geometry is very complicated. I offer here no clear explanation. The joint is a polycentric joint; I mean by this that the two shaft axes do not in general intersect. A fundamental geometrical fact is that an equilateral triangle joining the centres of the three balls on 2 can only move in such a way that its three apices remain within three planes which intersect in the same line (the central axis of 1) and are fixed in 1.

75. Of interest, perhaps, is the new joint at (*e*) in table 7.03. It is drawn more carefully in figure 7.15(*a*). Synthesising by the method of button pairs and directly from a special case of the regulus in figure 7.12(*e*), one can get the in-line arrangement of this joint. When the centre lines of the members are displaced from being collinear however, as has been done in figure 7.15(*a*), they become, in general, skew. The joint displays a seventh special 3-system of action screws and thus of motion screws as well (§7.71). The regulus of *n*-lines illustrated in figure 7.15(*b*) constitutes the pitch quadric for the system of action screws and this, as one can see, is not invariant as the joint engages in its action. The joint is, accordingly, a higher *f* 3 simple joint. It is, indeed, the middle joint appearing in photograph 7 and that physical model of it proves the joint to be remarkably smooth in both its action and motion. It could well operate as a plunging universal.

76. I should remark about the well known ball and slotted tube. Appearing at (*d*) in table 7.03 and drawn more carefully in figure 7.16(*a*), it is well known among the simple joints. Unless the ball stem is parallel

Figure 7.15(*a*) (§7.75) Another higher, simple *f* 3 joint. It is a 'plunging universal' and, when set with the central axes of its two members collinear, it displays third special 3-systems of screws. Otherwise it displays seventh special 3-systems, and one of these (the action system) is indicated by its pitch quadric in figure 7.15(*b*).

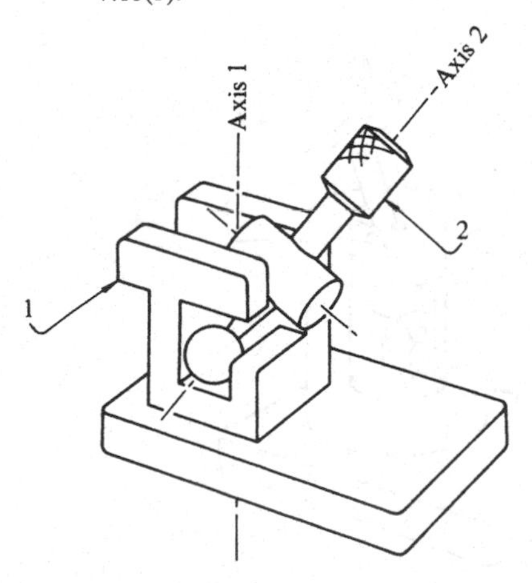

with the centre line of the tube (in which case the screw system at the joint is an eighth special 3-system), the system is a seventh special 3-system. If the stem is perpendicular to the centre line of the tube the system of action screws collapses into a fourth special 3-system which contains a whole planar field of $n$-lines. See figure 7.16($b$). In any event the screw system (and its reciprocal system) keeps changing as the joint engages in its action (and its motion), so the joint is a higher, $f3$ joint. Please note an alternative version of this joint which is drawn among the other things in figure 7.16($b$). Many joints of all kinds can be represented exclusively by balls, V-grooves, flats, etc., as this one has been, and many authors, for example Kraus (1952, 1954), make extensive use of this technique.

Figure 7.15($b$) (§ 7.75) The pitch quadric of $n$-lines for the system of action screws within the joint in figure 7.15($a$). Discovery of the corresponding sets of $r$-lines, and of the remainder of the reciprocal, seventh special 3-system of motion screws, is left as an exercise for the reader. There is no one line here containing screws of every pitch, as there was for example in figure 7.13($b$).

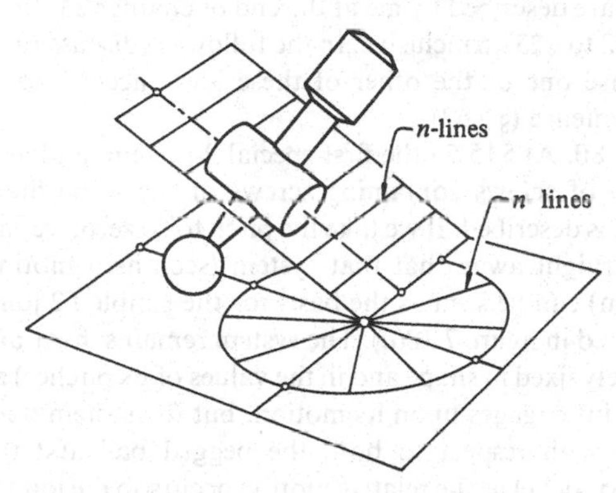


Figure 7.16($a$) (§ 7.76) Two general configurations of the ball and slotted tube, a higher, simple $f3$ joint. In general both systems of screws are seventh special 3-systems, and they change, but in special configurations the joint can display either an eighth or a fourth special 3-system. See figure 7.16($b$).

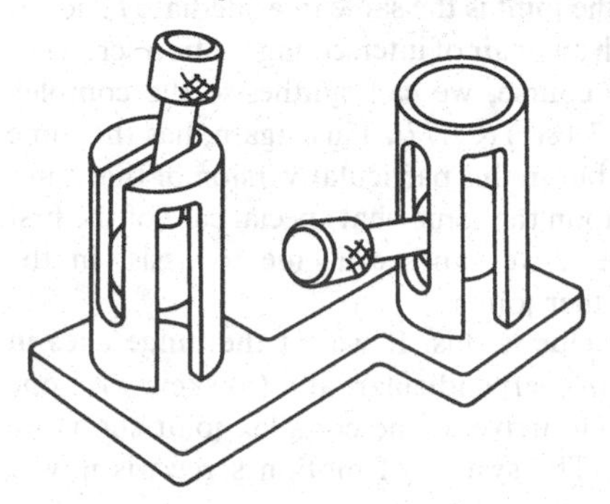

**The reciprocal and the complex $f3$ joints**

77. We have seen that, for each set of $n$-lines in a 3-system of action screws, there is a reciprocal set of $r$-lines in the 3-system of motion screws. This latter can be used as a basis for synthesising complex joints using hinges only. The hinges, of course, must be mounted in series (§ 3.41 *et seq.*, § 14.37 *et seq.*), but the principles of selection of the three hinge axes are substantially the same as they are for the three contact normals (§ 3.43). Take for example the reciprocal, complex, $f3$ joint which corresponds to the 3-ball, polycentric joint in figure 7.14($a$). A representative sample appears in figure 7.17. Note that, when in-line, the joint displays the same fourth special 3-system of screws, but that, when angled, the hinges no longer occupy parallel planes; go to ($c$) in table 7.03. I do not want to categorize endlessly here but the reader may try for an exercise the reciprocal complex joint for the simple joint in figure 7.15($a$). Refer for some parallel reading about 'reciprocal connections' to Hunt § 2.6.

78. On the question of complex joints in general I would like to make a few remarks. Please recall the contents of the passages from § 2.37 to § 2.41 inclusive and refer back to § 7.50. The situation in figure 7.07 is characteristic of the superimposition in series of two simple joints and their screw systems to make one complex joint and its screw system. Please explore by the principles outlined at § 3.09 the totality of the $r$-lines produced by 'adding' the two stars of $r$-lines in figure

Figure 7.16($b$) (§ 7.76) Reguli of $n$-lines for the ball and slotted tube in ($a$) its parallel, and ($b$) its perpendicular, configurations. At ($b$) I have drawn, not only the relevant regulus, but also an alternative version of the joint itself.

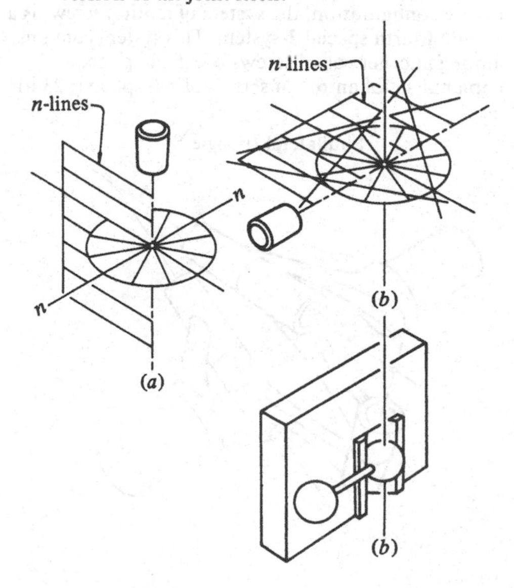

7.07; you will find a special linear complex of *r*-lines (§ 9.41). This should make it easier to see how it is in this particular case that 3 'plus' 3 is not 6 but 5, for the discovered linear complex of *r*-lines is the bare bones (the screws of zero pitch) of the particular 5-system of motion screws which is extant at the complex joint between the members 1 and 2 (§ 2.38). If you will think now, and still in figure 7.07, not in terms of the stars of *r*-lines there but in terms of the congruent stars of *n*-lines (§ 7.60), you will find that it is not the addition but the 'subtraction' of the two stars which is important. Please read at § 11.26 about *intersection*; it is explained there not in terms of stars but in terms of congruences. By this process – by the taking of an intersection between, or by a subtraction of, the two stars – you will discover the single action screw of zero pitch, the single *n*-line joining the centres of the two balls and common to both of the two stars, which constitutes the reciprocal 1-system of action screws at the complex joint. You will find, indeed, why it is in this particular case that 3 'minus' 3 is not zero but unity. Now study the joint in figure 2.02(*e*) or, better, the non-overconstrained joint in figure 1.03(*h*) and find that, when we have a parallel superimposition of two simple joints and their screw systems, the above procedures need to be reversed. These simple examples will go to show, I hope, that the mentioned matters of adding and subtracting screw systems in the arena of complex joints can become quite complicated, but not impossible. Refer to my discussion about the *f*-numbers at a complicated complex joint beginning at § 23.04. Please see that, although the joint there is complicated, a rational discussion of it and its instantaneous capacity for motion and action is a relatively straightforward matter. Refer also to [2]. Please observe that, *whereas addition in the above context is the nomination of all those screws which are linearly dependent upon all of the screws which are being added, subtraction is the nomination of all those screws which are common to both of the systems of screws being subtracted.*

Figure 7.17 (§ 7.77) This joint is a reciprocal of the 3-ball polycentric joint in figure 7.14(*a*). In its shown, in-line configuration, the system of motion screws is a certain fourth special 3-system. This system contains, among another $\infty^2$ of screws of infinite pitch, a coplanar set of an $\infty^2$ of screws of zero pitch (§ 23.61).

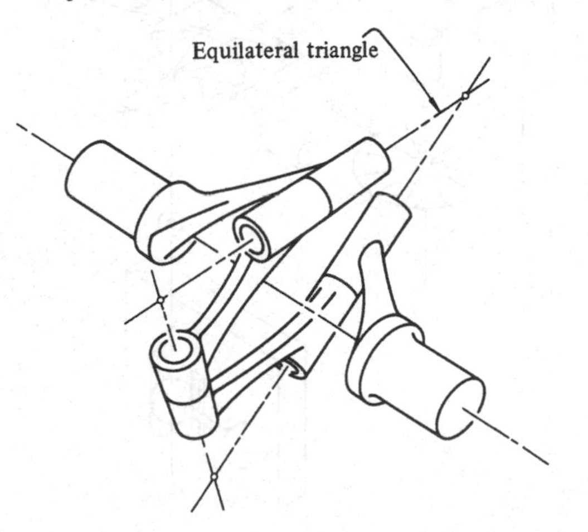


**The population of the simple *f* 2 joints**

79. An *f* 2 joint displays a 4-system of action screws which can always be seen as based upon the four contact normals at the four points of contact (§ 6.52). For the simple *f* 2 joints we could think about the ways to be special with four contact normals and thus about the various special forms of the 4-system (§ 23.71). Alternatively we could think in terms of the 2-system of motion screws, the cylindroid (chapter 15), and consider its general, special and degenerate forms. There are five special 2-systems, according to the categories of Hunt (1978), and these are described by me at the end of chapter 15, from § 15.50 to § 15.55 inclusive. There are, correspondingly, five special forms of the 4-system, and these are described by me at the end of chapter 23, from § 23.72 to § 23.76 inclusive. In the following discussion, I will use one or the other of these ideas according to convenience (§ 7.62).

80. At § 15.50 the first special 2-system, a planar pencil of screws containing screws of the same finite pitch, is described. If we take the pitch to be zero, we can see straight away that that system (seen as a motion system) can be seen as the basis for the simple *f* 2 joint depicted in figure 7.18(*a*). The system remains invariant (namely fixed in shape and in the values of its pitches) as the joint engages upon its motion, but the system does move with respect to both the pegged ball and the slotted socket as the relative motion occurs so the joint is not a lower but an intermediate *f* 2 joint (§ 7.15). Another joint with the same screw system is the one in figure 7.18(*b*). The only difference between this and the one at (*a*) is that its two 'defining' screws of zero pitch are not perpendicular; the capacity for relative motion is identical and the joint is the same intermediate *f* 2 joint. Beginning with any pair of intersecting motion screws of zero pitch, of course, we can synthesise the complex joint in figure 7.18(*c*) (§ 7.77). This, again, has the same screw system; but in this particular version of this joint, based as it is upon the somewhat special case of the first special 2-system, we can see a clue to assist in the discovery of other joints.

81. The clue is this. If we set the hinge axes in figure 7.18(*c*) (*a*) perpendicular, and (*b*) skew with one another, we can arrive at the complex joint shown in figure 7.19(*a*). The system of motion screws is now a

cylindroid of finite length, with the special feature that its two generators of zero pitch are at the two extremities, at $z = \pm B$ (§15.49). Next, by thinking in terms of Realeaux expansion and button pairs, we can soon produce the so called, toroidal joint which is pictured in figure 7.19(*b*). This joint is a well-known 'special' $f2$ joint, and yet, as we have seen, its system of motion screws is not one of Hunt's special 2-systems. The reason for this, of course, is that Hunt, with his carefully chosen classifications, does not regard the number zero as being qualitatively different from any other finite number. At my chapter 6 I have chosen to differ in this respect. Thus I found there my need for the term 'quasi-general'. For obvious and similar reasons I have found my need here for the term *quasi-special*; the caption to figure 7.19 is written accordingly.

82. At §15.53 one can see described the second special 2-system. One could choose there the finite pitch $p_\alpha$ to be zero and arrive at the simple $f2$ joint shown in figure 7.20(*a*). The reciprocal 4-system of action screws could be generated from a group of four contact normals arranged as follows: three are non-coplanar but parallel, while the fourth is set perpendicular to the three. Another version of the same joint is shown in figure 7.20(*b*); provided contact is maintained at all four points of contact, this joint is the same as the one in figure 7.20(*a*). See also figure 8.03 at §8.15. Also described at §15.53 is the third special 2-system. Seen as a system of motion screws for a joint, it is that system of screws which would underlie a joint with the capacity to restrict the motion of one of its members in such a way that it could only move in a planar manner with a fixed orientation with respect to the other member. I can not see a simple $f2$ joint for this. The only uncomplicated alternatives I can see are (*a*) the well-known complex, $f2$, Oldham joint or coupling, and (*b*) the well-known planar, double-parallelogram linkage used for example in draughting machines and adjustable desk lamps.

83. At §15.54 the fourth special 2-system is described and an account of its reciprocal system (the fourth special 4-system) will be found at §23.75. Thinking in terms of motion screws and the possibilities for a complex joint, we can immediately see that the complex joint in figure 7.21(*a*) displays a fourth special 2-system of motion screws. Note the angle $\zeta$ (§15.52) and locate the axes for $x$, $y$, and $z$. The pitch of the screw joint at $A$ may take any finite value, including zero, so the screw joint at $A$ might well be seen as a revolute. Both screw systems in figure 7.21(*a*) remain invariant as the joint engages in its motion, but please observe that they both

Figure 7.18 (§7.80) Various intermediate $f2$ joints. While the first two at (*a*) and (*b*) are simple joints, the third at (*c*) is complex. They all display, however, the same particular, first special 2-system of motion screws wherein the pitch that is characteristic of the planar pencil of screws of the same pitch is not simply finite but zero. Please imagine for the sake of an exercise that the joint at (*c*) might be equipped, not with a pair of screw joints of zero pitch, but with a pair of screw joints of the same non-zero pitch, either of the same or of opposite hand (§15.49, §15.50).

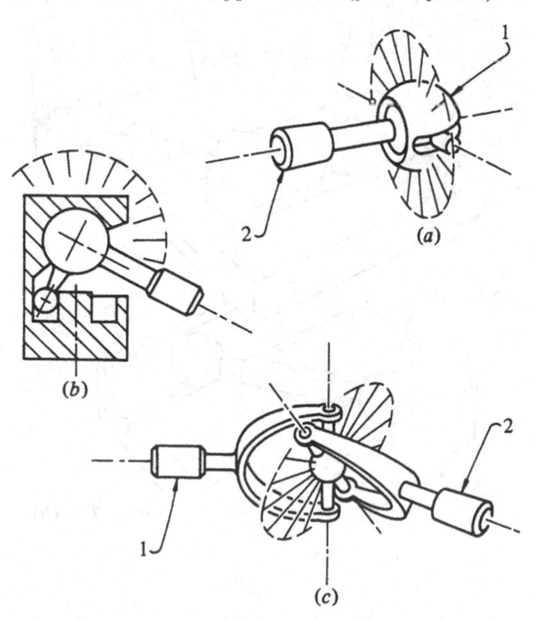


Figures 7.19 (§7.81) Development and details of the $f2$ toroidal joint. In the sense that its 2-system of motion screws (the cylindroid) has its two motion screws of zero pitch mutually perpendicular (§15.49), its system of motion screws is quasi-special. Please refer to the caption to the figures 7.22 for some further remarks about the mechanics of this ubiquitous joint.

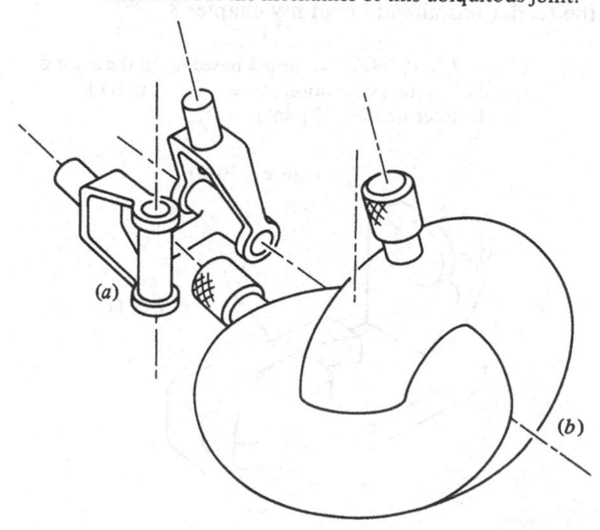

move (as a movement of the joint occurs) with respect to both of the links 1 and 2; the systems move by pure translation in the direction of the axis for $z$. The joint is accordingly an intermediate, complex, $f2$ joint. To see an intermediate, simple, $f2$ joint displaying the same fourth special 2-system of motion screws, one could set the pitch of the screw joint at $A$ to zero, then, by the method of button pairs, arrive at the joint in figure 7.21($b$). Please note in figure 7.21($b$) that the four contact normals intersect in pairs and that these pairs define planar pencils of action screws of zero pitch in two parallel planes. Thus the $\infty^2$ of screws of zero pitch in the $\infty^1$ of parallel planes of the fourth special 4-system are also defined (§ 23.75).

84. Now without loss of generality we can imagine the joint in figure 7.21($b$) to be constructed, not with two straight and parallel V-grooves, but with two straight and parallel tubes; refer to figures 8.05 and 8.06 for an impression of what I mean by this. If, next, and maintaining parallelism of the tubes, we allow the angle $\zeta$ in figure 7.21($b$) to diminish towards zero, we shall see that the tubes will coalesce in the limit into one straight tube. The double-ball-ended rod (the dumb-bell) will then be free to screw, within the single tube, with any pitch about the central axis of the tube. *With only one more step we can next arrive at the f2 cylindric joint, the C-pair, which is illustrated in table 7.01 and in figure 8.02.* It is fascinating to become aware in this way that this lower $f2$ joint, the $C$-pair of table 7.01, displays a fifth special 2-system of motion screws. The local origin of the fifth special 2-system is to be found, as we know (§ 15.55), at the extreme generator of an infinitely long cylindroid; there, as we also know, $\zeta$ is zero. Please refer to § 15.55, to § 23.76 and, for a fuller understanding of the motion and the action systems existing at this $f2$ cylindric joint, to § 8.13. I take it for granted here that the reader has already read my chapter 8.

85. Having looked in a cursory way at all of the special 2-systems and at some of the joints which either fitted them or were inspired by them, we might now ask this question: what are some of the simple, intermediate, $f2$ joints, if any exist, which spring from a *general* 2-system, from an ordinary, non-special cylindroid? Let us look first at figure 8.07. Here we have four contact normals with a bare minimum of specialty involved; two of the common normals intersect. Through the point of intersection of those two we can draw the unique line which cuts the other two (§ 11.20). Along this line lies one of the two available motion screws of zero pitch (§ 15.49). At the mentioned point of intersection there exists a planar pencil of $n$-lines (§ 3.09); this planar pencil is pierced by the two non-intersecting contact normals at two points and the line joining those two points is the seat for the other of the two available screws of zero pitch (§ 3.48, § 15.49). The two motion screws of zero pitch are, in general, skew but their locations in the fixed member 1 both change as the joint engages in a movement. The cylindroid of motion screws is, accordingly, not of a constant shape, and the joint in figure 8.07 is an ordinary, non-special, higher, $f2$ joint.

86. So let us go back to figure 7.21($b$) and, in order

Figure 7.21 (§ 7.83) A pair of mutually equivalent intermediate joints based upon the fourth special 2-system of motion screws. At ($b$) it is easy to see from the special arrangement of the contact normals there that the reciprocal system of action screws is the corresponding fourth special 4-system (§ 15.54, § 23.73).

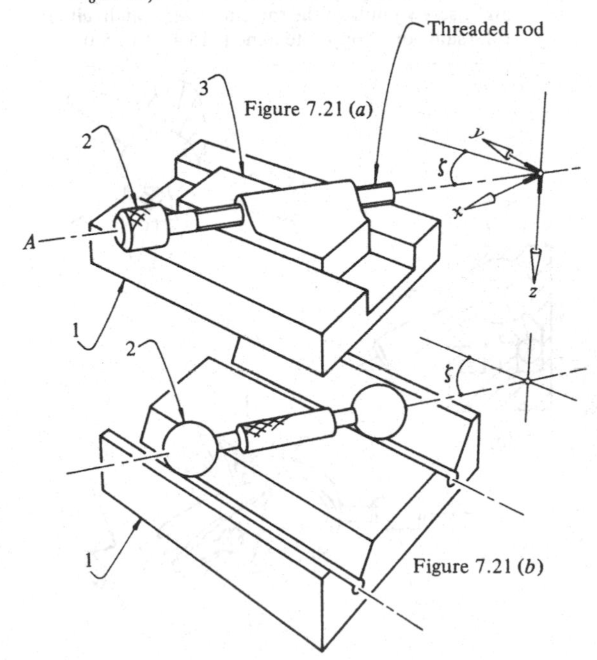


Figure 7.20 (§ 7.82) Two joints based upon the second special 2-system of motion screws. They are both simple, intermediate $f2$ joints.

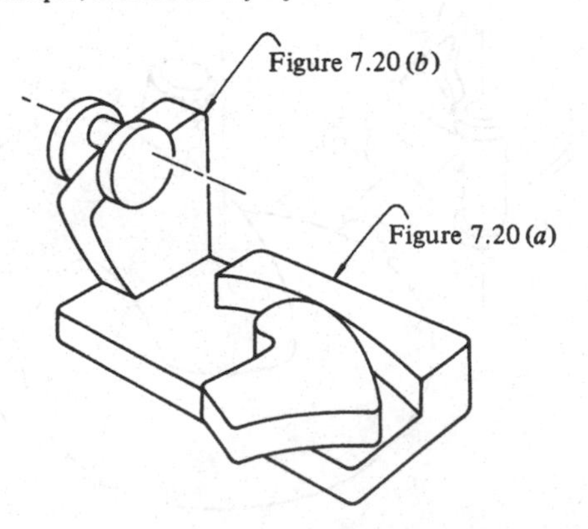

to cause there the parallel planar pencils of action screws (whose centres are at the centres of the balls) to intersect, let us successively 'bend' the whole bulk of the link 1 into the two curved, circular shapes shown in the figures 7.22. Notice in both cases in figure 7.22 that the straight V-grooves of figure 7.21(*a*) have been changed into curved, circular tubes. We can also see in both cases that the two motion screws of zero pitch continue to exist, first along the centre line of the dumb-bell 2 and second along the centre line *c–c* of the curved, circular member 1. The system of motion screws, which is accordingly an ordinary cylindroid, remains invariant now, provided the circular pieces of circular tubing remain coaxial. Both joints in figure 7.22 are intermediate, *f*2 joints. Notice now that we can go next to the figure 7.23(*a*) and then, if we wish, to figure 7.23(*b*) without losing the essential circumstance we have established, namely the invariance of the cylindroid of motion screws. In both of the figures 7.23 the centre lines of the tubes are helical and the helices are of the same pitch and coaxial, but in figure 7.23(*b*) the proportions have been chosen in such a way that the centre line *a–a* of the dumb-bell intersects the centre line *c–c* of the helices. Whereas the joint in figure 7.23(*a*) is a 'general', intermediate, *f*2 joint, the joint in figure 7.23(*b*) has, by virtue of its proportions, been allowed to collapse back into being the kind of joint which exhibits a first special 2-system of motion screws, a cylindroid of zero length. The latter, drawn small, belongs among the joints depicted in figure 7.18. *Please concentrate now upon the joint in figure 7.23(a); it is a representative example of the most general of the intermediate, f2 joints.* It will be mentioned again at §7.89.

Figure 7.22 (§7.86) Derived by distortion (Reuleaux expansion) from figure 7.21(*b*) the intermediate *f*2 joints at (*a*) and (*b*) display general 2-systems of motion screws namely general cylindroids of motion screws. Due to the circularity of the curved centre lines of the curved and coaxially arranged tubes the screw systems are invariant. Whereas at (*b*) the centre lines of the tubes reside in parallel planes and the cylindroid of motion screws has no special distinguishing characteristic, at (*a*) the centre lines of the tubes are coplanar and the two motion screws of zero pitch are mutually perpendicular (§7.81). The joint at (*a*), accordingly, is exactly equivalent with the toroidal joint already shown in figure 7.19(*b*).

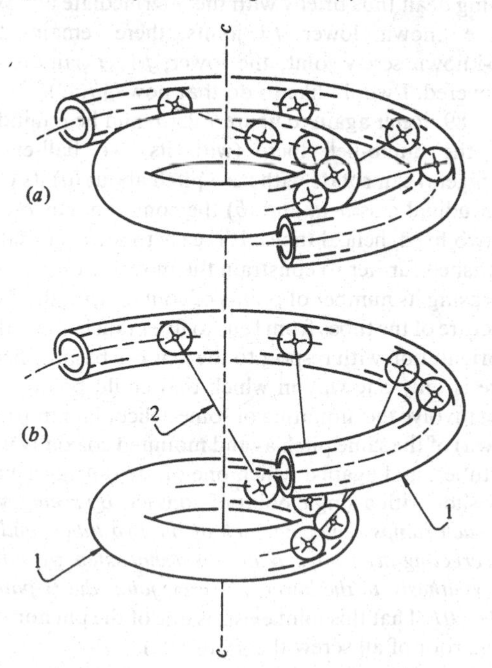


### The population of the simple *f*1 joints

87. Provided the contact normals are linearly independent, any five points of contact between a pair of bodies will define, with their contact normals, a linear complex (§11.30). The question arising for the lower and intermediate, simple *f*1 joints is this: in view of the fact that that complex will be the bare bones, the *n*-lines only (§7.62), of the 5-system of action screws extant, can that complex be arranged by means of 'button pair synthesis' to remain invariant? Refer to §6.59 *et seq.*, look at figure 3.07(*b*), figure 3.05, and, for the case of a linear complex of zero pitch, at figure 3.09(*a*) (§9.41). For methods of construction to find the central axis and the pitch of a complex defined by a set of five lines, please refer to §11.41 and, for a look at an irregular

Figure 7.23 (§7.86) At (*a*) we see a representative example of the most general of the intermediate *f*2 joints, while at (*b*) we see that the system of motion screws has collapsed again into a first special 2-system.

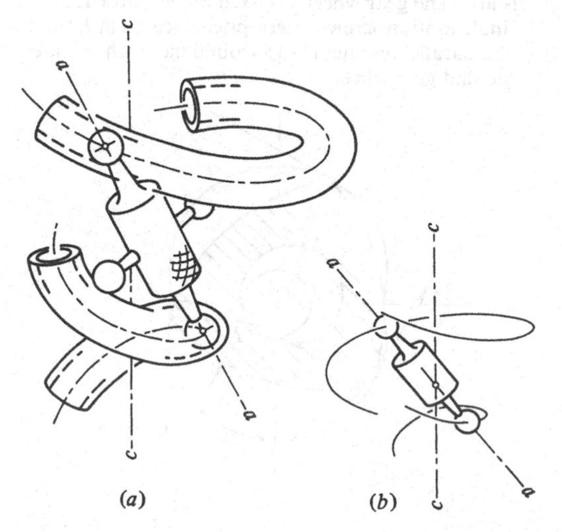

screw joint with five points of contact, go to §14.03 *et seq*. At §23.77 is my account of the first and only special 5-system. We will be concerned here (at §7.88) with a 5-system of action screws where the pitch $p_\alpha$ of the principal screw might be infinity. At §23.58, correspondingly, is my account of the first and only special 1-system. At §7.88 we will be thinking in terms of a 1-system of motion screws where again the pitch $p_\alpha$ of the principal screw might be infinity. We will be thinking in other words of a joint that might be able to sustain (that is, to transmit without any relative movement of its members) a pure couple applied anywhere and in any direction. We are about to consider some special and quasi-special sets of systems involving 1°F and a 5-system of action screws and it should be mentioned before we begin that, in accordance with the words 'in general' in my remarks at §7.03 and an already mentioned example at §7.63, we may actually find some motion screws, or a motion screw, which are, or is, collinear with a contact normal. Refer also to §23.71.

88. Now it is necessary here to see that the best known of the simple $f1$ joints, the ordinary hinge, the revolute, or the *R*-pair, does not derive from the first and only special 5-system of action screws. It derives from no special 5-system at all. It does however derive from that 5-system of action screws where the pitch $p_\alpha$ of the principal screw happens to be zero. The *R*-pair can sustain (that is, transmit without any relative motion of its parts) a wrench of zero pitch, namely a pure force, directed along (or to cut) its central axis. As the *R*-pair engages in its action the whole 5-system of action screws, that single infinity of coaxial complexes of screws of all pitches, not only remains invariant but remains fixed with respect to both of the links (§17.27). The *R*-pair is, accordingly, a lower, simple $f1$ joint. It is listed second in table 7.01. An example of an *intermediate*, simple $f1$ joint is the joint between the stator 1 and the rotor 2 of the Wankel engine. See figure 7.24. The evident central axis of the 5-system of action screws where $p_\alpha$ is zero and, of course, the central axis of the corresponding 1-system of motion screws, the single ISA of zero pitch, migrates as the joint engages in its motion in a parallel manner around the pitch cylinder of the fixed gear wheel 1. Refer to Reuleaux (1876) chapter 3, and Hunt (1978) §8.6, for a whole series of illuminating remarks about some examples of the kind of joint I have called here the intermediate $f1$ joint, where $p_\alpha$ is zero and the single ISA moves in a parallel manner. Another example is the joint I have already shown in figure 7.11 (§7.63); there the single ISA of the motion system moves in a spherical manner as the joint engages in its motion. I must say I have shown no example in this book of an intermediate, simple $f1$ joint where the mentioned ISA moves in a general, spatial manner. Nor have I shown an intermediate, simple $f1$ joint where the pitch of the single ISA is non-zero yet finite. When $p_\alpha$ is infinite however, we do get the first and only special 5-system of action screws. This can be set up with five contact normals put anywhere in five parallel planes. Thus we can easily arrive at the *P*-pair, the third of the lower, simple $f1$ joints which are listed in table 7.01. Having dealt thus briefly with the intermediate and two of the known, lower, $f1$ joints, there remains the well-known screw joint, the lower, $f1$ *H*-pair, to be discovered. I would like to do that now (§7.89).

Figure 7.24 (§7.88) Some details of the intermediate $f1$ joint in the Wankel engine. While the geometric central axis of the rotor 2 is at $C$, the instantaneous axis of rotation of the rotor with respect to the stator is at $I$. The gear wheel 1 is fixed in the stator 1, and the single motion screw of zero pitch, located at $I$, travels in a parallel manner (in 1) around the pitch cylinder of the said gear wheel.

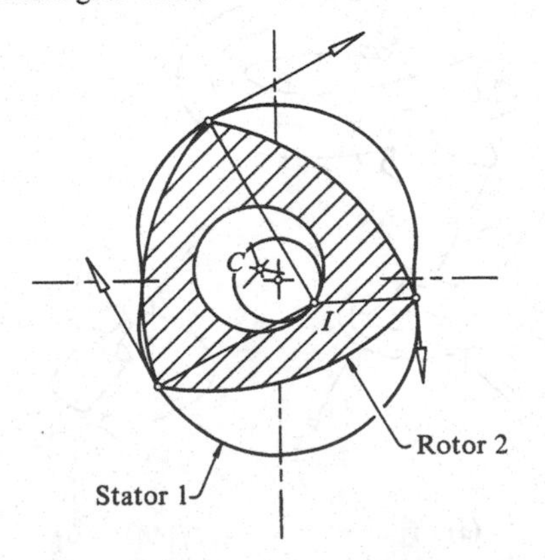


89. Refer again to figure 7.23($a$) and be reminded that the dumb-bell there (with its two ball-ended outriggers) can rotate with zero pitch about ($a$) its own longitudinal axis $a$–$a$, and ($b$) the common axis $c$–$c$ of the two fixed, helical tubes. It is easy to see next that, if we wished further to constrain the movable dumb-bell, increasing its number of points of contact with the fixed structure of the tubes from four to five in such a way that its orientation with respect to the axis $c$–$c$ became fixed, there is only one way in which that could be done. It must involve the mounting of some helicoidal ramp (not shown) of the same pitch as and mounted coaxially with the tubes, and against which one of the outrigger balls may slide with a single point of contact. *By seeing next two such ramps replacing each of the two tubes, and by thus erecting five ramps in all, it is a short step, by button pair synthesis, to the lower, f1 screw joint, the H-pair of table 7.01.* That this joint exists is one of the phenomena at the root of all screw theory (§0.05).

**Theorems of existence and nonexistence**

90. It is well known that Euclid showed there are no more than the known five, regular polyhedra. This theorem, that of the nonexistence of any other regular polyhedra, has been of continuing value to geometers, physicists, and engineers. In the field of mechanism, similarly, it would be useful to know, for example, whether my assertion at § 7.12 is sound. I said there that we believe no lower simple joints other than the known six listed in table 7.01 exist. I offer in the next six paragraphs reasoned support for that belief and, at § 7.97, treat of another important theorem of nonexistence which was first mentioned at § 7.11.

**The six (the only six) lower, simple joints**

91. In chapter 5 we saw that the characteristic thing about any relative motion between a pair of rigid bodies is its ISA, its instantaneous screw, or motion screw. If some particular screw system – either an action or a motion system – is to remain 'fixed' with respect to both of the links of some simple joint when the links are moving relatively to one another (§ 7.11), the system itself must be able to rotate or be twisted about some *axis* within itself without apparently altering its presented shape. I make the point for example that the quasi-general 5-system of § 6.25 could be rotated (or twisted with any pitch) about its central axis of axi-symmetry without our being able to detect the rotation or the twisting by occasionally examining, in a series of successive 'locations', the presented shape of the system. It is nonsense to speak about the rotation or the twisting of some such system about its own axis of axi-symmetry, in the same way that it is nonsense to speak about the rotation (about some axis through it) of some point.

92. It needs to be seen quite clearly that, if a joint is to be a lower joint by the definition offered at § 7.12 and clarified at § 7.15 and § 7.17, both of the screw systems at the joint – the action and the motion system, which are, as we know, mutually reciprocal systems (§ 23.03) – must exhibit that kind of symmetry (*a*) where an axis of axi-symmetry exists within which a motion screw or screws reside; (*b*) where, unless the pitch of one of those screws at the axis of axi-symmetry is zero, the axi-symmetry is such that the system has the same overall appearance at all neighborhoods along the axis of the axi-symmetry; and (*c*) where, if the pitch of a screw at the axis is zero, the system may be point-symmetric about a point in the axis of the axi-symmetry. It is easiest to think, first, in terms of the motion screws.

93. A single motion screw, namely a 1-system of motion screws, and its reciprocal 5-system of action screws are together axi-symmetric in the acceptable manner about the single motion screw; the combination can be of any pitch (§ 23.77). Given the pitches $p$ = zero, $p$ = non-zero yet finite and $p = \infty$, and given the method of button pairs, we can go without difficulty from the *n*-lines of a 5-system to the three *f* 1, lower joints in table 7.01. *We can see moreover that, except for the R-pair, the H-pair and the P-pair thus discovered, no other lower, simple f 1 joints can exist.*

94. Any pair of motion screws will determine a 2-system (§ 3.36, § 15.22). The only kind of 2-system of motion screws which is axi-symmetric in the acceptable manner is the fifth special 2-system, where $p_\alpha \rightarrow \infty$, $p_\beta = -k^2 p_\alpha$ and $k$ = unity (§ 15.55). The reciprocal 4-system of action screws is a fifth special 4-system and this, as we have seen (§ 6.29), not only carries the same single acceptable axis of axi-symmetry but is the only 4-system to carry such an axis. It is easy to isolate from among the screws of the said 4-system the special congruence of *n*-lines (or action screws of zero pitch) which is an axi-symmetric series of planar pencils of lines all normal to the line of the axi-symmetry. By the method of button pairs we can go next and without difficulty from that to the single, *f* 2 lower joint which appears in table 7.01. *We can see moreover that, except for the C-pair thus discovered, no other lower, simple f 2 joint can exist.*

95. Any three motion screws will determine a 3-system (§ 3.36, § 6C.01). A careful study of the ten special 3-systems, described at § 23.61 *et seq.*, will show that only two of them, the second and the fifth, exhibit the kind of axi-symmetry explained at § 7.92 as being suitable for founding a lower pair. As we know already (§ 7.73, § 10.71, § 23.23), the reciprocals of these 3-systems are the same systems – except for their handedness. So the question needs to be asked: among which of the reciprocals of these two systems can we find reguli of action screws of zero pitch, namely reguli of *n*-lines, which (with their button pairs) can be seen as the basis for a lower simple joint capable of continuity both of action and of motion? The second special 3-system, a star of screws of equal finite pitch becomes a star of *n*-lines if the equal principal pitches are not only equal but all zero; and thus we can go to the lower, simple, *S*-pair (§ 7.59). From the fifth special 3-system we can go, as we did at § 7.65, to the lower, simple, *E*-pair, where an $\infty^2$ of axes of axi-symmetry obtain; please be reminded of the item (*c*) at § 7.92. *We thus appear to have seen that, except for the S-pair and the E-pair thus discovered and already listed in table 7.01, there is no other lower simple f 3 joint which can exist.*

96. Any four motion screws will determine a 4-system. We have already seen, at § 7.94, but by way of action screws, that the only 4-system of motion screws to exhibit an axis of axi-symmetry is the fifth special 4-system. Its reciprocal here to be studied is the fifth special 2-system of action screws whose local *n*-lines (as distinct from the infinitely distant ones at the other end

of the cylindroid) inhabit only the end generator (§15.55). Now it is not possible to form an $f4$ joint of any kind, let alone a lower one, on two $n$-lines that are collinear, for such a synthesised joint could only be an $f5$ one. We know also that no $f5$ joint can be a lower joint (§7.48). It follows that no lower, simple $f4$ or $f5$ joint can exist. From this and the argument already pursued we can now conclude the following: *the six lower joints previously listed in table 7.01 are the only lower joints which can exist.*

**Another important theorem of non existence**

97. If a simple joint has a system of motion screws that is invariant, there are mechanical features of both elements at the joint which will have contributed to the invariance. Unless such an invariant system of motion screws is such that it might be suitable for a lower simple joint, and thus of such a sort that it has an axis or axes of axi-symmetry with an available screw or screws residing in that axis or those axes (§7.92), there will be at least a single infinity of other lines at various locations and in different directions which can act as seats for other motion screws within the system. Let the member 2 of some non-lower, simple joint with an invariant screw system make a movement about one of the generally available screws relative to the member 1. Because the element at 2 carries with it that feature which helped to determine the invariant system, it will (as it moves) carry the system with it. The system of screws will thus be movable with respect to the member 1. If the member 2 is thus free to make a movement relative to 1 and thus to carry the system with it, the system cannot be fixed with respect to 2. This is because the member 1 is next correspondingly free to make a movement relative to 2. Such a movement of the member 1 – it could be a different relative movement – would, in its turn, carry the mechanical feature of 1, and thus the invariant system, with it. It follows that, *if the screw system is invariant and if the simple joint is a non-lower joint, the screw system will never be seen to be fixed with respect to one member of the joint while being movable with respect to the other.* Accordingly we can say that, *if an invariant screw system at a simple joint is not fixed with respect to one of the members of the joint, the joint being an intermediate joint, the system will not be fixed with respect to the other member either.* We can say also, and conversely, that, *if the screw system at a simple joint is fixed with respect to one of the members of the joint, the joint being a lower joint, it will be fixed with respect to the other member too.*

**Some final remarks**

98. I am somewhat unenthusiastic about many parts of this chapter. I would like to feel more sure than I do that my arguments are everywhere sound. I do believe, however, that my proposal for a new category of joint, namely the intermediate (§7.15), will help towards a better understanding of the contact geometry at joints. I trust also that some further useful progress may now be made in the area of categorisation for design purposes. There is of course a bewildering variety of allegedly different kinds of joint, even among the lower and the intermediate, but many of these are, as we have seen, the same, at least from the point of view of screw theory. I have avoided the temptation to try to categorise all joints in all respects, choosing instead to comment upon the various possibilities. I hope that by so doing I have not disappointed the designer reader. There is much more to be done in this wide arena of contact geometry and I trust that in due course someone will have the new pleasure of doing it. Please refer again to my italicised remark at §7.36.

**Notes and references**

[1] Waldron, K.J., A method of studying joint geometry, *Mechanism and Machine Theory*, Vol. 7, No 3, p. 347–53, Pergamon 1972. This paper refers to an earlier work by the same author, On surface contact joints, *Report No 1971/AM/2*, School of Mechanical and Electrical Engineering, University of New South Wales, 1971. It is in this latter that Waldron claims to have shown that only the six known lower pairs exist. Refer §7.12.

[2] I mean by this at §7.15 that Waldron at reference [1] above makes it clear by implication that he defines a lower joint as a joint displaying surface contact only. When I use it again at §7.78, I mean by the same note that Waldron had already brushed with the matter of complex connectivities in his paper [2] already quoted at the end of chapter 1. Refer §7.15.

5. Most of the known shaft couplings for the transmission of rotary motion are [illegible] exhibiting two degrees of freedom. This one involving six revolutes and three spherical joints is a constant-velocity coupling

8. Most of the known shaft couplings for the transmission of rotary motion are complex joints exhibiting two degrees of freedom. This one, involving six revolutes and three spherical joints, is a constant-velocity coupling.

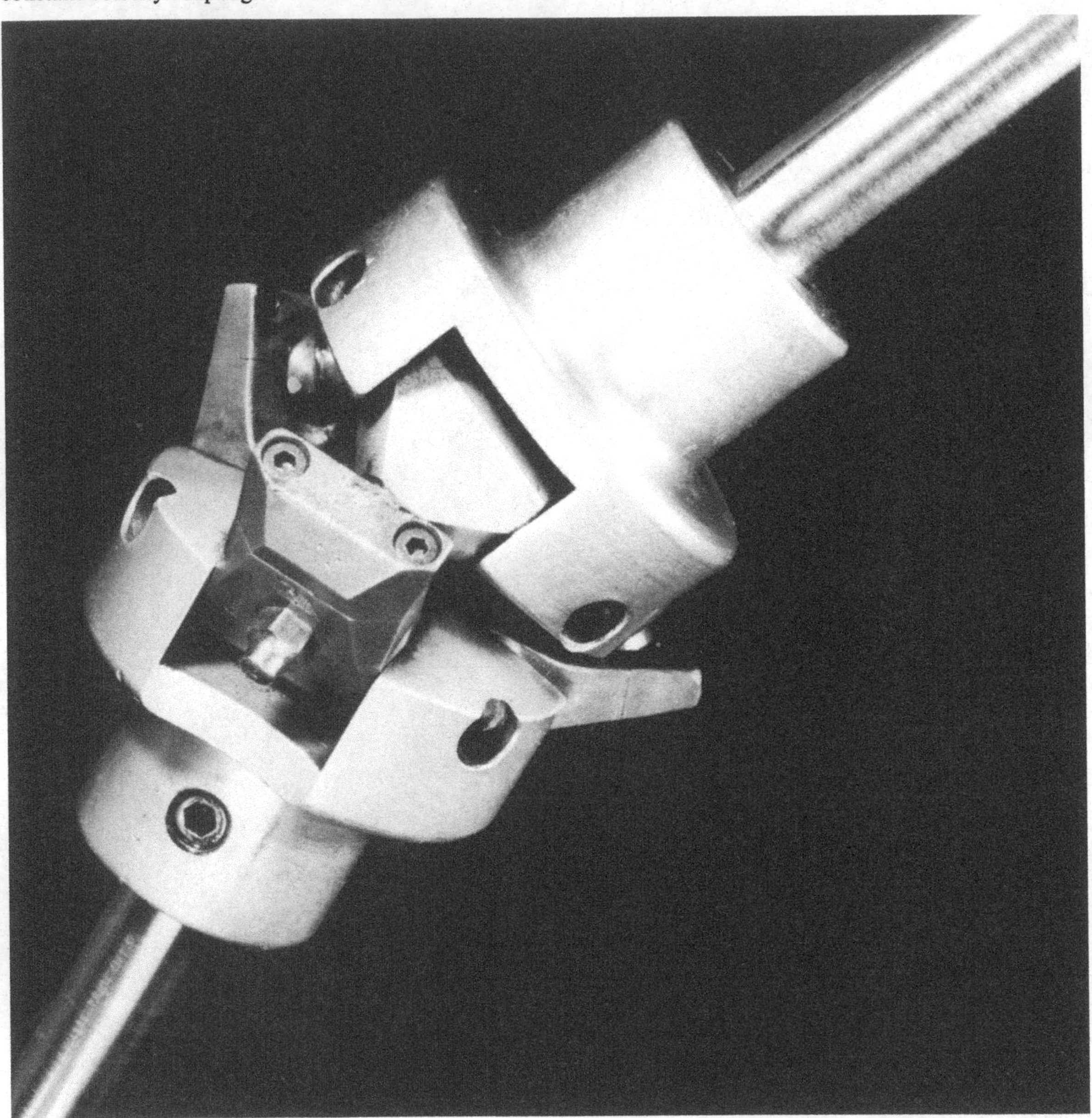

# Some elementary aspects of two degrees of freedom

### Some bases for an argument

01. For the reader already familiar with other discussions about the cylindroid (chapters 6, 15, 16 and elsewhere) it will be clear that a cylindroid is always lurking wherever there is a situation of two degrees of freedom – or wherever there are two forces or wrenches acting (§ 10.14). For the reader unaware as yet of the relevant generalities, the present chapter may be taken as a piece of purposely confusing, preliminary reading. It must be said in the kinematics (and the statics) of mechanism that the simpler a situation appears to be, the more baffling its deeper aspects often are. The science of the relative motion of rigid contacting bodies is bedevilled by degeneracies of its general geometry; these degeneracies, when taken in isolation, have a tendency to render the science a miscellany of unrelated facts and alleged separate theorems for which there appears to be no integrative binding. This chapter deals with some of that miscellany but with almost none of its binding; I shall be dealing here with the kinematics of only a few special cases of two degrees of freedom and I offer no conclusive results at the end of it.

02. However with no more than a primitive understanding of the relationship $c+f=6$, where $c$ is the number of *constraints* upon a body and $f$ is its number of *freedoms* (§ 1.52), we can begin a useful argument here about the matter of two degrees of freedom. We can set up for consideration a few (namely four) selected and 'practical looking' cases of a movable body 2 in direct physical contact with some other fixed body 1 at four points of contact. Then we discuss the four cases one by one. The setting up of the cases is done in the following four paragraphs, the separate discussions follow soon after, and the chapter ends with a brief mention of the generalities and thus of the ways that can be found out of the difficulties we have met on the way.

### The setting up of the cases for consideration

03. Please set up for consideration first, and see in advance in figure 8.01, the solid conical shaped body 2 with its spherical head centred at $A$. The head is secured to swivel within its mating spherical socket mounted in the body 1 and the conical piece has its circular base in contact with the flat platform of 1 at $B$. There are three points of contact within the spherical joint (§ 7.51) and a further one at $B$; these together make up the required total of four constraints.

04. Having so directly spoken about four constraints, why not set up next, in the mind's eye, the object in figure 8.02? It is a single rod-ended body 2 hanging freely from a fitted cylindrical hole in body 1. The four points of contact here are all inside the hole (§ 1.51) and the connection between the two bodies is a cylindric joint, a $C$-pair (§ 7.14).

05. In yet another case, employing this time another arrangment of the points of contact, let us look at the heavy pierced disc 2 in figure 8.03. Suspended upon a cylindrical peg in 1 which loosely fits the cylindrical hole in 2, the body 2 rests otherwise upon its three points of contact distributed across the flat surface of the supporting platform (§ 7.65). There is of course only one point of contact at the peg; the cylindrical peg can never be exactly square with the flat platform, nor exactly cylindrical for that matter (§ 6.05), so by virtue of gravity and of the pin's being loose in its hole the four points of contact are distributed as described.

06. Going next in figure 8.04 to another case – not this time of 'three plus one' constraints or of 'one plus three', but one of 'two plus two' – we might imagine the double ball-ended body 2 in contact with body 1 at four points, two upon each of the V-grooves or V-slots which, for the sake of generality, are arranged in such a way that the paths of the centres of the balls are neither parallel nor intersecting. Notice in the figure that body 2 has been shown to have some substance and be bent. It is not thereby some simple straight ball-ended rod, which may freely and irrelevantly spin about its own axis; the various possible locations in rotation of the body 2 are noticeable in ordinary practice and are significant.

### Some further remarks about the bases for argument

07. It would be wise for the sake of logical argument to bring in here from chapter 6 the following general idea: that a point of contact in this context, once made, cannot be unmade (§ 6.19, § 7.49). It will be convenient to imagine, without any loss of generality, that (*a*) there is no gravity; (*b*) the axiomatically irregular bodies are also suitably 'magnetic', so that they remain together at their points of contact; and (*c*) that at the points of contact there is no friction to impede any crosswise sliding which may occur.

08. Keeping in mind for consideration elsewhere (chapters 15, 23) the important fact that a body may possess 2°F by being connected to frame through a number of other bodies intervening along some open chain in various ways (§ 3.51), and that a body with 2°F may, indeed, be a link in the closed kinematic chain of a loop or of many loops (§ 2.39), let us consider now the 2°F of the bodies 2 in the array of simple two-body mechanisms of mobility 2 illustrated in the series of figures 8.01 to 8.04 inclusive (§ 1.34).

09. We shall examine the capacity for instantaneous motion of the bodies 2 in the series by asking on each occasion the following kinds of question: where are the ISAs in space (chapter 5) about which the body 2 may be screwed (or suffer a screwing, or be twisted) without losing contact at any of its points of contact at the instant and without 'digging in' at any of those points?; how many of such ISAs exist?; what are their respective pitches?; and in what pattern in space if any are their lines of action arranged? Such questions are asked, incidentally, throughout the argument of chapter 6, but here we are dealing, not with the whole range of freedoms from zero to six inclusive any one of which may be extant within a joint between two bodies, *but with a selected few of special cases, in the restricted field of 2°F only.*

### The first case (a) in figure 8.01

10. Taking the body 2 in figure 8.01 as an easy case to be dealt with first, we can see that by causing a rolling at *B* the body 2 can be screwed with zero pitch about the line *AB* (which line is marked *A–j* in the figure, the point *A* being at the centre of the ball). Alternatively by causing a sliding at *B*, given that the platform at *B* is horizontal, and nominating a vertical line through *A* marked *A–k* in the figure, we can see that the body 2 can be screwed with zero pitch about this line also. It is not very difficult next to see that by suitably combining sliding and rolling at *B* in various proportions at the instant, the body 2 can be caused to screw with zero pitch about any line of the whole planar pencil of lines (the whole 360° of them) defined by the lines *A–j* and *A–k*.

11. One main aspect of the matter in figure 8.01 is that the velocity $\mathbf{v}_{1A2}$ of the point $A_2$ in the body 2 must always be zero and it follows from this that all possible screw axes must pass through *A* and be of zero pitch (§ 5.24, § 7.61, § 12.07). There can in other words be no screw axes anywhere except through *A*, even with pitches other than zero. It should also be clear that it is not possible for the body 2 to screw with zero pitch about any axis through *A* which is not of the pencil *jAk*. If this were possible, the point $B_2$ at *B* would be either moving down into, or upwards away from, the surface of the platform there, and this behaviour at *B* (by definition of the action of the mechanism) is prohibited.

12. We see accordingly that the whole range of possibilities for instantaneous relative motion in figure 8.01 is summarised in the display of the single infinity of screw axes of zero pitch contained in the planar pencil *jAk*. Please keep this picture in mind for later reference, recognising meanwhile that we are dealing here with some kind of special case. Were the body 2 to be drawn with prongs at its points of contact, that is, were figure 8.01 to be drawn in the style of figure 8.08, there would be one prong at *B* and three prongs making contact inside the spherical socket at *A*. The contact normal at *B* would simply be normal to the platform there (§ 3.06) but the contact normals at the three points of contact within the spherical socket would be radially arranged and intersect at *A*; this latter is the special circumstance.

### The second case (b) in figure 8.02

13. Looking next at the body 2 in figure 8.02, which is perhaps another recruit for easy analysis by the same method, let us ask of this body the same set of

Figure 8.01 ((§ 8.10) A conical body 2 secured to swivel about a ball in a socket at *A* but constrained to roll or slide upon the flat surface of the body 1 at *B*.

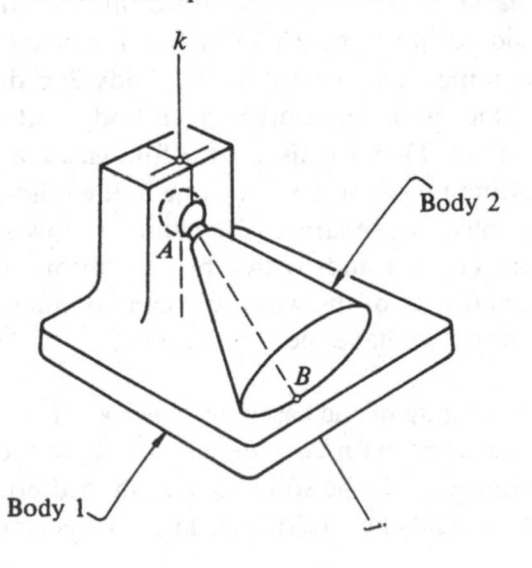

questions. In what ways (that is, about which screw axes and with what pitches) can it move instantaneously and relatively to its frame-body 1? Owing to the presence of the single cylindrical joint, whose mechanics is outlined elsewhere (§7.12, §7.84), and by virtue of ordinary inspection, we can see that the display of all possible screw axes for the instantaneous motion consists in this case of the following: (*a*) a set of screw axes of all pitches (ranging all the way from positive through zero to negative infinity) which are all collinear with the joint axis; these are there because the body 2 can clearly screw about this line with whatever pitch it pleases; (*b*) a set of screw axes of infinite pitch, one lying along each of the $\infty^2$ of lines in spaces which are parallel with the joint axis; this set is associated with the limiting possibility of there being pure axial sliding within the joint in the total absence of rotation; and (*c*) a set of screw axes of zero pitch which lie at right angles to the joint axis but infinitely far away from it; this latter set (which also relates to the same limiting possibility of pure sliding in the absence of rotation) is arranged as all of the tangents to all of the circles of infinite radius concentric with the joint-axis, spread out as they are along the whole of the infinite length of the joint.

14. One might complain with some justice here that, if there is some recognisable, regular arrangement in space of the sets of available screw axes with their various pitches in the general circumstance of 2°F, these current investigations are not having much success in finding that arrangement. But in figure 8.02 we have stumbled again upon a speciality and thereby upon another degeneracy of the general geometry. All four contact normals at the points of contact inside the cylindrical hole in figure 8.02 are radially arranged; they all cut the same line, the joint-axis, at right angles. It must be remembered also that, while a firm grip of the general geometry is essential for a proper understanding of the overall phenomena, it is precisely with these kinds of special case that the practical synthesiser is engaged in his day to day work with mechanism, and it is for this reason that I am moving so slowly in this area. We shall return again, here and elsewhere (chapter 6, 15, 23), to all of these deceptively simple special examples in 2°F, and when we do, there will be a much more powerful body of theory available to make some sense of them.

**The third case (c) in figure 8.03**

15. At the risk therefore of drifting even further away from the central theme of the main argument, let us look next at the pierced disc in figure 8.03. Take a line passing through the hidden point of contact *A* at the peg, which line is also drawn normal to the plane of the platform; the line in question is marked *j–j* in the figure. We see straight away that the body 2 can screw about this line *j–j* with zero pitch if no sliding is permitted at the point of contact *A*. If on the other hand another line *k–k* is chosen parallel with *j–j* but cutting also the contact normal *n–n* at *A*, it can be seen that the body 2 is also free to screw with zero pitch about that line, provided sliding is allowed to occur at *A*. Other lines parallel with *j–j* which are not in the plane defined by *j–j* and *k–k* are not available as ISAs for the body 2 because, unless the proposed line cuts the contact normal *n–A–n*, there will be (upon a movement) either a loss of contact at *A* or a digging in there, both of which are, by definition (§8.12), prohibited.

16. The line *k–k* is accordingly representative of a legitimate set of available screw axes of zero pitch, which is all of the lines parallel with *k–k* in the plane defined by *j–j* and the contact normal at *A*. We have therefore in figure 8.03 a whole planar array of available screw axes of zero pitch to be kept in mind, but we have some more as well.

17. If in figure 8.03 the disc were sliding bodily

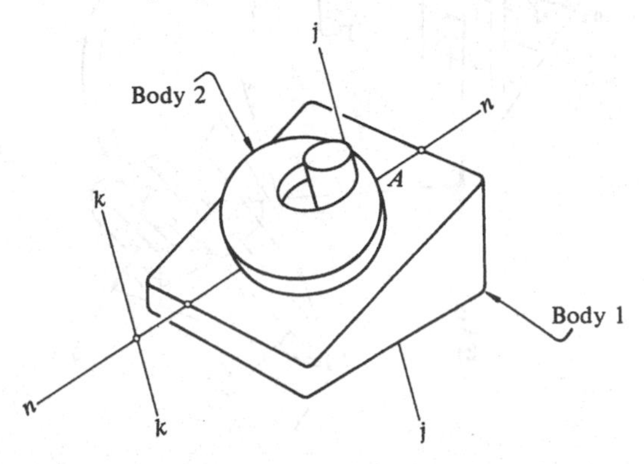

Figure 8.02 (§8.13) A cylindrical body 2 secured to turn or to slide within a fitted cylindrical hole cut in the body 1 (§7.84, §15.55, §23.76).

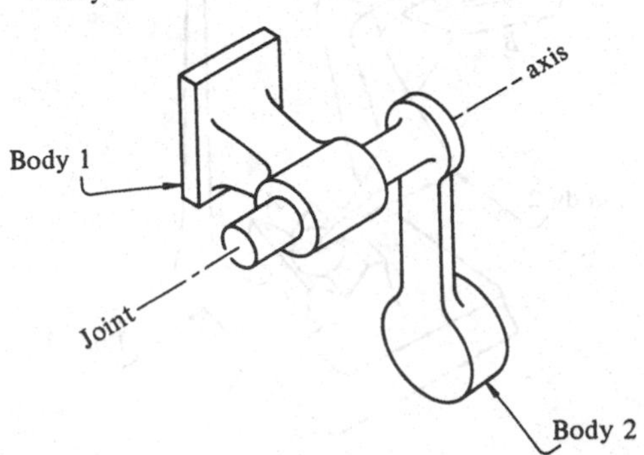

Figure 8.03 (§8.15) A pierced, disc-shaped body 2, hanging upon a cylindical peg. The peg protrudes from the sloping, flat surface of the upper side of body 1.

parallel with the face of the platform and in a direction perpendicular to the contact normal at $A$, this would be a legitimate instantaneous motion also. The whole $\infty^2$ of axes of infinite pitch parallel with the direction of motion must thus be seen to complete the total picture. These screw axes of infinite pitch are of course equivalent to the screw axes of zero pitch at the infinitely far distant ends of the already mentioned planar array. In any event, we might describe the particular speciality pervading figure 8.03 by pointing out that the three contact normals at the three points of contact upon the platform are parallel. The whole assembly is, thereby, a planar mechanism.

**The fourth case (d) in figure 8.04**

18. Going next to the fairly innocent looking apparatus in figure 8.04, in the hope that here at least we might escape the awkward clutches of infinity, let us see straight away that the body 2 can screw with zero pitch about the line $w$–$w$ which joins the centres $A$ and $B$ of the balls.

19. But can it screw with zero pitch about any other line? And can it, indeed, screw with any other pitch about this line or any other line? Putting the first of these questions another way, we could ask it as follows: were the balls with their respective centres $A$ and $B$ both to be moving with some pair of mutually consistent velocities $\mathbf{v}_{1A}$ and $\mathbf{v}_{1B}$ in the directions of the centre lines of their respective V-slots (take a glance at figure 5.03), would there be some line which could be drawn somewhere at right angles to both $\mathbf{v}_{1A}$ and $\mathbf{v}_{1B}$, in such a position that the shortest distances between $A$ and $B$ and the line are in proportion to the velocities $\mathbf{v}_{1A}$ and $\mathbf{v}_{1B}$? We shall see that there is such a line and that it is unique. The line in question can be seen as the line $A$–$x$ in figure 5.03, where the centres of the balls must be seen to be, not at $A$ and $B$, but at $B$ and $E$, and where, if $A$–$x$ is the ISA, the pitch of the screwing is clearly zero.

Figure 8.04 (§ 8.18) A double, ball-ended rod 2 in contact with a double, $V$-slotted body 1. There are four points of contact, two at each of the balls.

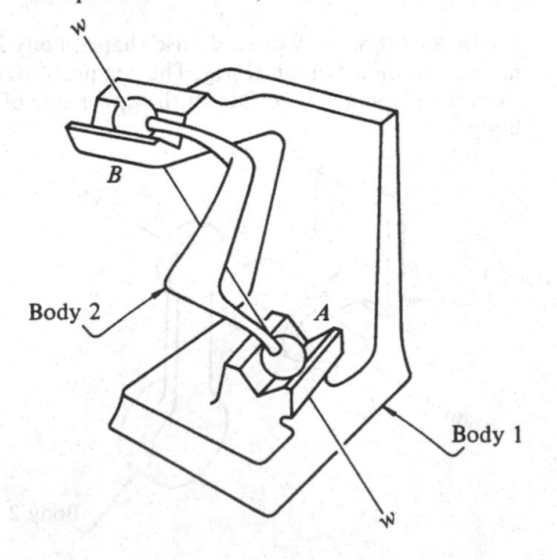


20. Whatever we choose the proposed velocities at $A$ and $B$ to be in figure 8.04, they will be obliged by virtue of rigidity to be in some fixed proportion to one another (§ 5.03). So without any loss of generality we can nominate some unit velocity $\mathbf{v}_{1A}$ at $A$, whereupon the corresponding velocity $\mathbf{v}_{1B}$ at $B$ will follow. Going now to figure 8.05, where such a pair of velocities is shown by vector arrows while their equal components along the rigid line $AB$ are shown by means of the thickened segments of line (§ 5.05), we see that the centre point $\Gamma$ along the line $AB$ has also been drawn (§ 21.03). The tips of the vectors representing the velocities of the points along the line $AB$ are also in a line as we know (§ 5.07, § 5.25), and the point $\Gamma$ is that point along the line where the velocity of the point is least (§ 21.03). The straight line $A\Gamma B$ in figure 8.05 represents body 2, and now we must do some geometrical reasoning to arrive at the foreshadowed conclusion, that *as well as the ISA of zero pitch which exists along the line AB itself, there is only one other ISA of zero pitch available*. The reasoning follows.

21. In figure 8.05 the line $\Gamma x$ is drawn perpendicular to the least velocity vector $\mathbf{v}_{1\Gamma}$ and perpendicular to the line $AB$; $\Gamma x$ is thus drawn normal to the plane

Figure 8.05 (§ 8.20) An analysis is here made to discover not only the first ISA of zero pitch $w_1$, but also the second $w_2$. The common perpendicular between the two lines cuts $w_1$ at $\Gamma$ and $w_2$ at $D$.

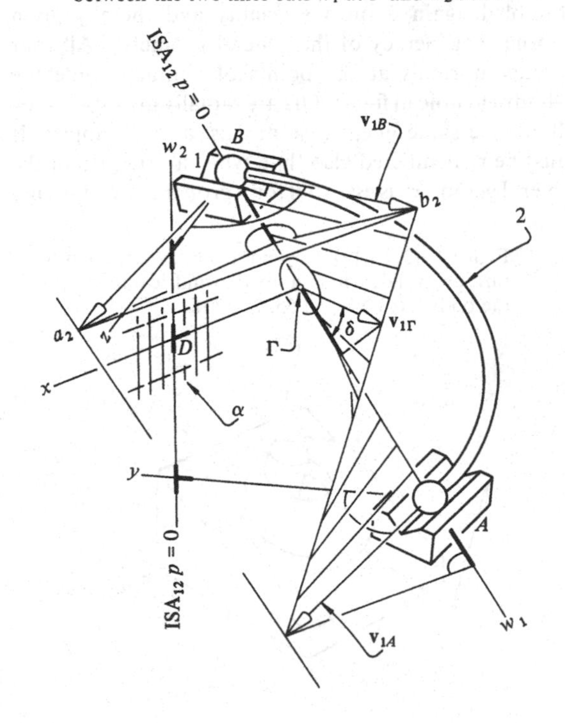

containing $\mathbf{v}_{1\Gamma}$ and the line $AB$. The plane in question is indicated by the circular 'disc' drawn at $\Gamma$ and the line $\Gamma x$ can be seen emerging forwards towards the left foreground of the picture. Now it is obvious, though shown elsewhere (§ 9.20), that no line anywhere in figure 8.05 can possibly be an ISA for the motion unless it cuts the line $\Gamma x$ somewhere at right angles. But it is also well known (§ 9.21 and please refer also to figure 21.01) that, parallel to the plane containing the vector $\mathbf{v}_{1\Gamma}$ and the line $\Gamma x$, there are planes which will contain the velocity vectors $\mathbf{v}_{1A}$ at $A$ and $\mathbf{v}_{1B}$ at $B$. It is of course normal to this set of parallel planes that any line for a possible ISA of zero pitch must lie. We have accordingly proven in figure 8.05 that any ISA of zero pitch (other than that one along the line $AB$ itself) must be somewhere among the planar array of parallel lines shown in the figure and designated $\alpha$; the plane of the array contains $\Gamma x$ and each line of the array is normal to the above-mentioned set of parallel planes.

22. The equal components along the line $AB$ of the vectors at $A$, $B$ and $\Gamma$ are shown, as has been said, by the thickened segments of line in figure 8.05. If we now pluck the vector at $A$ along with its component, slide the two of these bodily along $AB$ and plant them together again at $B$ (§ 5.31), we see immediately that the velocity difference there, the directed line $\mathbf{a}_2\mathbf{b}_2$ in the velocity polygon whose origin is at $B$ (§ 12.15), is perpendicular to $AB$. The directed line $\mathbf{a}_2\mathbf{b}_2$ must for all pitches be perpendicular to the ISA, but in this special case of zero pitch both $\mathbf{v}_{1A}$ and $\mathbf{v}_{1B}$ are also perpendicular to the sought-for ISA. It follows that $\mathbf{a}_2\mathbf{b}_2$ is parallel with $\Gamma x$; and it follows next (by ordinary Euclidean geometry) that the two lines $Ay$ and $Bz$, which are drawn perpendicular respectively to the velocity vectors at $A$ and $B$ and which are drawn in the previously-mentioned parallel planes respectively, intersect (at two different points) the same line of the display of parallel lines marked $\alpha$. This line, unique in the display, is shown in figure 8.05 to cut $\Gamma x$ in $D$. The line $AB$ and this new line which cuts $\Gamma x$ in $D$ have been designated in figure 8.05 $w_1$ and $w_2$ respectively; both of them will be seen to bear the legend $\mathrm{ISA}_{12}\, p=0$. *Except for the line AB, namely* $w_1$, *which was earlier called w–w and discussed as a special case, there is no line other than* $w_2$ *in the body 2 about which the body 2 can screw with a pitch of zero.*

23. Go back now to figures 3.10 and 3.11 and become aware that the two lines marked *r–r* in those figures, the two equivalent hinge axes, correspond exactly with the two lines which I have called here, in the particular case of the present apparatus, $w_1$ and $w_2$. Imagine here the two hinge axes at $w_1$ and $w_2$ series connected between the body 2 and its frame 1 (with the use, of course, of an intermediate body 3) and see that such an arrangement is kinematically equivalent at the instant to the original, 2-link apparatus shown in figure 8.04. In summary here we have found in figure 8.04 that there are two and only two ISAs about which the body 2 can screw with zero pitch with respect to the body 1. This discovery, as shown elsewhere (chapters 6, 15, 16, 23), is of quite profound significance.

24. At an earlier stage of the present argument the following question could have been asked: how can there be an ISA of zero pitch along the line $AB$ when some nominated unit velocity at $A$ is extant? One can reply that, having nominated some unit velocity at $A$, there will be a different angular velocity $\omega_{12}$ for each possible ISA which is chosen, and that, if the ISA is chosen to be along $AB$, the pitch will be zero because the angular velocity $\omega_{12}$ is infinite. This is simply another way of saying that, if the body 2 chooses to screw about the line $AB$, the pitch of the screwing will be zero and the linear velocities at $A$ and $B$ will be zero also.

**Intersecting complexes of right lines**

25. One more powerful way (among many others) to find the fact which has just been found, namely that there exist only two available screw axes of zero pitch for the body 2 in figure 8.04, is to argue in terms of contact normals drawn at the tips of prongs. Because it is instructive and because it will be used again in another place, this other way of argument will be pursued in the following three paragraphs. The discursive reader may

Figure 8.06 (§ 8.26) This figure shows a modified mechanism equivalent to that in figures 8.04 and 8.05. It shows also another construction for finding the two ISAs of zero pitch, $w_1$ and $w_2$.

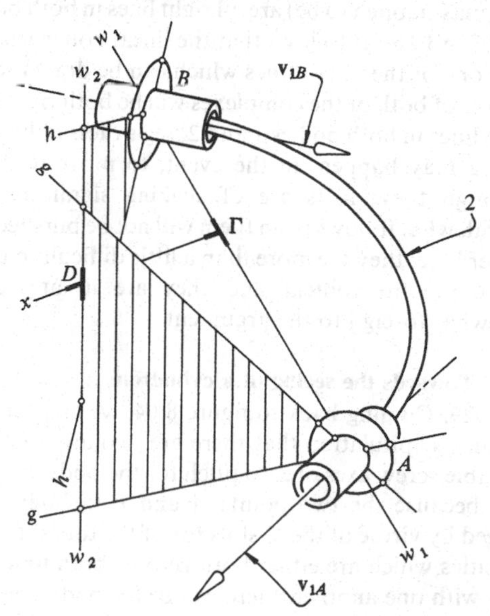

choose however to skip the third of these upon a first reading; the material is in a way parenthetical.

26. In figure 8.06 there is a redrawn version of figure 8.05 where the V-slots at $A$ and $B$ have been replaced by imaginary tubes which are fixed in the body 1. The tubes are fitted to match the balls and the body 2 is shown again as a double ball-ended, twisted, rigid rod. The balls are free to slide and rotate in the tubes but their centre points at $A$ and $B$ are confined to move in the same straight paths as before. The balls will touch their respective tubes at two points of contact only, the two points being somewhere around the equatorial 'circles of contact' in each tube.

27. At these two points on each ball one can imagine sharp prongs and through the tips of these the relevant contact normals can be drawn. The contact normals, two marked $A$–$g$ and two marked $B$–$h$ in figure 8.06, are radially arranged with respect to the tubes of course, and the 'flanges' upon the tubes have been drawn to help clarify this fact. Now the line along which the shaded plane $gAg$ cuts the shaded plane $hBh$ is the same available ISA of zero pitch as the one discovered laboriously above, and the point $\Gamma$ at the far foot of the common perpendicular $D\Gamma$ between this second available ISA of zero pitch and the first (which is along the line $AB$ itself) will turn out to be, of course, the same centre point $\Gamma$ of the line $AB$ in the body 2.

28. The quintessence of the matter in this particular approach is that the four lines $A$–$g$ and $B$–$h$ are all generators of two different, special linear complexes of zero pitch which have their respective central axes along the two ISAs of zero pitch in figure 8.06 (§ 9.41). Because they all cut the central axes of both of the complexes, all of the lines $A$–$g$ and $B$–$h$ (wherever the prongs in the tubes may happen to be) are all right lines in both bodies 1 and 2 (§ 3.16). It follows that the linear congruence of every one of the $\infty^2$ of lines which can be drawn to cut the axes of both of the complexes will be both right lines and $n$-lines in both bodies 1 and 2, regardless of how the body 2 may happen, in the event, to move (§ 11.40). Although these facts are of striking significance in general, what follows from them will not be pursued any further here; they are more than a little difficult to grasp in the present context and they are in any event somewhat foreign to the argument.

### Towards the seeing of a cylindroid

29. Coming back to figure 8.04, we can ($a$) state now in recapitulation that there are two and only two available screw axes of zero pitch for the body 2; ($b$) see that, because the two points $A$ and $B$ in body 2 are obliged by virtue of the V-slots (or of the tubes) to have velocities which are either both zero or both finite and skew with one another, there can be for body 2 neither screw axes at infinity nor any axes (either in the immediate neighborhood or infinitely far away) with pitches of infinity; and, accordingly, ($c$) see that, if there are any available screw axes other than the two of zero pitch (and there are, quite clearly), they must be somewhere in the immediate neighborhood and they must be of finite pitch.

30. Please note the remark at ($c$) above and ask the question: well, where are they then? This question (the answer is, upon the surface of a cylindroid) is the critical question asked at the beginning of chapter 16. Chapter 16 is a chapter similar in tenor to this one and I recommend it now as the next piece of preliminary reading.

31. In all of the four examples studied here (refer to figures 8.01 to 8.04 inclusive), our conclusions flowed from having first discovered either one or two of the axes of zero pitch. In each example there was at least a single infinity of other axes available for the screwing. If we could assume for the moment (it is, incidentally, a wrong assumption) that one will always find a pair of axes of zero pitch in a situation of 2 °F, we could observe with respect to those two axes ($a$) that in figure 8.01 they intersect; ($b$) that in figure 8.03 they were parallel, ($c$) that in figure 8.02 they were collinear (or there was only one of them), and ($d$) that it was only in figure 8.04 that they were neither intersecting nor parallel nor collinear. It was also in figure 8.04 that we found no axes at infinity nor any axes of infinite pitch. The unresolved aspects of figure 8.04, namely the questions surrounding the whereabouts of the other axes of other pitches, are pursued in greater detail in chapter 16, while elsewhere again (chapter 15) the significance and the geometry of the cylindroid is studied *ab initio*.

### Towards wider generality

32. It will be evident from the foregoing argument that, unless this matter of the whereabouts of the ISAs

Figure 8.07 (§ 8.32) A modification of the apparatus in figure 8.04; it is a move towards a wider generality. For a discussion of the actual $f2$ joint which is evident here, please go to § 7.85.

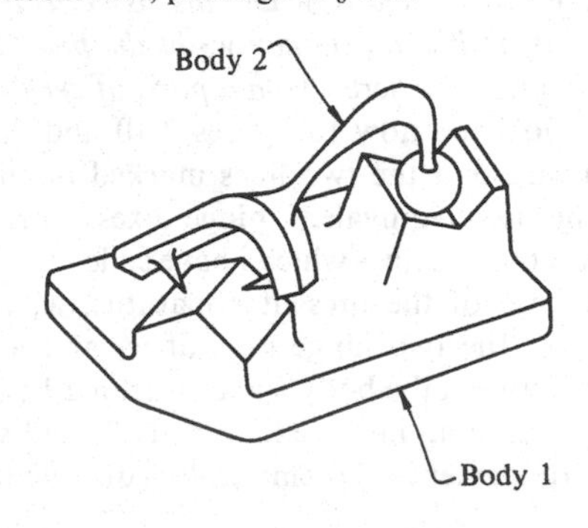

(or of the motion screws, §10.50) is tackled in general, little progress will be made. For more in the way of generality, and taking as a starting point the apparatus drawn in figure 8.04 (where there is the special circumstance that all four points of contact on body 1 remain fixed with respect to body 1 when pure rotation of body 2 occurs about the centre line *w–w* of the balls), we could produce as a variation the object in figure 8.07. In figure 8.07 two *prongs* on body 2 are resting respectively upon two intersecting planar surfaces on 1 in such a way that they irregularly straddle the ridge which can be seen there.

33. Beginning with figure 8.03 and generalising similarly, we could arrive at figure 8.08; but here we see immediately that any selected pair of non-parallel planes on 1 will always intersect along some line, along some ridge or V-groove somewhere; so, for the sake of generality, figure 8.08 might more effectively be drawn in the form of figure 8.09.

34. In figure 8.09 the four points of contact on the body 1 all move with respect to 1 whenever the body moves, but the four points of contact on the body 2 (which always remain at the tips of the prongs) do not move with respect to 2. This is a characteristic circumstance whenever sharp prongs and these flat facets are envisaged or employed (§7.20, §7.21). If we aim at a final generality here, there is no special reason for prongs on the body 2. Nor is there any special reason for the four flat facets on the body 1.

35. Please take a look, accordingly, at figure 8.10 and notice the *rounded protuberances* drawn upon both of the bodies there. These not only ensure that the required four points of contact can be made in the imagination, but also that they do move with respect to both bodies when relative motion occurs (§6.62). This nicety of the generality would be important if we were investigating the question of finite movements; but we are dealing in the present discussion with instantaneous motion only. We are dealing, in other words, with infinitesimal displacement only and its first derivative with respect to time, namely velocity, so the various points of contact in our presented (and projected) analyses are not required to move at all.

36. It is nevertheless true to say that, for a general consideration of the capacity for instantaneous relative motion of two directly contacting bodies 1 and 2 with two degrees of freedom, figure 8.10 with its evident points of contact, all four of which are unencumbered with specialities or ambiguities, is the most desirable. Such a device could be used to advantage in chapter 15. Refer to §15.46 and see that there, however, I have

Figure 8.08 (§8.33) Four points of contact between the bodies 1 and 2; the prongs and platforms here are irregularly arranged. Refer to §7.19 for prongs, platforms, and facets.

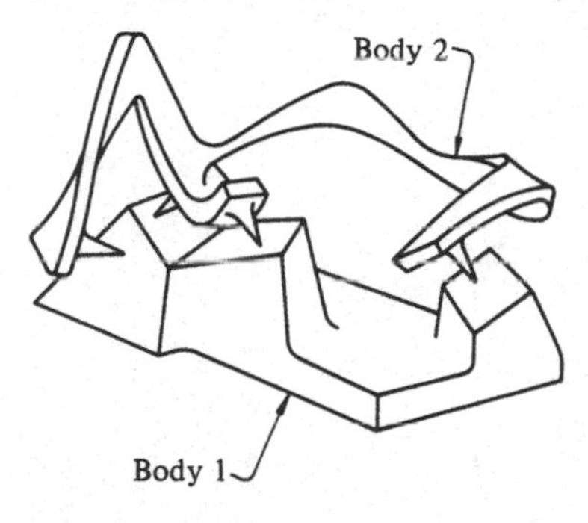


Figure 8.09 (§8.34) Another version of figure 8.08; the points of contact here can move upon the facets 1, but they must remain fixed with respect to 2; they remain fixed upon the tips of the four sharp prongs which may be seen. Refer to §7.19 for prongs, platforms and facets.

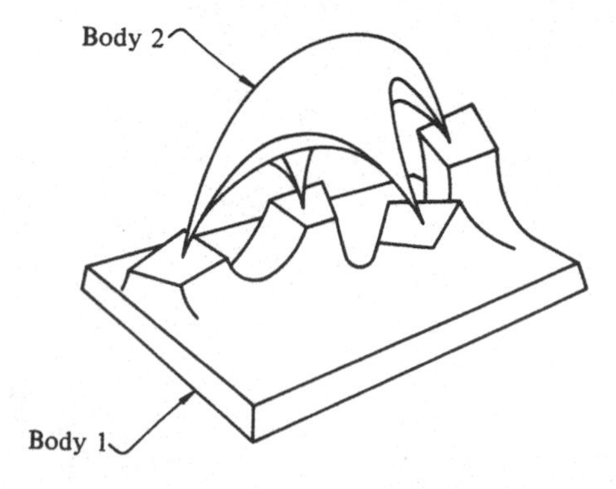


Figure 8.10 (§8.35) Two irregular, rigid bodies of general shape. They are in contact at four points. As the bodies move, all points of contact move with respect to both bodies, and we have here the most elegant of the generalities. See also figure 6.07 and refer to §6.62.

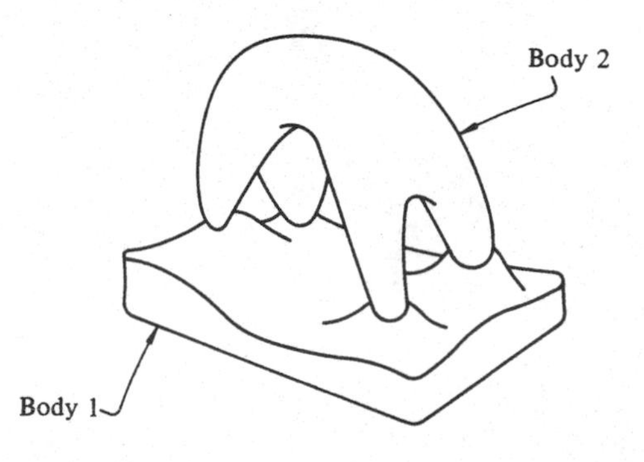

chosen for convenience to use figure 6.06. Beginning at § 15.46, and despite the specialities of figure 6.06, I show without any loss of generality that a complete cylindroid of screws will, as it were, spring into being as soon as a set of any four separated points of contact between a pair of rigid bodies is made. *Provided contact is not to be lost at any one of the four points, the cylindroid shows the layout in space at an instant of all of the ISAs or motion screws which are available.* Refer also to § 6.57; I have made exactly the same assertion in its context there.

9. The distribution about the motion (or the action) screw of the infinity cubed of right lines in a rigid body may be modelled by means of a wire framework. The twisted curves here are helices and the straight lines are the right lines.

# The linear complex of right lines in a moving body

## The right lines in a moving body

01. I began to explain in chapter 3 the importance of and the ubiquitous nature of that special kind of line in a moving body which I called there, for the want of a better name, *right line*. There and in other places the *n*-line and the contact normal are also mentioned (§ 3.01, § 3.05, § 3.06) and a distinction made between a line which is obliged by virtue of some constraint to be a right line in a body as soon as the body moves and other lines in the body which may or may not be right lines according to how the body moves. Because I am speaking here in chapter 9 about the kinematics of a body, I am speaking about the motional (not the actional) right lines within the body; the distinction between the two is explained at § 3.43.

02. In any event it is taken for granted here that the reader is aware of the practical significance of the right line in the kinematics of mechanism. The objective of this chapter is not to study the significance of right lines but to plot the overall distribution of them in a moving body or in a body which has the capacity to move with one degree of freedom only (§ 3.24). We are trying to discover, not only where the right lines in a body are, but also how many of them there are.

03. It should also be mentioned here (in reference to chapter 10) that, although this chapter 9 deals with motion and thus with the ISA and the linear complex of motional right lines surrounding it, it might just as well have been written in the field of *action* and thus in the field of the central axis of a wrench and the complex of actional right lines surrounding it (chapter 10). In that alternative circumstance, however, we would need to have a clear idea here of the physical significance of those straight lines in a body which join points in the body where the total *moment* vector there is perpendicular to the line. It might be better simply to argue that we are about to introduce via motion, and then to describe the geometry of, an abstract, geometrical thing known by the name *linear complex*. See § 10.09.

## The helicoidal field of velocity vectors

04. It can be shown that whenever a body is moving there is an axis unique at every instant called the instantaneous screw axis or the ISA about which the body is screwing with some pitch, and that surrounding the ISA there is a field of helices upon which is based a helicoidal velocity field; refer to the argument at chapter 5 leading to figure 5.09. Sometimes I use the word *twisting* synonymously with the word *screwing* as used above and, in chapter 10, I will often refer to the ISA as the *motion screw*. Thus we could also say here the body in figure 5.09 is *twisting at the instant about its motion screw*.

05. To describe this field we might imagine that all of the helices of the same pitch as that of the ISA are drawn concentric with the ISA. There must be in our imagination a helix at every radius from zero to infinity, and at each radius there must be an infinity of helices each displaced from its neighbor an infinitesimal distance in the axial direction. There is in this field of helices a double infinity, an $\infty^2$, of separate helices and together they fill the whole of space. Through each point in space one helix passes and along each helix there is a single infinity, an $\infty^1$, of points. These last two statements are consistent with the fact that in the whole of space there is an $\infty^3$ of points (§ 4.01).

06. In figure 9.01 there is drawn a selection of nine helices collected into three 'ribbons' of three. The ribbons are cylindrically arranged; the helices are of the same pitch, they are all right handed, they are coaxial, and together they are intended to convey an impression of the totality of the $\infty^2$ of helices under consideration. The *helix angles* $\theta$ of the helices are shown; they are all acute and are measured between the direction of the central axis of the helices and the directions of the tangents $T$ which are shown. Notice in passing that the helices of zero radius are of zero $\theta$ and are identical with the central axis, while the helices of infinite radius are sets of infinitely large circles set normal with the central axis. At § 9.15 it will be important to have understood the nature of these degeneracies.

07. Now we know (§ 10.47, § 10.49), and referring to figure 5.09, that the velocity at each point in a moving body is in the direction of the tangent to the helix at that point, and that together all of these $\infty^3$ of velocity vectors make up the helicoidal velocity field. So please see in figure 9.02 the following: *(a)* a single representative point $A$ in a body, which body is screwing about its ISA with some given (right handed) pitch; *(b)* the unique helix of that pitch passing through the point $A$; and (*c*) the relevant velocity vector **v** bound with its basepoint at $A$. In the figure there is a thing like the hub of a wheel at $A$, with its axis pivoted upon the velocity vector there.

**The wheel of right lines at a point**

08. Also in figure 9.02 is drawn the single infinity of lines through $A$, each line of which is perpendicular to the velocity vector **v**. This infinity of lines, which is a planar pencil of infinitely long lines (§ 3.08, § 11.03), is drawn as a truncated 'wheel'. Its 'spokes' represent the lines and its axle or axis is in the direction of the velocity vector. I now make the statement that, because each of the spokes of the *wheel* in figure 9.02 is a line containing a point whose velocity is perpendicular to the line, every one of them is a right line in the moving body. This statement relies upon and follows directly from the axiom of rigidity; if the component of the velocity along the line at one point in a line is zero then the same must apply for all points along the line (§ 3.04, § 5.03).

09. Through any point $A$ in the moving body, accordingly, there passes at least a single infinity of right lines, arranged in a planar pencil or as a *wheel of lines* as shown. The plane of the wheel is designated $\alpha$ in the figure. A first question now arises: are there any other lines through $A$ which could be right lines? To answer this question we can take any line through $A$ which is not in the plane $\alpha$, say the line marked $j$–$j$ in figure 9.02, and notice that the velocity at $A$ will then have some finite component along $j$–$j$. This means of course that the line $j$–$j$ and all lines like it cannot be right lines in the body.

10. There arises, as well, another question: is it possible that there are some lines in the plane $\alpha$, such as the line marked $k$–$k$ in the figure, which could be right lines? There are various ways of answering this question, all in the negative; one of them is as follows. If the line $k$–$k$ were a right line the velocities at both $R$ and $S$ (which must be perpendicular to $AR$ and $AS$ respectively because $AR$ and $AS$ are both themselves right lines) would have to be perpendicular to $k$–$k$ also. This means that, if $k$–$k$ were a right line, the velocities at $R$ and $S$ and indeed at all points across the plane $\alpha$ would be thereby normal to that plane. This is in clear contradiction to our accepted basis for argument, that the body is screwing about its ISA with a pitch. The proposition that lines like $k$–$k$ in the plane $\alpha$ might be right lines is, accordingly, nonsense.

11. Neither § 9.09 nor § 9.10 implies, incidentally, that right lines in the body cannot *intersect* the plane marked $\alpha$, for right lines can freely do so at all points other than at $A$. Indeed, unless they are parallel with it, all other right lines in the body will intersect the plane marked $\alpha$ somewhere. Having answered the above two questions in the negative, we can conclude that, *neither through A nor across the plane $\alpha$ of the wheel, are there*

Figure 9.01 (§ 9.06) Nine helices representing the whole field of the $\infty^2$ of coaxial helices of the same pitch that can be seen to surround an ISA. Note the helix angles $\theta$.

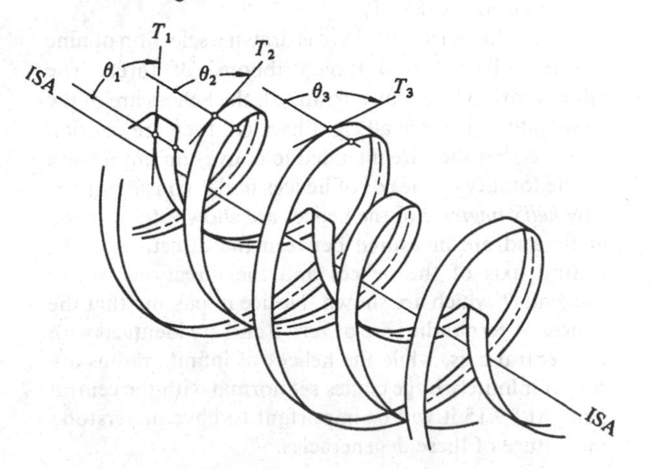


Figure 9.02 (§ 9.07) The linear velocity vector **v** at a point $A$ in a moving body is tangent to the relevant helix at $A$ whose central axis is at the ISA. A planar pencil of right lines, shown here truncated and called in the text a 'wheel' with radiating spokes, occurs in the plane at $A$ that is normal to **v**.

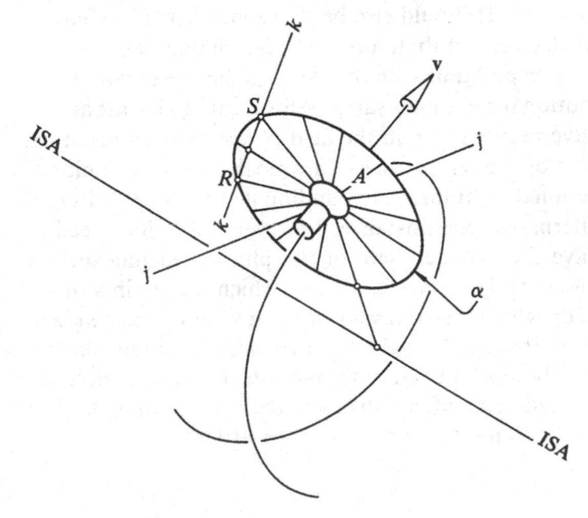

*any lines which are right lines in the body other than the spokes.*

**The layout in space of the right lines**

12. Try to imagine now the total layout in space of the whole set of the right lines. Let us allow first of all the wheel at $A$ in figure 9.02 with its infinitely long spokes to sweep through the whole of space (sweeping out right lines with its spokes as it goes) by sliding in the direction of its axis, tangentially along the curve of the helix, around and along in such a way that it 'nutates' as it sweeps its way from one end of the infinitely long helix to the other. Imagine next that this sweeping is allowed to repeat itself with the same wheel sliding around and along each of the other single infinity of helices of the same radius. This latter action could also be achieved in the imagination simply by sliding the whole picture produced from the first helix bodily along the ISA for one complete pitch distance. Alternatively the whole picture could be rotated about the ISA for one complete revolution.

13. Notice at this stage that there are no swept-out right lines cutting the ISA which are not perpendicular to the ISA. This must be so because there are as yet no spokes other than the 'helical–radial' spoke which can cut the ISA at all.

14. Remember next, however, that there are helices of all radii to be considered. For the smaller radii the imagined nutating wheel will sweep more broadside-on to the line of the ISA, while for the larger radii the wheel will be sweeping more edge-on to that line. In the limit at zero radius in our imagination the sweeping wheel will finally and simply be sliding along the ISA sweeping out some right lines already swept out. In the limit at infinite radius the wheel will be swinging around and along in such a way that all the right lines which almost (but never do) cut the ISA at angles other than a right angle will finally be swept out, and thus be included to complete the total picture.

**The question of the points A at infinity**

15. Despite my remarks at §9.06 it should be understood that, even in the long run when the wheel's axis is infinitely far away from the ISA (and thus at right angles to it), none of its spokes other than its helical–radial spoke can cut the ISA. As the helix radius approaches infinity, other spokes come closer and closer to cutting the ISA at angles other than a right angle, but none of them ever do. See figure 9.03. There, the wheel at centre $A$ is shown at some large radius $r$ of helix where the helix angle $\theta$ is almost a right angle. The helical–radial spoke $AO$ of the wheel cuts the ISA at O where an orthogonal set of axes $x$, $y$, $z$ is set up. Another spoke $AP$ inclined at some acute angle $\lambda$ to the helical–radial spoke is shown and it is seen to miss the ISA by the distance $Q$–$P$, which is measured in the $z$-direction. The angles $AOP$ and $OQP$ are right angles. Now $O$–$P=r\tan\lambda$, so $Q$–$P=r\tan\lambda\cos\theta$, but by definition $r=p\tan\theta$, where $p$ is the pitch, so $Q$–$P=p\sin\theta\tan\lambda$. When $r$ tends to infinity, $\theta$ tends to $\pi/2$ and $\sin\theta$ tends to unity; so in the limit when $r$ becomes infinity, $Q$–$P=p\tan\lambda$, which is independent of $r$. The intersection of the plane of the wheel with the $xy$-plane is the straight line O$P$, and (unless the pitch is zero) the distance $Q$–$P$ is never zero unless $\lambda$ is zero. *There are accordingly (unless the pitch is zero) no right lines in the body which cut the ISA at angles other than right angles.*

**The total number of right lines**

16. The foregoing description of the whole layout of the right lines rightly assumes that, because the point $A$ in figure 9.02 represents all points $A$ in the moving body, all right lines in the body are accounted for as soon as the planes $\alpha$ are counted. There are a single infinity of right lines in each plane $\alpha$ and there are an $\infty^3$ of the planes (one for each point $A$ in the body); but it does not follow from this that there is an $\infty^1$ multiplied by $\infty^3$, namely an $\infty^4$, of right lines in the body.

17. We know that the maximum number of straight lines which can be drawn in the whole of space is only $\infty^4$ (§4.09) so it is clear from this alone that there

Figure 9.03 (§9.15) This figure helps to show that, even at radius $r$ infinity, there are no spokes of any wheel other than the helical-radial spoke that can cut the central axis; no spokes cut the central axis at an angle other than a right angle. This property of the linear complex breaks down when the pitch of the complex becomes zero however (§9.41).

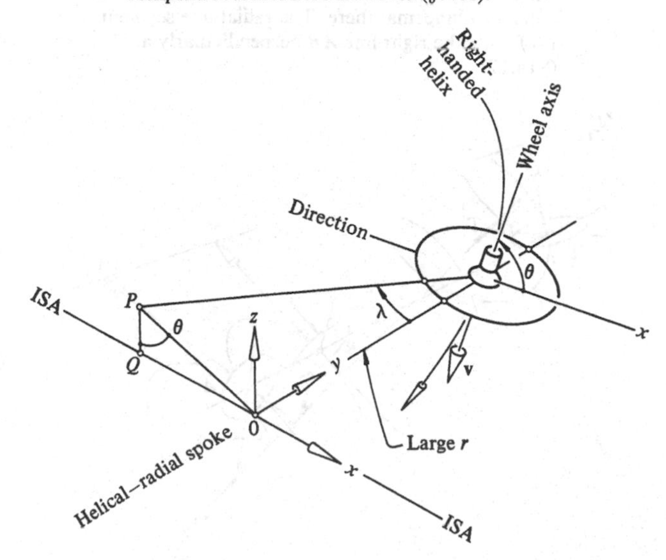

cannot be an $\infty^4$ of right lines in a moving body; there must be less than that.

18. Looking again at figure 9.02 we see that along any one of the right lines through $A$ in the plane $\alpha$ there will be some other point $B$ which might represent the whole single infinity of such points in that line, and that surrounding this $B$ there will be a plane $\beta$ containing the single infinity of right lines which pass through $B$. Refer to figure 9.04. The plane $\beta$ will contain the line $AB$; it is clear from this (because all of the infinity of planes $\beta$ will contain the same line $AB$) that each right line in the body, such as the right line $AB$, has been counted an infinity of times too often in the above false reckoning. There is thus in the body a total of $\infty^4$ divided by $\infty$, namely an $\infty^3$, of right lines; this lesser number makes much more sense in all the circumstances.

19. One of the circumstances is, of course, that each one of the $\infty^4$ of lines in a moving body will have some acute angle $\delta$ between itself and its least-velocity vector at its centre point Γ (§21.03). Appealing to the 'principle of the continuity of nature' (this is, admittedly, a dangerously unreliable principle to be using here), we can 'see clearly' that the angle $\delta$ has one chance in a single infinity of being a right angle. When the angle $\delta$ is a right angle, the line of course is a right line. So it follows from this argument also that there is only an $\infty^3$ of right lines in the moving body.

20. Figure 9.04 shows the mentioned right line AB (§9.18). It is important to see that this $AB$ in figure 9.04, being typical of the $\infty^3$ of right lines in the body, is not just any line $AB$ in the body – as was for example the line $AB$ joining the two arbitrarily chosen points $A$ and $B$ in figure 5.01. The line $AB$ in figure 9.04 is characterised by its special property that the wheels at $A$ and $B$ (and thus at all points along the line) rest with their axes perpendicular to, and thus with one of their spokes collinear with, the line $AB$ itself. Remember that starting from $A$ we spoke (at §9.18) about some other point $B$, *not somewhere along any line emanating from A, but somewhere along one of the spokes of the wheel at A.*

**All of the binormals to all of the helices**

21. The axes of the wheels of course represent the velocity vectors along the line $AB$ and in the special case where the velocity vectors are perpendicular to the line we can look along the line $AB$ (in the direction from $A$ to $B$) and see the velocity vectors displayed in true length; spread out as shown in the left hand part of figure 9.05. Compare here again figure 9.05 and figure 5.01. It has been assumed in figure 9.05 incidentally that $\mathbf{v}_B$ is smaller than $\mathbf{v}_A$ following the circumstance in figure 9.04 that the radius of the helix through $B$ is less than that through $A$. The tips of the vectors in figure 9.05 must be, and have been drawn, in a straight line. The fact established at §5.25, namely that the tips of the vectors for points along a straight line in a moving body must be in a straight line also, applies equally in this special case of the line being a right line.

22. Please see next in figures 9.04 and 9.05 that, at that unique point Γ along the line $AB$ where the line is closest to the ISA, there is for the line a least vector $\mathbf{v}_\Gamma$ (§21.03). It should also be clear that through Γ there

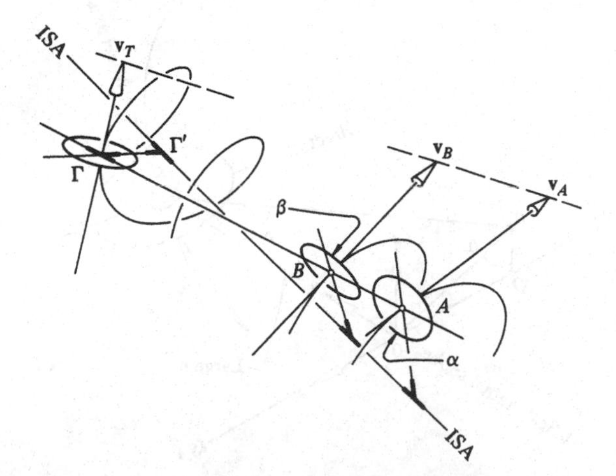


Figure 9.04 (§9.20) A typical right line $AB$ in a moving body. Its centre point Γ occurs at the point where $AB$, cutting the helix of least radius, cuts that helix as a binormal there. The radial line segment Γ′–Γ cuts the right line $AB$ perpendicularly at Γ (§14.17).

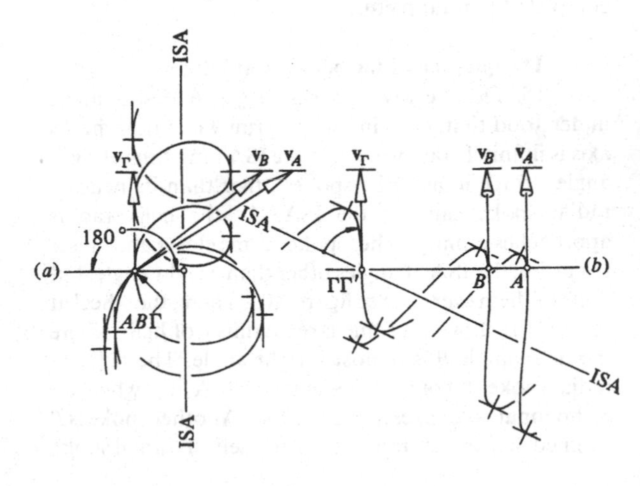


Figure 9.05 (§9.21) Two orthogonal projections of the right line $AB$ looking (*a*) along the line, and (*b*) perpendicularly across it and in the direction from Γ to Γ′. Note that the least velocity vector $\mathbf{v}_\Gamma$ occurs at the centre point Γ of the line (§9.31).

passes the unique helix of relevant pitch whose radius Γ-Γ′ is least for all points along the line. These phenomena can be seen in the model at the photograph 9.

23. At the right hand part of figure 9.05 there is drawn a projection taken in the direction ΓΓ′. Here we are looking along the common perpendicular ΓΓ′ between the right line *AB*Γ and the ISA, and we can see with the help of this view, (*a*) that all of the components of the velocity vectors along a right line are equal in the direction of the least velocity $\mathbf{v}_\Gamma$ for the line, and (*b*) that along any right line the velocity vectors must emanate from the line on one 'side' of the line only, namely from that side of the line indicated by the zone marked 180° in the figure. Photograph 3 shows a model of a right line complete with its display of velocity vectors at one side of itself only.

24. It is important now to notice that the right line *AB* is not simply any one of the single infinity of right lines perpendicular to the tangent to the helix at Γ; it is a special one of these, the *binormal* to the helix at Γ. Now the binormal at a point in a curve is perpendicular not only to the tangent to the curve at the point but also to the normal to the curve at the point (§ 5.52). In the case of a helix the normal at a point is collinear with the radius at the point (§ 5.54), so the binormal to a helix is not only perpendicular to the tangent but also tangent to the right circular cylinder upon which the helix itself is inscribed.

25. The point Γ is, as we have seen, unique in the right line *AB* and, in the context and in accordance with § 21.03, it may be called the centre-point of the right line. Thus we can see that if all of the right lines in a body with all of their centre-points Γ were drawn at an instant in space, we would see another, useful yet somewhat different, picture of the totality of the right lines. Refer to § 9.26.

26. The geometrical layout of the totality of the $\infty^3$ of right lines in a moving body may be described exhaustively yet much more simply thus: *it is that set of lines in a moving body (given the ISA and its pitch) which consists of the binormals drawn at every point of every one of the helices of that pitch which can be drawn coaxial with the ISA.*

#### Towards a visualisation of the set

27. Precisely the same set of lines was presented for the imagination, remember, by means of the nutating wheel at § 9.14 and it behoves me now to reconcile these two different views of the same set (§ 9.14, § 9.26). Geometrically speaking, and independently of how one chooses to see the set by means of the imagination, the set of the lines may be called by the name *linear complex*. Some of the other origins in mechanics of this particular set of lines in space are discussed in chapter 10 and some of its geometrical properties are mentioned in chapter 11. Here, however, I attempt merely to describe the linear complex. That I am speaking in terms of the right lines surrounding an ISA in a body is in some ways irrelevant, for the main matter at issue now is not so much the layout of the right lines in a moving body as the shape of the ubiquitous linear complex seen as a self-standing geometrical entity (§ 11.44). I wish to begin from the base line set at the end of the previous paragraph.

28. Figure 9.06 has been drawn to show a group of binormals representing the total set of the single infinity of binormals belonging to two equi-radial helices inscribed upon their right circular cylinder of relevant radius. By suitably repeating and superimposing in the mind's eye the contents of this picture, taking all of the helices of that particular radius, which are spread evenly along the cylinder as suggested in figure 9.01, we can arrive at figure 9.07. This shows a representative pair of the single infinity of right circular reguli, drawn upon their respective hyperboloids whose circular throats circumscribe the cylinder, and whose totality of generators constitute the $\infty^2$ of binormals to the mentioned set of coradial helices. The generators of the reguli are the $\infty^2$ of right lines which can be drawn tangential to the cylinder which is co-axial with the ISA at that particular radius (§ 9.24). In our example here the pitch of the screwing, or of the rate of twisting (§ 10.47), is right-handed, namely positive, so the reguli of which I speak are all left-handed reguli.

29. In order to visualise now the whole set of the $\infty^3$ of right lines in the moving body we might

Figures 9.06 and 9.07 (§ 9.28) all of the binormals to a pair of helices of the same radius have been drawn in figure 9.06, while at the figure 9.07 all of the generators of two circular reguli appear. Given the common, shown, circular cylinder against which all of the shown lines are tangential, the two figures are representative of the same set of an $\infty^2$ of lines (§ 6A.02).

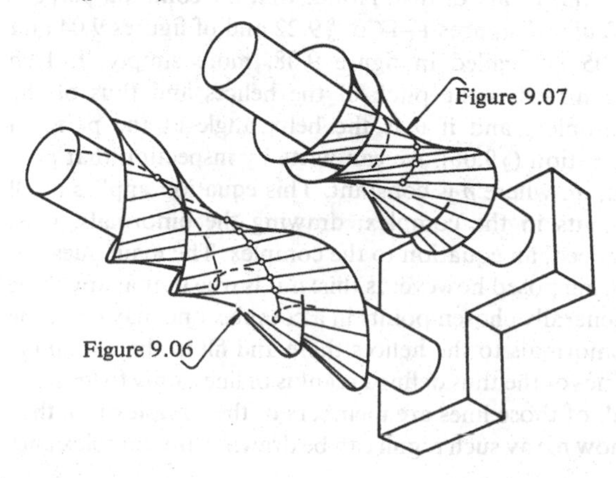

superimpose co-axially the whole series of cylinders represented by the cylinder shown in figures 9.06 and 9.07. The cylinders range from radius zero to radius infinity and each one must be imagined to carry its attendant set of the $\infty^2$ of reguli. Remember that as the radius increases the helix angle $\theta$ increases, so that as we approach infinity the generators of the reguli will become much more nearly parallel with the axis, while as we approach the radius zero the reguli will become much more 'twisted'. At radius infinity the reguli will all have become right circular cylinders congruent with one another, while at radius zero the reguli will have jointly become a set of parallel planar pencils with their respective generators all intersecting the ISA at right angles. All of the generators of these left handed reguli will constitute the total layout in the moving body of the $\infty^3$ of right lines. If the pitch of the screwing were left-handed, of course, the reguli would all be right-handed reguli.

30. In § 9.29 the linear complex is described in terms of right circular reguli, all of which reside upon right circular hyperboloids coaxial with the axis of the complex. About this description I wish to make two comments: (*a*) these reguli are not the only reguli which exist, for there are others, non-coaxial with the axis and non-circular; and (*b*) this method of visualising the complex, although useful, masks from immediate view what we know to be the polar planes, the wheels of § 9.14. I shall deal now with (*a*), then later, at § 9.45 *et seq.*, with (*b*).

### Other reguli of the complex

31. At § 3.62 I introduced an important theorem and I presented the theorem itself at § 3.64. The theorem refers to reguli of right lines in a moving body and it is directly relevant here. Take any three points in the moving body and draw through each one of them the binormal to the helix at that point. Refer to figure 9.08, where three such lines at the generally chosen points $\Gamma_1$, $\Gamma_2$ and $\Gamma_3$ are drawn. Notice that the common perpendicular distances $\Gamma$–$\Gamma'$ of § 9.22 and of figures 9.04 and 9.05 are called in figure 9.08, more simply, $r$. If $p$ (mm/rad) is the pitch of the helices and thus of the complex, and if $\theta$ is the helix angle at the point in question (§ 9.06), we can write by inspection that $r = p \tan\theta$, where $p$ is constant. This equation applies at all points in the complex; drawing the binormals, it is, indeed, an equation to the complex. The main question being posed however is this: if it is true that at any three generally chosen points in a complex one may draw the binormals to the helices there and fill in the remaining lines of the thus defined regulus of lines, only to find that all of those lines are members of the complex too, then how many such reguli can be drawn in the complex and,

Figure 9.08 (9.31) Three binormals to the helices drawn at three generally chosen points $\Gamma_1$, $\Gamma_2$, and $\Gamma_3$ within the complex. The complex is right-handed, and of pitch $p$. The binormals (and they are marked $b$–$b$) are lines of the complex, and the three mentioned points are the centre points of the lines. Note that the angles $\pi/2-\theta$ between the lines parallel with the central axis (the diameters) at the points and the binormals at the points are not the helix angles $\theta$, but the complements of them.

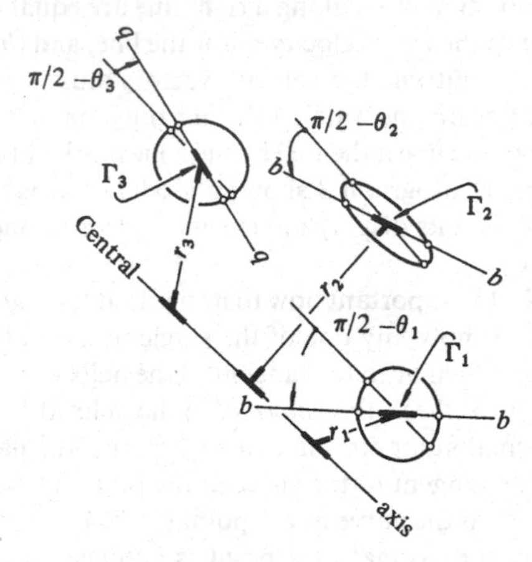


Figure 9.09 (§ 9.33) Any plane through the general point $P$ cutting the central axis of the complex at some angle in $Q$. A primary Cartesian triad $X$, $Y$, $Z$ at $P$ has $PX$ collinear with $QP$, and $PZ$ normal to the plane, while a secondary Cartesian triad $x$, $y$, $z$ at $P$ is rotated about $PZ$ through an angle $\phi$. At the inset we see an orthogonal view of the plane through $P$. Shown there is the throat ellipse of the single existing regulus that is rectilinearly arranged with respect to the axes $Px$ and $Py$. The regulus may be unreal (§ 9.34).

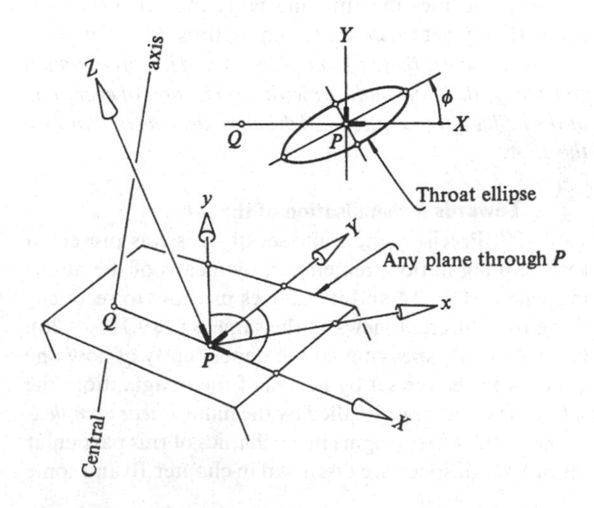

more importantly, *where are they*? Or, alternatively, how many complexes of a given pitch may be drawn to include a given set of three lines?

32. At §4.12 the reader will find, couched in the terms of the enumerative geometry there, an account of the relatively simple matter of finding the number of reguli which can be drawn (or can be seen to exist) within a given complex. The number is $\infty^6$. On the complicated and rather more difficult question of where these reguli are in the complex, I refer now to a recent work of Sticher [1] and briefly summarise its contents here.

33. Please refer to figure 9.09. It has been shown by Sticher that, at every point $P$ in a linear complex of lines, there is an $\infty^3$ of reguli whose hyperboloids are concentric there. This is consistent with the conclusion mentioned at §4.15 that there is an $\infty^6$ of reguli in the complex. There is no redundancy in the argument that an $\infty^3$ of concentric reguli at a point, taken in conjunction with an $\infty^3$ of points, leads to a total of an $\infty^6$ of reguli. Please be reminded that, while any hyperboloid has two reguli which may be inscribed upon it, a right and a left-handed regulus, each regulus which might be drawn in a complex, with or without its opposite-handed mate, will define its own hyperboloid. It is in any event true that the found number of reguli will be the same as the found number of hyperboloids, for, as we know, any finite multiple of $\infty^n$ is, quite simply, $\infty^n$ (§4.05).

34. But look again at $P$ in figure 9.09. Mount at $P$ any plane to contain $P$. The plane will cut the central axis of the complex at some point $Q$ obliquely. Mount at $P$ a fixed Cartesian frame $PXYZ$ such that $PX$ lies in the direction of OP, that $PY$ lies in the chosen plane and that $PZ$ emerges normal to the plane as shown. Next erect another movable set of axes Pxy, also in the chosen plane as shown, which can rotate about the axis $PZ$. Note the angle $\phi$. Now it can be shown that, for every value of $\phi$ in the chosen plane, there exists upon the axes $xy$, when taken as the principal axes, a single throat ellipse. This means that concentric at $P$ and coaxial with $PZ$, there is an $\infty^1$ of different hyperboloids. There is one regulus of lines from the complex drawn upon each one of these hyperboloids. With centre $P$ and in the chosen plane there is, as I have said, a single infinity of throat ellipses; they are all of different proportions and there is in general one *circular* throat among them. This means that, considering all of the $\infty^2$ of planes which can be drawn through $P$, there is an $\infty^2$ of circular hyperboloids (and thus circular reguli of the complex) concentric at $P$. Their axes form a star. We find however that not all of the throats are real. Real throats exist over limited ranges of $\phi$ only.

35. To begin to see more clearly the pattern of the hyperboloids within the complex, take now a special set of planes through $P$. Drop from $P$ onto the central axis of the complex a perpendicular $PO$ and mount upon this line the pencil of planes defined by it. Refer to figure 9.10($a$). There, four planes of the pencil are shown: plane 1 is normal to the central axis of the complex; plane 2 is normal to the velocity vector at $P$, namely normal to the helix through $P$; plane 3 contains the central axis of the complex; while plane 4 contains the velocity vector at $P$, that is, the plane 4 is tangent to the helix at $P$. Note next that the pitch of the ISA and thus of the complex has been drawn right-handed, namely positive. Now it is a feature of all of these planes of the mentioned special pencil that the ranges of $\phi$ for real throats extend only from zero to zero and from $\pi/2$ to $\pi/2$: there are, in other words, only two values of $\phi$, namely zero and $\pi/2$, for which real throats exist.

36. When these throats do exist – and that depends upon the particular plane of the pencil as we shall see – there is again an $\infty^1$ of them, and they are arranged as shown in figure 9.10($b$). Note that at each plane of the pencil where real throats exist there is one, and only one, circular throat. The next thing to see however is this: if the general plane 5 can be seen to be

Figure 9.10($a$) (§ 9.35) A pencil of planes whose common line of intersection cuts the central axis of the complex perpendicularly. Real throats exist within the ranges $\alpha$ (from 1 to 2) and $\beta$ (from 3 to 4). The plane 2 is the polar plane at $P$. Please note the location and the orientation of the radially opposite polar plane; it is centred at $P'$.

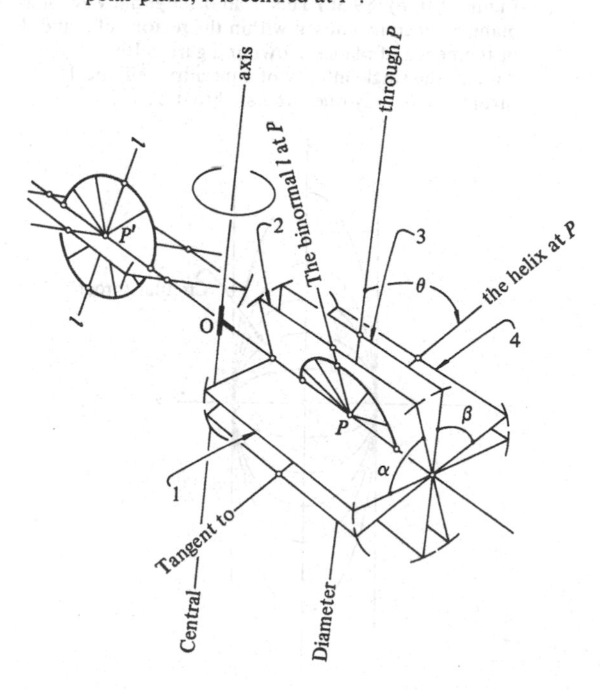

steadily rotating through the fixed planes 1, 2, 3, and 4 in that order, real throats will be seen to exist only within the ranges from 1 to 2, namely $\alpha$, and from 3 to 4, namely $\beta$. For this right-handed complex, the reguli upon the throats in $\alpha$ are left-handed reguli, while the reguli upon the throats in $\beta$ are right-handed. At each of the four ends of the two ranges $\alpha$ and $\beta$, namely at the four fixed planes, special conditions occur; I would now like to mention these.

37. At the beginning of $\alpha$ in plane 1 the radius of an infinitely large circular throat emerges; all lines of the regulus upon it are normal to plane 1 and parallel with the central axis of the complex; the regulus is an infinitely large circular cylinder concentric with the ISA. At the end of $\alpha$ in plane 2 we have arrived at a *polar plane* of the complex (§ 9.46, § 11.15); the circular throat has contracted to zero size and all lines of the regulus upon it reside within the plane 2; refer to figure 9.02. At the beginning of $\beta$ in plane 3 the circular throat emerges again with infinite radius, but here the regulus upon it resides within the plane; the regulus is a single infinity of lines tangent to the throat circle at infinite radius (§ 11.16). At the end of $\beta$ in plane 4 the circular throat has again contracted to zero size, but here the regulus upon it has collapsed into a single line; the line is marked $l$ in figure 9.10($a$). The line $l$ is a member of the regulus of lines in the polar plane (the plane 2) at $P$; *it is a member, in other words, of the planar pencil of lines in the polar plane at P*. See § 9.47.

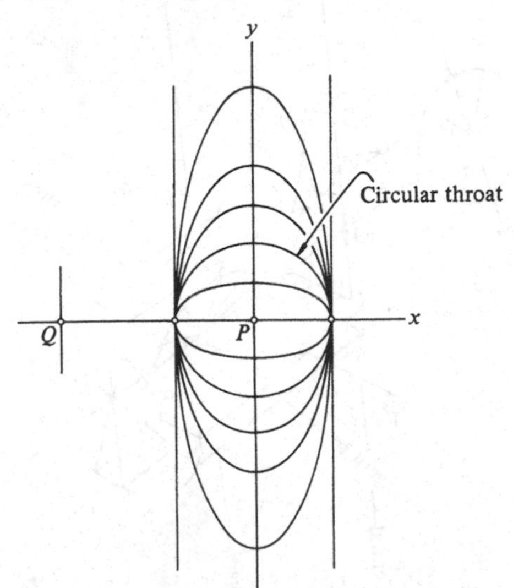


Figure 9.10($b$) (§ 9.36) This is an orthogonal view of a plane 5 generally chosen within the regions of $\alpha$ and $\beta$ of the pencil of planes shown in figure 9.10($a$). Among the single infinity of concentric, elliptical throats on 5, only one circular throat exists.

38. While looking at the planar pencil in the polar plane at $P$, look also at the 'opposing' point $P'$. This latter is the reflection of $P$ in the central axis. There is a planar pencil in the polar plane at $P'$ too; together these two planar pencils of lines (at $P$ and at $P'$) make up a regulus of a special kind. Refer to figures 6.05($d$) and 7.12($c$). This kind of regulus, whose hyperboloid consists of two intersecting planes, abounds in the linear complex, for, as will be seen, there will be one for every pair of points in space which contain any one line of the complex in common. There are an $\infty^5$ of such pairs of points.

39. It is interesting next to see what happens at points $P$ which are at zero radius $r$ from the axis of the complex, that is, at points $P$ which inhabit that axis. Let us note first of all in figure 9.10($a$) that, as $r$ tends to infinity, the ranges of $\alpha$ and $\beta$ both expand and, in the limit, they cover the whole of the pencil of planes on $OP$. In planes containing radial lines at least, there are no unreal throats at infinity. When $r=0$, however, $\alpha=0$ and $\beta=0$. This means that the only throats concentric at some point $P$ on the central axis are those, ($a$) within a plane containing $P$ which is *normal* to the central axis, and ($b$) within planes containing $P$ which *contain* the central axis. There are however an $\infty^2$ of throats of

Figure 9.11 (§ 9.39) Circular throats concentric at points $P$ upon the central axis of the complex. On the central axis it is only in planes such as the ones shown that throats exist. In each of these planes, however, there exists an $\infty^2$ of concentric, elliptical throats. In each plane a single infinity of the $\infty^2$ of elliptical throats are circular.

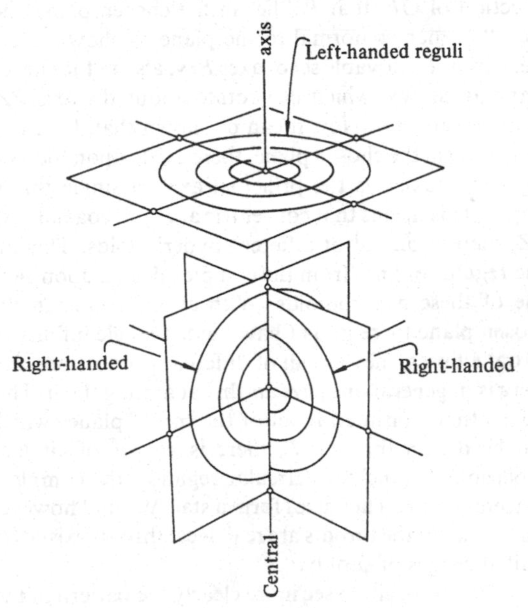

elliptical shape in each of these planes, including a single infinity of circular throats all concentric in each plane at $P$. Refer to figure 9.11. The figure shows only the circular throats in the planes, but it and the above results are consistent with the general statement made at §9.33 that there exists, concentric at every point $P$ in the whole of space an $\infty^3$ of reguli. Note the handedness of the different reguli mentioned in figure 9.11 and note that the left-handed ones are the ones we earlier saw in figure 9.07.

40. In summary we can say the following. There are an $\infty^6$ of reguli. There are an $\infty^5$ of circular reguli; indeed there are an $\infty^5$ of reguli of each prescribed throat proportions – recall §9.38 in this connection. Sticher says that, if $N_X$, $N_Y$ and $N_Z$ are the direction cosines of the central axis in the system, if $p$ is the pitch, if $r$ is the distance $Q$–$P$ in figure 9.09, and if the formula $ax^2+by^2-cz^2=1$, represents a regulus of the ensemble, then suitable general formulae for the reguli existing can be written as follows:

(1) $\pm N_Z\, c^{\frac{1}{2}} + (ab)^{\frac{1}{2}}(pN_Z - N_Y r) = 0$

(2) $\pm b\, c^{\frac{1}{2}}[N_Z\, r \sin\phi + p(N_X \cos\phi + N_Y \sin\phi)]$
$+ (ab)^{\frac{1}{2}}[N_X \cos\phi + N_Y \sin\phi] = 0$

(3) $\pm a\, c^{\frac{1}{2}}[N_Z\, r \cos\phi + p(-N_X \sin\phi + N_Y \cos\phi)]$
$+ (ab)^{\frac{1}{2}}[N_X \sin\phi - N_Y \cos\phi] = 0.$

Please note that Sticher's $a$, $b$, and $c$ here are the reciprocals of my $a$, $b$, and $c$ at §6C.09 and §11.13. Figure 9.12 is my sketch of a few of the reguli which exist. Please note (*a*) that the central axis of a complex cuts none of the reguli except through a generator perpendicularly (§9.15), and (*b*) that, if the former cuts the latter once, it cuts them twice (§11.20).

**When the pitch of the complex is zero or infinity**

41. Before I conclude this chapter an important remark should be made about the so called 'special' linear complex which results when the pitch becomes zero. This corresponds to the question in our context of what happens when the screwing about an ISA becomes mere turning or pure rotation, a common enough occurrence in the practice of mechanism. When the pitch becomes zero the helices all become sets of coaxial circles and all of the generators of the complex cut the central axis. It is only under this special circumstance of the pitch being zero that we find members of the complex cutting the central axis at angles other than a right angle (§9.15). *The special linear complex may be defined as that totality of lines in space which can be drawn to cut a given line; there are as might be expected an $\infty^3$ of such lines.*

42. In line with my opening remarks as §9.03 about motion on the one hand and action on the other in the mechanics of mechanism, let me remark here that, if a *force* were acting along its line of action, we could draw every line in space to cut that line of action and thus construct a special linear complex of lines. These lines would all be right lines in the body suffering the force; indeed they would be actional right lines (§3.43). At all points along the lines the *moment* there of the force would be perpendicular to the line. The analogous matter discussed at §9.41 is the fact that, at all points along the motional right lines, the velocity is perpendicular to the line. Note that here it would appear that, in the same way as linear velocity is consequent upon an angular velocity existing, moment is consequent upon a force. Refer to chapter 10.

43. What happens when the pitch of the complex becomes infinite is also interesting. All of the helices become straight lines parallel with the central axis and the $\infty^2$ of them fill the whole of space. The right lines, the generators of the complex, occupy planes normal to the central axis. There is an $\infty^2$ of lines in each plane and an $\infty^1$ of planes. It follows not surprisingly that there is in the linear complex of infinite pitch an $\infty^3$ of lines.

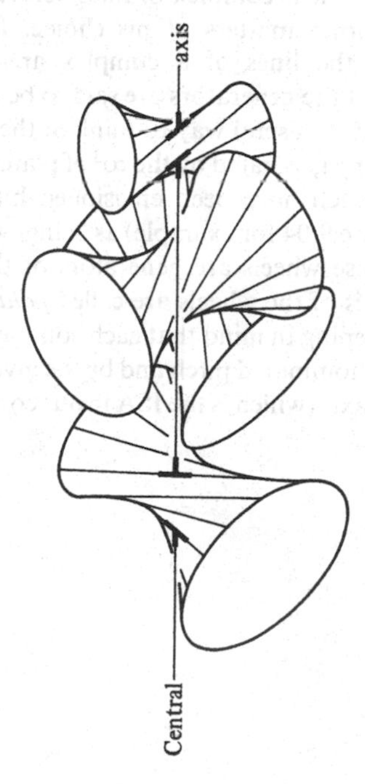


Figure 9.12 (§9.40) A very few of the $\infty^6$ of reguli that can be drawn using the generators of a single complex. Note that, if any one of the hyperboloidal surfaces cuts the central axis of the complex once, it cuts it twice, and that, at the points of intersection, the relevant generator of the regulus and the central axis of the complex intersect one another perpendicularly. In this sketch the complex is a right-handed one.

**Another interesting idea**

44. I would like to give here an answer to the question so often asked: seeing that we extract from the limited number of lines available in space, namely $\infty^4$, the very large number $\infty^3$ of lines to assemble any one complex, why not describe the complex not by the layout of those lines which are members of it but by the layout of those lines which are not? The answer is implicit in the somewhat confusing fact that when $\infty^3$ is subtracted from $\infty^4$ the remainder is still $\infty^4$: *there are very many more lines which are not members of a complex than the number which are, the ratio of non-members to members being infinity.*

**History and some geometry of the linear complex**

45. Having said all this and having come thus far, the extraordinary thing to find is that the pure mathematicians of more than a hundred years ago, generally uninterested in mechanism, knew this particular layout or arrangement or 'ensemble' of lines in Euclidean space, and called it the *linear line complex*, or linear complex for short. The word 'complex' refers to a single infinity (as distinct from $\infty^0$ or $\infty^2$) of lines passing through each point in space, the word 'line' refers to the fact that the lines are straight and not curved, while the word 'linear' refers to the algebraic formulation of the ensemble which obliges the single infinity of lines through each point in space to form not a cone of some shape but to be coplanar (§21.22).

46. It remains now to present here some basic terminology and some of the simpler geometrical properties of this linear complex of lines before the reader goes on to other matters of his choice. In accepted terminology, the lines of a complex are termed its *generators* and the generators are said to be *members* of the complex. One useful way to think of the complex is to see it as being generated by the $\infty^3$ of planar pencils of generators which have been envisioned hitherto (and shown in figure 9.04 for example) as being 'wheels'. The spokes of these wheels are generators of the complex and the planes of the wheels are called *polar planes*.

47. Keeping in mind that each complex is characterised by its nominated pitch and by the given direction of its central axis (which is the ISA in our context), it can be seen (with some difficulty) that a complex is wholly defined by the following simple statement: *through every point in space there must pass a complete polar plane of generators and no others, and in every plane in space there must reside a complete set of generators among which intersection occurs at one point only.*

48. The above statement infers, of course, (*a*) that no two polar planes can be coplanar, which can be understood from a consideration of the contents of figure 9.04; and conversely (*b*) that no two different generators from different polar planes can intersect, which can be seen from the fact that any generally nominated plane in space will always be cut by some unique helix at some unique point where the tangent to the helix is normal to the plane (§11.35). *Every generally nominated plane in space, in other words, is a polar plane at some point somewhere in any given complex.*

**Closure**

49. We have dealt here in a quite elementary way with one appearance only of the ubiquitous linear line complex. It has an origin also in the mechanics of groups of forces, statics, as Möbius saw (1837); and, if we wish, it can be seen to have an origin in the area of pure mathematics, as Klein explains (1908). It certainly has its place in the wide field of screw theory as applied to the elucidation and synthesis of the motion and action in machines and mechanism. Its main geometrical properties are therefore summarised at §11.30, its intersections with other complexes are discussed at §11.40 and its remarkably simple equation in Plücker's line coordinates is given at §11.44. One can look back over the nine chapters here contained, or forward to those appearing in volume 2, and one will see throughout that the linear line complex, the linear complex, or, more simply, the complex, figures largely everywhere. Please refer to chapter 10 and volume 2.

**Notes and references**

[1] This material arose in a private correspondence between myself and Dr Sticher during 1981. Except for the present condensed version of that material (which is written here by me) the work remains, as yet, unpublished. Refer §9.32.

10. Along the underside of the claw-arm of a crab there are five revolute joints in series. These and the requisite muscles connect six pieces of exoskeleton together to form an open articulated loop.

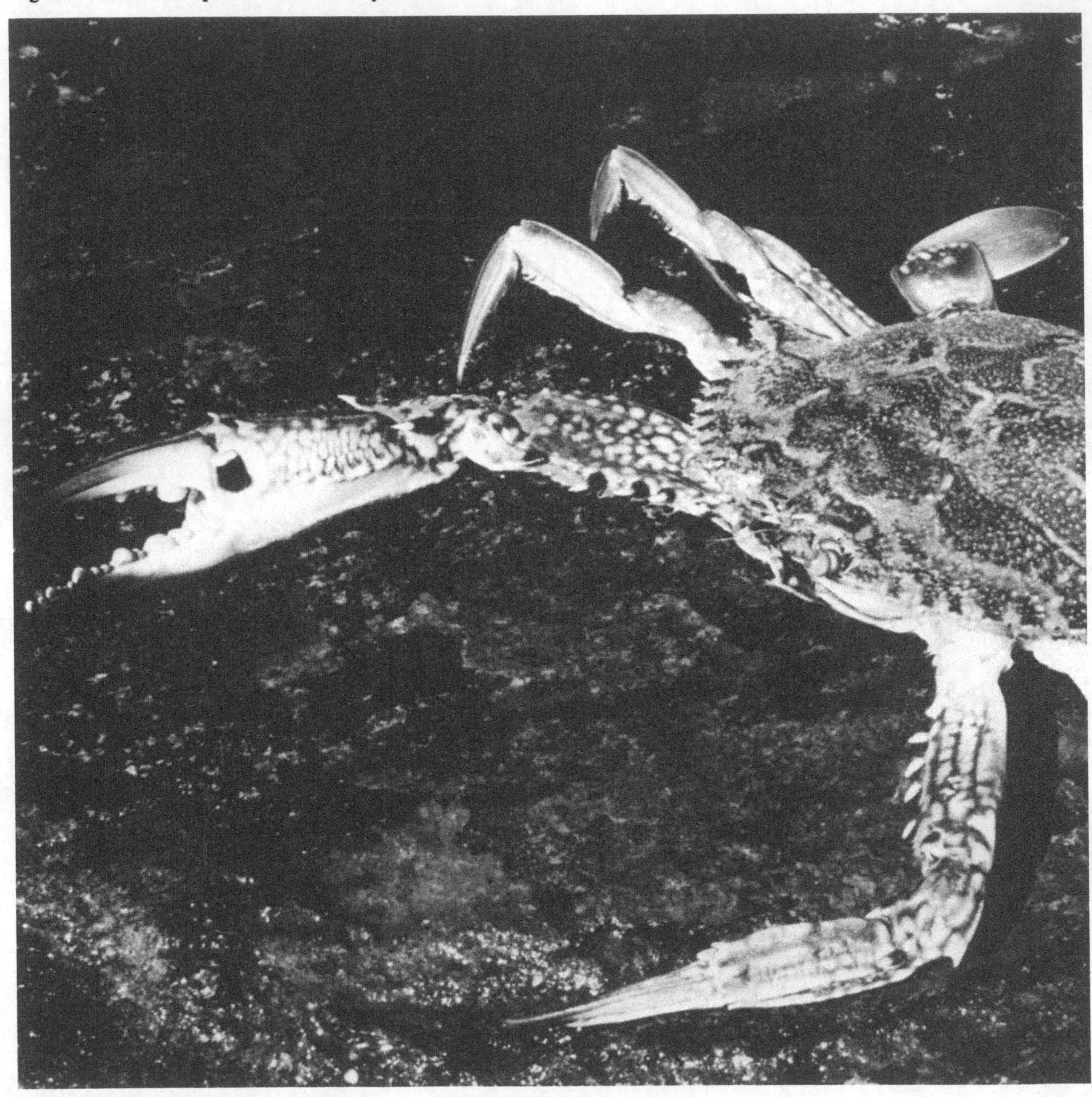

# 10 Line systems and the dual vectors in mechanics

## Opening remarks

01. In chapter 6 and in other parts of this book hitherto I have spoken about systems of ISAs or of screws, or more simply about screw systems. The origin in physical reality for most of these spoken remarks about screws was the capacity for instantaneous motion of some rigid body whose freedom to move at the instant was being restricted in some way. Relationships exist between these systems of screws about which small twists or rates of twisting of one body relative to another may occur, namely the systems of ISAs, and identical kinds of systems of screws about which wrenches and reaction wrenches between the same two bodies may act. An investigation of these two sets of systems of screws will reveal, at the end of this chapter 10, (*a*) an insight into the power expended in friction at working joints in mechanism, (*b*) an amplified meaning for the somewhat narrow term *joint* as defined for example at § 1.11, and (*c*) the beginnings of a method for calculating the forces at work at the joints of mechanism where mass and the consequent inertia of links is an important consideration.

02. With regard to (*b*) above I can mean by joint, as I shall show, the joint between for example the piston and the connecting rod of an engine designed for the transmission of power in the absence of loss, or the joint between for example a bulldozer blade and its one-off job where power is being releases in spurts, or the joint between a ploughing tool and its sod which is in a continuous, power-releasing action. Or I could mean the actuated joint or joints of an industrial manipulator arm which operates by virtue of a power-input only intermittently. I shall also begin to explain at the end of this chapter 10 the fundamental condition for the reciprocity of screws, a matter of pure geometry which can arise directly from the mechanics of joints in the absence of friction. Refer in advance to figures 10.14, 10.15, and 10.16. With regard to (*c*) above, an early reference might be made to § 10.59 *et seq.* The contents of these passages, along with the well-known principle of d'Alembert, are taken up and developed further in chapters 13 and 19.

03. Whereas a small twist of a body is a combination of a small angular displacement about an axis and a small linear displacement along it, a wrench upon a body is a combination of a force along a line of action and a collinear couple (§ 10.05, § 14.27). Small twists, or, if we wish, instantaneous rates of twisting (§ 5.50), and wrenches are both dual vector quantities (§ 10.50). They may both with physical meaning be added like to like (§ 15.11, § 19.20), and they may with physical meaning be multiplied with one another vectorially (§ 14.28, § 19.08). There are however not only these relationships between rates of twisting on the one hand and wrenches on the other, there is an analogy between the two. The analogy is important. Unless one sees for example that angular velocity $\omega$ is like force **F**, that rate of sliding $\tau$ is like couple **C**, and that linear velocity **v** is like moment **M**, one will not be able to see, in spatial mechanism and thus for example in robots and the like, that the product of wrench and the rate of twisting is power.

04. In the area of abstract vector analysis, any chosen screw of a chosen pitch may be ascribed a magnitude and a sense and it may thence be represented by a pair of vectors, namely by a dual vector. Dual vectors of this abstract kind are mere geometrical things: one of them might for example consist of a somehow-connected pair of measured and directed line segments, one standing for a unit rotation about the axis of the screw and the other for a pitch. The geometrical behaviour of any pair of abstract vectors may be arbitrarily defined: the vectors separately or the combination of the two may be caused to obey whatever rules for movement from place to place, for addition or for multiplication, we may wish. Usually however the rules are chosen in accordance with some perceived aspect (or imagined extension) of physical reality.

05. In the more demanding area of experimental mechanics on the other hand we conceive of actual physical quantities, such as angular velocity $\omega$ and force

**F**, and such as 'rate of sliding' $\tau$ (§ 5.50, § 10.45) and couple **C**. Accordingly and quite naturally we come to speak of: (*a*) *rates of twisting*, each of which involves an $\boldsymbol{\omega}$ and a $\tau$ and each of which is seen to occur upon or along, or 'about' a screw (chapter 5); and (*b*) *wrenches*, each of which involves an **F** and a **C**, and each of which is also seen to act upon or along or about a screw (chapters 10, 15). With an eye to correspondence with physical reality we next choose from existing material, or we write if necessary, sets of rules for the vector analyses. *It is very important to see in this physical arena that, unless the available or the written sets of rules accord with what we believe we see in the physical reality, we will reject the rules, for otherwise we would lose our hard-won, useful, and much-practised ability to predict.*

06. Against such a background and without further apology I shall speak in the first instance at § 10.19 *et seq.*, where some experiments are to be described, (*a*) of screws of zero pitch which might be seats for pure angular velocities $\boldsymbol{\omega}$, these latter being free to occur about the screws, and (*b*) of screws of zero pitch which might be seats for pure forces **F**, which latter will be said, not to occur about, but to act along, the screws. In the proposed experiments I shall thus confine myself in the first instance to single vector quantities. Please be aware that I have been employing here the words *occur and act*, and *upon, along, and about*, almost as though the grouped words were synonymous with one another. Notice also how the two words *rest* and *equilibrium* are analogously if not synonymously used in the following, duplex sentence. I shall be speaking soon of observing in physical reality (*a*) the addition to rest of a number of angular velocities imposed upon a body; and (*b*) the addition to equilibrium of a number of forces imposed upon a body. These experiments and associated observations will lead next to a consideration of wrenches and rates of twisting and of the question of power and, before this chapter is taken to its close, with some remarks about the matter of reciprocity (§ 10.62) and some notes about the overall significance of this for the mechanics of machinery (§ 10.63 *et seq.*), we shall have traversed already the area of the analogies to which I have alluded (§ 10.03, § 10.40). Refer in advance to figures 10.09 and 10.10.

### The linear line systems

07. Before describing those experiments, however, I wish to isolate in the field of Euclidean and pure projective geometry, and to present here for identification, a limited group of only three of the simpler of the so-called *line systems*. I wish to do this because these systems figure, as we shall see, in the experiments, and without some elementary understanding of the systems, the experiments will be spectacular but meaningless. The three systems, listed in rank order of advancing simplicity at § 10.09, might be called the *hierarchy of the linear line systems*. Each system of the hierarchy can be seen to be an infinitesimally small, nominated selection of lines extracted from all of the lines of its predecessor in the hierarchy, the first being extracted from that original system comprising the $\infty^4$ of lines which can be drawn in Euclidean space (§ 4.09).

08. The term *linear*, incidentally, derives from the linearity of the equations in homogeneous coordinates which describe the systems in the area of algebraic projective geometry; and I use the word *hierarchy* because the systems of lines are determined by a progressively increased number of restrictions being placed upon the lines by the number of equations they are obliged to satisfy. In Semple and Kneebone (1952) one can find a clear account of the broader aspects of these matters. One can find there too an explanation of what is understood in mathematical terms by another important matter physically at issue here: *the projective, or linear, dependence or independence of lines*. The adjectives projective and linear are exactly synonymous in the context of this book.

09. Very simply, the mentioned line systems are as follows: (*a*) there is a triple infinity, an $\infty^3$ of lines which constitute a linear line complex; it may be a general linear line complex or a special one (chapter 9); I shall call it simply a linear complex or even more simply a *complex*; (*b*) there is a double infinity, an $\infty^2$ of lines which constitute a linear line congruence (chapter 3); it may be either hyperbolic, parabolic, or elliptical (chapter 11); there are degenerate forms; I shall call it simply a linear congruence or even more simply a *congruence*; and (*c*) there is a single infinity, an $\infty^1$ of lines which constitutes a linear ruled surface, namely a regulus of lines (chapter 3); there are special and degenerate forms (chapter 7); and I shall call it simply a *regulus*.

10. In chapter 11 some useful geometrical properties of these three interrelated systems of the hierarchy are introduced and summarised, and in other chapters various examples of the systems may already have been seen (§ 10.13). It is important now, however, merely to know the following: *a complex, congruence, or regulus can be defined by a minimum of five, four, or three given lines respectively.*

11. In the same way as any three points define a circle of points and thus a circle, any three lines define a regulus of lines and thus an hyperboloid. In general the three lines defining the regulus will be skew, and the hyperboloid will have an elliptical throat. A good example of this can be seen in the figures 3.13. There the three contact normals at $A$ and $B$ and $C$ define the regulus of which they are members, and the regulus

(which is not shown) has generated the hyperboloid (which is shown). A congruence or a complex, incidentally, unlike a regulus which resides upon a surface, will always occupy the whole of space.

12. Once any one of the three systems has been defined in this way by its requisite minimum number of lines (§ 10.10), all lines of the system are, by definition, *projectively, or linearly dependent upon the given set*; and in the circumstances it will be clear that, before the given set of the three, the four, or the five lines can define its appropriate system, *all lines of the set must be projectively or linearly independent of one another*.

13. At § 10.10 I referred to various examples of the systems having been, perhaps, already seen. In chapter 9 we asked a simple question about the layout of the right lines in a moving body and, from elementary principles associated with rigidity and the ISA (chapter 5), we saw the emergence of the $\infty^3$ of lines of the linear complex. In chapter 3, at § 3.44 *et seq.*, we looked for the *n*-lines in a body with two degrees of freedom, and found another collection of an $\infty^2$ of lines which we saw to be a linear congruence. In chapter 3 and again in chapter 6 we saw a regulus of lines appearing when we asked the same or some similar questions about a state of three degrees of freedom. *These linear line systems are ubiquitous in the area of instantaneous kinematics; they are however equally ubiquitous in the area of statics, and this shall now be shown.*

**On the question of forces**

14. Take any two lines in space. They will in general be skew. Let there be acting along each of these lines a force of unspecified magnitude. Unless the two lines intersect or are parallel, or unless one of the forces is zero, the resultant of the two forces will never be expressible as a single force alone. The resultant will always be expressible however as a wrench. If we plotted the central axes and the pitches of the resultant wrenches for all possible pairs of magnitudes of the two forces, we would plot a cylindroid of screws. A similar set of facts applies, incidentally, for a pair of superimposed angular velocities: they will add, according to the chosen ratio of their magnitudes, to various rates of twisting, all upon the screws of a cylindroid. Refer in advance to the figures 15.01 and 15.02.

15. If the two lines intersect, however, or are parallel (in which case please see the point of intersection at infinity), the resultant, or the equilibrant, of the two forces is always expressible as a single force alone. Given that the magnitudes of the two intersecting forces may vary, the lines of action of all possible, single-force equilibrants make up a planar pencil of intersecting lines. Every line of the planar pencil is projectively dependent upon any chosen two of the lines, and the geometry of the circumstance is a matter of common knowledge. The direction of the equilibrant must accord with the polygon of forces which is a triangle (§ 10.18), and it has thus for a long time been seen that, *if any three forces are alone in equilibrium their lines of action will be coplanar and they will intersect at a common point*. There is an exact analogy for this in the field of kinematics: three angular velocities suffered by a body cannot add to rest unless, (*a*) they intersect at a common point, and (*b*) they are coplanar (§ 10.55, § 20.11).

16. Take now any *three* lines in space and be aware that, unless they intersect at a common point and are coplanar, or are parallel with one another and coplanar, they are projectively independent of one another. While remaining projectively independent they may indeed intersect at a common point and be non-coplanar (§ 7.60), they may intersect in pairs at two different points (§ 7.67), or they may intersect in pairs at three different points and be coplanar (§ 11.16), but in general the three taken lines will be both projectively independent and skew. Let there be acting along each of these lines a force of unspecified magnitude, and let us ask the new question now: *under what circumstances can these three forces be balanced by a single, fourth, equilibrating force?*

17. To answer the question we can argue as follows. Refer in advance to the figures 10.03 and 10.04. Draw any three transversals of the taken lines: draw in other words any three lines to cut each one of the taken three. Next project the whole system of the three forces orthogonally onto each of three planes put normal respectively to these transversals. We will see upon each of the projection planes a set of three intersecting component forces. Knowing that all projections of any balanced system of vectors is another balanced system, it will be clear that the three components at each projection plane can be equilibrated there only by means of another component in the plane whose line of action cuts the point of intersection of the existing three. It follows that any pure-force equilibrant of the three original forces must not only accord with the closed polygon of forces which is a skew quadrilateral, but also cut all three of the mentioned transversals. This means in effect, and please refer in advance to the figures 10.01 and 10.02, that, *if any four forces are in equilibrium, their lines of action will reside in space as the lines of some regulus*. Some further remarks about the geometry relevant here will be found at § 11.06.

**Some experiments with the balancing of forces**

18. That the polygon of vectors taken end to end of a balanced system of forces upon a body closes is a fact. It is in the present context an axiomatic fact, a matter for experiment. See Den Hartog (1948) for

example, where that author gives a brief account of the discovery by Stevin 400 years ago of the fact that any two intersecting forces are additive by means of the method of the parallelogram. He thus tells of the discovery in natural philosophy of the triangular polygon of forces for three forces in equilibrium upon a body, of how these three forces (in the absence of couples) are accordingly coplanar, and of how this discovery occurred only shortly before the time of Newton. The principle of Stevin can be extended, of course, to encompass balanced systems of non-intersecting forces and couples comprising four or more forces, and it is well known in all of these circumstances that, in the presence or the absence of couples, the many-sided, skew polygon of the force vectors always closes.

19. On the particular question of four balanced forces acting in the absence of couples upon a body, one might consult the work of Möbius (1837), find there at his §99 the whole gist of my § 10.17, and propose for oneself either or both of the following experiments. Refer also Ball (1900) § 185.

20. For the first experiment refer to figure 10.01. Employ gravity as one of the four forces by selecting some concentrated, heavy piece of metal, such as a solid brass ball. Drill the ball and erect about its sphere some asymmetrical but rigid framework, say of soft soldered brass wire. Next, by attaching a suitable balancing mass, return the centre of gravity of the constructed apparatus to the centre of the ball. Attach now three light, flexible strings at any three points among the framework and hang the whole piece up. Knowing the total weight acting downwards through the ball, and having measured the tensions acting in the three supporting strings, one can observe not only (*a*) all of those phenomena yet to be listed at § 10.22 but also, and most importantly in the present context, (*b*) *a set of four straight lines in space, namely the lines of action of the four balanced forces, each of which is projectively dependent upon the other three.*

21. For the second experiment refer to the figure 10.02. Although in mechanics on Earth it is never possible to ignore gravity, the above experiment could be repeated with a light frame and with four taut strings arranged across four suspended pulleys with clevises as shown. Construct such a light frame, arrange the clevises, and attach the four balancing masses, each of which should be relatively heavy and can be varied now. Note again (*a*) that the phenomena listed below at § 10.22 are all observable, and (*b*) that the projective dependence of the four lines mentioned at § 10.20 is in no way related to the special fact that gravity exists.

22. Refer to figures 10.03 and 10.04, and please

Figure 10.01 (§ 10.20) A fabricated, massive, rigid body 2, whose centre mass is at the centre of the bigger ball, is suspended against gravity by three light strings. In this case two of the strings are tied to the roof and one is tied to the floor. The four forces $\mathbf{F}_1$, $\mathbf{F}_2$, $\mathbf{F}_3$, and $\mathbf{G}$ (and these forces do not in general intersect) are a set of four forces in equilibrium.

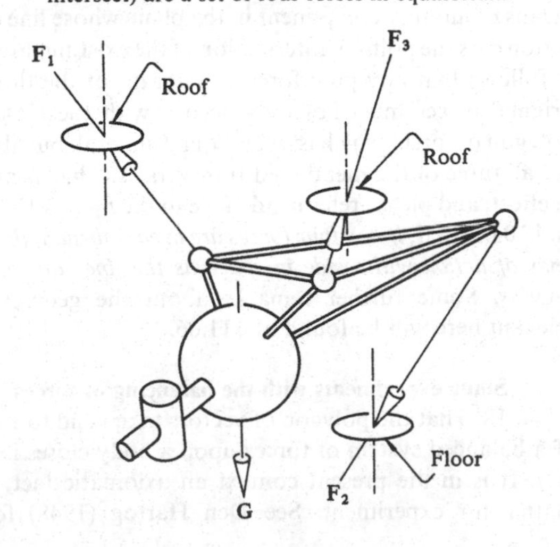


Figure 10.02 (§ 10.21) A fabricated, light, rigid body 2 is suspended by four light strings as shown. Such an apparatus will show that the four balanced tensions in the strings, when seen as four balanced forces upon the body, have their lines of action distributed in space as four generators of the same family of some hyperboloid. What makes this remarkable is that, while any three of the lines of action are enough to define the hyperboloidal surface (§ 10.11), the fourth line of action will be found, not to miss or to intersect that surface, but to lie automatically and exactly upon it (§ 11.05).

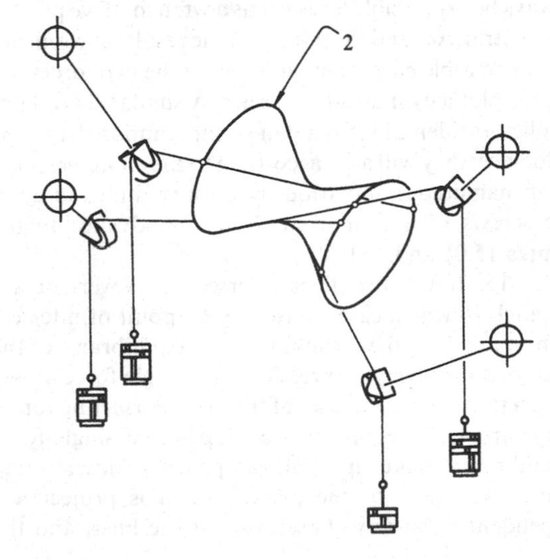

check the argument at § 10.17. It is not very difficult to observe by experiment with sets of four balanced forces that *(a) parallel with each of the four forces there is a line, a transversal, which cuts the lines if action of the other three; (b) independently of the order in which the vectors are taken, the polygon of forces which is a skew quadrilateral closes; (c) with a suitably chosen scale for the vectors, the polygon can be constructed wholly upon the hyperboloidal surface of the regulus along whose generators the forces lie; (d) when the polygon is thus drawn, its sides are alternatively along the lines of action of the forces and along the transversals of the complementary regulus; and (e) the polygon can be drawn to be either somewhere at the side of the hyperboloid as indicated in the figure 10.04(a) or wholly surrounding it as indicated in figure 10.04(b).* Figure 10.04(*b*) is extracted from reference [1].

23. For an example in ordinary engineering practice of the balancing of four pure forces upon a body in equilibrium, look at figure 10.05. There one can see a tractor and a trailed disc plough travelling at uniform speed across the land. Certain approximations have been made in the following explanation because there are, in fact, collinear couples in association with two of the resultant forces shown (§ 10.51), but roughly speaking we can say for the plough (*a*) that, while the pull **P** forwards and the gravity load **G** downwards upon it are skew, the two resultant forces, first of all the forces from the land upon the wheels, and secondly of all the forces from the ploughing upon the tools, **ΣW**and **ΣT** respectively, are also skew; yet (*b*) that the four forces together, **P**, **G**, **ΣW**, and **ΣT**, are in equilibrium with one another as shown. A similar analysis can be made for the tractor by selecting the following four groups of external forces acting: the pull rearwards **P**; the gravity load; the forces forwards, upwards, and sideways upon the driving wheels; the forces rearwards, upwards, and sideways upon the steering wheels. In this respect of the four forces, a simple sailing yacht can be likened to the plough. Because there is no plane of symmetry among the forces and because, therefore, the system of forces cannot be reduced to a system of only three forces, the mechanics of the equilibrium of the yacht in straight and steady sailing is also not well understood. Refer to item [2].

Figure 10.03 (§ 10.22) A view along one of the vectors of the polygon of forces for four forces in equilibrium. This could be the plan view of the polygon for the four forces in figure 10.01. While the located lines of action of the four forces are shown in the figure by means of the lighter lines, the polygon itself, a skew quadrilateral, is shown by means of heavier lines. Although the polygon is embedded into the actual layout of the actual forces acting here, it does not need to be so embedded; it may, of course, be drawn at any location in space and to any convenient scale.

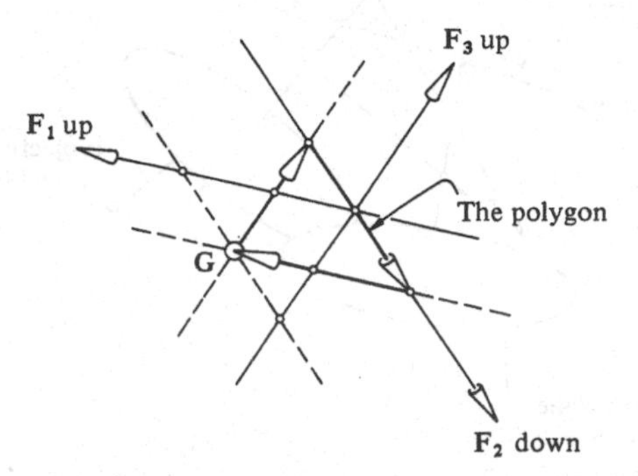


### Interlude

24. Be reminded that, with these various sets of four forces in equilibrium, we are looking at one of the origins in mechanics of the mathematical idea of the projective (or the linear) dependence of lines. To see another, apparently unrelated, origin of the idea, take a look at the kinematic phenomenon displayed in figure 7.08. The relevant text begins at § 7.51. That text says in effect that, without destroying one or other of the contacts at *A* or *B* or *C*, the contacting of a fourth prong at *D* is quite impossible. The reason resides in a special condition: *by virtue of a careful construction of the two pieces of the rigid-body apparatus, the incipient contact normal at D has been located, not to be just anywhere, but to be exactly along some line belonging to the regulus of lines already defined by the actual contact normals at A and B and C.*

25. These matters of physical fact are not being produced in a magical manner by some wizardry of mine: it is simply in the nature of nature that these things are true: I am trying to explain that there are to be found

Figures 10.04 (§ 10.22) These figures show that, for a given system of four forces in equilibrium, the polygon may with convenience be drawn, not only upon the actual surface of the defined hyperboloid, but also there in a number of different ways.

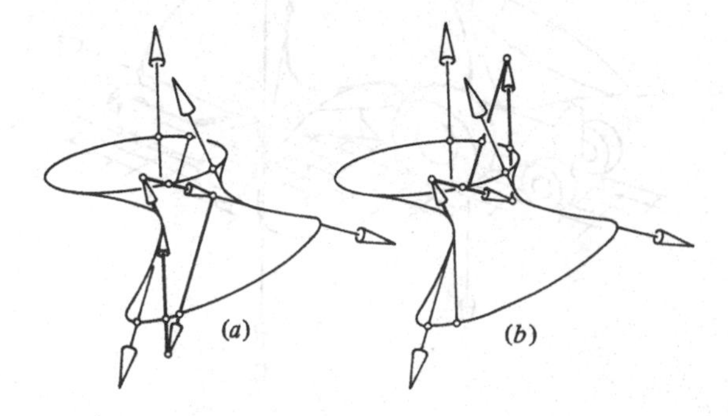

in physical reality facts which correspond exactly with certain aspects of our pure projective geometry.

**The cases of five, six, and more than six forces**

26. Let us look at the cases next of five and of six forces in equilibrium (§ 10.27, § 10.29). After similar argument and experiment in these more complicated cases, the following, observable facts of nature will emerge: *(a) when five forces are in equilibrium their lines of action will be members of the same linear congruence; (b) when six forces are in equilibrium their lines of action will be members of the same linear complex.*

27. To demonstrate by argument the first fact, namely that at (*a*) of the paragraph above, take any four forces $\mathbf{F}_1$, $\mathbf{F}_2$, etc. of variable magnitudes, and take them to be acting along a set of four given lines in space. Refer to the figure 10.06 where each one of the four lines is skew with the other three. In general the four taken forces will be unbalanced, first because they will not in general be arranged along the generators of a regulus, and secondly because their polygon of vectors will not be closing. Recall now the well-known theorem of projective geometry which states that, given four straight lines generally disposed in space, no more than two other straight lines can be drawn to cut the given four (§ 11.20). I have taken a case in figure 10.06 where the two lines, $w_1$ and $w_2$, are real (§ 11.21). Project in the manner of § 10.17 the system of the four forces onto the two shown projection planes put normal to $w_1$ and $w_2$ respectively. Thus find in figure 10.06 that any pure force equilibrant $\mathbf{F}_5$ of the four forces must cut both $w_1$ and $w_2$. This argument clearly demonstrates the first of the mentioned facts, namely the fact (*a*) at § 10.26, for the cases where the lines $w_1$ and $w_2$ can indeed be drawn. It is just as likely, however, that the two lines cannot be drawn: the two sets of intersections, and thus the two lines, may be either coalesced or both imaginary (§ 11.24).

28. If the lines $w_1$ and $w_2$ are imaginary, the equilibrating force will again be a member of the congruence defined by the lines of action of the existing four, but the congruence will be, not a hyperbolic congruence (§ 11.21), but an elliptic one (§ 11.24). Refer to § 15.48 and to § 23.35 *et seq.* for further information.

29. To contemplate the second fact, namely the fact (*b*) at § 10.26, one might consider the statics of a rigid body suspended – in stable equilibrium and in a parallel gravity field – by five taut strings. The equilibrium being stable, the centre of mass of the body is at some height above a selected datum level that is a local minimum. It follows that, in the event of a small displacement, the five points at the points of attachment of the strings upon the body and a sixth point at the centre of mass will all move in directions perpendicular to the forces acting there. The six linear velocities will be consistent with the unique ISA which must exist (§ 6.60), and the lines of action of the six forces – the five tensions in the five strings and the force of gravity – will reside as generators of the complex of motional right lines that surrounds the ISA (§ 3.43, § 9.27).

30. Referring again to Möbius (1837) we find that, for seven forces or more, no particular pattern must exist. This is analogous, incidentally, to the fact that in mechanisms consisting of a single closed loop of hinges

Figure 10.05 (§ 10.23) The two sets of four balanced forces on (*a*) the chassis of a trailed disc plough, and (*b*) the chassis of its tractor. Note the continuous straight line containing the two equal and opposite, pulling forces **P**.

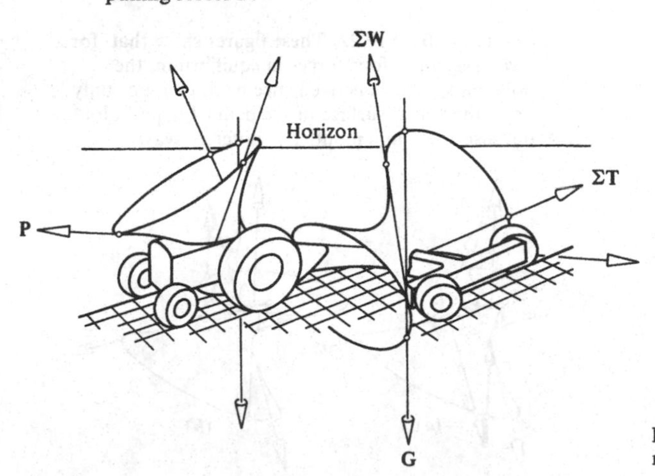


Figure 10.06 (§ 10.27) The lines of action of five forces in equilibrium must reside as generators of the same linear congruence.

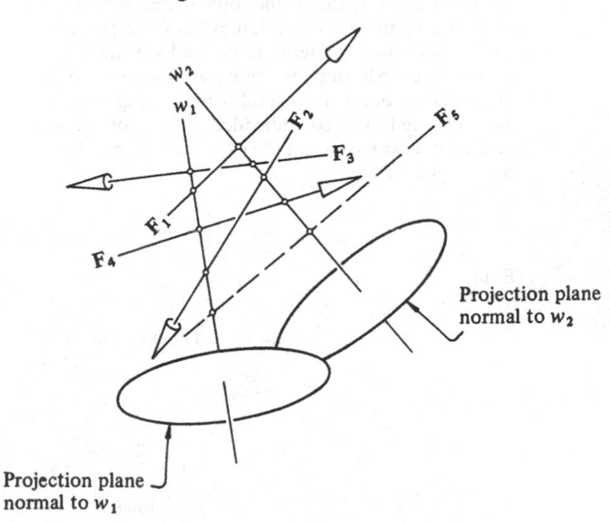

(§ 1.46), for all numbers of hinges from three to less than seven, special geometrical conditions are required to achieve an apparent mobility $M_a$ of unity (§ 10.56). It is only with seven hinges that we can set the hinges in whatever way we wish (§ 1.54). Refer also to chapter 20.

**Some notes about the concept of moment**

31. Conventionally in the teaching literature it is explained by diagrams such as the one reproduced in figure 10.07 that, whenever a force **F** acts within a Cartesian frame at a point *A* whose position vector is **r**, and along some line of action that neither passes through the origin nor is parallel with any one of the principal directions, there occurs at the origin a moment **M**. The moment vector is drawn to emerge from the origin in a direction normal to the plane containing **F** and the origin, and by the right-hand rule the vector is written, $\mathbf{M}=\mathbf{r}\times\mathbf{F}$. This moment at (or 'about') the origin of a force is almost always presented in the elementary books as a matter for mere definition. The various algebraic methods for determining the Cartesian components of **M** are next outlined, and the reader is encouraged in that way to learn, not so much what the moment of a force at a point might be or signify, but, more often, how most simply either to calculate its numerical value directly or to write the matrices for, and thus a computer programme for, its calculation.

32. This mere calculation of the moment is important. Experiment has repeatedly shown and we are thus convinced that for equilibrium among a number of forces, not only must the force polygon close, but a moment polygon (the moments being taken at any convenient point in the body) must also close. We say indeed, for equilibrium, (*a*) that $\Sigma\mathbf{F}$ must equal zero, and (*b*) that $\Sigma\mathbf{M}$ must equal zero.

33. The discovery of the first of these two necessary and sufficient conditions for the equilibrium of a body can be traced to Stevin (§ 10.18), but the final and total discovery of the second seems to have had a much more recent history. Because the discovery of this second condition is so recent, and because its application in current works of analysis is so convenient by way of matrices in the computer programming, we have tended to overlook, perhaps, the crucial importance in works of synthesis of the corresponding matters in line geometry.

34. *These matters deal, as we have seen directly and pictorially with the actual layout in space of the lines of action of the forces, and analogously as we shall see, with the layout in space of the axes of the angular velocities in a mechanism.* They are, accordingly and quite precisely, those matters which might be able to figure best in our attempts, unimpressive hitherto, at the synthesis of mechanism. Whenever the lines of action of the system of forces for some proposed device are already known, the checking of a moment polygon is a possible and a possibly useful thing to do. But if we do not yet know where the forces are, or have no clear idea yet of the motions of our intended mechanism, as in the case for example of some vaguely conceived robotic device, the construction of a moment polygon, or analogously, of a bound velocity polygon (§ 10.55), is not only impossible at such an early stage of design, but is also useless to try. There are other ways to exploit the concept of moment; and we shall be investigating them.

Figure 10.07 (§ 10.31) The moment at the origin of a force. We habitually write the cross product equation $\mathbf{M}=\mathbf{r}\times\mathbf{F}$ for this. Habitually also, we take the position vector **r** as being directed outwards from the origin, and we familiarly use the right-hand rule to check the relative orientations and senses of the various vectors.

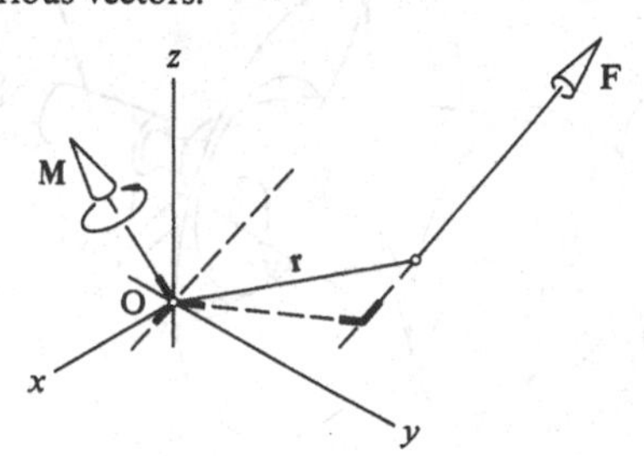


**The concept of couple**

35. It is well agreed that the only kind of 'action' which can act upon a body is a force or a combination of

Figure 10.08 (§ 10.35) The moment at the origin of a couple. The moments $\mathbf{M}_1$ and $\mathbf{M}_2$ at the origin are the respective moments there of the two equal and opposite forces $\mathbf{F}_1$ and $\mathbf{F}_2$ of the couple. The total moment **M** of a given couple **C** (and this applies at any point in space) has its magnitude and direction identical to that of the couple itself.

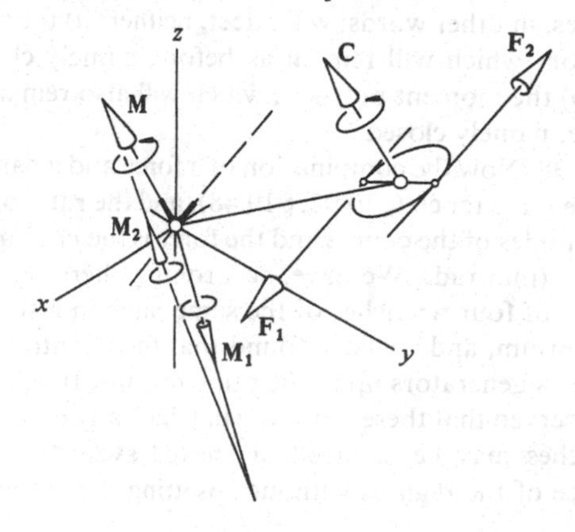

forces. It is argued however that forces often occur in special pairs wherein each force of the pair is equal to, parallel with, and opposite to the other. Such a pair of forces is called a *couple*. The magnitude and direction of the couple vector is held to be given by the vector equation, $\mathbf{C}=\mathbf{F}\times\mathbf{d}$, where $\mathbf{d}$ is the directed distance from one of the forces $\mathbf{F}$ to the other; and the couple vector $\mathbf{C}$ is held to be free (§ 10.43). The *moment* at a point of a couple, on the other hand, is naturally and quite simply the vector sum of the respective moments there of the two forces. See figure 10.08. Wherever the couple of forces comprising $\mathbf{F}$ and $\mathbf{F}$ might happen to be in the space, and however oriented its 'plane' might be, the resultant moment at the origin (or at any other chosen point) will always turn out to have the same direction and magnitude as the couple $\mathbf{C}$ itself. The moment $\mathbf{M}$ at a point of a couple $\mathbf{C}$, in other words, is simply $\mathbf{C}$. The reader might refer to Synge and Griffith (1949) for a parallel discussion of this most fundamental matter.

**Equilibrium among a selected group of wrenches**

36. We saw at § 10.22 that the lines of action of a set of forces balanced upon a body must be distributed in space in such a way that any fourth one of the lines will be a generator of the regulus defined by the other three. Let us take such a system of forces and be aware (*a*) that, from the force polygon which closes, the relative magnitudes of the forces can be known, and (*b*) that, moments being taken at any chosen point in space, the moment polygon will be known to close also.

37. Impose now upon the body four couples whose vectors are parallel respectively with the forces and in the same directions, and whose magnitudes are proportional respectively to the magnitudes of the forces. Because the directions and magnitudes of the couples in a 'couple polygon' will correspond with those of the forces, and because the force polygon closes, the sum of the moments of the couples at any chosen point in the body will be zero. The imposition of the four couples, in other words, will affect, neither (*a*) the force polygon, which will remain as before, namely closed, nor (*b*) the moment polygon, which will also remain as before, namely closed.

38. Now the combination of a force and a parallel couple is a wrench (§ 10.02, § 10.48), and the ratio of the magnitudes of the couple and the force is the *pitch of the wrench* (mm/rad). We have, accordingly, here set up a system of four wrenches of the same pitch in a state of equilibrium, and we have found that their central axes reside as generators upon the same regulus. It will next be observed that there is no way in which any one of the wrenches may be changed or shifted away from the surface of the regulus without upsetting the closure of one or other of the two closed polygons. We can conclude accordingly that, *if four wrenches of the same pitch are in equilibrium, not only will (a) the force polygon close and (b) a polygon of moments taken at any point close, but also (c) the screws of the wrenches will be distributed as generators of the same regulus.*

39. The matter being canvassed here is that whereas a line (in the sense of its being a member of a linear line system) is simply a line, the laws of nature, when looked at in terms of the dual vectors relating to the combination of force and couple, throw up for consideration linear systems not only of lines, but also of *screws*. These linear systems of screws involve the same powers of infinity of same-pitch screws as do the linear systems of mere (zero-pitch) lines, and the systems are indeed the same systems, namely the regulus, the congruence, and the complex (§ 10.09). I would like to remark that a similar demonstration might have been made here, not with forces and couples and thus with wrenches, but with angular velocities and rates of axial sliding and thus with instantaneous rates of twisting. Such analogous aspects of the matter (with their obvious relevance for the kinetostatics of mechanism) are to be dealt with next, and the reader might get ready now to think in terms of natural philosophy again.

**Some striking analogies**

40. If a body has an angular velocity $\boldsymbol{\omega}$ imposed upon it the effect at a point in the body is that a linear velocity $\mathbf{v}$ occurs. Refer to figure 10.09(*a*). The vector equation is written, $\mathbf{v}=\boldsymbol{\omega}\times\mathbf{r}$, where $\mathbf{r}$ is the directed distance to the point from the instantaneous axis of rotation (§ 5.45). Because we believe that we know what linear velocity is (§ 5.27), and are thereby ready to regard that vector quantity as a familiar entity, we say that the above stated, cross product equation is axiomatic: we say that it derives directly from our observation of

Figure 10.09(*a*) (§ 10.40) A body subjected to a pure angular velocity (a body that is purely rotating) displays a cylindricoidal distribution of linear velocity vectors, one such vector bound at each point within the body. We say that the equation $\mathbf{v}=\boldsymbol{\omega}\times\mathbf{r}$ is axiomatic in the circumstances, and we draw the linear velocity vectors accordingly.

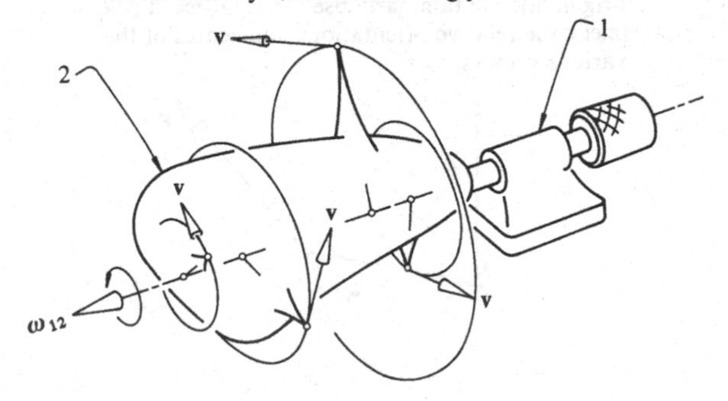

reality. *The cylindricoidal field of velocity vectors produced by the angular velocity in figure 10.09(a) fills the whole of space, and its central axis is the line of action, or the axis of rotation, of the angular velocity.*

41. If on the other hand a body has a force **F** imposed upon it the effect at a point is that a moment **M** occurs. The vector equation can be written, $\mathbf{M}=\mathbf{F}\times\mathbf{r}$, where **r** is the directed distance to the point from the line of action of the force (§ 10.31). Because we tend *not* to believe that we know what moment might be or signify, we are prone to regard it as a somewhat less familiar entity and we say more carefully on this occasion that the above stated, cross product equation is a matter of definition. We tend to say that the axiomatic truth in reality of the equation will emerge not now but later. *The cylindricoidal field of moment vectors produced by the force in figure 10.09(b) fills the whole of space, and its central axis is the line of action of the force.*

42. While exploring these and the following analogies we are speaking of course instantaneously. In the actual practice of terrestrial mechanics, where mass and gravity both exist, a body at rest must be supported somehow and thereby be in contact with other bodies, then next accelerated somehow, before an angular velocity $\omega$ can be caused to occur. If, also, a nominated single force **F** is to be suffered upon some body at rest, along some nominated line of action remote from its centre of gravity, that force must be the resultant of some other forces in combination with the force of gravity, and that also implies contact with other bodies. One must remember too that, as soon as a non-zero resultant force begins to act upon a body, the body accelerates. In this way the pattern of a system of applied forces and thus the resultant force upon a body will begin to change as soon as time begins to elapse.

Figure 10.09(*b*) (§ 10.41) A body subjected to a single force acting alone similarly displays a cylindricoidal distribution of moment vectors. Here we define moment by the equation $\mathbf{M}=\mathbf{r}\times\mathbf{F}$ (§ 10.31), or, if we measure the position vector *r* not from the point to the force but from the force to the point, $\mathbf{M}=\mathbf{F}\times\mathbf{r}$, and we draw the moment vectors accordingly.

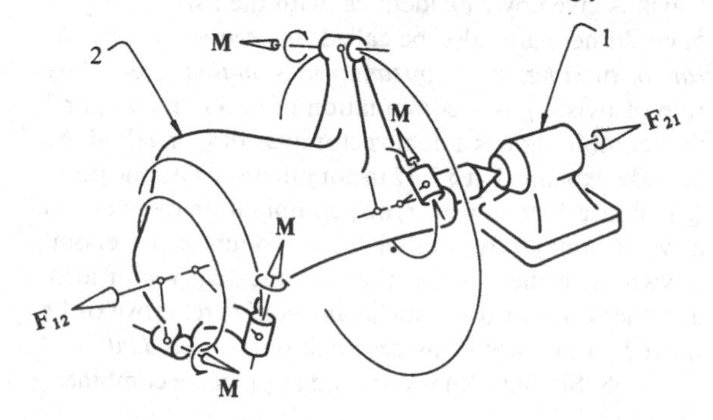


Similar remarks apply in the case of angular velocity (§ 10.54), and in the case of couple, next to be discussed.

43. If a pure couple **C** is applied to a body the immediate effect is that, at every point in the body, a moment **M** occurs. With gravity upon a free body counterbalanced or absent, a 'pure couple applicator' might be imagined to be a frictionless device like the one depicted in figure 10.10(*b*). There, a couple **C** is being applied at *A* along a flexible torque-transmitting cable and transferred via prongs at any pair of diametrically opposed castellations at the tubular protuberance there. The fact that a moment **M** immediately occurs at every point can be observed: at every point one can imagine a similar, cylindrical protuberance arranged with its axis parallel with that at *A*, and one can see that an equal and opposite 'pure couple reactor' could detect the moment there: it could do this by returning the body to equilibrium. In the sense that the applied couple **C** in figure 10.10(*b*) could be applied anywhere upon the body yet not affect the action or the argument, it can be said to be a free vector; but please refer in due course to my remarks at § 10.51 where I speak of the bound nature of a resultant moment vector which occurs at a nominated point within a body. Refer also § 23.27 and § 23.28. *The field of equal and parallel moment vectors produced by the couple in figure 10.10(b) fills the whole of space, and it has no central axis.*

44. If on the other hand we wish to impose at all points in a body some nominated linear velocity **v** at an instant, that is if we wish to cause a body to translate in the absence of rotation, we can do this by imposing a pair of equal, parallel, and opposite angular velocities $\omega$. Refer to figure 10.10(*a*). It is necessary incidentally that the two angular velocities be arranged, in the electrical sense of the following words, not in parallel with one another as the forces are in figure 10.10(*b*), but in series with one another as shown. A suitable device for applying the angular velocities is illustrated: parallel shafts with two pulleys of the same size are attached to the bodies 1 and 3 respectively, there being an intermediate body 2, and the connecting belt ensures that the two angular velocities applied to the body 3, namely $\omega_{12}$ and $\omega_{23}$, are not only parallel but also equal and opposite. Whatever the combination of the two equal, parallel, and opposite angular velocities may be called (and this question is not resolved as yet), it is, like couple **C**, a free vector; but (compare § 10.43) the velocity **v** at a nominated point in a twisting body at an instant is clearly a bound vector bound at its point. *The field of equal and parallel velocity vectors produced by the 'couple' of angular velocities in figure 10.10(a) fills the whole of space, and it has no central axis.*

45. In each of the figures 10.10 the direction of the field of equal and parallel vectors is normal to the plane

Figure 10.10(*a*) (§ 10.44) The body here (the body 3) is subjected to a couple of angular velocities, the pair of them being parallel, equal, and opposite in direction. The body displays a parallel field of equal, linear velocity vectors **v**. The vector value of the applied angular velocities, the applied rate of translation upon the body, is $\boldsymbol{\omega} \times \mathbf{d}$, namely **v**.

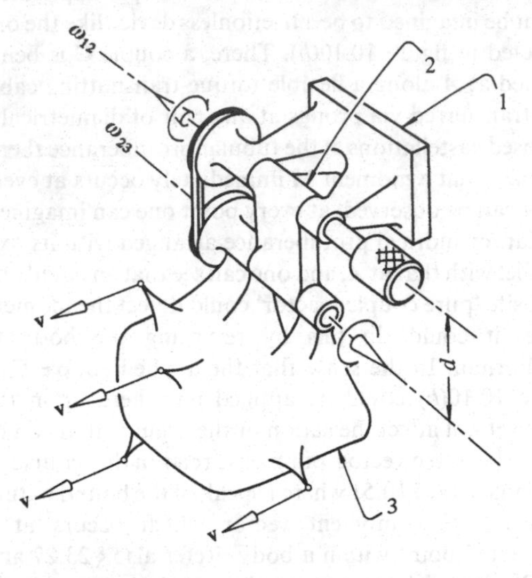


Figure 10.10(*b*) (§ 10.43) This body is subjected to a couple (a couple of forces). At each point in the body the couple produces a moment **M** in the direction of the couple and equal to it (§ 10.35). The vector value of the applied forces, the couple acting upon the body, is $\mathbf{F} \times \mathbf{d}$, namely **M**.

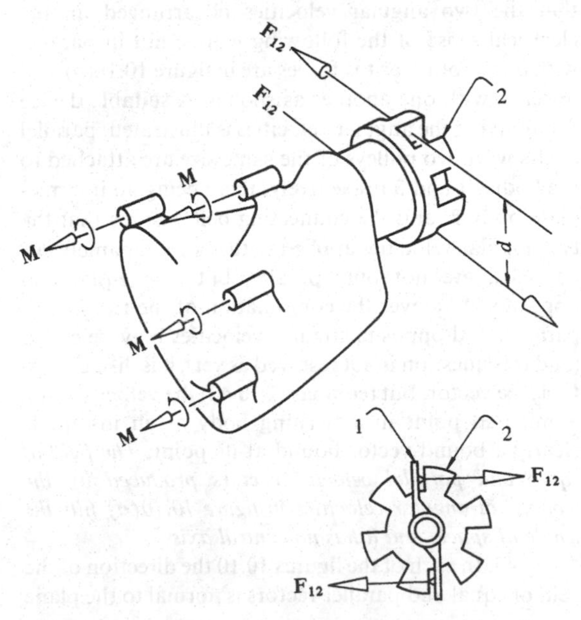


containing the 'couple', namely the pair of equal and opposite vectors imposed upon the body. In both figures the right-handed, corkscrew rule applies. In figure 10.10(*b*) each of the moment vectors **M** can be written, $\mathbf{M} = \mathbf{F} \times \mathbf{d}$; and the right-hand side of this equation is the applied couple **C**. In figure 10.10(*a*) each of the velocity vectors **v** can be written, $\mathbf{v} = \boldsymbol{\omega}_{23} \times \mathbf{d}$; but the right-hand side of this equation has no well-established name. We might call it however, and consistently in the context, *the applied rate of translation*, and we can use for it the symbol $\boldsymbol{\tau}$ (§ 5.49). Note that whereas **C** and $\boldsymbol{\tau}$ are both free vectors (as already explained), **M** and **v** are not.

46. Please also note that, whereas **C** and $\boldsymbol{\tau}$, and **F** and $\boldsymbol{\omega}$ from which they are derived respectively, are being seen as actions or events to be applied upon the whole of a body to initiate an action or an argument, the moments **M** and the velocities **v** are being seen as effects of these, occurring as a result, at points in the bodies. Please examine the question of whether or not these matters might be argued in the reverse order. Could for example a nominated linear velocity **v**, at a single point in a rigid body, call into action a predictable angular velocity $\boldsymbol{\omega}$ of the body as a whole, or call into action a pair of equal, parallel, and opposite angular velocities? This is an important question (in both fields), and the reader might refer to my remarks at § 5.28 *et seq*.

**Parallel assembly and the natures of the duals**

47. Next, and this is easy to see, if an angular velocity like the one in figure 10.09(*a*) were imposed upon an already translating body like the one in figure 10.10(*a*), in such a way that the imposed angular velocity was put parallel with the existing rate of translation, we would arrive immediately and quite simply at figure 5.09 of chapter 5. The central axis of the newly formed helicoidal field of resultant linear velocities **v** would be, in the event, exactly along the line of action of the imposed $\boldsymbol{\omega}$. The special velocity vectors **v** that were least in magnitude (and let us call these $\mathbf{v}^*$), could be written, $\mathbf{v}^* = \boldsymbol{\tau}$; and these would occur at that special single infinity of points that is distributed along the mentioned central axis. That axis, with its pitch which is given by $\tau/\omega$, identical with the ISA of chapter 5, could here and also be called, *the central axis of the rate of twisting, or the instantaneous motion screw*. Any rate of twisting is a combination of an $\boldsymbol{\omega}$ and a $\boldsymbol{\tau}$; and that combination is a dual vector quantity, or a dual. As already explained at § 5.47 in conjunction with the point *Q* in figure 5.09, however, any combination of an $\boldsymbol{\omega}$ and a **v**, meaningfully drawn at a common basepoint anywhere in the body, is the relevant dual vector also, and that dual vector, which can, as it were, move or be moved from place to place, I call the *motion dual*.

48. Similar remarks would apply if we combined,

again in a parallel manner, the force **F** of figure 10.09(*b*) with the couple **C** of figure 10.10(*b*). An axisymmetric field of resultant moment vectors **M** would spring into being, and that would be a helicoidal field as well. The central axis would be the line of action of the imposed force; and that axis, with its pitch, which is given by $C/F$, would be seen to be *the central axis of the wrench, or the instantaneous action screw*. At all points along that axis there would exist those moments **M** that were least in magnitude (and let us call them **M***), and these could be written, $\mathbf{M}^* = \mathbf{C}$. Any wrench is a combination of an **F** and a **C**; and that is another dual vector quantity, or a dual. But similarly also, any combination of an **F** and a **M**, meaningfully drawn at a common basepoint anywhere in the body, is the relevant dual vector also, and that dual vector, which can, as it were, move or be moved from place to place, I call the *action dual*.

**On the question of the terminology**

49. I wish to remark now and parenthetically about my choice for the terminology. In the above two paragraphs § 10.47 and § 10.48 I have spoken, first about (*a*) the parallel assembly of a motion dual, and next about (*b*) the parallel assembly of an action dual; and I have said that the two processes – they are of course intellectual processes – are analogous. Now I wish to make it clear that, although the actual terms *motion dual and motion screw*, and *action dual and action screw*, are here introduced by me, each of the concepts for which these terms stand have for a long time been well known. In 1865 Plücker [3] not only discovered the pair of vectors but also introduced his own word *Dyname* to describe the pair of vectors described at (*b*). More recently other writers in German for example Keler [4] have been using the newer word *Kinemate* to describe the pair of vectors described at (*a*). These two German words are difficult to use with meaning in the English language, and neither of them include in their spelling any sense of the two-ness of the vector pairs. Dimentberg, in an English translation of his original Russian (1968), uses the terms *force screw* and *velocity screw* to mean, in my terms, action dual at the action screw and motion dual at the motion screw respectively; he uses the terms *force motor* and *velocity motor* to refer to duals whenever they are expressed with their basepoints put remote from their central or their 'home' screws (see in advance figure 10.12 and refer to § 19.15); but these two sets of terms fail to accentuate, in my view, the essential fact, namely that it is angular velocity and not linear velocity which corresponds analogously with force; and 'force screw' and 'angular velocity screw' would be too cumbersome. At § 3.01 and § 3.43 I have used, out of a special need there, the contrived adjectives actional and motional; I hope that my explanation here will assist the reader to forgive me that. Study (1903), in his originative work, freely uses the form of words 'dual vector' when speaking of Plücker's *Dyname* and a suitable algebra for it (§ 19.13), and I copy, as Keler and others do, those words from him. Latin roots for the English words *action* and *motion*, incidentally, are *agere* to do, and *movere* to move, respectively.

**The general, oblique assembly of the duals**

50. We must look however at the general, oblique cases (and they are exactly analogous) where the superimposed parts of the two duals, namely the cylindrical or the first part, and the parallel or the second part, are not put parallel with one another. Refer to figure 10.11(*a*). This figure is a composite, drawn to illustrate both duals, both the rate of twisting and the wrench, and the argument that follows is written in such a way that it can apply to both. An angular velocity vector $\boldsymbol{\omega}$ (or a force vector **F**), free to be drawn anywhere along its line of action *b–b*, is joined at its basepoint *B* to a rate of translation vector $\boldsymbol{\tau}$ (or a couple vector **C**). This latter is drawn along the line *c–c* which, the vector being free in the manner explained at § 10.44 (or at § 10.43), may be drawn anywhere parallel with its own direction. The convenient and summarising combination of the two vectors drawn together at *B* can be called, in the two

Figures 10.11 (§ 10.50) The general, oblique assembly of the duals. We have here at (*a*), above, a figure to illustrate the argument, and we have in a separate figure at (*b*), below, an ordinary corkscrew with only one half of its handle.

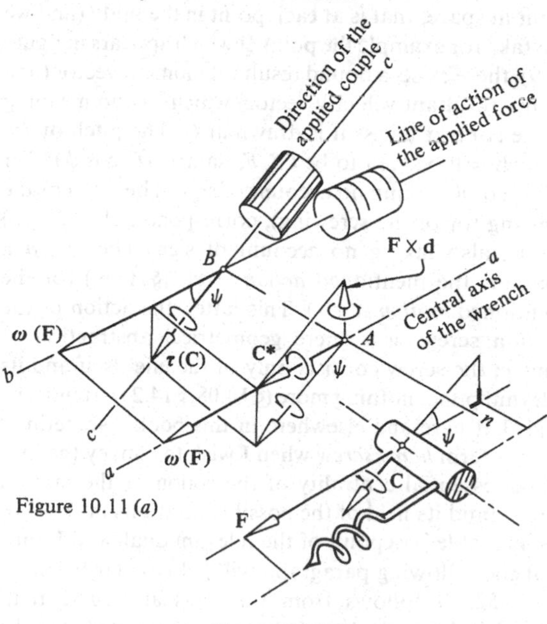


Figure 10.11 (*a*)

Figure 10.11 (*b*)

cases respectively, the relevant motion dual – and this is associated with its rate of twisting about its central axis or ISA or motion screw yet to be determined; and the relevant action dual – and this is associated with its wrench about its central line of action or its action screw which is also yet to be determined.

51. Not out of perversity but for the sake of the argument, I propose now to reverse the order in which I speak about the duals. In figure 10.11(*a*) the action dual (or the motion dual) represents out there at *B* the resultant wrench (or the resultant rate of twisting) of the system of vectors, and our job here is to locate the central axis of that wrench (or of that rate of twisting). That central axis is unique, and it must exist. Otherwise Poinsot [5] in 1803, and, correspondingly, Chasles [6] in 1830, were both wrong. There is along the line *B–n* (and this is drawn normal to the plane containing **F** and **C**) a unique point *A* distant *d* from *B*, where a moment due to the force, of magnitude *Fd*, counteracts the component in the opposite direction there of the moment due to the couple, of magnitude $C \sin \psi$. The particular component moment at *B* which is parallel with **F**, namely **M*** (and the magnitude of this may be written $M^* = C \cos \psi$) corresponds with that least couple which, collinear with a new force **F** along the axis *a–a*, will produce the same overall effect there as the inclined couple **C** did, together with the force **F** at *B*. An examination of the geometry will next reveal that *a–a* is the central axis of the resultant wrench (or of the rate of twisting). The axis *a–a* is surrounded by a helicoidal field of resultant moment vectors (or of resultant velocity vectors) that bears an exact resemblance to the layout in figure 5.09. At each point in space, that is at each point in the body (and we can take for example the point *Q* which appears in figure 5.09), there exists a bound resultant moment vector (or a bound resultant velocity vector) which has no meaning in the context unless it is drawn at *Q*. The pitch of the wrench will be seen to be $C^*/F$, namely $(C \cos \psi)/F$ or $(C/F) \cos \psi$, taking no account of sign. The pitch of the twisting (or of the screwing), correspondingly, is $(\tau/\omega) \cos \psi$, also taking no account of sign. The line *a–a* contains the mentioned *action screw* (§ 10.49) (or the mentioned *motion screw*). This latter, the action or the motion screw, is a mere geometrical abstraction of course; the screw consists only of the line itself and its relevant pitch, nothing more (§ 14.08, § 14.27). It must be said that here and elsewhere in this book I sometimes use the term *home screw* when I wish to convey the idea of the essential centrality of the action or the motion screw amid its field of the possible positions in space of the movable basepoint of the relevant dual, and I hope that the following paragraph will help to clarify this.

52. It follows from the facts at § 10.51 that, provided the angle $\psi$ of the associated couple (or of the linear velocity) is continually adjusted to suit, and provided the magnitude and the direction of the force (or the angular velocity) vector is not permitted to alter, the first part of any dual, the force (or the angular velocity) part, may be shifted about from place to place quite freely. Look at the first part, the force-part, of the action dual as it moves about and away from its central axis (or its home screw) in figure 10.12 while the overall effect of the dual remains the same; look also at the motion dual drawn at the point *Q* in figure 5.09, and peruse the relevant text at § 5.57. In the sense that the resultant force (or the resultant angular velocity) upon a body may be shifted about like this, invariant except for location, the force vector (or the angular velocity vector) is 'free' but, in the sense that there is always a unique, home-location for its action (or its occurrence) at the central axis of the dual, it must be seen as a so called *line vector* which is not free. The point at issue here is that, when looked at as an origin-line for the directed position vectors **r**, which latter abound in the vector algebra (§ 5.53), this origin-line itself is not free to move away from its unique, home-location at the central axis of its system; otherwise all of those vector equations involving **r** would have no meaning.

53. In figure 10.11(*b*) there is an ordinary cork-screw with only one half of its handle. This figure is intended to correspond exactly with figure 10.11(*a*). The figure may encourage the reader physically to feel the application of an action dual remote from its central axis when next he has occasion to use such a broken instrument. But here is a question: will it always be true in the real life that the central axes of the action dual

Figure 10.12 (§ 10.52) An action dual (**F**, **C**) moving about and away from its central axis or home screw, namely its action screw. Four alternative, parallel locations of the invariant force vector **F** are shown, and for each of these the accompanying, non-invariant, couple vector **C** is drawn. The least couple, called **C*** at § 10.51, occurs when **F** resides upon the action screw.

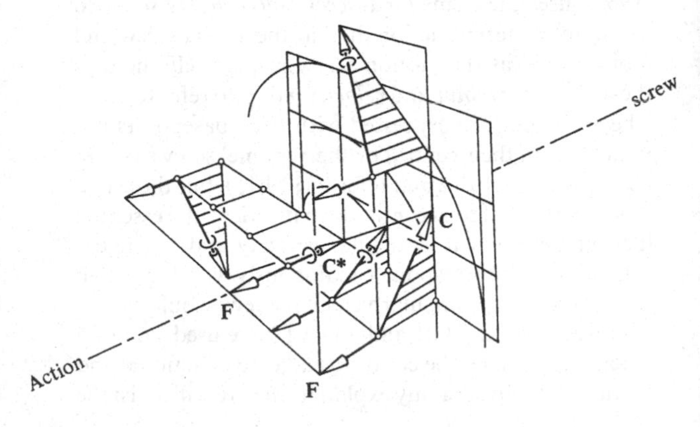

upon a corkscrew and the corresponding motion dual are collinear? And will the pitches be necessarily the same? Hint: think about the actual shape of the sharpened point; think about the distribution of the elementary normal reaction and friction forces acting from the material of the cork upon the working surfaces along the helical wire and at the tip of the instrument; think indeed about the action and the motion at the working edges of tools of all kinds (§ 10.01). *Become aware that, at joints in mechanism (even for example at an ordinary hinge where the motion dual is clear to see and where most action duals are quite clearly non-collinear), the central axes of the action dual and the motion dual are in general not collinear nor of equal pitch, that they are in general skew with one another and of different pitches.* Refer to § 10.59 *et seq.*, and to § 14.32 to § 14.34 inclusive, for some discussions about corkscrews and screwdrivers, and to the early parts of chapter 17.

**The addition to rest of rates of twisting**

54. If one were to wish to superimpose upon a body a number (say four) of rates of twisting, all of different pitches, one would need to build an apparatus such as the one depicted in figure 10.13. There four screw joints of different pitches are arranged in an open chain in series. One might, in practice, actually actuate (namely motorise) the four joints in figure 10.13, and make a kind of working manipulator arm, but please

Figure 10.13 (§ 10.54) Four screw joints (*H*) of unequal pitches arranged in series. We allow all points in the freely movable body 5 to become embedded in the body 1 (§ 1.06). We then allow the pitches at the joints to become first equal, than all zero. Thus we discover the condition for the full-cycle ability to move of the Bennett mechanism (§ 10.56, § 13.03, and chapter 20). At the position marked $A_1A_5$ there is a generally chosen pair of instantaneously coincident points. One point is *A* in the body 1, while the other is *A* in the body 5 (§ 1.06, § 1.13).

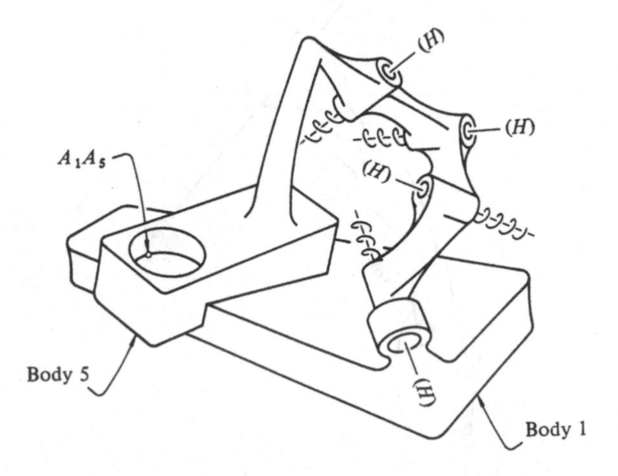


imagine here that the joints are not actuated. Notice that the terminal body 5 of the kinematic chain is freely interpenetrating the frame body 1 (§ 1.06). I now wish to ask whether, or under what circumstances, relative movement between the links might be possible if the body 5 became embedded within, that is if the body 5 became fixed within (and thus at rest within), the body 1?

55. Refer to § 10.36 *et seq.*, and contemplate the analogy. If the body 5 did become embedded in the body 1, it would mean that the four rates of twisting at the four screw joints must be 'adding to rest'. The total motion dual at the body 5, in other words, would be zero. This condition would require (*a*) that the sum of the angular velocities, $\omega_{12}+\omega_{23}+\omega_{34}+\omega_{45}$, be zero, *that is that the angular velocity polygon closes*; and (*b*) that the sum of the linear velocities at all points *A* instantaneously coincident at some representative point A in the four movable bodies, $\mathbf{v}_{1A2}+\mathbf{v}_{2A3}+\mathbf{v}_{3A4}+\mathbf{v}_{4A5}$, be zero too (§ 1.18), *that is that a bound velocity polygon at some representative point A closes as well.* This latter, the bound velocity polygon at *A*, is not the relative velocity polygon of chapter 12, which is drawn for a whole mechanism (§ 12.12). It is a velocity polygon taken at a point; and it is, of course, exactly analogous with the moment polygon taken at a point (§ 10.32). The above has shown that we may freely say, for rest, (*a*) that $\Sigma\omega$ must be zero, and (*b*) that $\Sigma\mathbf{v}$ must be zero too.

56. We could also say in figure 10.13, with 5 at rest in 1, that the four motion screws at the four screw joints must be members of the same 3-system: any one of the screws must, in the circumstances, be linearly dependent upon the other three (§ 3.36, § 6.49, § 23.23). For other readings about the processes of addition for rates of twisting, the reader may go to the robot arm at § 13.16 *et seq.*, to § 19.20, or to the latter parts of chapter 20 where three pure angular velocities are shown to be adding (by closed triangular polygon) to rest.

57. It will next be observed in figure 10.03 that, if the pitches of the four screw joints were equal (§ 10.36), the second condition (*b*), listed at the end of § 10.55 and enlarged upon at § 10.56, could be stated alternatively as follows: (*b*) *that the four joint axes be distributed as the lines of some regulus.* We would see in this way the formation of an overconstrained, 4*H*-loop of identical screw joints having a transient, apparent mobility $M_a$ of unity (§ 2.08). If the equal pitches of the four screw joints were next made equal to zero, that is if the screw joints were next made to be not only screw joints of the same pitch but simply hinges, we would find an overconstrained 4-link loop of *revolutes* of the same, transient, apparent mobility unity. Refer to figure 2.04 and notice there that, when the mechanism transitorily moves, the spherical joint at *B* operates in the same way as a revolute would, its effective axis through *B* becoming a

generator (of the same family as the other generators) of the regulus defined by the already established axes of the actual revolutes at *A*, *C* and *D*.

58. If, next, we were, with care, to proportion the links according to the formulae quoted and explained at § 20.53, *we would find that the italicised condition at (b) above would persist across all successive, finite movements.* We would in this way have arrived at the well known, overconstrained, yet mobile mechanism of Bennett: refer to figures 2.01(*j*), 20.25, and 20.26. This mechanism, founded upon the regulus as Bennett discovered, enjoys not only transient, but also *full-cycle*, apparent mobility of unity. This and other such matters are discussed in greater detail in chapter 20.

### When one or both of the duals or the power is zero

59. Each rigid body in the field of mechanics, and thus in the field of mechanism, suffers at an instant, (*a*) its total action dual, the resultant of all of the external forces and couples being imposed upon it, and (*b*) its total motion dual, the resultant of all of the angular velocities and rates of translation being imposed upon it. When the total action dual (which represents a wrench at its central action screw) is zero, the body is in equilibrium. When the total motion dual (which represents a rate of twisting at its central motion screw) is zero, the body is at rest. The existence of neither of these conditions implies the other, incidentally; the states of equilibrium and of rest can each obtain in the absence of the other. They must both be seen, of course, against some chosen frame of reference; this latter may or may not be fixed relative to Earth.

60. When either the action or the motion upon a body (briefly said) is zero, or when both are zero, the rate of working at the instant, namely the power, is zero. If the power is finite (and the power can be calculated in the manner of § 19.08) the body is accelerating, and its store of energy will be changing. This latter, like action and motion, needs to be measured and quoted against some frame of reference. If the body in question happens to be some rigid link of a mechanism moving in a steady cycle, its store of energy will be cyclically fluctuating: no link in steady cyclic motion *accumulates* its energy. It is accordingly argued that a machine in steady cyclic motion *transmits* its energy from input to output, losses of energy due to friction, or gains (§ 10.63), occurring only at the joints.

61. At any joint in mechanism – and a joint can only exist between a single pair of links (§ 1.11) – there are at an instant these two screws: (*a*) *the common action screw of the pair of equal and opposite wrenches acting at the joint*; in the case of a simple joint, a representative instance of these two wrenches is made up of all of the forces acting from one of the links upon the other at the points of contact at the joint; at each point of contact one will find, incidentally, not only the normal component of the force and its reaction force, which are collinear along the contact normal there, but also the friction component of the force and its reaction component force, which are collinear with the equal and opposite relative velocity vectors in the tangent plane; and (*b*) *the common motion screw of the pair of equal and opposite rates of twisting at the joint*; this latter, of course, is an entirely different screw; it is the screw of the relative ISA (chapter 5). I have spoken here about two screws, and upon each of them act a pair of equal and opposite duals, a pair of action duals upon the action screw, and a pair of motion duals upon the motion screw. I shall nevertheless refer from time to time simply to the action dual or to the motion dual which is acting at a joint. I do this for example at § 10.72. I do it again at § 17.03.

62. At any joint, in other words, there is (*a*) an action dual, and (*b*) a motion dual, each with its central screw (§ 10.59); the two screws, the action screw and the motion screw, are, in general, skew. In figure 10.14 I have sketched two screws, an action screw and a motion screw, which with their respective duals at a given instant, are together at work at a human shoulder joint.

Figure 10.14 (§ 10.62) Sketch of the motion screw and the action screw at a human shoulder joint. They are, at the instant, the seats for their respective duals. The duals are together at work at the instant, and the dot product of the two, suitably written and properly interpreted (§ 19.08, § 19.18), is the measure of power. Try for an exercise, and take for example the joint in figure 7.15(*a*), to see a pair of duals at work like this at the joint in the absence of friction.

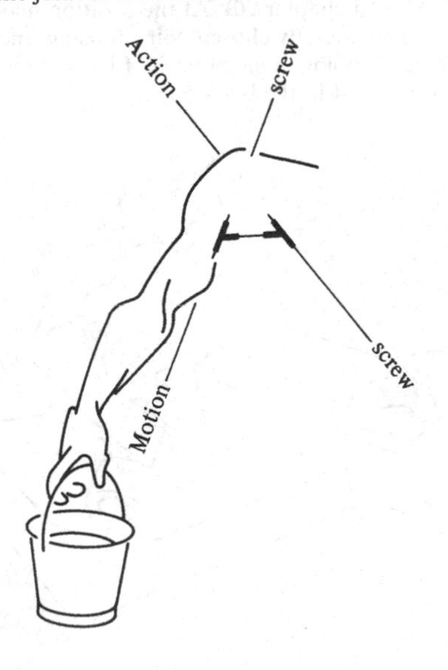

The screws are skew. They are not, however, reciprocal screws (§10.69, §14.25, §19.08, and below), for the power at the joint is not zero. Power at a joint (a scalar quantity) can be regarded either as the rate at which work is being done against an external load (and/or against friction) at the joint, or the rate at which energy is being usefully expended at the joint (§10.02). It should be mentioned here that Ball (1900) had already spoken in 1875 about the precise amount of work which was done by a wrench upon a screw, say $A$, of pitch $p_a$ in collaboration with a small twist about another screw, say $B$, of pitch $p_b$. Refer in advance to figure 19.01. Given that the distance apart of the two screws was $d$ and that the angle between them was $\phi$, he had arrived already at the multiplier $2Q=[(p_a+p_b)\cos\phi-d\sin\phi]$. The force $F$ of the wrench on the first screw (Ball used $\beta''$ for this) times the small angular displacement of the small twist on the second screw (Ball used $\alpha'$ for this) times the multiplier $2Q$ (Ball used the special symbol $\varpi$ for my $Q$) was shown by him to be the work done (§14.25 *et seq.*, §19.08 *et seq.*). Refer also to Ball §10 for the original argument. He had argued already, moreover, that when half the multiplier $2Q$ – its dimensions are those of length – was zero the work was zero and that the two screws were, by his definition, *reciprocal*. The point I am making here is that, at a joint in mechanism, the two screws implied by $A$ and $B$ above, which are or could be the two screws of Ball's argument, are not in general reciprocal. They are reciprocal only when the power expended at the joint and the power input there are either both zero or both equal. The first of these ideal conditions occurs in the absence of friction, while the second occurs only under other, special circumstances.

63. If, and I speak about machinery now, the joint is a mere *transmission joint*, an ordinary revolute in some spatial, working linkage for example, the power being lost at an instant (and this is usually minimised in good machine design) will be expended in wearing the mating surfaces at the joint and heating the lubricant there. If the transmission joint is held to be frictionless, the two screws (and these will be skew in general even if the linkage is a planar one) should be seen to be reciprocal. If the joint is an *actuated joint*, at the shaft of an engine for example or at the elbow of some robotic device where energy is applied by means of some electric or other motor, or at the above mentioned, human shoulder joint, the joint may well be a complex one, and in such cases analyses for power gain and friction loss might either be done directly or by considering separately the separate, simple joints. If on the other hand the joint is a *working joint*, such as that between a spade and its soil (§0.04, §10.01), between a disc and its sod (§10.23), a corkscrew and its cork (§10.53, §14.34), or a forging tool and its billet, the power being expended at any instant (and this may either be minimised or maximised according to the intended function of the joint) will be used in tearing, stirring, or penetrating some material, or distorting the shape of some workpiece, or in whatever other steady or non-steady work the joint has been designed to do. Refer in advance to the frictionless transmission joints at $A$, $B$ and $C$ in figure 10.15, and to the steadily working joint between the wheel and the ground in figure 10.16. These pictures are quoted just now to show some further examples here of action duals and motion duals at work in ordinary engineering practice. The first, in figure 10.15, is a case of zero friction and thus of uncluttered reciprocity at the relevant joints, while the second, in figure 10.16 is, quite clearly, a case where heavy work is being done.

### The lobster arm

64. Figure 10.15 shows a 'lobster arm'. The hinged arm is being subjected to a wrench at its 'gripper' (§2.62). From the fixed hinge axis $A$–$A$ down through $B$–$B$ and $C$–$C$ to the universal joint at $D$, the apparatus can be imagined to be a massless, frictionless, hinged manipulator in the absence of actuating power, or it might be the right arm of a light-weight crab, with weak muscles and poor purchase at its hinged elbows. The vertical line $z$–$z$ can move freely from place to place in the apparatus; the two horizontal ropes are very long. A wrench of force **F** and positive pitch $p_a$ is being applied downwards along this vertical line. By virtue of the arrangement of the three forces **F** upon 6 the wrench is acting via 5 at the gripper 4 at $D$–$D$. The three links 2, 3, and 4 of the arm, hinged together in series at $B$ and $C$, remain at rest and are in a state of stable equilibrium. Taking any one of the three hinges as the scene of the action (§10.60), or taking all three of the hinges together as a complex joint between the gripper 4 and the frame 1 §3.36, §6.49, §23.23), we see that any small movement which might occur at the joint or joints occurs at the expense of zero power. There being no friction at any one of the hinges and zero mass throughout, the dual part of the dual dot product of (*a*) the action dual along $z$–$z$, and (*b*) the motion dual at any one of the moving hinges or at any combination of the moving hinges, namely the power consumption upon a small displacement at a certain rate at any one or all of the hinges (§19.08, §19.18), is zero. The rate of working at any one of the three hinge axes of the wrench at the axis $z$–$z$ is zero in other words. Now as mentioned at §10.62 (and much more fully explained in chapter 19) the rate of working at any hinge is $F\omega[2Q]$, where **F** is the force of the wrench at the axis $z$–$z$ and where $\omega$ is the rate of twisting (the rate of pure rotation) at the relevant hinge. This requires for the hinges, each hinge having its radial

distance $d$ and its angle $\gamma$, all of which are shown in figure 10.15, that the expression $d \tan \gamma$ be a constant for the apparatus. It can be shown indeed, and it will be found by experiment, that, at each hinge, $d \tan \gamma = p_a$. It should be mentioned that the angles $\gamma$ here are the complements of the helix angles $\theta$ shown in figure 9.01 and mentioned for example at § 9.15. The distances $d$, also, are the distances $r$ of chapter 9 and of the passages § 11.44 *et seq.* Please refer in particular to § 6A.02 and to § 11.47. *It will be found by experiment, in other words, that the hinge axes of the apparatus will migrate to such locations in stable equilibrium that they will reside as generators of the same linear complex.* The axis of the complex is the axis of the wrench and the pitch of the complex is the pitch of the wrench, namely $p_a$.

65. Some further remarks are needed here however about this massless apparatus in figure 10.15. It will be noted that the zero-pitch screws at the hinges $D$–$D$ and $E$–$E$ are also reciprocal to the screw of the action dual whose pitch is $p_a$ (§ 14.29). If these hinges are to remain horizontal and to intersect upon the central axis of the action dual as shown in the figure, both of their radial distances $d$ must turn out to be zero and both of their angles $\gamma$ must turn out to be right angles. The cruciform piece 5, in other words, must achieve a central location with a horizontal orientation. *Unless in experimental practice the link 4 (or some other link) is bent to suit, however, this condition may never be achieved.* Why not? How many hinges in series, one might ask, are required before the wrench-applying, cruciform piece 5 can hang freely with its main axis vertical in the manner shown, and what, exactly and incidentally, are the conditions for stable as distinct from unstable equilibrium of the links?

Figure 10.15 (§ 10.64) The links are massless in this apparatus and there is no friction. Equilibrium prevails. An applied action dual, whose force is F and whose pitch is $p_a$, has its central axis (its home screw) along $z$–$z$. Each of the five motion duals, at home respectively at the five hinge axes, may in the event of a small displacement display an angular velocity $\omega$; but each of their pitches $p_b$ is and will remain zero. All of the screws of these zero-pitch motion duals, and all of the screws of all combinations of them (§ 15.02), are reciprocal to the screw of the action dual. We find, for the apparatus, (*a*) that $d \tan \gamma$ is a constant, and (*b*) that that constant is $p_a$.

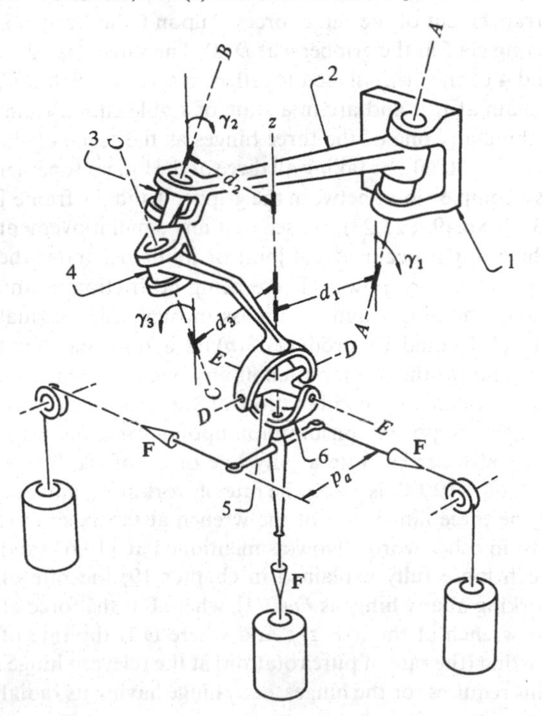


66. On beginning to ask the first of these questions, it turns out not to be a matter of the number of hinges at all: even with six (or even with more) hinges between the frame 1 and the cruciform piece 5, and with some nominated, central point of the link 5 at some stable position, the actual bulk of 5, namely the body 5, or a tick at the central point in the body 5 (§ 4.04), will not, most often, be free to orient itself in space to suit a vertically hanging rod 6. What we are doing here is (*a*) not only blundering into the complicated question of the 'zone of movement' of the body 5 (§ 4.04), but also (*b*) discovering some of the very real complexities within that zone. The question is not only whether or not a nominated central point within the body 5 can get to some nominated target position in space, but also whether or not the body 5 can freely orient itself about that position when the central point arrives. Two recent papers by Kumar and Waldron, [7] and [8], address themselves to some of the important questions at issue here. *It remains of course true however that, in whatever way the rod 6 hangs at the massless apparatus shown, the set of axial and radial reaction forces occurring at any one of the hinges will add to a reaction wrench exactly equal and opposite to, and collinear with, the wrench applied by the three hanging masses.* Refer to § 17.28.

### The rolling of a wheel

67. Look now at figure 10.16 where a steadily braked, obliquely rolling, vehicular wheel is shown. It could for example be one of the two front wheels of a steadily travelling road-grader, rolling in the direction $-x$. Let us study the action dual at this working joint between a wheel and its ground. One of the main purposes of the joint is to mobilise continuously from the ground and upon the wheel and thus upon the chassis (*a*) a rearwards component of force in the direction $+x$ to help to prevent the vehicle being driven

too fast by the driving wheels, (*b*) a sideways component of force in the direction $+y$ to help to counteract the side draft being suffered by the grading tool, and (*c*) a vertical component of force in the direction $+z$ to help to balance the force of gravity. The joint fulfils this purpose by continuously scuffing a rut at the ground, disturbing material there. This happens in conjunction with a continuously occurring though steadily changing elastic deformation at the tyre. In some arbitrarily nominated, horizontal, *xy*-plane at or near ground level, an origin of coordinates can be nominated. This latter I have put at the intersection of (*a*) the mentioned *xy*-plane, (*b*) the plane of the undistorted periphery of the tyre, and (*c*) the vertical plane through the axis of the wheel. The infinity of the elementary reaction forces across the area of the contact patch (§ 7.37), including not only the 'normal', but also the 'frictional' forces there, add to the reaction wrench represented by the force **F** and the couple **C** whose common axis is the central, or the home screw of the action dual (§ 10.59). The reaction wrench is shown emerging from the *xy*-plane at *J*. Because the wheel is in equilibrium,

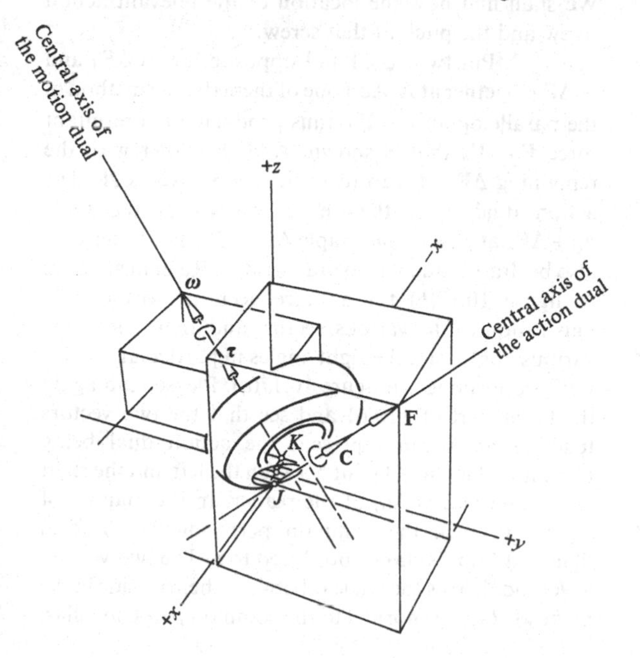


Figure 10.16 (§ 10.67) This is a braked, oblique-rolling, vehicular wheel. The central axis of the motion dual at the contact patch (namely at the joint between the wheel and the ground), parallel with the axle of the wheel, emerges from the ground at *K*. The central axis of the action dual at the same place (that is at the same joint) emerges from the ground at *J*. Both the motion screw and the action screw at the joint are of finite non-zero pitch. They are skew. They are however not reciprocal, for work is being done.

incidentally, the combined resultant of the axially and radially occurring reaction forces upon the wheel at the hub and the tangentially applied braking forces at the same place will form a wrench to balance exactly, and to be collinear with, the reaction wrench which is coming from the ground. The wheel itself (seen as a rigid body) is neither accelerating nor decelerating. Except for the small, hysteresis loss due to the flexing at the tyre, the wheel itself is not absorbing energy. The total action it suffers remains at zero (§ 10.57, § 10.58).

68. We are however dealing here, not with the wheel itself, but with the joint between the wheel and the ground (§ 10.59). Knowing the angular velocity $\boldsymbol{\omega}$ of the wheel with respect to the chassis and the direction of the angular velocity zero of the chassis with respect to the ground (the axis is at infinity), and by means of the first theorem of three axes outlines at § 13.12 *et seq.*, it will be possible to see that the central axis of the motion dual at the joint between the wheel and the ground will be somewhere as shown, parallel with the axle of the wheel and emerging from the *xy*-plane at *K*. The angular velocity of the wheel with respect to the ground is the same angular velocity $\boldsymbol{\omega}$ of the wheel with respect to the chassis, while $\boldsymbol{\tau}$, it will be found, will not be zero. It will be found, indeed, looking from the kinematical point of view and seeing the central portion of the wheel as a rigid body and allowing both $\boldsymbol{\omega}$ and the forward speed of the vehicle to vary, that the system of motion screws available for the ISA between the wheel and the ground is a fourth special 2-system (§ 15.02, § 15.54, § 23.57). The dual part of the dual dot product of the action dual and the motion dual at the joint, the 'product of action and motion' namely the power (§ 19.08, § 19.18), and, correspondingly, Ball's expression $2Q=[(p_a+p_b)\cos\phi-d\sin\phi]$, will not be zero. *The two screws at the joint are not reciprocal, for work is being done.* Given the necessary data, the rate of energy consumption namely the power loss at the joint, can be calculated easily. Please refer to chapter 19.

### The reciprocal systems at an f3 joint

69. Please imagine a knobbly, ice-encrusted pestle in your right hand inserted within a similarly encrusted mortar in your left. Let the handle of the pestle and the underside of the mortar be free of the slippery ice, so that a good grip of each of the pieces may be had. Let there next be three points of contact made between the frictionless, rigid, icy surfaces of the two pieces (§ 6.27, § 7.51). Try to imagine next that, in the absence of motion, a pure force is to be applied from right to left, and, accordingly, that a pure reaction force will respond from left to right. Refer to figure 10.17(*a*): the mortar is marked 1, the pestle 2, and the three points of contact *A*, *B*, and C between the two are clearly shown. We are

about to discuss the mechanics of a force closed, frictionless, $f3$ joint (§ 2.27).

70. The three reaction forces from the mortar upon the pestle are also shown. While their lines of action are naturally along the respective contact normals (§ 10.61), their senses – note the arrow-heads – must accord with the fact that, while compression is possible at the points of contact, tension, of course, is not. The three lines of action of the three reaction forces define a regulus and thus the hyperboloid which is shown (§ 10.10). Depending next upon the reader, for he is holding the apparatus, the relative magnitudes of the three reaction forces are now open to choice. Knowing the requirement that the pestle must be put into equilibrium by the application of a pure force (§ 10.69), and knowing the relevant laws of nature (§ 10.22), we see that the only pure forces which can be applied in the absence of motion must be applied from the reader's right hand upon the pestle (and thus another from his left hand upon the mortar) along lines of action somewhere among the shaded generators shown. Although there is thus only a limited zone upon the surface of the regulus within which the required pure force can be applied, and this limitation is due to the prohibition of tensions at the points of contact (§ 6.47), we see that all applicable pure forces, namely all applicable wrenches of zero pitch, are restricted to lines of action upon the surface of a specified regulus. This regulus is, as we shall see, the regulus of the pitch quadric of the two 3-systems of screws which are together relevant here (§ 10.74).

Figure 10.17($a$) (§ 10.69) Force relations within a general, frictionless, $f3$ joint. Points of contact between the links are at $A$ and $B$ and $C$. First a pure force $\mathbf{F}_Q$ along $q$–$q$ at $Q$ is seen to be transmitted by the joint. Then a wrench whose force is $\mathbf{F}_Q+\Delta\mathbf{F}_A$, and whose central axis is along $t$–$t$ at $T$, is seen to be transmitted by the joint. The 3-system of action screws within the joint and the 3-system of motion screws are seen to be reciprocal (§ 6.31).

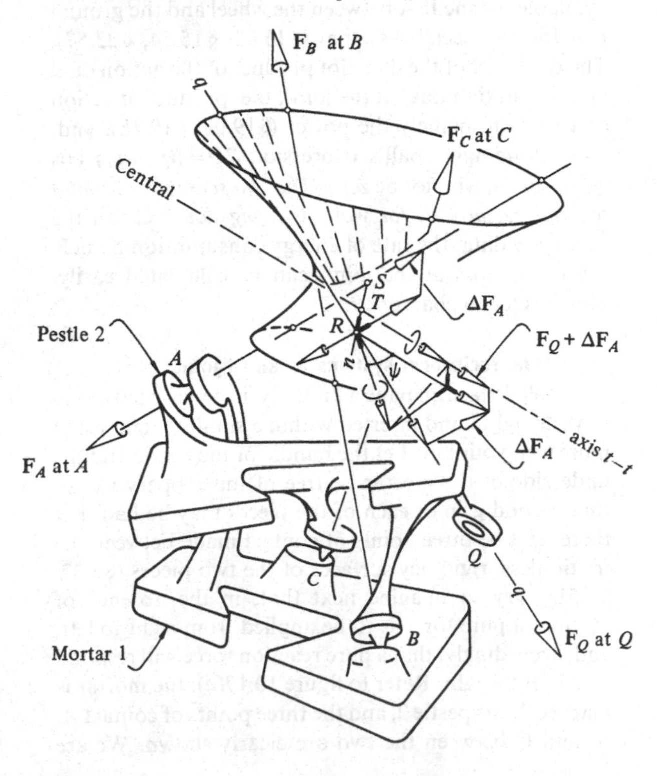


71. But can the joint sustain a wrench? And, if it can sustain a wrench of a certain pitch, can it sustain, as it can with a wrench of zero pitch (§ 10.70), a single infinity of others with the same pitch? For the sake of the argument, let the pure force $\mathbf{F}_Q$ of § 10.70 be acting along the line of action $q$–$q$; it may be convenient to imagine it to be acting upon the pestle 2 at the hook $Q$. All four forces – they are acting at $A$, $B$, $C$, and $Q$ respectively – are shown to be acting upon the pestle as I have said, not upon the mortar. This means that the shown $\mathbf{F}_Q$ is not the resultant of the other three forces, but their equilibrant. Construct now the common perpendicular $R$–$S$ between $q$–$q$ and the line of action of the force $\mathbf{F}_A$ at $A$, and let the length of $R$–$S$ be $2b$. Insert now at $S$ an extra force say $\Delta\mathbf{F}_A$ acting upon the pestle with the intention of decreasing the compressive force at $A$ while altering neither of the forces at $B$ and $C$. The vector sum of $\mathbf{F}_Q$ at $R$ and the shown $\Delta\mathbf{F}_A$ at $S$ (where the two vector parts of this sum are $2b$ apart) is now the required, new action dual upon the pestle from the reader's right hand. We shall find now the location of the relevant action screw and the pitch of that screw.

72. Put two equal and opposite forces $\Delta\mathbf{F}_A$ and $-\Delta\mathbf{F}_A$ together at $R$. Add one of them (by the method of the parallelogram) to $\mathbf{F}_Q$, thus producing the resultant force $\mathbf{F}_Q+\mathbf{F}_A$ that is shown. Add the other with the remaining $\Delta\mathbf{F}_A$ at $S$ to form the couple $\Delta\mathbf{F}_A\times 2\mathbf{b}$. The action dual upon the pestle consists of the force $\mathbf{F}_Q+\Delta\mathbf{F}_A$ at $R$, and the couple $\Delta\mathbf{F}_A\times 2\mathbf{b}$. The latter may also be drawn (and it is so drawn) at R. Refer in advance to figure 10.17($b$) for a more accurate, orthogonal illustration of the various vectors and angles here; the various angles and the right angles in particular are not well proportioned in figure 10.17($a$). Please read again the latter part of § 10.61 and see that the two vectors together at $R$ can represent the action dual being transmitted by the $f3$ joint between the left and the right hand (and vice versa) of the reader. In the manner of figure 10.11($a$), the common perpendicular $S$–$R$ in figure 10.17($a$) is shown produced to $n$. The two vectors at $R$ (and there is the angle $\psi$ between them) reside in the plane at $R$ that is normal to the common perpendicular.

Refer again to figure 10.11(*a*) and see that, at a distance *d* from *R* along the line *S–R–n* in figure 10.17(*a*), there is a point *T* where the force and the couple of the action dual are collinear (§ 10.51). The line *t–t* through *T* is the central axis, or, with its pitch, the action screw, of the action dual. Taking no account of sign, the pitch of the screw and thus the pitch of the wrench upon that screw is $|\Delta \mathbf{F}_A \times 2\mathbf{b}| \cos \psi / |\mathbf{F}_Q + \Delta \mathbf{F}_A|$; the pitch of the screw in figure 10.17(*a*), like the one in figure 10.11(*a*), is positive.

73. Become aware now that we have been dealing here with substantially the same problem as the problem of the motion screws discussed in conjunction with the 3-system at § 6.38. It is easy to show from the material of chapter 15 that the line *t–t*, the central axis of the newly applied equilibrating wrench, is a generator of the cylindroid determined by the two zero-pitch screws with which we have been dealing (§ 15.12). Both of them generators of the regulus already mentioned, they are (*a*) the one along the contact normal at *A*, and (*b*) the one along the line of action of the earlier applied, pure equilibrating force $\mathbf{F}_Q$ (§ 15.12). To present more clearly the relevant vectors and angles poorly proportioned in figure 10.17(*a*), figure 10.17(*b*) has been prepared. It is an orthogonal view of all the vectors, looking directly along the straight line *S–R–T*; we are looking from *R* through *T* to *S*. We note that, while the distance *S–R* is $2b$ and the distance *R–T* is $d$, $d = 2b \sin \psi$, and $z$, namely $(b+d)$, $= b \operatorname{cosec} 2\psi \sin 2\theta$. The symbols $z$, $b$, $\psi$, and $\theta$ all take here the meanings ascribed to them at § 15.05. See the equations (8) and (9) in figure 15.06.

Figure 10.17(*b*) (§ 10.73) An orthogonal view of the vectors in figure 10.17(*a*) taken in the direction of the common perpendicular *R–S*, The stippled triangle, with its angle $\psi$, is the triangle of moments occasioned by the shift of the force $\mathbf{F}_Q + \Delta \mathbf{F}_A$ from its original to its home location; refer to figure 10.12. This figure 10.17(*b*) shows more accurately the magnitudes and the angular relationships of the vectors. It reveals moreover the bald simplicity of what might otherwise be seen as a hopelessly complicated matter.

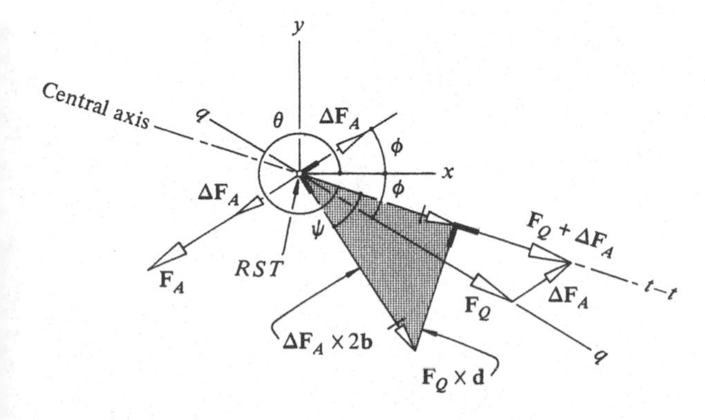


74. It should now be evident that, while the equilibrating force $\mathbf{F}_Q$ can be applied along any one of the shaded generators – indeed along any one of all of the generators if the *f*5 joints at *A*, *B*, and *C* were form closed in the manner of figure 7.06(*b*), there will be for the new wrench along *t–t* a whole single infinity of possible wrenches of the same pitch as the one we found. This $\infty^1$, in the manner of the 3-system explained in chapter 6, will reside as generators of another regulus concentric with the one already drawn upon the pitch quadric (§ 6.35, § 10.70). I have concluded this discussion by beginning to show, in a very preliminary way, and as an example of reciprocity, the following: *if there is a generalised, simple, f3 joint between a pair of bodies which determines a 3-system of motion screws in the manner explained at chapter 6, there is also a reciprocal 3-system of action screws existing; the latter is identical with the former except for the fact that the handednesses of the reguli upon the* $\infty^1$ *of hyperboloids are opposite.* Please refer to § 23.23 *et seq.*, but please read also § 10.75.

### Closure

75. I have introduced in this chapter a number of different, fundamental issues. I have talked about the hierarchy of the linear line systems, the natures of the two dual vectors (those for action and for motion), the separate concepts of equilibrium and of rest, the matters of friction and of power, and, in conclusion, I have brushed with the idea of the reciprocity not only of screws but also of screw systems. This latter arose while discussing (*a*) the lobster arm, (*b*) the rolling wheel, and (*c*) the three points of contact within the ice encrusted mortar and pestle. The discussion of the third of these, (*c*) at § 10.69, which was as I have said merely a batch of disguised remarks about the mechanics of the frictionless *f*3 joint, may have given rise to a wrong impression. It needs to be stressed at this juncture that it is only with this particular system, the 3-system with its 'knotty centrality' (§ 6.31), that any system itself and its reciprocal system are of the same 'shape'. With all other pairs of reciprocal systems – the ones and the fives and the twos and the fours, omitting the trivial pair, zero and six – this special circumstance does not obtain of course. Refer to the chapters 6, 14, and 23.

### Notes and references


[1] Phillips, J.R., A graphical method for skew forces and couples, *Australian Journal of Applied Science*, Vol. **6**, No 2, p. 131–48, CSIRO, 1955. Refer § 10.22.

[2] Phillips, J.R., *Skew forces and couples; the trailed disc plough; the oblique-rolling wheel*, doctoral dissertation, University of Melbourne, 1957. Refer § 10.23.

[3] Plücker, J., On a new geometry of space, *Philosophical Transactions*, Vol. **155**, p. 725–91, 1865. Fundamental

views regarding mechanics, *Philosophical Transactions*, Vol. **156**, p. 361–80, 1866. See also Plücker (1869), *Neue Geometrie des Raumes gegründet auf die Betrachtung der Geraden Linie als Raumelement*. Refer § 10.49.

[4] Keler, M., Die Verwendung dual-komplexer Grössen in Geometrie, Kinematik und Mechanik, *Feinwerktechnik*, Band **74,** Heft 1, p. 31–41, 1970. Kinematik und Statik einschliesslich Reibung in Mechanismen nullter Ordnung mit Schraubengrössen und dual-komplexen Vectoren, *Feinwerktechnik + Micronic,* Band **76**, Heft 1, p. 7–17, 1973. Refer § 10.49.

[5] Poinsot, Louis, Sur la composition des moments et la composition des aires, *J. Ecole Polytechnic*, Vol. **6,** p. 182–205, 1803. But see also Poinsot (1848), *Éléments de Statique*. This 9th edition of Poinsot's book contains (as do other editions which followed the first in 1806) an updated version of the above mentioned, original memoire. This 9th edition contains also, within the main text at Article 68, Poinsot's account of what we are now prepared to call the central axis of a wrench. For other historical notes on this [5], and the matter of Chasles [6], one could go to Hunt (1978), p. 49. Refer § 10.51.

[6] Chasles, Michel, 'Note sur les propriétés générales du système de deux corps . . ., *Bulletin des Sciences Mathématiques de Férrusac*, Vol. **14,** p. 321–6, 1830. Read Hunt (1978), p. 49, on Guilio Mozzi, who is said to have published much earlier, in 1763, well developed ideas about the instantaneous screw axis of a moving body. Whittaker (1927), discussing Chasles on page 4, also mentions this earlier writer Mozzi, and he speaks of him in similar vein. Refer § 10.51.

[7] Kumar, A. and Waldron, K.J., The dextrous workspace, ASME paper No 80-DET-108, Design Engineering Technical Conference, September 1980. Refer § 10.66.

[8] Kumar, A. and Waldron, K.J., The workspaces of a mechanical manipulator, *Journal of Mechanical Design*, Vol. **103,** (ASME), p. 665–72, July 1981. Refer § 10.66.

Freedom in machinery: Volume 2
*Screw theory exemplified*

This book explains what has become known as screw theory, and it deals by way of examples with the applications of this theory not only to the analysis and synthesis of mechanism in general, but also to the problems of real machine design. Because it is wholly three dimensional (and thus not easy to grasp when presented by means of mathematics alone), screw theory is presented here with the help of carefully drawn geometric figures which transport the reader directly into the three-dimensional domain. There are two important aspects of this book; it is firstly a fundamental work in the area of the kinetostatics of mechanism, which is a combination of the kinematics of the motions of and the statics of the forces between rigid bodies in contact. It is also a seminal work of importance for the mechanical design of robots, both the relatively simple robots of today and the much more versatile robots of the future.

# Contents

VOLUME TWO

# *Preface*

The index to this second (final) volume of *Freedom in Machinery* was already well enough written in 1983 to have its hundreds of items included among the overall index published with volume one in 1984. That latter, which might now be called the draft overall index at the end of volume one, was used by me in that volume for all backward and forward referencing. In 1984 there was thus a network of boundaries already established which delineated the already completed and the partly worked-out areas of subject matter which was simply waiting, it foolishly seemed to me, for its final filling out.

I met difficulties in some of those areas of course, and they were due to my growing perceptions which grew, in the intervening years, faster than I could follow with new clear words. The said set boundaries needed to be breached in a number of places, and many of the proposed paragraphs of roughly equal length were either enlarged or allowed, where possible, to proliferate. A somewhat extended index with a few unavoidable corrections is the result. The reader will find this revised index at the end of the now-completed work.

This second volume is very much the second part of a single work. Its numbered chapters are attached contiguously with those of volume one. The devices for referring from place to place both forwards and backwards throughout the work (§0.16) direct the reader in such ways that the closely related though sometimes apparently disparate subjects matter (which do go logically across the chapters) are better seen from the differing perspectives of one another.

My acknowledgements of March 1984 remain. Four years have passed (more or less safely), the complexities of the world's inter-territorial relationships continue, the relevance of machinery in all of this and the perceived importance of robots (and weapons) in the world of tomorrow remain disputed issues, but the IFToMM – as I said that I hoped that it might – does continue to flourish, and that in itself I suppose is a good thing to be glad about.

Jack Phillips, Sydney, March 1988.

11. It was arranged in this apparatus that four forces acting on a rigid body could adjust themselves to equilibrium. Four freely swinging metal rods revealed the locations of the forces. These were made the basis for the regulus of light wooden strips which may be seen.

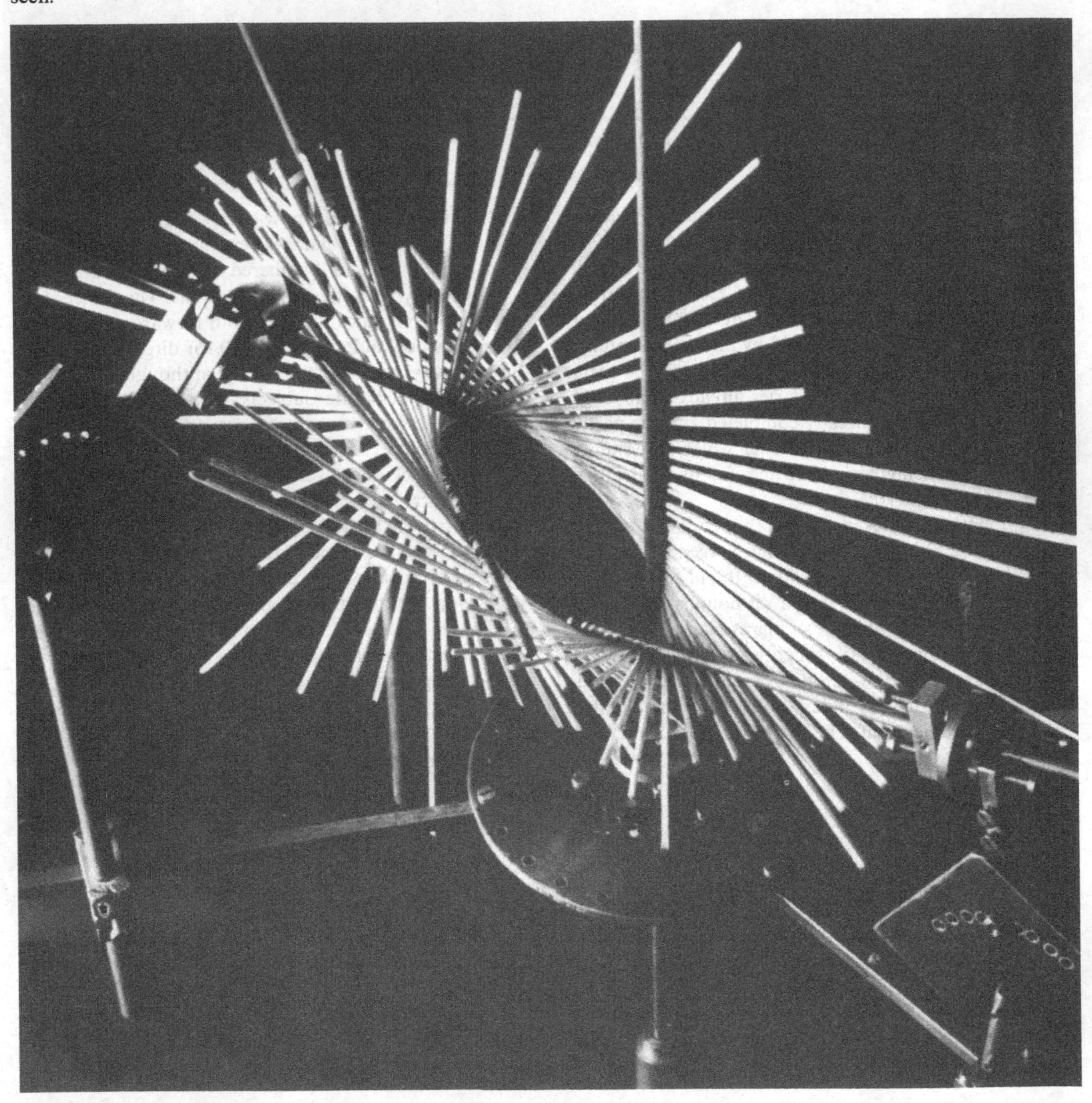

# 11 Geometrical properties of the linear line systems

**Opening remarks**

01. The linear line systems are brushed upon in chapter 3, mentioned in chapter 6, used in a primitive way in chapter 7, and better introduced among the kinetostatics of chapter 10. There is a relevant discussion beginning at § 10.07 and a useful summary at § 10.09. In this chapter I offer first some collected material about the properties of and the special and degenerate forms of the three mentioned linear line systems (§ 10.09). This earlier material is presented in a somewhat encyclopedic manner, dealing with the regulus first, the congruence next, and the complex last. Later, however, and beginning at § 11.43, I try to present a more ordered account of the linear line systems, employing as a vehicle for argument the Plücker line coordinates (§ 19.21). I trust that that material, presented in reversed order with the complex first and the regulus last, will be relevant to the overall question of freedom in machinery in a much more useful way.

02. The cylindroid, incidentally, although made up of lines, is made up of lines each with a pitch. The cylindroid is not one of the linear line systems; it is a system of screws; it is, indeed, the 2-system (§ 23.22). It belongs in chapter 15.

03. Any two lines generally disposed are linearly independent of one another: no other lines are linearly dependent upon the pair (§ 10.14). If the pair intersect, however, there springs into being a planar pencil of linearly dependent lines intersecting at the point of intersection and residing in the plane defined by the intersecting pair (§ 3.09, § 10.15, § 13.03). Each line of the pencil is then linearly dependent upon any other pair of lines of the pencil. The combination of the two original lines, the $\infty^0$ of them, has, upon intersection, degenerated into the $\infty^1$ of lines of the pencil. If the two lines are parallel, their point of intersection is at infinity, the said planar pencil resides in the plane defined by the two lines, and the lines of the pencil (now parallel and numbering not $\infty^1$ but $\infty^2$) occupy the whole of space. We shall see next that, whereas two lines can suddenly *bloom* in this way to become a planar pencil (or a field of parallel lines), a regulus – which is dependent not upon two but three linearly independent lines (§ 11.04) – can either collapse into a single line, become a planar pencil, or suddenly bloom into the $\infty^2$ of lines of a star.

**The regulus**

04. As mentioned elsewhere (mostly hitherto) any three linearly independent lines will define a general regulus; refer to § 6.32, § 6.34, and figure 6.05(a); but see also § 11.61 *et seq.* Any three randomly chosen lines will be linearly independent unless (*a*) they are coplanar and intersect at a single point, (*b*) they are coplanar and parallel, or (*c*) they are collinear. If the three lines are non-coplanar and intersect the regulus blooms into a star (§ 7.60, § 11.16); if the three lines are non-coplanar and become parallel the regulus blooms into an $\infty^2$ of parallel lines (§ 7.68, § 11.16); and if the three lines become collinear the regulus collapses into the single line of the collinearity (§ 9.37).

05. To see how three lines generally disposed (and thus in general skew with one another) actually define a regulus, one could proceed as follows. One could easily make a model of this. Erect the three skew lines in space and draw every line that can be drawn to cut all three of them. Draw in other words all of the *transversals* of the three lines. A single infinity of straight lines will thus appear, and these will delineate a surface – an hyperboloid. See for example figure 3.14. Now draw all of the straight lines in the surface of the hyperboloid that can be drawn to cut the already mentioned set of lines. These lines – another single infinity of them (and these of course are all transversals of the other set) – will constitute the regulus of lines defined by the three original lines (§ 6.33).

06. For every regulus, evidently, there is an opposite regulus. Alternatively we can say that, inscribed upon the surface of an hyperboloid, there are two reguli of lines, the right and the left handed regulus. These two are sometimes called the one and the other *family of generators* of the hyperboloid. It is important to see that every generator of one family cuts all

generators of the other. This in terms of the reguli can be said as follows: each line of a regulus and any line of the opposite regulus are coplanar. Please go to § 11.59, and look in connection with this at figure 11.13. There one may see that, parallel with each generator of a regulus, there is a generator (the so-called *conjugate generator*) which belongs to the opposite family. Any generator and its parallel, conjugate generator cut the *throat ellipse* at two points diametrically opposite. The throat ellipse is illustrated at figure 11.02; it appears there at the intersection of the hyperboloidal surface and the $xy$-plane.

07. There may be drawn in Euclidean space an $\infty^9$ of reguli (§ 4.13). There may be drawn in a linear line complex an $\infty^6$ of reguli (§ 4.15, § 9.33, § 9.40). Correspondingly there may be drawn in a linear line congruence an $\infty^3$ of reguli.

08. Any plane drawn normal to and at the mid point of a common perpendicular between a pair of lines of a regulus will pass through the central point of the regulus. This property of the regulus (or this theorem) may be used to locate the central point of a regulus whenever we are given only three of its generators (§ 6.34). A study of § 6C.10 will reveal that the theorem here stated may be seen as a special case of a much more general theorem. The general theorem relates, not simply to three defining lines and a single regulus of lines, but to three defining screws and a whole 3-system of screws. For a first look at the overall geometric relations within a 3-system, refer to § 6.31 *et seq.*

09. If short common perpendiculars are drawn between pairs of successively drawn and successively spaced lines of a regulus, the regulus will become ringed near its throat by a series of these common perpendiculars. The series will form a jagged band which, as the lines of the regulus (its generators) are taken closer and closer together, becomes in the limit a smooth curve. Such a curve (on this or any ruled surface) is called the *line of striction*. Refer to figure 7.09 for a picture of the line of striction surrounding a regulus, and to § 11.13 for some remarks about its algebra. The line of striction does not coincide with the ellipse in the plane of symmetry at the throat of the hyperboloidal surface; it is a twisted curve. However it does thus coincide if the regulus is a circular regulus (§ 7.58, § 11.15).

10. The common perpendiculars between all pairs of diametrically opposite generators of a regulus form a cylindroidal surface whose nodal line is the central axis of the hyperboloidal surface. Otherwise we can think in terms of erecting the common perpendiculars between the central axis and successive generators of the regulus. This alternative procedure will produce the same cylindroidal surface (§ 6.42).

11. If we drop the perpendiculars from a point $Q$ onto the generators of a regulus, we describe a cone of the fourth order – a quartic cone, whose vertex of course is at $Q$. If however the point $Q$ is taken anywhere within a certain closed region surrounding the centre of the regulus, the fourth order cone appears as two separate cones of the second order – two ordinary, elliptical, quadratic cones – which touch one another along a pair of intersecting lines. These two lines are the two (the only two) common perpendiculars between generators of the regulus which may be drawn to pass through $Q$ (§ 6.40). Refer to figure 11.12. A more thorough discussion of this particular property of the regulus is offered at § 11.57.

12. Some ways to construct a physical model of a regulus – with wire or string – might be mentioned. First the method already implied at § 11.05 might be used but, if one wishes to build a regulus of known proportions for some reason, that method would not be satisfactory. For a circular regulus one could use the well known method of a twisted series of originally parallel strings strung between a pair of circular plates, but that method produces with accuracy only a circular regulus (§ 11.15). It is possible, and geometrically acceptable, to proceed as follows: divide the peripheries of two elliptical plates at points which are equally spaced at angles $\phi$ according to the construction shown at figure 11.01; mount the planes of the two plates parallel with one another and normal to a central rod, adjusting the plates in such a way that the major axes of the ellipses are parallel also; next mount wires or strings between successively chosen points upon the peripheries, beginning with whichever pair will 'waist' the wires acceptably. One might also consider the fact that through any hyperboloid there is a series of parallel sections that are all circles; they are parallel either with the axis for $x$ or the axis for $y$. Refer to photograph 11; please see there the view of a regulus looking orthogonally along the central axis of its hyperboloidal surface.

13. The well known Cartesian, point-coordinate equation to the hyperboloidal surface of the regulus, with the central axis of the hyperboloid along the axis for $z$, and with the principal axes of the throat ellipse along

Figure 11.01 (§ 11.12) Construction for dividing the periphery of an elliptical plate.

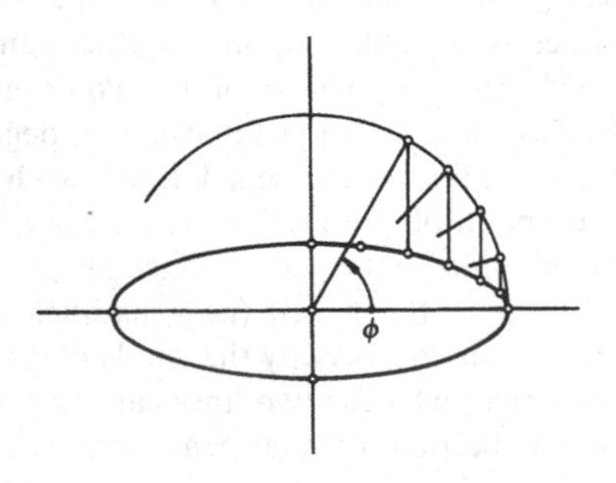

the axes for $x$ and $y$, runs as follows: $x^2/a + y^2/b - z^2/c = 1$. The equation is of the second degree, and the surface is of the second order, a quadric (§ 6.35). It is clear from the algebra that any straight line generally drawn in the Cartesian space can cut the surface no more than twice (§ 11.20). Figure 11.02 has been drawn to show why the coefficients $a$, $b$ and $c$ of the point-coordinate equation might be called the half-axes of the regulus (§ 6.45, § 6C.09, § 11.54). The hyperboloidal surface is indeed the only *ruled* quadric, and its two sets of rulings – the generators of the two superimposed reguli (§ 11.16) – can be exposed by the two pairs of the so-called *parametric equations.* These are explained for example in Sommerville (1934) and Hunt (1978), and one of the pair may be expanded and rewritten, following Bottema and Roth (1979), as follows:

$$(1) \quad bcx - ac\rho y + abz - abc\rho = 0$$
$$bc\rho x + acy - ab\rho z - abc = 0.$$

The other pair are similar. The equations (1) are the equations to two planes which intersect along a generator of the regulus, and, as will be seen, the setting of the parameter $\rho$ at all values will expose all of the separate generators of the regulus. Such equations are awkward for various reasons however, and I refer the reader to an explanation of the Plücker equations to a regulus, the so-called *line-coordinate equations,* at § 11.54. The line of striction is also an awkward matter algebraically. It is a part only of the curve of intersection of the hyperboloidal surface with a quartic, the so-called *striction surface.* This latter is symmetrical about all three principal planes of the hyperboloid, and its equation may be found at § 462 of Salmon (1915). The line of striction itself is not plane-symmetric about the principal planes as the striction surface is, but line-symmetric about the principal axes of the hyperboloid. Refer to § 11.09 and figure 7.09. Refer also to Hunt (1978) § 9.12 for a parallel discussion.

Figure 11.02 (§ 11.13) Showing the half-axes $a$, $b$ and $c$ of the hyperboloidal surface produced by the lines of a regulus. The half-axes determine the dimensions of the truncated elliptical cylinder shown, and this in turn determines the shape of the asymptotic cone.

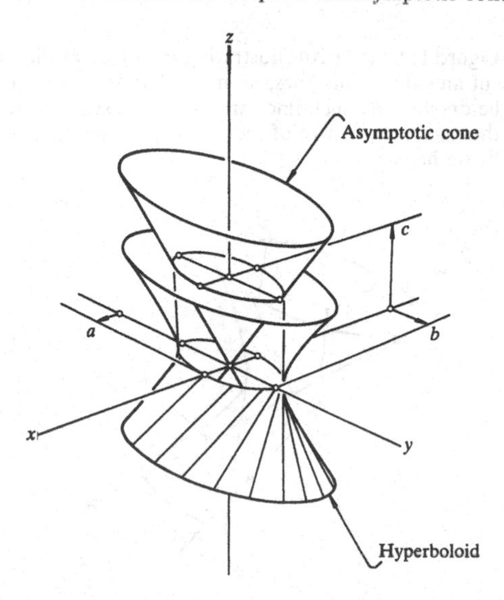


**Special and degenerate forms of the regulus**

14. I have variously and somewhat ambiguously used the words special and degenerate when mentioning some of the singular kinds of regulus which can exist (§ 11.03), and I wish to remain ambiguous still about the possible meanings of those two words. Here, however, I shall use the word *special* to describe a singular kind of regulus whenever the regulus continues to have its usual $\infty^1$ of generators, and I shall use the word *degenerate* whenever either (a) all of the generators have coalesced into a single line, namely an $\infty^0$ of lines, or (b) an $\infty^2$ of generators have suddenly sprung into being due to a blooming of the three defining lines, this latter occurring in the presence of a double linear interdependency (§ 11.03).

15. There follow now some of the ways in which a regulus may be special. We can have two intersecting planes with two polar points, as shown for example in figure 6.05(*d*) and described for example at § 6.47; it could be argued that the elliptical throat of the regulus there has degenerated into the straight line segment $Q'$-$Q''$ between the polar points; or it could be argued, as I have done at § 7.68 and in the figures 7.12, that the three defining lines of the regulus have been chosen in such a way that they intersect in a pair – either once or twice as shown in the figures 7.12(*c*); refer also to § 9.38. There can be two planes with only one polar point; see figure 7.12(*d*); in this case we might consider ourselves to be at one end only of an infinitely long, line-segmental throat (§ 23.67). If the coefficients $a$ and $b$ in equations (1) become equal (and please see figure 11.02), the throat ellipse of the two superimposed reguli will become a circle; in this case the hyperboloid is said to be an hyperboloid of revolution (§ 22.14); a circular regulus can be found for example at § 1.42. If the throat of a regulus has degenerated to a point, and the dimension $c$ has become zero also, the asymptotic cone may have collapsed to a plane and the regulus become a planar pencil; an example occurs at § 9.37. The generators of a regulus might reside in parallel planes; when this is so the regulus itself resides upon a parabolic hyperboloid; see figures 7.12(*g*) and 7.14, and refer for an example to § 7.74. Refer for other readings to chapters 5 and 6 of Salmon (1915), to Sommerville (1934), to chapter 9 of

Hunt (1978), and to my §11.51. At §11.51 Plücker's line-coordinate equations for a regulus are described and discussed in such a way that some of the special cases may be seen immediately.

16. Degenerate reguli occur when the three defining lines are non-coplanar and either intersect at a single point or are parallel (§11.03). If they intersect we get the star of lines; refer to figures 7.10 and 7.12(*a*). If they are parallel we get the parallel field of lines illustrated in figure 7.12(b) and mentioned for example at §23.65. If the three defining lines are coplanar they intersect in general in three places and they then define the $\infty^2$ of lines in that plane. If all of the three defining lines become collinear the generators of the regulus coalesce into the single line; an example of this occurs at §9.37. At §9.37 mention is made of all of the tangents drawn to a circle of infinite radius; at §22.55 I speak about all of the tangents drawn to a circle of finite radius; but in both of those cases (due to repeated interdependencies) the completed regulus is all of the $\infty^2$ of lines which can be drawn in the plane; an illustration appears in figure 7.12(*f*).

17. I have already said that I wish to avoid here the whole question of which areas in mathematics and in engineering practice the two words special and degenerate might cover, but I do wish to say this. Whatever the two words may mean separately, they can be taken usefully together in the following statements. If we are ever to understand and then to *manipulate* the geometry associated with freedom and constraint in working machinery we will need to know well, in a physical, visual way as well as mathematically, the special and degenerate forms of the geometry of the regulus. *The artistic, synthetical business of machine design can clearly be seen to deal, not only with the generalities of the geometry of the regulus, but also, and very much more importantly, with its special and degenerate forms.*

### The linear line congruence

18. In pure mathematical terms a congruence is, by definition, an ensemble (namely a layout) in Euclidean space of an $\infty^2$ of lines determined by two restrictive conditions. The conditions may be set for example as two algebraic equations written between the four parameters defining the line (§11.48, §19.27). It is subsequently argued from this definition that there will be some particular finite number of lines passing through each general point in a congruence and some other particular finite number of lines existing in each general plane. If $m$ and $n$ are those two finite numbers, respectively, they are called the *order* and the *class* of the congruence respectively; a congruence is said to be an $(m, n)$ congruence according to the values of $m$ and $n$. The congruence mentioned by implication at §6.44 in connection with the lines of the screws of the 3-system is for example a (3, 2) congruence (§6C.06).

19. Here, however, and springing from the natural philosophy of forces and angular velocities, we are dealing with a (1, 1) congruence – *the linear line congruence.* This latter is that ensemble of an $\infty^2$ of lines that is linearly dependent upon a given set of four mutually independent lines. It can be put for example this way: given a set of four unbalanced forces whose lines of action do not belong as generators of the same regulus and whose various magnitudes are open to choice, the $\infty^2$ of lines defind by the four lines of action are the lines along which a single pure resultant or an equilibrating force may act (§10.26).

20. I wish to demonstrate here the truth of a well known theorem. It deals with the number of straight lines (and these straight lines may be called *transversals* in the present context) that may be drawn to cut a set of four straight lines. Knowing that any line mentioned here is by implication a straight line, the theorem may be stated as follows: *if any four lines are drawn generally disposed in space, the number of lines that may be drawn to cut all four of them (namely the number of transversals that may be drawn) will not exceed two.* Let four given lines be generally disposed in space. Take any three of them and construct by the method of §11.05 the single infinity of their transversals. Construct thus a regulus of lines and thus the unique hyperboloid defined by the three taken lines. Refer to figure 11.03. The fourth line 4 may cut this constructed hyperboloid. If it does it will cut it twice however, and then no more than twice (§11.11). Figure 11.03 shows a fourth line 4 cutting the said hyperboloid twice; but this fourth line may, as can be imagined, either just touch the surface (at two points coalesced) or not cut it at all. The two lines shown upon the surface and

Figure 11.03 (§11.20) Illustrating the theorem that, to cut any given four lines, no more than two lines may be drawn. Any such lines are called transversals, and the maximum of two of them has been drawn at the figure here.

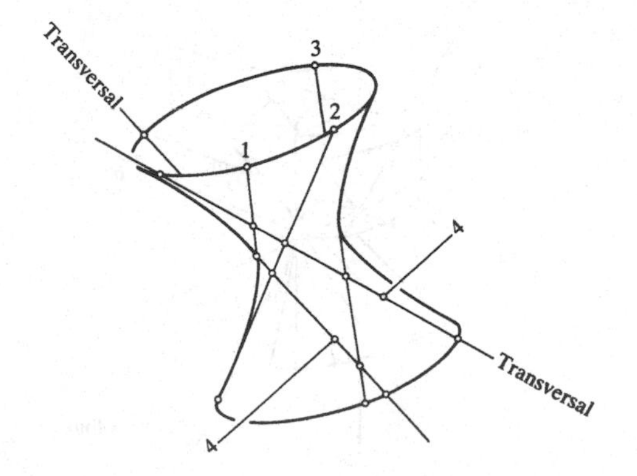

marked *transversal* in figure 11.03 are the two generators of the above mentioned regulus that are cut by the fourth line. They are, quite clearly, the two lines (and the only two) that may be drawn to cut all four of the four original lines.

21. The theorem at §11.20 is mentioned just here because it lies at the root of the linear line congruence. In chapter 10 I called the linear line congruence simply the congruence (§10.09), and so also do I call it here. According to whether the four defining lines of the congruence are cut by the maximum of two transversals, by only one, or by no transversals, the congruence becomes an *hyperbolic*, a *parabolic*, or an *elliptic* congruence. In the following paragraphs I try to describe the different geometrical appearances of these different kinds of congruence. In some of the following paragraphs, however, I speak openly and quite intuitively about what might be judged to happen at the infinitely far distant boundaries of Euclidean space, and I ask the mathematical reader to peruse in advance my §15.51. At §11.27 I begin to discuss the special cases and the degeneracies.

22. The first, the *hyperbolic congruence*, is the easiest. It is relatively easy to imagine all of the lines that may be drawn to cut a given pair of lines generally disposed, namely a given pair of lines that will be skew (§3.53, §6B.07, §7.27). While the two given lines are called the directrices of the congruence, the $\infty^2$ of lines that cuts the directrices is the ensemble which is called the congruence itself. One can readily see that through every general point in space only one line passes, and that in every general plane only one line exists. One can also see that, in zones remote from the zone of closest approach of the two directrices (I call this the central zone), space is filled with lines that are almost parallel. From any zone infinitely far away in a generally chosen direction, the lines will go 'radially' towards the central zone of the congruence, namely towards that neighbourhood where the directrices have their common perpendicular. Mount a sphere of infinite radius upon a centre within the central zone and see that there is upon its surface a kind of 'equatorial band' defined by the unique pair of planes containing the two directrices. In the neighbourhood of that band the lines through the zones are somewhat differently arranged. Try to imagine this. Perceive most importantly that, at any one of that $\infty^1$ of special points that exist along the two directrices, there is an $\infty^1$ of lines that passes through that point. Centred at each of these special points there is, indeed, a planar pencil of lines, and the plane of that pencil is defined by the special point itself and the other directrix.

23. To get a picture of the *parabolic congruence*, go first to figure 15.06 and consider, say, the axis $y$–$O$–$y$: at each of the single infinity of points along that axis, mount the planar pencil of lines that is (at the point) tangential to the surface of the cylindroid. The resulting ensemble of the $\infty^2$ of lines, each line cutting the axis $y-O-y$, will be a parabolic congruence. Refer to figure 11.04. Refer for background reading to §6B.14; read it and the material leading up to it. Go next to §15.57 *et seq.*, where a discussion involving the parabolic congruence and the relationship of one of its special cases with the $f2$ cylindric joint is pursued (§11.29). Note that in the parabolic congruence there is as before only one line through every generally chosen point in space and only one line in every generally chosen plane. Points along the central axis of the congruence, however, along the *coalesced directrices*, are special points; they are each at a confluence of a single infinity of lines. The following fact accounts no doubt for the name of this special congruence: the particular single infinity of lines of the parabolic congruence, each line of which is perpendicular to the central axis, is a regulus upon a parabolic hyperboloid. See figures 7.12(g) and 22.08.

24. The *elliptic congruence*, on the other hand, has no real directrices at all; there are no lines of special points at the centres of planar pencils in the congruence, and there is no series of special planes containing more than one line. At all generally chosen zones in space the zone is filled with lines that are nearly parallel. The overall layout of the lines can be glimpsed as follows. Refer to §6B.18. See how the elliptic congruence there described has a single central axis, is composed of an $\infty^1$ of coaxial, concentric reguli, and comprises an $\infty^2$ of lines that can be seen as being inscribed upon an $\infty^2$ of hyperboloidal surfaces which are all line-symmetric about the same axis and point-symmetric about the same, central point. The throat ellipses of all of the hyperboloidal surfaces reside concentric, coplanar, and similarly oriented in the central plane of symmetry which is normal to the central axis and, the bigger the

Figure 11.04 (§11.23) The planar pencils of lines of a pair of intersecting parabolic congruences (§6B.14, §15.60). The single infinity of the pencils are represented by only a few of them, and these are each illustrated stylistically by means of a circle.

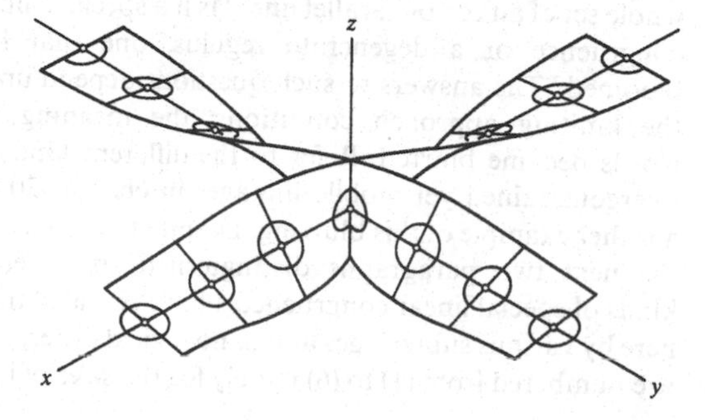

throats are, the more inclined the lines of the reguli are to the central axis of the congruence. It is instructive to scan the surface of the previously mentioned central sphere of infinite radius; in this way we can see the essential similarity of this the elliptic, the hyperbolic, and the (transitional) parabolic congruence. This elliptic congruence has like the others an equatorial band of special zones, the equator existing, as in the case of the others, in the central plane of symmetry normal to the central axis.

**Other remarks about the congruence**

25. We have said that any four linearly independent lines define a congruence. If the four lines have a pair of transversals (§11.20), these are the directrices of the congruence and, unless the two transversals are intersecting or parallel (§11.28), or coalesced (§11.29), the matter needs no further remark. If, however, there are no transversals (§11.24), a question arises: how can we construct the congruence given only the four lines? For an answer to this, please refer to §15.48 and §23.35.

26. Two linear line congruences intersect in two lines. What this means is that, if two congruences are seen to exist and be superimposed, there will be two lines in space common to them both. For more complicated line congruences which are nevertheless 'algebraic' (§11.18), there is a theorem due to Halphen [1] which states the following: if an $(n_1, m_1)$ line congruence intersects an $(n_2, m_2)$ line congruence, it does so in $n_1n_2 + m_1m_2$ lines. While speaking of intersections we might mention the following. We have said that there is an $\infty^3$ of reguli in a linear line congruence (§4.15, §11.07). In the case of a hyperbolic congruence it is interesting to note that, because each directrix of the congruence cuts all lines of the congruence, each one of the $\infty^3$ of reguli must reside on a hyperboloidal surface that contains the two directrices. The hyperboloidal surfaces, in other words, all intersect one another along the same two straight lines.

**Six special kinds of congruence**

27. The words special and degenerate are mutually ambiguous (§11.14). What should we think of the whole set of an $\infty^2$ of parallel lines? Is it a special elliptic congruence or a degenerate regulus, one that has bloomed? The answers to such questions depend upon the limiting approach conditions; the meanings of words become blurred. Refer to the different kinds of overconstrained yet mobile linkages in chapter 20 for another example of this blurring. Despite that, I have in the next two paragraphs distinguished six different kinds of special linear congruence. I have revealed them here by various subterfuges and in no logical order; they are numbered from (1) to (6) merely for the sake of later reference. The list is probably not exhaustive.

28. An interesting and important special hyperbolic congruence occurs when the two directrices intersect; *(1) this first kind of special congruence consists of all of the lines that may be drawn in the plane of the directrices, along with the whole star of lines at the point of intersection*; refer to §7.27; refer to §3.54 and consider what would happen to the equivalent joint made up of the four ball-ended rods if the rods all resided in the same plane; refer also to §14.45. Another occurs when the mentioned pair of directrices become parallel; *(2) this second kind of special congruence consists of all of the lines that may be drawn in the plane of the directrices along with all of the lines in space that may be drawn parallel to them.* Find another by going to figure 7.12; choose for example the picture (*e*) and see there a special biplanar regulus which occupies the pair of parallel planes defined by its three defining lines; choose now a fourth line in a new plane parallel with the existing planes and see that the now existing four lines together define a special congruence; *(3) this third kind of special congruence consists of planar sets of parallel lines, each set occupying one of a single infinity of parallel planes and arranged among themselves in such a way that the single infinity of lines that cut perpendicularly any one of the* $\infty^2$ *of common normals between the planes constitutes a rectilinear regulus upon a parabolic hyperboloid*; refer to figure 22.08 for a single example of such a regulus and to §23.68 for my meaning of the word rectilinear. Because it has no directrices it is evidently an elliptic congruence; refer to §23.73. See aso that, if this third kind of congruence be pierced by a line that is not normal to the planes, that line cuts a series of lines in the planes in such a way that the series is a non-rectangular regulus upon a parabolic hyperboloid. Going to the general elliptic congruence we can see that another special case occurs when the generators become parallel; *(4) this fourth kind of special congruence consists of a field of parallel lines and nothing more.* All four of these special congruences contain an $\infty^2$ of lines. Along with the two more outlined in §11.29, they figure among the five special 4-systems of screws which I describe at §23.71 *et seq.* Each of them might figure, of course, elsewhere as well.

29. At §11.23 I wrote an account of the parabolic congruence that occurs in connection with the cylindroid when two real directrices coalesce. A false impression will be given at a parabolic congruence that only one directrix exists, but there are two. The same false impression is given at another class of special hyperbolic congruence that is delineated when one of the two directrices is, in fact, at infinity. It is easy to see that, if one of the two directrices remains in the local neighbourhood while the other goes off to infinity, the

hyperbolic congruence will consist of a single infinity of planar pencils of lines arranged with their centre-points along the single available directrix and with their planes all parallel with one another. The angle between the planes of the pencils and the single available directrix will depend of course upon the orientation in space of the distant directrix. Indeed the angle between the planes of the pencils and the local directrix will be the angle between the directrices. So one could argue that, while another kind of special congruence occurs when that angle is (as it is in general) not a right angle, yet another occurs when it is. So we could be said to have (5) *a fifth kind of special congruence which consists of a single infinity of parallel planar pencils of lines arranged in their planes in such a way that their centre-points are in the same straight line, and* (6) *a sixth kind, a special case of the fifth, where the line of the centre-points is normal to the planes of the pencils.* A special congruence of this sixth kind (where all lines in all pencils contain screws of all pitches) occurs within the action system of the $f2$ cylindric joint (§23.76). It might also be arguable that these six different kinds of special congruence occur when certain pairs of special complexes intersect. Hunt (1978) discusses this and other such matters at his §11.5.

**The linear line complex**

30. The complex – the linear line complex – is determined, not by *three* or *four* lines as are the regulus and the linear line congruence, respectively, (§11.05, §11.09), but by *five* lines (§11.41). Any five linearly independent lines define a complex. What this means is that, if for example a system of five forces exists with a force along each of five given lines in such a way that no fifth line is a member of the congruence defined by the other four and that no four lines belong to the same regulus, then the $\infty^3$ of lines in space along which a single pure resultant or equilibrating force may act are the lines of the said complex (§10.29). For similar explanations in terms of angular velocities, please refer to §20.09, §20.23 and §20.24.

31. The complex – the linear line complex – is determined, not by *three* or *two* restrictive equations between the four parameters of the lines of the system as are the regulus and the linear congruence, respectively, (§11.54, §11.18), but by only *one* such restrictive equation. This is best explained, not here, but later. Please refer to §11.44, where the single, line-coordinate equation to a complex is discussed.

32. The whole of chapter 9 is devoted to the general layout in space of the lines of a complex. We glimpsed the complex there as we looked from a particular empirical point of view at the instantaneous motion of a moving rigid body. Except to repeat some of the main geometrical properties of the complex mentioned there, it would be duplicative to mention here the whole of chapter 9. Accordingly I simply list next the main, already mentioned properties.

33. The linear line complex is an axi-symmetric system of an $\infty^3$ of lines in space arranged about the axis of symmetry in such a way (*a*) that through any general point in space there passes an $\infty^1$ of lines in the form of a planar pencil, and (*b*) that in any general plane in space there exists an $\infty^1$ of lines intersecting at only one point. Accordingly there exists at each point in space a single, special plane containing lines of the complex; this plane is called *the polar plane at that point*. The axis of the complex is a special line, (*a*) because it is not itself a member of the complex unless the complex is a special complex (§9.41), and (*b*) because, at all points along it, the polar planes are normal to it and thus parallel with one another. The *special linear complex* (the complex of zero pitch) and another special one, the complex of infinite pitch, are mentioned at §9.41 and §9.43. Various implied or incidental appearances of the linear complex occur within the text at chapter 6 (§6.25, §6A.02), and another mention of the special linear complex is made at chapter 8 (§8.25, §8.28). Please be aware also that the linear complex figures importantly in connection with the overconstrained yet mobile 6-link loops of chapter 20. The number of lines in space that do not belong as members of a given complex is $\infty^4$ (§9.44).

**Other properties of the complex**

34. In chapters 9 and 10 first the complex and then it and the other line systems are examined from the empirical point of view. We ask questions, do experiments, and write down what is evident. In chapter 9 we are dealing with the linear complex of right lines that may be seen to exist among the helicoidal field of velocity vectors surrounding a motion screw. In chapter 10 we do, in effect, the same thing among the helicoidal field of moment vectors surrounding an action screw. In chapters 10 and 20 we deal in a parallel manner with the linear complex of possible pure force resultants or possible pure angular velocity resultants of a system of five linearly independent forces or five linearly independent angular velocities, respectively (§10.29, §20.26). At various places within chapters 6 and 10 and elsewhere we deal with complexes and other linear line systems of action or motion screws of the same finite pitch (§6A.02, §10.38, §10.56). A purely mathematical approach to the linear line complex may be seen at §11.44; there I speak about the Plücker line coordinates of a generator. From all of that experimental and mathematical work some of the following properties of a complex can be seen to grow.

35. We have said that at each point there is only one polar plane and that in each plane there is only one

polar point. An understanding of this can be strengthened by visualizing the following. In the whole of space there can be drawn an $\infty^3$ of planes (§4.11). Given the central axis of a proposed complex, we can see that every one of these planes will cut that central axis somewhere. Given next the pitch of the proposed complex, which is, incidentally, the pitch of its helices (§23.11), we can see that, if we go out from the central axis along the unique radial line which may be drawn in any one of the said planes, we will arrive at that unique point in the plane where a helix of the said pitch pierces the plane normally. The planar pencil of lines in the plane at the pierce-point, which is normal to the tangent to the helix there, is the unique planar pencil in that plane. *Any plane, in other words, is a polar plane at some point within a complex, and, correspondingly, any point is at the centre of some planar pencil.* One can see from the nature of the circumstance that no two planar pencils of a complex can be coplanar unless their centre-points are coincident; refer to figure 9.02. One can see conversely that, at no one point in a complex of non-zero pitch, can there exist more than one polar plane.

36. If we draw any line say $w_1$ through a complex the line will not, in general, be a member of the complex (§9.44). If we mount the polar planes of the complex at all points along such a line, these planes will intersect in another line $w_2$. If next we mount, conversely, the polar planes at all points along $w_2$, those planes will intersect in $w_1$. *Such pairs of lines in a complex are said to be mutually conjugate, or conjugate lines of the complex.* This 'conjugate lines' property of the complex can be demonstrated as follows. Refer to figure 11.05. Mount the polar planes at two points $A$ and $B$ on a first line $w_1$. Take two pairs of generators at $A$ and $B$ to intersect in two points as shown at $E$ and $D$. Establish the line $ED$ and call it $w_2$. Next take a third point $C$ in $w_1$. Now there must be a generator $EC$ existing at $E$ because of the polar plane at $E$ namely $EAB$, and there must be a generator $DC$ existing at $D$ because of the polar plane at $D$ namely $DAB$. The lines $CE$ and $CD$ are accordingly generators within the polar plane at $C$. This means that the polar plane at $C$ contains the line $w_2$, and thus that the property has been demonstrated. Please refer in advance to §11.39 where I mention that, if $w_1$ happens to be by chance, not just any line in space, but a line of the complex, $w_1$ and its conjugate $w_2$ will be collinear. *It is important to understand that each one of any pair of conjugate lines in a complex is not in general a line of the complex: each one of the pair of lines will be, in general, a non-member of the complex.*

37. The common perpendicular between a pair of conjugate lines cuts the central axis of the complex perpendicularly. To show this we can argue as follows. From say $B$ and $E$ in figure 11.05 drop the perpendiculars $BB'$ and $EE'$ onto the central axis. Because they are radial lines they are by definition generators of the complex and members of the planar pencils at $B$ and $E$ respectively (§9.15). Thus $BB'$ produced will cut $ED$, and $EE'$ produced will cut $AB$. This means (and please refer here to figure 11.06) that a perpendicular dropped onto the central axis from any point say $A_1$ on $w_1$, having its foot upon the central axis at $A'$, will go on to cut the conjugate line $w_2$ in an opposite point say $A_2$.

Figure 11.05 (§11.36) Illustrating a proof of that property of the complex concerning pairs of conjugate lines.

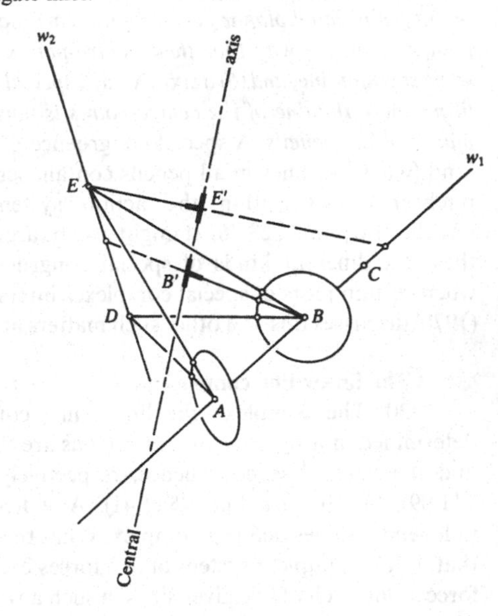


Figure 11.06 (§11.37) The common perpendicular distance $\Gamma_1$–$\Gamma_2$ between a pair of conjugate lines $w_1$ and $w_2$ within a linear complex.

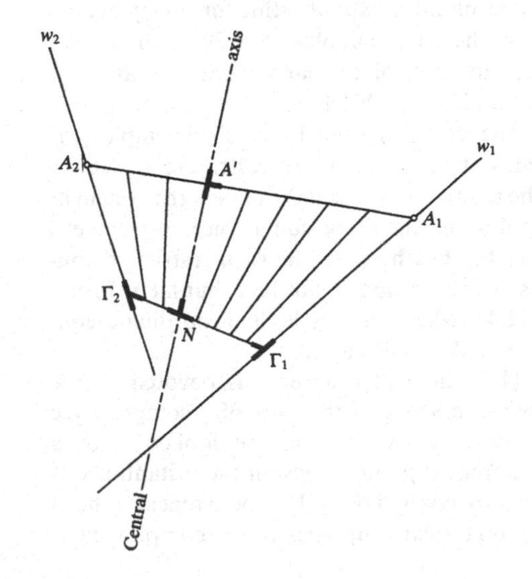

Refer to figure 11.06. It follows (*a*) that the common perpendicular $\Gamma_1\Gamma_2$ between a pair of conjugate lines $w_1$ and $w_2$ will cut the central axis perpendicularly, and (*b*) that given $w_1$ in a complex of unknown pitch, $w_2$ will be some generator of the parabolic hyperboloid $\Gamma_1\Gamma_2A_2A_1$; see figure 11.06. While the common perpendicular $\Gamma_1\Gamma_2$ cuts the central axis in $N$, $\Gamma_1$ and $\Gamma_2$ might be called the central points of the conjugate lines $w_1$ and $w_2$ respectively (§ 8.21, § 9.22, § 21.03). Although these quoted references all relate to the central points of motional lines (and the concept applies whether or not the lines be right lines), the same concept applies as we

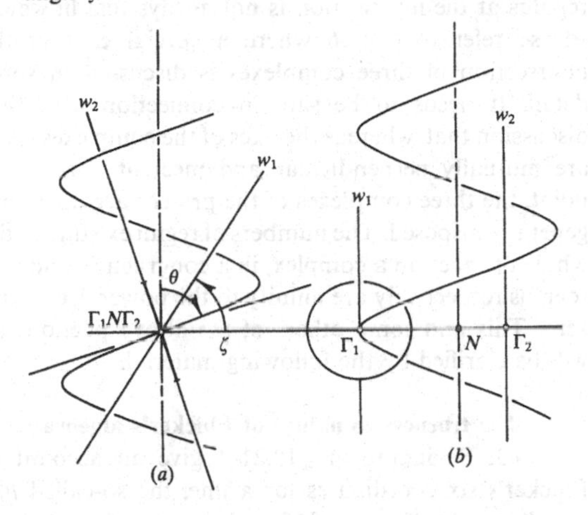


Figure 11.07 (§ 11.38) Showing two orthogonal views of a line $w_1$ and its conjugate $w_2$ set up within a linear complex. With all else fixed, we study the variation of the directed distance $N$–$\Gamma_2$ with the angle $\zeta$.

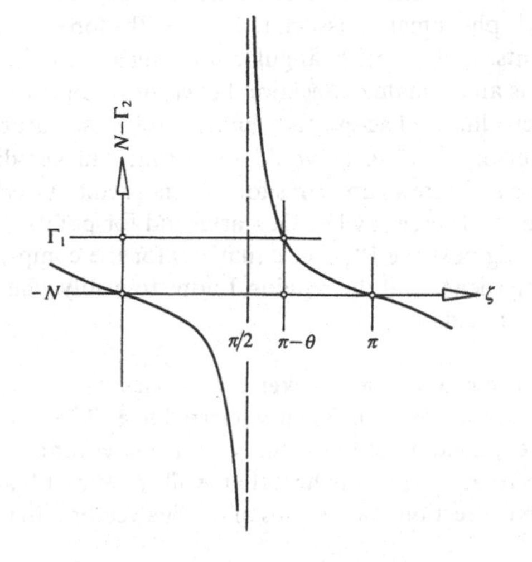


Figure 11.08 (§ 11.39) Showing the variation of the position of $\Gamma_2$ with $\zeta$.

know to actional lines as well (§ 3.01). Here, however, I have extended the central point, or the centre-point concept into the realm of this the abstract geometry.

38. It is interesting to take any line $w_1$ through a given complex, and to draw attention to the perpendicular distance $\Gamma_1$-$N$ between $w_1$ and the central axis. Refer to figure 11.07. We see in figure 11.07(*a*) a view of the said complex looking orthogonally along the line $\Gamma_1N$ and thus orthogonally at the central axis. The polar plane at $\Gamma_1$ appears as a mere line through $\Gamma_1$ in this view, and that line (shown thickened) must coincide there with the binormal to the helix at $\Gamma_1$. There is a point $\Gamma_2$ on the line $\Gamma_1N$ through which the conjugate line must pass (§ 11.37). This $\Gamma_2$ is shown in figure 11.07(*b*): the figure is another orthogonal view of the central axis looking, this time, at the common perpendicular $\Gamma_1N\Gamma_2$. Please see next that, independently of the angle of inclination $\zeta$ of the line $w_1$ at $\Gamma_1$ and of the consequent position of $\Gamma_2$ along the line $\Gamma_1N$, the conjugate line $w_2$, because it resides in the polar plane at $\Gamma_2$ (§ 11.36), will always appear in figure 11.07(*a*) at the fixed inclination shown. Allowing the directed distance $N$-$\Gamma_1$ to remain fixed and positive, let us now allow the angle $\zeta$ of $w_1$ to vary from zero through $\pi/2$ to $\pi$, and watch as we go the variation of the facing, directed distance $N$-$\Gamma_2$.

39. This paragraph is derived from a work of Lozzi [2]. Either by graphical construction or by algebra it is possible to show in the two views at figure 11.07 that the variation with $\zeta$ of the directed distance $N$-$\Gamma_2$ is such that $N\text{-}\Gamma_2 = -\tan\zeta$. Refer to figure 11.08. Remembering that $w_1$ is not a member of the complex unless it coincides with the binormal at $\Gamma_1$, this figure shows among other things (*a*) *that a line $w_1$ and its conjugate $w_2$ are not necessarily on opposite sides of the central axis, and* (*b*) *that the conjugate line for a line of the complex is the line itself.* The latter is a well known and important property, but the former is not so well known or understood. It will be made clear in the material that follows that this concept of the conjugality of two lines that may be drawn across a complex (the lines $w_1$ and $w_2$ being not themselves members of the complex) relates directly to the concept of the conjugality of two generators of the same pitch in a cylindroid (§ 15.49).

**On the intersections of complexes**

40. It is known that two complexes intersect in a congruence (§11.49). It is also known that the two directrices of the congruence of intersection, if real, are pairs of conjugate lines in both of the complexes. Refer to figure 11.09. This is a rough sketch of two complexes of generally chosen pitches generally disposed. Recall the helices of chapter 9 and see in the sketch that, at all general points in space (only one of which is shown),

there are two intersecting helices. See also that there are, at each general point, two intersecting planar pencils of generators each normal to its own helix. There is, accordingly, only one line through each general point which is common to both of the planar pencils there. This line is the single generator at that point of the congruence of intersection. If the directrices of the congruence are real there will be an $\infty^1$ of special points where the two helices are tangential with one another and the two polar planes coincide. Only two of these special points are shown in the sketch. *The special points are upon two straight lines; the lines are the directrices of the congruence of intersection; and the directrices both cut the common perpendicular between the central axes of the complexes perpendicularly.*

41. A single infinity of complexes can be drawn to contain a given congruence, and their central axes reside as the generators of a cylindroid (§ 6B.08). The relevant cylindroid, determined by its two screws of equal pitch at the two directrices of the congruence if the directrices are real (§ 15.46, § 15.47), carries screws which correspond to the pitches of the complexes. A complex having been thus erected, moreover, pairs of screws of equal pitch upon the cylindroid are not only pairs of conjugate screws of the cylindroid (§ 15.49) but also pairs of conjugate lines of the complex (§ 11.36). If the directrices of the given congruence are not real the cylindroid can be constructed as explained at § 15.48 and § 23.36. We can see from all of this that, if we have a congruence defined by a set of four linearly independent lines and a fifth line which is linearly independent of the already mentioned four, only one complex can be drawn to contain both the congruence and the fifth line (§ 11.30). A method for finding the central axis and the pitch of that complex follows: (*a*) take any set of four lines from the available five, construct the congruence defined by them (§ 11.20, § 11.22), erecting the cylindroid defined by its directrices or otherwise; (*b*) take another set of four lines and repeat the same performance; (*c*) the two cylindroids thus discovered will intersect along a common generator of the same pitch; the seat and the pitch of that generator are the central axis and the pitch respectively of the required complex (§ 7.87).

Figure 11.09 (§ 11.40) There are intersecting planar pencils of generators at all general points within an intersecting pair of linear complexes. Only one of these general points is shown here, but two special points are shown. These special points are representative of a single infinity of points – they are spread out along two straight lines – where the intersecting polar planes are in fact coplanar.

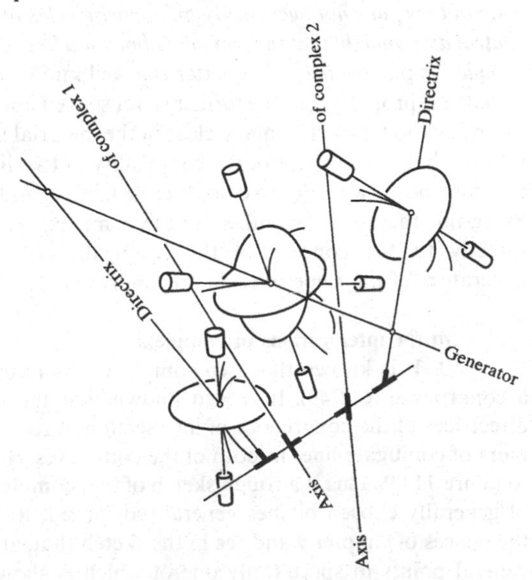


42. Three complexes intersect in a regulus. The regulus at the intersection is not always real however. Please refer to § 11.56 where a special case of the intersection of three complexes is discussed in some detail. It needs to be said in connection with that discussion that, whereas the axes of the complexes there are mutually perpendicular and meet at a common point, the three complexes of the present argument are generally disposed. The numbers of reguli existing in the whole of space, in a complex, in a congruence, and in a regulus respectively are infinity to the power 9, 6, 3 and zero. This and some others of the above phenomena will be clarified by the following material.

**The trueness to nature of Plücker's algebra**

43. Beginning at § 19.21 I give an account of Plücker's six coordinates for a line, the so-called *line coordinates* of a line. At § 19.28 I go on to describe and explain the six coordinates for a screw, the so-called *screw coordinates* of a screw, also due to Plücker. In chapter 10 accounts are given of the various observable natural phenomena associated (*a*) with forces and moments, and (*b*) with angular and linear velocities. There is an intimate connection between the algebra of Plücker's line and screw coordinates and these natural phenomena (§ 10.39). It would seem that Plücker did for lines and screws and thus for the linear line systems (of lines and screws) what Descartes did for points. By discussing next the Plücker equations for the complex, the congruence and the regulus, I hope to justify what I have just said.

**The complex in Plücker coordinates**

44. The reader is already referred to § 19.21 *et seq.* for an explanation of the Plücker coordinates for a line. Kindly recall by reading there that, while $L$, $M$ and $N$ are the three direction magnitudes of the line vector $\mathbf{L}$ in the

line, $P$, $Q$ and $R$ are the three direction magnitudes of the moment vector $\mathbf{L}_O$ at the origin (§ 19.24). I come now to the matter at issue here. If one were to try to locate in the Cartesian space a line ($L, M, N; P, Q, R$), with the simple restriction placed upon the line that

(2) $P=kL$

where $k$ is some constant, one would be trying to ensure that the magnitude of the moment about the $x$-axis of the line vector $\mathbf{L}$ in the line (namely $P$) is $k$ times the magnitude of the component of that vector in the direction $x$ (namely $kL$). Refer to figure 11.10($a$). Let us take any point say $A$ in the Cartesian space and imagine that a line segment $A$–$B$ (representing the line vector $\mathbf{L}$) is pivoted in a spherical manner at $A$. Let us try to locate the segment $A$–$B$ in space in such a way that the condition (2) is fulfilled.

45. In figure 11.10($a$) we have chosen a general position for the point $A$ and dropped from that point onto the $x$-axis a perpendicular $AH$. Please note the triad of axes at the top of the figure, and see that the line $HA$ is a generally chosen line radiating perpendicularly outwards from the $x$-axis. Looking now at the point $A$ it is clear from the figure that, with $\mathbf{L}$ in the special, thickened location shown by the numeral 1 (and regarding $\mathbf{L}$ now as a unit vector), $L=-\sin\theta$. Writing the radial distance $H$–$A$ as $r$, we can also write that $P=+r\cos\theta$. One can see at $A$ with the vector $\mathbf{L}$ located as shown by numeral 1 that, with a given $k$, we need only adjust the angle $\theta$ to ensure that the condition (2) is fulfilled. We have from above that $P/L=-r\cot\theta$, and we have that $P/L=k$, so $\theta$ must be such that $k=-r\cot\theta$. The question now is, with the angle $\theta$ at $A$ adjusted to suit (note the location of $A$–$B$ at 1), where else can the segment $A$–$B$ at $A$ be?

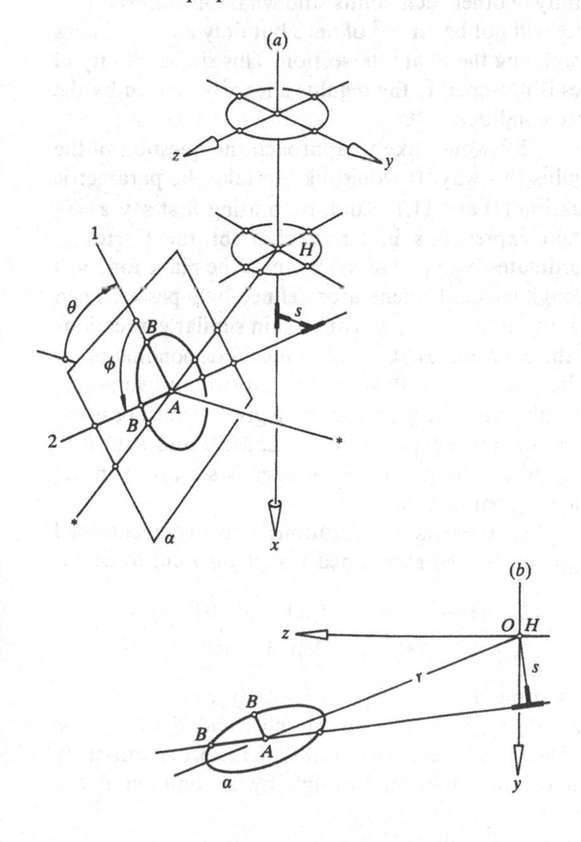


Figure 11.10 (§ 11.44) In panoramic view at ($a$) we locate the line vector $A$–$B$ in space to suit the condition $P=kL$. Please note at ($a$) that, if the axis for $x$ be vertical, the two lines marked by the asterisks are horizontal. In the orthogonal view at ($b$), which is taken looking along the $x$-axis, we see that the component $P$ of the moment at the origin is given by the projection of $A$–$B$ at 2 times the common perpendicular distance $s$.

46. Perceive that, if the vector $\mathbf{L}$ were not perpendicular to but collinear with the radial line $HA$, $L$ would be zero and so would be $P$; if $A$–$B$ at $A$ were collinear with $HA$, in other words, the condition (2) would be fulfilled in that location too. This gives the clue to a set of locations of the segment $A$–$B$ inclined at the variable angle $\phi$ in the plane marked $\alpha$. With $A$–$B$ in such a new location at $\phi$ (note the location of $A$–$B$ at 2), we can see quite easily that $L=-\cos\phi\cos\theta$. By taking a view orthogonally along the $x$-axis, see figure 11.10($b$), we can see moreover that, with $A$–$B$ at 2, $P=+r\cos\phi\cos\theta$. Thus we can see that, as before, $P/L=-r\cot\theta$. We have thus shown that, if $A$–$B$ at $A$ is anywhere at all in the plane marked $\alpha$, the condition (2) is fulfilled.

47. After having shown next – and this is easy to do – that the condition (2) will not be fulfilled at $A$ unless the segment $A$–$B$, spherically pivoted at $A$, is confined to the plane marked $\alpha$, and having seen that the whole of space is filled with an $\infty^3$ of points such as $A$, we can see straight away that the condition (2) is indeed the Plücker equation to a linear line complex, or, as we say more simply, to a complex (§ 10.09). The pitch of the complex can be seen to be $p=-k$. We know from §9.15 that the pitch $p$ can be written $r=p\tan\theta$, where $\theta$ is there (as it is here) the helix angle. The actual helix at $A$ in figure 11.10($a$) may be seen in the imagination to pierce the plane marked $\alpha$ (the polar plane at $A$) at $A$ orthogonally. As already shown at § 11.45, $\theta$ increases with $r$ in the same way as the helix angle $\theta$ in figure 9.01 does, according to the formula $k=-r\cot\theta$. *The complex defined by the single condition (2), namely $p=kL$, is the complex coaxial with the $x$-axis whose pitch is -k.*

### The congruence in Plücker coordinates

48. If we added to the condition (2) a second condition, namely that a similar relationship between $Q$

and $M$ should also exist, we would write a new set of conditions as follows:

$$(3) \quad P = k_1 L \quad Q = k_2 M$$

where $k_1$ and $k_2$ are different constants. Under these conditions (3) we would find that, at each point $A$ – and I refer again to figure 11.10 – only two isolated locations of the line segment $A$–$B$ would be found to be possible. The two locations would be found to be diametrically opposite in the plane marked $\alpha$; they would thus be found to delineate the same single line through $A$.

49. It may be seen this way. The condition $P = k_1 L$ will establish a polar plane at $A$ at a radial distance say $H$–$A$ from the $x$-axis, while the condition $Q = k_2 M$ will establish another polar plane at $A$ at a similar radial distance say $K$–$A$ from the $y$-axis. These two polar planes will intersect in a single line through $A$ (and thus in the two diametrically opposite, possible locations for $A$–$B$). Such lines, repeated at all of the $\infty^3$ of points such as $A$, would, after each line was traced this way a single infinity of times (§4.06), produce the $\infty^2$ of lines of a linear line congruence (§11.40). Please look again at the two intersecting polar planes in figure 11.09. *The congruence defined by the pair of conditions (3), namely $P = k_1 L$ and $Q = k_2 M$, is the congruence in which the two complexes $P = k_1 L$ and $Q = k_2 M$ intersect.*

50. It would appear that, unless the constants $k_1$ and $k_2$ are of opposite sign, the sets of helices of one of the complexes and the sets of helices of the other will nowhere meet in such a way that separate helices might be tangential with one another and the congruence will be an elliptic congruence. Perceive this by thinking about the two principal screws of a cylindroid; unless they are of opposite sign, no screws of zero pitch can exist in the cylindroid. We may say a similar thing (both forwards and in reverse) about the hyperbolic congruence. What are the corresponding conditions to be set for a parabolic congruence? Are they that the two axes of the complexes need to be collinear? It might be noted that a single infinity of complexes may be drawn to contain a given hyperbolic congruence. In such a case there is a cylindroid formed by the single infinity of the axes of the complexes. This is already mentioned at §11.41. We should also note however that the two complexes set here by the conditions (3) have their central axes intersecting one another perpendicularly. They are thereby at the centre of that cylindroid. Think about various sets of values for $k_1$ and $k_2$; when they are equal, for example, the length $2B$ of the cylindroid is zero and the single infinity of congruences collapses into only one special congruence (§11.28). The vaguely stated suppositions and unanswered questions of this paragraph might be taken by the reader as clear evidence (*a*) that the author has not thought very carefully about the details of the matter here, and (*b*) that he wishes for the sake of avoiding confusion to move on.

### The regulus in Plücker coordinates

51. We noted at §11.42 that three complexes intersect in a regulus. The regulus may or may not be real (§11.54). Here we shall look at the congruence defined by the two complexes at (3) being intersected by a third complex $R = k_3 N$. To do this we might first set out the following, alleged conditions for a regulus:

$$(4) \quad P = k_1 L \quad Q = k_2 M \quad R = k_3 N$$

Later we can check that these conditions (4) do, indeed, define a regulus; but first we might ruminate enumeratively as follows. If at each point $A$ in the Cartesian space there is only one line defined by the two conditions (3), there will be only one chance in a single infinity that the new polar plane at $A$ defined by $R = k_3 N$ will contain that line. What this means is that it is only at an $\infty^2$ of points in space that we might expect to find the three polar planes at $A$ intersecting in the same line. Along each such line of mutual intersection of the three polar planes at a single point $A$, however, there will be a single infinity of other such points; and what this means is that there will not be an $\infty^2$ of lines but only an $\infty^1$ of lines comprising the final intersection. This single infinity of lines is presumably the regulus allegedly defined by the three conditions (4).

52. I would like to approach the question of the regulus this way. It would like to take the parametric equation (1) at §11.13, and, by putting first say $z = 0$, obtain expressions in terms of $\rho$ for the Cartesian coordinates $(x_1, y_1, 0)$ of the point in the plane for $z = 0$ through which the generator defined by $\rho$ passes. Then by putting say $x = 0$, we can obtain similar expressions for the coordinates $(0, y_2, z_2)$ of the corresponding point in the plane for $x = 0$. With that data about two points through which the generator for a given $\rho$ must pass, we can write down expressions for $L, M, N$ and $P, Q, R$ in terms of $\rho$. The gist of the matter is set down in the following paragraph.

53. After the substitutions, the two mentioned points in the two mentioned planes turn out to be:

$$(5) \quad [2a\rho/(1+\rho^2), (b - b\rho^2)/(1+\rho^2), 0]$$
$$[0, (b + b\rho^2)/(1-\rho^2), 2c\rho/(1-\rho^2)].$$

By writing $L = x_2 - x_1$, $M = y_2 - y_1$, etc., and $P = y_1 z_2 - y_2 z_1$, $Q = z_1 x_2 - z_2 x_1$, etc., and these can be derived from the expressions (8) at §19.23, one can easily obtain, after division through by a common factor

(§ 19.24), the following, simple expressions for the Plücker coordinates of the generator:

$$
\begin{aligned}
(6)\quad L &= +a(1-\rho^2)\\
M &= -2b\rho\\
N &= -c(1+\rho^2)\\
P &= -bc(1-\rho^2)\\
Q &= +2ca\rho\\
R &= -ab(1+\rho^2).
\end{aligned}
$$

The same expressions, except for some differences in sign (?), are reported at page 326 of Bottema and Roth (1979). These authors are there discussing a special kind of motion they call Bennett motion (§ 20.53).

54. From the list of the Plücker coordinates at (6) we can easily see that the following three equations may be written:

$$
\begin{aligned}
(7)\quad P &= -(bc/a)L\\
Q &= -(ca/b)M\\
R &= +(ab/c)R.
\end{aligned}
$$

From these, in turn, it becomes obvious that the alleged conditions for a regulus quoted at (4) actually do define a regulus. It will also be seen that the three half-axes of the regulus determined by the conditions (4) – they are mentioned at § 11.13 – can be obtained directly from the following:

$$
\begin{aligned}
(8)\quad a^2 &= -k_2k_3\\
b^2 &= -k_3k_1\\
c^2 &= +k_1k_2.
\end{aligned}
$$

The question of sign is obviously important, for if the constants $k$ are not of suitable signs the regulus at (4) will be unreal. From (8) it can be seen that, unless the constants $k$ have such a set of signs that one of them is different from the other two, the regulus will be unreal.

55. We should look now to see how the generators of a regulus are formed as the parameter $\rho$ among the line coordinates at (6) is varied all the way from minus infinity through zero to plus infinity. In figure 11.11($a$) some lines of a particular regulus are shown upon the generated, hyperboloidal surface. I chose $k_1 = -0.5$, $k_2 = -3.0$, and $k_3 = +2.0$ to establish the example in figure 11.11($a$); these gave, from (8), $a = 2.45$, $b = 1.00$ and $c = 1.22$ approximately; these dimensions are those of the largest escribed rectangular box that is visible there and mounted with its far apex at the origin. The hyperboloid mounts itself symmetrically upon the Cartesian axes in the conventional manner; this is a consequence of the relationship established here between the given Plücker algebra and the conventional form of the point-coordinate equation quoted earlier at § 11.13. By using the first of the two sets of Cartesian coordinates at (5), we can see first of all that, as $\rho$ is allowed to vary, the point at which the thus nominated generator cuts the $xy$-plane traverses the whole curve of the throat ellipse in the manner shown. Selected points upon the throat ellipse are delineated and the generators may be calibrated in terms of $\rho$; see an orthogonal view of the throat ellipse in figure 11.11($b$). We can see from this calibration that, while the movement of the point in response to $\rho$ is symmetrical about the axis for $y$, it is not symmetrical about the axis for $x$.

56. It is easier now to observe the way in which the six Plücker coordinates listed at (6) define the whole of the shown regulus. By setting $\rho$ to the particular value $\rho = +0.5$, for example, we see that the particular generator marked $G$ is established by virtue of the following values:

$$
\begin{aligned}
(9a)\quad L &= +(0.75)a\\
M &= -(1.00)b\\
N &= -(1.25)c\\
(9b)\quad P &= -(0.75)bc\\
Q &= +(1.00)ca\\
R &= -(1.25)ab.
\end{aligned}
$$

Figure 11.11($a$) (§ 11.55) Showing how the generators of a regulus are determined by the Plücker equations (4), that the surface swept out by the radius vector $\mathbf{r}$ is an elliptical cone, that the throat ellipse in this particular case is in the $xy$-plane, and that the intersections of the hyperboloidal surface with the $yz$- and the $zx$-planes are hyperbolae. Notice that the shown branch of the hyperbola in the $yz$-plane cuts the shown throat ellipse perpendicularly at the vertex $V$ (of which there are four) and that this intersection relates to a similar detail at the single vertex $V$ of a parabolic hyperboloid in figure 23.04.

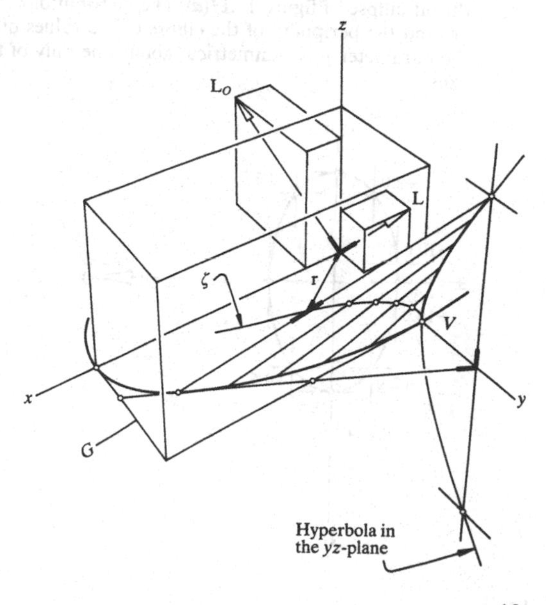

I have mounted at the origin in figure 11.11(*a*) not only the moment vector $\mathbf{L}_O$ (which belongs at that point) but also the line vector **L** which, more naturally, belongs in the line. Recognizing this and being aware that the figure is drawn somewhat approximately, please check for an exercise that the six values at (9*a*) and (9*b*) for the direction magnitudes of **L** and $\mathbf{L}_O$ (the six Plücker coordinates for the line *G*) are consistent with the way in which the regulus in figure 11.11(*a*) has been constructed; you might refer in so doing to § 19.24 and note that for the sake of clarity in the figure I have multiplied all six of the coordinates by $-0.3$. By putting the line vector **L** not in the line but at the origin I have shown more clearly, I hope, not only that we have here the geometrical behaviour of a dual vector (refer to figure 10.12), but also that the three vectors **L**, $\mathbf{L}_O$ and the radius vector **r** are mutually perpendicular. Please study the twisted curve marked $\zeta$ upon the hyperboloidal surface that contains the feet of the perpendiculars (namely the tips of the radius vectors **r**) dropped from the origin onto the lines of the regulus. The curve marks the intersection with the hyperboloidal surface of an elliptical cone whose vertex is at the origin. Already mentioned by implication at § 11.11, this $\zeta$ is mentioned again at § 11.58. For some examples of reguli presented in terms of their line-coordinate equations in this particular way, refer to Hunt (1978); find his equations (12.10) for all of the reguli of the general 3-system of screws, and his equations (12.15) for the single regulus of the pitch quadric. Whereas I use the symbol $p$ for the pitch of a screw, Hunt uses the symbol $h$.

Figure 11.11(*b*) (§ 11.55) An orthogonal view of the throat ellipse of figure 11.11(*a*). The distribution around the periphery of the ellipse of the values of the parameter $\rho$ is symmetrical about one only of the axes.

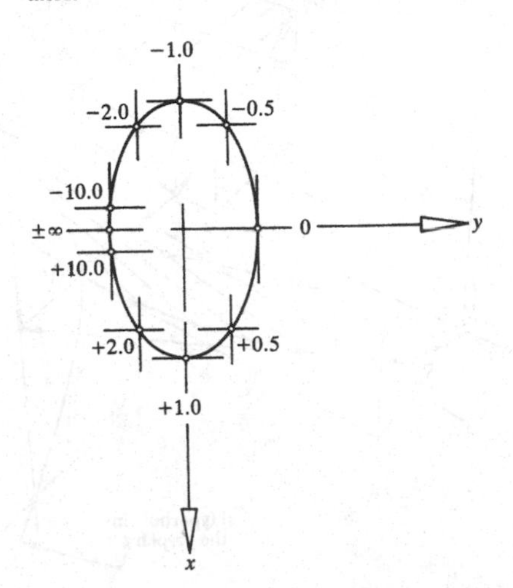


**Perpendiculars dropped to the lines of a regulus**

57. At § 11.11 it was pointed out in effect that, if we drew the common perpendicular between a selected pair of generators of a regulus and chose a point say $Q$ between those generators and upon that common perpendicular, we could then draw through $Q$ only one other common perpendicular between a pair of generators. We could, if we wished, next drop from that $Q$ the other perpendiculars (the non-common perpendiculars) onto the remaining generators, and thus construct the two elliptical cones (the special case of the general quartic cone) that was mentioned at § 11.11. Now it is not very easy to see (or to illustrate) how it is that the two elliptical cones, whose vertices are together at $Q$, actually touch one another along the two mentioned common perpendiculars. If we go to the recently employed figure 11.11(*a*), however, where a special case of the theorem outlined at § 11.11 is clearly exemplified, we can see that, with $Q$ at the origin $O$, the two common perpendiculars through Q are the minor and the major axes of the throat ellipse. We can see also by symmetry that the elliptical cone traced out by the radius vector **r** as it moves from $\rho = 0$ to $\rho = +1$ is wholly contained within the two planes $O$–$yz$ and $O$–$zx$. Part of the upper half of that cone is illustrated in figure 11.11(*a*), but the lower half, plane-symmetric with the upper half about the plane $O$–$xy$, and generated by the vector **r** as it moves from $\rho = \pm\infty$ to $\rho = -1$, is not shown there.

58. In figure 11.12 I have tried to make a sketch that will better illustrate the gist of the general theorem outlined at § 11.11 [3]. The point $Q$ has been chosen generally in figure 11.12, but within the central zone; the two elliptical cones are not of the same shape, and they touch in a manner unlike the manner of those symmetrically arranged in figure 11.11(*a*). The cones of course intersect with the hyperboloidal surface of the regulus; the thickened line, which traces the intersection around both upper sides of one of the cones and both lower sides of the other, is following the feet of all of the perpendiculars dropped from $Q$ onto the successive generators of the regulus. The twisted curve of intersection $\zeta$ at figure 11.11(*a*) is of course a special case of the twisted curve shown here.

59. It is instructive to consider the special case of the cone of perpendiculars dropped to the lines of a regulus when the said point $Q$ resides exactly upon the hyperboloidal surface. One effective way to consider that special case is to begin by taking a view of a general regulus looking directly along a generally chosen one of its generators. Refer to figure 11.13. The point marked $G$ in the figure is the said, generally chosen generator – this generator appears of course as a point. The point marked $G'$ on the other hand is representing that single, parallel generator of the opposite family through which

all of the other generators of the family in question must pass (§11.06). Looked at along a generator, a regulus always appears this way. It appears as an isolated point (the point $G$) and a single infinity of lines intersecting at another point (the point $G'$).

60. If now we put the point $Q$ upon the generator $G$ and drop the perpendiculars from $Q$ onto the other generators of the regulus in figure 11.13, we will construct an example of that special kind of cone that was mentioned at the beginning of the above paragraph. Note that the feet of the perpendiculars from $Q$ appear upon a circle. What this means is that, if the regulus is now cut by that unique circular cylinder defined by the pair of parallel generators $G$ and $G'$, these latter being conjugate generators set diametrically opposite upon the throat ellipse, the line of the intersection will be upon the surface of the constructed cone. The regulus, the cylinder, and the cone all intersect one another in the same closed curve. It may be a useful exercise for the reader to discover those particular circumstances if any that render this closed curve a plane ellipse. Is the curve, indeed, always an ellipse, and may it, under certain circumstances, even be a circle? It will be clear to begin with that the centre line of the cylinder will intersect the origin of coordinates and that the radius of the cylinder will be the perpendicular distance $r$ from the origin to the generator (§11.56).

Figure 11.12 (§11.58) Perpendiculars dropped from a point $Q$ onto the lines of a regulus generate a quartic cone. When $Q$ is within a certain central zone of the regulus, the cone breaks down into two quadratic cones that are tangential to one another along two lines.

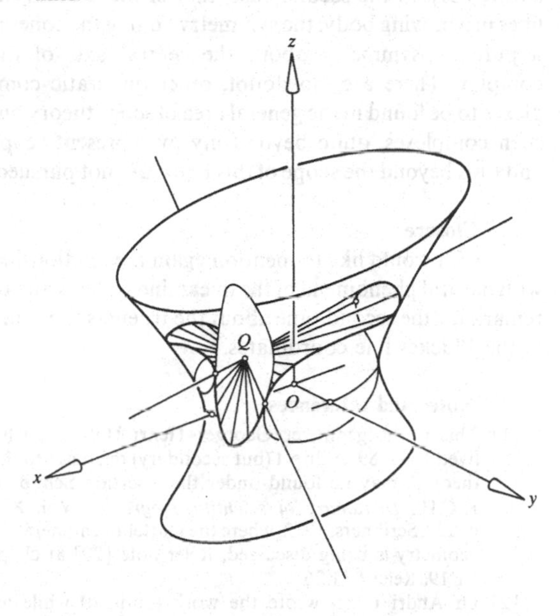


Figure 11.13 (§11.59) The appearance of a regulus looked at along a generally chosen one (namely $G$) of its generators. The figure also illustrates the argument concerning a point $Q$ upon the generator $G$ and the perpendiculars dropped from that point onto the generators.

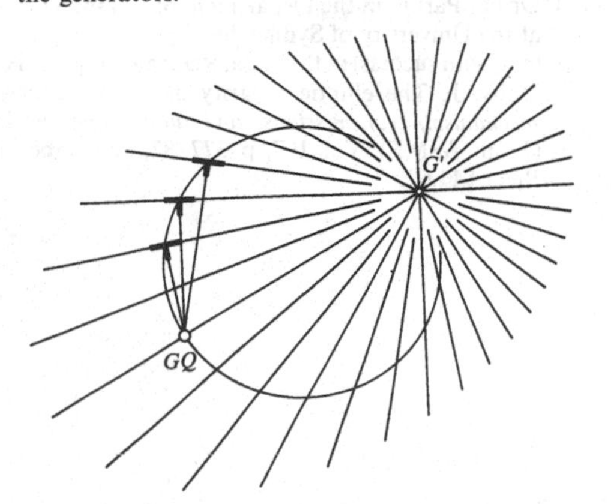


**The particular regulus defined by three lines**

61. I have offered in the present chapter a few theorems about the geometry of the regulus and the shape of its hyperboloidal surface. I have offered also some of the bare elements of its Plücker algebra. There is thus assembled just enough material for tackling the problem of how to locate the principal axes of, and how to determine the half-axis dimensions of, the particular regulus defined by a set of three given lines. A suggested method for doing this follows.

62. Refer first to §6.34 where another method for finding the centre of the regulus is given; this may be just as effective or even better; it depends upon the algebra. Take the three given lines and select for them a convenient origin of coordinates and a triad of Cartesian axes. Write the Plücker equations for the three lines and proceed – and I write now geometrically – as follows. Take any two of the lines and draw the unique transversal that is parallel with the third line. Repeat this with another two of the three lines, thus obtaining two sets of parallel lines. Erect between each pair of parallel lines that line that is equidistant between the two and parallel with them. These two lines will intersect at the central point of the regulus (§11.06). Position here a new, more convenient origin of coordinates. Erect now the third of the three transversals mentioned above, thus erecting, in all, three members of the opposite family of generators upon the as yet undiscovered hyperboloidal surface (§11.06). Any other transversal drawn across this opposite family will now locate a fourth line of the sought regulus. Erect this and, if you like, all of them, and thus obtain the whole of the required regulus that is now spherically pivoted, as it were, about its known

central point. Pluck the four line vectors **L** of the four found lines of the regulus and replant them at the origin. Refer for plucking and replanting to §5.31, and read about the spherical indicatrix at §19.40. Determine by algebraic metods the second order cone defined by the four vectors there. This cone is the asymptotic cone for the regulus. Note indeed, and this may be seen in figure 11.11(*a*), that the line vector **L** at the origin traces out this asymptotic cone. With the asymptotic cone thus established, the three principal axes and the half-axis dimensions *a*, *b* and *c* follow. One could also drop perpendiculars from the origin onto the four known lines of the regulus and thus locate the cone or cones of §11.57, but this would seem to be a more cumbersome method. Remember the circular cylinders of §11.60; they may offer an easier solution. There may be some other geometrical short-cut, but it is certainly true to say that, without some appropriate line-coordinate method (we have the Plücker method), the regulus is a very nasty thing to handle algebraically. Refer to Hunt (1978) page 307 for the following references: Grace (1902), Veblen and Young (1910), Somerville (1934), Semple and Roth (1949). Refer also to Hunt page 308 for some unproven algebraic manipulations. And refer to Jessop (1903) who gives at the end of his book some useful basic results. Refer to note [4].

**A note about the non-linear line systems**

63. Given the infinite variety of possibilities that are available for the writing of algebraic formulations to relate the Plücker coordinates with one another in sets of one, two, and three equations, there is, in the pure mathematical arena, an immense range of possibilities for the establishment of complicated kinds of line system that may be interesting. In the particular physical arena within which we are working here, however, the line systems other than the linear line systems ordinarily discovered appear to be comparatively simple ones. The circumstance is similar, perhaps, to the circumstance in the classical rigid body mechanics where we seldom seem to need more than the first and the second derivatives with respect to time of linear or angular displacement.

64. As well as the non-linear complexes mentioned by way of explanation at §9.45 (with an example at §6C.06), there are two examples to be found within this book of an occurrence within the explanatory text of a non-linear complex. At §6.29 the layout of the $\infty^3$ of screws of the 4-system is described, and at §21.22 the layout of the $\infty^3$ of helitangents within a moving rigid body is discovered. The system of lines in each of these cases is not a linear but a *quadratic complex*. It needs to be pointed out however that the two mentioned cases throw up quite different kinds of quadratic complex. In the first case, that of the lines of the screws of the general 4-system of screws, the quadratic cones of lines are arranged in such a way that there are symmetries of certain complicated kinds about a central plane and a rotary symmetry of the second order about a central axis (§6.29). In the second case, that of the helitangent lines in a moving body, the symmetry among the cones is a pure axi-symmetry about the central axis of the complex. There are, no doubt, other quadratic complexes to be found in the general area of screw theory but such complexes, quite beyond my own present scope and thus beyond the scope of this book, are not pursued.

**Closure**

65. I would like to mention again the relationship with natural phenomena of the linear line systems and to remark for the second time about the trueness to nature of the Plücker line coordinates.

**Notes and references**

[1] This is the geometer Georges–Henri Halphen, who lived 1844–89. A direct (but secondary) reference to the theorem may be found under the insertion Schubert, H.C.H., *Dictionary of scientific biography*, Vol. XII, p. 227, Scribners, 1972, where the matter of enumerative geometry is being discussed. Refer note [20] at chapter 19. Refer §11.26.

[2] Dr Andri Lozzi wrote the work (untitled) while an undergraduate in the Department of Mechanical Engineering at the University of Sydney in 1969. Refer §11.39.

[3] The truth of this theorem has been checked algebraically in a private correspondence between the writers K.H. Hunt and E.J.F. Primrose, 1982. Refer §11.58.

[4] Dr Ian Parkin in the Department of Computer Science at the University of Sydney has been working recently (and with success) in this area. Refer also Lipkin, H. and Duffy, J., The elliptic polarity of screws, *Journal of mechanism, transmissions, and automation in design*, (Trans ASME), Vol. 107, p. 377–87, September 1985. Refer §11.62.

1 Velocity polygons for [illegible] mechanism [illegible]
Fig. The model shows (a) a certain 3-bar
mechanism of mobility unity, (b) a velocity polygon
drawn for a certain configuration of that mechanism
and (c) another velocity polygon for the same
configuration of an inversion of that mechanism.

12. Velocity polygons for spatial mechanism may be drawn. This model shows (*a*) a certain 3-link mechanism of mobility unity, (*b*) a velocity polygon drawn for a certain configuration of that mechanism, and (*c*) another velocity polygon for the same configuration of an inversion of that mechanism.

# 12 The vector polygons for spatial mechanism

## Introduction

01. For any mechanism of mobility unity it is possible to draw the *angular velocity polygon*; refer to § 12.12. In a style that is similar to that which is well explained in the abundant literature on the kinematics of planar mechanism, it is also possible to draw the various *linear velocity polygons*. Such polygons may be used as a first device to help to locate the motion screws (the ISAs) among the links, to determine the velocities at the various points in the links, and to find the changing patterns of those velocities as the mechanism moves through its series of configurations. Corresponding remarks could well be made – and they will be made at § 12.04 – about the *force and moment polygons* that might be drawn. All of these polygons will change, of course, as the mechanism moves through its series of configurations, its cycles of motion and action.

02. I wish to make, here, two sets of remarks. First I wish to say that any vector polygon, awkward as it is to draw by hand in space, should not be seen as a useful device *per se*. Its main effectiveness in the areas of analysis and synthesis will reside in its capacity (*a*) to clarify the mind, and (*b*) to simplify the algebra. A polygon may help a creative worker to write the algebra of a mechanism. Alternatively, the geometrical aspects of a changing vector polygon might be calculated by computer and presented graphically (and rotating) upon the computer screen. As some chosen mechanism might in the same way be caused by computer-simulation to move through its cycles of motion and action, its various vector polygons might be examined for periods of rapid change. Thus a proposed real machine might be examined for mechanically dangerous aspects of its design. Secondly I wish to distinguish between that kind of graphics being discussed in this chapter and that kind of graphics produced by a programmed computer drawing upon its screen. Whereas the former is what might be called, I suppose, design to aid the computer, the latter is currently called (in many circles at least) computer aided design.

03. The reader might go now directly to § 12.34 to study in a preliminary way the three worked examples beginning there, or he may if he wishes read the intervening material first. The intervening material consists of (*a*) some further introductory remarks about the statics and dynamics of a link, (*b*) a survey of some of the relevant literature, (*c*) an outline of the nomenclature used by me for relative velocities and velocity-differences, and (*d*) a series of eight theorems relating to the construction of velocity polygons for spatial mechanism. These theorems are presented and proved in a relatively abstract way however, so it might be best to get some working impression first of the general area of investigation for which these theorems are intended. Please refer in advance to figures 12.03, 12.04 and 12.05.

04. Given the established literature about the forces sustained by a link of a slowly moving (or a massless) mechanism of mobility unity, these forces being due to the steady transmission of power through the mechanism, one might imagine that a composite force polygon for the statics of a spatial mechanism might be drawn. Looking at chapter 10 where force and angular velocity are discussed in relationship with one another, and knowing the relationships $\mathbf{v} = \boldsymbol{\omega} \times \mathbf{r}$ and $\mathbf{M} = \mathbf{F} \times \mathbf{r}$ which are also discussed in chapter 10 (§ 10.40, § 10.41), we might expect to find by analogy that moment polygons of various kinds will be significant also. The related matters of force and moment and angular and linear velocity are not well treated in the current literature. The following for example is not made clear. The static force and the angular velocity polygons for a mechanism will go together, analogously, and might be distinguished from the other polygons by means of the term *global*. Refer to § 19.40 for the origin of this term. Because the forces which need to be mobilized for accelerating the masses and the forces due to gravity upon the masses complicate the matter, any 'dynamic' force polygon (and its related moment polygons) will be more difficult to achieve than the static force polygon or the angular velocity polygon. All of the force and moment polygons will however be related somehow to the wrenches at work upon the action screws, and thus

to the real (or the global) and the dual (or the spatial) parts of the action duals respectively that actually exist at the instant within the working machinery (§12.54, §13.29, §19.62).

05. After a short survey of the literature beginning next (§12.06), I outline the mentioned scheme for the nomenclature (§12.08). Six simple theorems about the construction of velocity polygons follow, and in these I pay attention to the geometry of the velocity image (§12.14). Next a discussion of angular velocity, with two more theorems, is followed by the execution of three example-analyses in the kinematics of closed loops (§12.28). Next I speak about the statics of massless mechanism (or the dynamics of weightless machinery) and proceed to explain the construction of various force and moment polygons on the basis of given input and output data (§12.54). Lastly I take a brief look at the force polygons for the massive links in real, earth-bound machinery. This last material is an introduction to parts of chapter 13. Most of the purely kinematical content of the present chapter has already appeared in the conference literature [1] and been examined fairly thoroughly; but the content relating to statics and dynamics, much of which requires to be further clarified, is relatively new.

**On the corresponding matters of action and motion**

06. In this paragraph I write some scattered remarks about the matters of force and the force polygon in mechanism, leaving until the next some corresponding remarks about angular velocity. A massless frictionless mechanism of mobility unity can be seen as a statically determinate structure; a mutually consistent unequal pair of wrenches at the input and output links need to be applied however, and the consequent external reaction upon the frame requires consideration also (§17.21). Many modern authors – and one who gives a good account is R.L. Maxwell (1962) – proceed this way with planar mechanism. These authors write their arguments in the reference plane with wrenches that are either pure forces or pure couples or, sometimes, mutually perpendicular combinations of the two. I do the same, incidentally, at §17.02 *et seq.*, where a number of other relevant remarks appear. In the regular literature of planar mechanism static analyses are made by considering in turn the statics of each link of a mechanism; these are the analyses for transmission forces only. Then, sometimes, the equivalent of a *Maxwell–Cremona diagram* is drawn. This diagram, well known in the area of pin-jointed plane trusses, was developed by Cremona of Italy and Maxwell (Clerk Maxwell) of England, independently of one another about 1864. Den Hartog gives a simple and powerful account of it in his book (1948). Read also the well expressed ideas of Lamb (1912). But a Maxwell–Cremona diagram may be drawn for a spatial, massless statically determinate structure too; and I believe that a similar diagram may be drawn for a similarly idealized spatial mechanism of mobility unity. I make no remarks, just yet, about the nomenclature that might be used to identify the various apices in the relevant vector polygons (§12.67). Nor do I speak just yet about the systems of unbalanced forces involved in the dynamics of the corresponding real machinery. Real machinery is, of course, not only non-massless and thus a victim of the accelerating forces called into play upon its various links, but also the victim (most often) of a parallel gravity field (§12.62). In the realm of the kinematics, on the other hand, I wish to express the point of view that the diagram corresponding to the force polygon for a mechanism is not some one of the relevant linear velocity polygons as Maxwell claims and explains (1962), but the single relevant *angular velocity polygon*; see §12.28, §13.07, §13.09 and elsewhere in this and the next chapter.

07. In the kinematics of planar mechanism analytical studies have been made and techniques have been devised to discover the relative instantaneous centres or the poles of pairs of links. The geometrical relationships between these poles, their paths (polodes), and the shapes of the paths of points (coupler curves or point paths), have been explored and are comparatively well known. Some general methods and special techniques for path synthesis in planar mechanism are accordingly available: see for example Hall (1960), Erdman and Sandor (1984), and the journal literature. Not so well known, however, is the geometry of the motion screws (the ISAs) and of the paths of points in spatial mechanism. This is partly due to the fact that the ISAs in spatial mechanism are more difficult to envisage and to determine. Beyer (1963) has shown how certain of the ISAs can be determined, how certain portions of a velocity polygon for a spatial mechanism can be drawn, and how this and the angular velocity polygon can be used to advantage in quantitative analyses. Both Beyer and his younger colleagues appear to have been unaware of most of the theorems about to be outlined however. Neither he nor they (1963) dealt with more than three points in any one link and thus failed to draw the complete velocity image for any link (§12.14); none of them used the geometrical relationships of the cylindroid with which they might have determined some of the motion screws (§13.12); and none discussed the velocity-differences of pairs of instantaneously coincident points in their spatial mechanism (§12.09, §12.25). The literature in planar kinematics, on the other hand, deals extensively with this latter matter. The matter is dealt with well by Hartenberg and Denavit (1964) and, more recently, by Shigley and Uicker (1980). But the whole

question needs to be extended into three dimensions, and it is one of my aims in this chapter to try to do that. Let me repeat by way of final comment here that in a mechanism it is not the various linear velocity polygons that come from the instantaneous kinematics and relate with one another that are important from the kinetostatical point of view; it is the single angular velocity polygon that comes from the instantaneous kinematics and the single force polygon that comes from the instantaneous statics that correspond with one another and are analogous.

**Scheme of the nomenclature for velocities**

08. Links in the configuration diagram of a mechanism are designated by the numerals 1, 2, 3 etc., the initially 'fixed' link (if any) being assigned the numeral 1. Points that are nominated in the links of a mechanism are called by the letters $A$, $B$, $C$ etc., with a subscripted numeral appended where necessary to indicate the particular link to which the point belongs. Thus $B_2$ for example is a point $B$ fixed in link 2, while $B_3$ is an instantaneously coincident point $B$ fixed in link 3 (§ 1.13). In some areas of endeavour, however, where velocities are not at issue, we might depart from these conventions. In the field of finite screw displacements, for example, the symbols $A_1A_2$ etc. might stand for successive positions in space of an otherwise non-subscripted point $A$ (§ 19.30). In analyses to determine whether or not certain overconstrained linkages are mobile throughout a cycle, for another example, we might use letters (lower case letters) for the dimensions of the links and numbers to designate the joints (§ 20.29). Despite these departures (and in the context here of instantaneous motion) I go on.

09. The vector $\mathbf{v}_{2B}$ for example is the linear velocity relative to link 2 of point $B$, while the vector $\boldsymbol{\omega}_{34}$ for example is the angular velocity relative to link 3 of the link 4. Vectors such as $\mathbf{v}_{BC}$ are mentioned; these are not relative velocities in the ordinary way because a single point such as $B$ cannot be quoted as a frame of reference; any vector such as $\mathbf{v}_{BC}$ is a velocity-difference; for example $\mathbf{v}_{2C}-\mathbf{v}_{2B}=\mathbf{v}_{BC}$, or $\mathbf{v}_{3C}-\mathbf{v}_{3B}=\mathbf{v}_{BC}$. The vector $\mathbf{v}_{BC}$ may also be defined, however, as the linear velocity, relative to a frame of reference spherically pivoted at $B$ and keeping a fixed orientation with respect to the fixed link of the mechanism, of the point $C$. For a given configuration and a given rate of deformation (§ 12.28) vectors such as $\mathbf{v}_{2B}$ are independent of kinematic inversions but, unless the circumstances are special, vectors such as $\mathbf{v}_{BC}$ are not (§ 12.27).

10. Whereas non-coincident points in a configuration diagram are distinguished by different letters $A$, $B$, $C$ etc., points on different links that are instantaneously coincident are designated by the same letter $A_1$, $A_2$, $A_3$ etc. When a subscripted point, such as $E_5$ for example, is to be used as the second part of a subscript at a linear velocity which is quoted relative to some frame, say 6, the relative velocity is written, $\mathbf{v}_{6E5}$, with all of the subscripted material put together at the same level.

11. Points at the extremities of vectors in a linear velocity polygon are designated $a, b, c$ etc. according to whichever points in the mechanism $A, B, C$ etc. are involved. If a velocity polygon is drawn for a particular inversion of a mechanism where the link 2 for example is the fixed link, then the numeral 2 appears at the apex of the polygon. This particular apex, which is, incidentally, the velocity image of the fixed link, is called the *origin* of the polygon. The directed distances from point to point in the velocity polygon are referred to as **ab, bc, 2a, 2b,** etc. Whereas **ab** for example will represent a velocity-difference, **2a** will represent a relative velocity.

12. For a similar reason the apices of an angular velocity polygon are numbered 1, 2, 3 etc., the directed distances **12, 23** etc. representing the angular velocities $\boldsymbol{\omega}_{12}$, $\boldsymbol{\omega}_{23}$ etc. respectively. An angular velocity polygon is nearly always self explanatory, and nowhere in this book is the detailed geometry of such polygons discussed. Please refer to § 10.55, § 12.28, § 13.10, and § 15.20, however, for explanations of some of the phenomena. Following convention, all angular velocity vectors are directed according to the well known right hand rule, and the subscripts are always written in such a way that the first number refers to the frame of reference, namely the reference body, while the second refers to the rotating body (§ 3.52).

13. The term $\text{ISA}_{13}$ for example refers to the instantaneous screw axis for the motion relative to link 1 of link 3 (§ 5.44). Whereas the vector $\boldsymbol{\omega}_{13}$, mentioned above, is in a way a free vector which may be drawn anywhere parallel with the ISA (§ 10.52), the vector $\boldsymbol{\tau}_{13}$, which is the linear velocity of all of those points in link 3 that are residing at the instant in the $\text{ISA}_{13}$, is a bound vector whose location and direction is along the ISA. At any chosen configuration of a mechanism of mobility unity the location and pitch of an ISA are independent of prevailing velocities; an ISA, in other words, is a function of configuration only; an ISA may accordingly be determined as being either right or left handed and, where necessary, the handedness of an ISA may be indicated by the use of an appropriate device; such a device is used at figures 12.03, 12.04 etc., and at figure 15.03.

**Some remarks about the velocity image**

14. For any spatial mechanism of mobility unity, that is for any spatial mechanism constructed in such a way that each of its links is constrained to move with only one degree of freedom with respect to any other link

of the mechanism (§ 1.19), we may, for any chosen configuration of the mechanism, draw a linear velocity polygon. Knowing the idea of linear velocity polygon from the relevant literature in the field of planar mechanism, referring for example to Hartenberg and Denavit (1964) for a clear exposition of that elementary matter, the above statement follows naturally from the determinate nature of the whole circumstance. Once we have decided which link of the mechanism is for the meantime to be the 'fixed' link, or the *frame link*, of the mechanism, each other link will be seen to be suffering some spatial motion about its ISA relative to frame, and each of these links will accordingly have its velocity image in the relevant linear velocity polygon (§ 12.15). Smith [2] is credited by Ferguson [3] with having drawn the first velocity polygon for a planar mechanism, and Mehmke [4], according to Beyer (1953), was first to propose the idea of the velocity image of a link in a planar mechanism; and none of this requires further explanation. It is however the spatial analogue of Mehmke's image which is under consideration here, and all three theorems which now follow relate to it. I request the reader to look again at § 5.33.

### Three theorems about the velocity image

15. Refer to figure 12.01(*a*) and contemplate the truth of the following theorem: *theorem 1: although each link in the configuration diagram of a mechanism is suffering a twisting motion about its ISA relative to frame, its image in the velocity polygon will always be a plane figure.*

Figure 12.01(*a*) (§ 12.15) Local mechanics of the velocity image of a moving link. The plane quadrilateral *abcd*, which appears upon its projection plane set normal to the relevant ISA, is the velocity image of the tetrahedron *ABCD* embedded within the moving link. The shape of the image is the shape of, but not the same size as, the orthogonal projection onto the said plane of the said tetrahedron.

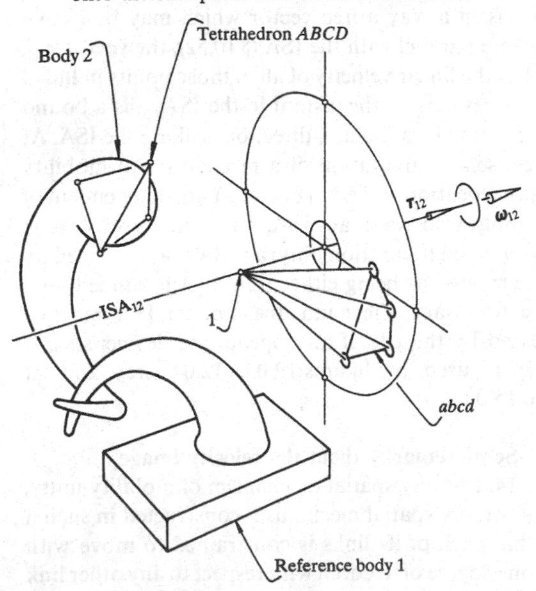


16. While the truth of this theorem is amply evident from the material of chapter 5, the following extra comments are offered nevertheless. Refer to figure 12.01(*a*). Consider the fixed link 1 and a floating link 2 of some mechanism of mobility unity and nominate in 2 any four non-coplanar points *A, B, C* and *D*. Erect within the body 2 in figure 12.01(*a*), in other words, any tetrahedron *ABCD*. It is now characteristic of the circumstance that $\mathbf{v}_{1A}$, $\mathbf{v}_{1B}$ etc. each have the same component $\tau_{12}$ in the direction of the ISA (§ 5.21). Plucked from the body 2 and replanted in the velocity polygon at origin 1, the relevant velocity vectors – they are the directed distances **1a**, **1b** etc. – are all drawn emerging from 1 at some known scale (§ 5.38). Because each of these has the same component $\tau_{12}$ in the direction of the ISA, the quadrilateral *abcd* (which forms the velocity image of link 2) can be none other than a plane figure.

17. Refer again to figure 12.01(*a*) and note there the truth of the following theorem also: *theorem 2: the shape of the velocity image of a link is the shape of the orthogonal projection of that link on a plane set normal to the direction of its instantaneous screw axis relative to frame.*

18. Take any point *Q* in link 2. It is characteristic of the circumstance that the component in the plane normal to ISA$_{12}$ of the velocity $\mathbf{v}_{1Q}$ is proportional to the distance from *Q* to the ISA. Remembering that each of the points *A, B, C* and *D* (like *Q*) suffers the same component $\tau_{12}$ of linear velocity in the direction of the

Figure 12.01(*b*) (§ 12.18) Illustrating the way in which the velocity image *abcd* upon the projection plane is rotated 90° away from the actual projection upon the projection plane of the embedded tetrahedron *ABCD*. The rotation occurs in the direction of the angular velocity of the link.

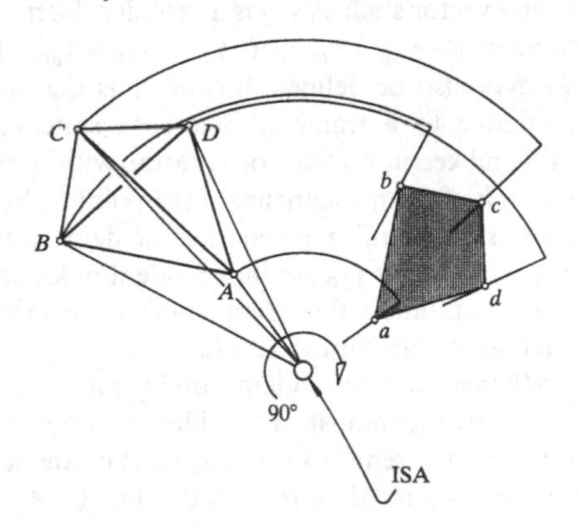

ISA, it may be seen that the velocity-differences $\mathbf{v}_{AB}$, $\mathbf{v}_{BC}$, etc. (represented by the directed distances **ab**, **bc**, etc. all drawn to the same scale in the velocity polygon) reside in planes normal to the ISA, and that the magnitudes of these velocity-differences are proportional to the orthogonal projections of $A$–$B$, $B$–$C$ etc. onto any chosen one of those planes. It follows that the plane quadrilateral *abcd* is geometrically similar to the orthogonal projection onto any plane normal to the $\mathrm{ISA}_{12}$ of the tetrahedron *ABCD*. Refer to figure 12.01(*b*).

19. In the combination of the figures 12.01(*a*) and 12.01(*b*) can be seen the gist of the following theorem: *theorem 3: there is a unique point in the plane of the image which represents the projection of the ISA; this point is at the foot of the normal dropped from the origin of the polygon onto the plane of the image.*

20. Each link in the configuration diagram is pierced by its ISA relative to frame along some certain line, and all points $S$ in the line have the same linear velocity $\tau$ in the direction of the ISA. It is therefore not surprising to find that the single point $s$ in the image is unique, and that the directed distance joining the origin of the polygon to $s$ is not only normal to the plane of the image but also in the same direction as the ISA. Please refer to §5.32.

**Three more theorems**

21. Refer again to the figure 12.01(*b*) and, in advance, to the linear velocity polygons drawn in figure 12.03. Perceive the following: *theorem 4: relative to its corresponding link in the configuration diagram each image is rotated through a right angle in the direction of the angular velocity of that link relative to frame; changing shape from configuration to configuration each image suffers, moreover, changes in its linear size in direct proportion to its changing angular velocity.*

22. As already shown, the directed distances **ab**, **bc**, etc. in the polygon are each at right angles to their corresponding projections of $A$–$B$, $B$–$C$, etc. onto the plane of the image (§12.18); and, by virtue of the nature of velocity-differences, **ab** is turned clockwise (or otherwise) away from $A$–$B$ when $\boldsymbol{\omega}_{12}$ is clockwise (or otherwise) when viewed in the direction of the vector according to the right hand rule. Because the component in the plane normal to the ISA of the velocity $\mathbf{v}_{1Q}$ is proportional not only to the distance of $Q$ from the ISA but also to the angular velocity $\boldsymbol{\omega}_{12}$ whose direction is parallel with the ISA (§5.45), it will moreover be seen that $\mathbf{v}_{AB}$, $\mathbf{v}_{BC}$, etc., and thus **ab**, **bc**, etc., vary with the magnitude of $\boldsymbol{\omega}$.

23. Go now to any one of the figures 12.03 to 12.05 inclusive and see there among the linear velocity polygons the truth of this fact: *theorem 5: each image is displaced away from the origin of its polygon a certain distance; this distance represents the linear velocity $\tau$ of all those points in the link that are residing instantaneously in its ISA relative to the chosen frame.*

24. This theorem is simply a corollary of theorem 3. For any one link, the directed distance drawn from the origin of the polygon to the point $s$ at the image of the link represents its vector $\tau$. When the magnitudes both of $\omega$ and $\tau$ become known for a certain link in a problem of analysis, it will be easy to determine the pitch of the ISA and, when the senses of $\omega$ and $\tau$ are compared, it will be easy to see whether the ISA is right or left handed (§5.50). Figure 12.02 is showing as circular discs the planes of three different velocity images in some velocity polygon. Each of them is at its own directed distance from the origin and these distances, $\mathbf{1s}_2$, $\mathbf{1s}_3$, etc., are shown. The directed distances $\mathbf{a}_2\mathbf{a}_3$ etc. in figure 12.02 are the velocity-differences $\mathbf{v}_{A2B3}$ etc. occurring at the instantaneously coincident points $A_2$, $A_3$, etc. (§12.10). The directed distances $\mathbf{a}_2\mathbf{b}_3$ etc. are the velocity-differences $\mathbf{v}_{A2B3}$ etc. between the points $A_2$, $B_3$, etc. which are not instantaneously coincident (§12.11).

25. Look at the equal and parallel pairs of lines $c_1c_3$ and $d_1d_3$ in both of the linear velocity polygons in figure 12.03. Look also at the non-parallel pairs of equal and parallel lines $c_1c_4$ and $d_1d_4$ in both of the linear polygons in figure 12.05, and become aware of the following important theorem: *theorem 6: the difference between the linear velocities relative to the same frame of any two instantaneously coincident points in a mechanism, namely the velocity-difference of those two points, is independent of the frame of reference chosen.*

26. This theorem is stating that a directed distance $\mathbf{q}_4\mathbf{q}_5$ for example, representing the velocity-difference $\mathbf{v}_{Q4Q5}$ in some velocity polygon, will have the same

Figure 12.02 (§12.24) Some of the relationships between the important vectors of a velocity polygon for a mechanism. This is a 1-polygon, drawn for the link 1 fixed, and the planes of the images of the links 2 etc. are shown normal to the directed distances $\mathbf{1s}_2$ etc. Note the relative velocities $\mathbf{1a}_2$ etc. Note also the two kinds of velocity difference $\mathbf{a}_2\mathbf{a}_3$ etc. and $\mathbf{a}_2\mathbf{b}_3$ etc.

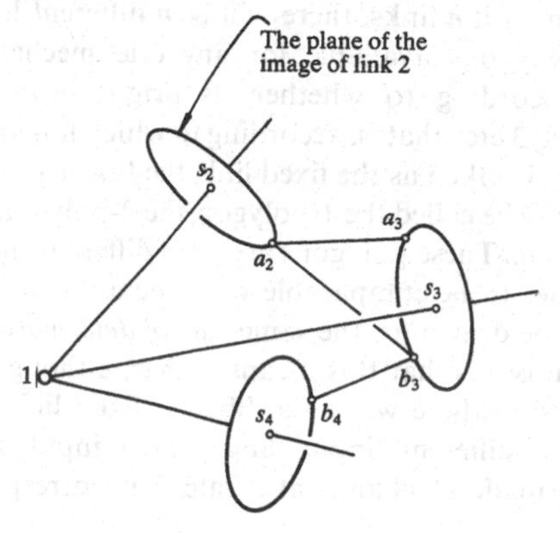

magnitude and direction in any other polygon drawn for the same configuration and the same rate of deformation of the same mechanism (§ 12.28): it is stating that $\mathbf{q_4q_5}$ is independent of which of the various polygons is drawn, there being as many different polygons as there are inversions of the mechanism (§ 12.28). It is important to be aware in this connection that, if two points $Q_4$ and $Q_5$ are instantaneously coincident, they are, by definition, at the intersection of their respective paths drawn relative to any link; and it is illuminating to note that the several pairs of paths drawn relative to the several links are different pairs of paths (§ 1.15).

27. To prove this theorem 6, take for the purpose of argument two points $Q_4$ and $R_5$ which are not coincident, and three different frames of reference 1, 2 and 3. The first two of the velocity-differences may be written as follows: $\mathbf{1r_5} - \mathbf{1q_4} = (\mathbf{q_4r_5})_1$; and $\mathbf{2r_5} - \mathbf{2q_4} = (\mathbf{q_4r_5})_2$. From the nature of relative velocities however we can write that $\mathbf{2r_5} = \mathbf{1r_5} + \mathbf{2r_1}$ and that $\mathbf{2q_4} = \mathbf{1q_4} + \mathbf{2q_1}$. Now $\mathbf{2r_1}$ does not equal $\mathbf{2q_1}$ unless $Q$ and $R$ are residing in some line parallel with $\mathrm{ISA}_{23}$. It follows that $(\mathbf{q_4r_5})_1$ does not equal $(\mathbf{q_4r_5})_2$ unless the mentioned condition is fulfilled. If the condition is fulfilled $(\mathbf{q_4r_5})_2$ does not equal $(\mathbf{q_4r_5})_3$ unless $Q$ and $R$ are residing in some line parallel with $\mathrm{ISA}_{23}$. But in general $\mathrm{ISA}_{12}$ and $\mathrm{ISA}_{23}$ are not parallel, so $\mathbf{q_4r_5}$ in one polygon is not equal to $\mathbf{q_4r_5}$ in another polygon unless, either the line joining $Q$ and $R$ and all the ISAs in the mechanism are parallel (in which case the motion is planar and $Q$ and $R$ are in effect coincident), or $Q$ and $R$ are coincident.

**On the instantaneous geometry of a mechanism**

28. In any mechanism of mobility unity the relative ISAs for the various pairs of links will be determined solely and uniquely by the configuration. If there are $n$ links of a mechanism there will be $^nC_2$ ISAs and, when the mobility is unity, it will always be possible to determine the locations of and the pitches of the ISAs for any chosen configuration. For each inversion of a mechanism of mobility unity a linear velocity polygon may be drawn and, because there are $n$ inversions of any mechanism with $n$ links, there will be $n$ different linear velocity polygons available for any one mechanism (§ 1.43). According to whether its origin bears the number 1, 2, 3 etc., that is, according to which link of the mechanism is taken as the fixed link, the linear velocity polygon may be called the 1-polygon, the 2-polygon, the 3-polygon etc. These polygons are all of different shapes but, in order to be comparable with one another, they should all be drawn for the same *rate of deformation* of the mechanism. What this means is that, although for each inversion there will be a different fixed link and, therefore, a different 'input' angle, each input angle should be made to change at a rate that corresponds with the chosen, unit rate of change of the original input angle. Compatible rates of change for the various input angles could of course be obtained directly from the angular velocity polygon. The directions of the edges of the angular velocity polygon will correspond, of course, with the directions of the corresponding ISAs, and thus with the directed distances $\mathbf{1s_2}$, $\mathbf{1s_3}$ etc. and $\mathbf{2s_3}$, $\mathbf{2s_4}$ etc. of the various linear velocity polygons. It is certainly clear that, for a given configuration of a mechanism, the configuration diagram, the linear velocity polygons, and the angular velocity polygon will all be closely related with one another. Similar remarks might expectedly be made, of course (§ 12.01), about the geometries of the force and the moment polygons for a mechanism. We might also say that, whereas the global polygons (the angular velocity and the static force polygons) are unique for a configuration, the others (the linear velocity and the moment polygons) are obviously not (§ 12.04, § 19.40). I should note here that, while static force polygons for mechanism involving gravity are somewhat of a problem, dynamic force polygons involving both gravity and inertia are doubly a problem (§ 12.62). Static force polygons may remain relevant upon inversion, but dynamic force polygons will obviously not. The effect of gravity upon machinery is always a complicating factor; but in high speed machinery the effects of gravity (in comparison with those of inertia) may become relatively unimportant. Think with respect to force and moment polygons in mechanism, however, about the effect of changing the direction of a parallel gravity field. And think with respect to forces in general about moving machinery in earth-orbit or moving machinery in interplanetary or interstellar space. Where are the inertial frames of reference? What exactly are they anyway? And how would we use them when we found the answers to these difficult questions? How might the dynamics of such machinery simplify or complicate design?

**Two theorems about the locations of the ISAs**

29. In some problems of analysis we go to one or other of the polygons for information. In others we go to the configuration diagram directly. The configuration diagram may be seen as a picture which, fixed at the instant, is determining the instantaneous relative motion of the links. Any three of the links of a mechanism (whether directly in contact with one another or not) may be seen as a group of three rigid but extendable bodies in relative spatial motion (§ 1.06).

30. The material which follows, here styled the theorems 7 and 8, is of quite deep significance for the whole of mechanism. It is presented here very briefly, however, and in a narrow context. No extensive proofs are offered. Please refer to chapter 13 for a more

thorough discussion of the matter which is there called, for the sake of a wider generality, the first of the two theorems of three axes (§ 13.12). Whereas the six theorems outlined above found their origin in [1], the two which follow (7 and 8) found theirs in [5].

31. Please think about the ways in which three angular velocities might add to zero and make thereby an angular velocity polygon that closes and that is, thereby, a plane triangular figure (§ 10.15, § 10.55, § 13.02, § 14.23, § 15.02, § 20.54). Think about three rigid bodies moving in any way relatively to one another, with or without constraints. If it is true (and it is by the nature of vector addition) that $\boldsymbol{\omega}_{12} + \boldsymbol{\omega}_{23} = \boldsymbol{\omega}_{13}$, then it is also true (and this by the same token) that $\boldsymbol{\omega}_{12} + \boldsymbol{\omega}_{23} + \boldsymbol{\omega}_{31} = 0$. Please read the relevant parts of chapter 15, and thus perceive the truth of the following theorem: *theorem 7: at any instant during the relative spatial motion of any three rigid links of a mechanism the three relative ISAs reside in parallel planes and have the same line as common perpendicular.*

32. Become aware of the geometry of the cylindroid by reading the relevant parts of chapters 13, 15 and 16, and thus perceive the truth of this last, the following theorem: *theorem 8: the three ISAs are, moreover, generators of a cylindroid, and this means that, if the locations of and the pitches of two of the ISAs are known, the cylindroid is determined and, if the mobility is unity, there is then a unique solution for the location of and the pitch of the third ISA.*

33. Now I wish to mention before I present some examples that, in mechanisms where there is a simple hinge at each extremity of some binary link (§ 1.38), the axes of those hinges are ISAs or motion screws of zero pitch which involve (*a*) the binary link itself, and (*b*) each of its contacting neighbours. Such pairs of ISAs or motion screws (which are special pairs of screws because in each case both of their pitches are zero) determine a cylindroid which moves unchanged and fixed upon the said binary link, the centre of the cylindroid being midway along the common perpendicular between the two hinge axes (§ 20.53). Please be aware that, whereas the mere superficial shape of a cylindroid may be determined by the nomination of any three lines that cut the same line as common perpendicular, a cylindroid in all of its aspects (the shape of its surface *and* the pitches of its generators) is wholly determined by any single pair of its screws (§ 15.49).

**Polygons drawn for angular and linear velocity**

34. In the following worked exercises in § 12.35 *et seq.* three different mechanisms are taken for the sake of demonstration. They have three, five, and four links respectively. Each has its kinematic chain in the form of a loop (§ 1.46), so they also have three, five and four joints respectively. Quoting the $f$-numbers at the successive joints (§ 1.26), and referring to figures 12.03, 12.04, and 12.05 in that order, the loop-order of the joints in the mechanisms are: (1) -1-4-2-; (2) -1-3-1-1-1-; (3) -1-3-2-1-. It will be seen that all three are comparatively simple loops, often employed in engineering practice, and that all three are of mobility unity according to the Kutzbach criterion (§ 1.34). In each case, $\Sigma f = 7$. In each of the worked exercises it is presumed that the dimensions of the mechanism and its configuration are known (§ 19.48), that a unit angular velocity $\boldsymbol{\omega}_{12}$ at input has been taken, and that at least one purpose of the investigation is to determine some certain ISA which is not obvious upon inspection. While sincere efforts have been made to avoid ambiguities and inconsistencies in the drawn figures, they should not be regarded as being more than mere sketches; their only purpose is to convey a picture of the 3-dimensional geometry; they are not drawn to scale. As already mentioned (§ 12.02), as soon as the essential geometry of a problem is solved and the corresponding algebra written (§ 19.47), analyses by means of digital computer will reveal wide ranges of numerical results accurately and quickly. It will be moreover conceivable, I hope, that iterative programmes based upon the algebra will often produce the optimum solution in problems involving the dimensional synthesis of linkages by trial (§ 19.66). Because it is so important not to be led astray at this juncture, please persue my remarks about the *closure equation* for a linkage, beginning at § 19.47.

**Worked exercise (1) in figure 12.03**

35. In this 3-link mechanism the object of the analysis is to locate the $\text{ISA}_{23}$ and to determine its pitch. Figure 12.03 shows as well as the mechanism itself two different linear velocity polygons and the angular velocity polygon all drawn for the reference configuration. An actual working model of the mechanism and wire models of the linear velocity polygons are illustrated at photograph 12.

36. Knowing $\mathbf{v}_{1B2}$ from the known $\boldsymbol{\omega}_{12}$, the directed distance $\mathbf{1b}_2$ may be drawn in the first velocity polygon whose origin is at 1. This is the 1-polygon. Knowing that the direction of the velocity-difference $\mathbf{b}_2\mathbf{b}_3$ is vertically downwards, that the direction of $\mathbf{c}_1\mathbf{c}_3$ is in the direction of $\boldsymbol{\omega}_{13}$, and that $\mathbf{c}_3\mathbf{b}_3$, which is the image of link 3, is in a plane normal to $\boldsymbol{\omega}_{13}$, the triangle $\mathbf{b}_2\mathbf{b}_3\mathbf{c}_3$ may be constructed. It can be seen that $\mathbf{c}_3\mathbf{b}_3$ is rotated through 90° away from $C_3$–$B_3$ in the direction of $\boldsymbol{\omega}_{13}$, and the ratio of $|\mathbf{c}_3\mathbf{b}_3|$ and $C$–$B$ gives the magnitude of $\boldsymbol{\omega}_{13}$.

37. Knowing that $\boldsymbol{\omega}_{12} + \boldsymbol{\omega}_{23} + \boldsymbol{\omega}_{31} = 0$, the complete angular velocity polygon which is a triangle may be drawn (§ 12.31). Thus $\boldsymbol{\omega}_{23}$ is determined.

Figure 12.03 (§ 12.35) The configuration diagram, two of the linear velocity polygons, and the angular velocity polygon drawn for the mechanism at worked exercise (1). The related parts of this figure are drawn by isometric projection, to scale, and they correspond with one another. Note that the vectors $\mathbf{c}_1\mathbf{c}_3$, $\mathbf{d}_1\mathbf{d}_3$ and $\mathbf{b}_2\mathbf{b}_3$ in the 1-polygon and their opposite numbers in the 2-polygon are, respectively, not only equal in magnitude but also parallel with one another. Notice that the three screw axes shown in the configuration diagram intersect the same line perpendicularly. Refer to photograph 12.

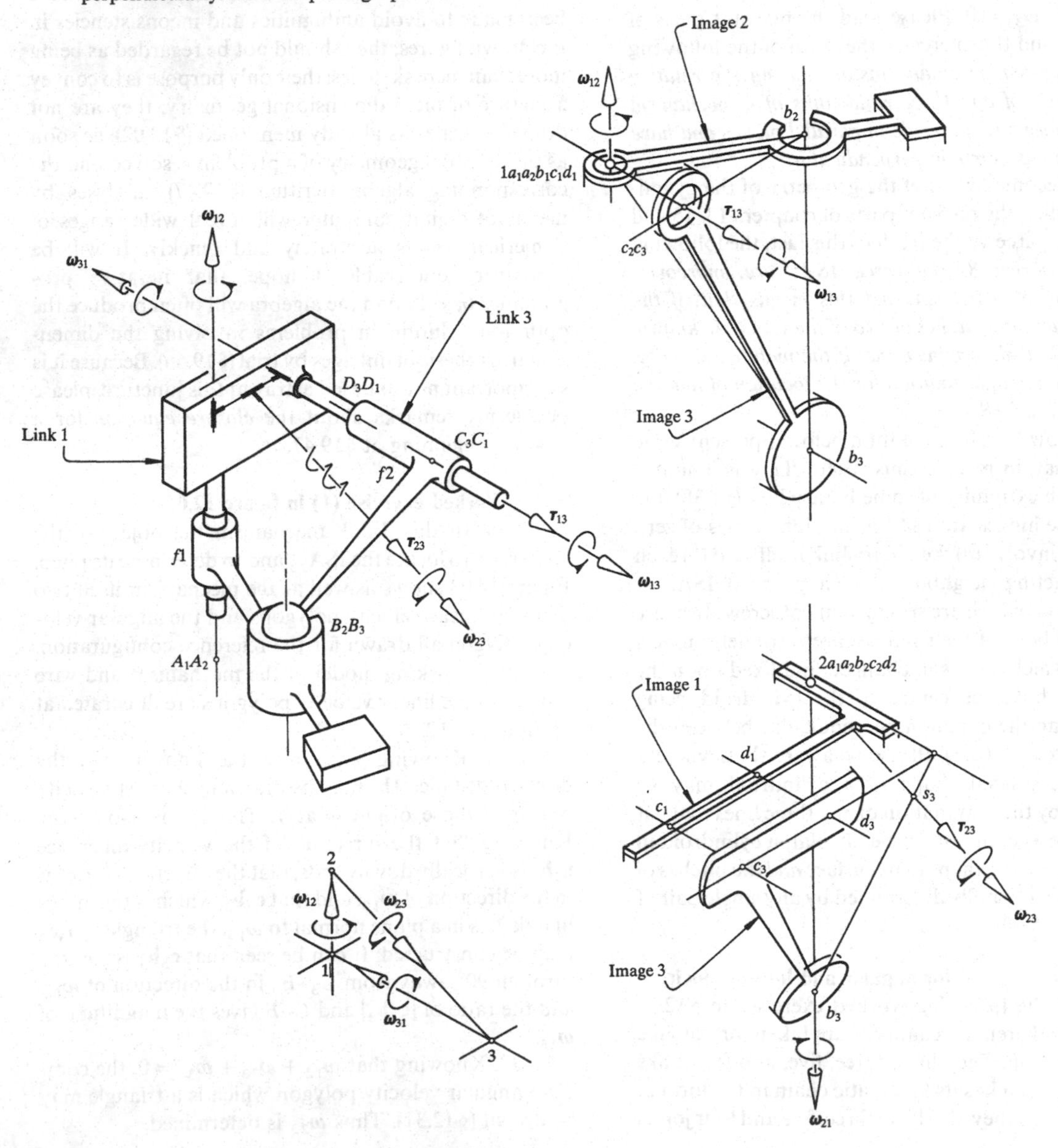

Figure 12.04 (§ 12.40) The configuration diagram, one of the linear velocity polygons, and the angular velocity polygon drawn for the mechanism at worked exercise (2). The related parts of this figure are based upon an isometric frame; they are nevertheless only approximately drawn. Notice that the three screw axes shown in the configuration diagram intersect the same line perpendicularly.

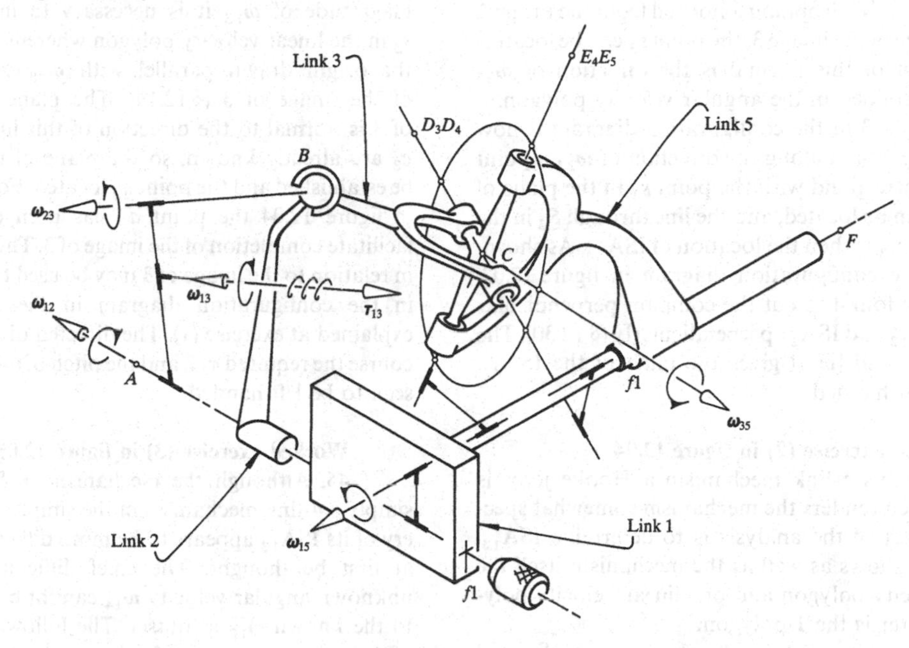


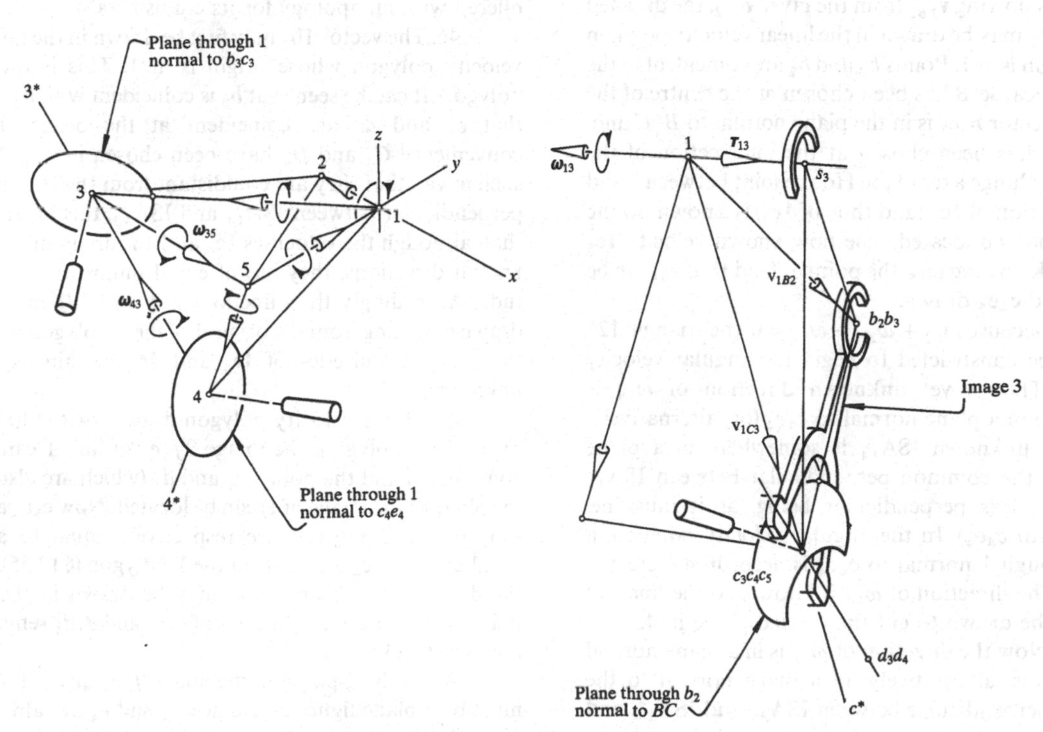

38. Taking 2 now as the fixed link, and knowing that $\boldsymbol{\omega}_{21} = -\boldsymbol{\omega}_{12}$, the image $a_1d_1c_1$ of link 1 may be drawn in the second velocity polygon whose origin is at 2. This is the 2-polygon. Knowing that $d_1d_3 = c_1c_3$, and that $(c_1c_3)_2 = (c_1c_3)_1$ already found, the points $c_3$ and $d_3$ may be located. Similarly $b_3$ may be found because $(b_2b_3)_2 = (b_2b_3)_1$. The triangle $b_3c_3d_3$ is the image of link 3. By dropping a normal from the origin 2 onto the plane of the image 3, the point $s_3$ can be located. The direction of this normal is the direction of $\boldsymbol{\omega}_{23}$ already determined in the angular velocity polygon.

39. If link 3 in the configuration diagram is now viewed in projection along the direction of $\boldsymbol{\omega}_{23}$ a point $S_3$ in 3 (to correspond with the point $s_3$ in the plane of the image) can be located, and the line through $S_3$ in the direction of $\boldsymbol{\omega}_{23}$ is then the location of ISA$_{23}$. As shown already at the configuration diagram in figure 12.03, ISA$_{23}$ will be found to cut the common perpendicular between ISA$_{12}$ and ISA$_{31}$ perpendicularly (§ 12.30). The ratio of $|\tau_{23}|$ and $|\boldsymbol{\omega}_{23}|$ gives the pitch of the ISA$_{23}$, which is right handed.

**Worked exercise (2) in figure 12.04**

40. In this 5-link mechanism a Hooke joint is involved which renders the mechanism somewhat special. One object of the analysis is to determine ISA$_{13}$. Figure 12.04 shows as well as the mechanism itself the angular velocity polygon and one linear velocity polygon. This latter is the 1-polygon.

41. Knowing $\mathbf{v}_{1B2}$ from the given $\boldsymbol{\omega}_{12}$, the directed distance $\mathbf{1b}_2$ may be drawn in the linear velocity polygon whose origin is at 1. Points $b_3$ and $b_2$ are coincident in the polygon because $B$ has been chosen at the centre of the ball. The vector $\mathbf{b}_3\mathbf{c}_3$ is in the plane normal to $B$–$C$ and, because $C$ has been chosen at the intersection of the intersecting hinge axes of the Hooke joint between 3 and 5, the direction of $\mathbf{1c}_5$ (and thus of $\mathbf{1c}_3$) is known, so the point $c_3$ may be located. The now known velocity $\mathbf{1c}_5$ gives $\boldsymbol{\omega}_{15}$. Knowing $\boldsymbol{\omega}_{15}$, the point $e_5$ (and thus $e_4$) can be located and $\mathbf{e}_4\mathbf{e}_5$ drawn.

42. Because $\boldsymbol{\omega}_{12} + \boldsymbol{\omega}_{25} + \boldsymbol{\omega}_{51} = 0$, the triangle 125 can next be constructed to begin the angular velocity polygon. The as yet unknown direction of $\boldsymbol{\omega}_{14}$ is somewhere in a plane normal to $c_4e_4$ (or, alternatively, the as yet unknown ISA$_{14}$ is somewhere in a plane normal to the common perpendicular between ISA$_{15}$ and ISA$_{45}$, this perpendicular being, as it must be, parallel with $c_4e_4$). In the angular velocity polygon, a plane through 1 normal to $c_4e_4$ is accordingly erected (§ 12.30). The direction of $\boldsymbol{\omega}_{54}$ is known, so the line 54* may next be drawn to cut the erected plane in 4.

43. Now the direction of $\boldsymbol{\omega}_{13}$ is in a plane normal to $b_3c_3$. It is, alternatively, in a plane normal to the common perpendicular between ISA$_{34}$ and ISA$_{14}$, and ISA$_{14}$ could be located if required by using the image $e_4c_4d_4$ later to be established. So, in the angular velocity polygon, a plane through 1 normal to $b_3c_3$ is erected, and the known direction 43* is drawn to cut this plane in 3. The angular velocity polygon is thus completed, and $\boldsymbol{\omega}_{13}$ becomes completely known.

44. To determine the location of ISA$_{13}$ and the magnitude of $\boldsymbol{\omega}_{13}$ it is necessary to find that point $s_3$ in the linear velocity polygon wherein a line through the origin drawn parallel with $\boldsymbol{\omega}_{13}$ cuts the plane of the image of 3 (§ 12.19). The plane of the image of 3 is normal to the direction of this line and $b_3$ and $c_3$ are already known, so the plane of the image can be established and the point $q_3$ located. For convenience in figure 12.04 the point $d_3$ has been established to facilitate completion of the image of 3. The position of $s_3$ in relation to the image of 3 may be used to locate ISA$_{13}$ in the configuration diagram in the same way as explained at exercise (1). The directed distance $\mathbf{1s}_3$ is of course the required $\tau_{13}$, and the pitch of the ISA$_{13}$ will be seen to be left handed.

**Worked exercise (3) in figure 12.05**

45. Although the mechanism in figure 12.05 is simply a 4-link mechanism of the simplest kind, discovery of its ISA$_{13}$ appears to be more difficult than might at first be thought. The chief difficulty is that the unknown angular velocity $\boldsymbol{\omega}_{14}$ cannot be related easily to the known $\boldsymbol{\omega}_{12}$ at outset. The following solution is offered with an apology for its clumsiness.

46. The vector $\mathbf{1b}_2$ may first be drawn in the linear velocity polygon whose origin is at 1. This is the 1-polygon. It can be seen that $b_3$ is coincident with $b_2$ and that $c_1$ and $d_1$ are coincident at the origin. For convenience $C_4$ and $D_4$ have been chosen in link 4 in such a way that they are equidistant from the common perpendicular between ISA$_{14}$ and ISA$_{43}$; this arranges that, although the velocities $\mathbf{1c}_4$ and $\mathbf{1d}_4$ are in different known directions, they are of equal unknown magnitude. Accordingly the directions $1c_4^*$ and $1d_4^*$ may be drawn emerging from the origin 1 of the 1-polygon while the exact magnitudes of $\mathbf{1c}_4$ and $\mathbf{1d}_4$ remain as yet unknown.

47. In the velocity polygon drawn for the link 2 fixed, the 2-polygon, the image $2f_1e_1$ of link 1 can be constructed and the points $c_1$ and $d_1$ (which are also in the plane of the image of 1) can be located. Now $\mathbf{c}_1\mathbf{c}_4$ and $\mathbf{d}_1\mathbf{d}_4$ in the 2-polygon are respectively equal to and parallel with $\mathbf{c}_1\mathbf{c}_4$ and $\mathbf{d}_1\mathbf{d}_4$ in the 1-polygon (§ 12.55), so the directions $c_1c_4^*$ and $d_1d_4^*$ may be drawn in the 2-polygon, the exact magnitudes of $c_1c_4$ and $d_1d_4$ remaining as yet unknown.

48. In the 2-polygon the image $f_4e_4c_4d_4$ of link 4 must be a plane figure and, since $f_4$ and $e_4$ are already

Figure 12.05 (§ 12.45) The configuration diagram, two of the linear velocity polygons, and the angular velocity polygon drawn for the mechanism at worked exercise (3). Here there are sketches only, drawn directly from the imagination. They suffer gross inaccuracies that the reader may care to elucidate.

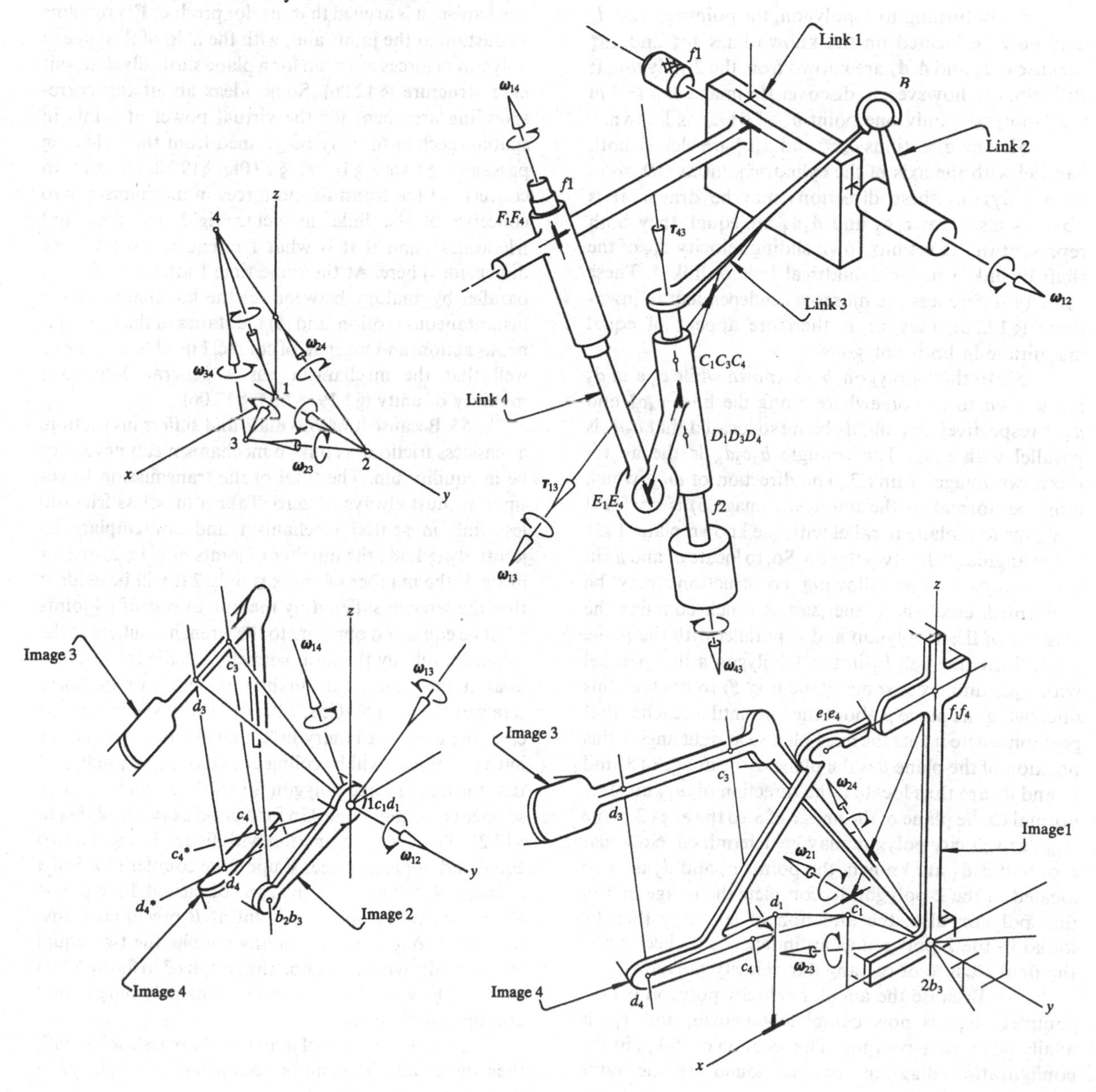

established (they are coincident with $f_1$ and $e_1$ respectively), the plane of the image of 4 must contain the line $f_4e_4$. All planes through this line are cut by the two known lines $c_1c_4^*$ and $d_1d_4^*$, but only one of the planes is so placed that $\mathbf{c}_1\mathbf{c}_4 = \mathbf{d}_1\mathbf{d}_4$. Thus $c_4$ and $d_4$ may be established, and the direction of $\boldsymbol{\omega}_{24}$ (which is normal to the plane of the image of 4) may be established also.

49. The triangle 124 of the angular velocity polygon may now be constructed because $\boldsymbol{\omega}_{14}$ is obvious. The direction of $\boldsymbol{\omega}_{43}$ is known so the line 43* may be drawn in the angular velocity polygon, but the exact location of the apex 3 on this line is as yet unknown.

50. Returning to 1-polygon, the points $c_4$ and $d_4$ may now be located on the known lines $1c_4^*$ and $1d_4^*$ because $\mathbf{c}_1\mathbf{c}_4$ and $\mathbf{d}_1\mathbf{d}_4$ are known from the 2-polygon. It still remains, however, to discover the image of link 3 in the 1-polygon, only one point $b_3$ of which is known.

51. The directions $c_4c_3^*$ and $d_4d_3^*$ are clearly both parallel with the axis of the cylindric joint at $CD$, so in both polygons these directions may be drawn. It is obvious also that $\mathbf{c}_4\mathbf{c}_3$ and $\mathbf{d}_4\mathbf{d}_3$ are equal: they both represent the as yet unknown sliding velocity $\boldsymbol{\tau}_{43}$ of the shaft of link 3 in the cylindrical hole of link 4. These velocity-differences are moreover independent of inversions (§12.25); they must therefore appear of equal magnitude in both polygons.

52. In the 1-polygon, $b_3$ is known while $c_3$ and $d_3$ are known to be somewhere along the lines $c_4c_3^*$ and $d_4d_3^*$ respectively, $c_3$ and $d_3$ being so located that $c_3d_3$ is parallel with $\mathbf{c}_4\mathbf{d}_4$. The triangle $b_3c_3d_3$ is the as yet unknown image of link 3. The direction of $\boldsymbol{\omega}_{13}$ (which must be normal to the unknown image 3) is confined however to a plane parallel with the known plane 143* in the angular velocity polygon. So, to locate $c_3$ and $d_3$ in the 1-polygon, the following constructions may be performed: erect the plane (say $\alpha$) which contains the origin 1 of the 1-polygon and is parallel with the plane 143*; draw through $b_3$ in the 1-polygon a line parallel with $c_4d_4$ and erect some plane (say $\beta$) to contain this line; swing the plane $\beta$ about the line until it reaches that position where it cuts the fixed plane $\alpha$ at right angles; this position of the plane $\beta$ is the plane of the image of 3, and $c_3$ and $d_3$ are thus located; the direction of $\boldsymbol{\omega}_{13}$ must be normal to the plane of the image of 3, so the apex 3 in the angular velocity polygon may be determined. Now that $\mathbf{c}_4\mathbf{c}_3$ and $\mathbf{d}_4\mathbf{d}_3$ are known, the points $c_3$ and $d_3$ may be located in the 2-polygon to complete the image of 3 in that polygon also; the direction of $\boldsymbol{\omega}_{23}$ may then be found in the 2-polygon as an independent check upon the final closure of the angular velocity polygon.

53. Because the angular velocity polygon is now complete $\boldsymbol{\omega}_{13}$ is now completely known, and $\boldsymbol{\tau}_{13}$ is available in the 1-polygon. The location of ISA$_{13}$ in the configuration diagram may be found by the same method as before – and the resulting ISA$_{13}$ will be found, as shown, to cut the common perpendicular between ISA$_{14}$ and ISA$_{43}$ (produced) perpendicularly. The pitch of the ISA$_{13}$, given by the ratio of $|\boldsymbol{\tau}_{13}|$ and $|\boldsymbol{\omega}_{13}|$, is right handed.

### On the force polygon for a massless mechanism

54. In elementary texts the so-called transmission forces in mechanism are calculated on the basis of applying (in the absence of mass and often of friction too) the principles of virtual work. Thus we derive the elementary method of virtual power: in studies in planar mechanism it is argued that the dot product $\mathbf{F}\cdot\mathbf{v}$ remains a constant at the joints and, with the help of this idea, a polygon of forces is drawn for a plane statically determinate structure (§12.06). Some ideas about the corresponding argument for the virtual power at joints in spatial mechanism may be gleaned from the following passages: §10.60, §14.28, §19.08, §19.28. A way to conceive of the transmission forces in mechanism is to conceive of the links as being rigid, massless and frictionless; and that is what I intend to do (in three dimensions) here. At the same time I intend to draw a parallel by analogy between (*a*) the kinematics of the instantaneous motion, and (*b*) the statics of the instantaneous action; and for that, of course, I need to assume as well that the mechanism has a general, Kutzbach mobility of unity (§1.19, §1.34, §12.06).

55. Because it has no mass and suffers no friction, a massless frictionless link in mechanism can never not be in equilibrium. The total of the transmission forces upon it must always be zero. Take a massless frictionless link in spatial mechanism and contemplate its jointivity (§1.38); the number of joints may be 2, or 3 or more. If the number of joints is only 2 it will be evident that the wrench suffered by the link at one of its joints must be equal and opposite to the wrench it suffers at the other. I imply by the term wrench here the total action dual at the joint, as it inhabits its action or its home screw at the joint (§10.59, §10.60). It will be clear in this case, the case of a binary link, that the two wrenches I have mentioned will be collinear and of equal pitch, and that the closed force polygon for the link will be a mere segment of straight line. Go for a good example of this to §17.23. There, in conjunction with figure 17.08, the two equal and opposite forces $\mathbf{F}$ upon the coupler link 3 of a massless *RSCR* are explained and discussed. In that case the nature of the spherical joint at $B$ precludes transmission through the link of any couple: the two equal and opposite wrenches upon the coupler 3 in figure 17.08 can only be wrenches of zero pitch, namely pure (equal and opposite) forces.

56. If the number of joints on the massless link is 3, then the equilibrium can be seen as being a balance of

*three* wrenches, one occurring at each of the three joints. In this case the closed force polygon for the massless link will be a triangle and, of necessity, the central axes of the three wrenches, namely the seats of the three action screws at the joints, will reside in parallel planes (§ 13.01). The three screws will moreover belong to the same cylindroid of screws, the same 2-system (§ 13.25). At this point we begin to see however that, whereas in an angular velocity polygon drawn for any three links in relative motion we could put numbers at the apices of the triangle, we cannot with any similar pretence at meaning do that here. If a number can be written at all, it is the single number of the link in question; it can be written not at an apex or upon an edge of the triangle of forces, but upon its *face*. I shall return to this matter of the faces within a composite polygon of forces later, at § 12.58, § 12.67, and again at § 13.26 and § 13.29.

57. If the number of joints upon the massless link were four there would be *four* action duals in operation and the closed force polygon for the link would be in general a skew quadrilateral. If the pitches of all four wrenches were zero, that is if for example the four joints were all spherical joints, the lines of action of the four forces, namely the seats of the action screws of zero pitch, would all belong to the same regulus (§ 10.19). The same would apply if the pitches of all four wrenches were, for some strange reason, equal (§ 10.38). In any event the four action screws would all belong to the same 3-system, and the polygon of forces would be, as I have said, a skew quadrilateral. Similar remarks could be made about corresponding cases where the jointivity of the link was 5, 6 etc.: the polygon for the link would be a skew pentagon, a skew hexagon, etc., and the screw systems involved would be the 4-system, the 5-system, etc. (§ 10.26).

58. Now how do we ascribe a number, the number of the link in question, to the 'face' of a skew, multi-sided figure? The number belongs to the whole of the relevant, completed skew polygon for the link. This question brings into prominence now, not the analogous nature of force and angular velocity, but the differences between the two. The main single difference touching us here is this: whereas we often freely conceive of the angular velocity of a link as being measured with respect to any chosen one of the other links of a mechanism, and can thus always draw the edges of an angular velocity polygon between its established apices wherever we wish and with meaning, we cannot always do this with or within a composite polygon of forces. Each force is meaningfully measured against the one and only inertial frame of reference for the mechanism, namely the fixed or the earth link. It is of course possible to construct a composite polygon, a spatial version of the Maxwell–Cremona diagram, but such a polygon cannot with clear meaning consist of a mere figure of apices complete with all triangular faces. Among the edges of the triangles and the tetrahedra, multi-sided skew force polygons (a separate one for each link) need to be delineated. Each edge may be ascribed however a letter to represent the joint (§ 1.11, § 1.13, § 12.67). There are moreover no letters or numbers which could with meaning be ascribed at the apices. The crux of the main matter at issue here has been mentioned or implied already, at chapter 3. *It is the fact that whereas a number of different angular velocities must be seen as being superimposed upon a body in series, a number of different forces upon a body must be seen as being superimposed in parallel.* Refer to figures 3.09(*a*) and 3.09(*b*). Refer also § 12.67, § 13.26.

**The case of the plough**

59. Refer to the plough at § 10.23, and see upon reflection there that the four wrenches in equilibrium belong to the same 3-system (§ 12.57). The plough is not a massless body, but a body (or a system of connected bodies) in equilibrium, and one of the four wrenches acting there is the pure force of gravity. Taking suitable account of the mentioned couples involved (§ 10.23, § 10.69 *et seq.*, and chapters 15 and 19), the hyperboloid shown will be seen to be the pitch quadric of the said 3-system. It will be seen moreover that the polygon of forces (the skew quadrilateral) may be drawn upon the pitch quadric in either of the two manners displayed in figure 10.04. Refer to photograph 19. This shows not only two of the ways but all of the ways in which the said force polygon may be drawn upon the said pitch quadric. Photograph 19 also shows a manoeuvre that may be used to change the two wrenches $\Sigma\hat{\mathbf{T}}$ and $\Sigma\hat{\mathbf{W}}$ into the two pure forces $\Sigma\mathbf{T}$ and $\Sigma\mathbf{W}$ which, when properly located along the common perpendicular between the said wrench axes, have the same resultant as the two wrenches themselves do (§ 15.18). Refer also to item [1] of the notes at the end of chapter 10.

**Planar analyses**

60. Here at the end of my brief remarks about transmission forces, and before I begin to mention the *dynamics* of a link, I would like to recommend the reader to Maxwell (1962). Other authors deal with the same material but Maxwell gives in his relevant chapters a thoroughgoing, logical, and readable account of (*a*) the static force analyses for finding the transmission forces at work in planar mechanism, and (*b*) the dynamic force analyses for finding the forces due to the presence of mass and gravity in the corresponding real earth-bound machinery. He discusses in two dimensions the nature of equilibrium, the origins of the various forces and couples, techniques for ignoring until later the masses of the links, the presence of friction, and the thus calculated

transmission forces. Next he deals, but again in two dimensions only, with classical particle and rigid-body mechanics. He deals with work, energy and impulse, and uses the results of these investigations to illuminate the dynamic analyses which may be made and which can help (at least in a rough way) to guide the practical business of planar machine design.

61. In the same vein I would like to mention Hirschhorn (1962, 1967). At chapter 9 of the first and at chapter 2 of the second of Hirschhorn's books one can find a clear account – again for planar motion only – of the various 'energy' methods which might be used to estimate the actual mechanical behaviour of a real machine where the masses and moments of inertia of its cyclically moving movable links are great, while the moments of inertia of the various wheels (at or geared to the input shaft) are small. Hirschhorn deals with the question of energy loss due to friction too, and he makes no bold assumption that the angular velocity at the input shaft of a working machine might ever remain a constant. The whole contents of these relevant chapters point to the many difficulties involved in making the so-called dynamic analyses of mechanism. It is partly for this reason that I have chosen not to dwell too much, in this book, on the matter of acceleration. Nor do I dwell upon the acceleration polygon which, drawn as we know for planar mechanism, might well be drawn for general spatial mechanism also. Until the matter of friction is adequately understood, especially the matter of static friction, which cannot be predicted accurately at reversals of direction of sliding within the joints of real machinery, not much progress towards a final accuracy can ever be made (§ 18.23). It must also be said at this juncture that, in any process of rational design, a designer should never know the masses and moments of inertia of his links until he knows the forces that are to be at work, and, conversely, he will never know the forces that are at work until he has proportioned the various links and chosen the various materials from which they might be made.

### The dynamics of a link

62. If it were possible by some analysis or measurement to determine at some certain configuration of a mechanism the linear acceleration of the centre of mass of a certain link of the mechanism, it would be possible to determine thereby the rate of change of the linear momentum of that link, and thus to calculate the vector sum $\Sigma\mathbf{F}$ of the forces that must be mobilized from somewhere to produce that rate of change (§ 10.32). Because in the normal course of events the mass $m$ of the link will remain constant, the rate of change $d(mv)/dt$ of the linear momentum may be written $ma$, where $a$ is the mentioned, linear acceleration. We could write (after Newton) that $\Sigma F = ma$, and thus calculate the said vector sum of the forces acting (§ 13.29, § 19.17).

63. If it were possible by means of some analysis to determine at some certain configuration of a mechanism the rate of change $d(I\omega)/dt$ of the angular momentum of some link of the mechanism, it would be possible to determine thereby the vector sum $\Sigma\mathbf{M}$ of the moments at the centre of mass of the forces that must be mobilized from somewhere to produce that rate of change (§ 10.32, § 13.08, § 19.17). In general, however, neither the matrix of inertia $I$ nor the angular velocity $\boldsymbol{\omega}$ of the link will be constant, so the problem here is not an easy one (§ 19.17). If the mentioned rate of change could be calculated or otherwise determined, we could write (after Euler) that $\Sigma M = d(I\omega)/dt$, and thus calculate the said vector sum of the moments (§ 19.17, § 19.62).

64. I do not wish to discuss here the manifest and manifold difficulties inherent in the practical business of doing what might be necessary to get, with any degree of accuracy, the $\Sigma\mathbf{F}$ and the $\Sigma\mathbf{M}$ of the above two paragraphs (§ 19.62). Here I wish to say simply this: if the mentioned $\Sigma\mathbf{F}$ and $\Sigma\mathbf{M}$ for a link could be determined, we would be in a position thus to know what might be called the 'd'Alembert' or the inertia wrench. This will be balanced by the forces and couples which occur (*a*) at the several joints upon the link where contact is made with neighbouring links, and (*b*) at the centre of mass of the link where the force of gravity acts. Putting the matter into different words we could say the following: *the several wrenches upon the link that are mobilized at*

Figure 12.06 (§ 12.65) This is a closed polygon of forces for the dynamics of a ternary link 2 having joints at *A*, *B* and *C* (§ 1.38). Note the inertia and the gravity loads $\mathbf{I}_2$ and $\mathbf{G}_2$. Superimposed there is a similar polygon for the contemporaneous dynamics of a ternary link 3 of the same mechanism. Link 3 has joints at *C*, *D* and *E*. Couples at the joints of the links, of course, do not figure in the force polygons; the pairs of forces comprising them cancel one another out.

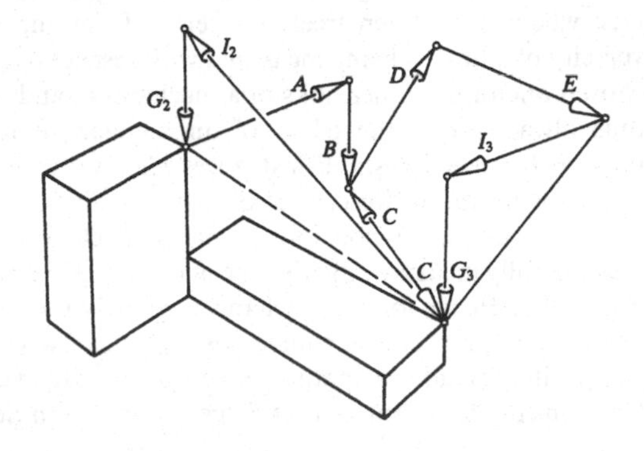

*the several joint elements at the link must balance the d'Alembert wrench plus the force of gravity.*

65. This means that, for the dynamics of any link at an instant, a force polygon can be seen to exist. Refer to figure 12.06. This is an imagined, composite, closed polygon of forces for the dynamics of two links where the jointivity of each is 3. The links, connected together at $C$, are $ABC$ (link 2) and $CDE$ (link 3). Figure 12.06 is a real (or a global) polygon dealing with the balance of forces. A dual (or a spatial) polygon would deal with the balance of moments. If $\mathbf{F}_A$, $\mathbf{F}_B$ and $\mathbf{F}_C$ are the vector sums of the forces upon link 2 at the several joints, namely the real or the global parts of the action duals upon that link at the several joints (§ 19.12), the closure of the polygon $ABCI_2G_2$ is a reflection of the fact that four wrenches are in equilibrium. The four wrenches are (*a*) the three wrenches $\hat{\mathbf{W}}_A$, $\hat{\mathbf{W}}_B$ and $\hat{\mathbf{W}}_C$ acting at the joints, and (*b*) the inertia wrench plus the force of gravity. Because the four wrenches are in equilibrium, incidentally, the screws of their respective central axes will all belong to the same 3-system (§ 10.55); and, for the same reason, the moments of the four wrenches taken at any chosen point in space – let us take for convenience the centre of mass of the link – will add to zero too. The sum of the moments due to the wrenches at the several joints, namely the vector sum of the dual (or spatial) parts of the action duals acting upon the link at the three joints (§ 19.14), will, when the dual parts are taken at the centre of mass, be the $\Sigma\mathbf{M}$ of § 12.63 (§ 19.17). The point here of course is that, when the moments (the dual parts of the duals) are thus taken at the centre of mass, the dual part at that point of the pure force of gravity is zero.

66. In the case of a binary link (and such a case is very common in machinery), we can easily think in terms of only three wrenches: (1) the mobilized reaction upon the link at the first of the two joint elements; (2) the mobilized reaction upon the link at the second of the two joint elements; and (3) the inertia wrench plus the force of gravity. At any instant these three wrenches will be in equilibrium, and from this we can see the following: (*a*) that the three forces lying along the central axes of their respective wrenches will reside in parallel planes; otherwise the force polygon could not close; (*b*) that the central screws of the three wrenches will belong to the same 2-system; they will belong in other words as generators of the same cylindroid; and (*c*), and this is a corollary of the others, that there will be a line in space, the nodal line of the cylindroid, that cuts all three wrench axes perpendicularly (§ 13.25). Knowing the natures of the joints (chapters 6 and 7), and knowing thereby the natures of the wrenches that can be transmitted at each joint in the absence or presence of motion (chapters 14 and 17), it should be possible in any determinate problem to use the algebra of the cylindroid to make some progress here (§ 15.24). In any real dynamical problem it will no doubt be necessary, as is usual, to winkle out the information about the wrenches bit by bit: there appears to be in this area (as there is in the whole wide field of mechanism and machines) no red-carpet route leading straight through to the answers we seek.

**Problems of symbolism in the polygons**

67. I would like to speak in conclusion now about the letters $A, B, C$ etc. and $I_2$ and $G_2$ etc. written at the edges of the polygons in figure 12.06. There was no real difficulty in the discussion beginning at § 12.65 when I formed the subscripted symbols $\mathbf{F}_A$, $\mathbf{F}_B$ etc. for the forces acting upon link 2 at the joints $A, B, C$ etc. The possibility that these subscripted letters might be inserted at the edges of the polygon had already been mooted at the end of § 12.58, and I have done precisely that in figure 12.06. But what should I do now in figure 12.06 about the equal and opposite force that is acting at the joint say $C$ *from* the said link 2 *upon* the next link 3? At § 17.02 I form the subscripted symbols $\mathbf{F}_{23}$ and $\mathbf{F}_{32}$ for example where the difficulty evident here is not obtruding. Such a system is no solution to the problem here, for in any composite Maxwell–Cremona diagram each force – and each force will be, in such a diagram, some *directed edge* of the composite polyhedron – would need to be known for example both as $\mathbf{F}_{23}$ and $\mathbf{F}_{32}$. For this and for other obvious reasons the numbers 1, 2, 3 etc. cannot appear with meanings at the apices of a composite force polygon. As explained already at § 12.58, it is at the *faces* of the various skew polygons that the numbers might be written. But the fact of the matter is of course that, unless these faces are broken into their subsidiary triangles (and this can always be done in a number of different ways), the faces remain skew faces.

68. We should note here that whereas angular velocity is a phenomenon that might occur between a pair of contacting or non-contacting links and in a reversible way, in the sense that we can write with meaning for example both $\boldsymbol{\omega}_{25}$ and $\boldsymbol{\omega}_{52}$ (§ 5.58) and put the numerals 2 and 5 at the relevant apices of the composite angular velocity polygon for a mechanism, force is a phenomenon that occurs (*a*) between a pair of links that must be in contact, and (*b*) in a reversible way *at a particular joint*. Refer again to my italicized remarks at the end of § 12.58.

69. On the matter of so-called forces such as $\mathbf{I}_2$ and $\mathbf{G}_2$ it must be said (*a*) that any given $\mathbf{I}$ is not a force at all, but merely the product of the mass of some link and the linear acceleration of its centre of mass, and (*b*) that any given $\mathbf{G}$, the force 'of gravity' upon some link, has no

reversed counterpart acting within the confines at least of the accelerating links of the mechanism. In any first try at a Maxwell–Cremona diagram to be drawn for the dynamics of some mechanism, these so-called forces will, for each link other than the frame link, have no equal and opposite force appearing. They might for that reason be caused to carry a subscripted *number*, the number ascribed to the link in question. The magnitude and direction of the parallel gravity field is of course an important factor in the dynamical behaviour of a majority of machines; gravity cannot be ignored.

70. I am thoroughly aware that the above remarks are confused. I am also aware that it is neither inevitable nor important that completed polyhedra to summarize the statical and dynamical forces at work within a moving mechanism must exist. It is not absolutely necessary to impose order where none exists. On the matter of moments and the possibility for moment polyhedra, I simply wish to say that analogous constructions corresponding with the linear velocity polyhedra of this chapter may be possible. If they are they will be useful. Please refer to §10.23, §13.26, §13.29.

### Notes and references


[1] Phillips, J.R., Determination of instantaneous screw axes in spatial mechanism, *Proceedings of the International Conference on Mechanisms and Machines, Varna, Bulgaria, 1965*, Vol. 1, p. 245–69, Mechanical and Electrotechnical Institute, Sofia 1965. Refer §12.05.

[2] Smith, R.H., A new graphic analysis of the kinematics of mechanisms, *Transactions of the Royal Society Edinburgh*, Vol. **32**, p. 507–17 and pl. 82, 1882–5. Refer §12.14.

[3] Ferguson, Eugene S., Kinematics of mechanism from the time of Watt, *US National Museum Bulletin 228*, Smithsonian Institution, Washington 1962. Refer §12.14.

[4] Mehmke, R., Über die Geschwindigkeiten beliebiger Ordung eines in seiner Ebene bewegten ähnlich veränderlichen Systems, *Civil Inginieur*, 1883. Refer §12.14.

[5] Phillips, J.R. and Hunt, K.H., On the theorem of three axes in the spatial motion of three bodies, *Australian journal of applied science*, Vol. **15**, p. 267–87, CSIRO, Melbourne 1964. Refer §12.30.

13. Somewhere through the teeth of this hypoid set passes the motion screw or the instantaneous screw axis for the relative motion of the teeth. It remains fixed in location with respect to frame and its pitch is constant. The line of this axis may be called with justification the pitch line.

# 13 On the two theorems of three axes

### Some elementary considerations

01. If three forces act upon a body – there being as well a number of couples (§ 10.35) – and the resulting effect is equilibrium, the lines of action of the three forces will reside in parallel planes. We saw an example of this at § 12.65. If the three forces are acting alone, however, the lines of action will reside in the same plane and they will intersect at a single point (§ 10.14).

02. If three angular velocities are imposed upon a body – there being as well a number of rates of translation (§ 10.45) – and the resulting effect is rest, the axes of the angular velocities will reside in parallel planes. If the three angular velocities are occurring alone, however, the axes will reside in the same plane and they will intersect at a single point (§ 10.14, § 20.54).

03. As soon as it is seen that any 'couple' of forces, namely any pair of equal, parallel, and opposite forces (§ 10.35), has no effect upon the closure of a force polygon, and that any closed force polygon with three sides can be none other than a plane figure namely a triangle, the self evident nature of the statements at § 13.01 becomes clear. It is true in any event that the contents of § 13.01 is common knowledge in the area of elementary statics. The contents of § 13.02, however, is not so well known or understood, and the reader is referred to § 10.55 and from there to § 10.32 for a clarification of these ideas. Please contemplate figure 10.13. If there were only three screw joints at the apparatus there, if they were all of zero pitch, namely hinges, and all transitorily operable, and if the bodies 5 and 1 remained relatively at rest, the three hinge axes would need to be coplanar and to intersect at a single point. Take a glance at the figures 20.27.

### On statics and kinematics

04. This chapter is dealing (*a*) with moving rigid bodies whose relative motions taken cyclically at the instant will always be adding to rest, and (*b*) with wrenches upon a single body whose respective actions at the instant are adding to equilibrium. The two circumstances here implied are regularly occurring together in the broad field of the kinetostatics of mechanism and the two are, as might be expected from the contents of chapter 10, very closely related with one another. We shall find once again in this general area of the relationships within linearly dependent sets of screws that the analogy across the dividing line between the statics and the kinematics of a body at an instant is complete (§ 10.57, § 10.59).

05. Various other remarks introductory of the main theme at issue here may be found elsewhere. See for example § 10.14, § 10.54, § 12.31, § 12.56, § 14.05 and § 15.03. Although those remarks refer as often as not to cases of three or more wrenches or rates of twisting upon their respective screws, the material of the present chapter converges very quickly upon the circumstance of there being, in a group, *only three, linearly dependent*, participating screws, and thus upon two analogous versions of a most important theorem.

06. In respect of the planar kinetostatics, it might be mentioned here that the well known theorem of Kennedy and Aronhold of about 1875 regarding the three collinear poles of any three laminae in instantaneous relative motion in the plane, and the well known theorem derived from Stevin of about 1620 regarding the equilibrium of three coplanar forces that are parallel, are each of them special cases of the more general theorem about to be put forward here. Unless that thoroughgoing analogy of chapter 10 between (*a*) the motions of, and (*b*) the actions upon, a rigid body at an instant is thoroughly gripped in the working imagination such apparently hazy matters as the similarity between these two analogous versions of the same theorem will never become completely clear however. So please be aware that at least a rough working knowledge of the contents of chapter 10 is being taken for granted here.

07. On the matter of relative motion I wish to mention that, if *n* rigid bodies are in relatively spatial motion, the cyclic sum of their relative angular velocities, $\boldsymbol{\omega}_{12} + \boldsymbol{\omega}_{23} + \cdots + \boldsymbol{\omega}_{(n-1)n} + \boldsymbol{\omega}_{n1}$, will always be zero. This is a matter for observation. It is a fact of nature. It is

inextricably bound up with our concept of angular velocity however, and thus with our Euclidean view of nature (§ 5.45, § 10.55). If there are associated with these angular velocities rates of axial sliding $\tau$, if in other words the bodies are instantaneously *twisting* with respect to one another (§ 5.47), then it can be said that the vector sum of all of the dual vectors for the instantaneous relative twistings taken cyclically around the bodies will be zero too (§ 10.55). Refer to [1].

08. Let me remark parenthetically that, in the case of a single, massive, accelerating link in mechanism, with its known mass and its known principal axes and values of inertia (§ 19.62), the sum of the following wrenches upon it will be zero: the wrenches applied to the link at its several joints (§ 10.59); a wrench of zero pitch in the direction of the fource of gravity and acting at the centre of gravity (namely the force of gravity); and the inertia wrench of d'Alembert calculated by application of Newton's and Euler's laws (§ 12.64, § 19.17). It is against the background of this remark, or against another and sometimes acceptable assumption about the masslessness of links in mechanism (§ 12.54), that I can speak of statics here in this broader context of dynamics. In any event it can be said that such a thus-taken static balance between the wrenches at a link in mechanism establishes itself at the same time as the relative motions between the links, and the point I am trying to make is that all planned analyses for these two disparate though related phenomena can always be taken hand in hand.

**Three rigid bodies in relative motion**

09. At any instant during the relative motion of three rigid bodies 1, 2 and 3, there will be three relative ISAs or motion screws. Using an established symbolism (§ 5.44, § 12.13), I shall call the screws $ISA_{12}$, $ISA_{23}$ and $ISA_{31}$. Now we know from the fact of nature mentioned at § 12.31 and again at § 13.07 that, independently of whether the bodies are mechanically connected or not, $\boldsymbol{\omega}_{12} + \boldsymbol{\omega}_{23} + \boldsymbol{\omega}_{31} = 0$. The angular velocity polygon drawn for this equation is a closed triangle and thus a planar figure. We can accordingly see straight away that, *at any instant during the relative motion of three rigid bodies, the three relative ISAs or motion screws reside in a set of parallel planes.*

10. Please commit to memory figure 10.13 and the gist of the text at § 10.54 and § 10.55. Now refer to figure 13.01 and see that, in the continued presence of motion of the bodies 2 and 3, the body 4 could be put to rest in body 1 provided (at the instant) the sum of the three motion duals at the three screw joints were zero (§ 10.49, § 10.55, § 13.07). Let it be so in figure 13.01; let 4 be 1, in other words, and let the resulting 3-link loop be transitorily mobile. From § 13.09 we can see that the axes of the three screw joints of the 3-link mechanism must now reside in parallel planes; at this stage, however, we can see no more than that regarding the required special configuration of the mechanism. With regard to its mobility we can see from § 1.34 that, unless it exhibits some very spectacular geometric specialities (§ 20.54), it can only be transitorily mobile. If the angular velocity and the pitch of the twisting at each of the screws $ISA_{12}$ and $ISA_{23}$ are given, however, we can write for the three bodies (which may, if we wish, now be considered free) that $\boldsymbol{\omega}_{13} = \boldsymbol{\omega}_{12} + \boldsymbol{\omega}_{23}$. Moreover if $Q$ be any point in the bodies (§ 12.10), we can write that $\mathbf{v}_{1Q3} = \mathbf{v}_{1Q2} + \mathbf{v}_{2Q3}$. These two equations are neither exotic nor extraordinary: they are well known and are often used in elementary applied mechanics. Together, however, and given two of the ISAs for any three bodies in relative motion, they can be used (as we shall see) to find the remaining ISA. Refer to both of the figures 13.02.

11. In figure 13.02($a$) we see three bodies in free relative motion. Although the bodies are not mechanically connected, let it be given that $ISA_{12}$ and $ISA_{23}$ are located as shown, and that the two pitches there are known. The common perpendicular between the two given screws has been drawn, and this has been labelled $p$–$p$ accordingly. It might be imagined next that two actual screw joints exist to match at the instant the two given screws. See figure 13.02($b$). The question is this: where is the remaining ISA or the motion screw $ISA_{31}$? Where is, in other words, and what is the pitch of the remaining imaginary screw joint that would complete the transitorily mobile 3-link loop discussed at § 13.12?

Figure 13.01 (§ 13.10) The idea of a bound velocity polygon at $Q$. This polygon applies whether or not the body 4 (which is shown here interpenetrating body 1) is fixed in 1. If body 4 is fixed in 1, and if motion within the linkage remains then possible, the motion duals at the three screw joints add to zero.

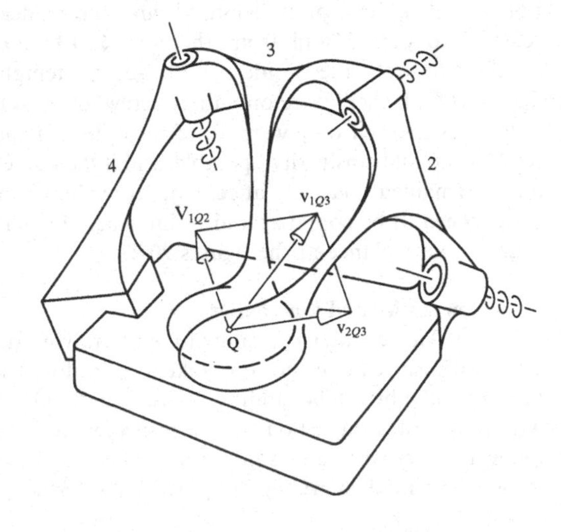

We can begin to answer this question by going to figure 15.05. There we see the mechanics of the vector addition of two action duals $\hat{\mathbf{W}}_1$ and $\hat{\mathbf{W}}_2$. Knowing that the same mechanics applies to motion duals, and with the help of figure 15.05, we may perceive, in the similar figure 13.02(*b*), *that the resultant motion at the instant resides upon a motion screw which cuts the common perpendicular p–p perpendicularly*. Please read chapter 15.

**The theorem of three axes in kinematics**

12. It follows from what has been said and from the contents of chapter 15 that the following theorem may be written down: *at any instant in the relative motion of three rigid bodies 1, 2 and 3 the lines of the three relative instantaneous screw axes $ISA_{12}$, $ISA_{23}$ and $ISA_{31}$ will be arranged in space in such a way that all three pairs of them will share the same line as common perpendicular; they will moreover be generators of the same cylindroid and the pitches of the ISAs will be appropriate to that.* A more sophisticated way of stating the same theorem would be to say this: *whenever we have three rigid bodies in relative motion the three motion screws at an instant will belong to the same cylindroid of screws.* It will of course be seen from the overall contents of this book that an even more concentrated way of saying the same theorem would be to state the following natural law: *any three linearly dependent screws will belong to the same 2-system.* The theorem may also be seen and derived as a special case of the mechanics of the screw triangle incidentally (§ 19.38, § 19.43); the screw triangle deals, however, not directly with the vector addition to rest of three instantaneous rates of twisting as this theorem does, but with the non-vector addition to rest of three finite screw displacements; the screw triangle is, accordingly, somewhat irrelevant here; but refer to [2].

13. The theorem is spectacular; and there are many methods available for the numerical calculation of

Figure 13.02(*a*) (§ 13.11) Three non-connected bodies moving relatively in space. At the motion screws $ISA_{12}$ and $ISA_{23}$ actual nuts and bolts might be imagined which may, with legitimacy and at the instant, exist.

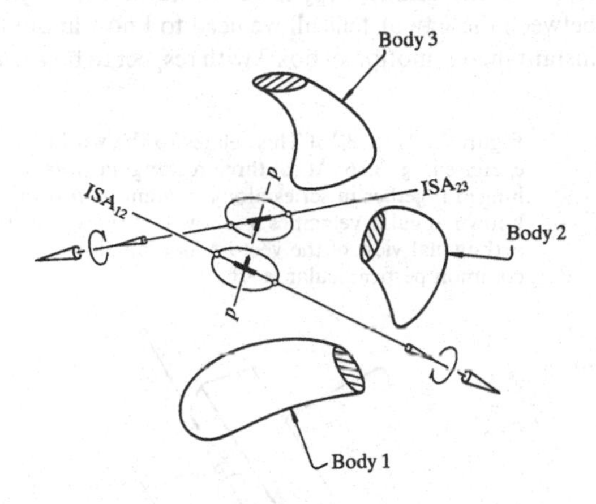


Figure 13.02(*b*) (§ 13.11) Two imagined screw joints located upon the motion screws for two of the three pairs of free bodies shown at figure 13.02(*a*).

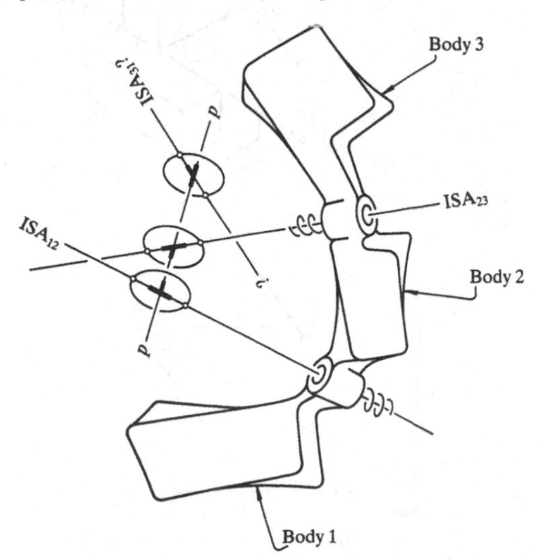


Figure 13.03 (§ 13.13) This figure is illustrating the theorem of three axes in kinematics. All three of the relevant motion screws are members of the same cylindroid of screws. Note the axes for *x*, *y* and *z*.

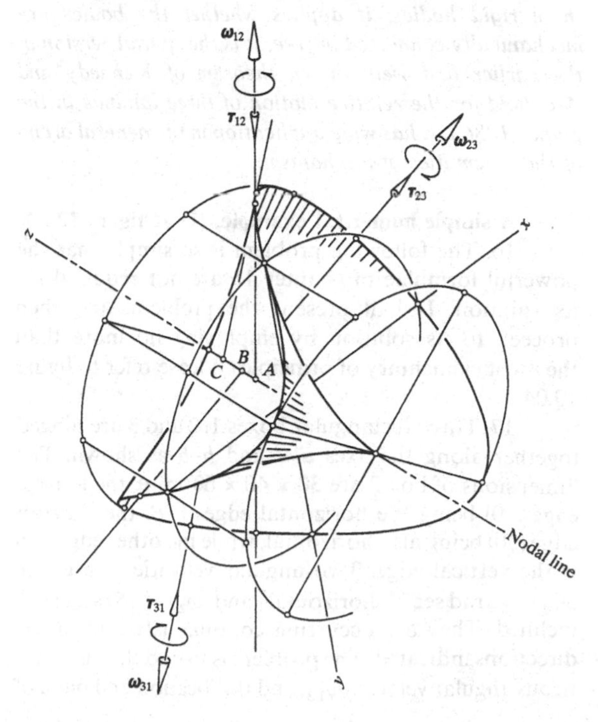

the components of practical problems. Depending upon the problem, these may be very simple (see the worked exercise for the special case at § 13.16), or they may be more complicated (see the worked exercise for the general case at § 13.20). In any event we may take it for granted that in an ordinary problem some sufficient items of data, say $ISA_{12}$ and $ISA_{23}$, are known. We may either (*a*) apply our geometrical knowledge directly, (*b*) go to the boxes of algebraic equations and criteria at figures 15.09 and 15.10, or (*c*) go to [2]. In these or other ways the relevant cylindroid may be determined and located upon its three Cartesian axes (§ 15.24). The unknown screw $ISA_{31}$ then lies as one of the generators of that cylindroid. Refer to figure 13.03.

14. Which generator it is, next, is determined by the ratio of the angular velocities involved (§ 10.14). By drawing the angular velocity polygon we can determine the direction $\theta_{31}$ of the angular velocity $\boldsymbol{\omega}_{31}$ and thus, by substitution into the now quantified expressions (12) and (13) in the box at figure 15.10, calculate the required $z_\theta$ and $p_\theta$ for the screw $ISA_{31}$. In this or a similar way, paying careful attention to sign, or graphically, using the circle diagram of figure 15.09, any determinate problem embraced by the theorem may be solved.

15. I have here called the theorem the theorem of three axes in kinematics [1]. There is a corresponding theorem in statics, but I have not yet stated it (§ 13.25). The following summarizing remarks are important: *the theorem of three axes in kinematics, stated at § 13.12, deals in the general area of the instantaneous relative motion of three rigid bodies; it applies whether the bodies are mechanically connected or free; it is the spatial version of the earlier and well known theorem of Kennedy and Aronhold for the relative motion of three laminae in the plane (1886); it has wide application in the general arena of the kinematics of mechanism.*

**A simple numerical example, (1) at figure 13.04**

16. The following problem is so simple that the powerful formulae of chapter 15 are not required for its solution. I shall present the problem, and then proceed to its solution by employing no more than the most elementary of principles. Please refer to figure 13.04.

17. Three rectangular boxes 1, 2 and 3 are hinged together along the axes *a–a* and *b–b* as shown. The dimensions of box 2 are 30 × 40 × 60 (mm), the longest edge (60) being the horizontal edge *A–B*, the shortest edge (30) being also horizontal, while the other edge (40) is the vertical edge. Two angular velocities are given, $\boldsymbol{\omega}_{12} = 12\,\text{rad sec}^{-1}$ horizontal, and $\boldsymbol{\omega}_{23} = 15\,\text{rad sec}^{-1}$ inclined. They are occurring continuously and in the directions indicated. The problem is to find the instantaneous angular velocity $\boldsymbol{\omega}_{13}$, and the location and pitch of the $ISA_{13}$. The reader may care to regard box 1 as being the fixed box, but that is not necessary.

18. We may shift the angular velocity vector $\boldsymbol{\omega}_{12}$ from its basepoint at *A* to a new basepoint at *B*, thus occasioning a linear velocity vector at *B* in the vertical direction. This occasioned vector is given by the cross product of the shifted angular velocity itself and the distance shifted. Its magnitude is 12 × 60, namely $720\,\text{mm sec}^{-1}$. We recognize this as the velocity relative to box 1 of point *B* in box 2; we write it $\mathbf{v}_{1B2}$ (§ 12.09). By vector addition at *B* we calculate by means of the parallelogram there that $\boldsymbol{\omega}_{13} = 24.19\,\text{rad sec}^{-1}$ and that this is inclined at angle $\psi = 60.26°$ to the vertical. See $\psi$ at figure 13.04(*b*). By virtue of the hinge along *b–b*, point *B* in box 3 is remaining coincident with point *B* in box 2 (§ 12.10). We see accordingly that $\mathbf{v}_{1B3} = 720\,\text{mm sec}^{-1}$ also. The dual vector at *B*, consisting now of the known $\boldsymbol{\omega}_{13}$ and the known $\mathbf{v}_{1B3}$ (and we know the angle $\psi$ between these two), tells all we need to know about the instantaneous motion of box 3 with respect to box 1. We

Figure 13.04 (§ 13.16) This relates to the worked exercise at § 13.16. At (*a*) three rectangular boxes are hinged together in series along *a–a* and *b–b* with known angular velocities as shown; at (*b*) we see an orthogonal view of the vectors looking along the common perpendicular *A–B*.

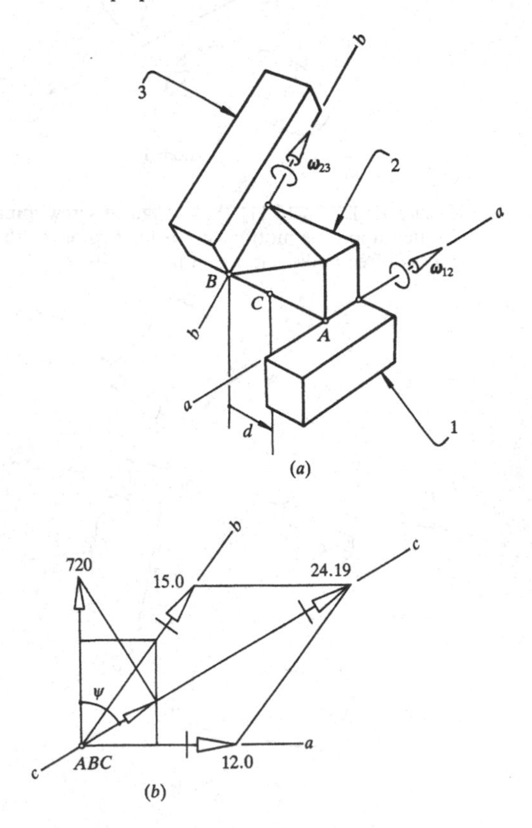

may next shift the basepoint of the dual vector at $B$ from $B$ to a new point $C$ along the edge $A$–$B$ by such a distance $d$ that the velocity vector occasioned by the shift is just enough to cause the resultant linear velocity at $C$ to be collinear with the angular velocity vector there. We write that $\boldsymbol{\omega}_{13} \times \mathbf{d} = \mathbf{v}_{1B3} \sin\psi$. Substituting the known values we find that $d = 25.85$ mm, and, by perceiving that the magnitude of the least velocity vector $\mathbf{v}_{1C3}$ at $C$ is given by $\mathbf{v}_{1B3} \cos\psi$, we find that to be 357.17 mm sec$^{-1}$. Next it is easy to calculate that the pitch of the twisting of box 3 with respect to box 1 is (357.17)/(24.19) namely 14.77 mm rad$^{-1}$.

19. This problem was a very simple one for two reasons, (*a*) because the common perpendicular distance between the motion screws of the two rates of twisting to be added was immediately obvious, and (*b*) because the pitches of the rates of twisting were both zero. Please note that, although the three boxes were hinged together in the manner shown, and thus that pairs of contiguous boxes were unable to slide with respect to one another, box 3 was found to be twisting (or screwing) and thus sliding with respect to box 1. This may be easier to understand when (*a*) one refers for the matter of clashing of links to §1.06 *et seq*., and (*b*) one sees that ISA$_{13}$, which lies along the line $c$–$c$ in figure 13.04(*b*), traces out a circular hyperboloid as the continuous, steady, relative motions occur; the axis of the hyperboloid is the hinge axis $a$–$a$. See also my remarks in connection with the hypoid gear set at §13.23, and note the material of chapter 22. This simple method for the working of the above exercise is a method that could be adapted and used, in general (and independently of the cylindroid), to prove the theorem under consideration here [1].

### A general numerical example, (2) at figure 13.05

20. Let it be given that three bodies 1, 2 and 3 are in relative motion, and that the motion screws ISA$_{12}$ and ISA$_{23}$ are known as follows: the pitches are $+50$ mm rad$^{-1}$ and $-20$ mm rad$^{-1}$ respectively, and the two screws or ISAs are displaced 30 mm apart perpendicularly and 60° apart right handedly. A first task is to locate the relevant cylindroid, and then, with the following information about the angular velocities, to locate the motion screw ISA$_{13}$ and the rate of twisting about that screw of the body 3 with respect to the body 1. The angular velocities $\boldsymbol{\omega}_{12}$ and $\boldsymbol{\omega}_{23}$ are 3.0 rad sec$^{-1}$ and 10.0 rad sec$^{-1}$ respectively, their directions being such that the two vector arrowheads of the angular velocity vectors straddle the acute (not the obtuse) angle between the screws. We are being required, in effect, to discover the instantaneous result of superimposing upon a body two different rates of twisting. The body might be the gripper of a robot; refer to figure 13.05(*a*). Refer to §15.23 for a strategy listed from (*a*) to (*f*) and to the boxes at figures 15.10 and 15.09 for the necessary equations and criteria.

21. Equation (11) in the box at figure 15.10 is based upon Pythagoras, and the question of sign is not important there. Substituting (*a*) the available data into that equation, we find that $2B$, the length of the cylindroid, is 87.94 mm; the circle of the circle diagram can thus be drawn using the value of $B$, namely 43.97 mm, as radius; see the circle diagram in figure 13.05(*b*). Substituting (*b*) the data about the given screws into criterion (15) we find that the expression there reduces to $-70 + 30 \cot 60°$, which is negative; the angle $2\theta_A$, next to be calculated, is accordingly in the first quadrant (not in the second). Going to equation (14) in the box at figure 15.10, we find (*c*) that $2B \sin 2\theta_A = 70.41$, that $\sin 2\theta_A = 0.8007$, that $2\theta_A = 53.20°$ in the first quadrant (not 126.8° in the second), and thus that $\theta_A = 26.60°$. From equation (9) in the box at figure 15.07, or directly from equation (14) as explained in the caption to figure 15.10, we find (*d*) that $z_A = -35.21$ mm, and thus that the system of axes illustrated in figure 15.06 is now located. See the circle diagram in figure 13.05(*b*) and its adjunct in figure 13.05(*c*). Substituting now into equations (12) and (13) in the box at figure 15.10 we find our expressions for

Figure 13.05(*a*) (§13.20) The gripper 3 of the robot has superimposed upon it the two different rates of twisting that are shown. They are additive by means of a method based upon the theorem of three axes in kinematics. A numerical result is calculated in the worked exercise beginning at §13.20.

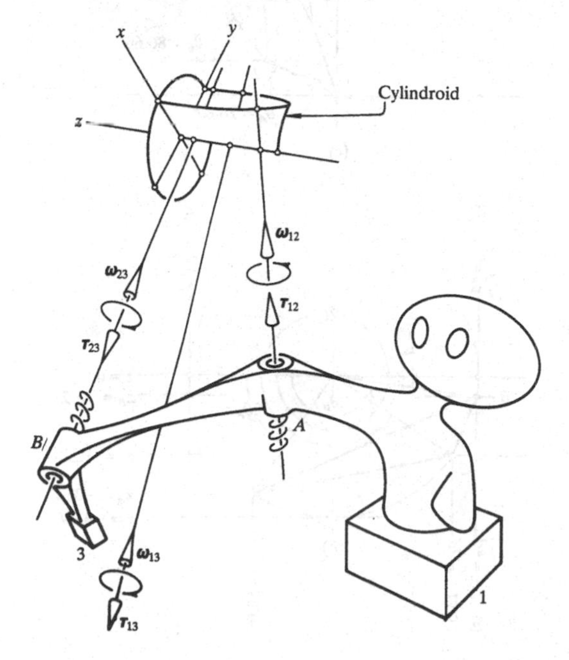

$z_\theta$ and $p_\theta$; they are (*e*) $z_\theta = -43.97 \sin 2\theta$, and (*f*) $p_\theta = 23.66 + 43.97 \cos 2\theta$; note that these are in fact equations (9) and (10) in the box at figure 15.07. It falls out now that $z_B = -5.21$ mm. It also falls out that, while $p_\alpha$ (namely $p_{max}$) $= +67.63$ mm rad$^{-1}$, $p_\beta$ (namely $p_{min}$) $= -20.31$ mm rad$^{-1}$; these latter are irrelevant to the limited problem at hand, but they help to complete the overall picture. Note the origin for $p$ duly indicated in figure 13.05(*b*).

22. Go now to the adjunct of the circle diagram illustrated in figure 15.09(*b*) and construct, not the force polygon which appears there, but the analogous, angular velocity polygon for the present problem; see figure 13.05(*c*). By simple trigonometry we find that, while $\omega_{31}$ is 11.97 rad sec$^{-1}$ as shown, its direction is given by $\theta$, namely $\theta_C$, which is 253.87°. Substituting this into the now-known expressions for $z_\theta$ and $p_\theta$, we arrive at the required result. It is that, while $z_\theta = -23.47$ mm, $p_\theta = -13.52$ mm rad$^{-1}$. The direction of the resultant rate of twisting of the gripper with respect to the robot body is opposite to that that is sketched in figure 13.05(*a*). These data could be used, of course, in the case of an actual robot arm, to determine whatever we might wish to know about the linear velocities of selected points in or upon the gripper. But refer also § 10.55, where an alternative method for linear velocity at points upon the gripper may be found. See also figure 13.01 directly.

Figure 13.05(*b*), (*c*) (§13.22) The circle diagram and other data drawn in conjunction with figure 13.05(*a*). At (*b*) the details of the cylindroid may be seen.

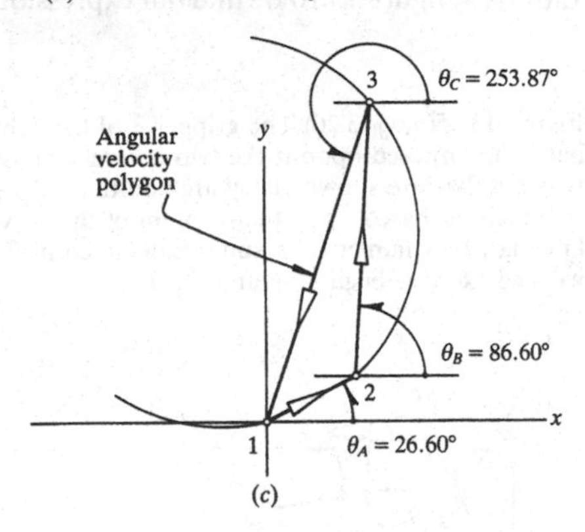


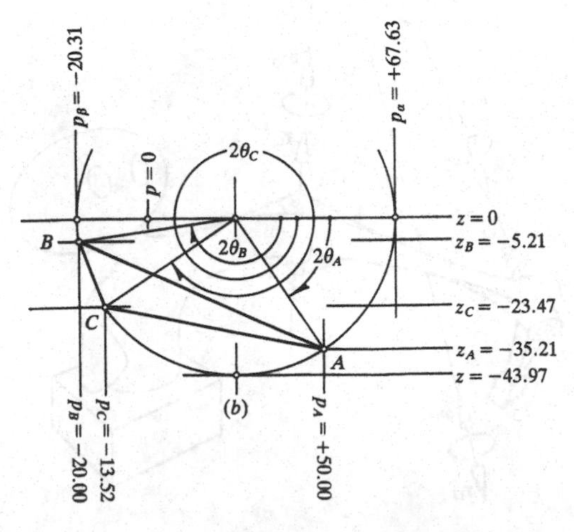


## Another numerical example, (3) at figure 13.06

23. I would like to deal with a typical hypoid gear set. There is a reference to this example at § 22.22. Refer to figure 13.06(*a*). The perpendicular distance between the shaft axes, the so-called centre distance $C$ (§ 22.07), is 40 mm. The numbers of teeth on the pinion 3 and the wheel 2 are 9 and 41 respectively. It is required for a check upon certain aspects of tooth behaviour to locate $ISA_{23}$. This is the ISA about which the pinion 3 will twist (or be screwing) with respect to the wheel 2. The actual existence of the $ISA_{23}$ may be confirmed by either imagining or actually doing the following experiment: fix the bare wheel flat upon its back upon a bench; put the bare pinion into proper mesh and, keeping it thus in contact, proceed to screw the pinion around and upon the wheel. You will notice by watching the motion or feeling the action or both (*a*) that $ISA_{23}$ resides in a vertical plane parallel to both of the shaft axes, (*b*) that the pitch of the screwing is not zero, and (*c*) that the axode traced in the wheel 2 by the moving $ISA_{23}$ is a circular hyperboloid. This hyperboloid will have its axis collinear with that of the wheel. Another circular hyperboloid will be traced by the $ISA_{23}$ in the pinion; this one (of smaller diameter) will reside with its axis collinear with that of the pinion. Refer to figure 22.12(*a*) for a clear picture of such a pair of hyperboloidal axodes; in that figure, however, the shaft axes of the two gear wheels are not perpendicular to one another as they are here. I wish to stress at this juncture that the pair of axodes for the relative motion here are determined solely by (*a*) the given perpendicular distance 40 mm, and (*b*) the tooth ratio namely the speed ratio 9/41 (namely 0.2195 approximately). The actual shapes of the machine-cut teeth and the complicated mechanics of their mutual meshing, although related, are both irrelevant to the fundamental question at issue here. The question is: where is the $ISA_{23}$ and what is its pitch?

24. This problem, like the simple problem of the three boxes (1) at figure 13.04, is also simple enough not to need the powerful general formulae of chapter 15. It is indeed an even simpler example of the same problem of two hinges in series. The two hinges are perpendicular in this particular case; the shaft angle $\Sigma$ is 90° (§ 22.07). But the problem itself is camouflaged here with new clothes. Refer to figure 7.19 and read the caption there. If the two free clevis-pieces in figure 7.19(*a*) were meshed together

Figure 13.06 (§13.24) These relate to the worked exercise beginning at §13.23. They show various views and aspects of the gears of a hypoid set. The revealed $ISA_{23}$ is the axis about which the pinion is twisting not with respect to the fixed frame but with respect to the rotating wheel. Refer to photograph 13.

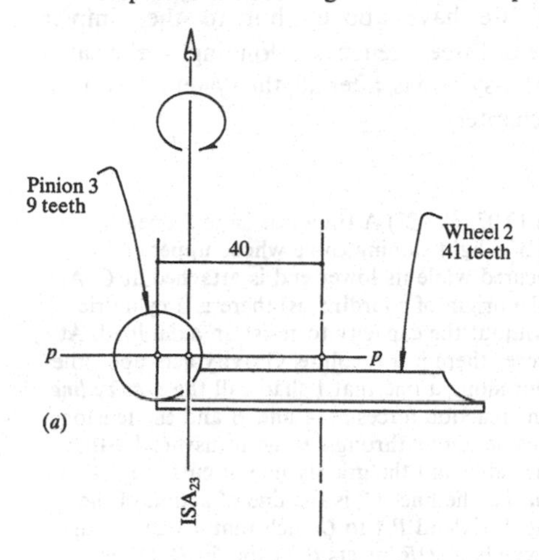


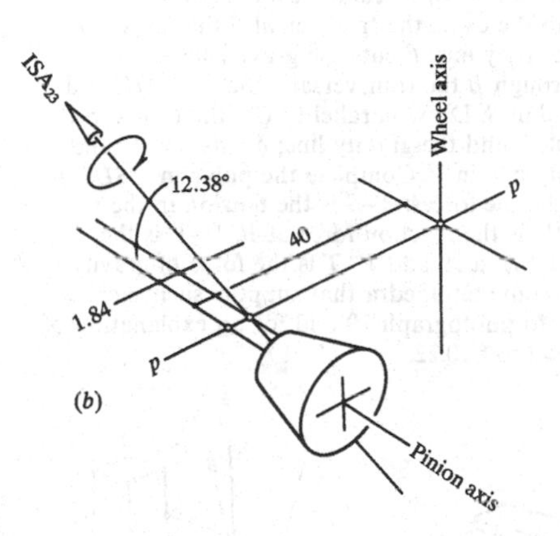


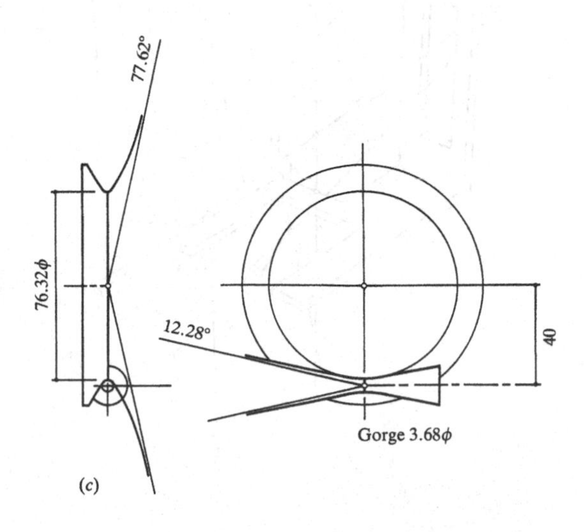


in a gear-like manner, that which is otherwise there a complex $f2$-joint would become the closed 3-link loop of figure 13.06(*a*). Given a single point of contact at the meshing, the mobility of the loop is unity. In table 22.01 at §22.24 the reader will find a number of special formulae for solution of the present problem; they are all derived in chapter 22 from the general formulae of chapter 15. They are not only directly applicable in chapter 22 and in the present problem (which is of course a gear problem), but also, indeed, in the simple problem (1) at figure 13.04. Notice in the present problem that, along the shaft axes, there are two known screws of equal pitch; they are moreover of zero pitch and mutually perpendicular; accordingly they are the extreme generators of the relevant cylindroid (§15.49); the length $2B$ of the cylindroid coincides exactly with the perpendicular distance between the shafts. Here we have $C = 40$ mm and $\Sigma = 90°$. With these data we may substitute directly into the equations (1), (2) and (3) at table 22.01 to obtain the following results: $\theta = 32.62°$, $r_3 = 1.84$ mm, and $r_2 = 38.16$ mm. The angle $\theta$, to be seen at figure 15.06 is measured right handedly from the positive $x$-axis of the cylindroid to the generator in question, so the angle between the horizontal plane in figure 13.06(*b*) and the $ISA_{23}$ is 45.00° minus 32.62° namely 12.38°. The radii $r_3$ and $r_2$ are the throat (or the so-called gorge) radii of the hyperboloidal axodes upon the pinion and the wheel respectively. The corresponding diameters are indicated in figure 13.06(*c*). The helix angle $\psi_3$ at the gorge of the axode at the pinion is 12.38°, and from equation (6) at table 22.01 the corresponding helix angle at the wheel is 77.62°. Note that whereas the tooth or the speed ratio (9/41) is 0.2195 the ratio of the gorge radii is not the same as this; it is 0.0482. Refer to §22.22 for some corresponding results in a general case and to §22.34 *et seq.* for an explanation of the fact that the actual radii of the real gear wheels in some real gear set will probably bear no obvious relation to these radii. Substituting $k = 9/41$ into equations (4) and (5) at table 22.01 we get the same values for the two radii as Konstantinov would (§22.26). Substituting our found values for $\psi_3$ and $\psi_2$ into equations (7) and (8) at the same table we find that they check with Steeds (§22.29).

### The theorem of three axes in statics

25. It also follows from what has been said and from the contents of chapter 15 that this theorem may be written down: *whenever three wrenches acting upon a body are in equilibrium, the lines of the three action screws will be arranged in space in such a way that all three pairs of them will share the same line as common perpendicular; the action screws will be moreover generators of the same cylindroid and their respective pitches will be appropriate*

*to that.* Another way to say the same theorem would be as follows: *whenever three wrenches are in equilibrium the three action screws will be members of the same cylindroid of screws.* A more concentrated way of saying the same theorem (and please compare this with what was said at §13.12) would be to say the following: *any three linearly dependent screws will belong to the same 2-system.*

26. It is necessary to comment here, as I did at §12.58, about the possible lack of a subscripted nomenclature for an action screw, say *s*, in the event of the body in question being a link in a mechanism – and this, in a sense, is always the event. We could insist that all mechanical contact between any single pair of links be regarded as a single joint (§1.11). In this way we could speak without ambiguity about the screw say $s_{23}$ or $s_{32}$ as being the action screw at the joint between the bodies 2 and 3. Similarly $\mathbf{F}_{23}$ and $\mathbf{F}_{32}$ could thus distinguish the force from 2 on 3 from the force from 3 on 2; refer for example to §17.23. But refer to §12.67 *et seq*, where the problems of such nomenclature are already discussed in some detail.

### The trapdoor problem and the 2-system

27. Before dealing with the mechanics of three wrenches by way of example, I would like to deal with the question of four. Refer to my remarks about the plough (§10.23, §12.59, §14.24). Refer to figure 10.16 and go from there to my description of a built laboratory machine for measuring by means of four adjustable wrenches in equilibrium the steadily acting wrench from the road upon a driven or braked obliquely rolling vehicular wheel [3]. Refer also to an earlier paper of mine [4] for a description and graphical solution of what I chose to call at the time the *trapdoor problem.* I gave the problem this easily remembered name because the problem seemed to be of some general significance. Figure 13.07 is a panoramic view reconstructed from a pair of orthogonal projections appearing in [4]. It shows superimposed (*a*) the said trapdoor, (*b*) the four pure forces which act upon it, and (*c*) one of the many four-sided skew force polygons which may be drawn; refer to the pair of figures 14.04 and photograph 19, and please read the caption at figure 13.07. Because the problem is statically determinate the polygon, whose geometry is based upon the 3-system and the principles of §10.22, may be constructed without difficulty. The polygon contains full information about all four of the four forces acting upon the trapdoor; its completed construction is in itself a solution to the problem. Everyone knows of course that such problems as these are solvable by the regular algebraic method of choosing an origin, summing to zero the resolved parts of all of the forces and the resolved parts of all of the moments at the origin in each of the three Cartesian directions, and solving the six resulting equations either by hand or by computer (§10.31, §10.32); but that is not the point (§10.33, §10.34). It is my intention here to mention that it may be convenient to relate, in such problems as these, not to the circumstance of four wrenches belonging to the same 3-system (as we have above), but to the simpler circumstance of three wrenches belonging to the same 2-system. The 2-system is, after all, the system at issue in the present chapter.

Figure 13.07 (§13.27) A trapdoor is held open by means of a light sloping cable whose upper end is well secured while its lower end is attached at *C*. At *A* (at the origin of coordinates) there is a cylindric joint without the capacity to resist an axial load. At *B* however there is a revolute. Gravity acts upon the trapdoor along a line that I shall call the *gravity line.* Find the reaction forces at *A* and *B* and the tension in the cable. Draw through *B* the transversal $\alpha$ that cuts the cable and the gravity line; it cuts the *yz*-plane in *P*. The line *AP* is the line of action of the force at *A*. Extend *PA* to *Q* such that a vertical line in the *xy*-plane, *QR*, meets the cable in *R*. Draw parallel to the cable the transversal $\beta$ that cuts *PQ* and the gravity line; $\beta$ cuts the gravity line in *T*. Draw through *B* the transversal $\gamma$ that cuts *QR* and $\beta$; $\gamma$ cuts $\beta$ in *S*. Draw parallel to *QP* the transversal $\delta$ that cuts $\gamma$ and the gravity line; $\delta$ cuts $\gamma$ in *U* and the gravity line in *V*. Complete the polygon *TSUVT* and ascribe the forces: *T–S* is the tension in the cable, *S–U* is the reaction force at *B*, *U–V* is the reaction force at *A*, and *V–T* is the force of gravity. Refer for some tetrahedra that support such skew polygons to photograph 19 and for an explanation of the method to §10.22.

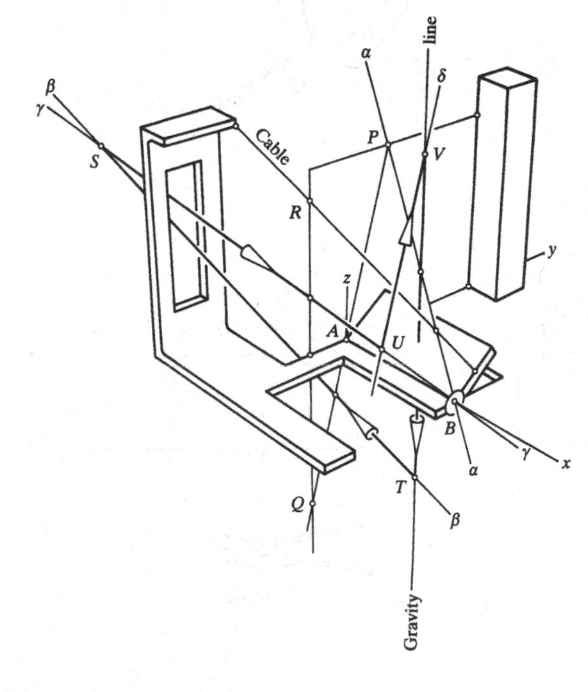

28. I confess however that I have found no easy manoeuvre for reconstructing the trapdoor problem in this particular way. It is true that if we had (or could find) full information about any two of the forces at the outset we could combine these two into their resultant wrench. The problem would then become a candidate for attack by the method of the cylindroid, for the application in other words of the relevant theorem of three axes. The trapdoor problem is a classical representative of its type, it is a simplification of the problems for example of the plough and of the yacht (§ 10.23). In the presence of four forces and the absence of symmetry in these problems, it is difficult for us to relate in a meaningful way to the nodal line of some convenient cylindroid.

**The dynamics of a link**

29. In §12.60 and §12.61 I wrote about dynamics in planar machinery and discussed the works of two authors. From § 12.61 to § 12.66 I branched into the arena of three dimensions and made some remarks about the balancing of the d'Alembert inertia wrench of an accelerating link by the force of gravity and the several other wrenches that must be mobilized from the neighbouring links to act upon the link in question at the relevant joints. There is accordingly no more for me to say at this juncture except that the theorem of three axes in statics has an obvious application now. If the number of wrenches upon a link can be reduced by manipulation to only three (and this is clearly possible in the simple case of a binary link), the axes of the three will reside in parallel planes, the relevant screws will reside upon the generators of a cylindroid, and problems will be soluble in terms of the theorem of three axes. I make this remark against an already expressed view that it will in the near future be possible, and maybe even usual in the processes of machine design, to converse with computers directly in terms of the relevant screws, to speak to and to get the required responses back from them in the same way, that is continuously and in the ordinary pictorial language of descriptive geometry.

**Closure**

30. I would like to close this chapter by saying again that, while all problems in the kinetostatics of rigid bodies fall somewhere into the wide arena of the screw systems, many of them fall directly into the limited arena of the 2-system, into the relatively simple mechanics of the cylindroid. I say again (and moreover) that, when the problem involves only three pure forces in equilibrium or only three pure angular velocities adding to rest, the relevant cylindroid becomes a mere planar pencil of screws of zero pitch. The problem then becomes the simplest of all problems in statics or kinematics; it becomes that of the three coplanar global vectors that are concurrent and the vector closure upon the planar page of a mere triangle.

**Notes and references**

[1] Phillips, J.R. and Hunt, K.H., On the theorem of three axes in the spatial motion of three bodies, *Australian Journal of applied science*, Vol. **15**, p. 267–87. CSIRO, Melbourne 1964. Refer § 13.19.

[2] The reader will find in the following papers not only that the various formulae by Bottema and Roth for the screw triangle may be reconciled with those of Hunt and myself for the cylindroid (§ 19.41, § 19.43), but also that new methods have become available for attacking more directly most of the kinds of problem that are dealt with here. Phillips, J.R. and Zhang, W.X., The screw triangle and the cylindroid, *Proceedings 7th worth congress IFToMM in Seville 1987*, p. 179–82, Pergamon, 1987. Zhang, W.X., Formulae for the connection of mechanics of finite and infinitesimal screw displacements and their use in rigid-body mechanics, *International journal of mechanical engineering education*, Vol. **16**, p. 107–18, 1988, I.Mech.E., London. Refer §13.12.

[3] Phillips, J.R., Brief description of the six-component oblique-rolling-wheel machine recently constructed and tested at the University of Western Australia, *Proceedings of the first international conference on the mechanics of soil vehicle systems at Torino Saint Vincent, June 1961*. Minerva Technica, Politechnico di Torino, 1961. Refer § 13.27.

[4] Phillips, J.R., A graphical method for skew forces and couples, *Australian journal of applied science*, Vol. 6, p. 131–48, CSIRO, Melbourne 1955. Refer §13.27.

14. This is a hand-held laboratory wrench-applicator. Built at the University of Sydney, it is used for the edification of one's kinesthetic sense. It can be used to apply a wrench of known pitch and, upon using the instrument upon some immobile body, one suffers the equal and opposite reaction wrench and, sensing this, transmits it to the floor.

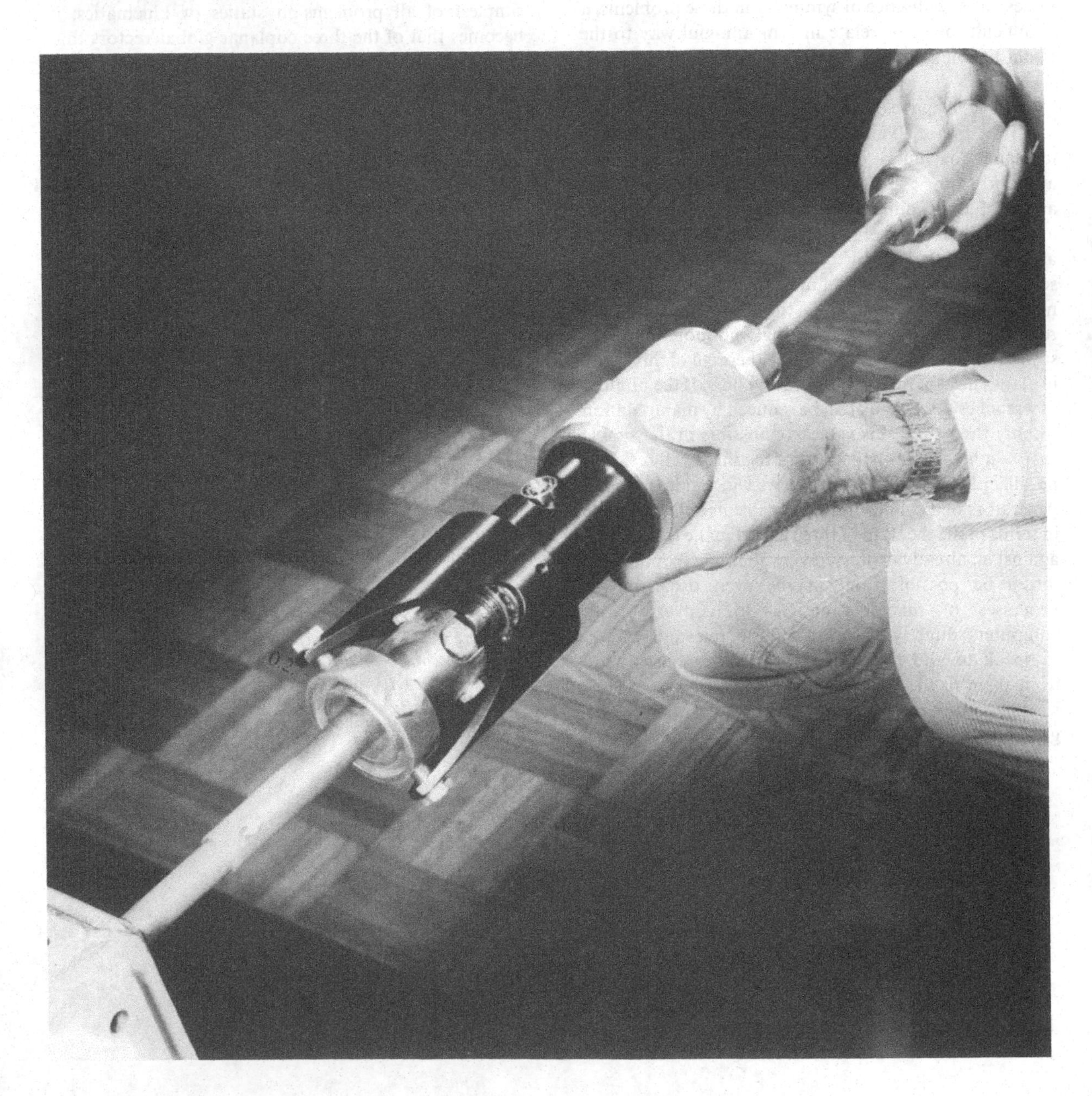

# 14 Some reciprocities across the middle number three

**Slidings and forces at joints**

01. At various places hitherto I have proposed the idea that, while relative motion between the rigid bodies of machinery exhibits itself in the linear velocities occurring within the tangent planes at the points of contact between the bodies, relative action between the bodies exhibits itself in the forces which also act at the points of contact between the bodies but in their cases to cut the tangent planes. While the velocities are called the rubbing or sliding velocities, the forces are called, in the absence of mass, the transmission forces. If the friction is zero (and throughout this chapter I will wish to assume, most often, that it is), the transmission forces will act normal to the tangent planes and thereby perpendicular to the sliding velocities. At both the input link or links and the output link or links of a working mechanism external, or applied forces act upon the links. These, like the transmission forces within the mechanism, must act at points of contact also. Accordingly we begin to see that, except for gravity, which can be avoided in a calculation by declaring the mechanism massless, and electrical forces, no forces can act upon a body unless some other body is there with which the body in question can interact. Bodies interact at joints (§ 1.11), and, as argued at chapter 6 and elsewhere, the number of points of contact at a joint not only varies from zero to six inclusively but also helps to characterize the joint.

**Slidings and forces at screw joints**

02. We know by experience and have found by logical argument in chapter 7 that one of the more singular among the joints – or at least the most often mentioned of them in screw theory – is the screw joint (§ 7.36). I propose therefore to begin a discussion here by adopting a screw joint of some general form, and by asking how the forces might be acting at the five points of contact within the joint when that joint is engaged upon its action. Refer to figure 14.01.

03. In the lower part of that figure I have assumed the absence of friction, and five forces are shown to be acting in five different directions along five imagined contact normals all cutting the surface of the bolt. The five points of contact between the bolt and the nut – the nut is not shown – have been chosen at random. I have done this because in practice the points of contact that occur depend upon the irregularities across the approximately helical surfaces of contact and the natures of the motion and the action (§ 1.49, § 6.03). I mean by the motion the relative motion at the instant between the bolt and its nut (this is a certain rate of twisting upon the known motion screw), and I mean by the action the location of the axis, the magnitude, and the pitch of the wrench that is being transmitted across the contact patch at the joint (§ 7.18, § 7.37).

04. Beginning at § 10.69 there is a discussion about a frictionless $f3$ joint between the two ice-encrusted parts of a mortar and pestle; and that discussion is wholly relevant here. Please read it with care and remain aware that, whereas the joint there has only three points of contact and is not form closed (§ 2.21), the joint here has five points of contact and is form closed. That we have five points of contact here means that the single motion screw comprising the 1-system of motion screws is unique at any one configuration of the joint (§ 6.49); and that the joint is a helical joint means that the single motion screw remains collinear with the axis of the joint and has for its pitch the pitch of the joint (§7.36, §7.87). That the joint is form closed here means that any combination of any of the forces of any magnitude that might be acting at the five contact normals shown – or at any other set of five contact normals which may be chosen at random from among the $\infty^3$ of them – can constitute the action. There is an $\infty^4$ of action screws in the 5-system of action screws available, and the geometry of this is discussed at § 23.12. Please note in reference to the title of the present chapter that, whereas at § 10.69 we were dealing with a 3–3 relationship (there being a 3-system of motion screws and a 3-system of action screws in figure 10.16), we are dealing here with a 1–5 relationship. Refer in advance to figure 14.08, and please read in due course §14.43.

05. It might be instructive in figure 14.01 first to

refer to §5.31 for an explanation of the plucking and replanting of vectors to suit the convenience of an argument, and then to view the force vectors of figure 14.01 plucked from their lines of action and replanted at the centre of a sphere as shown. The sphere at the upper part of figure 14.01 is the sphere of the spherical indicatrix (§19.40). One may contemplate there how it might be possible, by varying the magnitudes of the five pure forces, to vary the magnitude, the direction, the pitch and the location of the resultant wrench of them. One may also see how, by moving the points of contact about in the imagination and by reducing some of the forces to zero if one wishes, one can achieve those particular wrenches that can act exactly along the central axis of the joint; refer in advance to figure 14.02. One should also see in figure 14.01 that each of the five arbitrarily arranged contact normals there has its own unique shortest distance between itself and the central axis of the bolt, and that all of these distances, Γ-Γ′, are different. At each of the points Γ, however, only one of which is shown, we may erect a helix coaxial with the central axis of the bolt and perpendicular to the contact normal there and find that all of these helices are of the same pitch, namely the pitch of the thread. Be careful to note while checking this that the thread at the bolt is a 2-start thread. This matter of the helix at minimum radius is taken up again at §14.17.

Figure 14.01 (§14.02) Shown at (*a*) are five forces at the five points of contact within a frictionless screw joint that is loaded with a wrench; the thread is a 2-start thread; at (*b*) their sperical indicatrix may be seen.

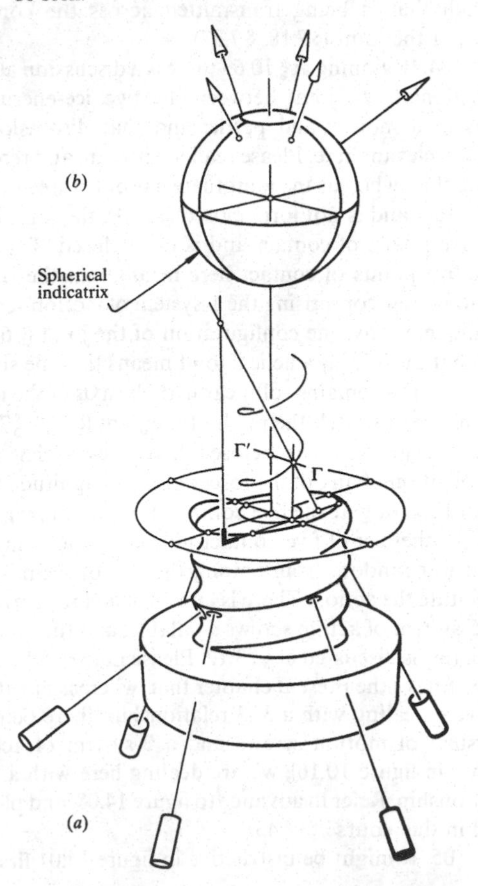


### When the two screws at a screw joint are collinear

06. In figure 14.02 only two pure forces are shown to be acting, at two points of contact only. The reader may care to see in figure 14.02 that the pure forces at the other three points of contact, wherever they may happen to be just now, happen, just now, to be zero. For the sake of convenience the thread is a 2-start, square thread (§14.17). The two forces **F** are equal, and they are diametrically opposite at equal radii $r$ from the central axis of the joint. The helix angle at $r$ is $\theta$. As before, there is no friction; so the forces **F** must be inclined to the central axis of the joint at angles $(\pi/2)$-$\theta$. It follows that the magnitude of the force of the wrench that is acting is $F_R = 2F \sin\theta$ while the magnitude of the collinear

Figure 14.02 (§14.06) This figure is illustrating the argument that shows that the pitch of the wrench sustainable axially by a frictionless threaded rod is minus that of the threaded rod itself.

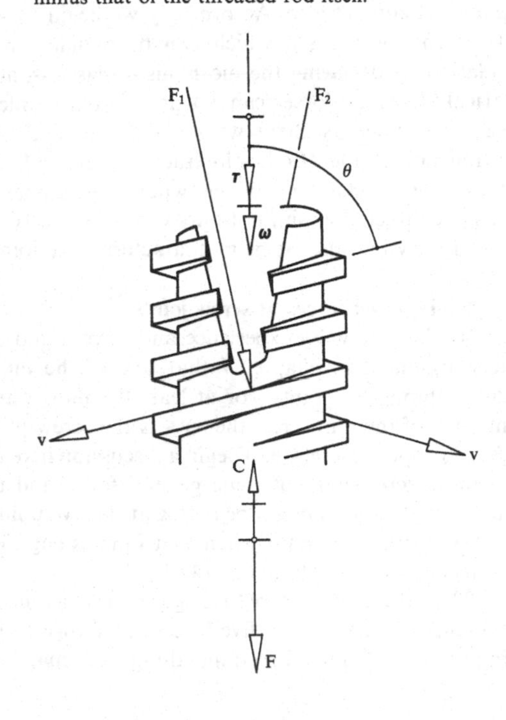

couple of the same wrench is $C_R = 2F\,r\cos\theta$. The central axis of the wrench coincides with the central axis of the joint, and the *pitch of the wrench*, or of the action (taking couple over force), is $C_R/F_R = -r\cot\theta$ (mm). Correspondingly the *pitch of the screw thread*, which is the pitch of the motion should it occur (taking linear over angular velocity), is $\tau/\omega = +r\cot\theta$ (mm rad$^{-1}$). I have been illustrating here now it is that, if an action transmitted by a frictionless screw joint happens to coincide with the axis of the joint, the pitch of the action screw can be none other than equal and opposite to the pitch of the motion screw.

07. The above is a most important concept for ordinary, engineering thinking, and, in the words of that activity, it can be said as follows: *if one grasps with frictionless fingers a threaded rod of pitch say* $+p$ (mm rad$^{-1}$), *and if one then applies a wrench along the rod, the pitch of the applied wrench can be none other than* $-p$ *(mm).*

08. I make a special point of saying this at the present stage for the following reason. Whereas Ball defines a *screw* (and so do I) as a line with a pitch, one must be very careful not to confuse that abstract idea 'screw' with the concrete idea 'threaded rod' (§14.07). Ball says in effect (and so do I) that, *whereas a motion screw is a line with a pitch which is the same as that of the rate of twisting that may occur about it, an action screw is a line with a pitch which is the same as that of the wrench that may act along it.*

**Let a screw joint at rest sustain a force**

09. If only one of the five points of contact within a screw joint transmits a non-zero force, or if any combination of non-zero forces at the points of contact add to a resultant that is a pure force (a force with no attached couple), the line of action of that force will be a member of the linear complex of the action screws of zero pitch. At this juncture the reader might be reminded of §6A.01 *et seq.*, go there to read again about the single infinity of coaxial complexes of screws of the same pitch of the 5-system, and become confused; he should remain aware (*a*) that I was discussing in chapter 6 motion systems, not action systems, (*b*) that the systems there were only quasi-general systems, and thus (*c*) that all screws of a given pitch resided (in chapter 6) upon a complex of that pitch. If (and I come back here) the pitch of the screw joint is $p_b$, the pitch of the complex of screws of zero pitch is $p_b$ also (§23.11). Because the pitch of this complex is not zero, there will be no lines of the complex cutting the central axis of the joint at angles other than a right angle (§9.15, §11.33). It follows that, although a screw joint can sustain a pure force provided it acts along one of the $\infty^3$ of members of the mentioned complex, it can sustain no pure force which cuts its central axis at an angle other than a right angle. These remarks are exemplified in the following paragraphs.

10. Refer to figure 14.03 which shows a fixed bolt 1 mounted horizontally, a frictionless, movable nut 2 upon the bolt, and a coaxially running ring 3 which is mounted by means of a frictionless $f1$ bearing to surround the nut. While the joint between the ring and the nut is an $f1$ revolute, the joint between the nut and the bolt is an $f1$ helical, or a screw joint (§1.25). Let the pitch of this latter be $p_b$ (mm rad$^{-1}$), and let the thread be right handed. Pivoted via a vertical pin to the ring 3 as shown, there is a massless clevis 4 and, acting along the central axis of this clevis in the first instance (and this is not shown), there is a force **F** due to tension in the light, loaded, flexible cable which can be seen. The nut will move along the thread in response to the force **F** until the angle $\phi$ becomes exactly a right angle. If next, however, the cable at the clevis were offset a distance $d$ from the axis of the clevis and thus from the axis of the screw joint (this is shown in figure 14.03), and if $\phi$ were the angle at which a massless and frictionless apparatus came to rest under those circumstances, the following relationship would obtain: $p = d\tan\phi$. This equation

Figure 14.03 (§14.10) Here a pure force **F** is applied to a frictionless massless nut. Note the dimension $d$ and the angle $\phi$, and that the pitch of the thread $p_b = d\tan\phi$.

(1) $p_b = d\tan\phi$

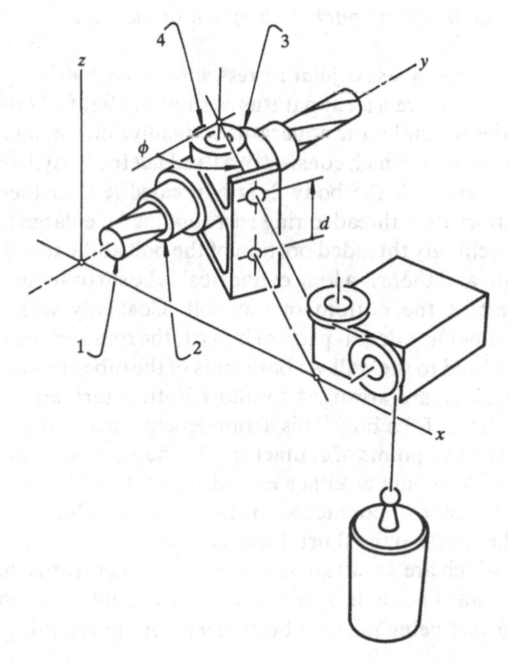

derives from elementary statics; it is listed (1) in the box at figure 14.03.

11. A sign convention for the apparatus, and thus for equation (1), is clearly necessary. What follows is consistent with Hunt (1978), and Hunt's convention is consistent here with Ball (1900). I say also, following both of them: *the pitch $p_b$ of the screw thread and of the motion screw is counted positive when the thread is right handed, and the angle $\phi$ is counted positive when measured from axis to axis right handedly about the distance d.*

12. It will be observed that this equation (1) describes the linear complex of action screws of zero pitch that was mentioned at § 14.09; but refer to § 21.28 and be aware that the angle $\phi$ here is not the helix angle $\theta$ there; the angles $\phi$ and $\theta$ are complementary. The pitch of the complex is $p_b$, and each of its $\infty^3$ of generators is a screw of zero pitch upon which a possible pure force **F** can act in the absence of motion at the frictionless screw joint. Please read § 9.30, then refer to figure 9.07 to gain an impression here of the layout of the linear complex of lines. Refer also to § 11.34 *et seq.*; those passages make some further remarks about the layout of the lines of a linear complex. Also relevant here is § 9.15 which deals, albeit by analogy, with those forces **F** which cut the central axis of the screw joint at no angles other than a right angle. The putting of $d$ to zero in (1) yields, as might be expected, $\phi = 90°$. In summary here I say this: all of the pure forces that, acting alone, can be sustained in the absence of motion at the screw joint – that is, all forces that are transmissible by the joint – will reside as members of a linear complex; *the complex is coaxial with the joint and its pitch is the pitch of the joint.*

**Let a screw joint at rest sustain a wrench**

13. Erect an apparatus as shown in figure 14.04. In order to render the apparatus statically determinate the screw joint, which consists of a fixed *nut* the body 1 and a movable *bolt* the body 2, is bifurcated (§ 2.22): there is a short wide threaded ring surrounding an enlarged and matchingly threaded portion of the bolt at the top of the bolt, and there is a long cylindrical tube surrounding the shaft at the bottom of the bolt. Coaxially with one another as integral parts of body 1, the ring and the tube are fixed to the wall. At both ends of the tube, frictionless bearings are arranged to allow both rotary and axial motion of the bolt. This arrangement ensures that four of the five points of contact within the joint occur at two double points at either end of that tube (§ 17.27). The fifth point of contact accordingly occurs alone, somewhere within the short threaded ring. The clearances (all of which are small) are so arranged in other words that it is always possible by means of a movement of the bolt – contact being made at both places within the tube, and in the absence of external load – to separate the threads of the joint from one another.

14. The rod $R1$ of the bolt is arranged to be vertical. A horizontal rod $R2$ is put through $R1$ and fixed to $R1$ as shown. Fixed in turn through $R2$ there is a third rod $R3$ set at right angles to $R2$ but at an angle of 40° to the vertical. The distance along $R2$ between the centre lines of the rods $R1$ and $R3$ is 100 mm. Set in a vertical plane which is parallel with the wall and at right angles to the rod $R3$, a fourth rod $R4$ is fixed. This rod, located centrally with 30 mm protruding on either side of the rod $R3$, is 60 mm long as measured between the points of attachment of the two horizontal cables which may be

Figure 14.04 (§ 14.13) An apparatus to demonstrate the fact that a given wrench applied to a movable threaded bolt, frictionless within its fixed nut, may cause no motion if the pitch of the thread is suitably chosen.

$d = 100$ mm
$\phi = 140°$ (or 320°)
$p_a = +60.0$ mm rad$^{-1}$
$p_b = -143.9$ mm rad$^{-1}$

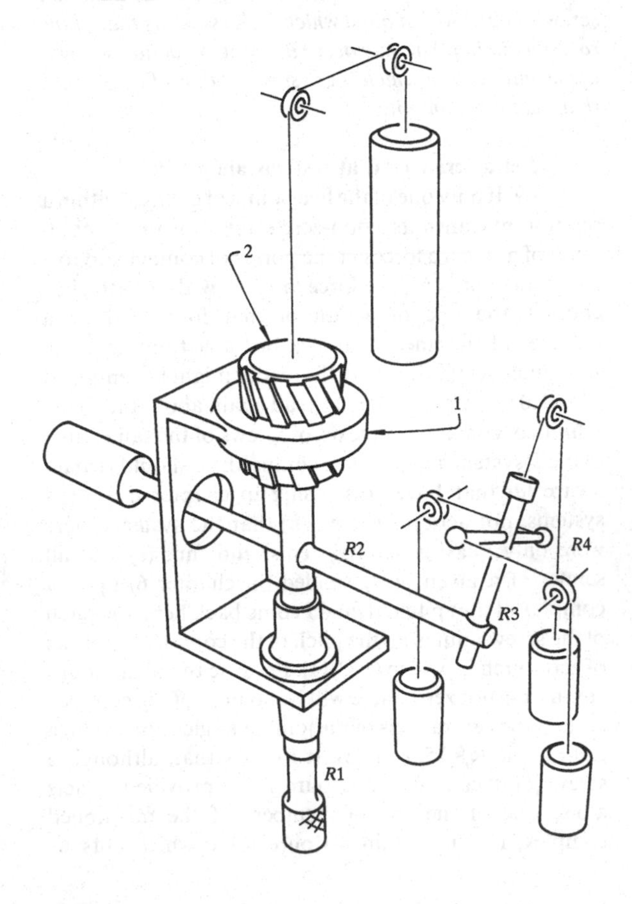

seen. Three pulleys, freely mounted on pins which are fixed in space with respect to the frame nut 1, carry flexible cables as shown and three dead weights are hanging down. The three masses are equal but, except in so far as they must be great enough to blot out the effects of friction in the cables and elsewhere, their actual magnitudes are not important. At the left hand end of $R2$ there is a counterweight arranged to bring the mass-centre of the body 2 onto the centre line of the rod $R1$ and, upwards and over a pulley at the top of the picture, runs a cable which is loaded to balance the total mass of the body 2. This device is simply another way of denying, in the statics here, gravity upon the bolt. Otherwise, of course, I could simply declare the mechanism massless as well as frictionless. Except in the event of an actual physical experiment, the devices used to balance mass and obviate the effects of friction here are unimportant.

15. We have an externally applied wrench upon the movable bolt now, whose central axis is along the centre line of the rod $R3$. It is a right handed wrench of pitch 60 mm, and the question I wish to ask is this: what must the pitch of the frictionless threads at the screw joint be before the joint can sustain this particular wrench without moving? What pitch of the frictionless threads surrounding $R1$, in other words, will permit this joint (in the absence of motion) to *transmit* to the wall this wrench which is acting upon the bolt.

16. One thing is clear: if the bolt is not to move there must be a reaction wrench from the nut to the bolt that is equal, opposite and collinear with the externally applied wrench (§ 10.59), and this reaction wrench must be mobilized from among the five pure forces at the five points of contact within the joint. We note that here of course four of the contact normals are coalesced in two pairs, both pairs being horizontal and intersecting the vertical centre line of the rod $R1$. The remaining contact normal, which is somewhere at the mating threads, is the only one that can have a vertical component.

17. Without loss of generality we can think of the thread as being a square thread; see the thread in figure 14.02. Imagine, in figure 14.04, a single reaction force to be occurring at radius $r$, in a direction normal to the helical surface of the square thread and acting at an arbitrarily chosen, single point of contact somewhere among the threads. That a square thread may be imagined without loss of generality can be seen from figure 14.01. Any one of the forces at the contact normals there will have a unique shortest distance between its line of action and the central axis of the joint, and that shortest distance could be the radius $r$ I have mentioned here. Refer to figure 9.04 and note the triad of right angles occurring exclusively at the special point Γ. Please contemplate also figure 3.04, and see there why it is irrelevant to worry in figure 14.04 where the real prong upon the real platform, namely the real prong upon the *facet* (§7.20), actually is.

18. Another thing that is clear is this: if we can establish the component of the total external force which is acting upwards along the rod $R1$, and the component of the total external couple which is acting parallel to and in conjunction with it, we shall be able to find the pitch of that component wrench, and thus the pitch of the threads that will be necessary to destroy it. I use the word destroy, in this context, after Ball (1900).

19. Figure 14.05 is a view of the relevant vectors looking from $R3$ towards $R1$ along $R2$. Notice that while the axis of the rod $R3$ carries the action screw, say $A$, whose pitch $p_a$ we know to be $+60$ mm, the axis of the rod $R1$ carries the motion screw, say $B$, whose pitch $p_b$ we do not yet know. Let us now shift the action dual (**F**, **C**) from its central axis, its home screw $A$, to a new location such that its basepoint is at the basepoint of the motion dual on $B$ (§ 10.51). Occasioned by that shift is the occasioned couple $\mathbf{F} \times \mathbf{d}$, where $\mathbf{d}$ is the directed distance we know to be 100 mm (§ 10.52). We now see from figure 14.05 that, while the component of the total

Figure 14.05 (§ 14.19) The forces and couples that are worklessly operating at the apparatus in figure 14.04.

$$(2) \quad p_a + p_b = d \tan \phi$$

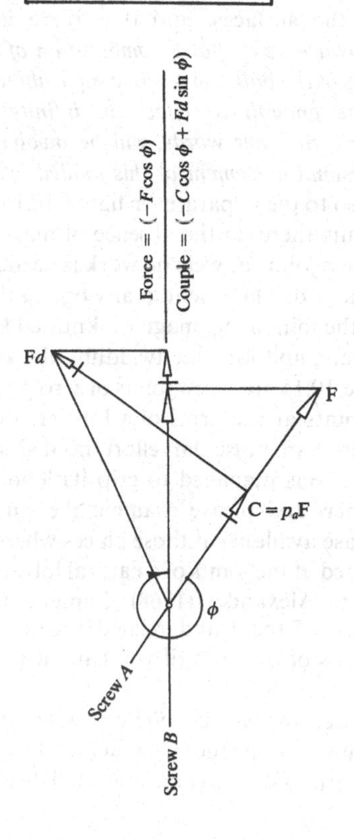

couple of the action dual along screw $B$ is $(-C\cos\phi + Fd\sin\phi)$, the corresponding component of the force is $(-F\cos\phi)$. The ratio of these two, which is the pitch of the component wrench on $B$, must be equal to minus the pitch of the screw thread (§ 14.07). So we can write, paying attention to sign (§ 14.11), that $(-C\cos\phi + Fd\sin\phi)/(-F\cos\phi) = -p_b$. But $\mathbf{C} = p_a\mathbf{F}$; we arrive accordingly at the simple relationship $(p_a + p_b) = d\tan\phi$. This last is significant: it is listed (2) in the box at figure 14.05. We have show that, before the apparatus in figure 14.04 can remain at rest, *the sum of the pitches of the screws at A and B must be equal to the common perpendicular distance between them times the tangent of the angle between them.*

**On the question now of action and motion together**

20. Substitution of the numerical data into equation (2) yields the result that the pitch of the motion screw, and thus of the thread at the screw joint in figure 14.04, needs to be $p_b = -143.9\ \text{mm rad}^{-1}$. See the data listed in the box at figure 14.04, and note that the thread at the screw joint, shown square there, is drawn left handed and accordingly. The reader is invited now to manipulate the apparatus manually: please take hold of the knurled knob. Be aware that all of the five transmission forces inside the joint between the nut 1 and the bolt 2 are normal to the surfaces at the points of contact between the surfaces, and that there is no friction. *Become aware next that a combination of zero force along the axis of the bolt and a zero couple about the same axis will be enough to effect an infinitesimal movement: no work, in other words, will be involved in making an infinitesimal movement at this loaded joint.*

21. Refer also to the apparatus in figure 10.15 and notice that all joints there, in the absence of mass and friction, are workless joints as well: no work is needed to effect a small angular displacement at any one of them: take for example the joint at $B$, imagine a knurled knob to be available there, and imagine twiddling the knob. The joints in figure 10.15 are screw joints of zero pitch of course, like the joints at the arms of a lobster. Please imagine the absence of muscular effort needed by a lobster whenever it has managed to grip its load in a satisfactory manner, and please examine the surprisingly small purchase available at those places where the tendons are attached at the joints of a natural lobster, or of a crab. Refer to Alexander (1968) chapter 2 for a relevant discussion of these and related phenomena, and study the joints of the crab (if you can) at photograph 10.

22. I have used the words *infinitesimal movement* above, not because the phenomena demand it, but because the apparatuses do. As soon as a small displacement occurs anywhere at the apparatuses in figures 14.04 and 10.15, the layout in space of the tensions in the cables alters, and the applied wrench alters accordingly. The equilibria are however stable rather than unstable and, left to their own resources after a small displacement at any one of the knobs, the apparatuses would oscillate. But the point I wish to make is this: if the wrench transmitted through the screw joint in figure 14.04 were somehow maintained constant, a continuous motion could be occurring at the joint while the rate at which any work was being done by the forces there would remain at zero. *The static balance between the mentioned forces and the external forces acting would continue to obtain, independently of the steady motion.*

23. To exemplify the above, one might think about the steady transmission of a constant wrench through a well lubricated gear-box involving a single set of skew gears. The assembled gear-box is simply a complex screw joint between the input and the output shafts of the box whose ISA from shaft to shaft might be the relevant motion screw (§ 22.16). By the same token, and according to how the actual box of the gear box might be mounted in practical reality (and I leave that for the reader to puzzle about), the screw at the central axis of the transmitted wrench – a different screw – is the relevant action screw. In the absence of friction the wrench upon the action screw continuously does no work in conjunction with the motion about the motion screw: no energy is expended at the mating teeth and no heat is generated. Equation (2) applies here because the question is, essentially, a question of statics in the absence of friction. Equation (2) applies independently of the continuous steady motion.

24. Refer now to figure 10.16 and to the few words of explanation at the caption there. In that case the transmitted wrench (the action dual) *is* doing work in conjunction with the motion dual. Energy is continuously being needed to displace the soil in the rut and flex the rubber of the rolling wheel and, while a consideration of the statics of the massless wheel will give the correct result that the wrench from the ground to the rubber of the wheel must equal the wrench from the chassis to the hub of the wheel, *the equation (2) at the joint between the wheel and the ground does not apply.* Looking at figure 10.05, incidentally, and given that the whole tractor-plough assembly is twisting about its screw of infinite pitch (the motion screw) which is everywhere parallel with the direction of the steady travel across the land, what might be the pitch and the location of the total wrench being transmitted from the land to the assembly and vice versa (*a*) in the total absence of friction and working at the towed and driving wheels and at the ploughing tools, and (*b*) in the presence of all of these? Refer to chapter 23.

**Ball's neglect of mechanism**

25. Ball in his book about screws (1900) never speaks about the concept of mechanism except that, at the beginning of his chapter 24 where he begins to deal with his theory of screw-chains, he mentions his idea of *mass-chain*. This in another terminology might well have gone by the name which others used at the time, in both the absence and presence of mass, *mechanism or open-chain mechanism*. Reuleaux (1875) and Kennedy (1886) were among a number of workers writing at the time of Ball; they dealt in mechanism and used the term quite naturally. Ball, however, appears not to have noticed mechanism. He had dealt thus far in his book (and he continued thus to deal in effect) with the dynamics of a single rigid body which was at rest. The wrenches acting upon the body and the constraints upon the body's motion were both seen by him as being as it were external to the body and therefore irrelevant, except of course for the fact that these 'applied' phenomena caused the body's behaviour at the instant. Presumably for this and other reasons connected with the development of his screw theory, Ball chose to speak in his book, not so much about mechanism and the steady occurrence of a varying wrench between a pair of bodies and a varying rate of the relative twisting of the bodies (in the way that I have done for the sake of arguing power consumption for example in chapter 10 and elsewhere), but more about the application of an *impulsive wrench* upon a single body, the constraints upon and thus the freedoms of that body, the body's mass, its various moments of inertia, and the resulting, instantaneously produced, *twist velocity* of the body.

26. Ball dealt, in other words, mostly in the dynamics of a rigid body at rest, considering the 'small oscillations' of the body. By speaking in terms of impulse, mass, moments of inertia, and instantaneously produced velocities both linear and angular, he avoided the concepts of linear and angular acceleration; he had no need for them. Nor did he deal with the practical mechanical matter of how the constraints are actually realized by links and joints in actual machinery, for that for him was also an irrelevancy. His treatment remains unique however. It is relevant (albeit primitively) to the instantaneous dynamics of a moving, accelerating, massive link constrained to traverse a prescribed track in mechanism (§1.20, §19.64); and it does, above all, continue to excite the reader. The fundamental aspect of his work that is relevant here, however, is not his dynamics of the rigid body at rest but his important idea, *the reciprocity of screws*.

**Ball's important idea, the reciprocity of screws**

27. The idea *screw* is a definition written by Ball. A screw by Ball's definition is a line with a pitch. The pitch is quoted by means of a single scalar parameter whose dimensions are L and whch can be measured, appropriately, mm rad$^{-1}$. The pitch is positive or negative by definition according to whether the wrench upon, or the small twist or the rate of twisting about, the screw in question is right handed or left handed respectively. One might expand on this to say – and I paraphrase Ball using my own words here – that Ball pronounces by definition and says the following. He says (*a*) that the central axis of an unbalanced system of forces acting upon a body, along and within which the resultant wrench (the combination of the force **F** and the collinear couple **C**) is acting, might be called a screw and that the pitch $p_a$ of that screw is the pitch of the wrench. He also says (*b*) that the instantaneous screw axis of the said body with respect to a fixed frame, along and within which the instantaneous time-rate of the twist or the 'twist velocity' of the body (the combination of the angular velocity $\boldsymbol{\omega}$ and the collinear linear velocity $\boldsymbol{\tau}$) is acting, might also be called a screw and that the pitch $p_b$ of that screw is the pitch of the time-rate at the instant of the twisting (or of the screwing).

28. Indeed Ball shows (and I paraphrase again) that, if F is the force of the wrench upon a body about a screw *A*, and if $\boldsymbol{\omega}$ is the angular velocity of a rate of twisting of the same body about some other screw *B*, the rate at which the wrench upon the screw *A* is doing work about the screw *B* is $F\omega[(p_a + p_b)\cos\phi - d\sin\phi]$, where $d$ is the shortest distance between the screws, and where $\phi$ is the angle between the screws measured according to the right hand rule. For reasons I try to make clear elsewhere (§19.08), Ball next introduces the devised quantity $(1/2)[(p_a + p_b)\cos\phi - d\sin\phi]$, and he calls this *the virtual coefficient*. He argues among other things that, when this coefficient (which involves nothing more than the geometrical data of the two screws and their relative locations in space) is zero, the mentioned rate of working is zero too; refer Ball §10. Other key references in Ball are §73 and §74 and, for a first impression of Ball's dynamics using the virtual coefficient, §90. The dimensions of the virtual coefficient are those of length L.

29. Ball accordingly argues that, if a wrench of pitch $p_a$ on a body, which is acting along some screw say *A* in the body, can make no contribution to the work that is done by a small twist of the body of pitch $p_b$, which may occur about some other screw say *B* in the body, the two screws – their pitches are by definition $p_a$ and $p_b$ respectively – must be related by the equation $(p_a + p_b)\cos\phi - d\sin\phi = 0$. This equation, listed (3) in the box at figure 14.06, will be seen to be, and not surprisingly, identical with equation (2).

30. Ball draws attention next to the 'symmetry' of the two terms $p_a$ and $p_b$ – appearing, as they do,

exchangeable in equation (3) – and he thus shows moreover that, if the applied wrench and the small twist were held to change places upon the screws, that is if a wrench of pitch $p_b$ were next held to be acting upon the screw $B$ while a rate of twisting of pitch $p_a$ were held to be occurring about the screw $A$, the wrench on screw $B$ would also, and again, be working worklessly in conjunction with the rate of twisting about the screw $A$. It is this latter, this reversibility of the argument, which led Ball to employ the term reciprocal in his context. Refer Ball § 20. He says (and I paraphrase again), *if two screws generally disposed accord with equation (3) – and this is a purely geometrical equation – they shall be said to be reciprocal.*

31. Two pieces of apparatus to demonstrate the idea of reciprocity are shown in figure 14.06. I trust that without explanation they are intelligible and that with appropriate arrangements made for the elimination of friction they can be constructed. Whereas numerically the pitch of the screws at $A$ in figure 14.06 is the same as that of the applied wrench in figure 14.04 (namely 60 mm rad$^{-1}$), and whereas the distance d is also the same (namely 100 mm), the angle $\phi$ is not the same; there is the matter of the handedness of $\phi$; whereas in figure 14.04 the angle $\phi$ was set at $-40°$, in both of the pieces in figure 14.06 it has the value $+40°$. The pitch of the screws at $B$ turn out to be, not $-143.9$ mm rad$^{-1}$ as before, but $+23.9$ mm rad$^{-1}$. Note the tommy bar of shorter length in the upper picture and the screw thread of shorter pitch in the lower picture. The numerical calculations for this result (omitted here) are an easy exercise for the reader. Refer to table 14.01.

### Congruent screws, the corkscrew and screwdriver

32. It is interesting to note in passing here that, while two reciprocal screws will render the virtual coefficient zero, two screws that are *congruent*, that is, collinear with one another and of the same pitch (say $p$), will render the coefficient $p$. Let the virtual coefficient be called $Q$, and note that the rate of working of a wrench of pitch $p$ upon a screw $A$, working in conjunction with a rate of twisting of pitch $p$ about the same screw $A$, is, according to §14.28, $P = 2Q\ F\omega$. If the two screws at a working joint are congruent – and an example here might be seen to be the joint between a correctly aligned 45° corkscrew and its cork, the corkscrew being operated properly (§ 14.34), and please refer in advance to figure 14.07 – the rate at which the work against friction is being done is $2p\ F\omega$. Please check this by thinking in terms separately to translations and rotations: the rate of working is the sum of $F\tau$ (namely $F\ p\omega$) and $C\omega$ (namely $pF\ \omega$); the rate of working is, accordingly, $2p\ F\omega$. Although Ball does not use the term congruent or any term like it, one can find the parallel reading in Ball at his § 33 and thereabouts.

33. In figure 14.07 a frictionless bolt in a fixed nut is loaded as shown with a wrench whose pitch $p$ is the same as the pitch $p$ of the bolt. Left to itself, and due to the unresisted wrench provided by the gravity forces **F** acting upon the three equal suspended masses, the body would begin to twist in a colckwise manner and continue thus to accelerate. If some agency such as the reader's own right hand, however, were grasping the handle shown, and causing an angular velocity $\boldsymbol{\omega}$ to be main-

Figure 14.06 (§ 14.31) Apparatuses to demonstrate the reciprocity of two screws. The apparatuses are frictionless and, apart from the hanging masses, they are massless. All four parts of the rigid rods containing the screws are horizontal while the two connecting pieces (each 100 mm long) are vertical. In both sets, the horizontal rods are inclined at 40° to one another.

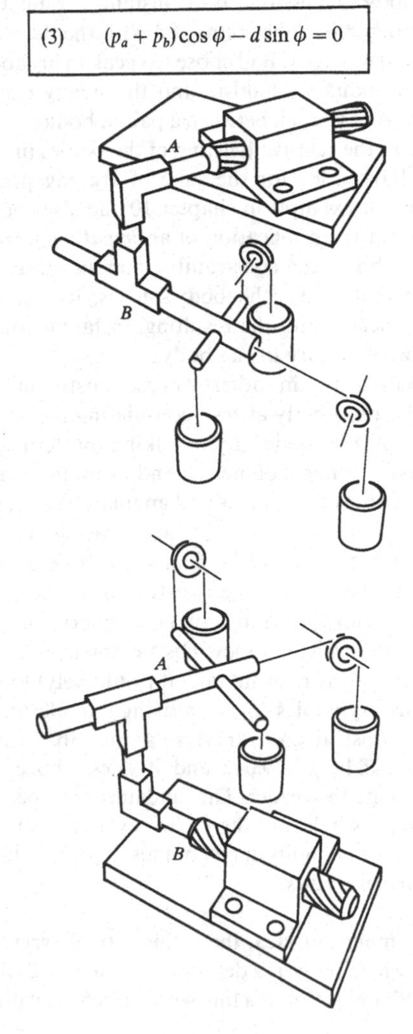

Table 14.01 (§14.31). *This table, which deals with some important special cases in the area of the reciprocity of screws, may be read either (a) as a geometrical exercise in conjunction with figure 14.06, or (b) algebraically with reference to equation (2) in the box at figure 14.05. One selects a box from the left hand column of boxes followed by a box from the top row; one then completes the sentence thus selected by reading the relevant box at intersection. The table – adapted from Hunt (1978), page 333 – is assembled here for the sake of general interest, but its importance as a source of constant reference will be made more evident at chapter 23.*

| | when the reciprocal screw is intersecting at some general angle, that is when $\phi \neq \pi/2$ and $d = 0$, | when the reciprocal screw is intersecting perpendicularly, that is when $\phi = \pi/2$ and $d = 0$, | when the reciprocal screw is parallel or collinear, that is when $\phi = 0$, | when the reciprocal screw is perpendicular but not intersecting, that is when $\phi = \pi/2$ and $d \neq 0$, |
|---|---|---|---|---|
| The pitch of a screw that is reciprocal to a given screw of pitch ZERO | is ZERO | may take ANY VALUE | is ZERO | is INFINITY |
| The pitch of a screw that is reciprocal to a given screw of pitch INFINITY* | is INFINITY | may take ANY VALUE | is INFINITY | may take ANY VALUE |
| | *A screw of infinite pitch having no definable axis is characterized by its direction only. Any single screw of infinite pitch might be seen to occupy accordingly all lines parallel to the said direction. It follows from this and the row of remarks above that the screws reciprocal to a single screw of infinite pitch comprise screws of all pitches in all lines in all planes normal to the direction of the single screw and screws of infinite pitch in all of the lines in space. Another way of saying this is to say that a frictionless prismatic joint has the capacity to transmit in the absence of motion all wrenches in all planes normal to the characteristic direction of the joint and all couples. | | | |
| The pitch of a screw that is reciprocal to a given screw of pitch FINITE NON-ZERO | is MINUS THE PITCH OF THE GIVEN SCREW | may take ANY VALUE | is MINUS THE PITCH OF THE GIVEN SCREW | is INFINITY |

tained in the direction shown, the rate of working at the reader's hand, namely the power supply to the apparatus, would be $P = 2p\ F\omega$. If the two pitches $p$ are measured in metres and metres rad$^{-1}$ respectively, the forces **F** in newtons, and the angular velocity $\boldsymbol{\omega}$ in rad sec$^{-1}$, the power $P$ is $2p\ F\omega$ watts. Note also (*a*) that the dimensions for power are $L^2M^2T^{-3}$, and (*b*) that, if there were friction present among the mating threads of the nut and bolt, we could not have written the result for power as we have written it. By distinguishing between internal and external forces acting within and upon the frictionless apparatus, please study the mechanics of the matter and discover the following: the result $P = 2p\ F\omega$ holds independently of the technique of the working hand; *whether a pure (turning) couple, or a pure (pulling) force, or some or any combination of these is applied at the handle, the rate of working, namely the power input there, is the same.*

34. Naturally in the present context one begins to wonder about such simple hand tools as the corkscrew and the screwdriver. But the mechanics of these is quite complicated. In the fairly unlikely event that the motion screw and the action screw at a working corkscrew are accurately collinear, they will in general be non-congruent. It is clear for example that, with a helical-wire-type corkscrew of a given pitch, the pitch of the resisting friction wrench varies with the helix radius. If the radius is small the wrench is almost a force – note that a hammer is used for a screwnail. If the radius is large the wrench is almost a couple – consider the action of a curved, helical needle, the bagneedle. It is for this reason that I mention the particular helix angle 45° at §14.32, and that I pursue no further here the fascinating mechanics of the corkscrew. With respect to the screwdriver and its slotted woodscrew, I ask the following questions here and leave the matter at that: (*a*) how might one calculate the pitch of the friction wrench acting from the wood upon the woodscrew; (*b*) what is the mechanics at the slot where the joint between the screwdriver and the screw, undamaged, may have the *f*-number 4 or 3, or even 2; (*c*) are there advantages in different shapes of slot, cruciform, staggered etc.; and (*d*) why, within limits, is a longer handle better than a shorter one? Refer to §10.53 where the above mentioned mechanical difficulties are, for the sake of simplicity, not mentioned.

Figure 14.07 (§14.33) A pair of congruent screws at a working joint. The pitch of the wrench is $p$ (m); the pitch of the frictionless thread is also $p$ (m rad$^{-1}$). Note that the thread is a 5-start thread. Independently of how the angular velocity $\boldsymbol{\omega}$ is being maintained, the power $P$ is given by $2p\ F\omega$ (watts).

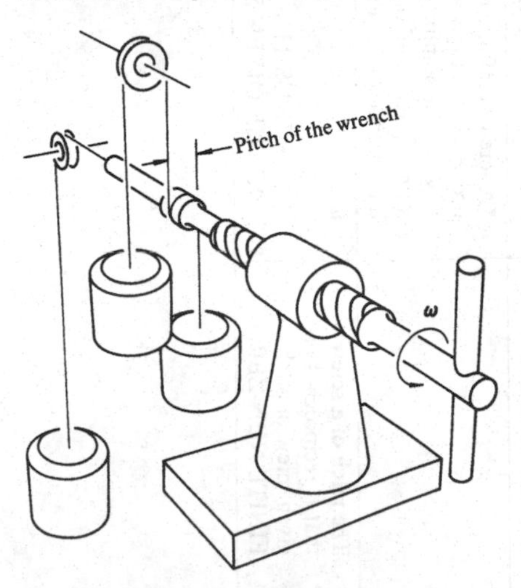


Figure 14.08 (§14.35) A graphic layout of the symmetrical relationships existing within the population of the seven screw systems. Reading the chart from side to side (not from top to bottom) we see for example that for each 2°F there is the corresponding 4°C; this means among other things that for each 2-system of motion screws for example there exists a reciprocal 4-system of action screws and vice versa. At the middle of the chart (at level 3–3) we find that the generally interwoven matters of freedom and constraint are thoroughly intertwined. Refer to §6.31, §10.75, and chapter 23.

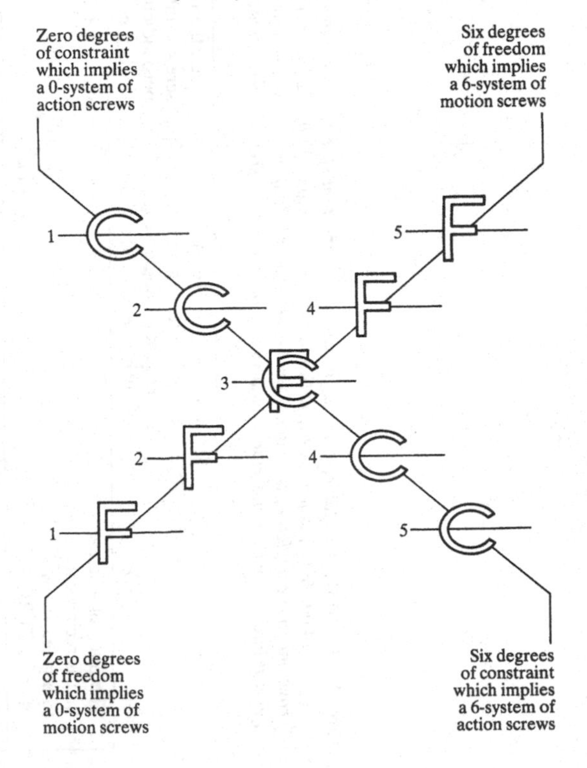

**Let two hinges in series sustain a force or forces**

35. Before beginning to read here please refer to figure 14.08; make a quick perusal of its contents and begin to contemplate the various aspects of its meaning. It comes from equations (2) and (3) which are identical that, if we have a pair of reciprocal screws and if both are of zero pitch, the pair will intersect. This may be exemplified by a single pure force being transmitted by a frictionless hinge in the absence of motion. The force must be arranged in such a way that its line of action is intersecting the axis of the hinge. Unless it is so arranged the force will have a moment about the axis, motion will occur, and that motion will not in practice be workless motion. But there is a question. In the absence of mass and friction, what can the work be done against? Or, alternatively, in the absence of mass and friction, can such a non-intersecting force exist? These are questions about the logic of 'virtual work' and they show that, in the absence of mass and friction, only certain configurations of any loaded, statically determinate mechanism are possible (§10.64, §14.47). Let now a massless body be connected to frame by two frictionless hinges in series as shown in figure 14.09. Let us imagine that the movable apparatus is to remain at rest. Next ask the question: along what lines of action can a force be applied to the body without moving the body? Leaving aside the stability or otherwise of the equilibrium to be established, we can see straight away the following: *any force on the body whose line of action cuts both of the hinge axes is a force that will produce no motion.*

Figure 14.09 (§14.37) A body 2 connected to frame 1 by means of two revolutes in series. Note the four reciprocal *n*-lines, each with its prong and facet (§7.19). The reader will not be surprised to find that this figure is, in all important respects, identical with figure 6.06.

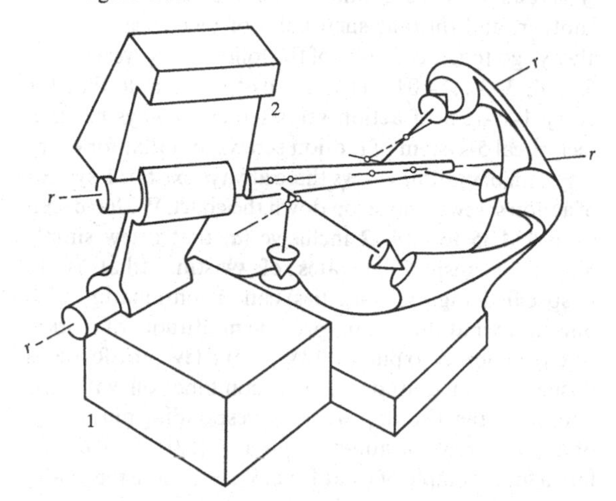


36. If an experiment is needed to show this fact, take a massive body whose centre of mass can be adjusted from place to place. Hang the body from the roof via two light, non-parallel hinges mounted in series and find that, wherever the centre of mass of the body may happen to be, a vertical line through that centre will always cut the two hinge axes. Such an experiment will work, incidentally, with any number of hinges, all such phenomena concerned with frictionless hinges in series being matters of simple statics. Note the lone hinge in figure 1.10(*c*) for example, and note that this hinge and others like it are, in the absence of friction and mass among the intermediate links of such suspended arrangements, *workless joints*. What this means is that the joint would respond with a small angular movement upon the application of a zero couple applied by means of a motor (an electric motor) at its axis. Please read §14.35 again and take a look at figure 6.06, noting the two directrices there.

37. In figure 14.09 the lines of action of four forces have been drawn – no more and no less than four. Two reasons for choosing this particular number of lines of action follow. If any fifth line of action were drawn it would be linearly dependent upon the other four, and any force along it could be reproduced by a combination of suitably chosen magnitudes of the forces along the other four (§10.27). If only three lines were drawn, that number would be less than the maximum number of linearly independent lines that could in the circumstances be drawn (§10.10). The suggestion is made in figure 14.09 that the four proposed forces between the body and frame could be provided at the four imagined frictionless prongs and facets that are drawn. Following the nomenclature of chapter 3, the axes of the forces are called *n*-lines in figure 14.09 and the axes of the hinges are called *r*-lines there. Notice that both of the hinges (and indeed all four of the prongs and facets) may operate as workless joints in the established circumstances.

38. We should see next that, whereas the two hinge axes or *r*-lines can provide a whole 2-system of motion screws when the two pure angular velocities there are taken and added in various proportions with one another (see chapter 15 and elsewhere), the four lines of action of the forces or the *n*-lines can provide a whole 4-system of action screws when the four pure forces there are taken and added in various proportions with one another (see chapter 6 and elsewhere). We begin to see in this way how a whole 2-system of motion screws can be reciprocal to a whole, related, 4-system of action screws. We see, indeed, how any chosen screw of the 2-system will be reciprocal, by equation (3), to any chosen screw of the reciprocal 4-system, and vice versa. *The whole 2-system of screws defined by the two r-lines at*

*the figure 14.09 is reciprocal to the whole 4-system of screws defined by the four n-lines there; and this can be said in reverse as well.* Refer to figure 14.08.

**Let two forces in parallel permit a hinge or hinges**

39. Let a massless body be connected to frame at two direct points of contact as shown for example in figures 6.04 and 15.01. Let the two contact normals there be noted and described as *n*-lines (§3.05). Refer to figure 14.10. Be aware now that, in the absence of other bodies, the frame can apply forces to the body only along the *n*-lines (§14.01). Let us ask the question: where are the axes of pure angular velocities (of the body relative to frame) that can exist at the instant without altering the forces? After consideration we can perceive the following: *any line that cuts both of the n-lines can be the axis of an imaginary hinge that could exist for the purpose of facilitating such an angular velocity.*

40. If an experiment is needed to show this fact, construct an apparatus as shown in figure 6.04. It should have at least one but preferably four hinges in series between the body and the frame as suggested in figure 14.10. Make all of the hinges capable of being immobilized at their reference configurations. On release of the hinges one by one, perceive that the infinitesimal motion of the body permitted by each is such that the tips of the prongs at the points of contact travel in directions parallel with the surfaces of the facets there: the tips will neither come away nor dig in at the facets (§7.21).

Figure 14.10 (§14.39) A body 2 connected to frame 1 by means of two points of contact (prongs and facets) in parallel. Note the four reciprocal *r*-lines, each with its revolute joint. Refer to figure 6.04.

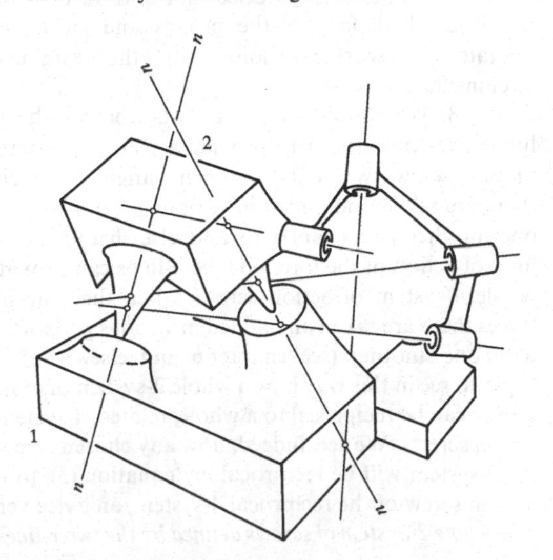


41. Providing now that only infinitesimal movements are permitted to occur, please notice (*a*) that all four hinges may be released and permitted to operate at the same time without loss of contact or digging in at the facets, and (*b*) that, whereas with only three hinges all possible combinations of movement are not possible, with five hinges one of that number would be superfluous for the mentioned purpose. It is for this reason that no more nor less than four hinges have been drawn in figure 14.10.

42. This §14.42 needs to be written now to correspond in a parallel way with §14.38. I leave that as an easy exercise for the reader. A more difficult exercise, however, is to ponder the general significance of the last eight paragraphs taken together and inclusively. Refer to figure 14.08 and go to chapter 23. Refer also to chapter 7 and be aware that the particular phenomena being discussed here are amply exemplified by the more general material discussed there that deals with the *f*2 and the *f*4 joints. On this overall matter of what might be called now the 2–4 (or the 4–2) reciprocity of the screw systems, the reader might care to go to Ball. Ball writes in his §123 and §124 some closely argued remarks; they tightly encapsulate large parts of my chapters 6, 14, 15 and 23.

**Reciprocity of systems and of joints**

43. Refer to figure 14.08. The chart deals with the crux of the present chapter, namely the correspondences and the symmetries among the systems of screws about the 'central' system namely the 3-system (§3.57, §6.31). The chart displays and is dealing with degrees of freedom F and degrees of constraint C, and we read the chart horizontally. Having established that individual screws may be reciprocal with one another in the manner described (§14.29 *et seq.*, figure 14.06, §19.08 *et seq.*), we are showing in the chart (*a*) that pairs of *systems of screws* may be found to be reciprocal with one another, and (*b*) that such pairs of reciprocal systems always go together in one of the following ways: 0–6, 1–5, 2–4, 3–3 (§14.04). This means for example that for every 1-system of action screws there always exists a reciprocal 5-system of motion screws and that for every 5-system of motion screws there always exists a 1-system of action screws, and so on down the chart. We have seen from §14.35 to §14.42 inclusive (*a*) that a few simple hinges of unspecified rates of twisting (that is, of unspecified angular velocities) can in combination with one another at the instant provide multitudes of motion screws of non-zero pitch, and (*b*) that a few pure forces of unspecified magnitudes can in combination with one another at the instant provide corresponding multitudes of action screws of non-zero pitch (§15.01, §15.02). To find a first example of what I speak about here the reader

may go to §6B.08. There the quasi-general 4-system of motion screws is being examined and we discover the reciprocal 2-system of action screws, the reciprocal cylindroid; in each of the systems there is a few, a sufficient number of screws of zero pitch which can be used to define the system; the few screws of zero pitch (four for the 4-system and two for the 2-system) do this by virtue of being able to be considered in combination with one another. Another example of a pair of reciprocal systems may be seen in figure 3.07(*b*). There the 5-system of action screws is defined by the five contact normals (each of which contains an action screw of zero pitch) while the reciprocal 1-system of motion screws consists of the single available ISA; refer to figure 3.05. The fact that this latter is not defined by means of a few (namely only one) screws of zero pitch is a reflection of the fact that the system is not in this case a quasi-general system; refer to chapter 23.

44. Please note that in this last example the reciprocal systems were the action system of one joint and the motion system of another and that the joints, moreover, were both simple joints; refer to §7.77. A third example of reciprocal systems may be seen in figure 20.22 (§20.48). While the relative motion of the links 1 and 4 of the Sarrus linkage shown is characterized by the first special 1-system of motion screws, and while this motion is reproducible by means of a single prismatic joint the action system for which is the reciprocal first special 5-system of action screws, the system defined by any five of the six hinges of the linkage is the same reciprocal first special 5-system; all six hinges of the linkage belong to this system and the system is a system of motion screws. Please go for a final example to the figures 7.14(*a*) and 7.17 (§7.77). There we see that the fourth special 3-system of action screws that characterises the joint in figure 7.14(*a*) is taken to be used as a motion system for the joint in figure 7.17; because they are both *f*3 joints the joints in this example display the same system of motion screws; they have in other words the same instantaneous capacity for relative motion. Like the last pair (the prismatic and the Sarrus) the pair may be said to be a pair of reciprocal joints. It will be clear to the reader that, whereas in some pairs of reciprocal joints there will be full-cycle reciprocity, in others the reciprocity will be only transitory; this matter is well presented, and discussed with the aid of other examples, by Hunt (1978).

**An appendix**

45. Please contemplate figures 3.11 and 3.12, pursue the associated text nearby, and see the general relevance of all that here. If in the apparatus in figure 3.11 the two hinge axes (the *r*-lines) were made to intersect, the joint between link 1 and link 4 would become a non-rectangular version of the Hooke joint like the one in figure 7.18(*c*), a regular rectangular version of which is shown for example in figure 2.01(*h*). My question is this: what would happen if we replaced the two hinges of a Hooke joint with four ball ended rods, all of which were inserted in the plane of the intersecting hinge axes namely the central plane of the cruciform piece? My answer is as follows: the congruence of lines defined by the pair of intersecting directrices (the *r*-lines in this case) would be a special one; there would be not only the $\infty^2$ of lines of the congruence (*n*-lines in this case) to be drawn in the plane of the directrices, but also another $\infty^2$ of them to be drawn to form a star of lines at the point of intersection, thus making the usual $\infty^2$ of them in all (§11.29); unless at least one of the ball ended rods were chosen from among the star of lines, and inserted to cut the point of intersection at some angle other than zero to the plane of the directrices, the substitute apparatus would collapse.

46. The system of motion screws available at a

Figure 14.11 (§14.47) A pair of Hooke joints hanging in equilibrium. The vertical load due to gravity upon the mass $M$ is collinear with one of the two reciprocal action screws of zero pitch. Please think however about the stability of the equilibrium (§18.19).

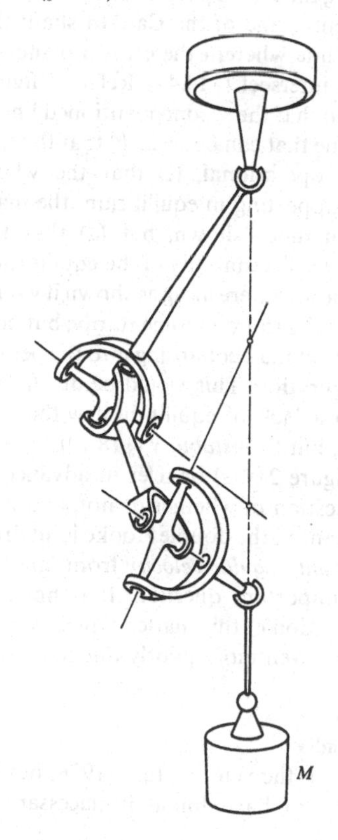

Hooke joint (where the freedom or the $f$-number is 2) is a first special 2-system (§15.50). The reciprocal system of action screws available at the same Hooke joint is, correspondingly, a first special 4-system (§23.72). In each of the two cases, moreover, the screw system is such that the two intersecting, principal screws of the system are not only of equal pitch but also of zero pitch: not only does $p_\beta = p_\alpha$ in a Hooke joint, but also $p_\beta = p_\alpha = 0$. This is the special circumstance that renders the Hooke joint not only highly special but also very useful (§1.47, §2.20).

47. In figures 2.01(*e*), 20.06(*b*) and 20.28 one may see the essential aspects of what might be called the double Hooke joint drive. The single, complex joint made up by the two Hooke joints is sometimes also called (quite loosely) a Cardan shaft or a Cardan coupling. Given this coupling as a complex joint in isolation, and referring to figure 14.11, we can see that, whereas the system of motion screws at the joint is a quite general 4-system, the reciprocal system of action screws is a quite general 2-system. This means of course that the layout in space of the lines along which the joint can sustain (or transmit) a wrench is a cylindroid. The cylindroid has two screws of zero pitch. These are the two action screws of zero pitch along which the joint can sustain a pure force. They are arranged as follows: one of them is the line joining the centre points of the cruciform pieces namely the centre line of the Cardan shaft; the other of them is the line wherein the central planes of the cruciform pieces intersect (§14.44). Refer to figure 14.11 and be aware (*a*) that the second mentioned line is the vertical broken line that can be seen, (*b*) that the two spherical joints are operational, (*c*) that the whole massless linkage is supporting in equilibrium the mass marked $M$ in the manner shown, but (*d*) that the equilibrium is unstable. The stability of the equilibrium is such that if the apparatus were hung as shown it would first remain in that configuration momentarily; but next it would slide away from that, collapsing into the better-known in-line configuration. This would be due (and I say it again) not to a lack of equilibrium within the apparatus as drawn, but to *instability* (§18.19).

48. Refer to figure 2.01(*e*) or refer in advance to figure 20.28. The question of whether or not, and if so under what circumstances, the double Hooke joint drive can transmit a *constant angular velocity* from input to output shaft is an important question. It is however irrelevant here. A mention of this matter, which is often poorly explained and even more poorly understood, is made at §20.56.

**Another appendix**

49. Somewhat in the vein of Hunt (1978) but for different reasons, I too have found it necessary to cogitate the design of a practical mechanical device for the steady application to a body of a wrench upon a given screw (§15.04). Please refer to Hunt §12.1 for his discussion about a *wrench support*. Contrary to Hunt in some respects, yet in parallel with him in others, I have conceived of this device as an operable hand tool, preferably adjustable. I have indeed built such a tool; it is however non-adjustable; refer to photograph 14. The device may be called a *wrench applicator*.

50. As soon as one sees that any nominated wrench can be transmitted through a minimum of five hinges in series – and here I appeal to a reciprocal version of figure 14.01, to the gist of figure 10.15, and please see also figure 3.09(*b*) where the five hinges transmitting a pure force (a wrench of zero pitch) are members of a special linear complex (§9.41, §11.33) – one sees how such a wrench applicator could be made. A wrench applicator could be made, indeed, from any string of five *screw* joints. Refer to figure 14.12. Given five such joints in generally chosen locations in space and connected together in series, there is only one action screw for the single wrench that could be transmitted

Figure 14.12 (§14.50) A wrench applicator across five screw joints which is almost certainly unstable. The joints are merely arranged as members of a 5-system of motion screws whose reciprocal 1-system is the required wrench. If the screw joints were all simply hinges, namely screw joints of zero pitch, the hinge axes would belong to the same linear complex whose pitch would be the pitch of the wrench.

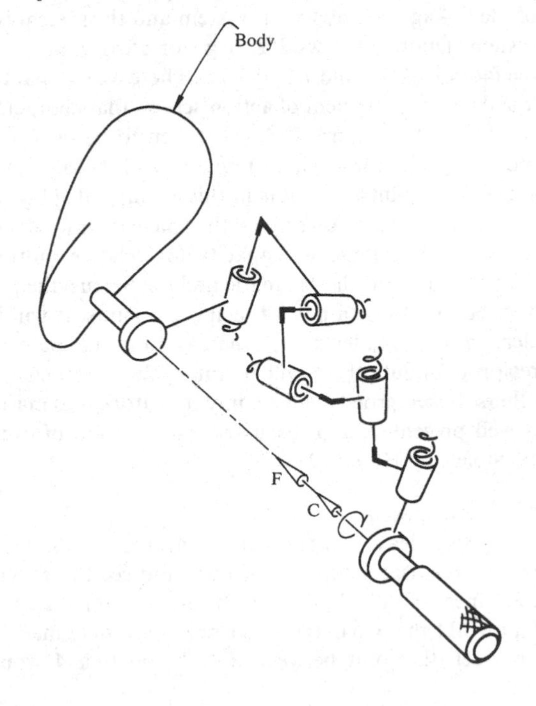

through them. This wrench occurs, of course, upon the single screw of the reciprocal 1-system which lies collinear with the axis of the thus defined 5-system. So we could, if we wish, set out the five screw joints in such a way that they become members of the system of screws reciprocal to that of the required wrench (§23.13), connect them together with links in series, arrange tightly restrictive stop surfaces at the joints which ring alarum bells upon a contact there, then try to apply the required wrench without ringing the bells.

Figure 14.13 (§14.51) Schematic layout of a stable wrench applicator. The mentioned gymbol rings which involve the four mentioned revolutes in series are at *G*1 and *G*2. The bearings indicated at *B*1 and *B*2 are combinations of journal and axial ball-bearings; these, along with the ball-bearing roller in its helical slot, make up the mentioned frictionless screw joint whose axis is collinear with the axis of the instrument. The grips at LH and RH are for the left and the right hands respectively. Refer to photograph 14 to see there a practical mechanical construction of this wrench applicator.

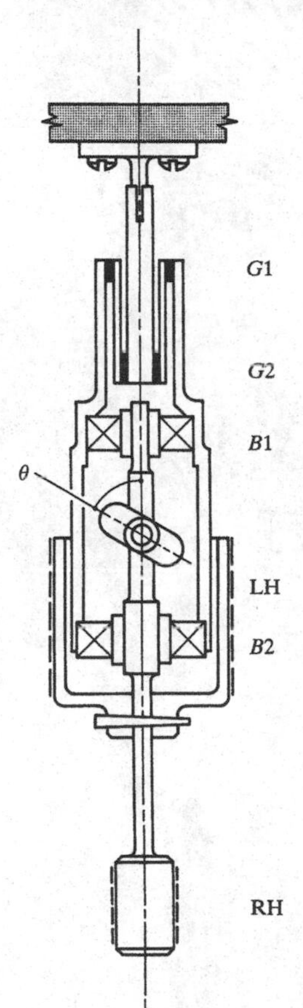


51. Under the circumstances of figure 14.12, however, the rig would be complicated to build and probably unstable (§18.19). It would be wiser and more convenient to design a simpler, stable arrangement. Progressive reduction of the possibilities into a more compact form produces in my mind the wrench applicator shown by the rough sketch in figure 14.13. Understand §14.07, construct a frictionless screw joint whose pitch is *minus* the pitch of the required wrench, attach a knurled knob or knobs concentric, and arrange four hinges to intersect the axis of the tool perpendicularly. Thus five screw joints, four of which are simply hinges for simplicity, define the 5-system of motion screws which is reciprocal to the 1-system of action screws namely the single screw of the required wrench (§23.13). Refer to photograph 14 and see there (with the aid of figure 14.13) how three rigid tubes, shown coaxial at the business end of the tool, are connected by means of two gymbol rings in series. Each of these gymbol rings involves two pin joints (revolutes), and these four revolutes, in series together with the screw joint (whose axis is at the axis of the instrument), make up the reciprocal complex joint. If the helix angle $\theta$ of the shown helical slot (in which a roller runs) were made to be adjustable, the wrench applicator of the photograph would be able to apply wrenches of all pitches. Please note the following: *if $\theta$ were set at 90° the pitch of the wrench would be zero and the wrench a pure force. If $\theta$ were set at 0° the pitch of the wrench would be infinite and the wrench a pure couple.*

### Conclusion

52. In this chapter I have not so much explained the idea of reciprocity but exemplified it, by discussing various unrelated pieces of machinery, by appealing to the muscle and the eye. I think it is not unduly anthropomorphic of me to argue that, whereas our kinesthetic sense relates to statics, our capacity to see relates to kinematics. Indeed it is very likely that unless we possessed those particular senses the sciences of statics and kinematics would not have been developed. Both of the sciences are a product partly of mathematics but mostly of man; and the art of machine design is a relevant activity.

15. This model of a cylindroid is drawn with embroidery silk and stretched within a black-painted steel tube. Made in Sydney, it is naturally thought by its makers to be the best. The cylindroid, at the heart of the screw system, is central to TMM.

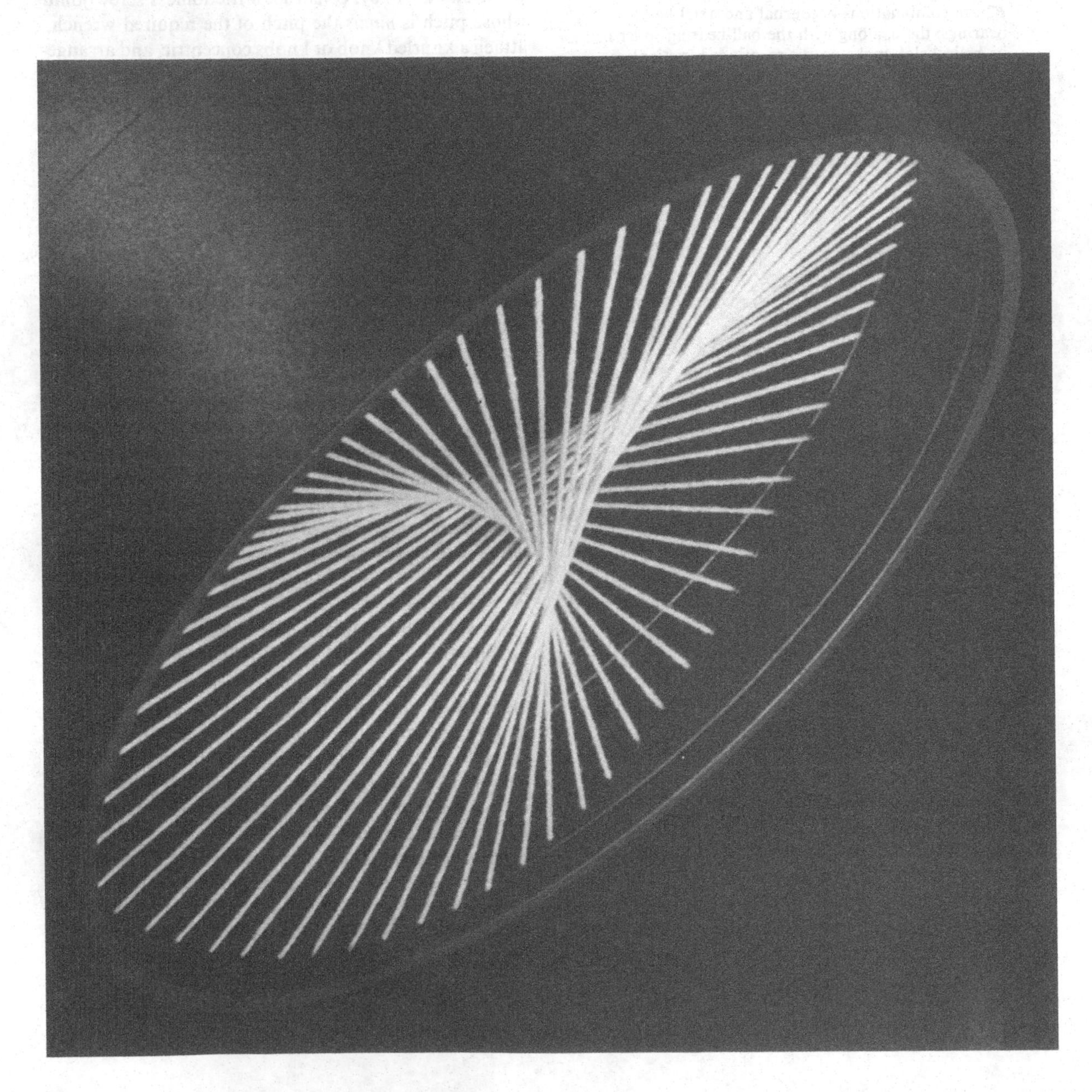

# 15 The generality and the geometry of the cylindroid

### Preliminary argument

01. Please peruse § 8.01 and § 16.01. Be aware too of § 15.49 before beginning here. Look now at figure 15.01 where two rigid, irregular, frictionless bodies are in contact with one another at two separated points. The two $n$-lines shown are the contact normals at the points of contact (§ 3.06), and along these lines pure forces between the bodies may act. By the term *pure force* I mean a wrench of zero pitch – a single, pure force uncluttered by any associated couple. Unless the $f5$ joints at the points of contact are either 'magnetic' (§ 8.07), or somehow form closed (§ 2.21, and please see figure 7.06), the forces between the bodies can only be compressive ones however. There is in any event the circumstance existing here where two forces of unspecified magnitudes might be applied along known lines of action between two bodies, and one begins to wonder about the following: *what are the various possibilities for the combined action of two such forces*?

02. Look next at figure 15.02 where a third body 3 is connected via two hinges mounted in series through a second, intermediate body 2 to a first or a frame body 1. The body 3 may rotate relative to 1, either separately or contemporaneously, about the two centre lines of the two hinges at the instant. The two rates of rotation $\boldsymbol{\omega}$ are quoted at the figure, each with its appropriate numbers subscripted; the numbers refer, first to the frame of reference, and next to the moving body (§ 12.12). Each of these, a rate of twisting of zero pitch, namely a pure angular velocity uncluttered by a linear velocity of axial sliding (§ 5.49), is open to choice. One begins to wonder, accordingly and correspondingly (§ 14.27), about the following: *what are the various possibilities at the instant for the combined effect of two such angular velocities*?

03. These questions, appearing ingenuous perhaps, first about an aspect of statics then suddenly about an aspect of kinematics, would not have been put by me in this particular way if it were not for the contents of chapter 10. At § 10.40 and § 10.41 the reader will see the beginnings of my argument about the analogy in natural philosophy which leads to the notions of *action dual* and *motion dual* in the general area of mechanics (§ 10.49). At § 10.37 the reader will see moreover my introduction of the general idea that it might be fruitful in our investigations to introduce and to consider, not only groups of pure forces (which are simply groups of wrenches of the same zero pitch), but also groups of *wrenches* of the same non-zero pitch. At § 10.57, correspondingly, I spoke about groups of *rates of twisting* of the same non-zero pitch, and there again the idea was found to be a fruitful one.

04. It is for similar reasons that I wish to begin here by introducing pairs of wrenches of the same pitch and pairs of rates of twisting of the same pitch. I wish to imagine just here, in other words, that wrench applicators on the one hand (§ 14.49), and ordinary helical joints which are rates-of-twisting applicators on the other (§ 1.11, § 1.25), come only in identical pairs. The advantage of this approach will become apparent soon.

### Take two fixed screws of equal pitch

05. Accordingly I ask the reader to modify in his imagination figures 15.01 and 15.02. Please replace in figure 15.01 the two points of contact and their contact normals with two correspondingly arranged and equally pitched wrench applicators (§ 14.49), and please replace in figure 15.02 the two ordinary hinges with two correspondingly arranged and equally pitched screw joints. One can thus provide a double basis in the imagination for figure 15.03. In figure 15.03 there is simply a pair of abstract screws; they are of the same pitch $p_\eta$. A screw has no particular direction of action or motion. The screws should be seen in figure 15.03 either as the seats for two wrenches of the same pitch $p_\eta$, or as the seats for two rates of twisting of the same pitch $p_\eta$ (§ 10.05). The screws are displaced apart a common-perpendicular distance $2b$ and they are angularly displaced left handedly along the common perpendicular by an angle $2\eta$. For reasons yet to be made apparent (§ 15.13), I shall call the infinitely long line of the common perpendicular the *nodal line*. Having noted the evident symmetry, I have chosen an orthogonal system

of Cartesian axes centrally and symmetrically located with its origin at $O$ as shown. I have chosen the axis for $x$ in such a way that the acute angle $\eta$, measured from that axis to either of the screws, is measured left handedly. Whatever the pair of screws or their relative layout, that can always be arranged. The axes for $x$, $y$ and $z$ make up a right handed system with the axis for $z$ taken along the nodal line. The somewhat contrived reason for the multipliers 2 will become apparent soon (§ 15.08).

**Speaking first in terms of wrenches**

06. Speaking first in terms of wrenches, and in the same terms as the question put in § 15.01, let us ask the following question: what are the various possibilities for the combined effect of two wrenches $\hat{\mathbf{W}}$ residing upon the two screws of equal pitch in figure 15.03? I wish to remark here that the circumflex above the $W$ is simply indicating that the symbol $\hat{\mathbf{W}}$ represents not a single vector but a dual vector (§ 19.14). Let us furnish the screws 1 and 2 with the forces and couples $\mathbf{F}_1$ and $\mathbf{F}_2$ and $\mathbf{C}_1$ and $\mathbf{C}_2$ respectively so that the wrenches there may be written $(\mathbf{F}_1, \mathbf{C}_1)$ and $(\mathbf{F}_2, \mathbf{C}_2)$ repsectively. Our general circumstance in figure 15.03 is now that the ratio of the magnitudes of the forces $\mathbf{F}$, which is the same as that of the corresponding magnitudes of the couples $\mathbf{C}$ because the pitches are the same, is open to choice. I shall adopt as positive those directions along the screws which enclose the acute angles $\eta$ between themselves and the positive direction $O$–$x$, and I shall write that $F_2 = kF_1$, the magnitude $F_1$ and the ratio $k$ both being free to vary all the way from negative to positive infinity. As already suggested, we are interes-

Figure 15.01 (§ 15.01) Two rigid, irregular, frictionless bodies in contact with one another at two points. Pure forces are transmissible between the bodies at the points of contact only, and the forces can act only along the contact normals there.

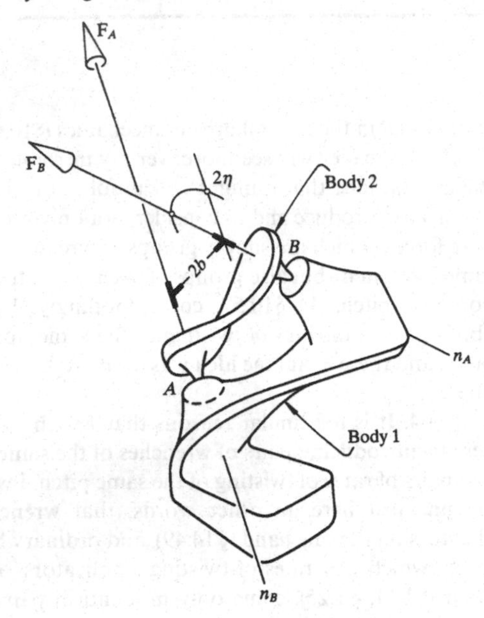


Figure 15.02 (§ 15.02) Two rigid bodies connected together with two hinges in series. Pure angular velocities between the bodies are possible only about the axes of the hinges, which are shown.

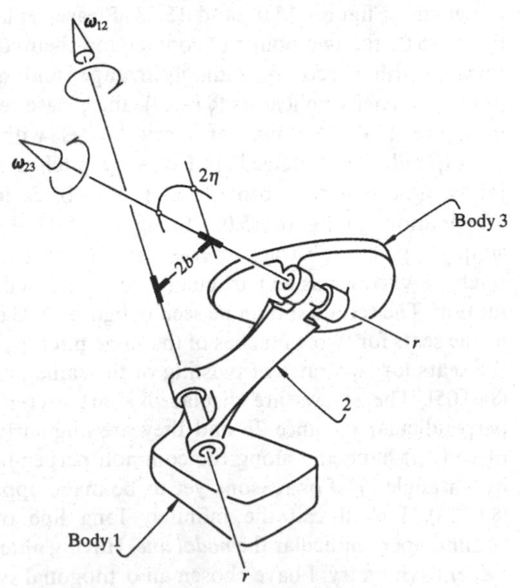


Figure 15.03 (§ 15.05) Two abstract screws of the same pitch $p_\eta$ established and ready to define their unique cylindroid of screws. Its geometric centre will be at the origin of coordinates and its nodal line will be along the axis of $z$ (§ 15.11).

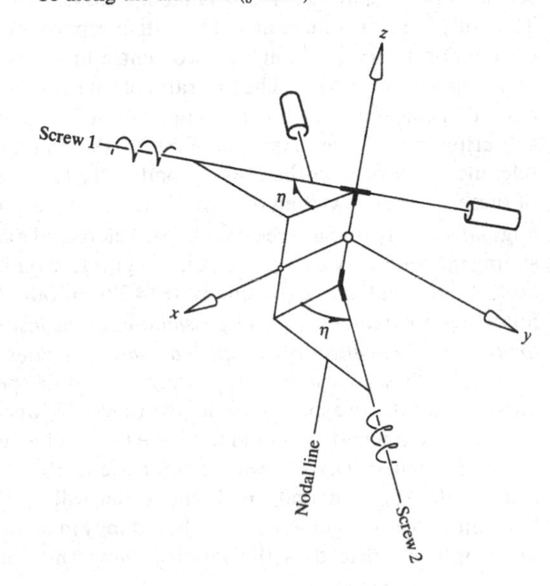

ted to plot the locus of the screws upon which all of the possible resultant wrenches lie (§ 10.14, § 15.01). Please remain aware, remembering chapter 10, that in this and the following paragraphs I could equally well have spoken, not as I have about these pairs of wrenches $\hat{\mathbf{W}}$, but about pairs of rates of twisting $\hat{\mathbf{I}}$.

07. At figure 15.04 two orthogonal views of figure 15.03 are drawn. They are taken against the planes $x$–$O$–$y$ and $y$–$O$–$z$ and they are shown in the usual relationship with one another. The forces and couples of two *equal* wrenches are shown. They are $\hat{\mathbf{W}}_1$ and $\hat{\mathbf{W}}_2$ seated upon the screws 1 and 2 respectively. The ratio $k$ in other words has been chosen to take the special value $k = +1$ in figure 15.04. By vector addition (§ 10.18), we can see that the sum of the equal forces $\mathbf{F}$ is in the direction $O$–$x$ with a magnitude $2F \cos \eta$, positive. By vector addition (§ 10.35), we can see that the sum of the equal couples $\mathbf{C}$ is in the direction $O$–$x$ with magnitude $2p_\eta F \cos \eta$, also positive. By looking however against the plane $y$–$O$–$z$ we can see that, whereas the components of the couple $\mathbf{C}$ in the direction $O$–$y$ cancel one another, the corresponding components of the forces $\mathbf{F}$ produce a couple in the direction $O$–$x$ with magnitude $2b\ F \sin \eta$, positive. Accordingly we can see, with the aid of §10.18 and symmetry, that, if the vector sum $\mathbf{F}_R$ of the forces resides in the axis $O$–$x$ with magnitude $(2F \cos \eta)$ positive, the vector sum $\mathbf{C}_R$ of the couples is parallel, and thus collinear if we wish, with the axis $O$–$x$, having a magnitude $(2p_\eta F \cos \eta + 2b\,F \sin \eta)$ also positive. The ratio obtained by division of these two magnitudes, taking couple over force (§ 10.48), I shall call $p_\alpha$. This is the pitch of the screw of the resultant wrench which acts about the axis $O$–$x$. An easily obtained expression for this $p_\alpha$ appears in the box at figure 15.04. It is listed there as equation (1).

08. Next by choosing $k$ to take the special value $k = -1$ it may be seen, by suitably modifying the layout in figure 15.04 and by a similar argument, that $p_\beta$, the pitch of the resultant screw about the axis $O$–$y$, may be given by equation (2). Knowing that the angle $\eta$ is always acute and measured left handedly from $O$–$x$ (§ 15.05), it may next be seen that, if $p_\eta$ is zero (which is the case as drawn at figures 15.01 and 15.02), $p_\alpha$ (always the maximum possible pitch among the resultant screws) will be right handed positive, while $p_\beta$ (always the minimum possible pitch among the screws) will be left handed negative. In general, however, the pitch $p_\eta$ of the two given screws may take any value, either positive or negative, so in general it may occur that $p_\alpha$ (the maximum) and $p_\beta$ (the minimum) are both negative together or both positive together. The signs of the pitches of the screws will be mentioned again at § 15.15.

09. What is important to see now, however, is this: *the difference between the two extreme values of pitch, $p_\alpha$ and $p_\beta$, which I shall call 2B, is independent of $p_\eta$.* By subtracting (2) from (1) in the box at figure 15.04 we can easily get the simple expression (3) for $2B$, which depends, as can be seen, solely upon $2b$ and $2\eta$. The difference $2B$ between the maximum and minimum pitches of the resultant wrenches depends wholly upon the geometrical layout in space of the two original, equally pitched screws in other words; it depends in no

Figure 15.04 (§15.07) This figure is helping to show that, for $k = +1$ and for $k = -1$ respectively, we get the equations (1) and (2) for the maximum and minimum pitches $p_\alpha$ and $p_\beta$. The difference between these pitches $p_\alpha - p_\beta$, which is the range of pitches available, is given by equation (3); it speaks about a length $2B$.

| | |
|---|---|
| (1) | $p_\alpha = p_\eta + b \tan \eta$ |
| (2) | $p_\beta = p_\eta - b \cot \eta$ |
| (3) | $2B = 2b \operatorname{cosec} 2\eta$ |

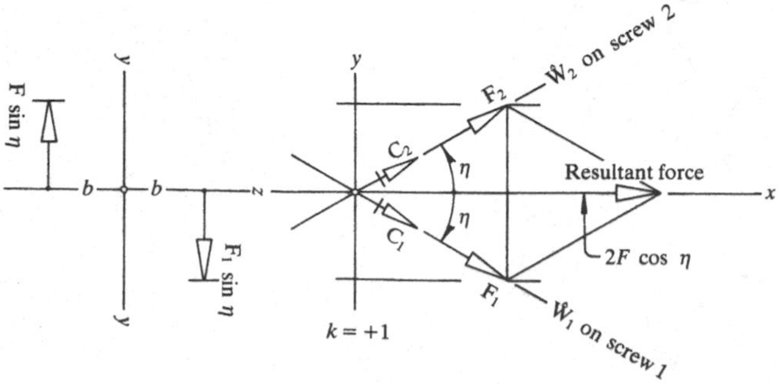

way upon the pitch $p_\eta$ of the screws. The units for $2B$ and thus for $B$ are the same as those for pitch, namely mm rad$^{-1}$, or, more simply, those for length, namely mm alone, and the special significance of this will become apparent soon.

**Towards the cylindroid of screws**

10. There are however resultant wrenches other than those two special ones we found at § 15.08, which two are able to act respectively about the axes $O$–$x$ and $O$–$y$ with pitches $p_\alpha$ and $p_\beta$. To find the others, let us vary now the magnitude of $\mathbf{F}_1$ and the ratio $k$, and let us watch as we do the behaviour of the general resultant wrench $\hat{\mathbf{W}}_\theta$ that occurs as the angle $\theta$ measured as shown in figure 15.05 moves out from the axis $O$–$x$. We have, in figure 15.05, $\hat{\mathbf{W}}_1$ made up of $\mathbf{F}_1$ and $\mathbf{C}_1$ upon the fixed screw of pitch $p_\eta$ at 1, and we have $\hat{\mathbf{W}}_2$ made up of $\mathbf{F}_2$ and $\mathbf{C}_2$ upon the fixed screw of pitch $p_\eta$ at 2. If we add vectorially, as we may (§ 10.52), the forces $\mathbf{F}$ by the method of the parallelogram in the orthogonal view at $x$–$O$–$y$, it will be clear first of all that for every value of $k$ there will be two values of $\theta$. They are (*a*) the value $\theta$ itself, which is shown at figure 15.05 (this $\theta$ I will later choose as the independent variable), and (*b*) the value $\theta + \pi$ which is also shown. This latter occurs, in the present reckoning, when both of the forces $\mathbf{F}$ are negative.

11. It is necessary now to demonstrate that, whatever the value of $k$ and thus of $\theta$, the central axis of the general resultant wrench $\hat{\mathbf{W}}_\theta$ will (*a*) intersect $O$–$z$ at figure 15.05, and (*b*) do so perpendicularly. I appeal first to § 10.18 to demonstrate that, because an equilibrating force $\mathbf{F}_E$ (not shown) will constitute the closing side of some triangular polygon of forces whose plane must be normal to $O$–$z$, the vector sum $\mathbf{F}_R$ of the forces must also act somewhere in some plane normal to $O$–$z$. Next I appeal to § 10.52 and to figure 10.12 to demonstrate that, if the two wrenches $\hat{\mathbf{W}}_1$ and $\hat{\mathbf{W}}_2$, both being action duals, are to have their axes shifted to such a new pair of intersecting locations that the two couples and the occasioned couples add up to a resultant couple collinear with the axis of the resultant force, the central axis of the resultant wrench must cut (perpendicularly) the common perpendicular between the axes of the two original wrenches. This common perpendicular is, precisely, the axis $O$–$z$ (the nodal line) in figure 15.05.

12. It should be clear by now that the central axis of the general resultant wrench $\hat{\mathbf{W}}_\theta$ will not reside in the

Figure 15.05 (§ 15.10) This figure is helping to show among other things that, for every value of $k$ between negative and positive infinity, there are two values of $\theta$ between zero and $2\pi$. Both $z_\theta$ and $p_\theta$ will be found to be sinusoidal functions not of $\theta$ but of $2\theta$.

| | |
|---|---|
| (4) | $F_2 \sin(\eta - \theta) = F_1 \sin(\eta + \theta)$ |
| (5) | $(b + z) F_2 \cos(\eta - \theta) = (b - z) F_1 \cos(\eta + \theta)$ |
| (6) | $F_R = F_2 \cos(\eta - \theta) + F_1 \cos(\eta + \theta)$ |
| (7) | $C_R = p_\eta F_2 \cos(\eta - \theta) + p_\eta F_1 \cos(\eta + \theta)$<br>$+ 2b\, F_2 \sin(\eta - \theta)$ |

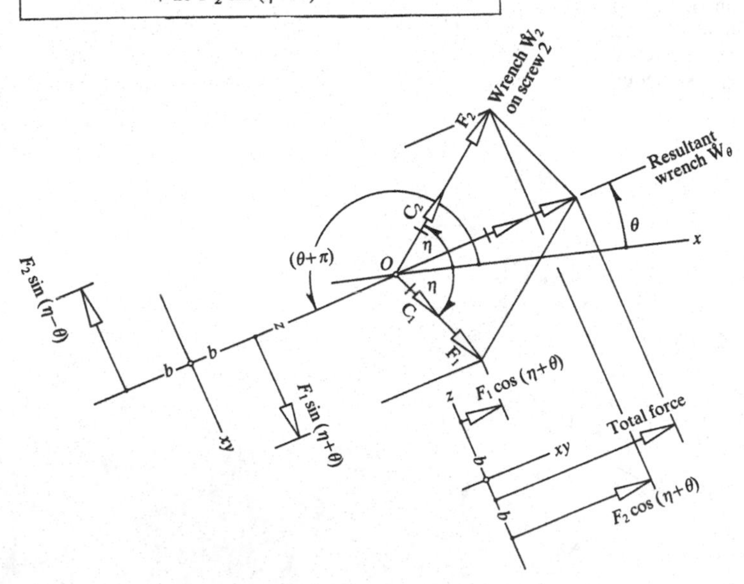

central plane $x$–$O$–$y$ of the figure (this may be taken as being horizontal in the figures), but be above or below it. Refer in advance to figure 15.06. Due to various phenomena better explained at chapter 10, the central axis of $\hat{\mathbf{W}}_\theta$ is displaced upwards from the origin $O$ by a distance $z_\theta$, this height above the horizontal being obtainable in terms of $\theta$ from figure 15.05. Resolving forces perpendicularly to the resultant in the view $x$–$O$–$y$ we get equation (4); by seeking the position of the vector sum $\mathbf{F}_R$ in the view $xy$–$O$–$z$, which is taken across $\mathbf{F}_R$, we get equation (5); by writing an expression for the vector sum of the forces we get equation (6); and by doing the same for the vector sum of the couples, taking account of the couple exposed by the view which is taken along $\mathbf{F}_R$, we get equation (7). Expanding and rearranging (5) and substituting (4), we get equation (8) which appears in the box at figure 15.06. By using (3) from the box at figure 15.04, however, we can rewrite equation (8) to get equation (9). *Equation (9) is important because it shows quite clearly that the displacement $z_\theta$ is a simple sinusoidal function of the angle $2\theta$, and that the range of variation of $z_\theta$, like that of $p_\theta$, is $2B$.* Notice that as $\theta$ (which is measured from $O$–$x$) takes off from zero in the positive direction $z_\theta$ takes off from zero in the negative direction. The negativity of this functional relationship between $\theta$ and $z_\theta$ has been a matter of choice for us; we have (already, you see) set down the inevitable result of our sign conventions here. Having decided that the nodal line shall lie along $O$–$z$ (that $O$–$z$ shall lie along the nodal line), there are only two reasonable ways for us to orient a right handed $O$–$xyz$ triad at the middle of the cylindroid. *We have chosen the way depicted in figure 15.06, not the other way.*

**Plotting the shape of the cylindroid**

13. Recapitulating we have shown here that, whenever two screws of the same pitch are taken and fixed in relation to one another by nominating the distance $2b$ and the angle $2\eta$ between them, we can write $z_\theta = -B\sin 2\theta$, where $B$ is a constant. These facts are summarized in equations (8) and (9) which appear in the box at figure 15.06. Plotting the shape of (9) in the Cartesian space established at figure 15.06, we must note first of all that this equation (9) is telling of a ruled surface which was called by Ball (1875) by the name *cylindroid*. In the Bibliographical Notes at the end of his more recent work (1900) Ball speaks of Plücker [1], Hamilton [2], and Battaglini [3] as having all seen in their respective areas of study various aspects of this surface before he himself independently did about 1870 presumably. As mentioned elsewhere Phillips and Hunt independently saw the same surface in 1962 (§16.32). Our other objective then, however, was to establish what has here become the kinematical part of my chapter 13; and this, as will be seen, deals with a related matter in the general area of mechanism and its analysis.

14. Reviewing now the plotted surface, we note in figure 15.06 and from the equation (9) that, while each infinitely long line of the surface intersects the $z$-axis at right angles and is thus parallel with the plane $x$–$O$–$y$, the completed area of the surface is wholly contained within the parallel planes $z = \pm B$. We see that, as $\theta$ increases from zero, $z_\theta$ decreases from zero (§15.12). Next we see that, when $\theta$ reaches $\pi/4$, $z_\theta$ reaches its minimum value at $z = -B$. As $\theta$ arrives to pass through $\pi/2$, $z_\theta$ returns to pass through zero again; and, as $\theta$ then increases further, $z_\theta$ increases into the positive region of $z$, moving towards its maximum value at $z = +B$. This maximum of $z_\theta$ occurs at $3\pi/4$, exactly $\pi/2$ of $\theta$ after occurrence of the minimum of $z_\theta$ at $\theta = \pi/4$. Figure 15.06 purports to show all this. In figure 15.06, however, I have followed the style of Dimentberg (1965) and the ruled surface of the cylindroid has been truncated to appear only within the annular space confined between the two coaxially drawn, right circular cylinders of arbitrary radii that can be seen. The figure reveals, accordingly, only that twisted ribbon of line segments which illustrates better than all of the whole lines would the shape of the surface we are studying. The *nodal line*, in figure 15.06, is that central line of the surface collinear with the axis for $z$; please refer to figures 6.05, 13.03 and 16.01 where the same surface, the cylindroid, has been drawn completely into its nodal line. The segment of the nodal line within the cylindroid itself – and that segment may itself be called the nodal line – is $2B$ long.

Figure 15.06 (§15.13) The locations that are possible for the central axis of the general resultant wrench $\hat{\mathbf{W}}_\theta$. This cylindroid of screws, defined by the two given screws at 1 and 2 which are of equal pitch $p_\eta$, is truncated in the figure by a pair of concentric, circular cylinders. This, done for the sake of clarity, is after the style of Dimentberg (1965). The angle $\theta$ is measured right handedly about the $z$-axis as shown.

| | |
|---|---|
| (8) | $z_\theta = -b\,\mathrm{cosec}\,2\eta\sin 2\theta$ |
| (9) | $z_\theta = -B\sin 2\theta$ |

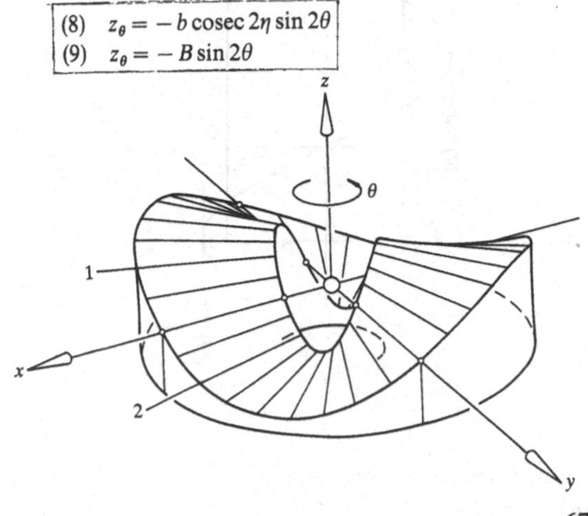

**On the distribution of pitch upon the cylindroid**

15. Having thus seen the shape of the ruled surface upon which all of the possible wrenches lie, and having seen that that shape is determined not by the actual wrenches residing upon the screws but by the *screws themselves* at 1 and 2 (which means that I might with equal effect have been using rates of twisting instead of wrenches since § 15.06), we can take a look now at the distribution of the pitch $p_\theta$ of the screws upon the cylindroid. Dividing (7) by (6), and using (4) to eliminate both $F_1$ and $F_2$ from the algebra resulting, we can get the expression (10) for $p_\theta$. Equations (9) and (10), for $z_\theta$ and $p_\theta$ respectively, are displayed in the box at figure 15.07.

16. Equation (10) shows that the pitch $p_\theta$ like the displacement $z_\theta$ is a simple sinusoidal function of $2\theta$. It shows also that $p_\theta$ like $z_\theta$ is a function which extends across the range of values $2B$ (§ 15.09). The plotted

Figure 15.07 (§ 15.15) Both $z_\theta$ and $p_\theta$ are sinusoidal functions of $2\theta$, and they both have the same range of values $2B$. The pitch $p_\theta$ lags the displacement $z_\theta$ by the phase angle $\pi/4$ however.

$$(9) \quad z_\theta = -B \sin 2\theta$$

$$(10) \quad p_\theta = +B \sin 2\theta + [p_\eta - b \cot 2\eta]$$

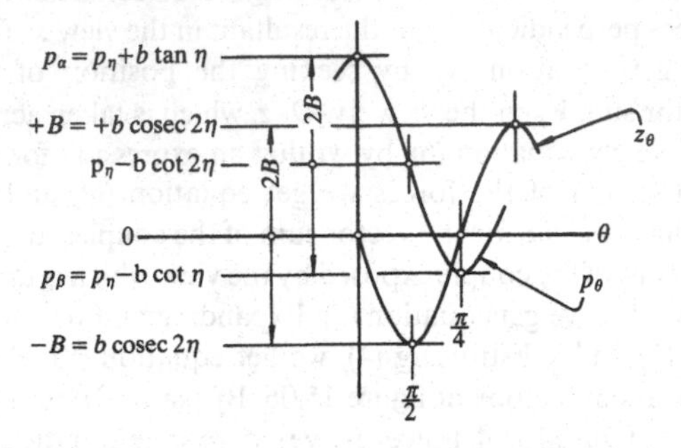


Figure 15.08 (§ 15.17) This figure (introducing the circle diagram) shows more clearly than figure 15.07 did that the range of the pitch $p_\theta$ is the same as the range of the displacement $z_\theta$ namely $2B$, and that $p_\theta$ always lags $z_\theta$ by exactly $\pi/4$. The radius of the circle shown is $B$.

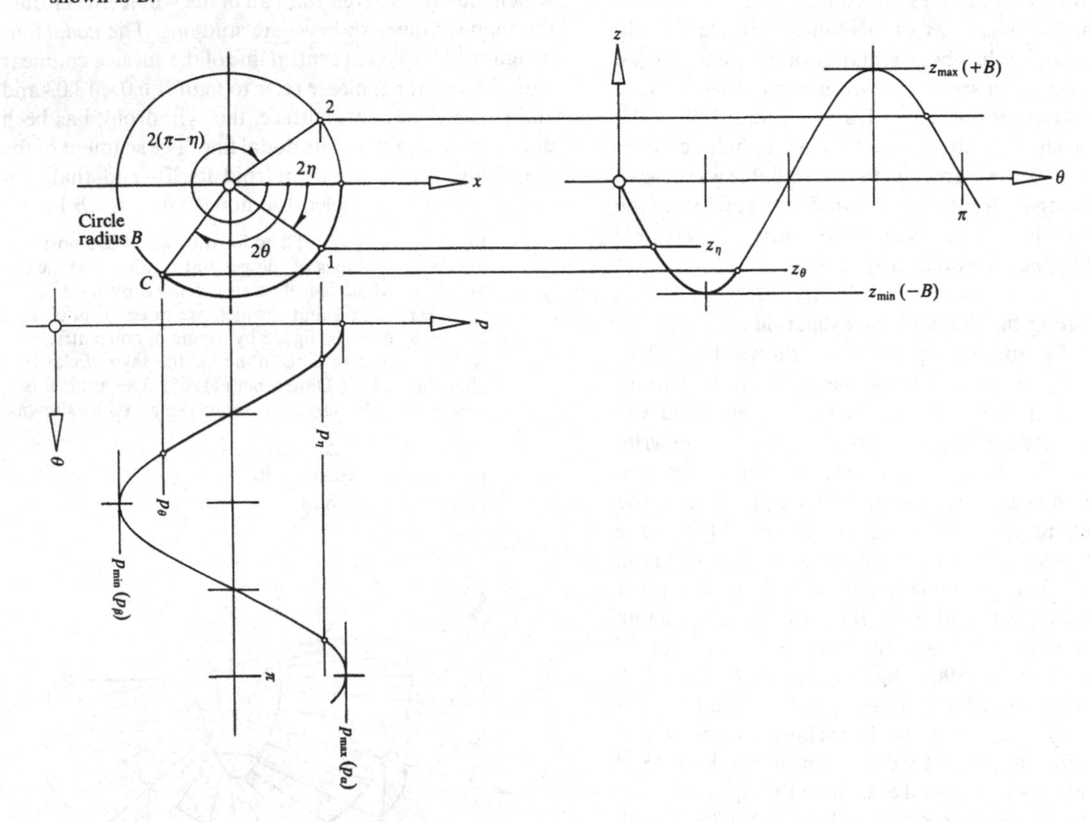

curves for $z_\theta$ and $p_\theta$ in figure 15.07 show these phenomena. The figure also shows that $p_\theta$ is lagging $z_\theta$ by the phase angle $\pi/4$ of $\theta$. This latter is the common property of all cylindroids, independently of the way of their definition (§ 15.49). Please notice also that, by virtue of the conventions adopted at § 15.05, $p_\alpha$ along $O$–$x$ will always be greater than $p_\beta$ along $O$–$y$. More importantly here, however, equation (10) is demonstrating the following: that the particular value $p_\eta$ chosen for the equal pitches of the two defining screws at 1 and 2 is irrelevant so far as the range of values $2B$ is concerned. To say it again it is showing that, whatever the mid-value of the pitches of the screws upon the cylindroid namely $(p_\eta - b \cot 2\eta)$ may be, *the numerical range of the values of the pitches is always the same as the length of the nodal line 2B.*

17. To show more clearly that the range of the pitch $p_\theta$ is the same as the range of the displacement $z_\theta$ namely $2B$, and that $p_\theta$ itself always lags $z_\theta$ by $\pi/4$, we can plot the so-called *circle diagram*, after Ball (1900). Refer to figure 15.08. The circle is of radius $B$ and within it a pivoted radius vector $O$–$C$ is imagined to have already rotated through the variable angle $2\theta$ from the crosswise origin-line $O$–$x$ as shown. Note that the circle diagram of figure 15.08 is a view of the cylindroid in figure 15.06 looked at from the underside; the crosswise origin-line $O$–$x$ of figure 15.08 is, indeed, the same $O$–$x$ as the $O$–$x$ of figure 15.06. By projecting the travelling point $C$ across the page for the variation of $z_\theta$ as shown, and by projecting the same point towards the bottom of the page and similarly for the variation of $p_\theta$, we can get the intelligible and useful chart which is the figure 15.08. Note that, while the origin for $z_\theta$ has been put at the mid-value of $z$ namely zero (§ 15.05), the origin for $p_\theta$ has been chosen arbitrarily in such a way that all of the pitches of the figure happen to be positive (§ 15.16). The two fixed points marked 1 and 2 upon the circle are representing the two fixed positions of the two screws of equal pitch $p_\eta$ in figure 15.03. Whereas the screw 1 was arrived at when $\theta$ was $\eta$, the screw 2 has yet to be arrived at; this will occur when $\theta$ becomes $\pi - \eta$. Refer to figure 15.06 to clarify the double-valued nature of the angle $\theta$ (§ 15.10).

### The cylindroid defined by any pair of screws

18. Next in this general argument it is of critical importance to show that, with no more than two wrench applicators of the same but optional pitch $p_\eta$, it will always be possible to generate with them in combination both of any pair of wrenches generally disposed, say $\hat{\mathbf{W}}_A$ and $\hat{\mathbf{W}}_B$. Let the points $A$ and $B$ be, respectively, at the two feet of the common perpendicular between the screws upon which the required wrenches lie, and be aware that freely to nominate the relative locations in space of the *screws* (*the screws alone*) of those two wrenches will involve the nomination of no more than the following: $(z_B - z_A)$, $(\theta_B - \theta_A)$. The distances $z$ are measured from the centre $O$ of the relevant cylindroid whose position along the line $AB$ has yet to be found; the angles $\theta$ are measured right handedly from the axis $O$–$x$ whose location in space is also yet to be found. We will have of course the pitches $p_A$ and $p_B$ of the two screws; we might note however that the actual numerical values of these are likely to be less important in the calculation than their difference $(p_B - p_A)$, provided however the value of $p_\eta$ can be chosen to lie between them. Firstly, and essentially, we are faced here with the problem of locating the unique cylindroid determined by any two of its own screws.

19. We can go to the circle diagram. Refer to figure 15.09. Knowing the directions for positive $z$ (from bottom to top) and for positive $p$ (from left to right) across the page, we can with the given data locate two representative points $A$ and $B$ on circle yet to be drawn, centre $O$,

Figure 15.09 (§ 15.19) This figure (a circle diagram) shows how any two screws generally disposed, at $A$ and $B$ with pitches $p_A$ and $p_B$ respectively, will define a cylindroid that can be located. The criterion (15) has been employed. Those geometrically inclined will notice that the stippled triangles have turned out to be similar.

(15) All given sign conventions having been obeyed, angle $2\theta_A$ is in the first or the second quadrant according to whether $[(p_B - p_A) + (z_B - z_A) \cot(\theta_B - \theta_A)]$ is negative or positive. See note [4].

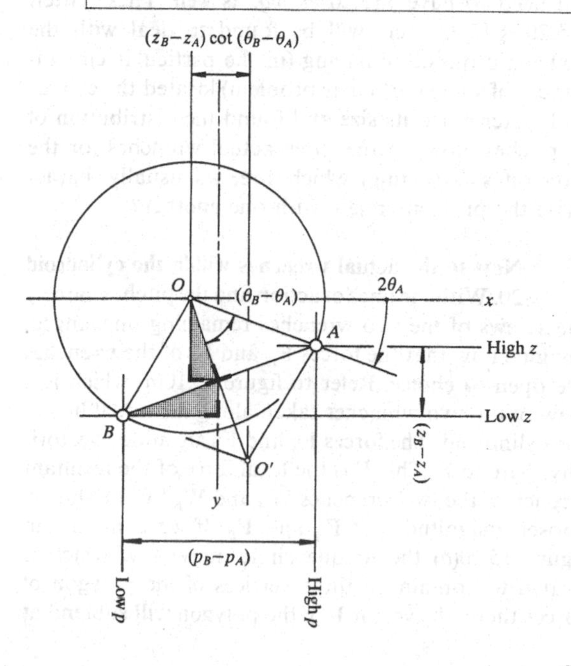

such that their distance apart horizontally is $(p_B - p_A)$ and that their distance apart vertically is $(z_B - z_A)$. We can next erect, upon the line segment $A$–$B$ as base, an isosceles triangle whose angle at the vertex is $2(\theta_B - \theta_A)$. This might seem at first to be straightforward, but a question obtrudes: do we mount the said triangle with its vertex $O$ above or below the line segment $A$–$B$? To distinguish correctly between the two possibilities at this stage of construction is important. The difficulty resides in the particular values that need to be given to certain inverse trigonometrical functions inherent (but not immediately apparent) in the geometry. These functions become apparent in the associated algebra however, and we can devise criteria. Please see the criterion (15) in the box at figure 15.09 and refer to note [4]. The criterion relates to the angle $2\theta_A$. We simply choose which of the above possibilities will cast $2\theta_A$ into the correct quadrant and carry on. Thus we locate at the mentioned vertex the central point $O$ of the circle diagram. The circle itself may now be drawn at centre $O$ to pass through $A$ and $B$ – its radius being, of course, $B$, where $2B$ is the length of the discovered cylindroid. Kindly distinguish here between the length $B$, which is the radius of the circle diagram and half-length of the cylindroid (§15.09), and the point $B$ which is mentioned here along with the point $A$; the two items $B$ are unrelated. *Note now that the cylindroid which has been drawn is unique: no other cylindroid may be drawn to contain the two given screws.* Go to §13.16 *et seq.* for some worked numerical exercises; go to §13.20 *et seq.* for a general numerical exercise using general data; but be aware that, to follow these exercises fully, the reader will need to have §15.20 *et seq*, as well. These latter, §15.20, §15.22 incl., will be found to deal with the next main question: having (in the particular circumstances of some particular problem) located the cylindroid, determined its size and found the distribution of its pitches, how do the three actual wrenches (or the three rates of twisting), which three will usually characterise the problem, relate with one another?

**Now to the actual wrenches within the cylindroid**

20. With the lines of action and the pitches namely the screws of the two wrenches remaining unchanged, imagine now that the forces $\mathbf{F}_A$ and $\mathbf{F}_B$ of the wrenches are open to choice. Refer to figure 15.10($a$) which is a view of the two wrenches taken along the nodal line of the cylindroid. The forces $\mathbf{F}_A$ and $\mathbf{F}_B$ are added vectorially there to $\mathbf{F}$. This $\mathbf{F}$ is the total force of the resultant wrench of the two wrenches $\hat{\mathbf{W}}_A$ and $\hat{\mathbf{W}}_B$ for the shown, chosen magnitudes of $\mathbf{F}_A$ and $\mathbf{F}_B$. If we construct in figure 15.10($a$) the unique circle marked $\alpha$, which is drawn to contain all three vertices of the polygon of forces there, the vector $\mathbf{F}$ of the polygon will subtend at the centre of that circle the angle $2(\theta_B - \theta_A)$. Note the magnitude of the angle $\theta$ that has occurred in figure 15.10($a$) and construct the corresponding angle $2\theta$ in the circle diagram at figure 15.10($b$). This will locate the point $C$ on the circle in figure 15.10($b$) that corresponds to that same $C$ of the rotating radius vector $O$–$C$ in figure 15.08. Observe that the stippled triangles in the figures 15.10 are similar; *$B$–$C$ is to $C$–$A$ in figure 15.10($b$), as $F_A$ is to $F_B$ in figure 15.10($a$).*

21. What this means is that, once we know the ratio of the magnitudes of the forces $\mathbf{F}_A$ and $\mathbf{F}_B$, namely the ratio of the magnitudes of the wrenches $\hat{\mathbf{W}}_A$ and $\hat{\mathbf{W}}_B$ because the pitches of these have remained unchanged, we can always construct in the circle diagram the radius vector $O$–$C$ and thus find not only ($a$) the angle $\theta$ but also ($b$) the pitch $p_\theta$ of the resultant wrench of the two wrenches and ($c$) its displacement $z_\theta$ from the centre of

Figure 15.10 (§15.20) The central axis of the circle diagram at ($b$) is set parallel with the nodal line of the cylindroid at ($a$). The trigonometry of the circle diagram displayed at figures 15.08, 15.09, and here at 15.10($b$) leads directly to the summarizing equations shown. The expression for $2B$ at (11) gives the total length of the cylindroid, while the expression for $2B \sin 2\theta_A$ at (14) gives the value not only of $\theta_A$ but also of $-2z_A$, equation (9), and thus the position of the centre of the cylindroid at the origin.

| | |
|---|---|
| (11) | $2B = [(z_B - z_A)^2 + (p_B - p_A)^2]^{1/2} \operatorname{cosec}(\theta_B - \theta_A)$ |
| (12) | $z_\theta = z_A + B(\sin 2\theta_A - \sin 2\theta)$ |
| (13) | $p_\theta = p_A - B(\cos 2\theta_A - \cos 2\theta)$ |
| (14) | $2B \sin 2\theta_A = (z_B - z_A) - (p_B - p_A)\cot(\theta_B - \theta_A)$ |

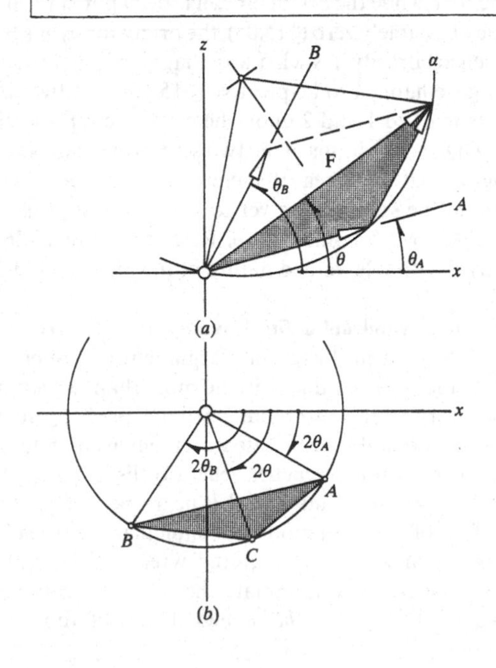

the cylindroid. We should be aware of course that whereas the cylindroid is determined by the screws of the two wrenches the resultant of the wrenches will reside upon some one screw of that cylindroid; refer to § 13.23 for a worked numerical exercise. It means moreover that we can always choose any pitch between $p_A$ and $p_B$ – or, indeed, between $p_\alpha$ and $p_\beta$ but, until we know the cylindroid itself, we do not know these two – and locate the angles $\theta = \eta$ and $\theta = \pi - \eta$, at which angles the resultant wrenches of this pitch lie. Refer to figure 15.08 to see the double values of these two angles drawn in connection with the two points marked 1 and 2 upon the circle there, and refer to figure 15.03 to see again the symmetrical arrangement about the axis $O$–$x$ of the two screws of equal pitch $p_\eta$. Indeed all pairs of screws of the same pitch upon a cylindroid are symmetrically arranged about the central point $O$, being line-reflected in both of the axes $x$–$O$–$x$ and $y$–$O$–$y$. Each screw upon the cylindroid exists as a member of such a pair, each screw of each pair is said to be *conjugate* with the other screw of that pair, and the screws of a pair are said to be *conjugate screws.*

22. Imagine now that the two points 1 and 2 at any suitable pitch $p_\eta$ are marked upon the circle in figure 15.10($b$), and recognize the following italicized fact. The fact itself is a paraphrasing of what was held to be necessary to show at the beginning of §15.18. *The cylindroid that is determined by any two screws of pitches $p_A$ and $p_B$ (generally disposed) is the same cylindroid as the one that is determined by any two screws of equal pitch which might be found upon it, the totality of the screws consisting of a single infinity of such pairs; each pair of screws of the same pitch is said to be a conjugate pair; the screws of the cylindroid range in pitch between the values $p_\alpha$ and $p_\beta$ inclusively.*

### A summary of the cylindroid formulae

23. Taking the above discussions in conjunction with the circle diagram which has appeared thrice and variously annotated in figures 15.08, 15.09 and 15.10 and by writing some simple trigonometry, we can arrive at the list of equations from (11) to (14) and the criterion (15) which appear in the boxes respectively at figures 15.10 and 15.09. Given a pair of screws generally disposed and nominated as outlined in §15.18 (the common perpendicular between the screws namely the nodal line of the cylindroid being put collinear with the $z$-axis), we can then use the listed formulae as follows: ($a$) equation (11) to determine $2B$ the length of the cylindroid; ($b$) criterion (15) to determine the status of $2\theta_A$, thus to determine the correct orientation (given the various sign conventions) of the $x$-axis; ($c$) equation (14) to determine $\theta_A$; ($d$) equation (14) again or equation (9) to determine $-B \sin 2\theta_A$ namely $z_A$ and thus to locate the centre of the cylindroid namely the origin of coordinates and thus the three Cartesian axes; ($e$) equation (12) to describe the shape of the cylindroid using only the cylindrical coordinates $z$ and $\theta$; and ($f$) equation (13) to discover the distribution of the pitches. For all determinate problems relating to known and unknown screws upon a single cylindroid, no data other than these should ever be required. Refer to §13.20 for a worked numerical exercise using general data; refer to the caption at figure 15.09 for some extra information about the criterion (15); and refer to the papers mentioned in note [28] of chapter 19.

### Applications of the cylindroid algebra

24. Whenever we have, in either of the arenas of action or of motion, two screws – and let me call them screw 12 and screw 23 – and we wish to find the screw 31, we can use the algebra of the cylindroid as outlined and summarised above. Specific examples of its useful application are explained at chapter 13 where the two theorems of three axes are outlined, at chapter 12 where certain matters concerning the statics and the dynamics of a massive link are considered, and at chapter 22 where the relative motions of meshing gears are studied. Please refer to these other chapters for further reading.

25. As well as the applications mentioned above we have the whole wide areas of freedom and constraint in mechanism, and of the general screw systems. At chapters 6, 7, 14, and 23 it will be seen that the so-called 2-system of screws is, in fact, simply and always, the cylindroid. An algebraic understanding of the geometrical properties of the cylindroid and of its relationships with other screw systems is therefore crucial for any calculations that may need to be made. I make a final remark here that, with suitable subroutines, it should not be impossible to speak to digital computers directly in terms of screws and thus in terms of the relationships within and the calculations of screw theory. This is discussed in greater detail at chapter 19.

### Cartesian analyses and the elliptical plane sections

26. From equation (9) we have that $z = -B \sin 2\theta$; and from figure 15.06 we can see that, with $z$ held constant, $y = x \tan\theta$. It is easy to eliminate $\theta$ from these two equations and thus to discover the Cartesian relationship $z(x^2 + y^2) + 2Bxy = 0$. This equation, listed (16) in the box at figure 15.11($a$), describes the surface of the cylindroid. Alone it describes the surface only of course, for it makes no mention of the pitches of the screws that are the generators of the cylindroid. Equation (16) might be useful somehow in the numerical analyses of mechanism because it is so simple. In the general arena of the mechanics of screws, however,

which is the arena of practicality for mechanism in my view, and thus the subject matter of this book, the cylindrical coordinates (namely $z$ and $\theta$) of the equations already offered would seem to be the more appropriate. For an equation to the cylindroid in terms of the Plücker line coordinates, please refer to §23.54.

27. Algebraically speaking, equation (16) is of degree *three*, and this means among other things, (*a*) that any straight line that cuts the surface once may cut it either never again or only twice more, (*b*) that any straight line that is tangent to the surface may cut it never again or only once more, and (*c*), in summary, that no straight line will cut the surface more than three times. This matter is taken up again at §15.36, there in connection with a theorem of importance which bears a relevance at §6B.17.

28. Another thing the algebraic degree of equation (16) means is that the curve of intersection between the surface of the cylindroid and any *plane* that cuts it will be a *cubic*. In Ball (1900) §160 the whole circumstance of the cubic of intersection is amply demonstrated by Cartesian methods. There are however special cases, and the following, explained at §15.29, is important. Refer to Ball §23 and to Hunt (1978) §12.2.1 for parallel discussions.

29. Please refer to figure 15.11(*a*). It will be clear that any intersecting plane that cuts a cylindroid along some generally chosen generator will be a tangent plane to the cylindroid at some one point along that generator. Let such a point of tangency – I choose my symbols in the figures 15.11 to follow Ball – be $L$, and let the mid point of the relevant generator (at the nodal line of the cylindroid) be $M$. In these special circumstances the infinitely long line $ML$ will be one part of the curve of intersection between an intersecting plane and the surface of the cylindroid. Imagine now that the plane is rotating about the line $ML$. Whereas the point $M$ will remain fixed in the line, the point $L$ will move either to or fro along it. *It is a fact that, wherever this rotating plane may be, it will be intersecting the cylindroid not only along the already mentioned generator of intersection but also in an ellipse.* The ellipse contains both of the points $M$ and $L$ and, $T$ being the point where the intersecting plane cuts the conjugate generator upon the cylindroid (§15.21), the ellipse of intersection contains $T$ also. *It is a fact that the angle $MLT$ will always be a right angle, that the major axis $F$–$G$ of the ellipse will always be parallel with $L$–$T$, and that the minor axis of the ellipse will always reside in the central, $xy$-plane of the cylindroid.* It is further a fact that, if $D$ and $E$ are the pair of points upon the ellipse where a pair of generators cut not only the ellipse itself but one another at the nodal line perpendicularly, the minor axis of the ellipse is $D$–$E$. The lines joining the pairs of points in which the ellipse is cut by two intersecting screws upon the cylindroid are all parallel with $D$–$E$, while the lines joining the pairs of points in which the ellipse is cut by two screws of equal pitch upon the cylindroid are all parallel with $F$–$G$. Ball's summary of these properties of the cylindroid appears at his §27, and my summary appears implicit in the figures 15.11(*a*) and 15.11(*b*).

30. Figure 15.11(*a*) shows another important aspect of the matter: viewed orthogonally in a direction parallel with the nodal line of the cylindroid, the elliptical plane section discussed at §15.29 will always appear circular. This is shown in the view $A$ in figure 15.11(*a*) where the cylindroid, which is truncated by two concentric cylinders and shaded in the manner of figure 15.06, is being viewed along its nodal line. Note the circle centre $C$, whose diameter $D$–$E$ is the true length of the minor axis of the ellipse. The projection onto the plane containing the generator of intersection and the axis $O$–$z$, namely the view $B$ in figure 15.11(*a*),

Figure 15.11(*a*) (§15.29) Any right circular cylinder containing the nodal line of a cylindroid intersects the surface of the cylindroid in a plane curve, namely an ellipse.

(16) $z(x^2+y^2)+2Bxy=0$

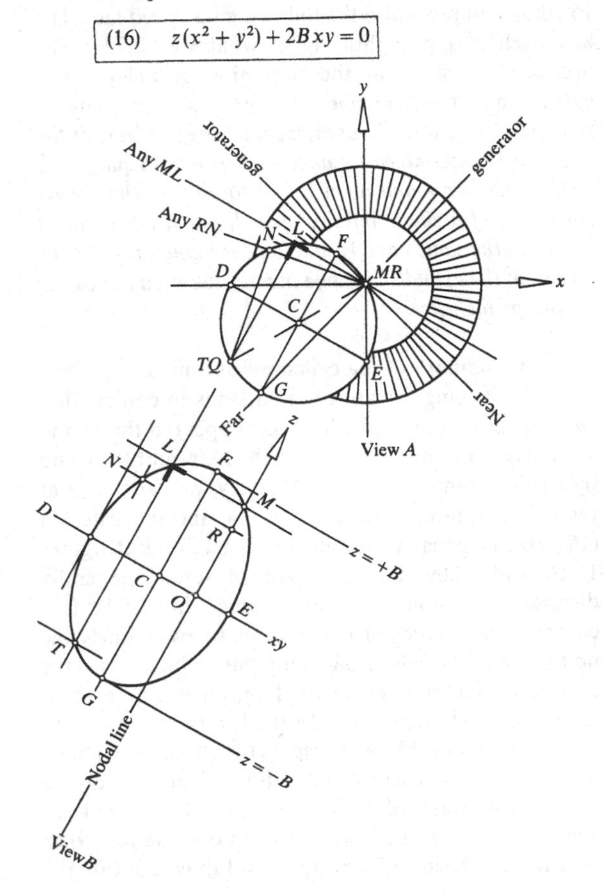

shows moreover that the major axis of the ellipse extends all the way from its one principal vertex $F$ on the extreme generator $z = +B$ to its other $G$ on the other extreme generator $z = -B$. The true length of the major axis $F$–$G$ is nowhere shown in figure 15.11(*a*); it appears foreshortened in both of the views; it is never less than $2B$ and, unless the intersecting plane includes the nodal line as well, in which case the ellipse has zero width, it is always greater than $2B$. These things can alternatively be said as follows: *if any circular cylinder parallel with the nodal line of a cylindroid cuts the cylindroid in such a way that the nodal line is contained at the surface of the cylinder, the section cut will be an ellipse upon the surface of the cylinder, the distance axially along the cylinder from one vertex of the ellipse to the other being 2B the length of the cylindroid.*

31. In the special case where a special generator-intersecting plane cuts, not simply any one of the generators, but an extreme generator (of which there are two, at $z = \pm B$), the special ellipse of intersection deserves a mention. Refer to figure 15.11(*b*). In the special event, $L$ is at $M$, $T$ is upon the other extreme generator, and the ellipse becomes symmetrically arranged with respect to the two branches of the cylindroid.

32. In figure 15.11(*b*) the point $T$ (and thus $G$) has been chosen upon the outer of the arbitrarily chosen truncating cylinders of the cylindroid, and a projection against the plane containing the extreme generator of intersection and the nodal line $z$–$O$–$z$ is also shown in the figure. The intersection of the cylindroid with its own truncating cylinder is as shown; it is, in this orthogonal projection, a *parabola*, with vertex at $G$.

Figure 15.11(*b*) (§ 15.31) When the plane of intersection contains one of the extreme generators, the cylinder is symmetrically located and the ellipse of intersection is symmetrically arranged with respect to the branches of the cylindroid.

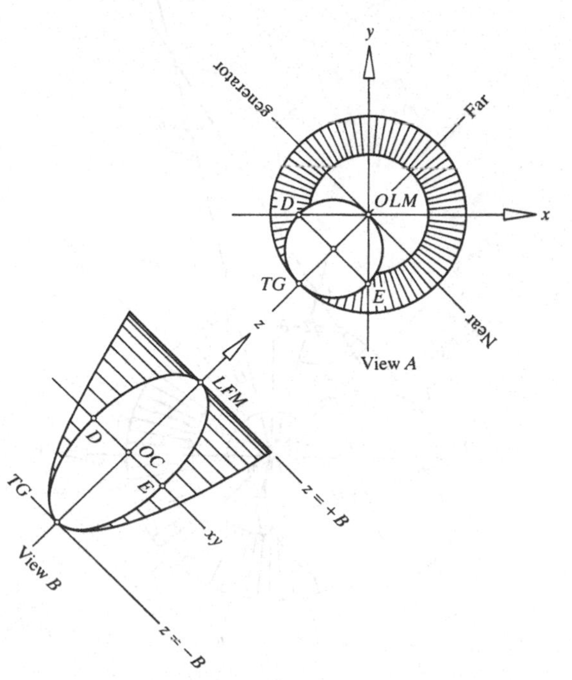


### A first look at the cones of reciprocal screws

33. Please go back to view $A$ in figure 15.11(*a*) and see that, if one looks along the nodal line of a cylindroid being intersected by a circular cylinder in the manner explained at § 15.30, one sees the nodal line as a point $M$ and the centre line of the cylinder as another point $C$. The ellipse of intersection appears as a circle in the view $A$, the centre of the ellipse is at $C$, and the straight line $M$–$C$–$T$ is a diameter of the circle. The right angle at $MLT$, which is not only a right angle in space (§ 15.29) but also a right angle in the view $A$ because all generators are parallel with the plane of the paper in that view, is clearly shown in figure 15.11(*a*); it subtends (as it always will) the diameter $M$–$T$ of the circle centre $C$.

34. Let there be erected, at every point upon the ellipse of intersection, a plane normal to the local generator there of the cylindroid. Were these planes to be drawn in the view $A$ in figure 15.11(*a*), they would all appear as lines; each line, originating at the circumference of the circle and being perpendicular to its local generator there, would intersect all of its companion

Figure 15.12 (§ 15.34) The single infinity of perpendiculars dropped from any point $Q$ in space onto the generators of a cylindroid form a cone. The cone is an ordinary quadric cone. It cuts the cylindroid in a plane curve which is, accordingly, an ellipse.

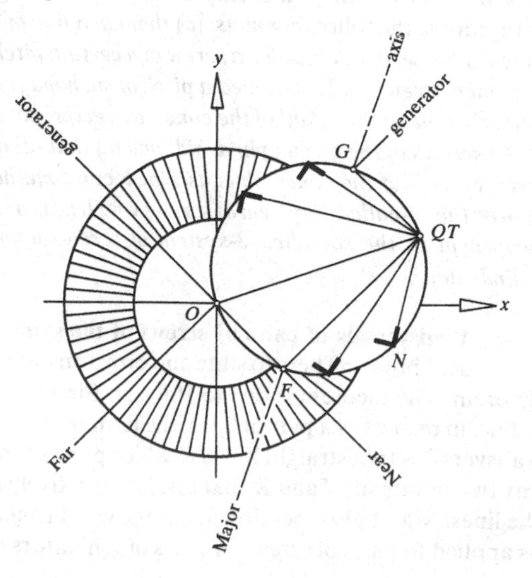

lines at $T$. This means that, through $T$ and parallel with the nodal line, there exists a straight line in which all of the mentioned planes intersect. I wish to nominate now a point $Q$ upon that line, and to drop from $Q$ onto each of the generators of the cylindroid a perpendicular such as the one that is already shown at the view A, namely $Q$–$N$. Please refer to figure 15.12, note the major axis $F$–$G$ of the ellipse of intersection with the cylindroid, and be reminded of the discussions beginning respectively at § 6.28 and § 6B.01. Figure 15.12 shows in effect that, *from any generally chosen point $Q$ in space, perpendiculars may be dropped to the generators of a cylindroid, and that a cone of lines can thus be nominated which intersects the cylindroid in a plane curve, namely an ellipse.* Please notice how it is that, if the point $Q$ is chosen at the surface of the cylindroid, the cone of lines becomes a planar pencil. Please notice also that the ellipse that appears in figure 6B.01 is not the ellipse of which I am speaking here. If you wish to see in chapter 6 the ellipse of which I am speaking here, trace from foot to foot of the perpendiculars dropped from $Q$ in figure 6.04. Please note however that the lettering there is not the same as it is here.

35. These cones – they are *quadric* cones, whose right sections are in general not circular but elliptical – are important for an understanding of screw theory. For a given cylindroid there is one cone for each point $Q$ in space, and thus there are for a given cylindroid and $\infty^3$ of them. Refer to § 6.27 *et seq.* where the nature of the quasi-general 4-system of motion screws in connection with two rigid bodies in contact at two points only is pursued, and to § 21.27, where the matter of these quadric cones of screws and the quadratic complex comprising the 4-system of screws is discussed against a different background. I wish now firstly and simply to present here, and then to exemplify in the following few paragraphs, the following facts: (*a*) *that each line of each cone can be seen as a seat for a screw of a certain pitch;* (*b*) *that each screw can be ascribed a pitch of such magnitude that all of the screws of all of the cones are reciprocal to all of the screws of the given cylindroid;* and (c) *that all of the screws on all of the cones, thus ascribed, can together be seen as the so-called 4-system of screws which is as a whole reciprocal to the so-called 2-system of screws upon the cylindroid.*

**Transversals of pairs of screws of the same pitch**

36. I break off here to state and prove an important theorem. The theorem is due to Plücker. First I wish to define, in respect of a pair of lines, the term *transversal*: a transversal is that straight link $JK$ which passes through any two points say $J$ and $K$ that reside respectively upon the lines. May I also mention again the word *conjugate*; as applied to pairs of screws or pairs of generators of the same pitch upon a cylindroid, it is defined at § 15.49 (§ 15.21). Plücker's theorem may now be stated as follows: *any transversal of a pair of conjugate generators upon a cylindroid will cut some third generator of the cylindroid perpendicularly.* Please recall § 15.27.

37. For a simple proof of this rather striking theorem, refer to figure 15.13 and note the two generators of equal pitch $p_\eta$; they are marked $O$–$J$ and $O$–$K$. They are parallel with the plane of the paper in figure 15.13 and they reside at the locations $z = +b$ and $z = -b$ respectively (§ 15.05). The generally chosen transversal $JK$ has been drawn and this is shown to cut a certain line $RN$ at $N$ perpendicularly; the line $RN$ (and thus the line segment $R$–$N$ within it) is at angle $\theta$ from $O$–$x$ and parallel with the $xy$-plane. I wish to hypothesize now that this line coincides with a generator of the cylindroid. I shall verify this by showing that the location of the segment $R$–$N$ is, in fact, $z = +B\sin 2\theta$ (§ 15.13). Examine the triangle $OJK$ and find by the sine rule that, if $O$–$J$ be unity, $K$–$J = (\sin 2\eta)/\cos(\theta+\eta)$. Correspondingly in the right-angled triangle $OJN$, find that $J$–$N = \sin(\theta-\eta)$. Multiply $2b$ by the ratio $(K$–$N)/(K$–$J)$,

Figure 15.13 (§ 15.37) This is illustrating a theorem of Plücker, that any transversal of a pair of conjugate generators of a cylindroid will cut some other generator of the same cylindroid perpendicularly.

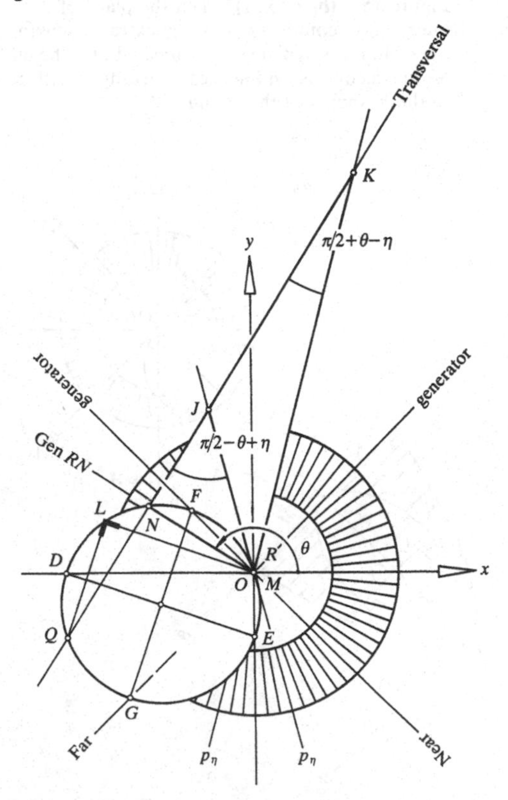

subtract $b$, and find that, for the segment $R-N$, $z = +B\sin 2\theta$. Thus or similarly the theorem can be proved by ordinary trigonometrical methods. Position a point $Q$, as shown, somewhere along the transversal $JN$ produced; we shall use this $Q$ later, at §15.44.

38. The converse of this important theorem cannot be stated as a mere reversal however. It must be stated as follows: *if a line is drawn to cut a generator of a cylindroid perpendicularly, and if it cuts another generator say G, it will also cut the one and only conjugate generator of the generator G.*

39. The subtlety here is that, if a line is drawn to cut a generator of a cylindroid perpendicularly (or, indeed, to cut the generator at any angle), it may not cut the surface of the cylindroid ever again (§15.27); it is a matter of whether or not the other two roots of the relevant cubic equation are real or imaginary. It should also be mentioned that, if the line drawn perpendicular to the generator is drawn in such a way that it cuts either of the two generators located in the $xy$-plane, namely the principal generators whose pitches are the maximum $p_\alpha$ and the minimum $p_\beta$ respectively, *the drawn line will be a tangent to the surface of the cylindroid there.* In either of these special cases the pair of conjugate generators is a collinear pair; and there exists a double point of intersection at the point of tangency (§15.27).

**Screws perpendicular to screws of the cylindroid**

40. Now it is a fact that any one screw of a cylindroid is linearly (or projectively) dependent upon any two screws of the same cylindroid (§10.14, §10.15, §15.22, and elsewhere). It is in particular a fact that any one screw of a cylindroid is linearly dependent upon any pair of conjugate screws of the same cylindroid (§15.05). Please read at this juncture the caption to figure 15.12. If two screws intersect one another perpendicularly they are reciprocal whatever their pitches are; see this in table 14.01 at §14.31. Any straight line that cuts a cylindroid once will, if it cuts it again, cut it twice more (§15.27). Refer to §15.38 and be reminded that, if a perpendicular drawn to some generator of a cylindroid cuts another generator whose pitch is $p_\eta$, it will also cut the conjugate generator whose pitch is also $p_\eta$. Go again to table 14.01 and find that, if a screw upon the said perpendicular is of pitch $-p_\eta$, it will be reciprocal not only to the screw of the generator across which the perpendicular was drawn but to the screws of the cut pair of conjugate generators as well. It will accordingly be reciprocal to all screws of the cylindroid.

41. This argument is not new. It came via the converse of Plücker's theorem (§15.38) from Plücker to Ball. The following §15.42, written to mimic Ball, is not only a paraphrasing of Ball's proof of Plücker's theorem, but an introduction to the actual distributions of the pitches of the reciprocal screws upon the cones of reciprocal screws. Each cone has for its vertex a different point $Q$ in space, and each generator of each cone carries a single screw of a certain pitch. I have already mentioned these cones preliminarily (§15.33).

42. Let the screw on $Q-N$ be reciprocal to the screw on $R-N$. Because the angle $QNR$ is a right angle, any newly ascribed screw on $Q-N$ may be of any pitch. Let the screw on $Q-N$ also be reciprocal however to the first-cut screw whose pitch is $+p_\eta$; its pitch must be, by equation (3) in the box at figure 14.06, $-p_\eta$. It follows next that, because the as yet uncut screw whose pitch is also $+p_\eta$ is linearly dependent upon the two screws of the cylindroid just mentioned, it must also be reciprocal to the new screw on $Q-N$. The two screws, accordingly, the one of pitch $-p_\eta$ on $Q-N$, and the as yet uncut screw of pitch $+p_\eta$ upon the cylindroid, must intersect. Refer to Ball §22.

**A second look at the cones of reciprocal screws**

43. Referring again to figures 15.12 and 15.13 one can now see that, keeping a fixed point $Q$, keeping the essential feature namely the transversal $KJN$ of figure 15.13, and allowing $N$ to migrate around the ellipse of intersection back and forth from its shown position and from screw to screw upon the cylindroid, the reciprocal screw on the moving generator $QN$ of the cone is free to vary continuously over all of the finite values of pitch between the limits $-p_\alpha$ and $-p_\beta$. These reciprocal screws are the type-$A$ screws of chapter 6, referred to first at §6B.07. They all cut the cylindroid not only once but twice more – not only at $N$, but at $J$ and $K$ as well. There are however two limits within which their continuous distribution across the surface of the cone is confined. The limits occur at those two positions of $N$ which cause the line $QN$ to be tangential to the surface of the cylindroid, the two points of tangency occurring at the two principal generators of the cylindroid which lie along the axes for $x$ and $y$ respectively (§15.39). At those limits, $J$ and $K$ coincide and, beyond the limits, $J$ and $K$ disappear. For points $N$ beyond the mentioned limits, any line $QN$ cuts the surface of the cylindroid only once, at $N$, and the reciprocal screws on the lines $QN$ are the type-$B$ screws of §6B.07. Refer to figure 15.13.

44. Recall that the general point $Q$ in the line $JKN$ produced was chosen at §15.37. This may be seen as the same point $Q$ that I previously chose and used in figure 15.12. Please see $Q$ in other words as the vertex of that cone of screws each screw of which cuts a generator of the cylindroid perpendicularly (§15.34). Where are the limits upon this cone within which the type-$A$ screws are confined? Refer to the figures 15.14($a$) ($b$). Take $Q$ in these figures as any point in space whose coordinates $x$ and $y$ are fixed but whose coordinate $z$ is (in the first

instance) both positive and greater than $B$; take $Q$ above the plane of the paper and above the shown near generator in other words; I have chosen to put $Q$ at $z = +3B$. In the figures 15.14 I have dropped the perpendiculars $Q$–$D$ and $Q$–$E$ downwards onto the axes for $x$ and $y$ respectively, that is, onto the two principal generators of the cylindroid in the $xy$-plane. I have also dropped the perpendiculars $Q$–$F$ and $Q$–$G$ downwards onto the near and the far generators respectively; these generators are at distances $B$ above and $B$ below that $xy$-plane respectively. It is of course not surprising to find in figure 15.14(*a*) that the mentioned points (all of which are shown) are in a circle (§ 15.30). This circle is the view in the direction z of the ellipse of intersection with the cylindroid of the cone of reciprocal screws at $Q$; see the ellipse itself in figure 15.14(*b*). To find the other ellipse of intersection, the one not with the cylindroid itself but with the $xy$-plane, I proceed as follows. I produce $Q$–$F$ to $F'$ where $(Q\text{–}F') = 2/3(Q\text{–}F)$ because, while $Q$ is $3B$ above the $xy$-plane, $F$ is only $B$ above it. I choose the point $G'$ on $Q$–$G$ such that $(Q\text{–}G') = 3/4(Q\text{–}G)$ because, while $Q$ is $3B$ above the $xy$-plane, $G$ is $B$ below it. I drop from $Q$ to $Q'$ in the $xy$-plane such that $Q$–$Q'$ is parallel with the nodal line. The five points shown by the larger blebs now define (indeed they over-define) the required ellipse. Its construction is not easy but there is this: its major axis is parallel with the far generator, and $Q'$ and $G'$ are symmetrically positioned with respect to that axis. Now the shown ellipse in the $xy$-plane cuts the axes for $x$ and $y$ in $U$ and $V$. $Q$–$U$ and $Q$–$V$ are the sought-for limits. The stippled sector of the cone shows the zone for the type-$A$ screws, and the thickened piece of arc from $U$ to $V$ corresponds with similar arcs in figure 6B.04. As $Q$ is moved upwards or downwards along a line parallel with the axis for $z$ in the figures 15.14, however, we find that the distribution of the type-$A$ screws changes. We find moreover that, if $Q$ is taken in the $xy$-plane, that is, if $Q$ is orthogonally opposite the centre of the cylindroid, half of the cone is type-$A$ while the other half is type-$B$. As $Q$ is taken further and further away from the $xy$-plane (in either direction), however, the area of type-$A$ screws diminishes. In the limit as $Q$ approaches an infinite distance away from the $xy$-plane, the type-$A$ screws disappear from the surfaces of the cones. Refer to figure

Figure 15.14(*b*) (§15.44) Check first in this figure that the cylindroid is occupying its conventional location within the Cartesian frame; its nodal line is along the $z$-axis, its principal generators are along the $x$ and the $y$-axes, its extreme generators (the near and the far) are located as shown and, as $\theta$ takes off from zero at the $x$-axis clockwise according to the right hand rule about $z$, $z$ takes off from zero in the negative direction. See the point $Q$ and check next that the right circular cylinder marked with the asterisks is the cylinder of § 15.30. The figure shows the cone of reciprocal screws at $Q$ cutting the $xy$-plane in the ellipse $EVDU$ of figure 15.14(*a*). The same cone cuts the cylindroid itself in the ellipse $FEGD$ which, as we know, is inscribed upon the said cylinder (§ 15.30). The type-$A$ screws of the cone at $Q$ occupy that portion of the cone which, stippled at figure 15.14(*a*), is delineated here $QVDU$.

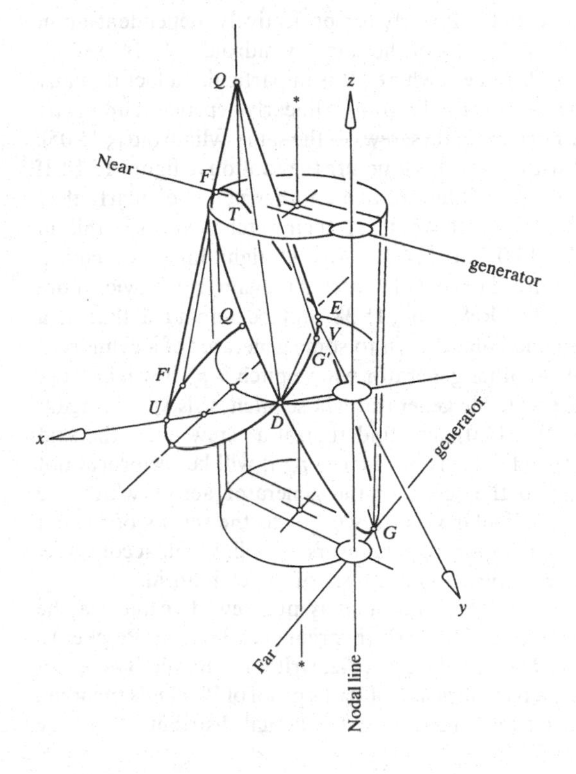


Figure 15.14(*a*) (§15.44) The distribution of the type-$A$ screws upon the surface of the cone at $Q$. The type-$A$ screws cover the stippled portion of the cone. This figure is an orthogonal projection along the $z$-axis of figure 15.14(*b*).

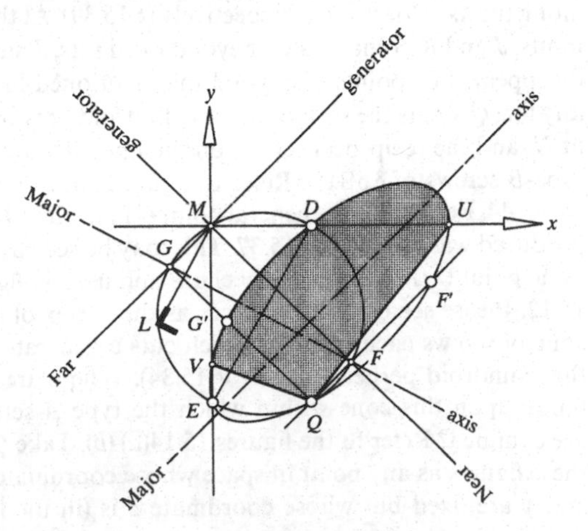

6B.01 and to the passages from §6B.15 to §6B.23 inclusive for related information, but please be aware that there the symbolism is different. The points $N$ and $L$ in figure 6B.01 are the points of tangency with the surface of the cylindroid; $QL$ and $QN$ in figure 6B.01 are the generators $QU$ and $QV$ in the figures 15.14.

45. All type-$A$ screws of a given pitch, there being one (and only one) at every point $Q$ in space, reside as generators of a hyperbolic linear congruence. All type-$B$ screws of a given pitch, there being, similarly, only one at every point $Q$ in space, reside as generators of an elliptic linear congruence. The totality of these congruences of screws of the same pitch constitutes the 4-system of screws which is reciprocal to the 2-system namely the cylindroid. Refer to the chapters 6 and 23 (§6.29, §6B.07 *et seq.*, §23.30).

**Four-point contact and the cylindroid**

46. In chapter 6 at §6.52 *et seq.* reference is made to the passages here; and here I refer to the passages there. Please see figure 6.06; there are four points of contact between the two bodies; at each point of contact the contact normal is drawn; and, in the figure there, the circumstance exists where two real transversals can be drawn to cut all four of the contact normals (§11.20). Now it may be seen by inspection that, if two revolutes were mounted at these transversals in the manner of figure 15.02, and in such a way that the body 3 in figure 15.02 became the body 2 in figure 6.06 (and refer to figure 3.11 to see the ideas there of the 'intermediate body' that needs to be inserted), any small movement at either or both of the revolutes would not disturb the nature of the contact at any one of the four points of contact; refer to figure 14.09. All four contacts would remain poorly conditioned ones (§7.26, §7.36). In the language of chapter 3 the four contact normals would remain $n$-lines in the body 3, while the two transversals, the axes of the revolutes, would remain $r$-lines in the same body.

47. Linearly dependent upon the two $r$-lines, however, and these are, in the language of chapter 6, two motion screws of zero pitch, there exists a whole cylindroid of motion screws of these zero and other, non-zero pitches, both positive and negative. That cylindroid may easily be located from the two $r$-lines by the methods outlined in the present chapter (§15.23). The paragraphs §14.35 to §14.44 inclusive might be of interest or found to be relevant here: what they say in effect is that, whenever there are two forces in parallel, or two hinges in series, there exists a cylindroid of screws.

48. If there were however no $r$-lines to be found in figure 6.06, that is if the two transversals there were unreal and could not be drawn, the existing cylindroid would be characterized by having its motion screws either all left handed or all right handed, and a pair of key screws upon the cylindroid could not be located quite so easily. In such a case we could take the contact normals two by two, erecting six cylindroids upon the six available pairs of contact normals. We could then peruse the six cylindroids of screws to find a set of four screws, each one from a different cylindroid, and all of the same pitch, that did have two transversals. The transversals would then be a pair of motion screws of the other, opposite hand, and we could proceed again as before (§3.54, §14.29, §15.23). Let me remark at this juncture that any four screws of whatever pitches determine a 4-system, and a more general problem that might present itself is the following: *given any four screws generally of different pitches, locate the central axis, the central point, and the principal screws of the thus nominated 4-system, and thus the nodal line and the central point of the reciprocal cylindroid.* This problem may be taken as an advanced exercise for the reader, or the reader may refer to Ball §24. Refer also §6B.08 and §23.36.

**Summary of some of the properties of the cylindroid**

49. The length $2B$ of a cylindroid namely the length of its nodal line (§15.14) is the same as the range $2B$ of its pitches (§15.09, §15.12). The extreme generators – those residing at the two ends of the cylindroid – are inclined at 45° to the axes for $x$ and $y$ respectively. Each parallel with the plane $x$–$O$–$y$, the extreme generators are perpendicular with one another (§15.14). The two principal generators occupy the axes for $x$ and $y$: they intersect perpendicularly at the centre $O$ of the cylindroid (§15.14). The maximum and the minimum pitches namely the pitches of the principal screws $p_\alpha$ and $p_\beta$ occur at the principal generators upon $O$-$x$ and $O$-$y$ respectively (§15.15). There are never more than two generators of the same pitch upon a cylindroid (§15.16). All pairs of generators of the same pitch are equally distributed on either side of the centre of the cylindroid (§15.16); such pairs are called conjugate pairs of generators (§15.21). Any two screws define a cylindroid; all screws of the cylindroid thus defined are linearly dependent upon the two defining screws (§15.18). There may or may not be a pair of generators of zero pitch; this depends upon whether or not the curve for $p_\theta$ cuts the axis for $\theta$ in figure 15.07; if it does not, the pitches of the generators are either all positive or all negative. The sum of the pitches of pairs of intersecting generators is a constant for any one cylindroid; this may be seen from the curves in figure 15.07 or easily derived algebraically (§15.16). Given that all generators of the cylindroid intersect the nodal line namely the axis for $z$ perpendicularly, an equation in cylindrical coordinates for

the shape of the surface is equation (9) in the box at figure 15.06. In terms of $\theta$, the pitch $p_\theta$ lags the displacement $z_\theta$ by $\pi/4$ always (§15.16). All of the above properties become evident upon inspection in figures 15.07 and 15.08. Any circular cylinder containing the nodal line of the cylindroid cuts the cylindroid in an ellipse (§15.30). Any straight line drawn to cut both of any pair of conjugate generators cuts some other generator perpendicularly (§15.36). If the angle between a pair of intersecting generators is $\zeta$, it is easy to show, figure 15.15, that $dp/dz$, the so-called pitch gradient, is given by $\cot\zeta$ (§15.54). The simplest equation in Cartesian coordinates (a point-coordinate equation) for the ruled surface of a cylindroid may be found at §15.26, and the simplest Plücker equation (a line-coordinate equation) may be found at §23.54. The mere shape of a cylindroid may be constructed by dropping from all points upon a given ellipse perpendiculars to any straight line that runs perpendicular to the minor axis of the ellipse and cuts the ellipse at one point only; this would appear to follow from the contents of §15.30. Any three lines drawn to cut a given line perpendicularly define the shape of a cylindroid (§12.33). The shape of a cylindroid may also be constructed by drawing the common perpendiculars between the central axis and all of the generators of a regulus (§6.42, §6C.06, §11.10). Refer also to Ball (1900) §27.

**The first of the five special 2-systems**

50. *The first special 2-system.* If two conjugate generators of a cylindroid intersect all of the generators become of that pitch and they all intersect at the same point; see equations (11) and (13) in the box at figure 15.10 (§15.23). The cylindroid becomes, in other words, a planar pencil of generators of equal pitch; the length $2B$ of the cylindroid becomes zero. I wish to remark here and elaborate later that this particular special cylindroid may be called in its wider context the first special 2-system of screws or, more briefly, the first special 2-system. This special system may moreover become, in a way, doubly special. If the screws are not merely of the same finite pitch but of zero pitch the system becomes a planar pencil either (*a*) of *n*-lines or (*b*) of *r*-lines. It then becomes of significance for example in (*a*) the conditions for equilibrium among a group of three forces (§7.32, §10.15), or (*b*) the synthesis of certain simple and complex $f2$-joints (§7.80, §8.12). The special and the doubly special cases may be seen if we wish as being in no way distinct from one another. Hunt (1978) who devised the currently accepted scheme for the classification of the special and degenerate screw systems took that view. For a note about my separate meanings for the two words special and degenerate, incidentally, please refer to §11.14. Remarks there about the regulus and the others of the hierarchy of the line systems apply also to the cylindroid and the others of the screw systems. Partly because I have no powerful inclinations to the contrary and partly because I agree with them, I accept the classifications of Hunt [5]. I use them here throughout (§23.57).

**On the difficulty at infinity**

51. In paragraphs to follow I deal with the remaining four of the special 2-systems. There is intuitive argument in those paragraphs, and I wish to try to explain here the different meanings I wish to attach there to the words *infinite* and *unrestricted.* Let us drop from a point say $Q$ in the local space a normal onto some plane that is infinitely far away. Suppose the foot of the normal to be unique upon the plane, find the relevant point, and call it $N$. Now go some finite distance away from $N$ to some other point say $M$ upon the plane. Join the points $Q$ and $M$ and perceive that the line $Q$–$M$ is also the normal at $Q$ to the plane at infinity. Perceive in other words that the foot of the said normal may be at any point in the plane at infinity that is no more than a finite distance from $N$ in any direction. The extent of the area containing possible points $M$ in the plane at infinity may not be infinite of course, for if $M$ were infinitely far from $N$ the line $Q$–$M$ could no longer be the normal. I accordingly wish to argue that the area containing possible points $M$ upon the plane at infinity is not infinitely large but merely *unrestricted.* Any dimension – length, area or volume – that might be said by me to be unrestricted in the present context should be seen by the reader as being very large (even as large as you wish), but not infinitely large, for that is larger than you could ever wish. It is true that if we argue with the projective geometers that a straight line is uniquely defined by two points and that no two lines can cut one another at more than one point (§19.26), we arrive at a different conclusion, namely that the said foot at $N$ is unique; but at least for my verbal decriptions of the special screw systems here and in chapter 23, I propose to ignore the conjecture. I wish to insist for the sake of useful argument in the paragraphs that follow that there is, in a given direction at infinity, not one point but an $\infty^2$ of them. Going to the regular literature to study the well known homogeneous coordinates $(x, y, z, w)$ for a point, we find that the coordinates $(x, y, z, 0)$ are said by the relevant writers there to nominate a *unique point,* set away at infinity in the precise direction $(x, y, z)$. And those writers argue that way for very good reason. But we may, if we wish, perceive the coordinates $(x, y, z, 0)$ as nominating not a point but a *plane at infinity,* set normal there to the said direction. And this plane should be imagined to be, I wish to argue, not infinite in extent but merely unrestricted in extent. Please refer for related

remarks and various other relevant matters to § 15.56, § 19.26, § 23.61, § 23.72.

### The pitch gradient

52. Since the length $2B$ is the sole measure of the size of a cylindroid (§ 16.33), it will be reasonable to state that a cylindroid becomes special whenever it becomes either zero in length (§ 15.50) or infinitely long. That is not the dead end of the matter however, for reasons now to be made clear. Refer to figure 15.15. Drawn there is a circle diagram for two generators $A$ and $B$ of a cylindroid at the same displacement $z_A$ from the origin. Note the angle $\zeta$; it is shown between the tangent to the circle at $A$ and the horizontal line segment $A$–$B$. Noting next the geometry of the figure, please see that $\zeta$ is the angle between the two intersecting generators of the real cylindroid. See next that, at $A$, $dp/dz = \pm\cot\zeta$ (17). As a check upon this please reconstruct the fact that, at the centre of the cylindroid where $\zeta$ is a right angle, $dp/dz$ is zero. Remember also that, at the ends of the cylindroid where $z = \pm B$, $\zeta$ is zero (or $\pi$), $z$ is stationary with respect to $\theta$, and $dp/dz$ is infinite. The dimensionless quantity $dp/dz$ is called the *pitch gradient*. In any cylindroid, $|z| = B\sin 2\theta$. Also, in any cylindroid, $\zeta = \pi/2 - 2\theta$. It follows that $|z| = B\cos\zeta$. We know already that $dp/dz = \cot\zeta$, so we can state that the angle $\zeta$ and the pitch gradient $dp/dz$ vary with $|z|$ as shown in figure 15.16. Go to equations (1) and (2); replace $b$ by $z$, $\eta$ by $\theta$, and $p_\eta$ by $p$; find by multiplication that $z^2 = (p_\alpha - p)(p - p_\beta)$ (18). Equations (17) and (18) appear together in a box at figure 15.16. It will be convenient to be able to express in terms of a ratio $k^2$ – where by definition $p_\beta = -k^2 p_\alpha$ (19) – the pitch gradient at those values of $z$ where generators of zero pitch occur. From equation (18) we can show, when $p=0$, (*a*) that $z^2 = -p_\alpha p_\beta$, and thus (*b*), using $k^2$, that $z = \pm kp_\alpha$ (20). After differentiating equation (18) we can also show that, when $p=0$, $dp/dz = \pm 2k/(1-k^2)$ (21); refer to figure 15.17. By solving equation (21) for $k$ it can moreover and conversely be shown that, when $p=0$, $k = (\pm 1 + \sin\zeta)/\cos\zeta$ (22). These expressions for our chosen definition of the positive constant $k^2$, for the value of $z$ in terms of $k$ and $p_\alpha$ when $p$ is zero, for the value of $dp/dz$ in terms of $k$ when $p$ is zero, and for the value of $k$ in terms of $\zeta$ when $p$ is zero, are displayed in the box at figure 15.17. They will be used to advantage later when some of the more obscure of the special screw systems will require to be elucidated (§ 15.55, § 23.67).

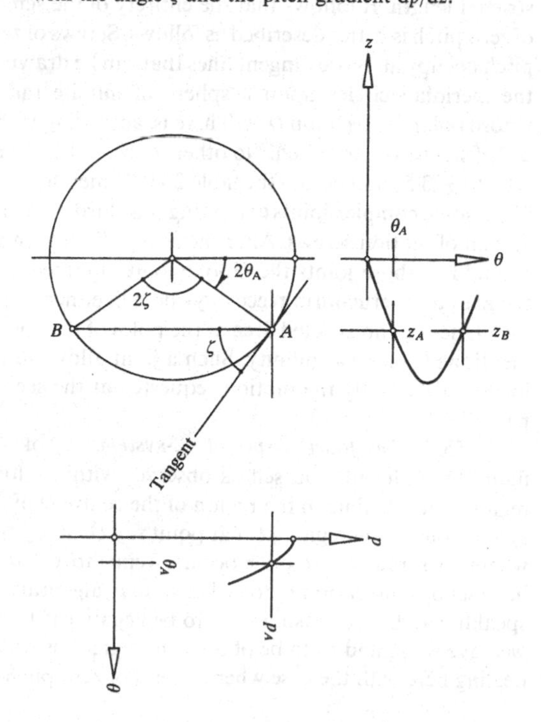


Figure 15.15 (§ 15.51) Circle diagram in connection with the angle $\zeta$ and the pitch gradient $dp/dz$.

### The remaining four of the five special 2-systems

53. *The second special 2-system.* Loot at figure 15.06, locate yourself as observer within the local region at $O$, and imagine the shape of the cylindroid when its length $2B$ and thus its half-length $B$ in each of the two directions $z$ have become infinite. The angle $\zeta$

Figure 15.16 (§ 15.52) Curves showing how the angle $\zeta$ and the pitch gradient $dp/dz$ vary with $|z|$ as we move outwards along the nodal line of a cylindroid from its centre $O$ to its extreme generator at $|z| = B$. Note that at $z = (1/\sqrt{2})B$, $\zeta = \pi/4$ and $dp/dz =$ unity.

| | |
|---|---|
| (17) | $dp/dz = \pm\cot\zeta$ |
| (18) | $z^2 = (p_\alpha - p)(p - p_\beta)$ |

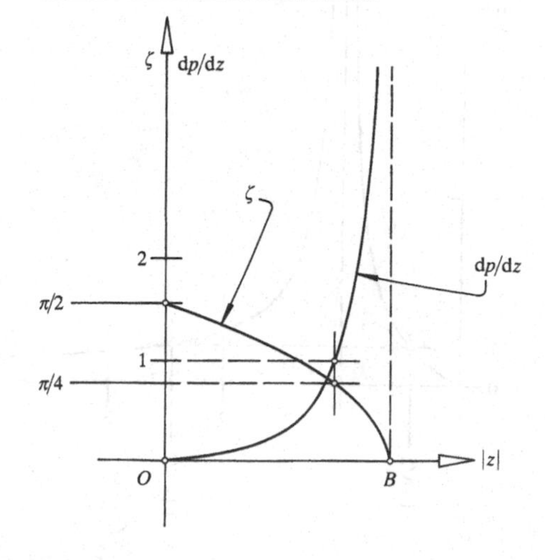

will be a right angle, and the algebraic difference (the range) of the principal pitches $p_\alpha$ of the screw along $O$–$x$ and $p_\beta$ of the screw along $O$–$y$ will be infinity (§15.16). Distinguish the two cases, (*a*) when $p_\alpha$ is finite while $p_\beta = -\infty$, and (*b*) when $p_\alpha = +\infty$ while $p_\beta = -\infty$. The second special 2-system occurs in case (*a*). We see in this case first an $\infty^1$ of parallel screws of some equal finite pitch $p_\alpha$ all parallel with $O$–$x$ and intersecting $O$–$z$ ($dp/dz = 0$). We see also an immense bundle of an $\infty^2$ of parallel screws of infinite pitch all parallel with $O$–$y$ and filling the local space. The said screws of infinite pitch are of course all infinitely long but the size (the diameter) of the bundle of them is merely unrestricted (§15.51). There will be also, and surrounding this bundle, infinitely far away and with each screw perpendicular to $O$–$y$, an $\infty^2$ of screws of zero pitch arranged to form an immense tunnel (§8.13). The diameter of this tunnel is infinite but its length is unrestricted. The tunnel is otherwise known as *the line* (*or the circle*) *at infinity*; refer to §19.26 for a discussion of the contradictions and the relevant projective geometry. The motion screws exhibited at the two joints in figure 7.20 (§7.82) both exemplify this second special 2-system.

54. *The third special 2-system.* This occurs in case (*b*) delineated above. We see in this case not only one but two immense bundles of an $\infty^2$ of parallel screws of infinite pitch. They interpenetrate one another perpendicularly. One is parallel with $O$–$x$ while the other is parallel with $O$–$y$. Each fills the unrestricted local space. There are planar pencils of screws of pitch infinity at every point in the unrestricted space however; these pencils, parallel with the $xy$-plane, are due to the fact that, at each point, the members of the pencil there are linearly dependent upon the pair of screws of pitch infinity that intersect at that point. It follows that we have an unrestricted array of parallel planes, each plane of which is normal to $O$–$z$, and each plane of which contains (in all of its $\infty^2$ of lines) its $\infty^2$ of screws of infinite pitch. There is, accordingly, this $\infty^3$ of screws of infinite pitch. It next becomes apparent that, in two planes of unrestricted area both normal to $O$–$z$ and infinitely far away from the local space, there are screws of pitch zero; these screws occupy all of the lines upon these two planes. But these are not all of the screws of zero pitch because, surrounding each of the single infinity of bundles of screws of infinite pitch, every one of which is pierced perpendicularly by $O$–$z$, there is a tunnel of screws of zero pitch like the tunnel described in §15.53. The axes of these tunnels are all in the $xy$-plane and the tunnels are of infinite diameter and of unrestricted length. It follows that the entirety of the screws of zero pitch is better described as follows. Screws of zero pitch occupy all those tangent lines that can be drawn to the meridian circles upon a sphere of infinite radius whose polar axis is upon $O$–$z$. There is, accordingly, this $\infty^3$ of screws of zero pitch. No other screws exist. Please refer to §23.57 and peruse the table 23.03. I mentioned in §7.82 some complex joints exhibiting this third special 2-system of motion screws. After the above discussion we can add to those joints the Hooke joint mounted with the axes of its cruciform piece anywhere in either one of the planes of unrestricted area at the poles of the above-mentioned sphere at infinity. Such a joint allows in the local space exactly the motion requested at the second part of §7.82.

55. *The fourth special 2-system.* Look at figure 15.06, locate yourself as observer within a local region, not this time in the region of the centre $O$ of the system, but in the region of that point say $O'$ along $O$–$z$ where the screw of zero pitch occurs. Remember that we have set by convention $p_\beta$ to be less than $p_\alpha$ algebraically speaking and, by choosing $-k^2$ to be negative (§15.52), we have set $p_\alpha$ and $p_\beta$ to be of opposite sign; thus we are dealing here with the case where screws of zero pitch do

Figure 15.17 (§15.52) This curve shows how the modulus of the pitch gradient at the screw of zero pitch varies with the value of $k$. Note that when $k$ has the special value unity the screw of zero pitch is at the extreme generator of the cylindroid and the pitch gradient at the screw of zero pitch is infinity.

| | |
|---|---|
| (19) | $p_\beta = -k^2 p_\alpha$ |
| (20) | $(z)_{p=0} = \pm k p_\alpha$ |
| (21) | $(dp/dz)_{p=0} = \pm 2k/(1-k^2)$ |
| (22) | $k = (1 \pm \sin\zeta_{p=0})/\cos\zeta_{p=0}$ |

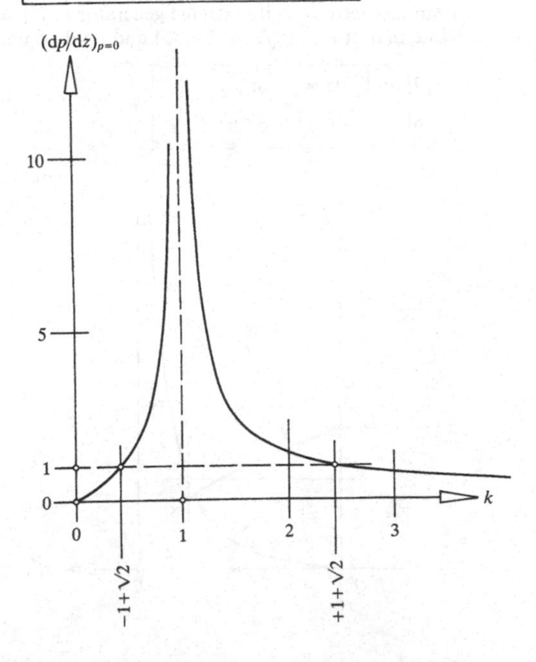

occur. Allow the constant $k$ to hold any value other than unity. You will be at $z = \pm kp_\alpha$, the angle $\zeta$ will be less than a right angle but not zero, and, according to which of the two screws of zero pitch we choose, $dp/dz$ will be $\pm 2k/(1-k^2)$ (§ 15.52). Now imagine the shape of the cylindroid when its length $2B$ becomes infinite. As it lengthens towards being infinitely long the scene in the immediate neighbourhood of $O'$ will remain substantially unchanged. We the observers are free to linger closely within the vicinity of $O'$ where the screw of zero pitch exists or we may with impunity migrate to neighbouring finite regions upon $O$–$z$ where the pitches of the screws might appear all finite positive or all finite negative. As $p_\alpha$ approaches infinity, and as $p_\beta$ varies accordingly ($p_\beta = -k^2 p_\alpha$), the pitch gradient along the whole of our finite space (which is tending now towards being unrestricted in length) will tend towards the constant value $\pm 2k/(1-k^2)$. We see accordingly that it is, under the circumstances, not important that our new origin $O'$ be exactly at the screw of zero pitch; you will find indeed that, when you derive from equation (18) the expression for $dp/dz$ in terms of $p$, it will, as $p \to +\infty$, arrive in the limit at the same $\pm 2k/(1-k^2)$ independently of $p$. So in the limit we shall see the following: *(a)* a single planar array of an $\infty^1$ of parallel screws of steadily varying pitch where $dp/dz = +2k/(1-k^2)$, each screw of which array is intersecting $O$–$z$ perpendicularly; *(b)* a bundle of unrestricted diameter of an $\infty^2$ of parallel screws of infinite pitch, also perpendicular to $O$–$z$ but penetrating the planar array of screws of varying pitch at the angle $\zeta$; and *(c)*, surrounding the mentioned bundle, a tunnel of screws of zero pitch (§ 15.53), the tunnel being of infinite diameter and of unrestricted length (§ 15.51). Refer to § 7.83 for an example of this system, and to § 10.68 for another. Go in advance to figure 23.04; the seventh special 3-system, which is explained nearby at § 23.67, incorporates this fourth special 2-system as a central component part.

56. *The fifth special 2-system.* This occurs at $z = \pm kp_\alpha$ when $p_\alpha \to +\infty$, $p_\beta = -k^2 p_\alpha$, and $k$ is unity. Look again at figure 15.06, locate yourself now at one end or the other of the cylindroid (where $|z| = B$), and repeat the exercise. On this last occasion, as the length $2B$ will tend to become infinite, we shall remain in a region where $\zeta$ is zero and where the pitch gradient $dp/dz$ is infinite. As $p_\alpha$ tends to plus infinity under these circumstances, $p_\beta$ will tend to minus infinity, and, in the limit, we shall see the following: *(a)* one single line, the extreme generator of the cylindroid at $|z| = B$, containing one screw of every pitch; *(b)* an $\infty^2$ of screws of infinite pitch parallel with this line and filling the whole of the local space; and *(c)*, surrounding the said line and the said parallel bundle of screws of infinite pitch, an $\infty^2$ of screws of zero pitch arranged to form an immense tunnel (§ 8.13, § 15.53), the diameter of the tunnel being infinite while its length is unrestricted (§ 15.51). This tunnel is, as it was before, the so called circle at infinity (§ 15.53, § 15.55, § 19.26). The fifth special 2-system is the last of the five special 2-systems. It is exemplified by the motion screws at the simple $f2$ cylindric joint or the $C$-pair (§ 7.84, § 7.94, § 8.14, § 15.57).

**Reciprocal screws at the screws of the cylindroid**

57. Unless the reader is especially interested in the obscurer aspects of the mechanics of the $f2$ cylindric joint (the $C$-pair) I would suggest that the whole of this passage, from § 15.57 to § 15.65 incl., be omitted upon a first reading. In discussing the quasi-general 4-system in § 6.26 *et seq.* and in appendix 6B at chapter 6, I began by explaining that all screws of a 4-system are reciprocal to all screws of its reciprocal cylindroid (its own reciprocal 2-system). At § 6B.07 I began to distinguish among the various screws of the 4-system, speaking about the type-$A$ and the type-$B$ screws. Whereas the type-$A$ screws cut the reciprocal cylindroid three times, twice by virtue of their being transversals of a conjugate pair upon the cylindroid (§ 6B.09) and once more, the type-$B$ screws cut the cylindroid only once. In the present chapter at § 15.36 *et seq.* and § 15.40 *et seq.*, I have further explained that whereas both types of screws cut some one screw of the cylindroid perpendicularly the type-$A$ screws are the ones that cut the conjugate pair. Refer to § 15.21. I have next established at § 15.42 that the reciprocal screw upon any transversal of a pair of conjugate screws of the cylindroid (namely any type-$A$ screw) has a pitch that is minus the pitch of the said pair. I reiterate here that any screw that is reciprocal to the screws of the cylindroid (whether it be a type-$A$ or a type-$B$ screw) will cut one generator of the cylindroid somewhere perpendicularly.

58. Taking now any generally chosen pair of conjugate screws upon a cylindroid (each one of which I will sometimes call the conjugate of the other), and taking the lines of these screws as a pair of directrices, it will be clear that the hyperbolic congruence of type-$A$ screws determined by the pair is a congruence of reciprocal screws of the same pitch. There is a single infinity of these congruences of reciprocal screws and each one of them is symmetrical about the centre of the cylindroid (§ 6.29). It is a fact incidentally that this single infinity of hyperbolic congruences of type-$A$ screws of the same pitch and the corresponding single infinity of the elliptic congruences of type-$B$ screws of the same pitch together make up the whole of the reciprocal 4-system (§ 6B.09).

59. What I wish to discuss here however is not that, but the pattern of the intersections of the reciprocal screws with each single screw of the cylindroid. Please go to § 11.22 to find that the lines of any hyperbolic

congruence through a given point on one of the directrices of that congruence constitute a planar pencil. The centres of the single infinity of planar pencils are in two sets upon their relevant directrices of course, but the planes of the pencils of each set are themselves arranged in such a way that they form an axial pencil of planes: all of the planes of the planar pencils whose centres are upon one of the directrices intersect in a single line, and that line is the other directrix.

60. At § 6B.14 I concluded an argument about the way in which a hyperbolic congruence of reciprocal type-*A* screws of the same pitch could be led towards a coalescence of its directrices at one or other of the principal axes of the cylindroid. I showed that a thus-changing hyperbolic congruence would become, in the limit, a parabolic congruence at the relevant principal axis (§ 11.29). Thus we saw the particular pattern of reciprocal type-*A* screws that cut each of the principal axes of the cylindroid. Each was a pattern of planar pencils lying along the relevant principal axis with one line of each of the pencils contained within the axis. The pencils themselves were angularly arranged in such a way that those lines of the pencils perpendicular to the axis formed the generators of a parabolic hyperboloid; refer to figure 11.04 (§ 11.23). There is, although it is drawn for another purpose, another picture of a parabolic hyperboloid in figure 22.08. The planes of the pencils along either one of the principal axes formed an axial pencil of planes as I have said, but in this special case (the case of the parabolic congruence) the axis of intersection of the planes was the principal axis itself. This meant that the centres of all of the planar pencils now lay along the same line. The pitches of the screws of the planar pencils were moreover equal to minus the pitch of the relevant principal screw of the cylindroid.

61. Consider next, however, each separate point (say *Q*) along the principal axes of the cylindroid. Try to see that there exists at each point *Q*, not only the above mentioned planar pencil of reciprocal screws of the same pitch, but another planar pencil (with its centre also at *Q*) of reciprocal screws of various pitches. This second pencil stands normal to the axis at *Q*. It stands there because it contains all those reciprocal screws at the point *Q* that are reciprocal screws by virtue of their cutting the principal screw of the cylindroid perpendicularly. With the exception of the particular pitch of the screws of the pencil described above, this second pencil at *Q* contains screws of all pitches and of both types. At all points *Q* along the principal axes of the cylindroid, accordingly, we have two concentric planar pencils of reciprocal screws orthogonally arranged. Please observe next that we are dealing here with special positions of the general point *Q* that was used for example at figures 6.04, 6B.01, 15.12, and elsewhere. The point *Q* was the vertex there of the cone of screws that exists at every point *Q* in the quadratic complex of lines that makes up the lines of the 4-system. Here we have taken the vertex *Q* of this general cone at various points along the principal axes of the reciprocal cylindroid. The cone at *Q* degenerates there into a pair of planes orthogonally arranged.

62. Let us go now to another (a general) generator of the cylindroid, and examine the pattern there of the screws reciprocal to its screws that intersect that generator. Spread along the length of a general generator there are (*a*) a pattern of the single infinity of the planar pencils of reciprocal screws of the same pitch whose centres lie on the line of the generator but whose planes intersect, not in the line of that generator, but in the line of its conjugate generator, and (*b*) a pattern of the $\infty^2$ of screws of all of the other pitches that are contained within a single infinity of planar pencils whose centres are on the line and whose planes (as before) are normal to the line. At each point along the general generator there are, accordingly, two planar pencils of reciprocal screws. The planes of these pairs of pencils of screws are, in general, not orthogonally arranged: they are inclined to one another at some angle other than a right angle.

63. Because for a general generator of the cylindroid its conjugate is inclined at some angle other than a right angle to it, the arrangement of the above mentioned planar pencils of screws upon the general generator is a 'twisted' arrangement: the planes of the pencils are not parallel; nor do they contain a set of parallel lines. If we go to the extreme generator, however, where the relevant conjugate (the other extreme generator at the other end of the cylindroid) is perpendicular to it, we see that the set of the planar pencils of screws of the same pitch is now configured in such a way that each plane contains one of a set of parallel lines. These lines are parallel with the conjugate generator and intersecting the generator in question perpendicularly. The set is no longer a twisted set; it is merely a 'bent' set, bent in a way to conform with the conjugate generator at the other end of the cylindroid. Here at the extreme generator the other set of planar pencils, the ones containing the screws of all pitches other than the one, remain as usual parallel with one another and normal to the generator.

64. Now perceive this: if the cylindroid is allowed to become of infinite length (§ 15.52), and if the 'end screw' is of finite pitch (§ 15.56), the patterns of the two sets of reciprocal screws intersecting the extreme generator will coalesce to become a single, double-set. The set of the pencils of screws belonging to the hyperbolic congruence of screws of the same pitch will become a parallel set and coincide with the set of the pencils of screws of all of the other pitches. All lines cutting the

extreme generator perpendicularly will accordingly contain reciprocal screws of all pitches. Because the conjugate generator (the other directrix of the hyperbolic congruence) is infinitely far away and perpendicularly arranged, all lines in all planes normal to the extreme generator will contain screws of infinite pitch.

65. Please refer to my account of the fifth special 4-system at § 23.76. Discover there the relevance of this discussion here to the question of the $f2$ cylindric joint.

### Closure

66. This discussion of the cylindroid, its geometry, and its degeneracies has been conducted by me in a way that differs from other treatments. Firstly there is my rigorous but contrived choice of the origin $O$ which has simplified the algebra of the cylindroid without destroying its applicability in design. Secondly I have tried throughout to accentuate the origins of the geometry in the empiricisms of natural philosophy, paying attention to the realities of statics and kinematics equally. Also, I believe, I have presented the theorems of Plücker and others regarding reciprocal screws and the elliptical plane sections through the cylindroid in a new and a simplified way. But lastly I wish to say this: I have dealt with the various degeneracies of the cylindroid, from § 15.50 onwards, in a somewhat unrigorous manner. I have offered a potted version only of the relevant contents of Hunt (1978), and for those readers whose inclinations extend towards the limit-arguments that are properly involved in the analyses here, and the possible use of the Plücker coordinates for a general argument, I recommend Hunt's original work. Please refer, however, to § 19.21 and § 23.47. A set of Plücker equations to the cylindroid may be found at § 23.54.

### Notes and references

[1] Julius Plücker, German, lived 1801–68. He worked successively in the fields of algebraic curves, physics, and line geometry. We see no reference to the cylindroid in his papers to the Royal Society mentioned in my note [3] of chapter 10, but we do find in the last of his books a succinct account of it; refer Plücker (1869) and scan the passages § 83 to § 87 inclusive. Plücker is discussing there the single infinity of linear complexes that intersect in a given linear congruence (§ 11.40, § 11.41), and he discovers there his *Linienfläche der dritten Ordnung* upon which the axes of the said infinity lie. This 'ruled surface of the third order' (which was called by Plücker by this and by no other name apparently) is of course the cylindroid. He deals there not only with the shape of the ruled surface of the cylindroid but also with the possible distributions of the pitches of the screws upon its generators, producing thereby the main gist of figures 9 and 10 of a much more recent paper – I refer to Phillips and Hunt quoted at note [1] of chapter 16. Be aware that the twisted curve of intersection of the cylindroid itself and the 4-lobed cylindrical surface, both of which surfaces are shown in the mentioned paper at figure 9, was known by Plücker and called by him *die characteristische Curve der Congruenz*. In a footnote on page 98 of the same book Plücker tells of his having built various models relating the cylindroid and of how these models afforded him 'great enlightenment'. Refer Ball (1900) page 517 for his remarks about Plücker and to page 520 of the same book for Ball's remarks about his own work. Be aware also that Felix Klein (then a student of Plücker) wrote about Ball's principle of reciprocity though under a different guise in 1869, refer Ball (1900) page 17; refer § 14.37. Klein in fact completed Plücker's book (1869) after the author's death in 1868. For a brief life of Plücker and a short account of his work go to *Dictionary of scientific biography*, Scribners, New York 1972; see also *Encyclopedia Brittanica*, 9th ed., 1875–89. Refer § 15.13.

[2] William Rowan Hamilton, Irish, lived 1805–65. He worked in the fields of mathematics, optics, and mechanics. His main claims to fame were his systems of rays (in optics), his general studies in mechanics, and his well celebrated quaternion. It was in his very early paper, First supplement to an essay on the theory of systems of rays, *Transactions of the Royal Irish Academy*, Vol. **16**, p. 4–62, 1830, according to Ball, that one finds his first discovery of the shape of the cylindroid; refer to Ball page 510 and to Ball on Ball page 520. See, for a brief life of Hamilton, *Dictionary of scientific biography*, Scribners, New York 1972. Refer § 15.13.

[3] Giuseppe Battaglini, Italian, lived 1826–94. He appears to have worked in the fields of statics and the kinematics of the rigid body, and as an editor. In the following paper Battaglini reportedly reveals in connection with statics the cylindroid. Battaglini, G., ⌊Nota sulla⌋ serie dei sistemi di forze, *Rendiconto accademia della scienze fisiche e matematiche di Napoli*, Vol. **8**, p. 87–94, 1869; refer also *Giornale di matematiche di Battaglini, Napoli*, Vol. **10**, p. 133–40, 1872; refer Ball (1900) pp. 20, 518, and, for Ball on Ball, to page 520. See, for a contemporary list of Battaglini's publications and some summaries of the above and other such papers, *Jahrbuch über die Fortschritte der Mathematik*, Band **2**, 1869–70. The reader may also see there summaries of the following pair of Battaglini's papers: (1) Nota sul movimento geometrico infinitesimo di un sistema rigido, and (2) Nota sul movimento geometrico finito di un sistema rigido: both of these appeared. *Rendiconti accademia della scienze fisiche e matematiche di Napoli*, in 1870; refer in connection with these to § 19.36 and note [20] of chapter 19. Refer § 15.13.

[4] I am indebted to Zhang Wen-Xiang of the People's Republic of China, Visiting Fellow at the University of Sydney, for having clarified the algebra of this criterion in 1985. Refer to the paper quoted in note [1] at chapter 16 and see the same criterion written otherwise there; it was equation (20). Refer § 15.19.

[5] Hunt claims (in a private correspondence at the time of writing, 1986) that his proposed system of classification for the special screw systems (1978) has now been found by himself and Christopher Gibson of Great Britain to be exhaustive. Together the writers claim to have found (after an excursion into projective 5-space) that no

existing one of the special screw systems remains unaccounted for. Hunt agrees in this correspondence that those special screw systems that I have dubbed here the doubly special systems (those involving the special finite number zero) require to be somehow separately listed; he speaks of sub-categories of the special screw systems in this regard. He speaks, moreover, of sub-categories of the general screw systems which might be held to occur when one or other of the principal pitches of those systems becomes zero, making various remarks about the self-reciprocity of certain systems and the fact that in most systems certain screws are members both of the system itself and its reciprocal system. Refer §15.50.

16. One needs to be able to see, in machinery, the wood for the trees. An important part of the education and experience of a person who tackles TMM is the development of this capacity. The geometry of the mechanism of machinery hides behind the distracting shapes and movements of the relatively moving real members.

# 16 The discovery in a mechanism of a cylindroid

**The cylindroid is ubiquitous in mechanism**

01. Following chapter 8, already recommended for an early reading, this chapter may be taken next; it may be seen, indeed, as a sequel to chapter 8. In chapter 15 I give an account from first principles of the way in which the cylindroid may be seen to derive from vector addition, either among forces (which occur in statics) or angular velocities (which occur in kinematics). The geometry of the cylindroid is studied there in some detail and its general significance is hinted at. In chapter 13 I discuss the summation to equilibrium of three wrenches in statics and the instantaneous relative motion of three rigid bodies in kinematics. I derive there the two theorems of three axes and explain how the cylindroid is inextricably involved in both of them. In chapter 8 I study some simple joints, each joint exhibiting two degrees of freedom. Once again the cylindroid appears in chapter 8, but there camouflaged in one or other of its special forms. There is however an example at § 8.18 (figure 8.04) where the elementary method of investigation being used there produces no result. In this chapter 16 I use a modification of that example to show that a general cylindroid of ordinary finite proportions can be seen to exist in a simple mechanism. The example will serve to introduce moreover the idea that the cylindroid is not only ubiquitous in mechanism but also somewhat elusive. The cylindroid is important in screw theory, and I wish to convince the reader that a person may effectively cope with it and its ramifications.

**We take an example**

02. Figure 8.05 in chapter 8 is derived from figure 8.04, and figure 8.05 is chosen now for a fresh beginning here. In this figure a rigid, double ball-ended rod 2 is shown making contact with another rigid, slotted body 1 at four points along two V-slots. The centres of the balls are at $A$ and $B$, and the centre point of the line $AB$ (which is not the half-way mark along the line segment $A$–$B$) is at $\Gamma$. The point $\Gamma$ is that unique point along the infinitely long line $AB$ in the body 2 whose linear velocity $\mathbf{v}_{1\Gamma}$ is least (§ 21.01 *et seq.*, and please see figure 21.01). It is explained in chapter 8 that in figure 8.05 there are two axes of zero pitch, or two $r$-lines (§ 3.39), about which the ball-ended rod 2 may screw with zero pitch (simply rotate) with respect to body 1. The two axes are marked $w_1$ and $w_2$ in figure 8.05, and they each carry the label $ISA_{12}$ $p=0$. It should not be surprising to find that the first of these two axes occurs along the line $AB$ itself.

03. The same two axes appear in figure 8.06 where they are marked again $w_1$ and $w_2$. In each of the figures 8.05 and 8.06 the common perpendicular distance $\Gamma$-$D$ between the two axes is drawn, and in both figures its straight-line extension into the foreground of the figure is marked with an $x$.

04. Please look now at the 4-member mechanism drawn in figure 16.01. This figure corresponds exactly with figures 8.05 and 8.06 so far as its layout and its overall dimensions are concerned. The earlier freedoms for the centres $A$ and $B$ of the balls to slide with their interrelated velocities along the centre lines of the slots or of the tubes, however, has been replaced in figure 16.01 by their respective freedoms to suffer interrelated velocities along the tangents to the circles of the cranks that are shown. Refer to figure 5.01 for an explanation of the interrelated velocities. The cranks in figure 16.01 carry matching sockets for the balls.

05. Following convention (§ 12.08) the links of the mechanism in figure 16.01 are numbered from 1 to 4 beginning with the frame link 1; so the coupler, the rod $A$–$B$, has the new number, 3. The letters $A$ and $B$ are retained for the centres of the balls, and the symbols $J$–$J$ and $K$–$K$ are used to signify the revolute axes at the output and input cranks respectively.

06. The mechanism in figure 16.01 is an RSSR, 4-link loop, and this arrangement abounds as we know in ordinary engineering practice; refer to figure 2.02(*d*); the corresponding text appears at § 2.39. I have effected the reference across from chapter 8 and thus the transformation from that other figure 8.05 to this figure 16.01 for three main reasons: (*a*) to show that the freedoms of the coupler $A$–$B$ in figure 16.01 are exactly represented by the identical freedoms of the twisted rigid rod in figures

8.05 and 8.06, and thus to make the important point that such equivalence of mechanism at an instant is always at our disposal for such manipulations as these; (*b*) to arrive at a suitable other example-mechanism for discussion without losing touch with material outlined in chapter 8; and (*c*) to allow this chapter 16 to stand alone as a piece that can be studied independently.

**Some parenthetical remarks about equivalence**

07. Before going on to pursue the main objective here, namely to look for all of the other axes of other pitches about which the coupler *A–B* might screw with respect to frame (and thus to find the cylindroid), let us take a closer look at the geometrical implications of the transformation of the figures (from 8.05 to 16.01) that has been effected.

08. In order that the possibilities for the instantaneous motion of the rod *A–B* remain unchanged, it was necessary only that the directions of the velocities at *A* and *B* remain unchanged. It follows that the revolute axes *J–J* and *K–K* could have been chosen by the writer to be anywhere in the *polar planes* which appear at *A* and *B* respectively. These planes are the flanges of the tubes in figure 8.06 that intersect in $w_2$, they are the planes that are normal at *A* and *B* to the velocities at *A* and *B*, and they are shown again in figure 16.02. Their fundamental significance is discussed elsewhere (§9.45 *et seq.*), but their significance in the present context will be clear. They are the planes in which (for equivalence) the revolute axes *J–J* and *K–K* must reside. This means, incidentally, that there is an $\infty^2$ of choices for the location of each one of those axes in figure 16.01 (§4.09). The said polar planes are also the planes *JJA* and *KKB*. They intersect in $w_2$, $w_2$ being the second ISA of zero pitch, namely the second *r*-line (§3.39), about which the coupler *A–B* (which is not a mere line but a rigid bulky link) may purely rotate relative to frame. An ordinary Euclidean analysis at any one of these related figures will reveal that, while $\mathbf{v}_{1A}$ and $\mathbf{v}_{1B}$ are both perpendicular to the line $w_2$, their respective distances from it are directly proportional to their magnitudes. I should say at this juncture that a mechanism may be equivalent to another so far as (*a*) velocities, (*b*) accelerations, (*c*) rates of change of acceleration, and (*d*) etc. are concerned. Refer to Hall (1960), chapter 5, for a clear explanation of this. In the present case, because the radii of curvature of the actual paths of the points *A* and *B* at the centres of the balls have not been matched in the equivalent mechanism, the equivalence at the instant extends to velocities only.

Figure 16.01 (§16.04) An *RSSR* 4-link loop. It is being used to show that, at this generally chosen one of its many configurations, there is a cylindroid of locations available for the $ISA_{13}$. The ball-ended coupler 3 has two degrees of freedom with respect to the frame 1. The coupler has, accordingly, at any generally chosen configuration of the mechanism, a single infinity of choices for its ISA with respect to frame.

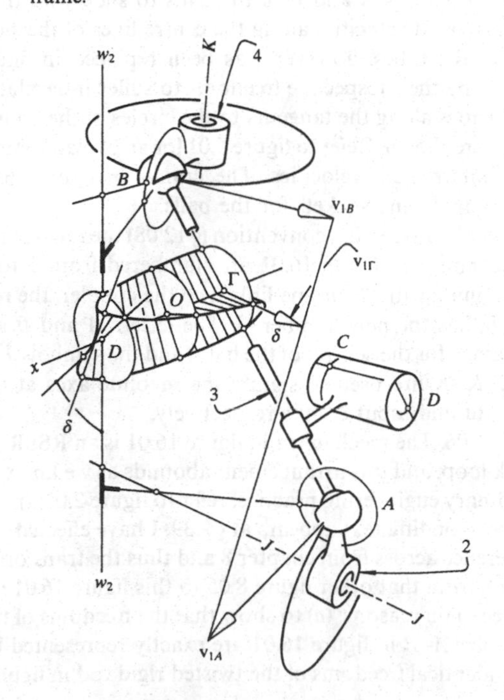


**Clarification of the question at issue**

09. In figure 16.01 the point Γ is the unique central point in the infinitely long line *AB* (§21.03). The line Γ–x has been drawn normal to the plane defined by (*a*) the vector $\mathbf{v}_{1\Gamma}$ at Γ, and (*b*) the line *AB* itself. It will be clear upon inspection that, if an $ISA_{13}$ of any finite pitch exists, it must cut Γ–*x* somewhere perpendicularly. It will also be clear from the contents of chapter 5 that other screw axes anywhere else are impossible. We note that both of the two axes already found, $w_1$ and $w_2$ (which are both of zero pitch), do cut Γ–*x* perpendicularly.

10. Figure 16.02 is another version of figure 16.01, redrawn to show more clearly the twisted shape of the existing display of velocity vectors along the line *AB*. This display shows that no possible ISA can be as far away as infinity along the line Γ–*x*. This means that all of the available axes, each of which must cut Γ–*x* perpendicularly somewhere, must be somehow distributed together within the finite immediate neighbourhood of the already discovered pair of axes of zero pitch.

11. To clarify the fact that the coupler *A–B* (namely the link 3 of the mechanism) has two degrees of freedom relative to its frame 1, and to accentuate its possible actual bulk in some real machine, it has been

equipped in figure 16.01 with the rigidly attached cylindrically shaped outrigger-piece that will be seen. The shape of the piece is arbitrarily chosen and the two points $C$ and $D$ which appear are embedded arbitrarily within it. Given the freedoms of the centres of the balls at $A$ and $B$ to move instantaneously and interrelatedly along their respective circular paths, and given the separate freedom of link 3 to rotate instantaneously and at the same time about its 'centre line' $AB$, we see that, as soon as we speak about some likely $\text{ISA}_{13}$ about which link 3 might screw with respect to frame 1, we speak not only about the likely velocities at $A$ and $B$ (which for the sake of the following argument are conveniently held to be fixed), but also about what happens at $C$ and $D$. And this, of course, is the crux of the present matter: while for each one of the various screw axes discussed in the following argument the instantaneous velocities at $A$ and $B$ are held to remain the same, the instantaneous velocities at $C$ and $D$ will of course be different. A useful thing to do in the imagination is to take hold of link 3 at its outrigger-piece and to ask one's self the double question: *about how many different axes can I personally cause this link to screw at the instant, and what is the layout in space of the whole set of these axes?* This question, or these questions, are the ones being asked and answered in this chapter.

Figure 16.02 (§ 16.10) This figure adds to the contents of figure 16.01. It shows how the straight-line array of velocity vectors along the line $AB$ of the coupler 3 is determined solely by the configuration of the mechanism (§ 5.25, § 21.01).

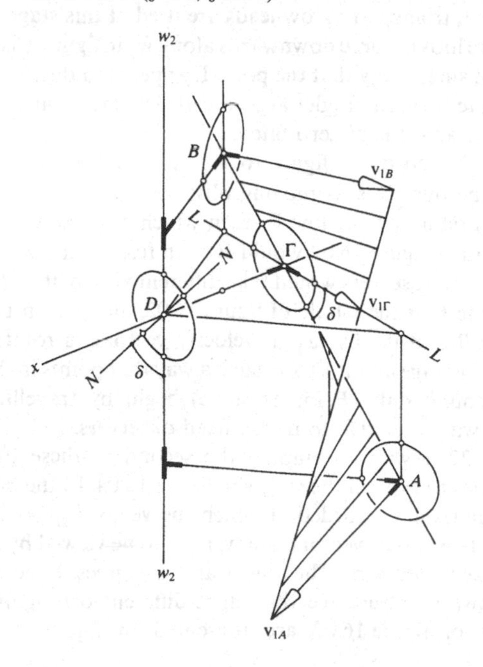


**Relevance here of the linear complex**

12. For the reader already familiar with the properties of the linear line complex (§ 11.30), the letters $w_1$ and $w_2$ will have a special significance. Thinking in terms of right lines and $n$-lines again (§ 3.06, and chapter 9), it will be seen that, while $A$ and $B$ might be two arbitrarily chosen points along some line $w_1$ that is drawn at random through some complex, the polar planes at $A$ and $B$ are intersecting along some other line $w_2$ of the same complex. Looked at in this way we see that the gist of our current enquiry can be stated as follows: we have this pair of conjugate lines $w_1$ and $w_2$ of some complex, but we don't know where the central axis of the complex is; we do know of course that the central axis must cut the common perpendicular between the conjugate lines somewhere perpendicularly (§ 11.37); but, in that the range of possible positions of the elusive central axis of this complex is the same as the range of possible positions here of an available screw axis of finite pitch, we know that already.

13. The above remarks do suggest however (*a*) that there is a symmetry about this question of where the available axes are, (*b*) that there is a single infinity of such axes (§ 4.01), *and* (*c*) *that the distribution of the axes, whatever it is, is line-symmetric about* $\Gamma$–$x$ *and point-symmetric about the half-way mark O along the common perpendicular distance* $\Gamma$–$D$ *between the pair of axes* $w_1$ *and* $w_2$.

**More early hints about the nature of the symmetry**

14. The components of the velocities $\mathbf{v}_{1A}$ and $\mathbf{v}_{1B}$ in the direction of the least velocity for the line $AB$, namely in the direction of $\mathbf{v}_{1\Gamma}$, are equal to $\mathbf{v}_{1\Gamma}$ (§ 21.05); so it is quite clear that there is an available screw axis of non-zero pitch through the point $\Gamma$ and in the direction of the velocity $\mathbf{v}_{1\Gamma}$. It is marked $L\Gamma L$ in figure 16.02. The pitch of this axis is left handed, namely negative.

15. Notice next in figure 16.02 that the common perpendicular $NDN$ between the two intersecting lines $BD$ and $AD$ at $D$ is perpendicular to $AB$ and thereby normal to the plane $BAD$, and that the components of $\mathbf{v}_{1A}$ and $\mathbf{v}_{1B}$ in the direction of this normal $NDN$ are equal to one another and to the component of $\mathbf{v}_{1\Gamma}$ in the same direction (§ 21.03). Notice also that, when the velocity vectors at $A$ and $B$ are projected onto the plane $BAD$, their components there are perpendicular respectively to the radii $DA$ and $DB$, and that the tips of all of the projected vectors along the line $AB$ are still in a straight line after projection upon that plane. It follows that there is along $NDN$ another available screw axis of non-zero pitch whose pitch is also left handed, negative.

16. Notice next that the angles $\delta$ appearing in figure 16.02 are equal, and that (as shown in figure 16.03

which is an orthogonal view of all of the newly found axes looking along $D$–$\Gamma$) the angles $\delta$ are displaced exactly 90° apart. Given the data summarized in figure 16.03, the reader should next be able to show (and this is an exercise for the reader) that the pitches of the two left handed axes along $L\Gamma L$ and $NDN$ are equal.

17. Before going on to clarify the whereabouts of the remaining single infinity of the axes, it would be wise to pause for reflection here about the nature of the symmetry emerging. Refer to figure 16.02. Notice that at each end of the line segment $D$–$\Gamma$ (where the discs are drawn) there is a pair of axes. At each disc there is one axis of zero pitch and one of non-zero pitch, and at the discs the same included angle $\delta$ obtains. The pairs of axes at the discs with their equal included angles $\delta$ are displaced right handedly along $D$–$\Gamma$ exactly 90° apart. However, while the two axes of zero pitch are displaced $(90+\delta)°$ apart, the two axes of non-zero pitch are displaced $(90-\delta)°$ apart. This information is made clear in figure 16.03.

18. The information is confusing, but it does leave the possibility that, if we take into consideration merely the lines of action of the axes irrespectively of their pitches, there are planes of symmetry at $s$–$s$ and $s$–$s$ as shown in figure 16.03. These planes both contain $\Gamma$–$x$, and they are at right angles to one another. We could go

Figure 16.03 (§ 16.16) Looking orthogonally along the line $DM\Gamma$ we see this view of the already discovered, few possible locations of the $ISA_{13}$. These will be seen next to be a few locations only of a whole cylindroid of possible locations, there being a single infinity of them. The line $DM\Gamma$ is the nodal line of the cylindroid. The balls in the tubes appearing at $A$ and $B$ correspond with those that appear in figure 8.06.

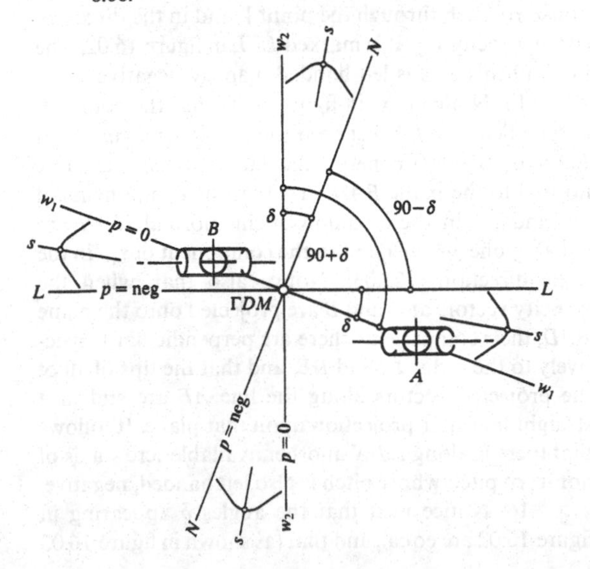


further and suppose that, if the display of available axes is continuous (as it is for example wholly in figure 8.01 and partly in figure 8.03), we might find at least some of the available axes outside the confines of the line segment $D$–$\Gamma$.

19. It remains now to take a look at this whole question in one or another of a number of ways. A direct physical way will be to look, by progressively taken orthogonal projections in a series of radial directions across $\Gamma$–$x$, at the display of velocity vectors along $AB$. If there is an available screw axis in the direction of the projection it will be clear to see. Important matters of symmetry will also present themselves for easy viewing, and we shall see, indeed, that there is an available screw axis which is easy to see in every possible direction of projection.

**Rotation about the nodal line**

20. Look top left at figure 16.04 and please see (at 1) an orthogonal view of the line segment $\Gamma$–$D$; it is shown thickened. Reading downwards through the various stages of the argument which are consecutively numbered in figure 16.04, see also that the same segment $\Gamma$–$D$ appears at every stage. The vector $\mathbf{v}_{1\Gamma}$ (marked $TL$ at 1) also appears as a thickened segment of line at every stage; it carries no arrowhead. At stage 1 $\mathbf{v}_{1\Gamma}$ is itself being viewed orthogonally; it appears in the plane of the paper there; $TL$ stands for true length. The vectors $\mathbf{v}_{1A}$ and $\mathbf{v}_{1B}$ also appear with their own true lengths at this stage 1 because the relevant ISA, which resides in the line $w_2$ and appears as the double-circled dot at $D$ in figure 16.04, is of zero pitch. It is for this reason that the unusual triangular arrowheads are used at this stage 1. We are looking here downwards along $w_2$ in figure 16.02 and in such a way that the point $\Gamma$ appears to the left of $D$. Note the right angles at $A$ and $B$. The ISA is at $D$ at stage 1 and it is of zero pitch.

21. Next in figure 16.04 we could either (*a*) imagine ourselves as moving observers to be rotating head-first about the line $\Gamma$–$x$, in which case the whole content of figure 16.02 would remain fixed while we as travelling observers would take different view of it, or (*b*) imagine that the whole of figure 16.02, along with the line $AB$ and its display of velocity vectors, is rotated stage by stage about $\Gamma$–$x$ in such a way that points in the foreground (take $B$ for example) begin by travelling downwards relative to us the fixed observers.

22. I wish to employ the second of these two schemes to explain the stages in figure 16.04. In the box the angles are recorded at which the vector $\mathbf{v}_{1\Gamma}$, along with $A$–$B$ and its vector display, is set to be viewed by us the fixed observers. The angles are in degrees. Reading downwards please see that eight different orthogonal views of figure 16.02 are presented in figure 16.04.

Dimensions have been chosen so as to reveal unambiguously the phenomena, and the drawings have been made to scale. The angles of rotation set for the viewings accord with the various critical aspects of the matter revealed in figure 16.03 and mentioned previously. Please note the *clock* which may be seen, and be aware

Figure 16.04 (§16.20) Successive views of the straight-line array of velocity vectors taken along the successively chosen ISAs. The relevant ISA as generator will be seen to be tracing out the mentioned cylindroid.

| | |
|---|---|
| 1 | $\theta$ |
| 2 | $\delta/2$ |
| 3 | $\delta$ |
| 4 | $\delta/2+45$ |
| 5 | 90 |
| 6 | $\delta/2+90$ |
| 7 | $\delta+90$ |
| 8 | $\delta/2+135$ |

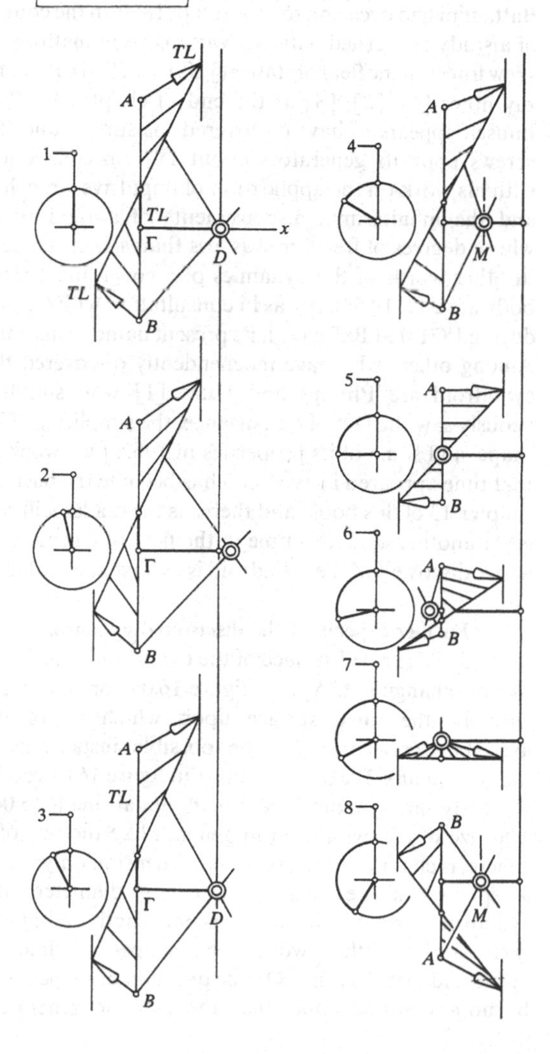


that the angles listed in the box correspond with the settings of the clock.

23. Leaving aside stage 2 for the moment, and looking next at stage 3 in figure 16.04, which occurs at the critical angle $\delta$ on the clock, we see that we are looking at $D$ along the line $NDN$. Along $NDN$ is the left handed ISA which was previously found in figure 16.02 and already recorded in figure 16.03; it is shown with the double-circled dot at $D$. At this stage 3 the segment $A$–$B$ appears in true length, and the vector $\mathbf{v}_{1\Gamma}$ is foreshortened in the proportion $\cos\delta$. All three vectors are emerging upwards out of the paper and this fact, along with the clockwise rotation looking downwards, reveals the left handedness of the ISA at $D$.

24. Going back to stage 2 we now see that, following earlier suppositions about the symmetry which are summarized by the lines $s$–$s$ and $s$–$s$ in figure 16.03 (§16.27), the critical angle $\delta/2$ has been chosen for this first rotation from stage 1 to stage 2. At stage 2 neither the vector $\mathbf{v}_{1\Gamma}$ nor the segment $A$–$B$ is seen in true length, and the perpendiculars to the projected vectors at $A$ and $B$ cut the line $\Gamma$–$x$ at a point outside the segment $\Gamma$–$D$, beyond $D$. It is not very easy to see why this should be so, but contemplation of a wire model built in three dimensions or a careful construction by drawing will show, not only that the ISA for stage 2 is in the position shown by the double-circled dot, but also that for no other position between the original position and stage 3 is the ISA any further to the right than it is here.

25. To shorten this long story somewhat, the reader might cast his eye now quickly across the whole range of movement of the double-circled dot in figure 16.04 as the viewing of $A$–$B$ and its display changes from stage to stage. Notice that there is an even oscillation of the dot about the central point $O$, that the amplitude is somewhat greater than the half-length of $\Gamma$–$D$, and that, for continuous rotation on the clock, the movement appears to be quite smooth as the dot goes from one end of its travel to the other and back. One could construct at this stage a physical wire model of the figure 16.02, cast its shadow orthogonally upon a wall, and watch the moving shadow as the model is appropriately and steadily rotated.

26. Continuing, however, note that at stage 4 the segment $A$–$B$ is again foreshortened, with $B$ moving upwards now and into the background. The clock at stage 4 has been set at $\delta+\frac{1}{2}(90-\delta)$, namely $\delta/2+45$, and the ISA that can be seen is exactly at $O$. The pitch here is maximum left handed, namely minimum negative. Moving next across the stages 5 and 6 where the pitch is still left handed though diminishing in magnitude, and where the figures are self explanatory, we arrive at stage 7. At this stage we are looking directly

along $A$–$B$ (from $A$ to $B$), and the pitch of the screwing is zero again. The clock is set at $\delta + 90$ exactly. All of this accords with the early information summarised in figure 16.03.

27. At stage 8 (refer stage 4) the clock has been put at $(90 + \delta) + \frac{1}{2}(90 - \delta)$, namely $\delta/2 + 135$, which is exactly 90 in advance of stage 4. Notice that, although the geometry is different, with $B$ emerging into the foreground now while $A$ is beginning to recede, and with the projected length of the segment $A$–$B$ being quite different from what it was at 4, the double-circled dot is again exactly at $O$. The pitch of the screwing will be seen at stage 8 to be at its maximum right handed namely maximum positive. However its maximum positive magnitude here is not equal to, but somewhat less than, its maximum negative magnitude at stage 4.

28. Stage 9 (which is not drawn) could have been set at 180 on the clock, whereupon a situation exactly symmetrical with stage 1 would have been revealed; $A$ and $B$ would have changed places, but the geometry would be the same; we would look upwards along the line $w_2$ not downwards along it, and the pitch of course would have diminished to zero again.

**The sinusoidal nature of the discovered display**

29. Having thus checked upon the various key stages in the oscillatory movement of the double-circled dot in figure 16.04, it is not very difficult next to discover (by construction, by algebra, or otherwise) that, for a constant speed of rotation on the clock, the oscillation is exactly simple harmonic. It is also easy to see that, for each complete revolution of the clock, the available ISA goes from left to right and back again (from one end of its spectrum of possible positions to the other end and back again) not once but twice.

30. In this way there is traced out not once but twice during each revolution of the clock the remarkable ruled surface which is shown pictorially in figure 16.01; please refer also to figures 13.03 and 15.06. Only some of its straight-line generators are drawn, and the surface is suitably truncated at the figure by a right circular cylinder whose axis is upon $\Gamma$–$D$. Otherwise the single infinity of infinitely long generators would fill the whole page with lines. In the whole of Euclidean space, of course, the ruled surface is not unlike a flat plane (with a slight twiddle at the middle); it cuts the whole of the space into two halves. Refer to photograph 15. The surface is, indeed, the expected cylindroid, and some more of its main features are soon to be described (§ 16.33).

31. It should be mentioned here that an important one of its main features, that there is a similar, simple harmonic variation in the pitch of the possible screwing, can also be seen with the help of figure 16.04. The amplitude of the variation of the pitch is the same as the amplitude of the variation in position of the ISA along the axis of the cylindroid. Should the half-length of the cylindroid, its amplitude, be $B$(mm) in other words, its length would be $2B$(mm), and the algebraic difference between the maximum and the minimum pitches would also be $2B$(mm rad$^{-1}$). This and other important aspects of the cylindroid are discussed more fully in chapter 15.

**A potted history of the cylindroid**

32. According to Ball (1900) the shape and some of the properties of this renowned surface were first described by Hamilton in 1830 in connection with certain systems of reflected rays in optics. It was later found and described in detail by Plücker in 1868 in the context of some pure mathematical work in the field of the linear line complex (§ 9.30, § 11.30, § 11.44), and the surface became known for a time as Plücker's *conoid.* A year later and independently, again according to Ball, Battaglini had occasion to report it in 1869, in the course of a study concerned with the various combinations of skew forces in the field of statics (§ 10.14, § 15.01). Refer to my notes [1], [2], [3] at the end of chapter 15. Ball himself appears to have discovered the surface and the screws upon its generators about 1870 in connection with his work on the application of impulsive wrenches and the infinitesimal displacements of a rigid body whose degrees of freedom was less than six, in the field in other words of the dynamics of a constrained rigid body at rest (§ 19.64). It was in consultation with Cayley during 1871 that Ball gave it its present name, *cylindroid.* Among others who have independently discovered the cylindroid are Phillips and Hunt [1] who simultaneously saw the fact of its existence, the simplicity of its shape, and some of its properties in 1962. Our work of that time appeared in 1964 and has become the basis of chapter 13 of this book, and there, as the reader will see, is yet another area, this time in the field of mechanism, where discovery of the cylindroid is almost unavoidable.

**Further aspects of the discovered cylindroid**

33. The ruled surface of the cylindroid traced out by the changing $ISA_{23}$ in figure 16.04, or, more accurately, the ruled surface upon which all of the available screw axes for the possible instantaneous motions of link 3 relative to link 1 in figure 16.01 reside, is clearly line-symmetric about its nodal line (§ 15.04). The size of the cylindroid in general, like the size of a sphere, is characterized by its one dimension only; while a sphere's single dimension is its diameter, the cylindroid's single dimension is its length. Whatever their size, in other words, the shapes of different cylindroids are the same. On the question of shape it will be noticed for example that the extreme generators

(those at the two far ends of the cylindroid) are perpendicular with one another, while at the mid-point $O$ another pair of generators s intersect perpendicularly. These and other aspects of the cylindroid in general are the subjects of study in chapter 15. Please see the main properties of the cylindroid summarized at § 15.49, and be aware that most of the matters mentioned there are important in screw theory and sophisticated machine practice.

34. Even more important, perhaps, in the field of ordinary machine practice, are the geometrical properties of the cylindroid when that surface becomes, as it does in various ways, special or degenerate. If the reader will now return to chapter 8 and pursue again the various arguments presented there he will, I think, begin to suspect the following: (*a*) that in figure 8.01 the cylindroid is of zero length, having become a simple planar pencil of generators all of zero pitch; (*b*) that in figure 8.04 the cylindroid exists in the ordinary way as we have discovered it here to exist, in no way special or degenerate; (*c*) that in figure 8.03 the cylindroid is infinitely long; and (*d*) that in figure 8.02 (which illustrates, of course, the well known $f2$ cylindric joint) something doubly extraordinary has happened. The cylindroid has in figure 8.02 not only collapsed inwards somehow towards some central axis (which is, I hasten to hint, not the nodal line), but also exploded outwards somehow towards infinity (§ 15.56). Please go to the index, item Cylindric (*C*) joint, and be guided there in your reading about this last case (*d*). The ordinary $f2$ cylindric joint might appear to be the simplest and oldest of the six lower joints, in the same way as the lathe that can make it might be seen to be the simplest and oldest of the machine tools. It will however be found to be, not only the simplest of them all, but also, by far, the most intriguing of them all (§ 8.01, § 15.57).

35. Apart from the kinds of circumstance already mentioned here and in other chapters where the cylindroid is seen to exist in connection (*a*) with two degrees of freedom (chapters 8, 14 and 16), (*b*) with the addition of wrenches with one another (chapters 13 and 15), and (*c*) either with the free relative motion in space of any three rigid bodies or the equilibrium among a set of three wrenches that add to zero (chapter 13), there are other circumstances as well. The cylindroid can be made to figure usefully in our thinking even though the mechanical motion being considered is entirely constrained or even overconstrained. Take a look for example at (*d*) aspects of the famous mechanism of Bennett (§ 20.52), and (*e*) the relevance of the cylindroid in gear sets (§ 22.16, § 22.20). It is not unreasonable to argue that the phenomenon of the cylindroid is at the root of all mechanism. It is at the root, too, of the three central systems of screws, namely (*a*) the 2-system which consists indeed of the cylindroid itself, (*b*) the 3-system which is the central one of these three, and (*c*) the 4-system. Refer to chapters 6, 14 and 23.

## Notes and references

[1] Phillips, J.R. and Hunt, K.H., On the theorem of three axes in the spatial motion of three bodies, *Australian journal of applied science*, Vol. **15**, p. 267–87, CSIRO, Melbourne 1964. Refer to § 16.32.

**17. Here is a revolute joint being dislocated by a wrench. The action screw of the wrench is reciprocal to the motion screw of the joint of course, for otherwise a movement would occur.**

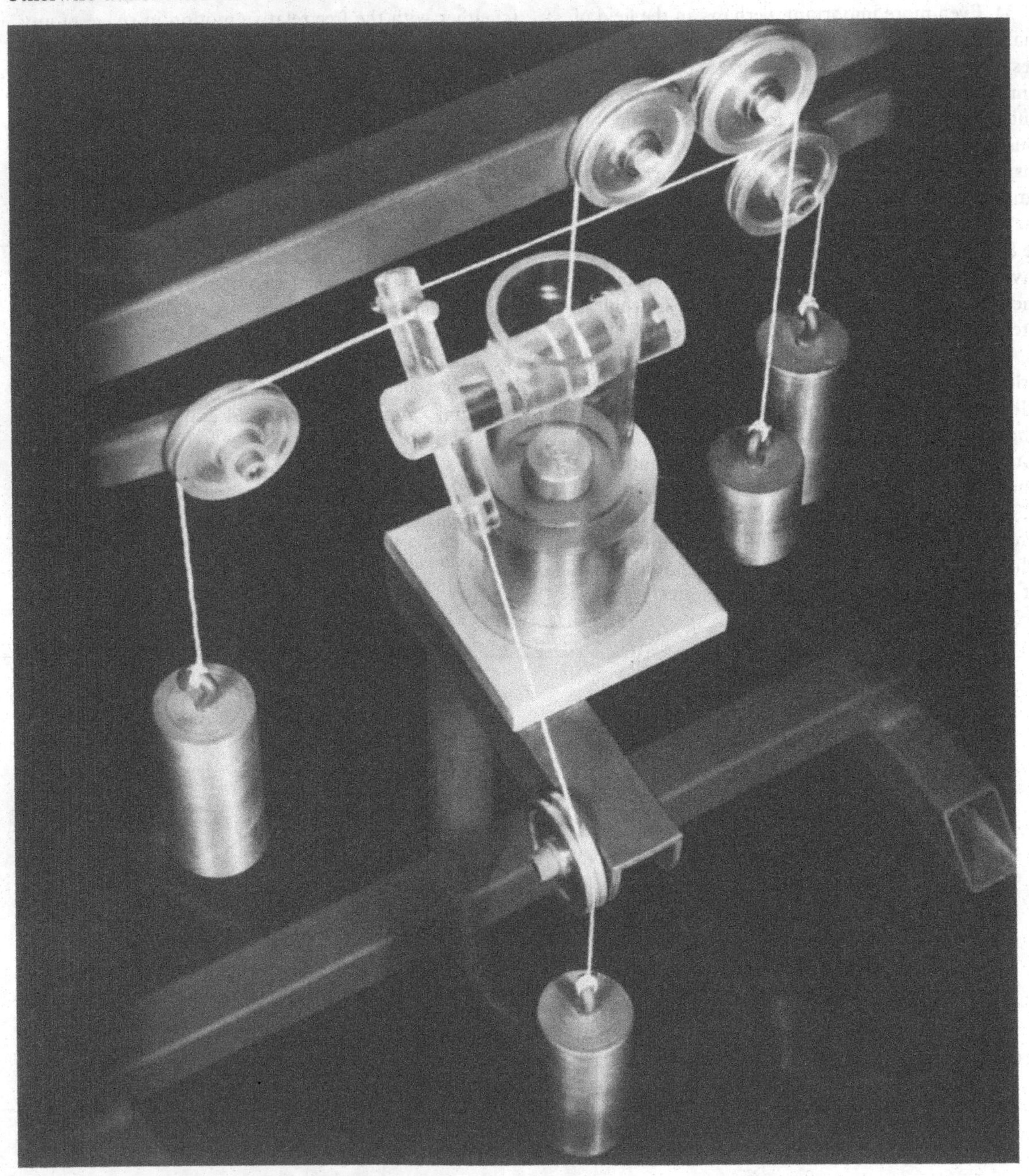

# 17 Action, motion, clearances and backlash

## Introduction

01. The combined occurrence of the action dual and the motion dual at a joint in working machinery is discussed in general at § 10.59. Examples of this are illustrated in figures 10.14 and 10.15. That the dual dot product of the two duals is the power loss at the joint is mooted at § 10.59 *et seq.*, and mathematical methods for calculating such things are outlined at § 19.18 and § 19.28. See chapter 14 and photograph 17 for reminders that, at any simple joint in the general arena of spatial mechanism, a pair of equal and opposite wrenches will be acting, Either one of these or both of them may be seen as the action dual being transmitted by the joint. In the absence of friction at a joint, the screws of the action and the motion duals will be reciprocal. So the study of action and motion at the joints of mechanism in general has these kinds of opportunities to begin. It will be instructive upon beginning here, however, to study the highly special mechanics of a well known special joint – let us take for example the simple revolute joint – when that joint is at work in the special arena of planar mechanism. We will study the joint within the context of the well known planar, 4-link, 4*R*-loop (§ 2.35, § 2.36). We will do it, moreover, with and without clearance at the joint (and thus with and without the likelihood of backlash there), and with and without friction. Having thus explored the somewhat uncertain realm of the kinetostatics of joints in planar mechanism briefly, having studied, also, the behaviour of the planar 4*R*-loop as a whole, we return to more familiar ground. We return to ordinary Euclidean 3-space at § 17.21.

## The frictionless revolute in planar mechanism

02. Refer to figure 17.01(*a*). There the well known, planar, 4*R*-loop is illustrated. Its input (an externally driven shaft resisted by internal forces) is at *A* while its output (an internally driven shaft resisted by external forces) is at *D* (§ 2.35). We will argue for the sake of simplicity that the moving links are *massless*. This means that no accelerating forces upon the links are necessary and that no gravity forces are acting. Alternatively we could argue that the mechanism, although massive, is working *slowly* upon a smooth horizontal table, the gravity forces becoming thereby irrelevant. There are no clearances or friction as yet. With these idealized circumstances established, the coupler 3 of the mechanism is in direct compression with two equal, opposite, and collinear forces $\mathbf{F}_{23}$ and $\mathbf{F}_{43}$ acting upon it as shown. The first number among the subscripts here refers to the 'body of origin' of the force, while the second refers to the body in question, namely the 'free' body suffering the force (§ 12.28, § 13.26). Analogously we would write $\boldsymbol{\omega}_{23}$ and $\boldsymbol{\omega}_{43}$ whenever we wished to quote the angular velocities relative to 2 and to 4 respectively of link 3 (§ 12.12). Link 3, although moving relative to 1 and, accordingly, not at rest relative to that frame, is however in equilibrium relative to that frame. It is suffering the two mentioned forces only, and they add to zero (§ 10.57).

03. Look in figure 17.01(*a*) at joint *C*. The bearing belongs to the coupler 3, while the pin (or the journal) belongs to link 4. There is a force $\mathbf{F}_{34}$ acting from 3 upon 4 and, of course, an equal and opposite force $\mathbf{F}_{43}$. In the absence of radial clearance these two are distributed forces acting across the interface between the elements at the joint. The common point of application of their respective resultants is at *Q*. *The seen common line of action of the two pure forces acting alone at Q is the central axis of the action dual (or of the action duals) at the joint.* There is no couple **C** associated with this vector sum of the forces acting from one link upon the other at the joint (§ 10.59). The pitch of the wrench along the central axis of the action dual is zero in other words. This means that the pitch of the screw at the central axis upon which the wrench and its reaction wrench reside is also zero.

04. Look next in figure 17.01(*a*) at the equal and opposite angular velocities $\boldsymbol{\omega}_{34}$ and $\boldsymbol{\omega}_{43}$ occurring at joint *C*. In the absence of radial clearance these angular velocities reside upon their common line of occurrence through the geometric axis *R* of the cylindrical interfac-

ing surfaces at the joint, the line of occurrence being normal to the plane of the paper. *The seen common line of occurrence of the two angular velocities acting alone at R is the central axis of the motion dual* (*or of the motion duals*) *at the joint.* There are no rates of translation $\tau$ associated with these angular velocities (§ 10.49). The pitches of the motion duals are zero. The pitch of the rate of twisting of either one of the links relative to the other is zero in other words. This means that the pitch of the screw at the central axis upon which the equal and opposite rates of twisting reside is also zero.

05. The respective central axes of (*a*) the zero-pitch action dual $\mathbf{F}_{34}$, and (*b*) the zero-pitch motion dual $\boldsymbol{\omega}_{34}$, intersect one another at right angles. The moment $\mathbf{M}$ of the force $\mathbf{F}_{34}$ at the point anywhere upon the axis of the motion dual is perpendicular to that axis (§10.31); it has no component in the direction $\boldsymbol{\omega}_{34}$; so the work being done by the force per unit time (namely the power being lost to friction at the joint) is zero (§ 10.62). *The screws of both duals are of zero pitch; they intersect one another perpendicularly; they are a pair of reciprocal screws, not because they intersect, but because the power is zero.* Refer to table 14.01 at § 14.31. Refer to § 19.28.

Figure 17.01 (§ 17.02) Layout of the forces and linear velocities in the planar 4*R*-loop with and without clearances and with and without friction.

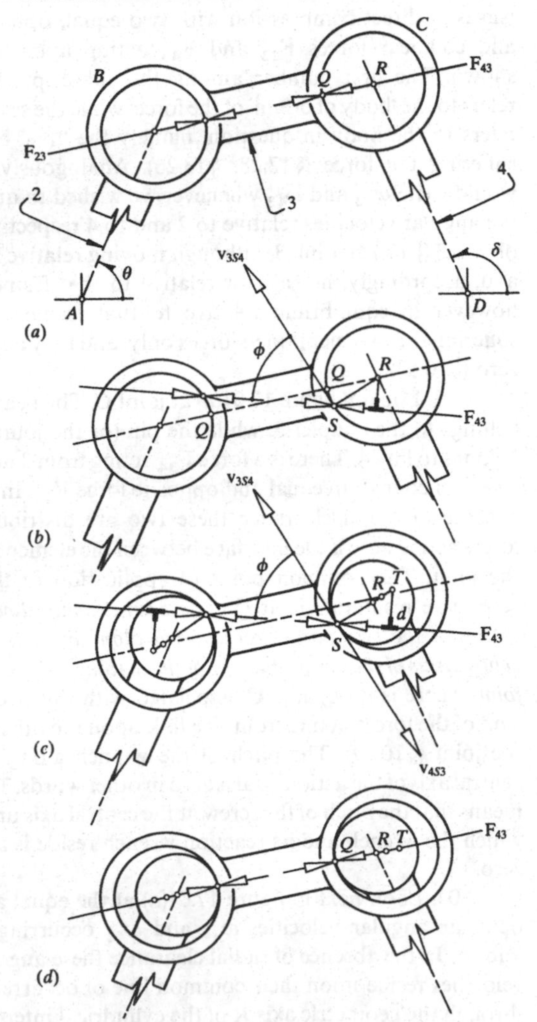


**Introduction of some friction and clearances**

06. In figure 17.10(*b*) let there be, now, at both *B* and *C* some friction. It is well argued in the established literature – and I take for example Hirschhorn (1962, 1967) – that the resultant line of action of the acting forces will have moved from *Q* to some new location through *S* as shown. A new angle $\phi$ (now less than a right angle) will have developed between the direction of the total force $\mathbf{F}$ at *S* and the direction of the rubbing velocity there. That the rubbing velocity $\mathbf{v}_{3S4}$ (§ 12.10) is in the direction shown is consistent with the relevant motion screw, the $\text{ISA}_{34}$, being normal to the plane of the paper at *R*, the pitch of the screwing at *R* being zero, and the angular velocity $\boldsymbol{\omega}_{34}$ being clockwise (§ 5.44, § 17.05). The power, or the rate at which the work against friction is being done, is $P = \mathbf{F} \cdot \mathbf{v}$; that is, $P = Fv\cos\phi$; and that is, $P = F\omega d$, where $d$ is the radius of the so-called *friction circle* at the revolute joint, and where $\boldsymbol{\omega}$ is the relevant angular velocity (§ 19.08). *The line of action of the resultant force and the axis of the angular velocity no longer intersect; the two screws mentioned at § 10.05 are no longer reciprocal, not because they no longer intersect, but because the power loss is no longer zero.* Refer to § 10.62 and again to table 14.01 at § 14.31.

07. In figure 17.01(*c*) let there be now at both *B* and *C* not only friction but also a radial clearance, allowing backlash to occur. Due to friction the point of contact which was at *Q* will be at *S*; how far it will have moved from *Q* to *S* depends upon the severity of the friction; note the angle $\phi$. Due to clearance the motion screw $\text{ISA}_{34}$, which was at *R* at the centre of the bearing circle, will be now at some other point say *T* along the line *SR*; how far it will have moved from *R* to *T* and in what direction depends not upon the clearance but upon the time-rate of change of the angular velocity. The motion screw, normal to the plane of the paper, reveals itself as a point of course, the *pole* as we know it in planar kinematics. The central axes of (*a*) the zero-pitch action dual through *S*, and (*b*) the zero-pitch motion dual through *T* are now no longer intersecting. They are skew. *The screws of both duals are of zero pitch; they are perpendicular yet skew with one another; they are not a pair of reciprocal screws, not because they are skew, but because the power is not zero.* Refer to table 14.01.

08. The power being lost at *C*, in any one of the figures in figure 17.01, may be written more precisely

$P = F_{34}v_{3S4} \cos \phi$, and that is, $P = F_{34}\omega_{34}d$. Power is a scalar, so the question of which duals of the equal and opposite pairs of duals in each of the central axes might be taken for the calculation is a nonsensical question (§ 10.59). With force expressed in newtons, angle in radians, distance in metres, and time in seconds, the calculated power will find itself expressed in watts.

09. In figure 17.01($d$) the presence of friction has been removed but the clearances remain. The angle $\phi$ is again 90°, the radius (or the distance) $d$ and thus the power loss is again zero, and the two screws of zero pitch (passing through $Q$ and $T$ respectively) intersect again. *The two intersecting screws are reciprocal; it is however very important to be aware that the two reciprocal screws at a frictionless joint in mechanism do not in general intersect; in general they are skew; also, in general, their pitches are not zero.* Refer to chapter 10 and to § 17.26 *et seq.*

### Backlash in the planar 4*R*-loop

10. Before going on to examine the general question at issue in this chapter, namely the question of action duals and motion duals and backlash in spatial loops in the absence of friction, it will be useful to look at the question of backlash in the planar 4*R*-loop. Refer again to figure 17.01. We will assume that all four hinged joints of the linkage have equal radial clearances $\Delta r$ and, as before, that axial clearances are absent, so that the axes of the bearings and those of the pins remain parallel at all times. Take now the linkage as set up in figure 17.02($a$) where the four pins are central in their bearings, where the angle $\theta$ at the input link 2 is nominated, and where the angle $\delta$ at the output link 4 is determined by $\theta$ and the fixed link-lengths $a$, $b$, $c$ and $d$. For zero backlash, there are well known equations that relate all of the relevant variables throughout a cycle of the motion (§ 19.47, § 19.48).

11. Ignoring gravity and friction we see that should a pure clockwise couple $\mathbf{C}_{12}$ of unit magnitude be applied to the input link 2 as shown in figure 17.02($b$), and should the output link 4 be held at its orientation $\delta$ by a pure counterclockwise couple $\mathbf{C}_{14}$ of the necessary magnitude, a clockwise rotation of 2 through an angle $\Delta_1\theta$ takes place before the backlash in that direction is taken up. Because the coupler can sustain no bending, and because there are no forces associated with the couples applied, the distorted configuration of the linkage exhibits dislocations at each of its four joints in directions parallel with the centre line of the coupler. The four pairs of reaction forces **F** at the four joints are equal, and, as noted by the subscripts, the unequal reaction couples applied to the input and output links have originated in the frame. The frame 1 accordingly suffers not only its pair of reaction forces **F**, which together constitute a counterclockwise couple, but also the unequal pair of reaction couples $\mathrm{C}_{21}$ and $\mathrm{C}_{41}$. The sum of these three couples acting upon the frame is zero.

12. In figure 17.02($c$) a unit couple has been applied at 2 in the opposite direction namely counterclockwise, and a clockwise reaction couple of the necessary magnitude is maintaining the orientation $\delta$ at 4 as before. With the same overall conditions obtaining (absence of gravity and of friction) a counterclockwise rotation $\Delta_2\theta$ of 2 has occurred to take up the backlash in this opposite direction. The two angles $\Delta_2\theta$ and $\Delta_1\theta$ are in general not equal incidentally, but we may define for convenience now the total angular backlash $\Delta\theta$ at $\theta$ to be $\Delta\theta = |\Delta_1\theta| + |\Delta_2\theta|$. Be aware not only that $\Delta\theta$ will

Figure 17.02 (§17.10) This figure helps to show that, for a radial clearance $\Delta r$ at all four joints, and for a combination of clockwise and counterclockwise couple at the relevant input link, we can quote the two difference backlashes $\Delta\theta$ at $A$ and $\Delta\delta$ at $D$ by means of equations (1) and (2).

(1) $\Delta\theta = |\Delta_1\theta| + |\Delta_2\theta|$

(2) $\Delta\delta = |\Delta_1\delta| + |\Delta_2\delta|$

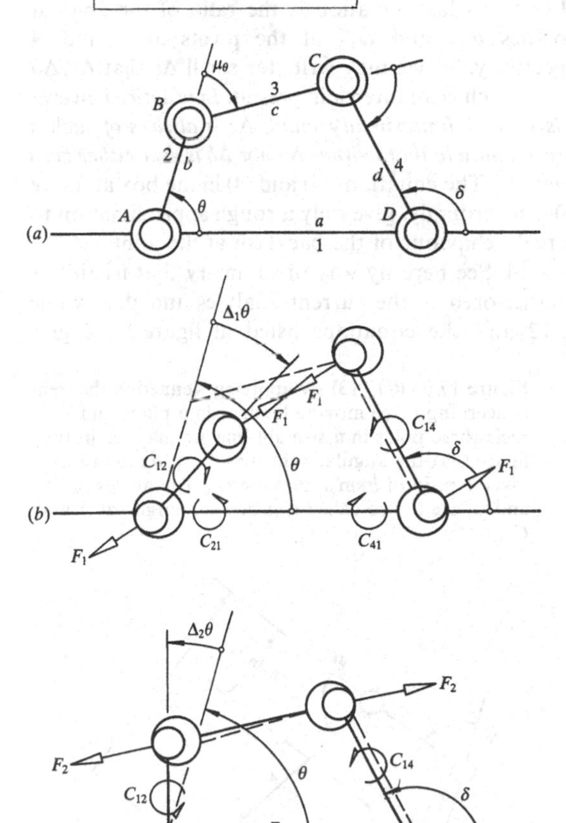

vary somehow with $\theta$, but also that it will be very large whenever the acute transmission angle $\mu_\theta$ at $B$ is very small. Reversing now the direction of the argument, driving at each direction of $\delta$ and resisting on each occasion at $\theta$, we see that a total angular backlash $\Delta\delta$ at output can similarly be nominated, $\Delta\delta = |\Delta_1\delta| + |\Delta_2\delta|$. For any given $\theta$ (or for any given $\delta$), *the total backlash $\Delta\delta$ at output does not, in general, equal the total backlash $\Delta\theta$ at input*. See equations (1) and (2) in the box at figure 17.02.

13. Please make a careful study now of the dislocations illustrated in figure 17.02(*b*). Refer to the related figures 17.04(*a*) and 17.04(*b*). These latter are not explained as yet, but they can be used to discover that, in the limit as $\Delta r \to 0$, $\Delta_1\theta = 4\Delta r/(b \sin \mu_\theta)$. From figure 17.02(*c*) the same expression (again for small $\Delta r$) may be found for $\Delta_2\theta$; $\Delta_2\theta = 4\Delta r/(b \sin \mu_\theta)$. We may accordingly write for small $\Delta r$ that $\Delta\theta = 8\Delta r/(b \sin \mu_\theta)$. Similarly a similar expression may be written for $\Delta\delta$. See equations (3) and (4) in the box at figure 17.04. By dividing these equations we may write (and this is again for small $\Delta r$) that $\Delta\theta/\Delta\delta = (d \sin \mu_\delta)/(b \sin \mu_\theta)$. But by Kennedy's theorem in figure 17.03 (§ 13.06) it will be seen that the RHS of this last equation is the ratio of the angular velocities $\boldsymbol{\omega}_{14}$ and $\boldsymbol{\omega}_{12}$ at the pivots at $D$ and $A$ respectively. So we may write for small $\Delta r$ that $\Delta\theta/\Delta\delta = \dot{\delta}/\dot{\theta}$, which is an interesting result. *In practice however $\Delta r$ is never infinitesimally small; $\Delta r$ is always of such a finite magnitude that neither $\Delta\theta$ nor $\Delta\delta$ is ever either zero or infinity.* The equations (3) and (4) in the box at figure 17.04, accordingly, give only a rough approximation to the real behaviour of the backlash at the joints.

14. See here by way of summary that friction is being ignored in the current analyses and that, while § 17.12 and the equations listed at figure 17.02 give accurately the backlashes in the abscence of friction, § 17.13 and the equations listed at figure 17.04 give more simply but only approximately the said backlashes. Please notice also by way of transition that, to calculate accurately the values throughout a cycle of $\Delta\theta$ and $\Delta\delta$ for a finite $\Delta r$ we could use the following idea [1].

15. Refer to the upper part of figure 17.04(*a*) and notice that, if each link of the 4*R*-loop *ABCD* is replaced by a thin rod of length equal to the nominal centre-to-centre length of that link, and if the four thin rods are then joined at their extremities by short flexible strings of length $\Delta r$, the resulting planar device would (upon a smooth table) be wholly equivalent in its geometrical action to the original 4*R*-loop with its finite radial

Figure 17.03 (§ 17.13) By applying Kennedy's theorem concerning three moving laminae in a plane and their three poles in a straight line, we may see in this figure that the angular velocities $\dot{\theta}$ and $\dot{\delta}$ are in the inverse ratio of $b \sin \mu_\theta$ and $d \sin \mu_\delta$. The angles $\mu_\theta$ and $\mu_\delta$ are the so-called transmission angles at $B$ and $C$.

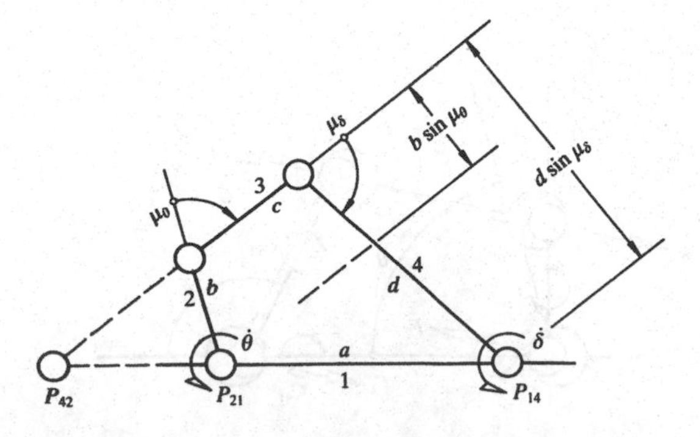


Figure 17.04(*a*) (§ 17.14) The two parts of this figure relate directly to the two parts (*a*) and (*b*) of figure 17.02. The radial clearances $\Delta r$ at the four joints of a planar 4-link loop are replaced by four flexible inelastic strings of length $\Delta r$. These are then seen to join the end-points of rigid rods whose lengths are the nominal lengths of the original links. While the top picture shows the initial configuration of the linkage before the clearances are taken up, the lower picture shows the dislocations after the clearances are taken up and the imagined short coupler whose length is $(c - 4\Delta r)$. To save space, the imagined long coupler, implicit in figure 17.02(*c*), is not shown explicitly, either there or here.

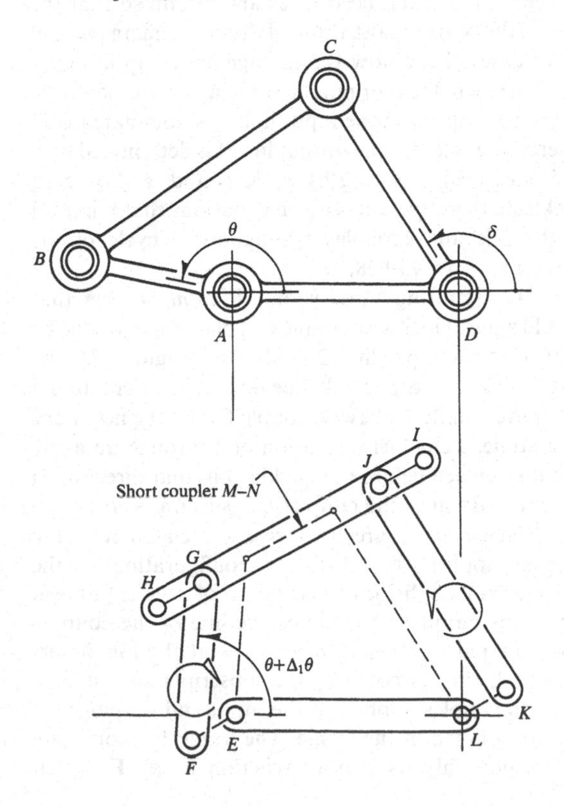

clearances $\Delta r$. Refer to the lower part of figure 17.04(*a*) and to the whole of figure 17.04(*b*) and see that these purport to show in their different ways the mentioned rods and strings in action, that relevant points in the actual linkage *ABCD* above have been relettered below, and that the lengths $\Delta r$, namely *E–F*, *G–H* etc., have been for the sake of clarity exaggerated. The arrangement of rods and strings is being subjected to a driving clockwise couple at input and a resisting counterclockwise couple at output in such a way that $\delta$ remains unchanged as before while the link 2, namely *F–H*, is displaced through $\Delta_1\theta$; the four short strings *E–F*, *G–H* etc. have taken up their parallel locations and are all in tension. The mentioned figures both correspond in other words with figure 17.02(*b*). See next in the lower part of figure 17.04(*a*) an imagined 4*R*-loop *ENML* without clearances, where *E–M* and *N–L* are equal and parallel respectively with *F–H* and *J–K* as shown. Notice in this newly imagined linkage, which has what I have called in the figure a *short coupler* of length $(c-4\Delta r)$, that the functional relationship between its $\delta$ at output and its $\theta$ at input is the same relationship which would obtain in practice between $\delta$ and $\theta$ in the original linkage with clearances when either *A–B* or *C–D* is being driven in such a way that the coupler *B–C* is in compression. An

Figure 17.04(*b*) (§ 17.14) A panoramic view purporting to show more clearly the attachments of the strings and the dislocations in the lower part of figure 17.04(*b*). A clockwise couple is being applied at the input link 2 (which appears in the middle foreground) and the four short strings are all in tension.

| | |
|---|---|
| (3) | $\Delta\theta \simeq 8\Delta r/(b \sin \mu_\theta)$ |
| (4) | $\Delta\delta \simeq 8\Delta r/(d \sin \mu_\delta)$ |

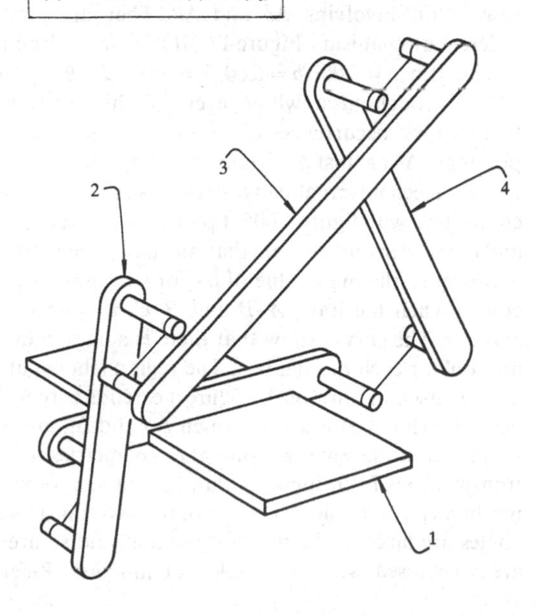


exactly similar argument will lead to the notion that a *long coupler* of length $(c+4\Delta r)$ in another newly imagined linkage with no clearances will provide the same functional relationship between its $\delta$ and $\theta$ as the one that obtains in the original linkage with clearances whenever the manner of driving is such that the coupler *B–C* is in tension. The concepts (*a*) of the short coupler, and (*b*) of the long coupler (the long coupler is not explained by means of figures here), both need to be digested by the reader.

**Some numerical examples within the planar 4*R*-loop**

16. In the following examples we take three planar linkages, all 4*R*-loops like the one that is shown in figure 2.01(*m*). Their groups of four links are of the following sets of lengths: (1) $a, b, (c-4\Delta r), d$; (2) $a, b, c, d$; and (3) $a, b, (c+4\Delta r), d$. We choose some numerical values and calculate for each of these linkages (each with zero clearances) the relationship between $\delta$ and $\theta$ for a cycle of the motion. We use for this work the well known closure equation of Freudenstein (§ 19.48). We get for each set of numerical values (that is for each real mechanism with its clearances) a set of three curves as shown for example in figure 17.05(*a*), and note them correspondingly (1), (2) (not marked in the figures), and (3). Figure 17.05 is drawn from numerical results computed for the crank rocker linkage $a=9.00$, $b=4.00$, $c=6.00$, $d=8.00$, with $\Delta r=0.05$. The linkage itself is drawn to scale in figure 17.05(*b*). Notice firstly that, with *A–B* driving counterclockwise and $\theta$ increasing from zero in figure 17.05(*b*), the real mechanism with its clearances will follow the thickened curve through the two blebs appearing in figure 17.05(*a*). The slope of this curve is discontinuous at the blebs. After descending to its minimum the curve jumps in transition at the critical value $\delta_1$ of $\delta$ from the short-coupler curve (1) across to the bleb on the long-coupler curve (3). During this jump the resisting couple at the output link *C–D* is presumed to be zero. We can argue geometrically that, between $\theta_1$ and $\theta_2$ of $\theta$, the output link *C–D* remains at rest at the value $\delta_1$ of $\delta$, having fallen short as it were of its nominal limit $\delta_2$. A similar jump occurs at the critical value $\delta_3$, $\delta$ having failed to reach its nominal limit $\delta_4$, and a similar irregularity occurs there with $\theta$.

17. Figure 17.05(*a*) also shows a computed plot of $\Delta\theta$ against $\theta$, the accurate curve for $\Delta\theta$ being superimposed upon the approximate curve for $\Delta\theta$ (§ 17.14). Please note that the range of $\theta$ goes only from zero to 180° or so; the curves are not complete. Here it can be seen by way of example, however, how the approximate equation in the box at figure 17.03 is very close to reality at all configurations of the linkage except where the linkage is at or near the occurrence of the value zero of $\mu_\theta$

Figure 17.05 (§17.17) At (*a*) and drawn thick is a portion of the computed input–output curve $\delta$ versus $\theta$ for the chosen crank-rocker mechanism with its chosen clearances and, against the same axis for $\theta$, is plotted the backlash $\Delta\theta$. Both curves are taken across the first of two bad patches for $\theta$; this patch extends from $\theta_3$ to $\theta_4$. All marked angles are in degrees. At (*b*) a polar plot against $\delta$ of the backlash $\Delta\delta$ is constructed. Certain key configurations of the linkage itself are also shown at (*b*).

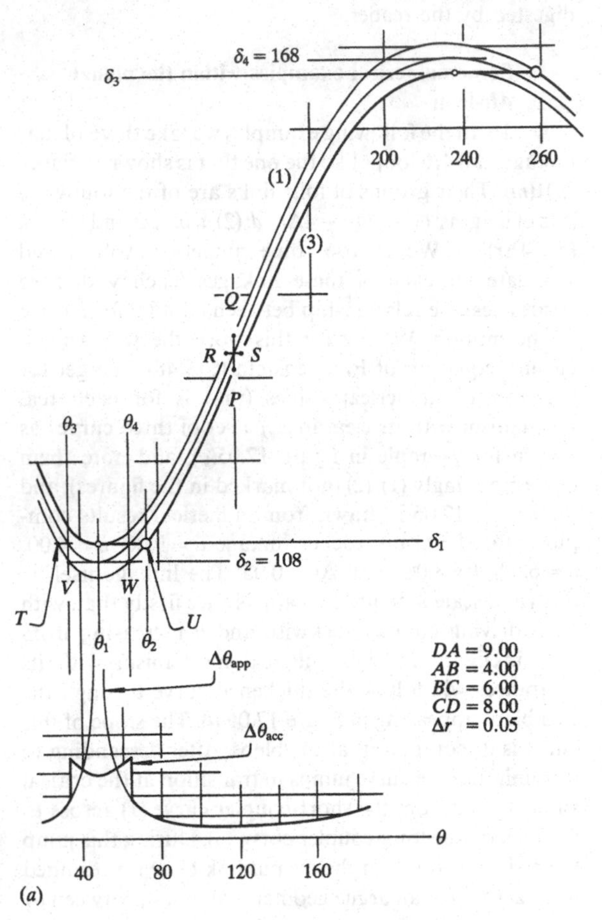


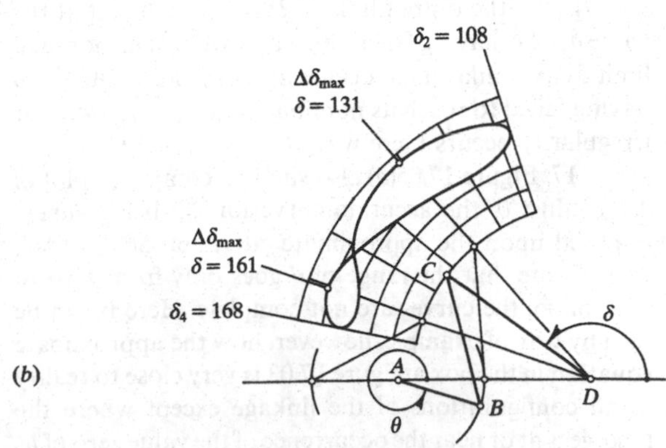


(§18.08). At the critical values $\theta_3$ and $\theta_4$ of $\theta$, which straddle the configuration of zero $\mu_\theta$, finite jumps occur in $\Delta\theta$ and, between these jumps, the curve for $\Delta\theta$ dips to a local minimum as shown. At the general value $\theta$ shown, $\Delta\delta$ is $P$–$Q$ while $\Delta\theta$ is $R$–$S$. At $\theta_3$ and $\theta_4$, as shown, $\Delta\theta$ is $T$–$U$, while the local minimum value of $\Delta\theta$ between the peaks at $\theta_3$ and $\theta_4$ is $V$–$W$. These displayed phenomena are typical of a transmission angle zero wherever it may occur and in whatever kind of planar 4*R*-linkage. Notice however that there is no corresponding occurrence of a zero $\mu_\delta$ in this particular example; we find only that $\Delta\delta$ arrives at a local maximum whenever $\mu_\delta$ arrives at a local minimum, that is, whenever $\theta$ arrives at zero or 180°. The backlash $\Delta\delta$ is plotted radially outwards from the arc traced by the pin $C$ in the changing configuration diagram shown in figure 17.05(*b*).

18. Figure 17.06 deals with results computed for an inversion of the same linkage where $a = 8.00$, $b = 9.00$, $c = 4.00$, $d = 6.00$, and $\Delta r = 0.05$ as before. At figure 17.06(*a*) the loop-like relationship between $\delta$ and $\theta$ for this double-rocker linkage is shown for short (1), nominal (2), and long (3) couplers. In figure 17.06(*b*) critical values of $\delta$ and $\theta$ are shown by means of a changing configuration diagram for the linkage. In figure 17.06(*c*) the backlash $\Delta\theta$ is plotted radially outwards against $\theta$ on the arc traced out by the crank-pin $B$. The approximate curve for small $\Delta r$ – refer to equation (3) in the box at figure 17.03 – is again superimposed upon the accurate plot. The characteristic departures from the approximate of the actual $\Delta\theta$ are clearly shown; these occur at the two occurrences of a zero $\mu_\theta$.

19. The separated figures 17.07(*a*) and 17.07(*b*) show plots involving $\Delta\delta$ and $\Delta\theta$. They are for two different mechanisms. Figure 17.07(*a*) is for the drag-link mechanism $a = 4.00$, $b = 6.00$, $c = 8.00$, $d = 9.00$, $\Delta r = 0.05$. Throughout a whole cycle of this mechanism there are no occurrences of zero transmission angle. I plot there $\Delta\delta$ against $\Delta\theta$. Figure 17.07(*b*) is for the crank rocker mechanism already drawn and discussed in connection with figure 17.05. I plot there $\Delta\theta$ against $\Delta\delta$; and the reader will see there that $\Delta\theta$ does indeed grow on occasion to the high value of 88° or so. This occurs of course when the links $A$–$B$ and $B$–$C$ lie atop of one another. The curves show that there is a 'region of safe motion' for each mechanism. The region (shown at the linkage itself in both of the figures) extends across that neighbourhood of $\theta$ within which $\Delta\delta$ and $\Delta\theta$ are both small. In these safe regions the geometries of the transmissions from input to output and vice versa are mechanically reliable. The sums of the two transmission angles are large in the relevant regions, and so are the areas enclosed within the relevant linkages. Refer to

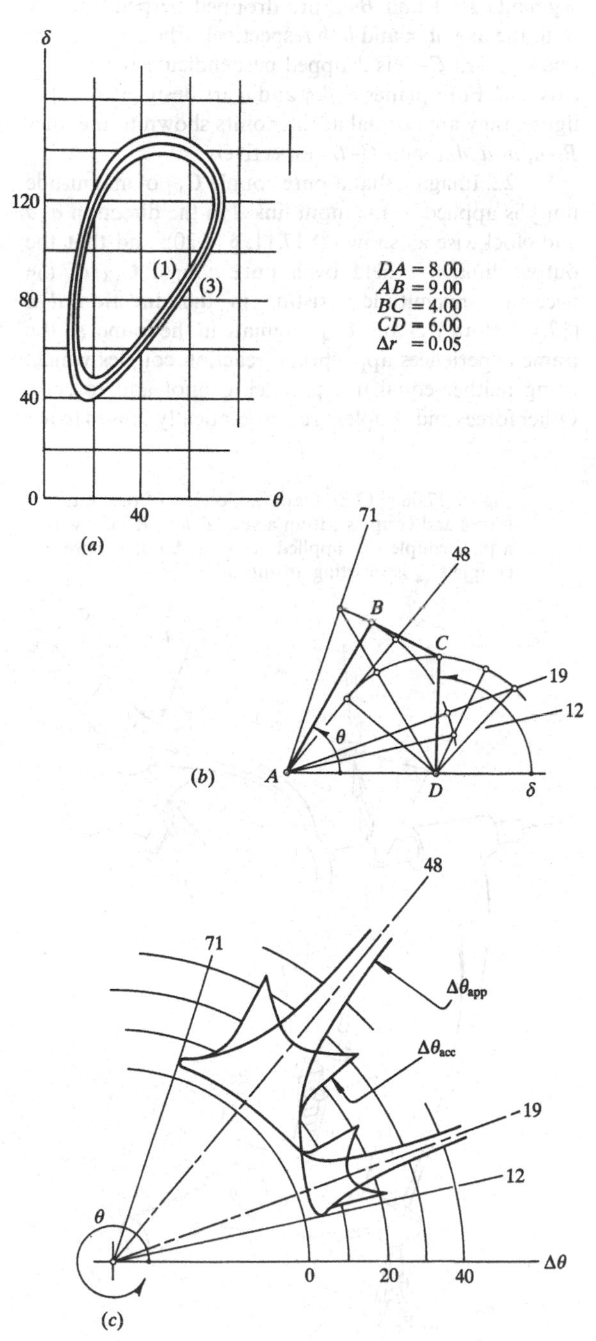


Figure 17.06 (§17.18) At (*a*) is shown the three computed input–output curves ($\delta$ versus $\theta$) for the three versions of the chosen planar double-rocker mechanism given its chosen clearances. Refer to figure 2.03 for the loop-shaped curve which is characteristic of all such double-rocker arrangements, whether planar or spatial. At (*b*) the linkage itself appears in certain key configurations, and at (*c*) we see the backlash $\Delta\theta$ at input plotted in a polar manner against the input angle $\theta$.

§18.08 where I mention some of the difficulties met in defining transmission angle and where I refer to the earlier geometrical works of Alt and Volmer (§2.42).

**Some important tentative conclusions**

20. The foregoing material is not especially spectacular and it is, quite probably, not even entirely new. It does however give a glimpse of the difficulties that might be encountered in the analysis of spatial mechanism where clearances, and thereby the tendency to backlash at its joints, exist. It shows moreover some of the possible ways for clarifying that general and, apparently, much more difficult question. It will already be clear for example (*a*) that the application of a pure couple at an arbitrarily chosen input link is an arbitrarily chosen method of loading which has, perhaps, no special claim

Figure 17.07(*a*) (§17.19) This shows the chosen planar drag-link mechanism in a number of key configurations along with a Cartesian plot of the backlash $\Delta\delta$ against the backlash $\Delta\theta$. This curve, calibrated in terms of input angle $\theta$, shows very clearly the safest range of $\theta$ for accurate transmission of angular displacement.

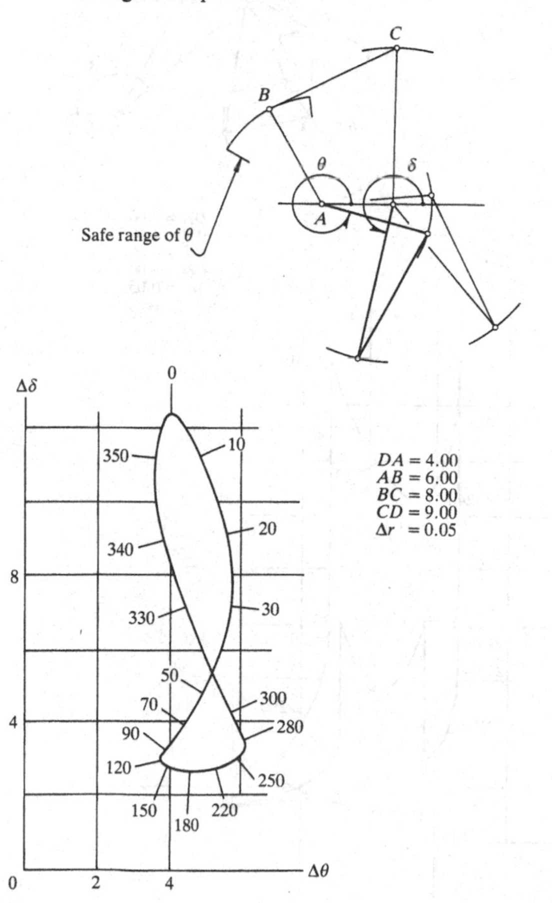

to merit over other methods of loading, (*b*) that any definition of backlash will depend upon the method of loading, (*c*) that for any kinematical purpose a static force analysis will always need to be made of a linkage, for otherwise the dislocations at its joints will not have been determined, (*d*) that no static force analysis can be made unless the mobility is unity, (*e*) that the whole of the current argument is purely geometrical, taking no account of friction, mass, or inertia, and (*f*) that complicated phenomena are sure to occur at what I have called, in various places and somewhat vaguely, the change points (§ 18.16). The following, limited discussion is of a 4-link spatial loop of mobility unity, the linkage *RSCR*.

Figure 17.07(*b*) (§ 17.19) Refer to figure 17.05. Here we see the Cartesian plot of $\Delta\delta$ against $\Delta\theta$ for the crank-rocker mechanism of figure 17.05. Note once again that this curve for a mechanism shows very clearly the safe and the dangerous ranges of $\theta$ for the accurate transmission of angular displacement.

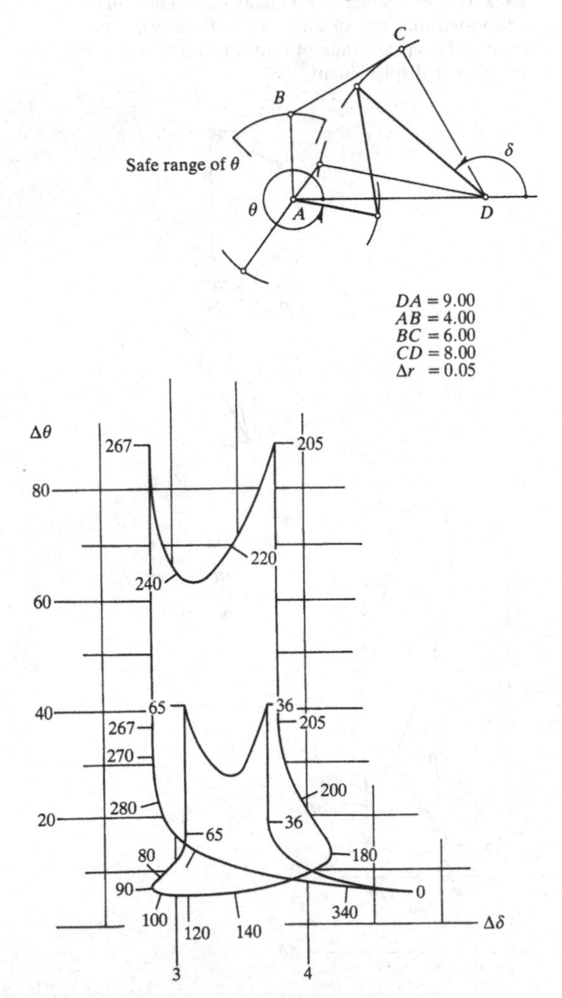


### A static analysis of a spatial 4-link loop

21. At figure 17.08 there is a frame link 1 which is not shown, an input link 2 with a hinge to frame on the axis *a–a* and a ball whose centre is at *B*, a coupler link 3 extending from the socket of the ball to the barrel of a cylindric joint on the axis *c–c*, and an output link 4 from the shaft of the cylindric joint to a second hinge to frame on the axis *d–d*. The loop is an *RSCR*. From *B* the line segments *B–A* and *B–C* are dropped perpendicularly onto the axes *a–a* and *b–b* respectively, then from *C* the line segment *C–D* is dropped perpendicularly onto the axis *d–d*. Four planes $\alpha$, $\beta$, $\gamma$ and $\delta$ are designated in the figure; they are normal at the points shown to the lines *B–A*, *a–a*, *d–d* and *C–D* respectively.

22. Imagine that a pure couple $\mathbf{C}_{12}$ of magnitude unity is applied to the input link 2 in the direction *a–a* and clockwise as shown (§ 17.11, § 17.20), and that the output link 4 is held by a pure couple $\mathbf{C}_{14}$ of the necessary magnitude resisting in the direction *d–d* (§ 7.12). Both $\mathbf{C}_{12}$ and $\mathbf{C}_{14}$ originate in the frame, so the frame experiences appropriate reaction couples which, being neither equal nor parallel, cannot add to zero. Other forces and couples are automatically caused to act

Figure 17.08 (§ 17.21) Panoramic view of the static forces and couples within a spatial *RSCR* loop with a pure couple $\mathbf{C}_{12}$ applied at input. Another pure couple $\mathbf{C}_{14}$ is resisting at output.

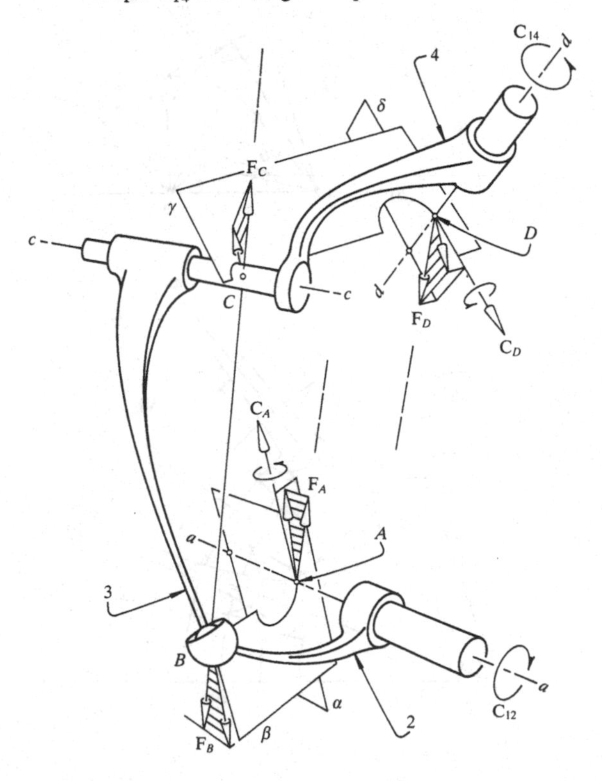

upon the frame however (§ 17.25, § 23.28), and these will be seen to effect the overall balance.

23. Independently of what may be written about action and action duals in chapters 6, 7, 10, 13 and 23 we can see from first principles here the natures of the joints at $B$ and $c$–$c$. The ball and socket at $B$ can transmit only a wrench of zero pitch namely a pure force which passes directly through the point $B$ at the centre of the ball. We can see also that the cylindric joint at $c$–$c$ can transmit only a wrench whose central axis is perpendicular to $c$–$c$ and passing through $c$–$c$. It accordingly becomes clear that a pure force reaction $\mathbf{F}_{32}$ on link 2 is mobilized from 3 at the point $B$; this force, shown in figure 17.08, is given the name $\mathbf{F}_B$ there. The position and direction of $\mathbf{F}_B$ must be such that it is collinear with the line $C$–$B$. Its magnitude however is determined by the capacity of its equal and opposite reaction force $\mathbf{F}_A$ (acting on 2 from 1 anywhere through the axis $a$–$a$ of the hinge but for convenience shown to be acting through $A$) to combine with $\mathbf{F}_B$ to produce a reaction couple on 2 whose component in the direction $a$–$a$ will balance the unit couple applied. In the figure the equal and opposite components of $\mathbf{F}_B$ and $\mathbf{F}_A$ in the plane $\beta$ are shown. The product of the magnitude of these components and the distance between their lines of action must be unity. The equal and opposite components in the direction $a$–$a$ of $\mathbf{F}_B$ and $\mathbf{F}_A$ however are not balanced by the unit couple $\mathbf{C}_{12}$. They are balanced separately by the couple $\mathbf{C}_A$ which is mobilized from the frame link 1 at hinge $a$–$a$ and can be nominated to act (along with $\mathbf{F}_A$) on link 2 at $A$. The action dual at A consists of the combination of $\mathbf{F}_A$ and $\mathbf{C}_A$. It acts from 1 upon 2 and it, along with its equal and opposite action dual which acts in reaction from 2 upon 1, constitutes the total 'action' being transmitted by the hinge $a$–$a$ (§ 10.52). Please refer to § 12.67 and § 13.26 for the confusing mixture of subscripts that may be used for the vectors **F** and **C**.

24. This 'transmission of an action' through a frictionless joint in spatial mechanism is associated with the whole question of reciprocal screw systems (chapters 14 and 23) and cannot be discussed in detail here. It must be said in explanation however that any frictionless joint has the capacity to transmit a limited range of wrenches without suffering relative motion of its links, and it will be found in the present example that the action dual at $A$, when expressed as a wrench upon its home screw, will belong to that limited range of wrenches which can be transmitted in the absence of friction by the hinge at $a$–$a$; compare the passages beginning at § 10.69, § 14.27 and § 17.27. I now return to remarks I made at § 2.02 and mention here that the presence of friction at the cylindric joint on $c$–$c$ would clearly affect the behaviour of the present apparatus; the eccentricity of the loading at that joint (seen in relation to the location and the length of the actual barrel which is drawn there) may cause jamming or judder to occur; for a short discussion of these phenomena please refer to § 18.11 through to § 18.15 inclusive.

25. Continuing to argue in figure 17.08 (and in the absence of friction), we arrive at the action dual at D which is shown and which is acting from the frame upon link 4. It consists of ($a$) the force $\mathbf{F}_D$ which is equal and opposite to $\mathbf{F}_A$, and ($b$) the couple $\mathbf{C}_D$ which balances the components in the direction $d$–$d$ of the equal and opposite forces $\mathbf{F}_C$ and $\mathbf{F}_D$. Again it will be found that this action dual at $D$ when expressed as a wrench will be acting upon a screw (one of the limited number of action screws) that is reciprocal to the single screw of zero pitch (the single motion screw) about which the relative rotation at $d$–$d$ takes place. The reaction action duals acting upon the frame at $A$ and $D$ which are unequal and the reaction couples $\mathbf{C}_{21}$.and $\mathbf{C}_{41}$ which are also unequal add collectively to zero. These latter reaction couples, which must be sustained to course at the frame, were mentioned by implication previously (§ 17.22).

**Spatial dislocations at the separate joints**

26. By using a simple example that was not only simple but also highly idealized, I have thus shown how a static analysis might be made of a movable spatial mechanism. I loaded a simple loop of mobility unity with a pure unit driving couple and an unequal non-parallel resisting couple, the necessary balancing couple was mobilized from the frame, and I assumed the absence of friction. I would like to suggest next what kinds of calculation might be made to discover the dislocations occurring at the separate joints with clearances that are caused by the transmission through those joints of the given discovered wrenches. Apart from remarking that White's idea of the flexible strings [1] can be extended into three dimensions with suitable modifications – there being frictionless sliding rings upon rods and frictionless sliding plates upon planes as well as ordinary knots at the extremities of the strings to equivalate the various joints – I shall do no more than make a brief analysis here, first of the forces at the points of contact within a revolute joint transmitting a wrench, and next of the dislocations permitted by small clearances within that and other joints (§ 17.29).

27. Figure 17.09($a$) shows a revolute joint of the kind where two double points of contact are made between the journal and the bearing and one single point of contact is made between one end of the bearing-piece and the face of a thrust-piece fixed upon the journal (§ 1.49, § 1.53). There exists of course the other kind of revolute joint where these two numbers (the two and the one) are reversed. A wrench of magnitude $F$ and of pitch $p$, namely the combination of a force **F** and a couple $p\mathbf{F}$,

is applied from outside and is acting as shown upon the movable bearing. An equal and opposite reaction wrench is mobilised from the fixed frame and is acting from the fixed pin upon the bearing to hold it stationary. Refer to photograph 17. The relevant dimensions $e$, $f$, $g$, $q$ and $\phi$ are known. The three expected reaction forces $R_\alpha$, $R_\beta$ and $R_\gamma$ are shown acting in their known directions at their respective points of contact, and the object of the exercise is to determine them. This involves the calculation not only of their magnitudes (their directions being known), but also their locations; we require to know the angles $\alpha$, $\beta$ and $\gamma$ that are shown. Note that the angles $\alpha$ and $\beta$ will be in general different. By means of an apparatus such as the one that is shown in photograph 17 one may solve this problem for at least one case experimentally; by providing a small radial clearance and a large axial one, one may ensure that the revolute is of the type mentioned; and, by using a trace of oil that may escape, the points of contact within the joint may be shown up clearly by the wedge-shaped meniscuses that will form.

Figure 17.09($a$) (§17.27) A wrench is here sustained by a revolute joint and the five reaction forces at the five points of contact are shown. Two pairs of the five points of contact are coalesced, at $A$ and $B$ respectively.

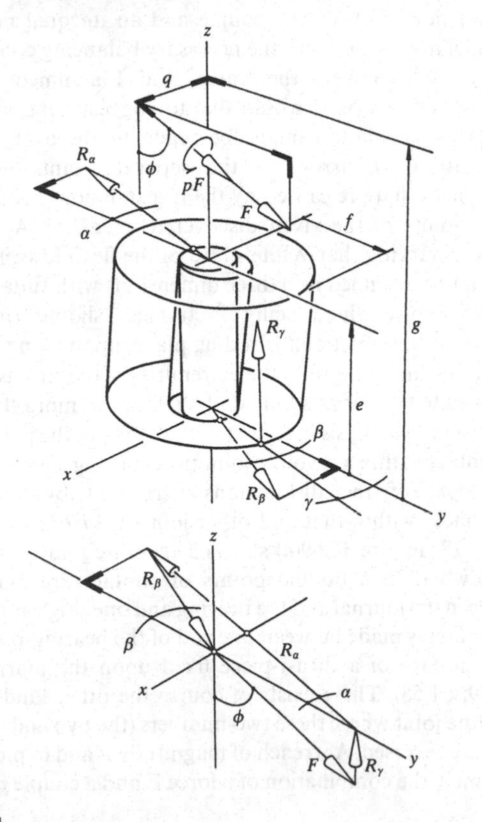


28. By taking moments about the $z$-axis in figure 17.09($a$), we obtain the relationship $p = q\tan\phi$. This is the fundamental condition that needs to be fulfilled before the wrench upon its action screw and the rate of pure rotation upon its motion screw can be reciprocal (§14.19, §14.29). Refer also §10.64. It is moreover the fundamental condition that sets the circumstance here that, having freely chosen a line in space for the central axis of the applied wrench, its pitch will be determined; refer to appendix 6A at chapter 6. By resolving forces and taking moments in each of the three directions in figure 17.09($a$), by simplification, and by writing equation (10), which can be seen by inspection on looking against the $xy$-plane, we arrive at the following five equations:

$$R_\beta \sin\beta - R_\alpha \sin\alpha = 0 \quad (6)$$

$$R_\beta \cos\beta - R_\alpha \cos\alpha = -F\sin\phi \quad (7)$$

$$eR_\alpha \cos\alpha + fF\cos\phi\cos\gamma = gF\sin\phi \quad (8)$$

$$eR_\alpha \sin\alpha + fF\cos\phi\sin\gamma = qF\sec\phi \quad (9)$$

$$2\gamma = \alpha + \beta. \quad (10)$$

These are not easy to solve for the five unknowns unless the semi-graphical method due to Sticher [2] is used. Refer to figure 17.09($b$). The point $(x, y)$ on the circle must be chosen by trial until the angles $\beta$ are equal, whereupon the solutions for $\alpha$, $\beta$, $\gamma$, $R_\alpha$ and $R_\beta$ are as shown. Iterative methods could also be used to solve the equations numerically.

Figure 17.09($b$) (§17.28) This figure illustrates the semi-graphical solution for the five equations which are numbered in the text from (6) to (10); the equations are derived from the apparatus in figure 17.09($a$). The angle $\gamma$ is altered until the angles $\beta$ become equal.

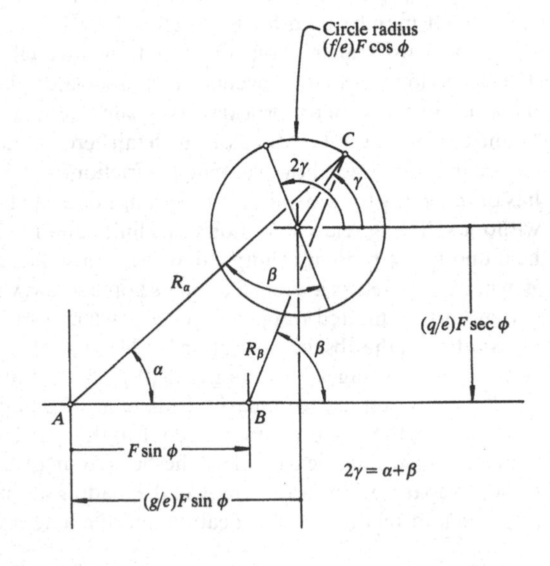

### Accumulation of the dislocations around a loop

29. The reader will have seen by now that the statical analysis of a $4R$-mechanism of mobility unity, or what we might have called at the time *the statics of a loop*, which began at § 17.21 dealt only with an example; generalisation of the method was left by implication as an exercise for the reader. Similarly I say that the study of the points of contact within the revolute joint, or what we might have called at the time *the statics of a joint*, which began at § 17.26 dealt only with an example; the reader will have seen by now that he will have to have and to do his own examples.

30. Generalizing our studies in such a way that we begin to consider the statics not only of *loops* in general but also of *joints* in general, we might get to the stage where the combination of those two might be seen by us as the central study. Next we could introduce the occurrence of clearances (whose dimensions of course would be small compared with those of the links) and thus begin to study the dislocations at the joints. We could then call the combined study, for the want of a better name, *dislocation accumulation.*

31. I would like to project what I have said in this chapter forward into the material of chapter 18. There are many matters mooted here that might be discussed there. Can we explain for example that the total dislocation in a driven closed loop of mobility unity will be very great both at and near those special configurations of the loop where an extra transient mobility suddenly presents itself? Please think about sogginess and the so-called singular configurations in robot arms where difficulties of various kinds inexplicably occur (§ 2.46), and refer to chapter 18.

### Disclaimer

32. The very wide question of clearances and backlash has been treated here by me in a desultory way. I have simply picked out a few spectacular matters and played with a few elementary ideas. Dynamics has been ignored, and none of the work is conclusive.

### Notes and references

[1] The ideas outlined in § 17.15 are due to Kim White; see Preface. Refer § 17.14.

[2] The sèmi-graphical solution is due to Friedjof Sticher; see Preface. Refer § 17.28.

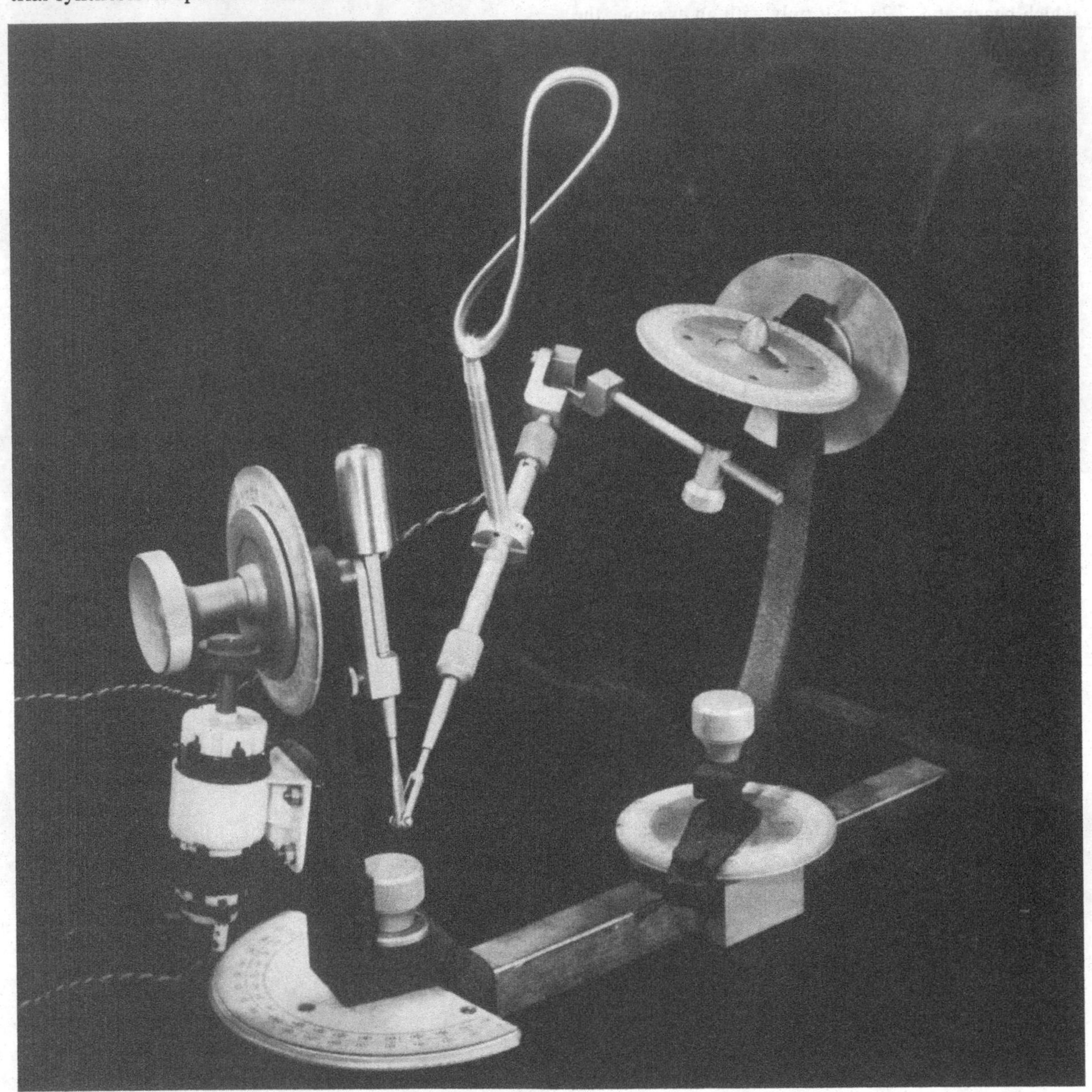

18. A lit coupler point of this adjustable 4-link machine is tracing its path in the fixed space. Such adjustable apparatuses are often valuable in the first trial syntheses of spatial mechanism.

# 18 Singular events in the cycles of motion

### Where to begin

01. As I address myself to the title of this chapter and begin to write, I am appalled at the overall complexity and the nebulous nature of the matter. However it did seem to me as I wrote the title, long ago, that any book seriously pretending to discuss the nature of machinery and its mechanism must deal somehow (and thus somewhere) with the phenomena I have in mind. The phenomena are the unwanted knocking and chattering noises, the sudden changes of pace, the momentary uncertainties, and the regularly occurring catastrophic events associated with the taking up of clearances. Look for example at the discontinuities in the otherwise smooth coupler curve traced out by the lamp filament in photograph 18. When observing such phenomena in a piece of real machinery we argue most often that the piece is faulty, that it suffers somehow important mistakes in its design.

02. Let me begin with a rough list of some events in the cycles of motion that might be considered to be (in the context here) singular. Here is the list: (*a*) the arrival of a link at some limit of its travel; (*b*) the occurrence of a maximum value of the linear acceleration of the centre of mass of some link or the angular acceleration of some link; (*c*) the occurrence of a minimum value of the transmission angle at some joint; (*d*) the reversal of the direction of rotation or of axial sliding at some cylindric joint; (*e*) the arrival at any such condition where judder for a time, or jamming might occur due to friction; (*f*) the arrival at parallelism, collinearity or coplanarity of two or more revolute axes; (*g*) the passing through of any such geometric speciality in the configuration of the links and joints; (*h*) the sudden occurrence of a new contact between a pair of links with a catastrophic change in certain velocities; (*j*) the reversal in direction of the radial or axial or other component of force or couple being sustained at some joint namely the occurrence of backlash at some joint; (*k*) the arrival at some configuration where what I shall call the stability of the linkage might be called into question; (*l*) the actual physical collision of the bounded rigid bodies of some two of the relatively moving links; (*m*) some relevant screw system suddenly becomes reciprocal to some other, or becomes special.

03. I would like to examine now the somewhat loosely chosen items (*a*) to (*m*) above. I would like to discover, by rejection of the irrelevant, which fundamental matters need to be considered here.

### The limits to the travel of some link

04. Take the item (*a*). What I have in mind here is the behaviour of a link such as link 4 in figure 1.11(*a*). Its behaviour is discussed at § 1.45 and illustrated at the inset to the figure there: twice in a cycle, and whatever the mode, the otherwise continuous rotation at the output revolute *D* is seen to reverse its direction. Think now what the geometrical situation with regard to motion might be, first at a general configuration of this mechanism, and next at the instant of such a reversal. Refer here to figure 10.13 (§ 10.54, § 10.55). Refer also to figure 13.01 (§ 13.09, § 13.10).

05. In any general configuration in figure 1.11(*a*) the four motion screws at the four joints will not only pass along the axes of the revolutes or pass through the centres of the balls as the case may be, but also they will be of zero pitch. This is due, quite simply, to the natures of the joints. Referring first to § 7.80 and then to § 10.56, we can see that the motion screw of zero pitch at *C* must be so chosen from among its single infinity of possibilities that the hyperboloidal surface of the regulus defined by it and the motion screws of zero pitch at *A* and *D* contains the centre of the ball at *B*. All four screws of zero pitch, in other words, will belong to the pitch quadric of the same 3-system. This applies of course at the instants of reversal also. At those two configurations of the mechanism, however, the angular velocity $\boldsymbol{\omega}_{14}$ at *D* will be zero, the centre of the ball at *C* will be stationary with respect to frame, *and the three motion screws at A and B and C will be coplanar and intersect.* This follows from the fact that, when $\boldsymbol{\omega}_{14}$ is zero, the sum of the three angular velocities $\boldsymbol{\omega}_{12}$, $\boldsymbol{\omega}_{23}$, and $\boldsymbol{\omega}_{34}$ must be zero too (§ 13.03). Refer to figure 18.01.

06. What we have seen here is that, at the two instants of reversal of the angular velocity $\boldsymbol{\omega}_{14}$, namely at the two limits to the travel of the output link in figure 1.11(*a*), the 3-system of screws defined by all four motion screws at the joint-axes of the mechanism becomes a somewhat special one; refer to figure 7.12(*c*). It is special in the sense that $p_\beta$ has become zero and thus that the pitch quadric has become two intersecting planes (§7.70, §11.15). One plane and one of the polar points namely $Q'$ are defined by the three coplanar intersecting screws of zero pitch, while the other polar point $Q''$ is at the intersection with the said plane of the non-operating but nevertheless existing motion screw of zero pitch at $D$; refer to figure 18.01. Seeing the matter another way we can say that, at the two limits to the travel of the output link 4 in figure 18.01, the link 4 is stationary with respect to 1 and is accordingly, and as it were, fixed to it; the motion screws at the three joints $A$, $B$ and $C$ belong then to the same first special 2-system (§15.50, §20.54). It should be possible to see from this elementary example that, whenever we speak of the limit to the travel of some link and mean by that the reduction to zero of an angular velocity at some relevant revolute or other joint, we speak indeed of the likely occurrence of some special system of motion screws being defined by the motion screws at the other joints. The whole question of travel limits, in other words, is inextricably involved with the question of the occurrence of special screw systems among the motion screws.

Figure 18.01 (§18.04) Here the coupler 3 of a moving spatial crank-rocker is momentarily in the plane defined by its crank 2 and its input shaft. In this configuration of the mechanism there is spherical motion among the links 1, 2 and 3; the three shown angular velocities are coplanar; and the link 4, at one of its travel limits, is at rest.

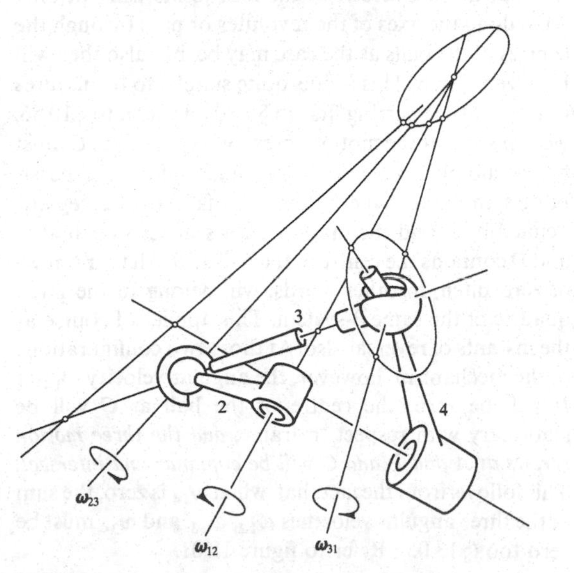


**On the question of the dynamics of a link**

07. Take next item (*b*) at §18.02 and wonder whether the matters there are relevant to what we ought to be discussing here. In the ordinary case of a quantity being at a local maximum or minimum in its variation with time or input crank-angle, there is no point in calling the occasion singular. It is not true – even in two, but especially in three dimensions – that any maximum or minimum among the combination of forces and couples that constitutes the d'Alembert wrench upon a link will reflect itself in corresponding maxima or minima at the supporting joints. Nor is it true that a maximum value of the angular acceleration of a link coincides with a maximum value of the d'Alembert couple at the centre of mass or with any other maximum value of force and couple at the supporting joints. It follows, it seems to me, that, except for its use in determining what the forces in a piece of machinery actually are or might be, any domestic aspect of the instantaneous dynamics of some particular link is not only irrelevant, but also to some extent alien here. I discuss the dynamics of a link at §12.61 *et seq.*, §13.29 *et seq.*, §19.17, and §19.62 *et seq.*

**The transmission angle at some joint**

08. Refer to item (*c*) at §18.02. Refer to §2.43. In the special planar mechanics of the planar 4*R*-loop we have the concept of *transmission angle*. This angle occurs in the reference plane at each of the revolute joints remote from frame. Refer to Hartenberg and Denavit (1964) and see both §2.10 and §10.11 of that book for a well developed, balanced discussion of this important angle. Refer to Volmer *et al.* (1968) and see both §11.3 and §12.3 of that book for a full account of the original German work; it came from Alt, Volmer, Bock, and others. The angle involved in German is *Übertragungs-winkel*. We know from experience that this angle, as Hartenberg and Denavit remark, relates to how well a mechanism might 'run' when still in its kinematical or skeletal form upon the drawing board – when, as one might say, the mechanism was massless. Please digest the gist of chapter 17: I discuss there the statics of, and the dislocations due to clearances in, frictionless massless planar mechanism; I discuss the elementary idea of the friction circle at revolute joints in planar mechanism; I mention the matter of transmission angle (§17.12, §17.17 etc.); and I make a poorly argued extension of these ideas into the realm of real spatial machinery (§17.21 *et seq.*). In order now to introduce here the influence of friction, refer to figure 17.09(*a*) and imagine the possibilities for the static balance of the two equal and opposite wrenches at this particular joint (*a*) when

the joint is frictionless (as it is in chapter 17), and (*b*) when it is not. Under any circumstances the range of the possible wrenches that can be transmitted is very large (there being an $\infty^4$ of reciprocal action screws and, accordingly, an $\infty^5$ of possible wrenches that will in the absence of friction cause no motion at the joint), but in the presence of friction the range of the possible wrenches is vastly greater than this. I would venture to say (and I cautiously quote the last sentence of §4.11 in support of my argument here) that, *even though there is an* $\infty^6$ *of wrenches that will cause the joint to move in the presence of friction, there is, equally, an* $\infty^6$ *of wrenches that will not*. Although an amplification of this statement (and thus, I hope, a clarification of it) is made at §18.10, it will be useful if the reader will, unaided, begin to digest its meaning here.

09. Well, is there a transmission angle at the joints of spatial mechanism? At least in so far as the statics is concerned (and here I ignore the matter of clearances and the questions of dislocation), I think that this is a non-question, a question without meaning. Transmission angle, defined as it is most often in the realm of the planar 4*R*-loop, is the angle between the centre lines of one or other of the two pairs of connected movable links. For this definition to make any sense in general the links must be binary links and the joints revolutes; for otherwise (*a*) the links would have no definable centre lines in the accepted sense, and (*b*) we could not think clearly enough about the forces in the links. The first of these latter are the equal and opposite force upon the 'floating' link, the coupler link of the planar 4*R*-loop, whose collinear lines of action are known because of the absence of couples acting upon that particular link. The second of them are the equal and opposite forces upon the relevant input or output (the unloaded) link whose lines of action are parallel but not collinear unless a small transmission angle and the friction at the relevant joint have combined to make them so. If the transmission angle at the relevant joint (in this massless, planar, unloaded mechanism) is small enough and/or the friction there is large enough, the joint will jam. The joint will be transmitting then a force (a wrench of zero pitch) in the absence of motion. But due to friction the line of action of the force (the action screw of zero pitch) and the axis of the angular velocity (the motion screw of zero pitch) do not intersect. They are moreover not reciprocal, not because they do not intersect, but because the joint is not frictionless (§17.06).

10. Taking these elementary ideas into the realm of three dimensions and considering for example the revolute joint in figure 17.09(*a*), we can see that in the presence of friction the joint may be transmitting in the absence of motion a wrench upon an action screw which is not reciprocal to the motion screw. We can see however that, in this absence of motion in the presence of friction, this action screw of the transmitted wrench is related to (or is a distortion of) the transmitted wrench in the absence of motion and of friction. For every single action screw reciprocal to the single available motion screw there is the discoverable set of the five points of contact (§17.28); but for every set of the five points of contact there is not one but a single infinity of action screws which in the presence of friction can be transmitted in the absence of motion by the joint. This justifies the italicized statement at the end of §18.08. We see accordingly that 'transmission angle' is a kind of myth. The question at issue here relates not to some mere angle but to the problem of friction and jamming at joints in general (§18.11). It is not so much a matter of some easily definable transmission angle, but rather a matter of the action and motion screws, frictional phenomena at the determined points of contact within the relevant joint, and reciprocal systems of screws in operation first in the absence of but then in the presence of friction. The above remarks are relevant to a literature emerging just now which is beginning to deal effectively with the complicated phenomena associated with the gripping of rigid bodies by multi-fingered robots; refer to the notes from [1] to [4] at the end of chapter 6; these notes refer to some of the earlier works.

**Friction, chatter, and jamming**

11. Take item (*d*) at §18.02 and think about Coulomb friction. It might be imagined that the two mentioned velocities are never simultaneously zero, so the scene could be a benign one so far as likely trouble with slip-stick friction (and chatter or judder) is concerned; look for example at the *f*2 joint in figure 1.11(*b*). One can see on the other hand from this example that there might be problems in planar machinery where, upon reversal of motion at the oscillating revolutes and prismatics that abound, the complicated phenomena associated with static friction might obtrude. In the absence of a rubbing velocity diminishing to zero (§14.01), there may well be no trouble with static friction at a joint. But of key importance in machinery are the phenomena of chatter (or judder) and jamming, and the question I wish to ask is this: what are the main characteristics of these phenomena?

12. Consider item (*e*) at §18.02. Refer to figure 18.02 and see there an incipient jamming at a joint in planar mechanism. A prismatic joint consists of a fixed rectangular shaft and a movable rectangular collar. The shaft and the collar and the collar's attachments are mounted horizontally as shown in elevation at (*a*) and in plan at (*b*). Ignoring the effect of clearances the only possible relative motion of the parts is pure translation in the prismatic direction. The collar is about to sustain

a force **F** in the prismatic direction at the hook at $A$, and the location of the hook upon the collar is laterally adjustable. In this simplified analysis I wish for the sake of simplicity to let the coefficients of static and kinematic friction be identical. Let the force be applied. If sliding becomes imminent (or is already occurring slowly), both reaction forces upon the collar – they act at $B$ and $C$ – will be inclined as shown at their equal friction angles $\phi$; the lines of action of these two intersect at $Q$. Let the hook be adjusted to let the line of action of **F** pass exactly through $Q$. The three forces upon the collar will then be in equilibrium and it could be said that sliding is imminent or already occurring steadily; see the composite closed force polygon at (*b*) showing the applied force **F** and the well-understood components **N** and $\mu$**N** of the reaction forces.

13. If however $A$ were moved and the line of action of the force **F** were set to pass a distance say $d$ to the left of $Q$ (but still to the right of $B$), the action of the apparatus would be different. An apparently smooth translational motion might occur, but I wish to say that it would (if it did) occur as the result of a fast-changing intermittency. Initial application of the force **F** would cause a small clockwise shift of the collar to effect the contacts at $B$ and $C$. Subsequent translational motion in the prismatic direction would mobilize the same forces as before at $B$ and $C$ at the same angles $\phi$ but, due to the new geometry, these two would meet no adequate response in **F** to achieve an equilibrium. What would happen is that the overall resultant of the forces acting, a counterclockwise couple **F** × **d**, would next be impressed upon the collar. The points of contact at $B$ and $C$ would be released, the forces there would disappear, and the cycle of operation would begin again. Thus the collar would *judder* as it went, the frequency depending among other things upon the length and breadth of the collar and the relevant clearances.

Figure 18.02 (§ 18.12) An apparatus for studying the mechanics of plane jamming. Here the rectangular collar is incipiently slipping with the three forces upon it steady in equilibrium. Go to § 18.12 for (*a*) accelerated motion with **F** pulling at the left of $Q$, and (*b*) jamming with **F** pulling at the right of $Q$.

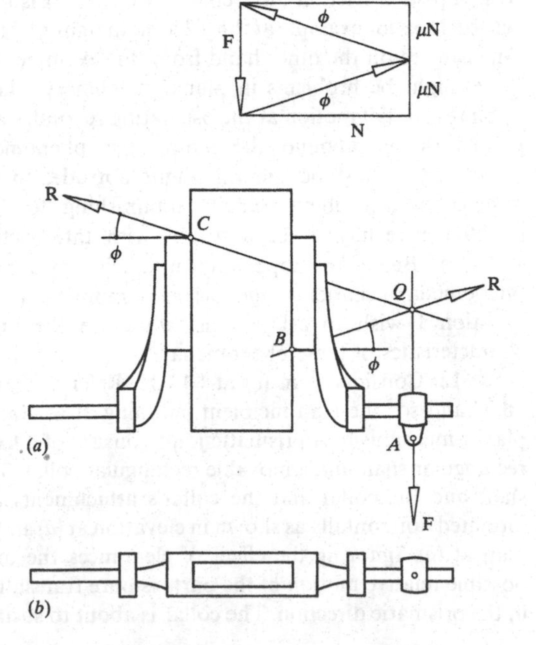


14. If next the line of action of the applied force **F** were set to pass at a distance say $d$ to the right of $Q$, the apparatus would immediately *jam*. Upon inception of the action the growing reacton forces being mobilized at $B$ and $C$ (inclined as yet at angles that are growing but less than $\phi$) would soon intersect in the line of action of **F**. The frictional components of 'these two forces (both of them less than their maxima $\mu$**N**) would not have grown to their full values before that intersection – and thus the possibility of an overall equilibrium – had occurred. Due to frictional forces not yet fully mobilized the apparatus would become immobile. It would moreover remain so whatever the increasing magnitude of the applied force **F** might happen to be. Indeed the greater that force the more stuck the collar would become. There would be, of course, local failures of the material due to compression at the points (or lines or Hertzian areas) of contact at $B$ and $C$, and thus an increased $\mu$ as well (§ 1.49).

15. I depart from the plane to speak in general now and say the following. In the jamming discussed above an external wrench was applied to a joint in the presence of friction. Normal reaction forces **N** acting along the contact normals at the points of contact within the joint were mobilized in response and, given the friction coefficients $\mu$, frictional reaction forces were mobilized as well (§ 10.61). Before all of the frictional reaction forces reached their maxima $\mu$**N**, however, the externally applied wrench was balanced by the reaction forces and thus destroyed. Had the applied wrench not been balanced and thus destroyed in this way, the joint would have operated (or been, as we say, *actuated*); relative motion would have occurred. It appears that the question to be asked at a likely jamming at a spatial joint is this: *can the reaction wrench mobilized within the joint in the absence of friction grow in the presence of friction to balance the applied wrench and thus destroy it, or will it fail to do so*? Refer to chapter 14 and see for example that the joints in the apparatus in figure 10.15 are, by the arguments of this book, jammed. If there is no friction they are not only jammed but workless and they can without energy be actuated at the instant by

actuating motors (or muscles) or otherwise. If there is friction they are not workless and energy is required to actuate any one of the joints. But if they can be actuated, how can they be jammed? An answer to this could come as follows: *that this or that joint is jammed depends not upon the joint but upon the conditions of its loading.* We know that any jammed joint can be unjammed if we change the conditions of its loading; we can always arrange to apply to the joint overall, with the help of a hammer maybe, a congruent wrench (§ 14.32) or something near to it. To jam a frictionless joint we apply a reciprocal wrench – or, in the presence of friction, something near to it. And whether in general the joint elements of a jammed joint become damaged or not depends solely upon the geometry of the joint, the nature of the materials in contact, and the details of the external applied wrench. Refer to photograph 14 and reflect there and elsewhere about the ways in which the action and the motion duals present themselves in joints and on how these ways are modified by the absence and the presence of friction. Refer to Obwovoriole; see note [3] at the end of chapter 6; it is my impression that that author tried to grapple the matters at issue here, failing, in three dimensions, effectively to do so. Refer to my few notes on the assembly of a peg in a hole at §.18.24. See also Whitney, Whitney and others, and Gustavson [1].

**Change points**

16. Refer to items (*f*) and (*g*) in § 18.02. No serious student of such phenomena should miss the chapters 12 and 13 of Hunt (1978). They deal with the screw systems and their special cases (chapter 12) and some applications of these in the design of spatial mechanism (chapter 13). For reasons he explains very carefully, Hunt avoids using the somewhat discredited, poorly defined terms *limit position* and *change point* when dealing with the matters under consideration in his chapter 13. He proposes there instead, and clarifies the separate meanings of, his new terms (*a*) *stationary configuration* which refers to the configuration of a mechanism at a stage in its cycle where some freedom at a joint is inactive, and (*b*) *uncertainty configuration* which refers to the configuration of a mechanism at some stage in its cycle where no single input can be used to remove the mechanism from the said configuration reliably. But the idea of the possibility of switching (or of *changing*) from one mode to another at certain configurations of a mechanism (§2.42) is not heavily stressed by Hunt, and I wish for my own reasons here to stress precisely that.

17. The term *change point* refers as we know, not to some actual point in a linkage, but to some particular configuration of the linkage. I wish to defend the term nevertheless; for it describes so well a recognizable group of related phenomena. Whenever a linkage in motion adopts a configuration where it has the capacity to change (or switch) from one of its modes of operation across to another (§ 1.45, § 2.57), the screw systems involved suddenly change. At a change point precisely, the screw systems obtaining become different from those that obtain on either side of the change point. This applies even in the likely event of sogginess (§ 2.46); for while sogginess might blur the actual mechanical activity at a change point, it renders the change point itself (a geometrically defined phenomenon) no less sharp. Whatever the merits or demerits are of the actual term, we are driven to the conclusion here that the passing through of a change point brings into play special screw systems that are not apparent in the otherwise general operation of the linkage. It follows again that we have here a range of phenomena most easily explicable in terms of screw theory.

**Clearances, backlash and vibration**

18. Refer to items (*h*) and (*j*) in § 18.02. If errors in manufacture of necessary small clearances at joints are not great likely dangerous events such as the passing through of a change point may be interpreted in terms of screw theory (§ 18.16); but if the clearances at joints are so great that the possible backlash is commensurate with the dimensions of the linkage itself the poorly built mechanism in question might better be seen as being some other mechanism, a lock or a trap for example, where the cycle of movement involves the ordered, intended, intermittent breaking of old joints and the subsequent making of new ones (§ 2.53). In a middle event, where the clearances although considerable have not entirely destroyed the accuracy of the intended relative motions of the links (and here I infer by the word intended a designer), the problem is bedevilled by the dynamic behaviour of rigid bodies in free flight, for the links of loose machinery do fly freely as they adjust and respond to the forces at work. Except for making my few remarks about static dislocation in chapter 17, I wish to stay away from such free-flying links. Nor do I wish to become engaged with questions of vibration and wear. Except to observe that these are important problems for the machine designer, and that they do obtrude within each one of the subjects matter being dealt with here, I am obliged somewhat ingenuously to say that they are beyond the scope of this book.

**The stability of a linkage**

19. Take the item (*k*) in § 18.02. It mentions the matter of stability. We enter here an arena in the theory of machines that is neither well delineated nor well studied. There is a paucity of literature that deals not merely with the statics of the systems of forces that act

upon mobile linkages in equilibrium, but with the stability of such systems. Machine designers do not understand very well the stability of asymmetrical trailed implements such as the multi-furrow disc plough (I tried to understand at note [2] of chapter 10); the relevant engineers may or may not understand the stability of the flying kite (or of the ocean-going paravane); and the designers of long trains of hitched-together rubber-tyred vehicles must surely worry about the mechanism necessary to achieve an accurate follow-on. In all such complicated devices there are certain wrenches due to the behaviour of disrupted earth or flowing fluid or other frictional phenomena that change with the changing circumstances; and there are certain others (pure forces) that change, not by changing their whole lines of action, but merely by virtue of their *fixed, spherical-jointed points of application at the linkages*. The point of attachment of the pull-rod at a plough, connected as it is with the plough's steering mechanism (an added complication), is of vital importance in the matter of stability. So too is the point of attachment of the string that is knotted at the underside of the flying kite. Go to figure 14.11 and ask the question: why is the linkage there (which is in equilibrium) unstable? What is the exact mechanics of its collapse when its moves (as it will) from the shown configuration to another, stable one? Why indeed do certain structures in the course of construction – or even after construction – buckle and fall down? What do we know about spatial linkages employing springs? What is the exact mechanics of the stability of the front wheels and the steering mechanism of a motor car (or of a bicycle)? These questions are easy to find and easy to ask, but not easy to answer. Look at the apparatus in figure 10.15 and consider the stability: find yourself thinking first in terms of screws, screw theory and reciprocity (because you are invited thus to think), and then in terms of what happens in the event of small displacements at the joints. Certainly for planar cases (but less certainly for spatial ones) there are established methods for such problems; they employ the somewhat split-off, elementary concept of the virtual work done by a force applied to a point throughout the length of some small displacement (§19.10). Refer to Meriam (1975) for some excellent reading in this particular area. It is hard to deny, however, that the more wholistic idea of the virtual work that might be done by a wrench upon a screw, applied not to a single particle but to a whole rigid body and throughout a small twist of that body about some other screw, might have an important impact here (§19.08). The stability of linkages is a new area for the application of screw theory.

**The collision of links**

20. Take item (*l*) in §18.02. Be aware that the matter is mentioned at §1.07. It is also mentioned (but by omission) at §10.54, §13.10, §20.33, §22.34, §22.43, §22.62 and elsewhere. In cyclically operating machinery where the mobility is unity or less, one could deal with this phenomenon in an ordinary brutal way by (*a*) chopping off offending bits of material from the troublesome links, and (*b*) adding to the thus weakened links other non-offending bits of material to account for the choppings off. In doing this one might be persuaded to keep in mind, if one could, the respective *tracks* with respect to frame of the links (§1.20). But read also the latter half of §1.06 and see that, if the relevant pair of links were both in motion with respect to frame, the question of synchronism would obtrude; the materials of the two links may of course occupy the same zones in the fixed space, *but they may not simultaneously do so*. There would be, as well, other difficulties. These would be (*a*) how, actually, to plot the tracks of the two links, and (*b*) what to do about concavities or actual holes in the links. One would discover in other words that there is more to the idea of track (and especially the idea of *two intersecting tracks*) than meets the eye.

21. Here I wish to mention Ganter [2]. Ganter has shown (*a*) how to deal to some extent with the problems associated with concavities or holes in links, (*b*) how to plot the outer enveloping surface of the track of the so-called *convex hull* of a link, and (*c*) how the technique of inversion can obviate the mentioned problems associated with synchronism. With respect to items (*b*) and (*c*) I wish to make the following remarks; they are written to accentuate the fundamental role of screw theory in this new area of investigation. To arrive at and to interpret the actual track of a rigid body (that is, the actual collection of the separate paths of all of the points embedded within the physical boundary of the body) is a much more difficult thing to do than to arrive at the mere outside envelope of the track. But to arrive at the outside envelope is itself a challenging task, not easily achieved without a flow of data about the ISA of the body with respect to frame as the body moves. Given such data, one might (*a*) choose an instant and look orthogonally in the direction of the ISA at the *silhouette* of the convex hull, this being a twisted curve in space which, at the instant, is the locus of points upon the convex hull whose linear velocities are tangential to the hull, (*b*) plot that locus in the fixed space, and (*c*) move to the next instant and repeat. Ganter indeed does exactly this, but he plots in this way the envelopes of the tracks, not of two moving links in some spatial closed-loop mechanism, but of two moving rigid bodies both of whose relative motions in the fixed space he already knows. Having done this he finds that his envelopes intersect. Anxious to show that this intersection does not necessarily mean that the bodies actually collide, Ganter

then inverts his mechanism (§1.47). He holds one (the first) of his moving bodies fixed and, keeping the same relative motion between the bodies, he plots the track of the second body with respect to the first. The envelope of the track of the second body does not now intersect the physical boundary of the fixed body. He concludes from this of course that the two bodies do not in fact collide.

### Special screw systems

22. Take the item (*m*) in §18.02. This item was planted there of course by me; but have the arguments above been too facile and thus quite false? There may be other matters needing our attention here, and there may be other methods. But on the whole we have shown I reckon that screws, the screw systems in general, and the special screw systems in particular may be seen as a kind of network for relating all of the relevant matters under consideration here.

### Some disorganized left-over ideas

23. I make the statements here as they come to mind; and I ask the questions not to be rhetorical but to wonder. Can we think about singularities not simply in terms of links and the points in links but also in terms of joints? The collision of links and the engagement of accidental stop surfaces are two aspects of the same phenomenon however. On the matter of change points refer to the 5-link loop at figure 1.11(*c*) and consider the possibilities for intersecting tracks of the ball; what connections might be here with figure 20.09? On the points in a link, two separated single infinitudes of points are enjoying singular experiences at the instant; there are two straight lines of points along the relevant motion and action screws; the points are suffering either the same minimum velocity or the same minimum moment. Instantaneous screw systems can be added and subtracted of course (§7.78), and we have the phenomenon of transitory mobility (§2.08); so why not reexamine §7.36? Can we distinguish more clearly between poorly and well conditioned contact to consider better what happens in machinery? Carefully define the word *active* that was used at §6.63 and at the caption to figure 7.13(*a*); then discuss that definition in the context here. But are the singularities of a mechanism required and achieved by the designer and thus good or are they not and thus bad? Consider a key; it turns in its lock; then suddenly it slips in; but consider too the restrictive title of this book and limit the scope of that discussion. Think about those machines that are carefully designed to avoid what might be called *reversal friction*. See for example the crank-slider in figure 18.03 and read the caption there (*a*) for clues about the intended meaning of these words, and (*b*) for some unrelated remarks about the link-lengths and offsets of chapter 20 [3]. Overconstrained yet mobile mechanism is both the friend and the bane of man; freedom without constraint is paradoxical; unrestricted freedom for our machinery is inadvisable unless we wish to weaken it; go to the middle ground; understand the yin and the yang of freedom and constraint.

Figure 18.03 (§18.23) The angular velocity $\omega_{12}$ of the input crank 2 about its axis $AB$ is constant. The oscillating block 4 connects to the fixed frame 1 through a cylindric joint. At the limits of its linear travel 4 is not translating but rotating and, at the limits to its angular travel, it is not rotating but translating. Let $R$ be a point at the contact patch of the said cylindric joint and note that the magnitude of the rubbing velocity $\mathbf{v}_{1R4}$ at $R$ is the same for all points $R$. This never diminishes to zero. If incidentally $S$ were a point at the axis of the said joint, all points $S_4$ would (with all of their equal velocities $\mathbf{v}_{1s4}$ collinear there) oscillate, but in some non-sinusoidal manner. Whereas $B$–$C$ (say $r$) and $C$–$D$ (say $c$) are the lengths of the links 2 and 3 respectively, $A$–$B$ (say $f$) is not the length of link 1 (which is zero) but its offset (§20.29). If independently of the fixed angle $\zeta$ the special condition obtains that $c^2 = f^2 + r^2$, the points $S_4$ oscillate with exact simple harmonic motion [3].

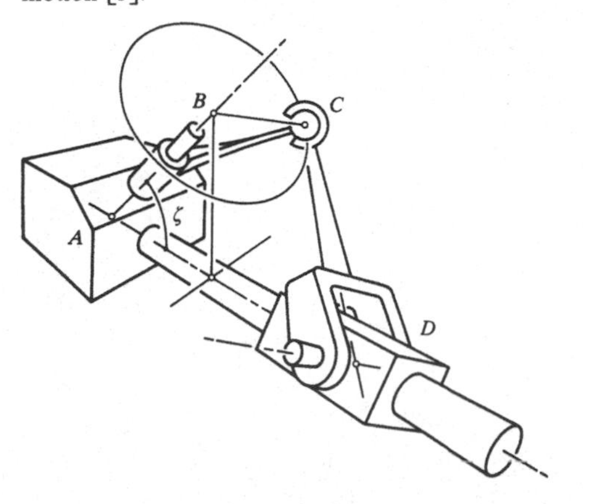


24. Many of the problems of industrial assembly (the peg in a hole for example) are now solved in the sense that the relevant processes can be effected; force and moment sensors and compliant wrists at the ends of robot arms are used to monitor and guide mechanical activity. In terms however of the actual forces at the frictionful points of contact and the possibilities for the actual motions at the instant, the phenomena are by no means adequately explained. What is the exact mechanics for example of the multi-fingered robotic grippers being developed now?

### Notes and references

[1] Whitney, D.E., Quasi-static assembly of compliantly supported rigid parts, *Journal of dynamic systems,*

*measurements, and control*, Vol. **104**, p. 65–77, ASME, March 1982. Whitney, D.E., Gustavson, R.E. and Hennesey, M.P., Designing Chamfers, *The international journal of robotics research*, Vol. **2**, No. 4, p. 3–18, MIT, 1983. Gustavson, R.E., A theory for the three-dimensional mating of chamfered cylindrical parts, ASME paper No. 84-DET-115, *Design engineering technical conference, ASME, Cambridge, Mass., October 1984*.

[2] Ganter, Mark, A., *Dynamic collision detection using kinematics and solid modelling techniques*, doctoral dissertation at The University of Wisconsin Madison, 1985. Refer §18.21.

[3] Sherwood, A.A., The mechanical genertion of simple harmonic motion by three-dimensional linkages, *Australian journal of applied science*, Vol. **9**, p. 96–104, CSIRO, Melbourne 1958. Sherwood, A.A., The dynamics of the harmonic slider-crank mechanism, *Journal of mechanisms*, Vol. **1**, p. 203–8, Pergamon, 1966. Read also Hunt (1978), Chapter 10. Refer §18.23.

19. Given four forces in equilibrium the polygon of forces may be drawn conveniently in a number of different ways. The polygons of forces nestle with one another in the manner here shown. This model relates to the plough and some couples are involved.

# 19 Fundamental relations and some algebraic methods

## Vectors and the motor dual vector

01. We believe we see in the arena we call reality various mechanical quantities which we identify, then try to quantify. We look in the area of statics and instantaneous kinematics and we see for example force, displacement, velocity, and acceleration. In the wider field of applied mechanics we see for example quantities such as mass, momentum, energy and so on. It might be said that in every case we fashion (or select) the quantity in question because we believe it to be, in some way, not only the most fundamental in a scientific sense, but also the most useful from a practical point of view.

02. A unique and a somewhat different aspect of nature is the discovered phenomenon of number. Without that special idea of number we cannot measure any of the quantities. Measurement is done in terms of the units of the quantities of course, and the whole matter of the units is another, related area of study. In any event we find by experiment that there are laws of number and that, in certain areas, these laws work to our total satisfaction. They are the laws of addition, multiplication and so on of the scalars, and the business of applying these laws to the scalars we call arithmetic.

03. Among the various quantities mentioned above however there are many like force, the displacement of a point, the linear velocity of a point, and the angular velocity of a rigid body, which cannot be fully quantified or described by a number only; and some of these are the well known, single vector quantities which are well introduced in almost every text on the subject of mechanics. It is argued (*a*) that a single vector quantity requires not only its magnitude but also its direction to be given before it can be known, and (*b*) that the quantity in question must be additive by the parallelogram law both commutatively and associatively before it can be called by the name vector.

04. It is well understood and accepted that, before we could successfully add, subtract, and multiply the single vector quantities with one another, laws other than the simple arithmetic laws of number had to be discovered and applied. It was found in the course of scientific endeavour that, unless the correct laws were applied, the results of our wrong mathematics would not accord with what we saw. It was found by experiment, in other words, what the correct laws for the manipulation of the vectors had to be, and the business of applying these laws to the vectors of nature we now call vector analysis.

05. In vector analysis we find for example that, whereas the various polygon laws for addition and subtraction of vectors apply among vectors of the same kind, the various laws of multiplication apply, most often, among vectors of different kinds. Polygon addition and subtraction can be used among groups of vectors comprising only forces for example, or only angular velocities; it makes no sense to try to add together or to subtract from one another forces and angular velocities; we do not do that. We do not do it, not simply because we choose not to do it, but because we know that such a practice would not accord with anything we see in objective reality. Similarly we know that, whereas the scalar quantity power and the vector quantities force and linear velocity are connected in particle mechanics by the scalar, or dot product equation $P = \mathbf{F} \cdot \mathbf{v}$, the vector quantities linear velocity, angular velocity, and distance from the ISA are connected in rigid-body mechanics by the vector, or cross product equation $\mathbf{v}_{\text{circ}} = \boldsymbol{\omega} \times \mathbf{r}$ (§ 5.45, § 5.47). These phenomena are rooted in nature, we reckon, and the corresponding vector algebra arises directly out of it. The phenomena are not produced by the algebra in other words. The phenomena display themselves; and we try with our algebra to describe the phenomena. The relevant algebra is well understood by us and is often and well introduced in the regular literature.

06. It is not so often or well introduced, however, that there are different kinds of vector and that these behave in different ways (§ 5.27). *Nor is it well understood or presented that, in rigid-body mechanics, certain kinds of vectors appear to go in pairs, and that these pairs are the quantities or the entities we have chosen to call by the name dual vector, or dual.* Please refer to § 10.03, § 10.47,

and § 10.48. In chapter 10 and in other parts of this book we have asked and answered the following kinds of question. Is the angular velocity $\boldsymbol{\omega}$ of a rigid body a 'line' vector or is it not? If it is a line vector, confined to its 'line of action' (the ISA or the motion screw), why can it be quoted as being anywhere in its rotating body, and thus be spoken of (as it often is) as being 'free'? And what do we mean when we say that the linear velocity of a point in a rigid body is a 'bound' vector in the sense that it must always be drawn to emanate from the point whose linear velocity it describes? Please go for an answer to § 5.57 and, at the risk of some confusion, to § 5.31. What do we mean, first in states of equilibrium and then in stability studies, by the 'point of application' of a force? How is this last idea compatible with the idea of a force being a line vector just like angular velocity? And what do we mean conceptually when we say that the moment at a point of a force – another bound vector – is $\mathbf{F} \times \mathbf{r}$, where $\mathbf{r}$ is the directed distance from the force to the point? Please go for answers to § 10.14 *et seq.*, § 10.31 *et seq.*, and § 15.01 *et seq.* We have asked and answered such questions hitherto, and we have come to an understanding of the following: *that in the kinetostatics (the kinematics and the statics) of rigid bodies there are delineated two dual-vector quantities: they are, respectively, the rate of twisting and the wrench.*

07. Looking again at figures 5.09 and 10.12, please notice next that the two dual vectors I have mentioned, the rate of twisting and the wrench, geometrically within the body and at the instant, behave in exactly the same way. I mean by this that they both exhibit the same kind of line-symmetric, helicoidal field of bound vectors surrounding a central axis, the central axis being at the motion screw or at the action screw as the case may be. And here I wish to make a remark. There lurks in many of us the vague supposition that there are – or that there might be – other dual vectors or duals which behave in the same way. We may have in mind for example the combination of angular and linear acceleration (§ 19.17), or that combination of rotation and translation which is the finite screw displacement of a rigid body (§ 0.07, § 19.30). Neither of these, however, behaves as the rate of twisting and the wrench do. The field of the bound, linear acceleration vectors in an accelerating, rigid body is not even axi-symmetric. It is not helicoidal. The field has a central point (namely that unique point in the body which is possessed of zero linear acceleration); and the 'ellipsoidal' field of the bound vectors is point-symmetric in certain respects about that point. Refer to Sticher [1] and Konstantinov *et al.* [2]. Refer also Bottema and Roth (1979) § 6.12. In the case of the finite screw displacement (this will be dealt with later at § 19.30), it is not even acceptable to call the combination of the finite rotation and the finite translation by the name dual vector, because, as we know, this particular combination does not, as the rate of twisting and the wrench do, obey the commutative and associative laws of vector addition (§ 19.03). Refer however to Keler [19]. The point I am wishing to make, right here, is this. *There is no guarantee that all pairs of vectors which may be seen or spoken of as dual vectors will behave as the rate of twisting and the wrench do; the rate of twisting and the wrench can be seen as belonging to a special class of dual vector which is characterized by its possession of the 'helicoidal' property outlined above.* It was only dual vectors like these, incidentally, which were the ones manipulated and called, under certain circumstances, by Brand (1947), *motors.* I will return to motors, mentioning some of the various relevant workers and discussing the so-called *motor algebra*, at § 19.14. In the meantime, however, and whenever appropriate, I propose to distinguish those particular dual vectors which, like the rate of twisting and the wrench, exhibit this helicoidal property, by giving them a special name. I shall call them *motor dual vectors.*

**Units, and the dot product of action and motion**

08. Except for having preliminarily broached the idea of the dot product of two motor dual vectors at § 14.27 *et seq.*, the questions of work and power have not been properly canvassed yet, and I wish to try to do that now. I rely here on material developed in chapters 3, 5, and 10, and I take it for granted that terms and concepts such as dual vector, action dual, motion dual, basepoint, action screw, motion screw, motor etc. are well understood. They derive, as we have seen, from those kinds of quantities like wrench upon a body and rate of twisting of a body which clearly require, not only two vectors to describe them, but also the motor method of treatment for their manipulation. Central to the idea of the motor dual vector is the *screw*, and I wish to say the following. Refer to figure 19.01. Let there be in a body two screws $A$ and $B$ whose pitches are $p_a$ and $p_b$ metres per radian respectively. Let their perpendicular distance apart be $d$ metres, and their angular displacement from one another by the right hand rule be $\phi$ radians. Let a wrench upon the body of magnitude one newton be acting along the screw at $A$. The pitch of the wrench will be $p_a$ and the magnitude of its least couple will be $p_a$ newton metres (§ 14.08). Let this wrench be acting in conjunction with a rate of twisting of the body of magnitude one radian per second about the other screw at $B$. The pitch of the rate of twisting will be $p_b$ and the magnitude of its least velocity will be $p_b$ metres per second. The combined effect of the two phenomena, namely the wrench which might be called the *action* and the rate of twisting which might be called the *motion*, can be seen as a kind of dot product of

the two. It is, as we shall see, the *rate of working*, $F\omega[(p_a + p_b)\cos\phi - d\sin\phi]$ watts; and the other name for this of course is power (§ 19.28).

09. To prove this fact, namely that the power is given by the expression quoted above, let us argue the matter simply as follows. Go again to figure 19.01 and see there (*a*) a wrench ($\mathbf{F}$, $p_a\mathbf{F}$) on a screw *A* of pitch $p_a$, (*b*) a rate of twisting ($\boldsymbol{\omega}$, $p_b\boldsymbol{\omega}$) on a screw *B* of pitch $p_b$, and (*c*) the common perpendicular distance *d* and the right handed angle $\phi$ between the screws. Paying attention to the screw *B*, and considering first 'force times linear velocity along the screw', and next 'torque times angular velocity about the screw', we can write that

$$P = (F\cos\phi)p_b\omega + (p_aF\cos\phi - Fd\sin\phi)\omega$$

$$(1)\quad = F\omega p_a\cos\phi + F\omega p_b\cos\phi - F\omega d\sin\phi$$

$$(2)\quad = F\omega[(p_a + p_b)\cos\phi - d\sin\phi],$$

where the expression between the square brackets is equal to twice Ball's virtual coefficient (§ 14.28). Ball represents this important quantity – its dimensions are that of length – by the special symbol $\tilde{\omega}$, but I prefer to represent it by the symbol *Q* (§ 14.28, and Ball § 10). Whereas, also, $\mathbf{F}$ and $\boldsymbol{\omega}$, in Ball's terminology, are the intensity $\alpha''$ of the wrench and the twist velocity $\dot{\beta}$ of the twisting motion respectively (Ball § 7), I call them both, and more simply, the *magnitudes* of the respective duals.

Figure 19.01 (§ 19.09) A wrench of magnitude *F* and pitch $p_a$ is acting here upon a body while the body is suffering a rate of twisting of magnitude $\omega$ and pitch $p_b$. The duals are being dot multiplied to obtain an expression for power.

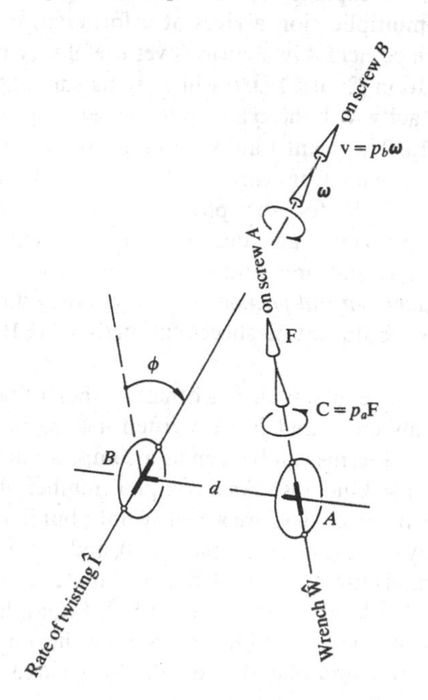


So we can write that

$$(3)\quad P = 2QF\omega,$$

where *F* and $\omega$ are the magnitudes of the action and the motion duals respectively, and where *Q* is the virtual coefficient. When *Q* is zero, incidentally, the power is zero because the wrench is doing zero work. Under that special circumstance the screws are said to be reciprocal (§ 14.25 *et seq.*, and Ball § 20), but the matter of reciprocity, otherwise important, is not the matter under observation here (§ 19.28).

10. We should pause here to examine the ways in which the above equations behave when (*a*) a pure force and a linear velocity only, or (*b*) a pure couple and an angular velocity only, are involved. Such special circumstances are often met in the simpler problems in the mechanics of machinery (see below), and the corresponding expressions for power are well known: $\mathbf{F}\cdot\mathbf{v}$ may be written $Fv\cos\phi$, and $\mathbf{C}\cdot\boldsymbol{\omega}$ may be written $C\omega\cos\phi$. When the couple of the wrench is zero the pitch $p_a$ is zero, and when the angular velocity of the rate of twisting is zero the pitch $p_b$ is infinite. Substituting $\omega = 0$, $p_a = 0$, and $p_b = \infty$ into equation (1) we get: $P = \text{zero} + F(\text{zero} \times \infty)\cos\phi - \text{zero}$. It is not very difficult to see from this that, in the limit, as both *C* and $\omega$ tend to zero,

$$(4)\quad P = Fv\cos\phi.$$

Correspondingly it can be seen that, as both *F* and *v* tend to zero, the power, in the limit, is given by

$$(5)\quad P = C\omega\cos\phi.$$

Please notice (*a*) that, in both of these special circumstances, the distance *d* is irrelevant, and (*b*) that this fact corresponds with our experience with nature and machinery. Please notice that, *whereas equation (4) relates at an instant to the power loss due to friction at a prismatic joint when that joint is transmitting in the presence of motion a pure force obliquely applied, equation (5) relates at an instant and similarly to a revolute joint when that joint is transmitting a pure couple obliquely applied.* Refer to § 14.32; there I deal with another special case where the two screws at *A* and *B* are not only collinear but also of equal pitch; $d = 0$, $\phi = 0$, $p_a = p_b = p$; *Q* becomes the pitch *p*, and the power *P* becomes $P = 2pF\omega$. Other special cases I leave as exercises for the reader. Refer to § 19.28.

**The need for axes and an origin of coordinates**

11. Please go to figures 5.09 and 10.12 and be reminded that they are showing how first a motion dual and next an action dual may be *shifted*. Without loss of meaning the duals are shifted in those pictures away from their respective central axes – the axes are the motion screw and the action screw respectively – simply

by the shifting in each case of the basepoint accordingly. Please also see figures 10.07 and 10.08 where, in effect, a shift of a wrench has been effected; in these pictures a wrench – first of pitch zero and next of pitch infinity – has had its basepoint shifted from the central axis to a new position, namely the arbitrarily positioned origin of the given Cartesian coordinate system. Look also at figure 19.01 and at the argument in § 19.09 and see there that we shifted the basepoint of the wrench (**F**, $p_a$**F**) from point *A* on screw *A* to point *B* for the sake of convenience. There are, indeed, many examples in this book of our having shifted the basepoint of a dual vector from one position to another for convenience. In each case, of course, the second-quoted or the 'dual' part of the dual vector (namely the couple or the linear velocity part) was changed appropriately whenever the basepoint of the dual was radially shifted away from the central axis.

12. Now in any design or analytical work it is necessary somehow to keep one's feet on the ground, and, given that computers work the way they do, the most convenient ground is still a Cartesian coordinate system. Go to figure 19.01 again and imagine that, at some third point *O*, a triad of axes *O–xyz* has been established. It may have been established for the calculation of the location of the rigid body involved, or for the linear velocities of certain points, or for many other reasons as well as the reason given there – the calculations of power. Under such circumstances it may have been convenient – for the sake of our own thinking or for the computer or for both – to shift not only one but both of the basepoints of the duals to some other one point, to that remote origin of coordinates *O*. It is due to circumstances such as that that we begin to hear, in the literature, about such things as the following: dual vectors at the origin, the spherical indicatrix there, the real (or the global) parts and the dual (or the spatial) parts of the dual vectors (§ 19.14, § 19.40) and so on. It might be mentioned parenthetically that, as shown in figure 10.12 for example, the real part of the dual vector – the dual vector consists there of the force **F** and the couple C – is the force **F**, which is invariant, while the dual part is the couple **C**. The dual part of a dual vector is not invariant; it changes according to the position chosen for the common basepoint. Thus too do we hear about quaternions, dual quaternions, motors, etc. (§ 19.15), about the six Plücker coordinates for a line (§ 19.21) and for a screw (§ 19.28), and about other such useful devices for nominating the locations of the various relevant lines, screws, wrenches and rates of twisting in the Cartesian space. There is a wide literature of the mathematical devices used for locating bodies, lines and points in the moving mechanism of machinery, and an excellent summary of much of that can be found in Rooney [3] [4]. One might also refer to the final chapter of Bottema and Roth (1979) where a similar summary is correspondingly, though in a much more general manner, made. Lebedev (1966) is also interesting, for there one can find a long account of the various methods used by kinematicians in the wide field of TMM. In all of those summaries, however, there is a recourse at all times to some convenient set of axes having an origin of coordinates.

**Use of the complex multiplier epsilon**

13. In those fields of natural philosophy suited to it the complex multiplier i (where $i^2 = -1$) can be used to advantage. By choosing an origin of coordinates, by following certain rules and without our caring how, we can get answers to certain physical questions which continue to satisfy. Examples occur in the fields for example of alternating electric current and in the analysis of planar mechanism. Refer for the latter to Hartenberg and Denavit (1964). In the area of dual vectors on the other hand – and one should refer for the early beginnings here to Clifford [5], Everett [6], Kotelnikov [7], Hamilton (1899), Study (1903), and Brand (1947) – we find that the choice of an origin and the introduction of the complex multiplier $\varepsilon$ (where $\varepsilon^2 = 0$) can be useful. Among the few contemporary workers in this special field with certain kinds of vector is Keler [8] [9] [10]. His writings (*a*) of a dot-like product of a pair of suitably chosen duals, and (*b*) of the sum of a pair of duals of the same kind, are suitable for consideration and presentation here; see § 19.18, § 19.20. Keler's explanations are interesting because (*a*) his dot multiplication arrives at a formulation for power which coincides in a curious yet useful way with that of equation (2) at § 19.09, while (*b*) his summation coincides exactly with the cylindroid geometry outlined in chapter 15. The point I am wishing to make is that, despite the bewildering variety of vector and other methods available for the physical and numerical solution of problems, and despite their apparent differences in style and presentation, *they all relate in the end to the same natural phenomena.* Unless they did, of course, they would not be successful methods (§ 10.05, § 19.05).

14. Keler explains (using other symbols) that, if two real numbers *a* and *b* are written $a + \varepsilon b$, such a combination of terms can be seen as a complex number of a certain new kind, just like a complex number of the well known, older kind where we write not $\varepsilon$ but i. Keler quotes Study (1903) and Blaschke (1924), and it must be said that Brand (1947) deals with the same material in his well known book. Refer also Everett [6]. Brand, however, forbears to discuss the mechanics before he outlines his mathematics, and blankly calls the whole area, as I

have mentioned before, *motor algebra*. Certain laws of addition and multiplication can be made to apply, certain extensions into the field of vectors can be made, and certain conventions adopted. A dual vector can be written for example $\hat{\mathbf{W}}$ where the circumflex indicates duality. We can write for a wrench at its central axis, at its action screw, $\hat{\mathbf{W}} = \mathbf{F} + \varepsilon\mathbf{C}$. Correspondingly we can write for a rate of twisting at its ISA, at its motion screw, $\hat{\mathbf{I}} = \boldsymbol{\omega} + \varepsilon\boldsymbol{\tau}$. Quoting these things, not at the relevant *home screw but at the origin, we can write* $\hat{\mathbf{W}} = \mathbf{F}_O + \varepsilon\mathbf{M}_O$ and $\hat{\mathbf{I}} = \boldsymbol{\omega}_O + \varepsilon\mathbf{v}_O$; refer to figure 19.02. We say that $\mathbf{M}_O$, the moment at the origin of the wrench, is the moment at the origin of the force $\mathbf{F}$ plus the moment at the origin of the couple $\mathbf{C}$ (§ 10.31, § 10.35). We say correspondingly that $\mathbf{v}_O$, the linear velocity at the origin, is, like the moment, obtained in two parts and similarly (§ 5.47). The vectors $\mathbf{M}_O$ and $\mathbf{v}_O$ are both bound vectors; they exist and are bound at that particular point in the moving body which is, at the instant, at the origin. Whereas the terms $\mathbf{F}_O$ and $\boldsymbol{\omega}_O$ are known as the real (or the global) parts, the terms $\varepsilon\mathbf{M}_O$ and $\varepsilon\mathbf{v}_O$ are known as the dual (or the spatial) parts of the dual vectors.

Figure 19.02 (§ 19.14) Dual vectors at the origin. Please note the angle $\psi$ at figure 10.11(*a*); it is the same $\psi$ as the one shown here. Note also the figures 10.12 and 5.09; they show in turn the essential natures of the helicoidal fields of (*a*) the moment vectors, and (*b*) the velocity vectors surrounding their respective central screws.

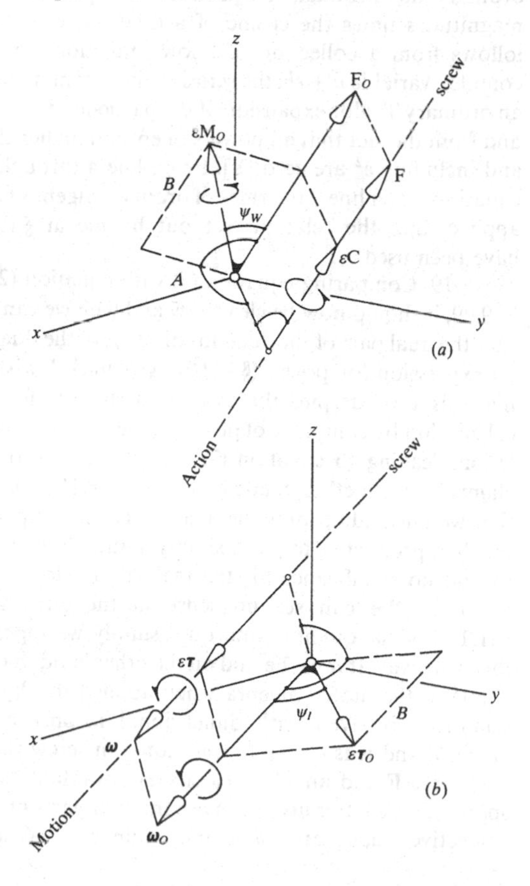


15. In figure 19.02(*a*) we see a dual vector $\hat{\mathbf{W}}$ expressed with its basepoint at an origin of coordinates. In figure 19.02(*b*) we see a dual vector $\hat{\mathbf{I}}$ expressed with its basepoint also at an origin of coordinates. Following the terminology of Brand – and of Waldron [1] at § 7.12 – these pairs of vectors at the origin may be called *motors*, and, as already explained (§ 10.49), Dimentberg for example calls them *force motor* and *velocity motor* respectively. Dimentberg claims at the same time (and after Clifford), incidentally, that the word *mo'tor* is a combination of the words moment and vector. If we imagine, however, all four of the single vectors $\mathbf{F}_O$, $\mathbf{M}_O$, $\boldsymbol{\omega}_O$, and $\mathbf{v}_O$, all nominated for a given point in the same rigid body, all erected in other words at the same origin, we will see how Keler sets up, in his somewhat complicated diagrams, a pair of dual vectors ready for a dot multiplication to obtain power. Refer to figure 19.03. See there that a triad of axes with a fixed orientation in space has been set with its origin at a point $O$ in the mentioned moving body, and that the four mentioned vectors have been mounted at that point ready for consideration.

16. It should be remembered in the context of this book that the power being mentioned above might be the rate at which the resultant of the unbalanced forces $\hat{\mathbf{W}}$ on some rigid link is doing work upon the link as the link twists (or screws) at the rate $\hat{\mathbf{I}}$ about its ISA with

Figure 19.03 (§ 19.15) Two dual vectors at an origin. The origin here is at a point fixed in the moving body, while each of the axes $x$, $y$ and $z$ has its direction fixed in space.

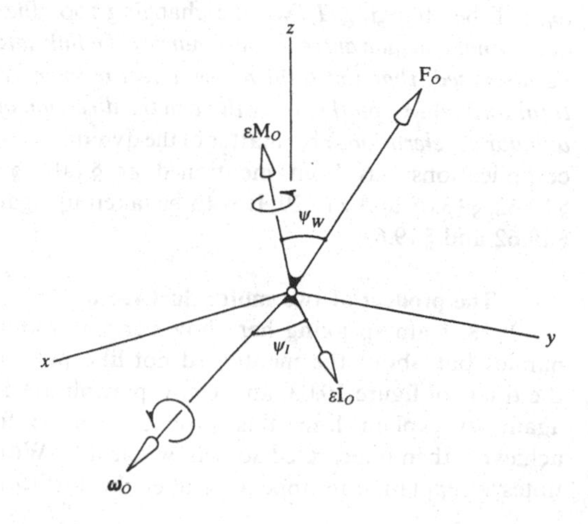

respect to frame. If the link is part of some frictionless machine where the motion is cyclic, the mentioned power is equal to the instantaneous rate of increase (either positive or negative) of the link's total energy. In such a cyclically operating machine, of course, the total energy of each link varies, as the movements do, in a regular periodic manner, no link ever accumulating energy. Refer to the passages about work and energy beginning at § 10.59.

### Interlude, Newton and Euler

17. If, incidentally, the origin in figure 19.03 had been chosen, not to be just anywhere, but at the centre of mass of the rigid body, and here I digress to make an early remark about this particular and most important matter (§ 19.62), *the vector representing the time rate of change of the linear momentum of the body would be identical with that of the total external force* $\mathbf{F}_O$ *at O, and the vector representing the time rate of change of the angular momentum of the body would be identical with that of the total external moment (or torque)* $M_O$ *at O.* These remarks derive from Newton's second law for the motion of a massive particle (1687); they are a flat statement of Euler's two laws for the motion of a massive, rigid body [11]. Whereas in our cases however the mass m of a link in mechanism will be remaining constant as the linear momentum $mv_O$ might be undergoing change due to a changing $\mathbf{v}_O$, and whereas this will always ensure that the total force $\mathbf{F}_O$ on a link will be in the direction of the linear acceleration of the centre of mass of that link, the relevant inertia matrix $I_O$ of the link will *not* in general be remaining constant as the angular momentum $I_O\boldsymbol{\omega}_O$ might be undergoing change due to a changing angular orientation of the link or to a changing direction of $\boldsymbol{\omega}_O$. There are two aspects to be noticed here: (*a*) the inertia matrix $I_O$, a tensor, will have no particular direction through $O$ – it has, in general, no definable axis there; and (*b*), whenever the direction of $\boldsymbol{\omega}_O$ will be changing, $I_O$ will be changing too. *There is accordingly no guarantee in the dynamics of a link, later to be discussed, that the total moment vector (namely the total torque)* $M_O$ *on the link will be in the direction of the angular acceleration.* This matter of the dynamics and its complications has been mentioned at § 1.02, § 1.03, § 12.62, § 13.08 and § 13.29. It is to be taken up again at § 19.62 and § 19.64.

### The product of two motor dual vectors

18. I am speaking here however not about dynamics but about the mentioned dot-like product of the duals of figure 19.03, and I now paraphrase Keler again to explain how this product can be firstly achieved, then interpreted somehow sensibly. With my notes in explanation appearing after the forthcoming algebra, Keler's argument runs as follows:

$$\hat{\mathbf{W}}\cdot\hat{\mathbf{I}} = (F+\varepsilon M)(\mathbf{a}_o+\varepsilon\mathbf{a}_1)(\omega+\varepsilon v)(\mathbf{b}_o+\varepsilon\mathbf{b}_1)$$
$$= \mathbf{F}_O\boldsymbol{\omega}_O + \varepsilon[\mathbf{F}_O v_O + \boldsymbol{\omega}_O\mathbf{M}_O]$$
$$(6)\qquad = (F+\varepsilon M)(\omega+\varepsilon v)\cos(\phi+\varepsilon d)$$
$$= (F+\varepsilon M)(\omega+\varepsilon v)[\cos\phi - \varepsilon d\sin\phi]$$
$$= F\omega\cos\phi + \varepsilon[M\omega\cos\phi + Fv\cos\phi - F\omega d\sin\phi]$$
$$= F\omega\cos\phi + \varepsilon[(M\omega+Fv)\cos\phi - F\omega d\sin\phi]$$
$$(7)\qquad = F\omega\cos\phi + \varepsilon F\omega[(p_a+p_b)\cos\phi - d\sin\phi].$$

The expressions $(\mathbf{a}_0+\varepsilon\mathbf{a}_1)$ and $(\mathbf{b}_0+\varepsilon\mathbf{b}_1)$ in line 1 above are dual unit vectors; these are carefully defined in the relevant literature [8]. Line 2 follows from the fact that all powers of epsilon higher than and including $\varepsilon^2$ are zero. Equation (6) on line 3 follows from the mechanics of the algebraic device which gets us away from the origin and back to the two home screws (§ 10.49); it introduces $d$, the shortest distance between the two home screws, and $\phi+\varepsilon d$, the dual angle between them (§ 19.33). Note that at (6) we have the external form of an ordinary dot product; the product of a pair of dual magnitudes times the cosine of a dual angle. Line 4 follows from a collection of known functions of the complex variable $a+\varepsilon b$; the series derives from making an ordinary Taylor expansion of the particular function and from the fact that all powers of epsilon higher than and including $\varepsilon^2$ are zero [8]. From line 4 through to equation (7) on line 7 the rules of ordinary algebra have applied, and the relations set out by me at § 19.08 have been used.

19. Comparing equation (7) with equation (2) at § 19.09, it should now be clear how and why we can say that the dual part of the dual dot product of the duals is an expression for power (§ 10.60). A remark I wish to make is this: despite the evident usefulness for the calculation by computer of power in the various formulations leading to equation (7), and this is one of the claims for the method made by Keler in his [8], the fact that we must admit only the dual or the spatial part of the dual product is, to put it simply, untidy. We appear to find no significance for the real or the global part $F\omega\cos\phi$. We can see no sense in the dimensions $MLT^{-3}$ of that quantity, and, quite simply, we appear to throw it away. It must be said on the other hand that the whole of the motor algebra when applied to physical quantities is riven with dimensional inconsistencies anyway, and it is certainly well noted in screw theory that force $\mathbf{F}$ and angular velocity $\boldsymbol{\omega}$ powerfully correspond. These latter are the invariant, real parts of their respective duals, and $\phi$ is the ordinary, real angle

between the relevant screws. So well we might ask: is there a meaning for $F\omega \cos \phi$, or, alternatively (and this is the putting of another question), does $F\omega \cos \phi$ *need* to have a meaning? Refer to § 19.28.

**Addition of two motor dual vectors**

20. I am still speaking here in terms of the dual complex algebra of Keler or the motor algebra of Brand (1947); they are the same. Relying to some extent on Brand again, but referring to the better developed and diagrammatic explanations of Keler, we can see that the process of motor dual vector addition is an algebraic process which coincides exactly with the seen geometry of the cylindroid (§ 13.13, § 15.20). Referring to Keler's figures 6(*a*) and 6(*b*) in [8], and rewriting his symbolism into mine, he says that $\hat{\mathbf{I}}_{13} = \hat{\mathbf{I}}_{12} + \hat{\mathbf{I}}_{23}$. Or, to put it cyclically and in terms of the dual unit vectors $\hat{\mathbf{a}}$, $\hat{\mathbf{b}}$ and $\hat{\mathbf{c}}$ in the directions of the three screws, he says that $I_{12}\hat{\mathbf{a}} + I_{23}\hat{\mathbf{b}} + I_{31}\hat{\mathbf{c}} = 0$. Keler next multiplies this equation by $\hat{\mathbf{b}}$, and by using the fact that $\hat{\mathbf{b}} \times \hat{\mathbf{b}} = 0$, he writes that $I_{12}(\hat{\mathbf{a}} \times \hat{\mathbf{b}}) + I_{31}(\hat{\mathbf{c}} \times \hat{\mathbf{b}}) = 0$. He concludes from this that the three screws must all cut the same line in space perpendicularly, and proceeds to construct next a dual vector triangle. This, he finds, must close. From the closure of the dual triangle, this six-sided figure in this particular case having three of its alternate sides collinear (§ 19.43), he derives a dual version of the ordinary sine rule for real angles in a real triangle, and from that deduces that not only an angular velocity polygon in one plane but also a linear velocity polygon in another may be drawn. The latter polygon helps to determine the distances between the screws along the mentioned line in space, the mentioned line being, of course, (*a*) the common, common perpendicular between the three sets of two screws, (*b*) the three collinear sides of the dual vector triangle, and (*c*), as may be seen from my remarks at § 19.43, the nodal line of the relevant cylindroid. *Although Keler never mentions the fact (never mentioning, indeed, the cylindroid), it is clear from the geometry of his diagrams that his algebraic method for the addition or subtraction of two motor dual vectors of the same kind bears its expected, direct relationship with the geometry of the cylindroid.* Refer to § 19.28.

**Motor algebra and the Plücker coordinates for a line**

21. Having studied the idea of motor, namely the idea of a dual vector (or of a dual) expressed at a basepoint somewhere remote from its home screw (§ 10.49), we can study now the relationship between the algebra of motors namely the motor algebra and the Plücker coordinates for a line. It should be mentioned by way of clarification here that a line in the present context may be seen as a screw of zero pitch and that, at § 19.28 I look again (and more incisively) at the mechanics of screws in the light of the Plücker coordinates for a screw. I intend to proceed from the particular to the general in other words; so the reader may go to § 19.28 directly, paying attention there to the entirety of that long paragraph or to the latter parts of it. In any event a few useful references are as follows: Plücker [12], Jessop (1903), Maxwell (1946, 1951), Woo and Freudenstein [13], Hunt (1978), and [14].

22. So let us study first the coordinates for a line. They are also known more simply as line coordinates. Refer to figure 19.04(*a*). If we wished to locate a line $\ell$, free as yet, in a Cartesian frame $O$–$xyz$, we could proceed as follows. We could select in the line $\ell$ a line segment $A$–$B$ of unit length, then erect upon that segment as diagonal a rectangular box as shown. We could next determine the shape of the box by choosing only two – any two would be enough – of its three dimensions, say $l$ and $m$ as shown. Next, by aligning $l$ parallel with $O$–$x$ and $m$ parallel with $O$–$y$, we could fix the orientation of the line in the Cartesian space. We could then say that the ratios $l$/unity and $m$/unity were two of the three *direction cosines* of the line $\ell$, the third, $n$/unity, being determined by the other two. Kindly recall here that, if $\alpha_x$, $\alpha_x$ and $\alpha_z$ are the angles respectively between the directions of $O$–$x$, $O$–$y$ and $O$–$z$ and the line $\ell$, then $l, m$ and $n$ are the cosines of those angles respectively (§ 19.25). To locate the line, we

Figure 19.04(*a*) (§ 19.22) The location in space of a line $\ell$ by means of only two of its direction cosines $l$ and $m$ and the radii $a_x$ and $a_y$ of two circular cylinders it must touch. The other direction cosine $n$ and the third cylinder radius $r_z$ are both determined by the choice of the four parameters. Note that the line $\ell$ is here shown passing beneath the cylinder on $O$–$y$; two of the three points of contact between the line and the cylinders are thereby hidden.

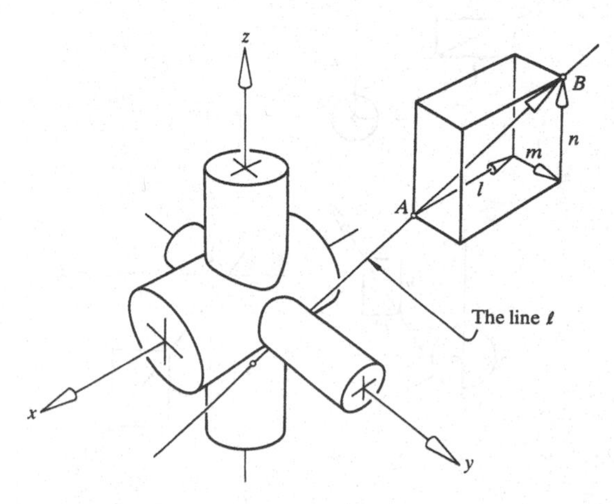

could erect two circular cylinders of radii say $a_x$ and $a_y$ coaxial with the axes $O$–$x$ and $O$–$y$ respectively, then move the already oriented line until it touched the two cylinders as shown. Leaving aside just now the obvious questions of sign (and of handedness) – of whether we touch the cylinders at their fronts or at their backs or by way of some combination of these – please see that the fixing of the two radii $a_x$ and $a_y$ have been sufficient to locate the line $\ell$. *Having once chosen the unit length A–B, we have needed only four length parameters to locate the line,* $l, m, a_x$, *and* $a_y$. We note of course that a third radius $a_z$ of a circular cylinder coaxial on $O$–$z$ has been determined by the prevailing circumstances.

23. Allow a unit vector say **L** to exist in the line, extending say from a basepoint at $A$ to a tip (or an endpoint) at $B$. Refer to figure 10.07 where the moment at some chosen origin of a force (or, alternatively, the linear velocity at some chosen origin of an angular velocity) is shown. Now in the same right handed manner let us mount at the origin in figure 19.04(*a*) a moment $\mathbf{L}_O$ normal to the plane containing the now located line $\ell$ (which itself contains the unit vector **L**) and the origin. Referring to the orthogonal views in figure 19.04(*b*) we can see that, if $l_O$, $m_O$ and $n_O$ are the magnitudes of the components of this moment vector $\mathbf{L}_O$ along the axes $O$–$x$, $O$–$y$ and $O$–$z$ respectively,

Figure 19.04(*b*) (§ 19.23) Three orthogonal views of the figure 19.04(*a*). They are each taken towards the origin looking along the axes respectively for $x$, $y$ and $z$. The various dimensions that correspond within the three views are interrelated, of course, not only by virtue of the geometry of projection but also by virtue of the redundancies. Note that the magnitudes of the moments $\mathbf{l}_O$, $\mathbf{m}_O$ and $\mathbf{n}_O$ about the axes for $x$, $y$ and $z$, respectively, are obtained by multiplying the diagonal of the relevant rectangle by the radius of the relevant circle.

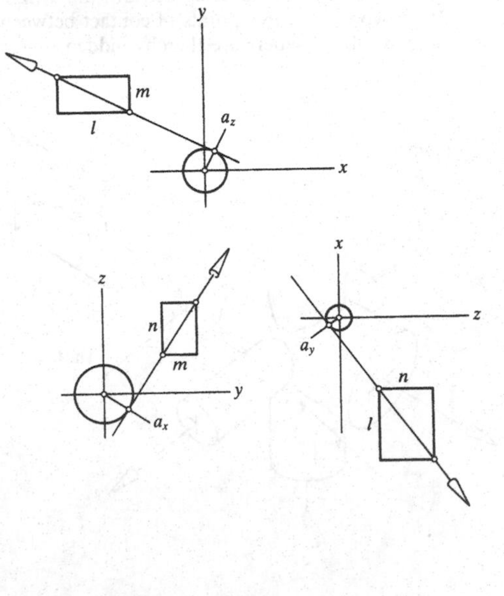

(8) $$l_O = a_x(m^2+n^2)^{1/2}$$
$$m_O = a_y(n^2+l^2)^{1/2}$$
$$n_O = a_z(l^2+m^2)^{1/2}$$

See now, paying due regard to sign and to handedness, that the six parameters $l, m, n, l_O, m_O$ and $n_O$ may be used to locate the line $\ell$ definitively. These are the six Plücker coordinates for the line. They are commonly written $(l, m, n;\ l_O, m_O, n_O)$. Perceive clearly that only four of them – two from each group – are sufficient for the purpose, the remaining two being in fact redundant. The six are also commonly written $(l, m, n;\ p, q, r)$. Since I have, however, already used the symbol $p$ for pitch elsewhere in this book, and used the symbol $r$ for both the distance from an origin to a line and the distance from a screw to a point, I wish to avoid from here onwards in this chapter the otherwise convenient lower-case letters $p$, $q$ and $r$ for the second three of the six Plücker coordinates.

### Homogeneity of the Plücker coordinates for a line

24. The six Plücker coordinates display an important property. Because we can multiply all six of them through by any non-zero scalar (say $k$) and still get the same result namely the same line $\ell$, the six coordinates are said to be *homogeneous*. Thus we can use for the same line $\ell$, having abandoned the need for a unit length $A$–$B$, the new formulation $(L, M, N;\ P, Q, R)$. This series of six new numbers – let me call them the *new coordinates* – we can get by simply multiplying through. There are, however, less limiting relationships among such a set of new coordinates. The relationships reduce not to only four but to only five the apparent number of six open choices for the new coordinates to nominate a line. The relationships are as follows: (*a*) the six new coordinates must be such that the two vectors **L** and $\mathbf{L}_O$ they specify must be perpendicular, that is that the dot product of the two vectors must be zero; (*b*) the square root of the sum of the squares of $P$, $Q$ and $R$ must be equal to the magnitude of the moment vector. Taken together these become the same thing as saying (*c*) that, when the six new coordinates are divided again by the non-zero scalar $k$ in such a way that the line vector **L** becomes unity again (whereupon the set of coordinates might be said to have been *normalized*), the magnitude of the moment vector at the origin must be equal to the length of the perpendicular dropped from the origin onto the line. These interrelated conditions can be set down more succinctly in the form of only one condition, namely

(9) $$LP + MQ + NR = 0.$$

This single limiting condition (9) is called the *orthogonality condition* or the *quadratic identity*; it must always apply among a set of line coordinates.

25. The gist of figure 19.05 comes directly from Rooney [4]; paper [4] quotes in turn Rooney [28]; Rooney claims in these two works that, while we know on the one hand that

$$(10)\quad l = \cos\alpha_x,\quad m = \cos\alpha_y,\quad n = \cos\alpha_z,$$

we can find on the other that

$$(11)\quad l_O = -a_x \sin\alpha_x,\quad m_O = -a_y \sin\alpha_y,\quad n_O = -a_z \sin\alpha_z.$$

Please see in figures 19.05 that, while the radii $a_x$, $a_y$ and $a_z$ are, respectively, the common perpendiculars drawn between the axes $O$–$x$, $O$–$y$ and $O$–$z$ and the line in question namely the line $\ell$ for which the Plücker coordinates are written, the angles $\alpha_x$, $\alpha_y$ and $\alpha_z$ are the angle between the three mentioned directed axes and the directed line (§ 19.22). For an explanation of the idea of the dual angle $(\alpha, a)$ between a pair of skew lines, please go to § 19.33. It might be observed here incidentally that, if, with these data, an expression were written for the shortest distance between the origin and the line $l$, the expression would reduce to the square root of the sum of the squares of $l_O$, $m_O$ and $n_O$; for, after all, if the vector $\mathbf{L}$ in the line $\ell$ is unity, unity times this distance must be equal to the magnitude of $\mathbf{L}_O$.

26. An advantage of the homogeneity of the six coordinates for a line is that the distance from the origin to a line with a fixed orientation in space can be varied all the way from zero to infinity without employing the actual number $\infty$ at all. If we chose the direction magnitudes $L$, $M$ and $N$ for some line and we found that, given its distance from the origin, its coordinates were $(L, M, N; P, Q, R)$, the coordinates $(L, M, N; 0, 0, 0)$ would set the same line parallel with itself and passing through the origin, while the coordinates $(0, 0, 0; P, Q, R)$ would set the same line parallel with itself and at infinity. In the latter case, however, there is more to it than that. Whereas the coordinates $(L, M, N; 0, 0, 0)$ unequivocally set the direction of the line vector $\mathbf{L}$, the coordinates $(0, 0, 0; P, Q, R)$ do not set this direction; they set only the direction of the moment vector $\mathbf{L}_O$. In the limit, in other words, as $L$, $M$ and $N$ tend to zero (and these should be seen to keep their proportionality as they go), the direction of the perpendicular $\mathbf{r}$ from the origin to the line becomes indeterminate. As the line $\ell$ recedes to infinity, in other words, it becomes in the limit a *circle* at infinity. This circle, this so-called 'line at infinity', straight, of course, yet curved into a circle of infinite radius, clearly resides in that particular plane through the origin which is normal to the moment vector $\mathbf{L}_O$. Refer to § 8.13, § 15.53, § 15.56 etc. where I speak, in connection with real mechanical joints and the screw systems they display, about a tunnel of unbounded length whose diameter is however infinite (§ 15.51). This tunnel is imagined there and elsewhere to be formed by screws of zero pitch located at infinity; the screws, with each screw perpendicular to the axis of the tunnel, are seen to be tangent to circles of infinite radius; the planes of the circles are normal to the axis of the tunnel and the centres of the circles are, if not upon, then within some finite distance from the said same axis. With this difficult matter illuminated here in the comfortable glow of homogeneity, and keeping in mind the implications at infinity of the idea that *no two lines can intersect at more than one point*, I invite the reader now to ponder the one and only line (or circle if you wish) at infinity. Its Plücker coordinates are $(0, 0, 0; P, Q, R)$.

27. It is worthwhile to notice that, whereas the well known four homogeneous point coordinates for a point $(x, y, z, w)$ have one redundant parameter, thus reducing the parameters open to choice for a certain point to only three, there being an $\infty^3$ of points in space (§ 4.01), the six homogeneous line coordinates for a line $(l, m, n; l_O, m_O, n_O)$ have two redundant parameters, thus reducing the parameters open to choice for a certain line in space to only four; and there are, as we know, an $\infty^4$ of lines in space (§ 4.10). But let me repeat that, whereas only four of the lower-case (the original or normalized) line coordinates are enough to determine a line, only four of the upper-case (the new) line coordinates are not enough to determine the line; we need five; refer to figure 19.06 and read the explanation at its

Figure 19.05 (§ 19.25) Geometrical relations between the angles $\alpha$ and the radii $a$ of the three circular cylinders shown at figure 19.04(*a*). The angles $\alpha$ are the real or the global parts of the dual angles $(\alpha, a)$ that obtain.

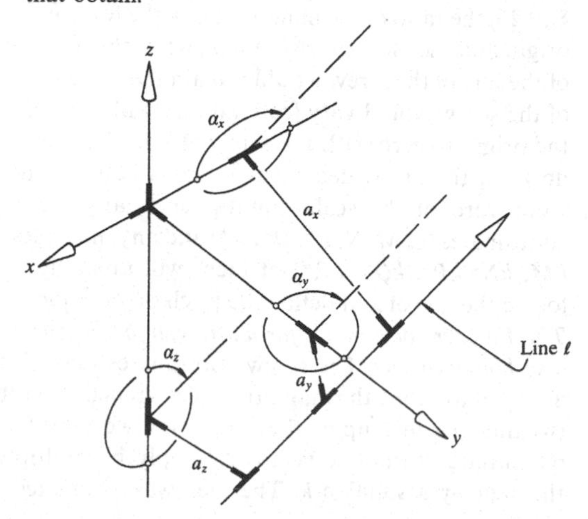

caption. The remarkable simplicity of the Plücker line coordinates may be seen when we write equations for the three linear line systems so evident in nature – the complex, the congruence and the regulus. That these systems of lines are so evident in nature is explained in chapter 10, and an elementary account of the mentioned equations is given in chapter 11, beginning at § 11.43. Application of the Plücker line coordinates in the field of screws and screw systems is also remarkably precise – at least preliminarily. Please refer to my limited remarks in chapter 23, beginning at § 23.47. But go now to § 19.28 for an explanation of the so-called screw coordinates for a screw that were drawn again to our attention recently by Woo and Freudenstein [12] [13].

**The Plücker screw coordinates and the mutual moment of two screws**

28. Referring now to § 19.14 we can see that if the line vector **L** in the line $\ell$ were say a pure force **F**, and if the moment $\mathbf{L}_O$ at the origin were the pure moment say $\mathbf{M}_O$ of that force, we could write for the motor at the origin, $\mathbf{F}_O + \varepsilon\mathbf{M}_O$. The two sets of the three components of the magnitudes of $\mathbf{F}_O$ and $\varepsilon\mathbf{M}_O$ at the origin would simply be then, as we have seen before, the six Plücker coordinates for the line. Equally we could say that the six elements $(L, M, N; P, Q, R)$ are the six coordinates of the force, the magnitude of the force becoming known from

Figure 19.06 (§ 19.27) It should be clear from this figure that the mere choosing of say $L$, $M$, $P$ and $Q$ will not be enough to locate the line $\ell$. Although the vectors $\mathbf{L}$ and $\mathbf{L}_O$ must be mutually perpendicular at the origin as shown, this condition (the orthogonality condition) is not enough to determine $N$ and $R$. To locate the line $\ell$ we must nominate not only four but at least five of the six upper-case (the new) Plücker coordinates.

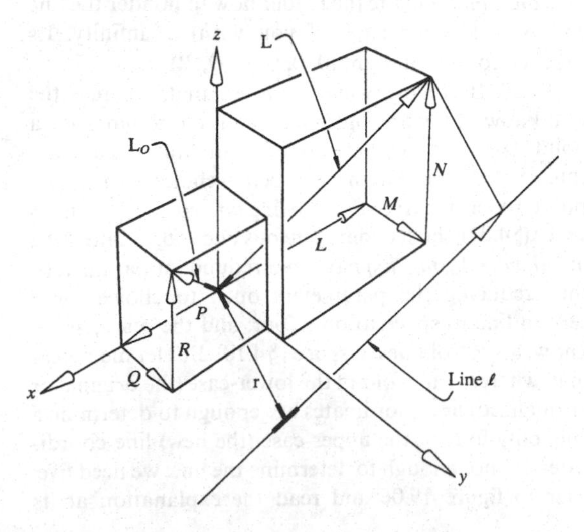


the magnitudes of $L$, $M$ and $N$. We see accordingly that the idea of motor on the one hand and that of the six Plücker coordinates on the other are closely related. Parenthetically I wish to say this: whereas in the motor algebra the relevant workers might change the symbol **F** to $\mathbf{F}_O$ whenever an **F** (the invariant part of an action dual) is shifted from its line of action or its home screw to the origin, I have not done this with the similarly invariant line vector **L** – either in this paragraph or in figure 19.06. The trouble is, here in this chapter, that my discussion in encumbered with a double standard for the nomenclature. In trying to keep faith with regular usages in two diverse aspects of the literature, I am running the risk of not reflecting properly either of them; refer to Rooney [4]. Suppose now that there were in the line, as well as the force **F**, a collinear couple **C**. Together the two would comprise a wrench $\hat{\mathbf{W}} = \mathbf{F} + \varepsilon\mathbf{C}$. The motor at the origin expressing this wrench could be written $\hat{\mathbf{W}}_O = \mathbf{F}_O + \varepsilon\mathbf{M}_O$, where the vector $\varepsilon\mathbf{M}_O$ exhibits not only one but both of the two components explained at § 19.14 and shown in both of the figures 19.02; the components are marked there $A$ and $B$; while $A$ has the magnitude of $\mathbf{F} \times \mathbf{r}$ (where **r** is the distance from the axis of the screw to the origin), $B$ has the magnitude of **C**. It should be clear from the figures – and the following was well understood by Plücker – that, if we wrote the components of the magnitude of $\mathbf{F}_O$ and the components of the magnitude of $\varepsilon\mathbf{M}_O$ in the same way as we write the six Plücker components for a line, we would, as it were, be writing the six Plücker-like coordinates for the relevant *screw*, the pitch of the screw being the pitch of the wrench. Let us write for these components $(L, M, N; P^*, Q^*, R^*)$; I follow Hunt (1978). On doing this, however, we find (as Plücker did) that the property held by the Plücker coordinates for a line, namely that property outlined at § 19.26 involving the oriented line at the origin and the similarly oriented line at infinity, does not obtain. For if we varied (as we did for the mere line at § 19.26) the ratio of the magnitudes of the force $\mathbf{F}_O$ at the origin and the moment $\varepsilon\mathbf{M}_O$ there, while the orientation of the line of the screw would remain the same the pitch of the screw would vary from infinity with the screw at the origin to zero with the screw at infinity. We do see however that, provided we avoid (as we did before) the value zero of the scalar multiplier $k$, any set of six coordinates $(L, M, N; P^*, Q^*, R^*)$ and any multiples $(kL, kM, kN; kP^*, kQ^*, kR^*)$ of them will unambiguously locate the line of and define the pitch of a unique screw. *The Plücker coordinates for a screw, in other words, are also homogeneous.* The screw coordinates may also be adjusted to define the magnitude of a wrench or a rate of twisting residing upon the screw, for we may fix the magnitude of the line vector **L** simply by multiplying through by a suitable $k$. The idea was originated and examined by Plücker [12]; it was recently reexamined by Woo and Freudenstein [13] and by Hunt (1978). Let me expand a little here and, in the steps of the mentioned workers, write down some of the well established results. I wish to write first (*a*) an expression for the pitch $p$ of a given screw, and (*b*) the line coordinates for the line of a given screw. Knowing that the pitch of the screw is $C/F$ (or $v/\omega$), and referring to the essential geometry of the figures 19.02 (or of the many other such figures in this book), we can show quite easily (*a*) that the pitch $p = lP^* + mQ^* + nR^*$, and (*b*) that the line coordinates for the line of the screw are ($L, M, N; P^* - pL, Q^* - pM, R^* - pN$). Another spectacular result, taken directly from Plücker [12], item 2, but modified for presentation here, is as follows: the resultant of a number of different forces ($L, M, N; P, Q, R$) is ($\Sigma L, \Sigma M, \Sigma N; \Sigma P, \Sigma Q, \Sigma R$). If, by chance, the quadratic identity applies, that is if $\Sigma L \Sigma P + \Sigma M \Sigma Q + \Sigma N \Sigma R = 0$, the resultant is a pure force. If not, and this will be the case in general, the screw coordinates ($\Sigma L, \Sigma M, \Sigma N; \Sigma P, \Sigma Q, \Sigma R$) will nominate the resultant wrench; this may be seen to be in the form of its equivalent *dyname* at the origin [12] (§ 10.49). As explained above any set of screw coordinates may be adjusted to refer directly to a wrench (or a rate of twisting), and we find that the resultant of a number of different wrenches or rates of twisting ($L, M, N; P^*, Q^*, R^*$) may be written ($\Sigma L, \Sigma M, \Sigma N; \Sigma P^*, \Sigma Q^*, \Sigma R^*$). This spectacular result follows also of course directly from the fundamental principles of kinetostatics (§ 10.32, § 10.55). *Thus with screw coordinates we can write down the whole of the central gist of the kinetostatics of a rigid body in a single linear algebraic statement.* It follows more clearly than ever now that in the context of screws a line is simply a screw of zero pitch. I wish to change the subject now and refer to the *mutual moment*, not only of a pair of lines (which is relatively well known), but also to that of a pair of screws. Given two lines with a distance $d$ and an angle $\phi$ between them, we may say (and we do) that the mutual moment of the two lines is $d \sin \phi$; this is clearly a measure of the capacity of a unit force in either one of the lines to do work on a body free to rotate about the other; it is also clear from this (and useful to know) that, if the mutual moment is zero, the lines will either intersect or be parallel; refer to Jessop (1903). But Jessop also explained in his book that the mutual moment $d \sin \phi$ of two different lines ($L, M, N; P, Q, R$) may be written $L_1P_2 + M_1Q_2 + N_1R_2 + L_2P_1 + M_2Q_1 + N_2R_1$; and this, in the context here, is of course most interesting. *It can however and moreover be shown that, if we choose to define the mutual moment of two different screws* ($L, M, N; P^*, Q^*, R^*$) *by the same formula, that is by the same summation of six terms* $L_1P^*_2$ *etc. and* $L_2P^*_1$ *etc., the two screws will, in the event of their mutual moment being zero, be reciprocal*; refer to Lipkin and Duffy [14] and Parkin [14]. The importance of this result cannot be overemphasized. The mutual moment of two screws is clearly a measure of the capacity of a unit wrench in either one of the screws to do work on a body free to twist about the other. The reader may care to reconcile the underlying algebra here with that of Ball (§ 19.09, § 14.25). *The reader will find that the mutual moment of two screws, as expressed above in terms of the Plücker coordinates, is identical with* $[(p_a + p_b)\cos\phi - d \sin\phi]$, *namely with Ball's 2Q.* The reader will find as well however that, whereas I spoke at § 19.08 of power being this 'kind of dot product' of the two dual vectors, and whereas I spoke when speaking about the dual vector algebra at § 19.19 of the dismissal by mere banishment of the unwanted real part of the said dot product as being (to say the least) 'untidy' in the circumstances, the formulation here can also not be seen to be an ordinary dot product of an ordered pair of six-vectors. In the mutual moment as written above there is a switching over of the groups of elements of the two six-vectors; the direction-parts and the moment-parts are interwoven. The root phenomena associated with this may be found however by making a careful study of the physics outlined at § 19.10. It may thus be seen moreover that the symbol 'dot' might best be avoided when writing vectorially about the mutual moment, that some new symbol for this admittedly other kind of product might better be used; refer to Parkin [14]. Finally I refer to the last-cited papers at [14]; these and a few others of similar kind are beginning to show that the Plücker screw coordinates for relative rates of screwing in robot arms admirably lend themselves to the formation of new kinds of matrices. These are (6 × 6) matrices involving various rows and columns made up of the screw coordinates. They are helping to solve directly the many problems associated with the sudden gain or loss of freedoms at the end-effectors of robot arms due to the so-called *singularities* – the geometrical specialities of chapter 2, the problems of § 4.04 made manifest, and the linear interdependencies of the screws of chapter 20 – that accidentally occur in the arm-configurations (§ 19.55).

**The exterior calculus of Grassmann**

29. Let me repeat that we need to deal algebraically not only with sets of lines (which may be seen as screws of zero pitch) but also with sets of dual vectors and thus with the sets of screws that abound in kinetostatics. From § 19.13 to § 19.20 inclusive and in § 19.28 I have discussed the motor representation and the screw coordinates that may each be used to deal with these. But I wish to mention now in addition to those devices the extraordinary vector method *Ausdehnungslehre* (extension theory), or the so-called *exterior calculus* of Grassmann. It was founded by Grassmann in

1844 and reported then; his collected works (1896) (1904) (1911) are available. The following few remarks about what appears to be a very small part of these works are ingenuously extracted by me (and ruthlessly paraphrased here) from recent papers by Browne and Pengilley of Melbourne [15]. The authors explain among other things the following: that a force is not satisfactorily expressed by a free vector; that such deficiencies in contemporary practice may be put right by the use of 'verbal appendages' to the mathematical statements; that for example one might say that a force acts not simply along its line of action but through some point; that the exterior calculus is capable of representing a force in magnitude, direction, sense, and its line of action; that points **p** in the exterior calculus are vector-like elements (but not vectors) which, when taken together one by one, may be 'extended' by means of their exterior product into first the line defined by a pair of them and then the plane defined by that line and another one of them, and so on; that *exteriorness* is the geometric equivalent of linear independence; that if two points are coincident their exterior product is zero; that the sum of a vector and a point is a point, and that this leads to the difference of two points being a vector; that a force **F** is denoted by the exterior product of the force vector **f** and any point **p** in its line of action; *that any system of forces may be represented by a single bound vector and a so-called bivector, the first being the vector sum of the forces of the system acting at an arbitrarily chosen point, while the second is the sum of the moments at that point of the same forces.* Be aware however that in [15] we are led to believe that the opposite obtains in the realm of kinematics. The authors speak about the 'velocity' of a rigid body. Included within this concept is (*a*) the linear velocity of a point within the body, and (*b*) the angular velocity of the body. *Whereas next this velocity of a point in the body appears to be seen by the authors as the bound vector* (*like force*), *the angular velocity turns out to be, and is seen to be, not the bound vector as one might expect, but the bivector* (*like moment*). The papers [15] are short, and parts of them are obscure to me, but the working reader in English might care to find in [15] a useful introduction to the exterior calculus of Grassmann.

**The finite screw displacement of a body or of a line**

30. A theorem said by some to be due to Chasles [16], which deals with the finite screw displacement of a rigid body, is written about and proven by many authors. The theorem states, in effect, that we can always get from any given location 1 of a rigid body to any other given location 2 by means of the combination of (*a*) a single pure rotation $\phi$ of the body about an axis which is unique, and (*b*) a single pure translation $d$ of the body in the direction of that axis. If the two movements are imagined to occur together the whole displacement may be seen as a *finite screw displacement*; we may write the displacement ($\phi$, $d$). Given the two locations of the body, say 1 and 2, one may find the mentioned unique axis by following these brief hints: (*a*) refer in advance to figure 19.07; (*b*) remove by pure translation the body $ABC$ from its location $A_2B_2C_2$ to another location such that $A_2$ goes directly to $A_1$; call this intermediate location $A_1B'C'$; it is shown in the figure but, except for the point $A_1$, it is not so named; (*c*) erect through $A_1$ a line parallel to the common perpendicular between the straight lines $B_1B'$ and $C_1C'$; this will give the required direction (but not the location) of the sought-for axis; (*d*) take now an orthogonal view of the whole figure by looking along this newly discovered direction and locate the mentioned axis (marked FSA in the figure) by means of the method well known for dealing with the finite displacement, by pure rotation about a point, of a lamina $AB$, $BC$ or $CA$ in the plane. Hunt (1978) makes a nice point at his § 3.12, paying attention to the fact that, in getting from the one given location of the body to the other, the *pitch* of the relevant finite screw displacement (or of the twist) may take a series of different values. The value taken, as he points out, depends upon the number of complete turns in excess of the necessary minimum which might be undertaken by the body as it moves from start to finish; the pitch of the twist may be written $(\phi + 2n\pi)/d$ where, usually, $n$ will be put to zero. It must moreover be said (and I say it now) that the pitch of the screw displacement will depend upon whether we go from the first to the second location of the rigid body right handedly through the angle $\phi$ or left handedly through the angle $2\pi$-$\phi$; in many cases it might be wise to adopt a convention about this; refer to note [17] at § 19.32. The *location* of the axis of the twist is however unique according to the theorem, and I shall call that axis *the finite screw axis or FSA*.

31. In the area of TMM the concept of the FSA is an important one for at least two reasons. First it can be seen that rigid links in mechanism and objects moved by robots move (or are moved) from place to place in space; quite naturally, we wish to study such gross movements (§ 2.58, § 2.60). Next it can be seen that, as a geometric device for going in the mathematics from one joint-axis to the next in a closed loop mechanism, the idea of the FSA can be employed; we often employ it thus when we try to write a closure equation for the loop (§ 2.43, § 7.10, § 19.47). I try to explain in the next paragraph how these two, quite different ways of application of the FSA are related.

32. Refer to figure 19.07 where a rigid body is shown in two locations 1 and 2 (§ 12.08). Between the two locations the body has suffered a finite twist that

may be written ($\phi$, $d$). Fixed in the body there is a point $A$; it is a generally chosen point; but with the body in location 1 it is called $A_1$, in location 2 $A_2$, and so on (§ 12.08). Ignoring just yet the shown intermediate location of the body, notice that a line in the body drawn from $A_1$ to cut the FSA perpendicularly at $O_1$ can be seen to have undergone a finite screw displacement (from $O_1$–$A_1$ to $O_2$–$A_2$), that is, the same as that of the body namely ($\phi$, $d$). It should be clear from this that, if we have two lines fixed in the same rigid link (they may be the seats for the motion screws at successive joints in some closed loop mechanism), the combination of the angle say $\alpha$ between them and the perpendicular distance say a between them, namely ($\alpha$, $a$), can be seen as a finite screw displacement of a line. This latter FSA from one location to another of a mere line, however, is not unique; the ubiquitous cylindroid is found to be involved; see notes [17] [47]. The FSA for a given pair of locations of the same line segment or for a given pair of locations of the same whole line is unique only if the axis of the FSA is, as it has been here, already nominated. That complicated matter notwithstanding, the existence of two lines fixed within a rigid link can be seen if we wish, not only as a screw displacement between two locations of the same line, but also as a geometrical representation in space of the bounded body of that link. Please recall here the idea *tick* discussed at § 14.04. See with the two lines of the tick that, whereas the real part of the dual angle between the two lines is finite in value, the dual part of the dual angle between them is zero.

Figure 19.07 (§ 19.32) The finite twist or the finite screw displacement ($\phi$, $d$) of a body from an original location 1 to a new location 2. Given only the locations 1 and 2 of the body the shown unique axis of the finite displacement (which is called by me the finite screw axis or the FSA) can always somehow be found (§ 19.32). Although the figure suggests that the following movements might take place separately and in the reverse order, the body may be seen (*a*) to rotate through the finite angle $\phi$ about the FSA, while (*b*) it translates at the same time through the finite distance $d$ in the direction of the FSA. The overall screw displacement, which is shown here by the drawn helical track of the body, is also shown as being achievable in two other separated stages: (*a*) a pure rotation through $\phi$ about an axis parallel with the FSA passing through some generally chosen point $A$ in the body (the point being marked here $A_1$ in location 1), and (*b*) a pure translation of the body in such a direction that the point $A_1$ at location 1 goes directly to $A_2$ at location 2.

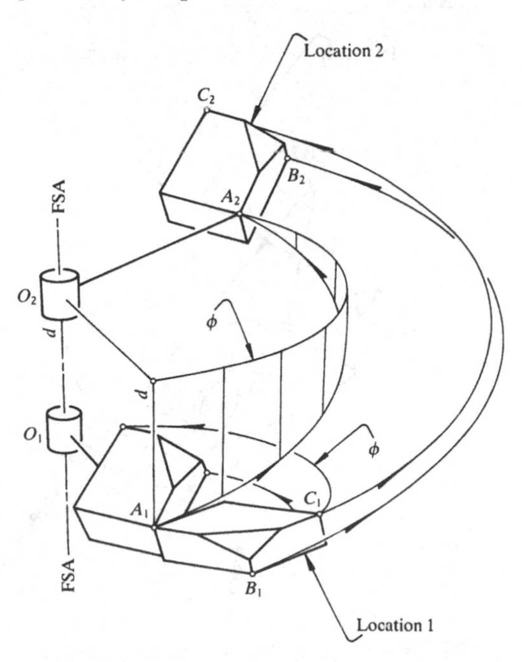


33. In an appropriate algebra, the dual-complex algebra as many might call it (§ 19.36), combinations such as ($\alpha$, $a$) may be written $\alpha + \varepsilon a$, and referred to as the *dual angle* or the *dual angular displacement* between the two lines or axes or screws. Similarly combinations such as ($\phi$, $d$) may be written $\phi + \varepsilon d$, and such combinations may be called the dual angular displacement from one location to another of a rigid body. *Neither combination, ($\alpha$, $a$) nor ($\phi$, $d$), may be called with acceptability dual vector however, because they do not commute; nor may either be seen as a motor because, as we shall see, the field of the linear displacement vectors is not helicoidal and the motor algebra cannot apply.*

34. We have already seen the point $A_2$ in the body in the body's location 2 in figure 19.07. I wish to take two steps. This is the first step: draw through $A_1$ and parallel to the FSA the intermediate axis of rotation and execute the finite angular displacement $\phi$ to bring the body to its shown intermediate location; see figure 19.07. See also figure 5.09 and note there the point $Q$. Read also § 5.57. Now for the second step: translating the body without rotation, take the point $A_1$ to its final position $A_2$ and note that the angle $O_1A_1A_2$ is not a right angle. Please think about this and perceive that, *unless the screw displacement from 1 to 2 is an infinitesimal one, the angle $O_1A_1A_2$ will never be a right angle.* This is the fundamental reason (*a*) for the fact that the distribution of the bound finite displacement vectors within a moved (or about to be moved) body is not helicoidal, and (*b*) for the consequent fact that the ordinary vector laws for commutative addition cannot be applied to finite screw displacements.

35. What this means is that the end result achieved by successively applying finite screw displacements to a body will depend upon the order in which the displacements are applied (§ 2.60). It means that the motor

algebra associated with the motor dual vectors is inapplicable here. It means also in mechanism that, if we wish to write closure equations for the various closed-loop linkages, as Duffy (1980) and others have done, some other algebra for the general spatial displacements of bodies and lines needs to be found. Before I go on to mention, later, some of the *matrix methods* used for this work, I wish to mention next, and under a fresh heading, that branch of the literature which concerns itself in a direct, dual-angle way with the finite screw displacements of a rigid body and the composition (namely the addition) of such displacements. The literature appears to have originated with Kotelnikov. I have noted him already [7].

**Rotations, reflections, and the screw triangle**

36. Studying the more recent literature, Dimentberg (1965), Roth [18], Bottema and Roth (1979), Keler [19], and Duffy (1980), one begins to see (from the works of Dimentberg particularly) that Kotelnikov in Kazan in 1895 and Study in Dresden (1903) independently laid the foundation for what was to become the dual or the dual-complex algebra for the analysis and composition of finite screw displacements. At the same time Roth [18] and Bottema and Roth (1979) collectively draw attention to the breathtakingly concise paper of Georges-Henri Halphen of 1882 [20]. This paper explains and encapsulates in four short pages the essential geometry of what we know today as the screw triangle (§ 19.38).

37. One fundamental fact at the root of the dual algebra for the manipulation of finite screw displacements is the following: *the finite screw displacement* ($\phi$, $d$) *of a rigid body from one location to another can always be duplicated by two successive pure rotations.* The two pure rotations are each half-revolutions about an axis, there being two axes, and the two half-revolutions are often called, in the literature, line-reflections in, or reflections about, the relevant axis. Refer to figure 19.08($a$). Both reflection axes must cut the FSA perpendicularly, the angle between them must be $\phi/2$, and the distance between them must be $d/2$. Except for the question of handedness (and here we note that the displacement from one to the other of the two axes must be of the same hand as the original screw displacement of the body), there is no other requirement. There is, accordingly, an $\infty^2$ of different ways for choosing the first of the two reflection axes. Having chosen the first, however, the second is already determined. In figure 19.08($a$) I have tried to illustrate in a single picture the gist of these successive half-rotations about, or this double reflection in, a pair of axes. Note the line segments $O_1$–$A_1$ and $O_2$–$A_2$, both perpendicular to the FSA; they are carried over from figure 19.07. I leave the required proof of the proposition as an easy exercise for the reader – view the scene, first along the FSA, and then orthogonally across the FSA [20].

38. Now Roth gives, in the first of his [18], not only a clear explanation of this double-reflection phenomenon, but also an explanation of what he calls *the screw triangle.* It is also in this paper that Roth remarks incidentally that the geometry of the matter comes from Halphen. When we go to Dimentberg (1965) we find a similar account of the screw triangle, but the idea is there said to be due to Kotelnikov. I should mention here that, although the screw triangle deals with the composition of two finite screw displacements to produce a resultant, namely a third finite screw displacement (§ 19.39), *it is not a vector polygon*; refer to figure 19.08($b$). It is in fact a diagram of the relative locations in the fixed space (or in the moving body, as we shall see) of the three relevant FSAs. Please study carefully the caption at figure 19.08($b$). Now by virtue of inversion (§ 1.47) there must

Figure 19.08($a$) (§ 19.37) A tick (§ 4.04) embedded at $A$ in a rigid body is marked $A_1$ to represent the body in some original location 1. The tick (and thus the body itself) is transported from location $A_1$ to a new location $A_2$ via an intermediate location $A^*$ by means of two successive half-rotations about two rotation axes which, inclined to one another at $\phi/2$, are $d/2$ apart. Call the common perpendicular between the two rotation axes $p$–$p$. Each half-rotation may be seen as a line-reflection of the tick in the relevant rotation axis; the axes are accordingly also called reflection axes. It is easy to show that the body has suffered, overall, a finite screw displacement from 1 to 2 of ($\phi$, $d$) about $p$–$p$. The distance $O_1$–$O_2$ is $d$.

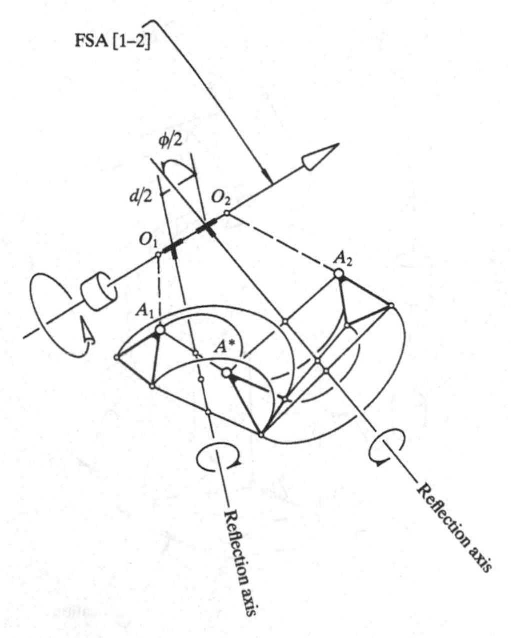

exist not only one but two screw triangles, one drawn in the fixed space (as this one is) and another drawn in the moving body. It is however not well explained in the cited literature either (*a*) that both of the said screw triangles do indeed exist, or (*b*) (and this is the more intriguing), *that they are identical.* They work in conjunction with one another like this: upon completion of any one of the three finite screwings (at which completion, of course, we have the beginning of the next), a relevant pair of equal common perpendiculars of the two screw triangles become collinear, coalesced; the two FSAs – the recently operated one and the next one, both of which must appear at this juncture as continuous lines in both the fixed space (namely the fixed body) and the movable body – are represented by abutting pairs of sides of the two screw triangles; the members of each abutting pair are collinear, abutted at the two feet of the said coalesced common perpendiculars. In the event of insufficient imagination (it was the event with me), the reader might construct wire models of two generally chosen but identical screw triangles and perceive thereby the truth of what I have been saying here. The screw triangle is clearly related to the dual vector triangle and the polygons of Keler [8] mentioned at § 19.20; those polygons relate to the vector addition of infinitesimal displacements however; this means of course that the screw triangle is related to the cylindroid; refer in advance to § 19.43 where the geometry of the cylindroid (which relates to rates of screwing and thus by implication to infinitesimal screw displacements) is shown to be a special case of that of the screw triangle. For some early remarks about the nature of the screw triangle (or the dual triangle as the following two workers called it) refer to the doctoral theses of Keler [21] and Yang [22].

39. In the matter of the screw triangle Roth began to examine, in 1967, the spatial analogue of the Burmester geometry for the 3, 4, and 5-location theory for a movable lamina in the plane. He thus examined among other things the geometrical circumstances surrounding three successive locations in space of a movable rigid body. This led to the relative locations of the three available FSAs (§ 19.30) and these Roth discussed, naturally, in a cyclic manner; he spoke, in other words, of three finite screw displacements of an originally located body that returned the body to its original location. He used the facts outlined at § 19.37. Roth was thus able not only to construct his screw triangle but also to write a suitable algebra for it (§ 19.41). The screw triangle itself, the closed spatial hexagon appearing in figure 19.08(*b*), consists of the three FSAs and the three common perpendiculars drawn between the pairs of them; it may also be seen to consist of course of the three reflection axes and the common perpendiculars (the FSAs) drawn between the pairs of them. The three initially mentioned common perpendiculars here (when taken in successively chosen pairs) turn out to be sets of suitable half-rotation or reflection axes for stepping in a cyclic manner from each of the three locations of the body to the next [20]. I have tried in figure 19.08(*b*) to illustrate in a single picture the whole geometrical gist of the screw triangle; I have tried to show, among other things, how, as well as the three locations of the thrice screwed body, there is the common thrice reflected body at the 'centre' of the triangle. If the three FSAs be taken to be the 'sides' of the screw triangle and the three common perpendiculars to be the 'apices' of it, the construction may be seen to represent the ordered 'triangular'

Figure 19.08(*b*) (§ 19.39) This picture is my attempt to illustrate what Halphen explained in his [20], namely (*a*) that a series of three successive sets of two reflections of a rigid body in pairs of reflection axes taken cyclically from a set of any three reflection axes will return the body to its original location, (*b*) that the three intermediate locations represented by the single location $A^*$ at figure 19.08(*a*) will coincide to form a single 'central' location of the body, (*c*) that the three common perpendiculars represented by the single common perpendicular *p–p* at figure 19.08(*a*) may be seen as the three FSAs about which the body may be seen, successively, thrice to have screwed, and (*d*) that we thus have enough information to understand how any two successive finite screw displacements may be compounded into a single resultant finite screw displacement. The skew hexagon clearly delineated here (with a right angle at each of its six vertices) is the figure that was found by Roth and called by him the screw triangle.

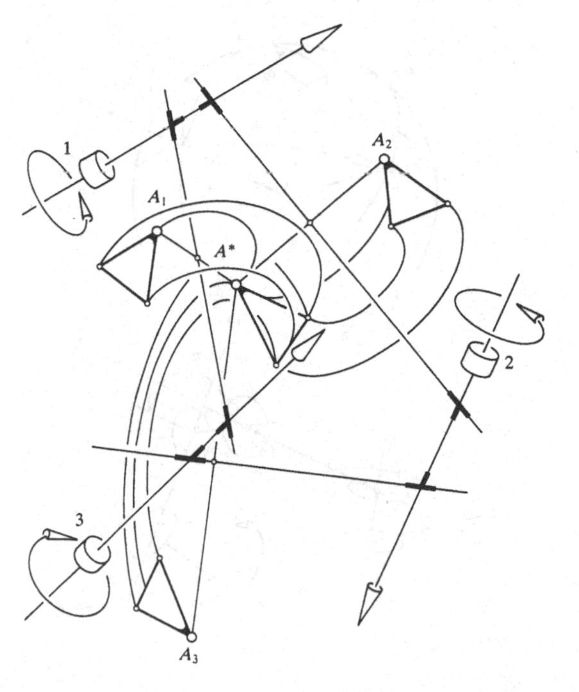

addition of two finite screw displacements of a rigid body to produce the resultant screw displacement.

**The spherical indicatrix**

40. The idea of the spherical indicatrix goes back for a long time (to the time of Gauss, some say), and Bennett for example uses it. And please refer to my few remarks about the matter at § 0.11. The idea is, essentially, a simple one. If there exists some group or set of lines in space in some organized relationship with one another, organized for example as the joint axes of some spatial mechanism or as the members of some mathematically determined rules surface, one might construct and examine the relevant spherical indicatrix as follows: (*a*) draw to pass through a single point all lines that are parallel to lines of the group or set, (*b*) describe a sphere of unit radius about the said point, and (*c*) examine the pattern of traces upon the sphere of the said lines. Find an example drawn in figure 14.01(*b*). Find another suggested in figure 11.02 where lines through the origin in the figure and drawn parallel to the generators of the regulus will form the asymptotic cone. Go to Hunt (1978) to find a fine picture of the spherical indicatrix for (or of the axes of the revolute joints of) the Bennett mechanism. I next present, after Rooney, and thus after Yang [22], my figure 19.09(*a*). It shows at the bottom three skew lines or screws 1, 2 and 3 in space, with the dual angles $(\alpha_{12} + \varepsilon a_{12})$, $(\alpha_{23} + \varepsilon a_{23})$, and $(\alpha_{31} + \varepsilon a_{31})$ between them; it shows at the top the corresponding spherical indicatrix. Because the radius of the sphere is unity, the real angles $\alpha_{12}$, $\alpha_{23}$ and $\alpha_{31}$ appear, not only as the angles between the corresponding edges in the spherical construction, but also as the corresponding sides of the spherical triangle. The external angles $\delta_1$, $\delta_2$ and $\delta_3$ at the apices of the spherical triangle correspond with the twist angles (the angles between successive common perpendiculars) upon the lines 1, 2 and 3. But as Duffy explains at the end of his § 2.2 on spherical trigonometry (1980) there will be for a given dual triangle as many spherical indicatrices as there are interpretations for the directions of the lines or screws and the handednesses of the twist angles between the successive common perpendiculars; for a given set of three radial lines drawn to cut the surface of a sphere at six points, there are three planes intersecting at the centre of the sphere and thus eight different spherical triangles that may be drawn. The figures 19.09 are relevant to my remark at § 19.20 where I spoke of Keler's dual vector triangle; in that special triangle in [8], however, the common perpendiculars $a_{12}$ etc., were parallel. Figure 19.09(*a*) is relevant too for explaining Keler's terminology. Keler calls the real and the dual parts of a dual vector the *global* and the *spatial* parts respectively (§ 19.19), for it is a fact, and this is explained in Keler [8], that, if each of the three 'sides' of the general spatial triangle is a dual vector, the real or the global parts of the dual vectors transfer to the sphere of the spherical indicatrix, while the dual or the spatial parts do not. The idea of the spherical indicatrix is at the root of the methods employed by a number of writers. Keler [8] [9] [10], Yang [25] [26], Sandor and Bisshop [27], and Waldron [36] all try to solve the ongoing problems of spatial mechanism by making use of it in one way or another. Please be aware that, whereas Keler speaks about the dual complex algebra and dual vectors, whereas Yang speaks about dual-number quaternion algebra and the screw calculus, whereas Sandor and Bisshop speak about stretch-rotation tensors, and whereas Waldron speaks about motors, the algebraic

Figure 19.09(*a*) (§ 19.40) Three screws in space, their common perpendiculars taken in cyclic order, and the relevant spherical indicatrix (after Rooney). Have I adopted here a consistent scheme for the directions of the angles $\alpha$ and $\delta$ and the distances $a$? In what circumstances might such matters matter? Refer to § 19.38, § 19.40, to Duffy, and to note [17] where I mention the handedness of finite screw displacements. Notice the skew hexagon involving six right angles in the lower part of the figure (which might, according to circumstance, be called the screw or the relevant dual triangle) and be aware that the magnitudes of all of the angles there correspond with those of their counterparts above.

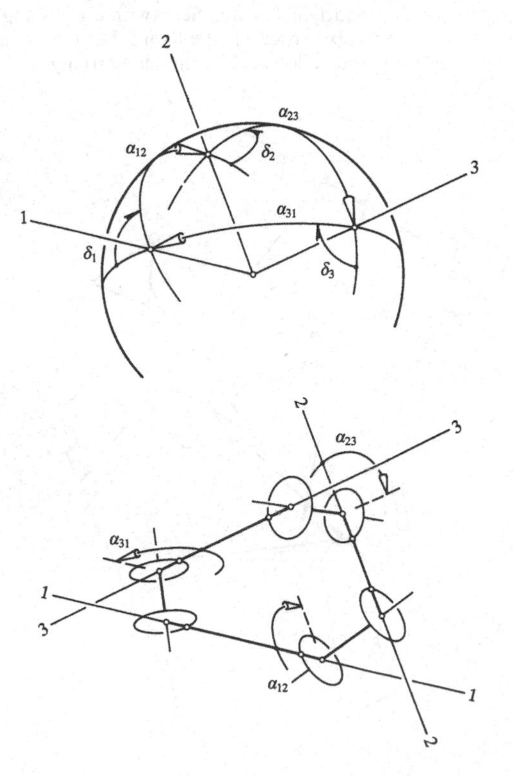

methods of the mentioned writers are from a fundamental point of view substantially the same. Waldron [36] explains in his § 2.7 that, if the spherical linkage derived by means of the indicatrix from a spatial linkage of lower joints is not mobile, then the spatial linkage itself is either not mobile or contains permanently locked joints; he explains that the prismatic joints of a linkage will appear in its spherical counterpart as permanently locked joints. Duffy (1980) uses the spherical indicatrix continuously to sort out the geometrical concepts in his important book (§ 19.59).

**The screw triangle and the cylindroid**

41. One of my references for this paragraph is Bottema and Roth (1979) chapter 4. Notice first that the bare bones of the screw triangle of Roth, the spatial triangle of Yang, and the dual vector triangle of Keler are the same [20]. Refer to figure 19.08(*b*) and arrange your perceptions to see the FSAs there to be joined by the common perpendiculars between them; these latter are the reflection axes $a_1$, $a_2$ etc., or the normals as Roth also calls them. Perceive at the same time that the lines 1, 2 and 3 in figure 19.09(*a*) are joined by the common perpendiculars between them. Go next to figure 19.09(*b*), which is somewhat approximately drawn, and see there a composite spherical indicatrix constructed (*a*) for the FSAs of the screw triangle in figure 19.08(*b*), and (*b*) for the reflection axes of the same screw triangle. Following a tradition that may be found in the general literature surrounding the ideas of Burmester (1888), let us call the

Figure 19.09(*b*) (§ 19.41) This is a composite spherical indicatrix drawn for the FSAs of the screw triangle in figure 19.08(*b*) and for the common perpendiculars (the reflection axes) of the same skew triangle. Following a tradition found generally in the literature, and after Rooney, I have called the stippled triangle 123 (which comes from the FSAs) the *pole* triangle and the other, 1′2′3′ (which comes from the reflection axes), the *polar* triangle. As an exercise for the reader I ask the same question as I did at the previous figure: are the various edges here drawn in their proper cyclic order?

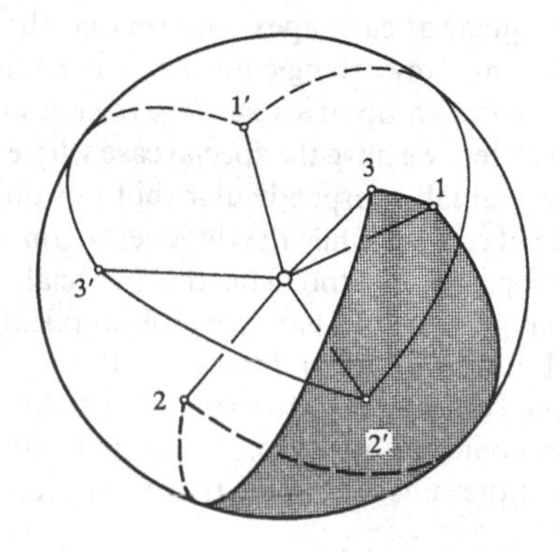


stippled triangle 123 which comes from the FSAs the *pole triangle* and the triangle 1′2′3′ which comes from the reflection axes the *polar triangle*. Kindly note and keep in mind for later that the edge 3′ is normal to the plane defined by the edges 1 and 2, that the edge 2′ is normal to the plane defined by the edges 3 and 1, and so on cyclically around the set. Now by employing the cosine rule and the sine rule for spherical triangles in the pole triangle 123, and by projecting the six straight sides of the screw triangle onto the axis of the FSA [3–1], we can write, after Roth,

$$(13)\quad \cos(\phi_{31}/2) = \cos(\phi_{12}/2)\cos(\phi_{23}/2) - \cos\alpha_2 \sin(\phi_{12}/2)\sin(\phi_{23}/2)$$

$$(14)\quad \sin(\phi_{31}/2) = \sin(\phi_{12}/2)[\sin\alpha_2/\sin\alpha_3]$$

$$(15)\quad (d_{31}/2) = (d_{12}/2)\cos\alpha_1 + (d_{23}/2)\cos\alpha_3 - a_2 \sin\alpha_1 \sin(\phi_{12}/2)$$

These three equations (along with their cyclically obtainable variants) are the bases upon which Bottema and Roth write the rest of their algebra to find the FSA [1–3] given an FSA [1–2] and an FSA [2–3]. They need to determine, however, not only the direction, magnitude and pitch of the resultant FSA [1–3], but also its *location*; and to do this they depart quite soon from trigonometrical methods; they nominate position vectors for crucial points within the figure and apply the vector algebra. Please refer to the book for details; the details do lead (it must be said) to awkward algebraic statements; refer to [28].

42. One should not be surprised to find that Bottema and Roth (1978), Dimentberg (1965), Keler [19] [21] and Yang [22] [25] are all involved in their respective algebras of finite displacements with the tangents of half angles. Indeed Keler writes in his recent [19] a so-called dual-vector algebra for the representation of finite displacements; it employs what Keler calls 'dual-vector half-tangents'. The method involves the taking of reflections not in certain lines but in certain planes. Keler's use of the word vector in this arena is misplaced however (§ 19.03). Keler connives with Dimentberg in the matter but fails to be convincing; he distorts the clear behaviour of the *instantaneous* dual vector as described in § 19.07 above and distorts his mathematics to fit. Vectors (and especially dual vectors) are useful for the statics and instantaneous kinematics of mechanism, but the dual angular displacements $\phi + \varepsilon d$ and $\alpha + \varepsilon a$ do not lend themselves to vector addition – either for bodies as they are displaced from location to location, or as we jump in the mathematics from joint to joint. It is true as well that for numerical solutions we must go in the long run to Euler angles and matrix transformations of the fixed and the changing variables and so on, so we could refer directly now to the matrix

methods of Uicker and the others (§ 19.52). Whatever the mathematics, however, the geometry of the screw triangle is such that it displays a number of special cases; refer to Bottema and Roth (1979), chapter 4, § 13 for a listing (but no discussion) of these. I intend to enlarge now on what I believe to be the most important of these special cases.

43. I ask this: *what happens to the screw triangle (both geometrically and algebraically) when the finite screw displacements become infinitesimal*? The three reflection axes in figure 19.08(*b*) will coincide. The three FSAs will thus share a single common perpendicular and themselves reside separately in a set of three parallel planes. Any central point $A^*$, reflected outwards thrice, will reflect on each occasion into itself precisely; this means that the three successive locations of the rigid body will collapse into the same (central) location; this may also be seen with the help of figure 19.08(*b*). I remind the reader here of my own chapters 13 and 15 about the theorems of three axes and the cylindroid. At the same time I remind myself of some perceptive remarks of Rooney [28] which helped me to clarify the following. Go to the spherical indicatrix in figure 19.09(*b*) and be reminded that the directions of the radial lines there have been drawn to correspond with the directions of the relevant lines in figure 19.08(*b*). Please refer here to note [46]. Keeping fixed, now, the mentioned right-angular relationships between the lines and planes at the centre of the sphere (§ 19.41), alter the figure by pulling together in the mind's eye the three edges 1′, 2′ and 3′ until they become collinear, that is, until the polar triangle 1′ 2′ 3′ is diminished to a point. The edges 1, 2, and 3 of the pole triangle have no alternative now than to become coplanar; the pole triangle becomes a great circle of the sphere and it becomes clear in the new circumstances that the three FSAs of the screw triangle must reside in parallel planes. Also by putting to zero the relevant variables in Bottema and Roth (1979) one may arrive at the results of Phillips and Hunt and thus at the formulae of chapter 15; refer to Phillips and Zhang [28]. *Thus we reduce ourselves to the geometry of the addition of infinitesimal displacements, namely to that of the composition of rates of screwing (chapter 13), and thus of course to the geometry of the cylindroid.* The material of this paragraph, better developed in Phillips and Zhang [28], is, I believe, a key to the unification of the kinematics of mechanism. We deal in two disparate areas; we have on the one hand studies in the gross movements of bodies, and we have on the other studies in instantaneous motion. Even though they should for the sake of better machine design be better related (§ 0.18), the two areas are not well connected in the literature. Having begun to make the connection here however and in this way, having taken the finite down to the infinitesimal in this particular study, we have an intellectual leap that needs to be made. *The leap here is to see that a circumstance (a) whereby three suitably chosen successive infinitesimal screw displacements of a single rigid body return the body to its initial location (§ 19.43), and a circumstance (b) wherein the instantaneous relative motion of three rigid bodies is represented by the layout at the instant of the three relative motion screws (§ 13.15), are the same. One needs to see that, whereas in (a) one might imagine a central location and three infinitesimally displaced yet superimposed locations of the same rigid body, in (b) one might imagine some chosen bounded volume in space being occupied at the same time by three separate moving rigid bodies interpenetrating one another.*

**Some other special cases of the screw triangle**

44. Also with the spherical indicatrix in figure 19.09(*b*) we could do this: we could pull the edges 1, 2 and 3 of the pole triangle together into collinearity. In this case the pole triangle becomes a point while the polar triangle becomes a great circle of the sphere and we produce the spherical indicatrix which relates to the Burmester geometry of three finitely separated locations of a lamina moving in the plane. The planar pole triangle of Burmester is (as we know) of finite size; it is not in general a point but, because the actual FSAs are parallel (all normal to the plane of the paper), its counterpart, the spherical pole triangle upon the sphere of unit radius, is indeed a point. In the relevant special screw triangle the reflection axes are not necessarily coplanar; they may reside separately in three parallel planes. This simply means that the Burmester geometry is applicable also to the addition of finite screw displacements of finite pitch provided the axes of these displacements are parallel. I make the same remark as I did before. We have here the possibility of welding together disparate parts of the science. Let me also mention the special case that can be hinted at by pulling the spherical indicatrix into such a configuration that the pole and the polar triangles (both spherical triangles upon the unit sphere) become congruent. In this case we find (*a*) that the triangles must both be right-angular at each apex, and (*b*) that the six sides of the relevant screw triangle must be so arranged that they may be drawn upon six successive edges of a rectangular box. Here we have the special case where the three FSAs are mutually perpendicular (but not mutually intersecting of course), while the three reflection axes are mutually perpendicular too. This third special case of the screw triangle may well have a simple application in industry. I am not pretending that the abovementioned special cases of the screw triangle are exhaustive or that they could not have been discovered without using the idea of the spherical indicatrix. I am pretend-

ing however that the spherical indicatrix is more than a mere handy device; it and related developments might well help to change the face of kinetostatics (§ 19.45).

**Kotelnikov's principle of transference**

45. This principle (which employs the spherical indicatrix) is of much wider significance than the mere spherical indicatrix itself which has, here, led up to it. Dimentberg (1965) gives a history of the development of the idea mentioning a period of cooperation between Kotelnikov [7] of Kazan and Study (1902) of Dresden. In 1975 Rooney [4] mentions the matter and gives a formulation there which I have paraphrased here as follows: all valid laws and formulae relating to a spherical configuration (involving intersecting lines and real angles) are also valid when applied to an equivalent spatial configuration of skew lines if each real angle $\phi$ or $\alpha$ in the original formulae are replaced by the corresponding dual angles $\phi + \varepsilon d$ or $\alpha + \varepsilon a$. In a somewhat earlier publication [29] in which he proves the principle, and which arose no doubt from [30], Rooney explains as follows: 'The principle of transference... enables one to formulate the laws describing spatial linkages in terms of the simpler laws of the equivalent spherical linkages.' Duffy (1980) uses the principle of transference extensively in his analyses of spatial linkages of mobility unity with lower joints or pairs, and (following a clear account of the bases of the algebra of dual numbers) he describes the principle (at his § 2.8) as follows: 'A trigonometrical or loop equation for a spherical $n$-gon is identical to a dual angle trigonometrical equation for the equivalent spatial $n$-gon.' He goes on to say that 'any dual angle trigonometrical equation can be expanded into a primary equation which is identical to the spherical loop equation, and a secondary or spatial loop equation which relates'... the twist angles, the joint angles, the link lengths, and the offsets (of the linkage). Refer to § 19.59 for some further remarks about the methods of Duffy.

**Interlude, multiple loops**

46. I ask the reader to keep in mind now whatever impressions he may have of the spherical indicatrix and the principle of transference discussed above and to contemplate at the same time the mechanism shown in figure 19.10. The mechanism may be seen as the complex joint between its frame member 1 and its movable member 2 (§ 2.40). Although the mobility of the mechanism is 2 (§ 1.34), the connectivity $C_{12}$ of the mentioned complex joint is unity (§ 2.37). This means that, with the possibility of a single angular input displacement say $\theta$ inserted at $A$ or $D$ (but not at $E$?), the link 2 is constrained to travel a unique track in 1 (§ 1.20). Three separate paths of that track (two of them circles) are mentioned in the caption. Another way to contemplate the mechanism is to perceive it first as an *RSSR* loop with one point namely $F$ of its coupler constrained to travel the surface of a sphere. The line segment $E$–$F$ forms the radius of this sphere. Compare the mechanism here with the one shown in figure 3.01 incidentally; in the one shown there the centre of the mentioned sphere is set to remain infinitely far away along the line $EQ$. The mechanism in figure 19.10 is not a single-loop but a multi-loop mechanism. It or something like it, however, may provide a simple solution for certain practical problems. Consider for example the shoulder joint of a one-armed industrial manipulator and the combination of strength (the capacity to transmit great wrenches) and the absence of backlash due to clearances that is required. As you read the remarks that follow, please remember (*a*) that such mechanisms as the one displayed in figure 19.10 are available, but (*b*) that, unless their motions can be analysed easily by applicable algebraic methods, they will find no great favour in the open field of engineering practice.

**To write a closure equation**

47. The analysis of a mechanism cannot begin unless full initial information is available about the

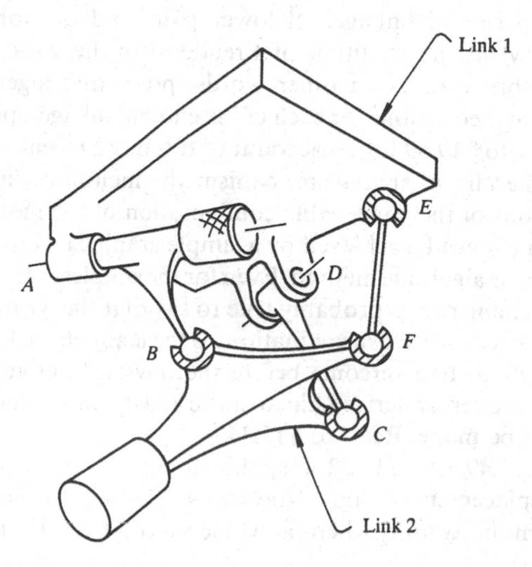


Figure 19.10 (§ 19.46) A multi-loop mechanism where the connectivity between the frame link 1 and the floating link 2 is unity. Each point in the link 2 has a predetermined path in link 1, and all of the paths together make up the track of 2 in 1 (§ 1.20). The paths of the centres of the sockets at $B$ and $C$ are arcs of circles, the circles being coaxial with $A$–$A$ and $D$–$D$ respectively. The path of the centre of the socket at $F$ is a twisted curve upon a sphere at centre $E$ whose radius is $E$–$F$. These three paths, those of $B$, $C$ and $F$ in 1, are wholly representative of the track of 2 in 1.

relative locations in space of the several links of the mechanism for every configuration of the cycle. As soon as that information is available, it becomes a relatively simple matter to determine the velocities and accelerations of key points fixed in the several links and thus to understand the relative motions. A closure equation (or equations) for a mechanism is an algebraic or other statement about the mentioned initial information; it deals with the geometry of the assembly of the given rigid links into the actual mechanism and thus, in a fundamental way, with the matter of displacement.

48. In those textbooks where the mechanics of planar machinery is discussed solely with the aid of configuration, velocity, and acceleration diagrams, the matter of the closure equations for a mechanism is seldom mentioned. The reason, of course, is that the closure equations are effectively solved in those books and works before they are ever written. The changing configuration of the planar mechanism in question, and thus the successive locations in the plane of its several movable links, is discovered by simple graphical layout before the rest of the analysis begins. The same is true of my chapter 12; there I simply draw the mechanism concerned and then, in the absence of any values for the dimensions of the links, simply pronounce that the mechanism is closed. It is worthwhile here to study the well celebrated algebraic analysis of the planar 4*R*-loop that was made by Freudenstein [23]. This begins with an algebraic equation for the mechanism which may be written (and manipulated and understood) without a great deal of difficulty. Various accounts of this closure equation for the planar 4*R*-loop (the originating paper [23] having appeared in 1955) are given in the modern texts, for example Hall (1960), Hartenberg and Denavit (1964), Hunt (1978), and Duffy (1980). Duffy (1980) not only mentions Freudenstein [23] and repeats his work but also deals with a host of other linkages (all single-loop spatial linkages of lower pairs and of mobility unity) in the tradition of Freudenstein; he writes the algebra directly in other words, presenting algebraic closure equations for each of his chosen linkages; please refer to § 19.59 for an account of this more recent work. Generally in spatial mechanism the mentioned initial layout of the changeable configuration of the mechanism cannot be achieved by a simple graphical construction or algebraic method. Even for the simplest of spatial mechanism it is probably true to say that the writing of the necessary closure equations is the main obstacle that needs to be overcome before the obviously useful but subsequently derived kinetostatic and dynamic analyses can be made. Refer to § 12.34.

49. At § 19.32 I spoke about the finite screw displacement of a line. Now in a single-loop mechanism of mobility unity where, as we step successively from link to link and we find that each link in the kinematic chain has its single motion screw with respect to its neighbouring link, we can, if we wish, represent the links themselves with the pairs of motion screws which are, as it were, and at the instant, attached to the links. If, moreover, the joints of the chain are lower simple joints, the pairs of screws can be seen as mere pairs of lines, and these pairs of lines can then be seen to remain rigidly attached to their respective links as the mechanism moves (§ 7.12). Thus we begin to see how closure equations of a different kind can be written for many of the simpler linkages. I refer here to the matrix-kind of closure equations to be dealt with next.

50. A set of coordinate axes – a triad, or a tick (§ 4.04) – may be established in each of the links of the loop. This may be done in a multiplicity of different ways (§ 19.52), but one well known way is as follows. Going around the loop, each of the locations of the successive seats of the motion screws (namely the lines fixed in the links) are employed as the $z$-axis of a triad. We next erect the common perpendiculars between the successive pairs of these $z$-axes and, by deciding that the $x$-axes will reside thereon, establish an origin of coordinates in each link. The $y$-axes are last located (according to standard convention) in such a way that all of the triads are right handed. Refer in advance to figure 19.11; the figure is not yet fully explained; it will give a rough impression nevertheless. Having identified the necessary parameters (the so called fixed parameters which depend upon the dimensions of the rigid links and the variable parameters which depend upon the configuration of the mechanism), coordinate transformations can next be written for all successive pairs of adjacent triads. As we step from link to link around the loop, we say – and here we write a matrix equation – that the product of all of the transformations (one for each of the links of the loop) will result in the unit transformation. If the triads are carefully located, and if the transformations are expressible (as they nearly always are) by means of matrices, we can hope to arrange next that each element of each transformation matrix will be some simple function of the parameters already mentioned. The parameters involved in the matrix $A_i$, for example, might be those describing the location of the $(i+1)$th triad relative to the $(i)$th triad. Upon expansion, the above mentioned matrix equation will produce a number of complicated simultaneous equations involving transcendental terms and all of the variables. In order to arrive in due course at useful closure equations for the linkage (the linkage under consideration being of mobility $M$) we need to extract from those equations explicit expressions for each single 'output' variable parameter in terms of each set of $M$ 'input' variable parameters and the fixed parameters. Some workers, concerned as they are with

existence criteria for the overconstrained yet mobile loops, are more interested in linkage geometry than in input–output relatioships; and their problems in this respect are somewhat different; please refer to chapter 20. In any event the final extraction from the simultaneous linear equations of the required expression or expressions to clarify the closure of a closed mechanism constitutes at least one of the major difficulties. Please be aware here that even the open loop of an industrial manipulator arm becomes a closed loop as soon as the gripper is seen to be grasping in a specified way a stationary object (§ 2.61, § 19.55).

51. During the last 20 years or so a bewildering variety of different strategies has become available for this work and there is, accordingly, a bewildering variety of authors and methods spread throughout the recent literature. The variety exhibits itself both in the transformations themselves and in the strategies employed for solving the resultant sets of simultaneous equations. Firstly there are some fairly straightforward methods associated with direct, non-screw displacements by means of translation and rotation matrices (§ 19.50, § 19.52). There is the method of Sticher and related methods (§ 19.56). Next there are methods employing dual numbers and dual quaternions (and among these is the 'dual calculus' of Yang) to achieve the various direct screw transformations (§ 19.57). Related with these are the various methods involving a compact tensor notation to achieve 'stretch rotation' effects (§ 19.57). There is discussion in the literature on the differences between ordinary point-coordinate transformations and rigid-body transformations (§ 19.58). And there are, as I have mentioned, the more recently developed (and quite different) methods of Duffy (§ 19.59). Almost without exception the abovementioned methods employ, in the long run, the writing of sets of $(3 \times 3)$ or $(4 \times 4)$ matrices of one kind or another. This results in the need to solve (by iterative or other methods) the many sets of simultaneous equations which invariably arise and so, to this extent at least, the methods are similar. Rooney in [4], incidentally, draws an important distinction between those matrix methods used for spatial displacements involving point-transformation devices, and those involving line-transformation devices. And there do exist, as we know, plane-transformation devices, but these latter are not much used in the current literature. I shall deal here first with a well known and a well tried method that is based upon point transformations (§ 19.53), and next I shall deal with the others (§ 19.56 *et seq.*). I try in what follows to give a brief but intelligible account of only some of the main schools of the general endeavour. I have used Uicker, Denavit and Hartenberg [24], Yang and Freudenstein [25], Yang [26], Sandor and Bisshop [27], Rooney [4], Duffy and Rooney [30], Sticher [31], Gal [32], Suh and Radcliffe (1978), Duffy (1980), Paul (1981) and a host of other references. Despite the many references, and yet perhaps because of them, the following account of the various methods is a somewhat limited one.

**A brief account of the matrix methods**

52. One of the earliest methods for stepping from link to link around the loop of a closed-loop spatial linkage to write the closure equations, to make a *displacement analysis* for the linkage, appeared in the journal literature in 1964; it was the work of Uicker, Denavit and Hartenberg [24]; refer also to Hartenberg and Denavit (1964). The bare bones of the beginning of the method are summarized in figure 19.11, the gist of which figure is extracted from the mentioned works. It shows the joint axis (or the motion screw) $O_i - z_i$ for the $(i-1)$th and the $(i)$th link of a kinematic chain, the joint being any kind of simple joint, and the chain being a single loop of mobility unity. It is easier, however, to think of all of the joints as being in the first instance either revolute, cylindric or helical joints. The common perpendicular is drawn between the joint axes (or the screws) $O_i - z_i$ and $O_{i+1} - z_{i+1}$, and the length of this, the *link length* of the $(i)$th link, is called $a_i$. Notice next the angle $\alpha_i$ between the mentioned joint axes; this is the *twist angle* (or the skew angle) of the $(i)$th link. Provided the joints are $R$, $C$ or $H$ (§ 7.12), the parameters $\alpha_i$ and $a_i$ are fixed dimensions characteristic of the $(i)$th link.

Figure 19.11 (§ 19.52) This figure (after Denavit) shows one way of defining the fixed and the variable parameters of the $(i)$th link of a generalized single-loop linkage. The variables relate to the matrix method of analysis outlined at § 19.52.

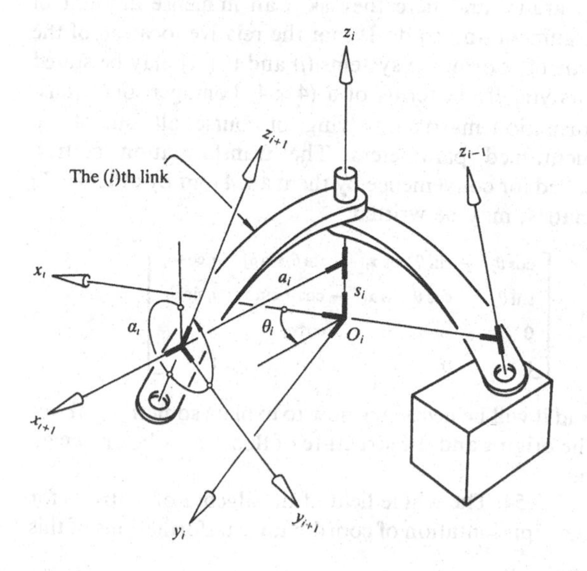

Observing next that the position of the origin $O_i$ is determined in the same way as that of the origin $O_{i+1}$, namely by the construction of a common perpendicular, notice that the parameters $\theta_i$, the *joint-rotation angles*, and $s_i$, the *offsets*, are variable dimensions that depend in general upon the configuration of the linkage (§ 20.29). Having seen this, please see next that the four parameters $\alpha_i$, $a_i$, $\theta_i$, and $s_i$ are defining the relative locations of the successive triads at $O_i$ and $O_{i+1}$, namely the successive systems of Cartesian point coordinates. It is legitimate now to think of the triad upon the origin $O_i$ as being fixed in the ($i$)th link and the triad upon the origin $O_{i+1}$ as being fixed in the $(i+1)$th link and so on throughout the whole of the linkage or loop. Using the usual convention for numbering the links of a loop (§ 12.08), we see that an origin $O_1$ and its three axes for $x_1$, $y_1$ and $z_1$ may be seen as being the first triad fixed in the first (namely the fixed) link of the linkage.

53. Now the purpose of figure 19.11 is to lay the foundation for displacement analyses (or closure equations) of the closed loop linkages that can be analyzed. Not all linkages can be analyzed by the mentioned methods but, if the linkage is a simple loop, and if the joints are lower joints (§ 7.12), there will be a good chance that at least the overall matrix equation can be written fairly easily and some chance that it may be solved for the closure equations explicitly. Joints that are not revolutes can often be seen as a combination of revolutes and prismatics (always, in fact, if they are lower joints other than screws). For example a cylindric joint may be seen as a revolute in series with a prismatic whose direction of sliding is parallel with the axis of the revolute, and a ball and socket may be seen as three revolutes in series whose axes intersect at the centre of the sphere. Having said all this the authors of [24] go on to argue (and here they skip an immense amount of argument and so do I) that the relative location of the pair of coordinate systems ($i$) and $(i+1)$ may be stated analytically in terms of a $(4 \times 4)$ homogeneous transformation matrix involving, of course, all four of the mentioned parameters. The transformation matrix, called for convenience by them an $A$ (but by others a $T$) matrix, may be written,

$$A_i = \begin{bmatrix} \cos\theta_i & -\sin\theta_i\cos\alpha_i & \sin\theta_i\sin\alpha_i & a_i\cos\theta_i \\ \sin\theta_i & \cos\theta_i\cos\alpha_i & -\cos\theta_i\sin\alpha_i & a_i\sin\theta_i \\ 0 & \sin\alpha_i & \cos\alpha_i & s_i \\ 0 & 0 & 0 & 1 \end{bmatrix};$$

and it will be necessary now to explain something about the origins and the structure of this matrix before we go on.

54. The whole field of the algebra of matrices for the representation of coordinate transformations of this kind is a subject within itself, and it is, moreover, quite beyond the scope of this book. A few remarks about the nature of this particular matrix and some useful references may be in order however. I could mention to begin that while transformation matrices like this one are called A (or T by some), the so-called rigid-body transformation matrices, which are in a sense opposite (§ 19.58), are sometimes called $R$. The adjective *homogeneous* derives from the fact that, while the vector from the origin to a point in a coordinate frame may be written in terms of the four homogeneous point coordinates of the point in question namely $(x, y, z, 1)$, the direction of that vector may be written $(x, y, z, 0)$, these latter four being the homogeneous point coordinates of the relevant point at infinity (§ 19.27). Homogeneity means that not only the separate elements of such a vector but the separate elements of a whole matrix derived from a set of such vectors may be multiplied by a common non-zero scalar without affecting the represented transformation. See figure 19.11 and notice in the matrix above that, whereas the first three columns relate to the directions of axes referred from the old to the new Cartesian frame, the fourth relates to the translation of the origin. Notice also that, within the $(4 \times 4)$ at top left, there is a $(3 \times 3)$ which relates to the two rotations $\theta_i$ and $\alpha_i$ about the relevant axes of the Cartesian frame. At section 7.1 of Rooney [4] (and elsewhere) one can find accounts of these and other such matters argued from first principles, and in [4] one can find that somewhat similar matrices were used in the early seventies by workers in the field of computer graphics. The whole of chapters 1 and 2 of Richard Paul (1981) is given over to accounts of these particular matrices, and Paul claims there that their use is becoming, just now, almost universal in the field of robot control. Minkov-Petrov (1985) claims to be able to separate the fixed and the changing variables of the linkages in such a way that the above kinds of matrix algebra are simplified. He claims moreover that, under the usual special circumstances of common engineering practice (parallelisms, intersections and perpendicularities among the various joint axes), the matrix algebra for computer use may be even further simplified.

55. The matrices are, in any event, multiplied together around the single closed loop of the analyzable. mechanisms, and the displacement or closure equations already mentioned in § 19.50 emerge; all of this is well explained in [24]. In manipulators and robot arms the matrices may be multiplied along the open arm, but by relating in the long run one end of the open arm to the other, the equations may again be written in this closed form (§ 2.61). In the language of those who deal with industrial manipulators, the process of multiplying the matrices along the open arm to the gripper is called the

'forward kinematics' of the open arm; knowing the fixed dimensions of the connected links of the arm, and given a set of joint-angles for the joints therein, this process will locate the gripper in the coordinate system of the frame, or, as they say, in the 'world' coordinates of the robot. If the gripper is seen (or held) to be gripping its manipulated body in some particular location in these world coordinates (§ 1.17), and a corresponding configuration of the linkage of the arm is not yet known, a closure equation for the completed *loop* may be written (§ 2.61); and this may subsequently be solved. This process is called the 'inverse kinematics' of the manipulator arm, and the matrix formulations thus produced often spawn a multiplicity of sets of solutions for the required joint-angles. This relates directly of course to the question of the *modes* of a linkage (§ 1.44). These modes, and the possible intervening change points or uncertainty configurations and the sogginesses that may occur in certain configurations (§ 2.46), constitute one of the major difficulties in the overall design and control of robots; for accuracy in the absence of backlash is almost always required. Please be aware that we are discussing here only the kinematics of linkages (open or closed). Uicker and others go on to the statics and dynamics of the separate links in later works, but these are not due to be discussed by me until § 19.62; see my note [8] of chapter 0.

56. As a variant of the above I wish to mention the method due to Sticher [32]. A modified form was used by Gal [33] and was studied by Gal [34]. The method uses not a common perpendicular between joint-axes but a line segment which may be drawn directly from any chosen joint-centre to the next chosen joint-centre. The line segment gives a straight translation of origin, and three successive orientations of the second joint-axes are taken by Euler angles. Sets of simple non-homogeneous $(3 \times 1)$ and $(3 \times 3)$ point-coordinate transformation matrices appear [32], and certain advantages are claimed [34].

57. There are other ways of stepping (or of jumping) from link to link to write the transformation matrices, and according to the ways of jumping the resulting matrices differ. Here I promised to mention the biquaternion due to Clifford (see note [1] at the end of chapter 0) and the dual-quaternion method of Yang [22] [25] [26]. Yang develops his method through the work of Hamilton (1899), introduces the idea of a *dual-quaternion* for the jumping job, again produces matrices for multiplication around the loop and deals, as the others do, with simple loops employing lower joints. In [26] we find an account of Yang's *dual calculus*; he claims there (yet not convincingly in my view) that this can cope with matters not only of kinematics but of the statics and dynamics of a link in the absence of friction. Related with [25] we find that Sandor and Bisshop [27] jump from link to link by means of a *stretch rotation tensor*, producing yet another kind of transformation matrix at each jump. The matrices are however again multiplied around the loop to produce the final matrix equation for the loop, and this final equation needs to be, as usual, expanded into a series of simultaneous equations for solution.

58. There is an important difference, pointed out by some, between the above mentioned point-coordinate transformation matrices and the somewhat more easily understood *rigid-body transformation matrices*; refer to Suh and Radcliffe (1978). With the former (as we have seen) we position various single points of a rigid body within a first Cartesian frame and note their different sets of coordinates with respect to some second Cartesian frame which is displaced somehow with respect to the first. But with the latter we hold a single Cartesian frame fixed in an initial location while we watch the rigid body as it moves from location to location within that frame. Whereas with the former in other words we move the frame of reference, with the latter we move the rigid body. It should not be difficult to see with the help of such imaginative devices as that of kinematic inversion, however, that we can go mathematically from either of these circumstances to the other; the different matrices must be, indeed, related with one another. With respect to the particular and somewhat special *finite screw* displacement of a rigid body (§ 19.30), which clearly lends itself to treatment by a suitable rigid-body transformation matrix, the reader will find such a suitable matrix explained in Suh and Radcliffe (1978) § 3.6. The authors also explain, at their § 3.9, how to decompose the resultant screw-displacement matrix of a number of successive screw-displacement matrices into the elements (the linear displacement and the rotational twist) of the final screw displacement itself. Also in connection with the mathematics of the finite screw displacement I wish to mention here the recent idea *Drehzeiger* (turn-pointer) due to Hiller [35].

**Duffy, his methods, and the 7*R*-loop**

59. I wish to discuss the works of Duffy (1980). In the same way as many others do he constructs a *spatial polygon* to represent the layout of a chosen linkage. The polygon consists as usual of link lengths and common-perpendicular offsets. Duffy's overall strategy in the kinematic analyses of linkages is next based not upon matrix transformations but upon the spherical indicatrix, Kotelnikov's principle of transference and, most importantly, the decomposition of the spatial polygon into component spatial triangles [30]. Duffy's objective is not to write a series of matrix equations for multiplication around the loop which will lead in turn to

sets of simultaneous equations for solution by trial (§ 19.52), but to write algebraic closure equations for the linkage directly. Corresponding to the actual spatial polygon there will be, by virtue of the spherical indicatrix (§ 19.40), a *spherical polygon* for the mechanism. Using the general trigonometrical properties of the spherical triangle (which he develops fully and delineates carefully in his book), he writes on the basis of the spherical polygon his *trigonometrical equations for the loop*. Then he dualises those equations (§ 19.45) to generate *additional equations for the loop*. Next, with tangent half-angle substitutions for sine and cosine, he transforms this collected material into pure algebraic equations so as to perform more easily the elimination of unwanted variables. He takes clever advantage of established methods for doing all of this. Now an input variable may be for example the crank angle at some clearly defined input crank, while an output variable may be the joint angle or linear displacement at one of a number of intermediate joints or the crank angle for example at some clearly defined output crank. For each one of the relevant input–output relationships Duffy searches next for solutions, not to a set of simultaneous equations with transcendental terms, but to a *polynomial in a single variable*. For each specific loop he tackles, he tries to express each output variable as an explicit algebraic function of the input variable, reducing the apparent degree of the function to its absolute minimum and thus eliminating (or trying to eliminate) all extraneous roots. Reflecting the multiplicity of the modes of assembly of the linkage (§ 1.44, § 2.57, § 19.48), these polynomials each have multiple roots of course. There will be an unreal root whenever the links of the loop are so proportioned that the loop is physically unable to close in that particular mode. To separate the real from the unreal and to solve for the real roots numerically is another problem; but with the polynomials reduced in degree and written down, Duffy's main aim is already achieved. Using his declared new methods he finds spectacular success. He is able to write the algebra for a host of listed simple loops of mobility unity with lower joints or pairs. Quite naturally he is keenly interested in the general 7*R*-loop. His solution to this – it is unique at the time of writing but it has, as he says, some faults to be cleared – appears not in the book (1980) but in the journal literature [37]. To obtain some idea of the complexity of the 7*R* and its associated algebra, the reader might care to peruse not only [37] but also some of the international literature that preceded it. I would like to mention in particular the very early paper of Urquart [38] and the doctoral thesis of Albela [39]. Waldron wrote a Book Review on Duffy [40]. In more recent times Wohlhart has produced a most interesting pair of linkages. He has shown that if the link lengths of the 7*R* are all made zero, and if the offsets are all made equal, we get right hand and left hand versions [41]; and Wohlhart claims that these linkages and certain other special cases of the 7*R* manage to elude the wide net cast by Duffy. By way of transition comment I wish to repeat that all these displacement analyses including (*a*) those of Hartenberg, Denavit, Uicker, etc., and (*b*) those of Duffy deal almost exclusively with relatively simple loops using only lower joints. I say this despite the wide range of their scope and the difficulties and complexities of the algebra. So a few inconclusive remarks about the much more complicated kinematics of the general higher joints might be in order here; refer to § 19.60.

**On sliding contact at rounded protuberances**

60. This matter was mentioned at § 6.21, § 6.62 and § 8.35. At § 6.21 and § 6.62 there were references forward to this paragraph. And it should be understood that what we are discussing here is the question of the general higher joints. Refer to figures 6.07 and 8.10 and perceive there what I mean by rounded protuberances. It should be clear that in general, as two rigid bodies in contact at some number of points of contact less than six move with respect to one another, the points of contact move across the surfaces of the contacting bodies and trace paths thereon with respect to both. In discussing the path of such a point of contact we are not discussing the path traced out in a fixed space by some point fixed or embedded within some moving body (§ 1.18), as the point $F_2$ for example is embedded at the tip of the sharp prong on body 2 in figure 6.07. We are discussing the path upon the surface of a body of some point of contact with another body, and the point under these circumstances is not fixed or embedded in either of the bodies. Some simple aspects of the mechanics of this matter (presented in connection with some planar mechanical joints) were studied by Reuleaux (1875). A more general investigation (but again in the realm of two dimensions only) was written by Rosenauer and Willis (1952); the material appears in their book under the general heading of the Euler-Savary equation. Begin reading about the idea of *envelopes* at their § 14, and study their figures 69 and 70. More recently Dijksman (1976) has written a clear account of envelope theory; his work applies also in the realm of the planar kinematics of two laminae and may be found in section 2.2 of his book.

61. Sliding contacts at fully rounded, or ordinary spatial protuberances will be seen to be occurring continuously in all joints, even in those that are specially designed to be lower joints; all we need to do is observe the macroscopic phenomena (§ 7.36). But in a much more blatant way they occur in the general higher joints such as those that are commonly found in the wide field of biomechanism. Take for example the joint between the

human jaw and its skull. Please read § 2.64 *et seq.* about the mechanics of biomechanical joints and recognize that, given the presence of ligaments and muscle control, the number of points of contact at such joints may well be less (from time to time) than five; please read § 3.27 *et seq.*, § 6.58 *et seq.* and § 7.87 *et seq.* about the question of five points of contact between a single pair of bodies nevertheless. This latter will reveal the possibility that a perfectly general force-closed higher joint might be machined; the joint might consist of two rigid undulating surfaces designed to make contact at five sliding points of contact; the shapes of the two elements of the joint might be defined numerically. Such a joint with all five pairs of points in contact would be an *f* 1 joint. Given such a joint, how might we be able to calculate the axodes and thus the track of one element with respect to the other? We could (*a*) locate somehow an initial set of the points of contact, (*b*) determine somehow the slopes of the tangent planes at the points of contact (refer to photograph 6), (*c*) locate the five common normals, (*d*) locate the unique ISA, (*e*) make a small movement, (*f*) relocate the points of contact and repeat the cycle of calculation, and (*g*) refer to § 18.21 for calculation of the track. Or, alternatively (and this might be more difficult), given a required track, how might we determine firstly the axodes, and secondly some suitable shapes for the elements of the joint? This reverse process might be seen to be a synthesis by means of button pairs (chapter 7). There would be no unique solution; and there would be considerations of friction.

**The machine dynamics of Uicker and Chace**

62. Please go to the Index and peruse there the listings under Dynamics. I wish to deal here with the works of Uicker [42] and Chace [43]. It was during the recent early sixties that these separate American workers established the bases for their respective methods for the analysis of the dynamics of massive machinery (§ 0.13). When I employ the word *massive* here I simply mean that for these workers machinery is non-massless. Both methods of analysis are essentially Cartesian in character. I mean by this that in each case all vector quantities such as force, moment, angular velocity, linear velocity and the accelerations associated with the links and points within the links are split into the usual two ($x$, $y$) or three ($x$, $y$, $z$) components for calculation. This, of course, is not quite the case with Ball. Uicker and Chace have both developed their strategies during the intervening twenty years and the results of their works are presented now in the form of tightly packaged computer programmes designed for use by others. There is IMP (Integrated Mechanisms Program) due to Uicker which can deal with spatial mechanism, and there is DRAM (Dynamic Response of Articulated Machinery) due to Chace. Uicker and Chace are not the only workers presently in the field of the dynamics of mechanism, but they were the pioneers; and they are, still, the two best known of them. It must be understood that the details of the methods are much more complicated than they might at first appear from the following simple explanation.

63. Both Uicker and Chace see the whole of their linkage (whatever the linkage may be) as a series of mechanically connected bodies and, by writing Lagrangian equations according to the number of independent inputs required, arrive at sets of differential equations accordingly. Whereas Uicker writes the necessary number of algebraic equations to describe the configuration of the mechanism (the closure equations), Chace does not do this. Chace interrupts the linkage at a convenient one of its joints and nominates there the unknown reaction force and couple vectors acting. This means that, whereas Chace writes a larger number of differential equations involving various inertia and acceleration terms and a smaller number of other differential equations after differentiating the data, which equations relocate the elements of the interrupted joint, Uicker has a smaller number of differential equations of the same or a similar kind and a larger number of algebraic equations associated with the closure. Next, by means of iterations involving successive small increments of time and reconstructions of the configuration after the calculated small displacement of the whole mechanism, both authors can arrive at an overall answer to the following question: how does some input angle theta (along with its theta-dot) vary with time as the mechanism moves? Both authors have the capacity to introduce in their analyses not Coulomb but viscous friction at the joints, they both become involved in the long run with matrix methods of calculation, and they are both able on the basis of these exact dynamic analyses to answer (*a*) the so-called dynamic constraint force problem and (*b*) the so-called dynamic time response problem. They both claim, in other words, to be able to solve Wittenbauer problems of the first and the second kind (§ 1.03). Writers in the field of robotics – and I take for example Richard Paul (1981) – explain that we cannot yet do the necessary calculations by these or similar methods for the dynamics of robots in real time. Workers in the field of robots sometimes pre-do dynamic calculations on the basis of certain simplifying assumptions and build the results of these into the control system. Nor can the robot workers know, of course, what the operating (applied) forces might be upon the robot at any one time, and they cannot know the effect of other external irregularities such as accidental collisions. For an excellent summary and a comparison in mathematical and numerical terms of the overall strategies of Uicker

and Chace and others for the dynamics of machinery, refer to Burton Paul [44]. I venture to say that this paper is a first piece of required reading for any person wishing to pursue the dynamics of machinery seriously, and without hesitation I say that the above few remarks about the works of Uicker and Chace are adequate only for the limited purposes of this book.

**Machine dynamics, Ball**

64. If the reader will go now to chapters 12 and 13 he will find expressed there some views about dynamics of my own (§ 12.60 *et seq.*, § 13.08, § 13.29). In effect I was asking there this: can we not programme the computers to listen easily to input data in terms of screws, to solve the given problems in dynamics in terms of screws, and, moreover, to speak the answers back in terms of screws? One thinks of course of Ball. So go to Ball. Find first that his Introduction deals exclusively with (*a*) the geometrical theorem of Chasles (§ 0.07, § 10.51, § 19.30) which he proves in a very careful way by means of projective geometry, and (*b*) the theorem of Poinsot (§ 0.08, § 10.51, § 10.53) which he proves in such a short way that the theorem is made to appear to be some mathematical proposition rather than a physical law. His chapter 1 introduces basic concepts and terminology. Ball deals there not with wrenches and rates of twisting as I have done throughout this book, but with wrenches, impulsive wrenches, and the infinitesimal screw displacements of a rigid body namely twists. This fundamental aspect of his work and especially his idiosyncratic terminology need to be understood. One needs to wonder, indeed, as one reads Ball, that if we have (as we do) both instantaneous and non-instantaneous kinematics, namely the kinematics (*a*) of infinitesimal motion, and (*b*) of finite movements (§ 19.41 *et seq.*), what then is non-instantaneous statics? Towards the end of his § 13 Ball explains the very severe restriction he puts upon the area of his investigations. He says there that he will deal in his book only with small screw displacements namely infinitesimal twists of a single rigid irregular body being disturbed from rest by the application of an impulsive wrench. In this special case of course (where there is no initial angular velocity of the body) the relevant inertia matrices are not complicated with gyroscopic terms. Ball's actual dynamical considerations are confined to his chapters 7 and 21. Before attempting chapter 7, however, one must refer to his chapter 4 to understand there (*a*) his general concept of the six screw coordinates of a screw, (*b*) his idea of a set of six co-reciprocal screws, and (*c*) his set of canonical co-reciprocals. In chapter 7 (at § 79) we find his first equation of motion. This equation relates to a perfectly free body which suffers an impulsive wrench upon a certain screw. In the general course of events such a body would begin to twist about some other screw, but in this particular case the screw is collinear with one of the principal axes of inertia of the body, the pitch of the screw (and thus of the impulsive wrench) is equal to the radius of gyration of the body about that axis, and it is shown that the body begins to twist (with an angular velocity $\dot{\omega}_1$) not about some other screw but about that screw. It is shown that $\dot{\omega}_1 = (e/aM)\,\omega_1''$, where $e\omega_1''$ is the magnitude of the impulsive wrench, $e$ is the very short time, $a$ is the radius of gyration, $M$ is the mass of the body, and $\omega_1''$ is the magnitude of the very great wrench. The magnitude of the impulsive wrench $e\omega_1''$ is subsequently called by Ball $\omega_1'''$; its dimensions are $LMT^{-1}$. From this early important result Ball goes on to study the effect of given impulsive wrenches upon bodies that are constrained in various ways. At his § 90 he deals in particular with a body constrained to enjoy one degree of freedom only. He applies there to a body of mass $M$ which is free to twist exclusively about a certain screw $\alpha$ an impulsive wrench $\eta'''$ which acts about another screw $\eta$. He finds that the resulting angular velocity $\dot{\alpha}$ is given by $\dot{\alpha} = (\eta'''\varpi_{\eta\alpha})/(u_\alpha^2 M)$, where $u_\alpha$ is a contrived radius of gyration that takes account not only of the rotation about but also of the translation along the ISA (Ball § 79) and where $\varpi_{\eta\alpha}$ is the virtual coefficient for the two screws $\eta$ and $\alpha$. The dimensions of the virtual coefficient are L (§ 14.28). For further exploratory reading in Ball the reader may go to his chapter 21. Here the scope of the argument is widened to admit the possibilities (*a*) that the body might be entirely free, and (*b*) that the impulsive wrench might be caused to act about any screw. This along with the earlier work sounds interesting and promising, but remember that Ball's body is always initially at rest, so that his formulae are not sufficiently general to be applied to a link that is already moving. Remember too that in any piece of machinery where the mechanism is of mobility unity each massive link is constrained to enjoy one degree of freedom only. At this juncture I wish to mention the work of Lovell [45]. Lovell argues in his thesis that Ball's incisive contributions (his ideas comprising and surrounding the principal screws of inertia of a body) are not only elegant but also useful in a fundamental way for the analysis of moving machinery. When these and Ball's other ideas are sufficiently well developed, he says, they can be applied with success to the dynamics of a link.

**A note about the future technology**

65. Much of the present chapter has been written as though all kinds of spatial arrangements of mechanism are easily realisable in the actual machinery of ordinary engineering practice. We know on the other hand that the machinery of the present day – even the openly spatial machinery of robots and the like – is based upon relatively simple special cases of the more

general kinematic chains. There is much in the way of parallelism, coaxiality and orthogonality among the joint axes; and the machinery is generally machinable by ordinary machining methods. We are waiting in the new arena of spatial machinery not only for more sophisticated algebraic analyses and for faster methods of computer control, but also for some of the kinds of things that might be represented by the following: (*a*) methods of fabrication or of casting (and of machining) for the actual links of spatial machinery which easily allow the introduction of compound angles, threads of variable pitch, known common-perpendicular distances and other practical difficulties; (*b*) methods of construction of joints and methods of lubrication that encourage use of the special and various intermediate and higher joints. Above all we need to know new kinds of machine tool; these will truly allow the practical design and actual manufacture of odd-shaped three-dimensional links with two, three and four joints. And some of these joints may (in the not too distant future) be of quite new kinds; please refer to chapter 7.

**Some remarks about synthesis and design**

66. The building blocks for the synthesis of linkages are links and joints (§ 1.10, § 1.11). Because useful mechanism consists in the main in loops (§ 1.15), the joining up of links into loops by joints, or the joining up of joints into loops by links (§ 1.38), is the main business of the synthesis of mechanism and thus of machine design. Once the topology of the mechanism has been decided (and thus the state of its freedom and constraint) or, in other words, once its so-called type-synthesis has been achieved (the mechanics of this activity is quite obscure, rooted as it is in the whole question of creativity), we can get on with the much milder, mathematical business of optimising the dimensions of the synthesised linkage. We can decide the actual shapes of the links then, and check the distribution of mass. We do this to try to suit whatever our requirements are of the real machine. All of this latter design activity, often called – yet wrongly in my view – syntheses, might be seen to consist very often of a mere reversal, a turning inside out of the analyses outlined in this chapter. But even the methods of optimization are not mere reversals of analysis. It is not possible for us as designers, equipped as we are with our mathematical methods, to arrive without insight and imagination at radical original syntheses. It is surely absurd to say that machinery is a mere mathematical problem waiting for solution. It is a wide field of endeavour for us, a cultural activity. There is and there will continue to be endless scope for the synthesis of new mechanism in the art of machine design.

**Notes and references**


[1] Sticher, F.C.O., *An investigation into three dimensional motion*, undergraduate dissertation, Department of Mechanical Engineering, University of Sydney, 1965. Refer § 19.07.

[2] Konstantinov, M.S., Genova, P.I., and Topencharov, V.V., Acceleration in the spatial motion of the rigid body (in Bulgarian), *Bulletin of the Higher Institute of Mechanical and Electrical Engineering, Sofia*, volume for 1960 or thereabouts, p. 95–103, date uncertain. Refer § 19.07.

[3] Rooney, Joseph, A survey of representation of spatial rotation about a fixed point, *Environment and planning B*, Vol. **4**, p. 185–210, Pion, 1977. Refer § 19.12.

[4] Rooney, Joseph, A comparison of representations of general spatial screw displacement, *Environment and planning B*, Vol. **5**, p. 45–88, Pion, 1978. Refer § 19.12, § 19.45, § 19.51.

[5] Clifford, W.K., Preliminary sketch of biquaternions, *Proceedings of the London Mathematical Society*, Vol. **4**, Nos. 64, 65, p. 381–95, 1873. Refer § 19.13.

[6] Everett, J.D., On a new method in statics and kinematics, *Messenger of mathematics*, Vol. **45**, p. 36–7, 1895. Refer § 19.13.

[7] Kotelnikov, A.P. I take some of the information about this obscure author from Dimentberg (1968) and some from elsewhere. Please refer to note [6] of chapter 0 for the titles of works and be aware that they first appeared, reputedly, in *The Annals of the Imperial University of Kazan*, 1895. Refer § 19.13, § 19.34, § 19.39, § 19.45.

[8] Keler, Max., Die Verwendung dual-complexer Grössen in Geometrie, Kinematik und Mechanik, *Feinwerktechnik*, Band **74**, Heft 1, p. 31–41, 1970. Refer § 19.13, § 19.20.

[9] Keler, Max., Analyse und Synthese räumliche, sphärische und ebene Getriebe in dual-komplexer Darstellung, *Feinwerktechnik*, Band **74**, Heft 8, p. 341–51, 1970. Refer § 19.13.

[10] Keler, Max., Kinematik und Statik einschliesslich Reibung in Mechanismen nullter Ordnung mit Schraubengrössen und dual-komplexer Vektoren, *Feinwerktechnik + micronic*, Band **76**, Heft 1, p. 7–17, Carl Hanser Verlag, München 1973. Refer § 19.13.

[11] Truesdell, C., Whence the law of moment of momentum? This is an essay from a collection of Truesdell's work entitled *Essays in the history of mechanics*, p. 239–71, Springer-Verlag, Berlin 1968. Refer § 19.17.

[12] Plücker, Julius. There are two well known papers in English by this author. They are quoted in note [3] of chapter 10. The second, of 1866, reveals the origin not only of the Plücker coordinates for a line but also of the so-called screw coordinates for a screw. See also there and listed in my Bibliography the details of Plücker's book (in German); see note [1] of chapter 15. Refer § 19.21, § 19.28.

[13] Woo, L. and Freudenstein, F., Application of line geometry to theoretical kinematics and the kinematic analysis of mechanical systems, *Journal of mechanisms*, Vol. **5**, p. 417–60, Pergamon, 1970. Refer § 19.21, § 19.28.

[14] Lipkin, H. and Duffy, J., The elliptical polarity of screws, *Journal of mechanisms, transmissions, and automation in design*, Vol. **107**, (Trans ASME Series R), p. 377–87, September 1985. Two papers by Parkin, I.A., (1) Coordinate transformations of screws with applications to screw systems and finite twists, (2) Location of the central geometries of the general screw systems, *Mechanism and*

*machine theory*, in press 1988. Two papers by Hunt, K.H., Special configurations of robot arms via screw theory (parts 1 and 2), *Robotica*, Vol. **4**, p. 171–9, 1986, Vol. **5**, p. 17–22, 1987, Cambridge University Press. Refer §19.28.

[15] Browne, J.M. and Pengilley, C.J., Applications of the exterior calculus to Newton's law, *Proceedings of the 6th World Congress IFToMM in New Delhi 1983*, p. 1238–41, Wiley Eastern, 1983. This paper is one of a number by these authors. They grew from the doctoral dissertation of Browne which makes a detailed study of Grassmann's work. See also Grassmann, *Dictionary of scientific biography*, Volume **XV** (Supp. 1), p. 192–99, Scribners, New York 1978. Ball (1900) also mentions Grassmann; see his few remarks among his Bibliographical Notes, p. 515. Refer §19.29.

[16] Hunt says (and so does Whittaker) that it is not from Chasles but from an earlier worker (§0.07). Nevertheless I quote Chasles, M., Note sur les propriétés générales du systeme de deux corps semblables entr'eux et places d'une maniere quelconque dans l'espace; et sur le deplacement fini ou infiniment petit d'un corps solide libre, *Bulletin universal des sciences et de l'industrie, Section 1, Bulletin des sciences mathématiques, etc.*, Vol. **14**, p. 321–6, Paris, 1830. Refer §19.30.

[17] We must distinguish between the finite screw displacement of a line segment, or of a *point-line* as it was called by Sticher (references to follow), and a whole unbroken line. Sticher, F.C.O., Line system geometry applied to closures of spatial mechanisms, *Proceedings 5th World Congress IFToMM in Montreal 1979*, p. 489–93, ASME, 1979. Also Sticher, F.C.O., Geometric and algebraic study of the RRRRl zero link-length point-line system, *Mechanism and machine theory*, Vol. **19**, p. 337–47, 1984, Pergamon 1985. Sticher has shown in a recent private communication and in a simple way (*a*) that the locus of the FSA in the case of a point-line is distributed upon the shape of a cylindroid, but (*b*) that the pitches of the FSAs upon the cylindroidal shape are distributed in a non-sinusoidal way; he has shown that there is always one but only one FSA of zero pitch upon the cylindroidal shape and, taking the finite screw displacements as being all right handed (§19.30), that there is always one but only one local maximum of the pitch. However the first of the following authors (references to follow) arrived at some of the same conclusions in 1973, while the second of them (two other authors appearing in the same issue of the same journal) simply mention the cylindroid but mysteriously. Bottema, O., On a set of displacements in space, *Journal of engineering for industry*, Vol. **95**, (Trans ASME Series B Vol. **95**), p. 451–4, May 1973. Tsai L.W. and Roth, B., Incompletely specified screw displacements: geometry and spatial linkage synthesis, *ibid.*, same volume, same date, p. 603–11. Refer §19.32. Refer also [47].

[18] Two papers. Roth, B., On the screw axes and other special lines associated with spatial displacement of a rigid body, *Journal of engineering for industry*, Vol. **89**, (Trans ASME Series B Vol. **89**), p. 102–10, February 1967. Also Roth, B., The kinematics of motion through finitely separated positions, *Journal of Applied Mechanics*, Vol. **34**, (Trans ASME Series E Vol. **89**), p. 591–8, September 1967. Refer §19.36.

[19] Keler, M., Dual-vector half-tangents for the representation of the finite motion of rigid bodies, *Environment and planning B*, Vol. **6**, p. 403–12, 1979. Refer §19.36, §19.41.

[20] Halphen, Sur la theorie du deplacement, *Nouvelles annales de mathématiques*, 3rd Series, Vol. **1**, p. 296–9, Paris, 1882. The author of this paper, despite his designation M. Halphen (the M. stands for Monsieur), is the well known George-Henri Halphen of my note [1] of chapter 11. He explains and encapsulates without diagrams the whole of the bare bones of what Roth calls the screw triangle in less than four small pages of text. The reader might check also the works of Battaglini mentioned in note [1] of chapter 15. Refer §19.36.

[21] Keler, Max., *Analyse und Synthese der Raumkurbelgetriebe mittels Raumliniengeometrie und dualer Grössen*, doctoral dissertation, TH-München, 1958. Refer §19.38.

[22] Yang, A.T., *Application of quaternion algebra and dual numbers to the analysis of spatial mechanisms*, doctoral dissertation, Columbia University, New York, 1963. Refer §19.38.

[23] Freudenstein, Ferdinand, Approximate synthesis of four-bar linkages, *Trans ASME*, Vol. **77**, p. 853–61, 1955. Refer §19.48.

[24] Uicker, J.J. Jr, Denavit, J. and Hartenberg, R.S., An iterative method for the displacement analysis of spatial mechanisms, *Journal of applied mechanics*, Vol. **31**, (Trans ASME Series E Vol. **86**), p. 309–14, June 1964. Refer §19.51.

[25] Yang, A.T. and Freudenstein, F., Application of dual-number quaternion algebra to the analysis of spatial mechanisms, *Journal of applied mechanics*, Vol. **31**, (Trans ASME Series E Vol. **86**), p. 300–8, June 1964. Refer §19.38, §19.51.

[26] Yang, A.T., Calculus of screws. This is an essay within the anthology, *Basic questions of design theory*, Ed. W.R. Spillers, p. 265–81, Elsevier, New York 1974. Refer §19.38, §19.42.

[27] Sandor, G.N. and Bisshop, K.E., On a general method of spatial kinematic synthesis by means of stretch-rotation tensor, *Journal of engineering for industry*, Vol. **91**, (Trans ASME Series B Vol. **91**), p. 115–20, February 1969. Refer §19.51.

[28] The material of §19.43 springs from remarks made by Joseph Rooney in private discussion with me in 1981. The idea of manipulating the spherical indicatrix as described is due to him, the final italicized passage at the end of §19.43 is due to me, while the mathematical analyses outlined in the following papers are due to Zhang Wen-Xiang. Phillips, J.R. and Zhang, W.X., On the screw triangle and the cylindroid, *Seventh world congress IFToMM in Seville 1987*, p. 179–82, Pergamon, 1987. Zhang, W.X., Formulae for the connection of mechanics of finite and infinitesimal screw displacements and their use in rigid-body mechanics, *International journal of mechanical engineering education*, Vol. **16**, p. 107–18, I. Mech. E., London 1988. Refer §19.43.

[29] Rooney, J., On the principle of transference, *Proceedings of the 4th World Congress IFToMM in Newcastle-upon-Tyne 1975*, p. 1088–94, I. Mech. E., London 1975. Refer §19.45, §19.51.

[30] Rooney, Joseph, *A unified theory for the analysis of spatial mechanisms based on spherical trigonometry*, doctoral dissertation, Liverpool Polytechnic, January 1974. See also Duffy, J. and Rooney, J., A foundation for a unified theory of analysis of spatial mechanisms,

ASME Paper No. 74-DET-70, p. 1–6, 1974. Refer § 19.45.

[31] Gal, J.A., *Survey of closure equations*, undergraduate dissertation, Department of Mechanical Engineering, University of Sydney, 1970. Refer § 19.51.

[32] Sticher, F.C.O., Use of conics in studying the characteristics of some spatial mechanisms, *Journal of mechanisms*, Vol. **6**, p. 303–39, Pergamon, 1971. Refer § 19.56.

[33] Gal, J.A., *Kinematics of mechanical manipulators*, doctoral dissertation, University of Sydney, 1983. Refer § 19.56.

[34] Gal, J.A., On the use of physical link parameters in the algebraic description of mechanical manipulators, *Proceedings of the ICAR in Tokyo 1985*, p. 267–74, IFS Pub's. and North Holland/Elsevier, 1985. Refer § 19.56.

[35] Manfred Hiller and Christoph Woernle, A unified presentation of spatial displacements, *Mechanism and machine theory*, Vol. **19**, p. 477–86, 1984, Pergamon, 1985. In this paper these writers translate their original term *Drehzeiger* into 'screw displacement pair'. Refer § 19.58.

[36] Waldron, K.J., *The mobility of linkages*, doctoral dissertation, Stanford University, 1969. Refer § 19.40.

[37] Duffy, J. and Crane, C., A displacement analysis of the general spatial 7-link, 7R mechanism, *Mechanism and machine theory*, Vol. **15**, p. 153–69, Pergamon, 1980. Refer § 19.59.

[38] Urquart, P., Analysis of the seven-bar chain, *Journal of mechanisms*. Vol. **2**. p. 77–83, Pergamon, 1967. Refer § 19.59.

[39] Albela, Hiram, *Displacement analysis of the n-bar, single-loop, spatial linkage; application to the 7R single-degree-of-freedom spatial mechanism*, doctoral dissertation, Technion, Hiafa, 1976. Refer § 19.59.

[40] Book Review, Waldron on Duffy, *Mechanism and machine theory*, Vol. **18**, p. 193, Pergamon, 1983. Refer § 19.40, § 19.59.

[41] Wohlhart, Karl, Die homogene othogonale und nicht versetzte Zwanglaufkette 7*R* und ihr Spiegelbild, *Proceedings 6th World Congress IFToMM in New Delhi 1983*, Wiley Eastern, New Delhi 1983, p. 272–6. Refer § 19.59.

[42] Uicker, J.J., Denavit, J., and Hartenberg, R.S., An iterative method for the displacement analysis of spatial mechanisms, *Journal of applied mechanics*, Vol. **31**, (Trans ASME Series E Vol. 86), p. 309–14, June 1964. Uicker, J.J. Jr, Dynamic behaviour of spatial linkages, *Journal of engineering for industry*, Vol. **91**, (Trans ASME Series B Vol. **91**), p. 251–65, 1969. Sheth, P.N. and Uicker, J.J. Jr, IMP (Integrated Mechanisms Program): a computer-aided design analysis system for mechanisms and linkages, *Journal of engineering for industry*, Vol. **94**, (Trans ASME Series B Vol. **94**), p. 454–64, 1972. Uicker, J.J. Jr, *Users' guide for IMP (Integrated Mechanisms Program): a problem oriented language for the computer-aided design and analysis of mechanisms*, University of Wisconsin, Madison, 1973. Refer also note [8] of chapter 0. Refer § 19.63.

[43] Chace, M., Vector analysis of linkages, *Journal of engineering for industry*, Vol. **85**, (Trans ASME Series B Vol. **85**), p. 289–97, August 1963. Chace, M.A. and Smith, D.A., DAMN – digital computer program for the dynamic analysis of generalized mechanical systems, *SAE Transactions*, Vol. **80**, p. 969–83, 1971. Smith, D.A., Chace, M.A. and Rubens, A.C., The automatic generation of a mathematical model for machinery systems, ASME Paper 72-Mech-31, 1972. Chace, M.A. and Sheth, P.N., Adaption of computer techniques to the design of mechanical dynamic machinery, ASME Paper 73-DET-58, 1973. Chace, M., *Dram and Adams*, Mechanical Dynamics Inc., 1103 So. University Avenue, Ann Arbor, Michigan, USA. Refer § 19.63.

[44] Paul, Burton, Analytic dynamics of mechanisms – a computer oriented overview, *Mechanism and machine theory*, Vol. **10**, No. 6, p. 481–507, Pergamon, 1975. Refer § 19.63.

[45] Lovell, Gilbert, *A new approach to motor calculus and rigid body dynamics with applications to serial open-loop chains*, doctoral dissertation, University of Florida, 1986. Refer § 19.64.

[46] I have made a mistake at § 6.41. The geometry there is the same as it is here; the last three lines of § 6.41 are wrong; the facts as stated there apply only if the mentioned triplet of motion screws (and thus the triplet of nodal lines) are mutually perpendicular; this will not occur at a general point in a general 3-system; it will however occur as a special case at the origin, namely at the centre of the point-symmetry of the system.

[47] Ian Parkin of [14] has shown more recently that, if the pitch of a finite screw displacement be defined not $d/\phi$ but $(d/2)/\tan(\phi/2)$, Sticher's distribution of finite pitch mentioned at [17], which is smooth except that it suffers a sudden change of sign at that single special generator of the cylindroid where the angle of rotation $\phi$ is exactly $\pi$, becomes replaced by a distribution which is exactly sinusoidal. Parkin argues from this that his new definition has implications for linearizing the mathematics of the combinations of finite screw displacements. Parkin's new definition reduces in the special case of instantaneous motion to the well accepted rate of change $dd/d\phi$ of course; and that sits comfortably with the contents of my chapter 5. Refer to [17] and thus back to § 19.32.

20. The two reguli of the hinge axes of two Bennetts back to back. The reguli intersect along a pair of mutually shared generators. This model, of brass wire and balsa wood, is elucidating the geometry.

# 20 The special geometry of some overconstrained loops

## The need for overconstraint

01. In heavily loaded machinery we often intentionally overconstrain the mechanism. Such overconstraint ensures a multiplicity of paths for the transmission to frame of external wrenches applied upon the moving links. The distribution of these paths depends upon the compliances of the links in the same way as it does in statically indeterminate structures. The overconstraint will reveal itself by a surprisingly low $M$ obtained from the general mobility criterion (1) of § 1.34. Consider the bulldozers and front-end loaders whose heavy, eccentrically applied and varying external loads at the blade or bucket are discussed in chapter 10 (§ 10.02, § 10.63); we duplicate the planar linkages on either side of the transporting tractor and join the corresponding links together rigidly. In such machinery we often find wide (laterally wide) and striking examples of the well known overconstrained planar 4*R*-loop. This overconstrained loop among others is often used in situations where loads are heavy, eccentrically applied, or otherwise demanding upon the flexibility of the machinery. The effect of overconstraint in the continuing presence of the capacity to move of a mechanism is to increase strength and stiffness, to increase the difference between the lower Kutzbach mobility $M$ of the mechanism and its higher apparent mobility $M_a$, and to increase the need for accuracy of construction. One may look for other examples of overconstraint among the multiplicity of devices used in the internal construction of folding cardboard boxes, for the opening and closing of pop-up greeting cards, and in the operation of various cunning kitchen devices. One may also look (in advance) at § 20.16.

## Mixed mechanism

02. I wish to start here by shifting the argument away from the rigid crooked linkages of chapter 1 – all of which obey the general mobility criterion (1) of § 1.34 (§ 1.05, § 1.59, § 2.01) – into the new arena of the overconstrained yet mobile, non-crooked linkages of the present chapter. I wish to begin with an example of what I meant when I spoke at § 1.36 about the difficulties met in 'mixed' mechanism. The example I take comes from Waldron [1], but the term *mixed* comes from Boden [2]. I think incidentally that the term mixed betrays an outdated belief, namely that the whole of mechanism can be divided into two areas: into planar mechanism, which can be understood, and 'spatial' mechanism which is mysterious. I consider the term mixed to be outdated because of the growing strength of screw theory. The elucidation of the general screw systems and their special and degenerate forms has rendered that particular term, applied as it is to the mixing of planar and spatial mechanism, almost meaningless.

03. Despite that, please look at figure 20.01(*a*) and see there that the planar 5-link loop of revolutes *ABCDE* can be seen. See also that the link 4 has a peg at $F$ engaging in a circular slot on centre $G$ and cut into the material of the frame link 1. See that the 5-link multi-loop mechanism in figure 20.01(*a*) is a planar mechanism with six joints. Using the special criterion mentioned at § 2.36 which uses the multiplier 3, and knowing that the slotted hinge at $F$ is $f2$, we can find that the Grübler mobility say $M_g$ of the planar mechanism in figure 20.01(*a*) is unity. It is mobile and, in terms of planar kinematics, just constrained. The link 4 can rotate at the instant about the pole $P_{14}$ which is marked in the figure, and there is no evident difficulty.

04. Look however at figure 20.01(*b*). A ball-ended rod $G$–$F$ is inserted between the coupler 4 and the frame 1 in the manner shown. The overall dimensions are such that an orthogonal view of the new, mixed mechanism in figure 20.01(*b*) will reveal the original dimensions displayed in 20.01(*a*). In figure 20.01(*b*) the centre of the ball at $G$ remains in a fixed position while the centre of the ball at $F$ is obliged to travel in a particular circle. The circle exists where the sphere at centre $G$ and of radius $G$–$F$ is cut by the unique plane at $F$ normal to the five hinge axes. So far as the five flat links are concerned, the two mechanisms are kinematically equivalent, but – and here we can see an advantage – the mixed mechanism with its set of long hinges and its strut link $G$–$F$

could be built very stiff and thus strong against external loads applied in the transverse direction.

05. Uncouple the ball and socket at $F$ and see that we have remaining an overconstrained $5R$-loop (§ 2.10). It has an apparent mobility $M_a$ of 2. Couple-up again at $F$ and, ignoring the freedom of the rod $G$–$F$ irrelevantly to rotate (§ 1.37), observe that the overall, apparent mobility of the apparatus is unity. If we uncouple the hinge at $C$, however, we reveal the 4-link loop $GFDE$, whose general, Kutzbach mobility $M$ is $+2$. If we couple-up again at $C$, we (*a*) lock in a measure of overconstraint at the 5-link loop $ABCDE$, while (*b*) permitting a measure of underconstraint in the 4-link loop $GFDE$. The Kutzbach mobility $M$ for the whole apparatus in figure 20.01(*b*) is $M = 6(6 - 7 - 1) + 11$, namely *minus one*; and my point is that this bald result, taken alone, is useless.

06. The above, somewhat ill disciplined discussion has pointed up, I hope, that a mere unthinking application of the formulae of chapter 1 will not solve all of the even simple problems of practical machine design (§ 2.14, § 2.27). It should be evident also that a better way to go about the matter in figure 20.01(*b*) might be to examine somehow the special system of motion screws available at the instant for the link 4 which is a floating link. Notice incidentally the nature of the complex joint between this floating link 4 and the frame 1; it has three branches. The special system in this case, and this can be seen upon inspection in figure 20.01(*a*), is a 1-system where the pitch of the single screw is zero [1]. It must be observed in general, however, that such links as this link 4, having no direct contact with frame and thus no cover under the quasi-general methods of chapter 6, can present difficulties for which we have at the present time no effective answer. But refer to Baker [3] and Davies [4] to be mentioned later. Look also at figure 19.10 and, in the absence of a Kutzbach analysis for the mobility of that whole mechanism, ruminate there about the connectivity $C_{12}$ with frame link 1 of the thrice supported, floating link 2 (§ 1.31).

Figure 20.01 (§ 20.02) The two mechanisms here are equivalent in that their projected instantaneous motions could be said to be the same. Whereas at (*a*) the mechanism is planar, it is at (*b*) however mixed, and there are at (*b*) certain special difficulties in applying the Kutzbach criterion for mobility.

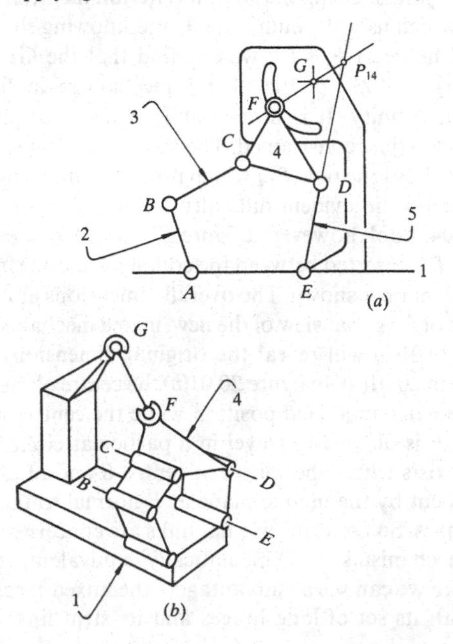


### Single-loop as distinct from multi-loop mechanism

07. We do appear to have no direct or easy way, in other words, for finding the connectivity with frame of such links as 4 in figure 20.01(*b*), or for finding the precise screw systems available to links such as 4 for motion at the instant. A reader might well conclude from what I have said that the methodology of the application of screw theory to closed multi-loop mechanism is not sufficiently well developed; and in my view the reader would be right; but the reader might care to read again in this respect Waldron [1], and refer to the much more recent works by Baker [3] and Davies [4]. These latter are seminal works of importance in the new literature concerning freedom and constraint in what Davies has begun to call *mechanical networks*. Davies uses the concepts of degrees of mobility $F$ and degrees of redundancy $R$, and relates the two in multi-loop mechanism and in a kinetostatic manner by means of wrenches and rates of twisting at the so-called passive and at the redundant joints. There does exist, however, a wide range of overconstrained yet mobile *single loops* which employ moreover lower joints. Various analyses of these relatively simple loops have produced some intelligible though fragmentary theorems. I discuss these next from § 20.08 to § 20.10 inclusive.

08. At this juncture it might be wise to recall from chapter 2 that the overconstrained single-loop mechanisms whose apparent mobilities $M_a$ are nevertheless unity will always fall into one or other of these two categories: (*a*) those that exhibit their extra mobilities $M_s$ only transitorily (take for example the mechanism in figure 2.04 and see there that the motion will often be 'soggy' as described and explained at § 2.46); and (*b*) those mechanisms that exhibit their extra mobilities $M_s$ openly and continuously throughout the full cycles of their movement. But the determination of whether or not a mobile mechanism is continuously mobile presents difficulties not only in the cases of multi-loop mechan-

isms but in the cases of isolated single loops as well. A relevant exercise for the reader is as follows. Within the mechanism in figure 20.02 the Kutzbach mobility $M$ is $-2$ but, due to the parallelism of the centre lines of the revolutes $A$ and $C$ and the coplanarity of all three link-lengths (§ 20.02, § 19.52), the apparent mobility $M_a$ is zero. Is this apparent mobility full-cycle or only transitory? And how does this 3-link loop compare with the ones that are shown in figure 20.27? I think that the reader may find the apparent simplicity of these questions quite complicated. Refer to chapter 23.

**The screw systems within linkages**

09. We saw in chapters 12 and 13 that the three action screws of any three wrenches in equilibrium or the three motion screws of any three rates of twisting that add to rest belong to the same 2-system, namely to the same cylindroid of screws. We saw in chapter 10 that the four action screws of four wrenches in equilibrium belong to the same 3-system of action screws (§ 10.36 *et seq.*, § 10.69 *et seq.*); and we saw correspondingly that a closed loop of four screw joints that was transitorily mobile could be represented at the instant by its four screws all belonging to the same 3-system of motion screws (§ 10.54 *et seq.*). In such ways and otherwise we can be led to make the following statement about the kinematics of a simple loop: it is a series of facts – and these relate to our ideas about the dependence and independence of screws – that, *if there are 3, 4, 5, or 6 screw joints in a simple loop, and if the loop is transitorily mobile* (*or mobile throughout a full cycle of its motion*), *the motion screws at the screw joints at the instant* (*or continuously*) *are members of the same 2-system, 3-system, 4-system, or 5-system of motion screws respectively; in special cases they may be members respectively of even lower systems of motion screws.*

Figure 20.02 (§ 20.08) This is a 3-link loop with geometrical specialities which is, nevertheless, immobile. The geometrical specialities have increased its apparent mobility from $-2$ to zero, and the question is this: is this a full-cycle mobility or is it only transitory?

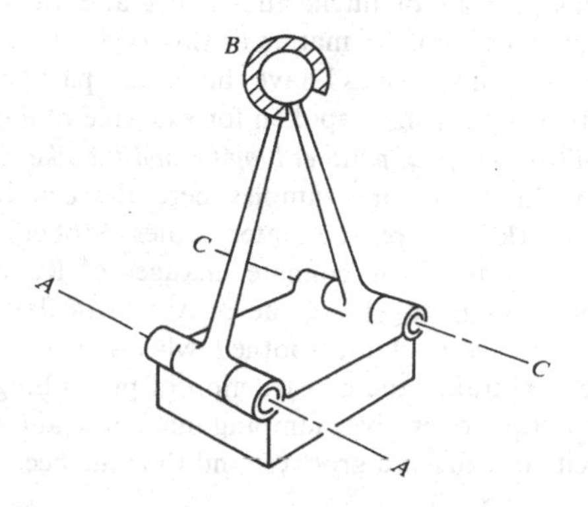


10. These facts have led to a recent literature about the existence or otherwise of overconstrained yet mobile loops containing all kinds of lower pairs (§ 7.12). As well as the zero-pitch helicals (namely revolutes) that might be expected to abound, sphericals, cylindrics, helicals of all pitches and prismatics abound in this difficult literature. And the relevant screw systems, of course, are often special or degenerate. Refer for example Waldron [5] and Hunt [6]. In answer to the question often asked in the mentioned literature of whether the linkages found to be transitorily mobile are mobile for their full cycles, however, the mentioned authors always speak – or they spoke at the time – in terms of the symmetry of the particular linkage if any, or they considered some other special aspect of the overall geometry. Except for the algebraic methods of Baker and Waldron [7] who began to write some closure equations for some of the linkages with non-parallel helical joints at that time, there seems to have been no way, until quite recently, for checking this question of full-cycle mobility if the linkage in question was asymmetrical in all respects or otherwise of a general arrangement. This question of the full-cycle mobility of asymmetrical overconstrained linkages involving all of the lower pairs has been, and continues to be, a difficult one [4].

11. There is however another branch of the literature dealing with a series of often symmetrical, single loops of an even simpler kind; *these single loops involve revolutes only*; and the main matter at issue in this literature is the already mentioned question of the presence or otherwise of full-cycle mobility. Spread over at least a hundred years, Sarrus, Bricard, Bennett, Delassus, Myard, Goldberg, Voinea, Atanasiu, Dimentberg, Yoslovich, Waldron, Baker, Schatz, and Yu are some of the contributors here, and it is a summary of the results of these workers (separately referenced later) that forms the bulk of the chapter here. For the sake of a general presentation – and such a presentation is due to follow soon – the mentioned linkages may be grouped into (*a*) the 7*R*-loop which is mobile throughout its cycle with $M = 1$, (*b*) the particular 6*R*, 5*R*, and 4*R*-loops which have full-cycle apparent mobility $M_a = 1$, and (*c*) the trivial mobile 3*R*-loops which also demand a mention (§ 20.54). It is convenient to mention the 3*R*-loops because they exemplify the special relationship that must exist between any three angular velocities that add to rest (§ 10.15). Beginning at § 20.17 I deal with aspects of all of these, and for each aspect I deal with the linkages concerned in the mentioned descending order, descending by the number of links.

**Linkages that are replicated linkages**

12. Before going on with that, however, we might look first at the following possible way for classifying into a single special group some of the more complicated of the overconstrained yet mobile linkages. It is possible (and probably useful) to argue that there is a whole class of commonly occurring, overconstrained linkages in engineering practice that depend for their capacity to execute a full cycle of motion, not upon the special peculiarities of their line geometry, but upon the mere fact that they consist in the *replication* of some simpler linkage. I wish to explain in the following paragraphs what I mean by this; and I wish to remark again before I begin that, whereas each single mechanism may be well enough constructed to deal with elastic and other distortions of its frame and its other links, each replicated linkage is, by virtue of the replication, stiff in its overall construction and thereby strong against expected heavy loads.

13. Figure 20.03(*a*) is a schematic drawing of the central balls and one branch only of the three branches of the constrant velocity coupling or CV joint appearing in photograph 8. A check with the general mobility criterion of § 1.34 will reveal that the coupling of photograph 8 has the necessary and sufficient freedoms to operate satisfactorily. I wish to mention here (and parenthetically) that, in the absence of the central assembly of the two balls in contact within the closed tube, this coupling is capable not only of continuing to function as a CV joint but also of plunging; this latter facility removes the need for a prismatic joint (a spline) in the intermediate shaft of a double CV-joint assembly where the driving and the driven shafts are free to move relatively. Well, were we to manufacture couplings of this kind with not only three but with *four or more* parallel loops (we might do this for the sake of improving torque carrying capacity or for some other practical reason), we would render the mechanism overconstrained by virtue of replication. Other examples of the same phenomenon might be found in the kinds of constant velocity coupling that operate with many balls imprisoned by means of crossed or crossing slots. The Rzeppa, and the more recent Demag coupling, are examples. Refer to figure 20.03(*b*). In the case of the Demag coupling, only partly illustrated there with two shaft bearings fixed as shown and one loop only of its mechanism extant, the general mobility is unity. In practice, however, many more loops and balls than one are required; an angle-bisecting cage for the balls is also required. The mechanism of the completed coupling is grossly overconstrained, it requires great accuracy for its construction, and suffers certain defects due to its overconstraint. A deeper discussion of the mechanics of this particular coupling may be found at reference [8].

Figure 20.03 (§ 20.13) Some examples of replicated linkages. They are replicated in the sense that a basic linkage in each case is duplicated, triplicated, or otherwise multiplied for the sake of load distribution, symmetry, or strength.

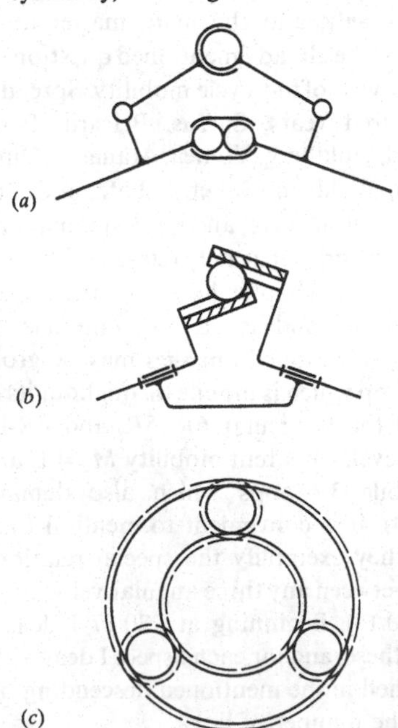


14. For another example of what I mean by the idea of replication, please refer to figure 20.03(*c*). With one planet only in operation this multiple gear set – looked at from the narrow, purely kinematic point of view – will operate satisfactorily. With three planets installed to share the load, however, we find (*a*) that only certain ratios of the tooth numbers may be employed, (*b*) that great accuracy must be attained to ensure an easy assembly in the presence of small clearances, and (*c*) that, due to the evident overconstraint, the planets do not in fact share the load at all, until a combination of clearances, wear, and elasticity modifies the difficulty.

15. Running through the scientific literature there is a diffuse thread of intellectual effort and thus of written comment on the matter of this *replication.* In somewhat the same vein as I have chosen this particular word, other writers have spoken for example of *duplication, mirror imaging, parallel looping and the like.* The reader might recall for examples here the classical, duplicated 4*R*-loop of the letter scales (Roberval's linkage), the sets of the cognate linkages of Roberts (imagined running together), the provision of flexible roots for the teeth of the toothed wheels in certain epicyclic gear trains, the construction of piano hinges, certain multiply operative stamping machines, and the multiplicity of balls and grooves (and thus the need for

loose-fitting cages) in the constant velocity couplings. I quote Davies [4] again, who has, most recently, vigorously tackled in a kinetostatic manner his own questions regarding *multiple parallel loops*.

**The advantages of overconstraint**

16. I wish to take up again the theme I put down at the end of § 20.01. Whereas a mechanism that exactly fits the Kutzbach criterion will most often be chosen wherever continuous functioning of the machine is required against a background of rigidity or permanent accidental damage among its links (§ 1.59), a mechanism that is overconstrained may well be the best choice in problems of machine design when great and variable loads must be sustained by means of mass and compliance, and when the maintenance of mechanical accuracy is important. As an example of the first alternative take the linkages in a flexible glider which transmit the inputs from the joy-stick to the corresponding outputs at the control surfaces in contact with the slipstream. Consider next however the need for an overconstrained yet mobile single-loop or multi-loop arrangement of links at the strong shoulder-joint of some futuristic industrial manipulator arm; refer to figure 19.10 and see there a mechanism that might be seen to be too underconstrained for such a purpose. Think too about machine tools. In machine tools we need accuracy; this is obtainable by careful machining of the machine tool. In machine tools we need a resistance to resonance with high frequency vibration; we get this by designing the links (the tables, the carriages etc.) to be large and thereby heavy. But also in machine tools we need flexibility. We need this to ensure the proper closure under gravity of the accurately cut prismatic joints (the slideways, the driveways etc.); the separate elements of these are integral with the massive links. It is therefore clear that the machine tool depends for its successful operation upon the flexibility of its links. Overconstraint in heavy machinery is very much more the rule than it is the exception; it is an advantage, often, not a disadvantage; and our theories of mechanism must be able to cope with this. Now it may be argued that the remainder of this chapter is irrelevant to the real problems of machining and the manufacturing industry. It may on the other hand be argued, however, that the machine tools, manipulators and robots of today are hopelessly ill developed mechanically in the light of what there is to follow.

**Connectivities within the $7R$- and the $6R$-loops**

17. To begin a study of all of the mobile simple loops of revolutes of apparent mobility unity one might look at photograph 1. Let us look at this 7-link, hinged, closed-loop linkage (the $7R$-loop) and think about its mobility. Let us do this, however, not by means of the already established general criterion (§ 1.34), but in terms of the $r$-lines and the $n$-lines that may be drawn (§ 3.43). Refer to figure 20.04. Notice there that the link 4 of the $7R$-loop has been broken into two pieces, the piece $4L$ at the left, and the piece $4R$ at the right.

18. Consider the open linkage 1–2–3–$4L$. A whole regulus of $n$-lines may be drawn to cut the three $r$-lines (the three hinge axes) that are evident. Any separated three of these are linearly independent of one another (§ 10.12); any three lines are enough to define a regulus (§ 11.04); in this context $n$-lines may be seen as prospective contact normals with imagined sliding button pairs (§ 7.19). So let us take any three of the $n$-lines, mount upon $4L$ and 1, respectively, three imagined prongs and three imagined facets (§ 7.20), and perceive that the three $n$-lines characterize the capacity for motion of the half-link $4L$ as having three degrees of freedom only, relative to 1.

19. Consider next the open linkage 1–7–6–5–$4R$. Here only two $n$-lines may be drawn to cut the four $r$-lines that are evident (§ 11.20). These two $n$-lines may be seen in the context as prospective contact normals with sliding button pairs. The two may thus be seen to characterize the capacity for motion of the half-link $4R$ as having four degrees of freedom only, relative again to 1.

20. If now the two half-links $4L$ and $4R$ are joined together, each will be constrained, not only by its own imaginary set of contact normals and prongs and facets, but by the other's set as well. It is thereby clear that in

Figure 20.04 (§ 20.17) A general $7R$ loop broken for the sake of argument at link 4. It is argued that this link 4 (or, indeed, any chosen one of the six movable links) enjoys only 1°F with respect to frame.

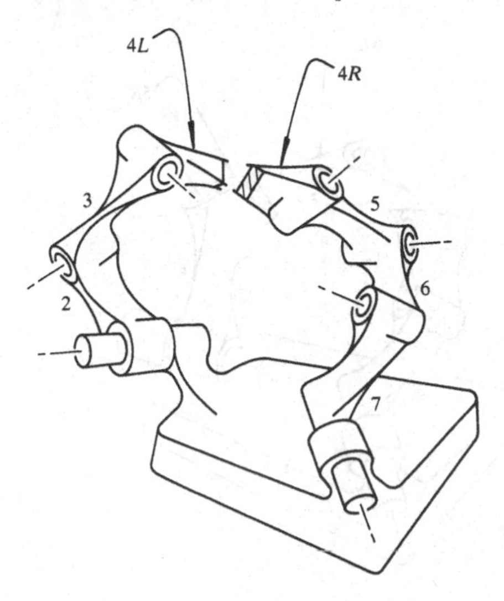

general the unbroken link 4 of the $7R$-loop is encumbered with *five* independent contact normals, and thus with five imagined sets of prongs and facets (§ 3.29). The five $n$-lines will characterize the capacity for motion of the reconstructed link 4 as having one degree of freedom only, with respect of course to 1.

21. Notice that we have found here and coincidentally a method for detecting the connectivity $C_{14}$ between a nominated pair, namely 1 and 4, of non-contacting links (§ 1.31, § 2.39); each link of the chosen pair is mechanically remote from the other. These ideas are derived from Waldron [1] and they relate to my remarks at § 20.07.

22. One may use a similar argument to show why it is in general that any link (say the link 5) of a $6R$-loop cannot move. Refer to figure 20.05. Look at the capacity for motion relative to frame 1 of the half-link $5L$. The open linkage on the left, connecting $5L$ to frame, gives (through the four hinge axes that are evident) only two contact normals with imagined sliding button pairs. Look at the capacity for motion relative to frame 1 of the half-link $5R$. The open linkage on the right, connecting $5R$ to frame, gives (through the two hinge axes that are evident) an hyperbolic congruence namely an $\infty^2$ of contact normals with imagined sliding button pairs; four of these are enough. We see that a reconstructed link 5 will be supported by *six* independent contact normals and imagined sliding button pairs. We see accordingly that the link 5 is (in general) left with zero degrees of freedom. It cannot move.

Figure 20.05 (§ 20.19) A general $6R$ loop broken for the sake of argument at link 5. It is argued that this link 5 (or, indeed, any chosen one of the five non-frame links) is, in general, immobile with respect to frame.

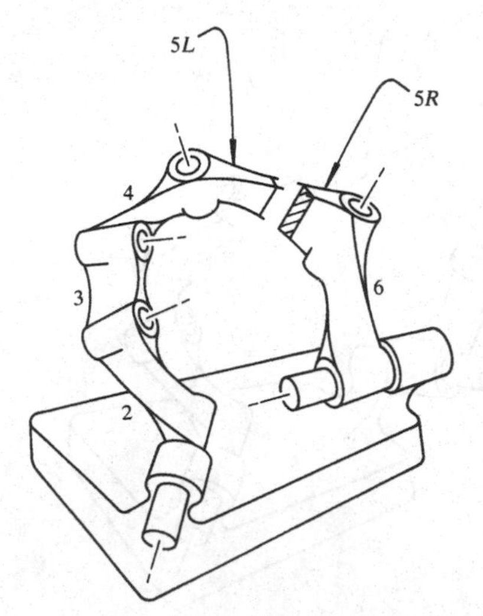


## A few special linkages reported in 1954

23. An interesting question arises now however. By arranging the $r$-lines and the $n$-lines in some particular way, might it be possible to render say a $6R$-loop transitorily mobile, or mobile, perhaps, throughout a full cycle of its motion? Refer to figure 20.06($a$). If for example all six hinge axes of a $6R$-loop are, in some particular configuration, constructed to intersect the same straight line, it will be clear from § 14.35 and from the principles of reciprocity that the loop will be transitorily mobile. The loop will have, at that configuration at least, an apparent mobility $M_a$ of unity. As motion occurs, however, the configuration will change (in general) in such a way that the hinge axes no longer cut the one straight line. The mechanism will, accordingly, not function on either side of the particular configuration – provided, of course, clearances at the joints and flexibilities of the links are absent, the presence of either of which would permit the phenomenon of sogginess to be apparent (§ 2.46, § 20.08). If for another example the hinge axes cutting the line in figure 20.06($a$) are reconstructed to be grouped in such a way that two groups of three successive hinge axes each intersect at a common point, and this is shown in figure 20.06($b$), we shall find by experiment that the linkage (which is overconstrained of course) is mobile throughout the full cycle of its motion. Refer in this connection to the apparatus in figure 2.01($e$). Such facts as these were no doubt known to Altmann [9] who reported in 1954 his workable linkage shown in figure 20.07($a$). I should mention that figure 20.07($a$) is

Figure 20.06 (§ 20.23) Two simple examples of $6R$ loops which are, by virtue of special geometrical conditions at their shown configurations, transitorily mobile. Whether these loops are continuously mobile throughout whole cycles of their motion is of course another question (§ 20.32, § 20.55).

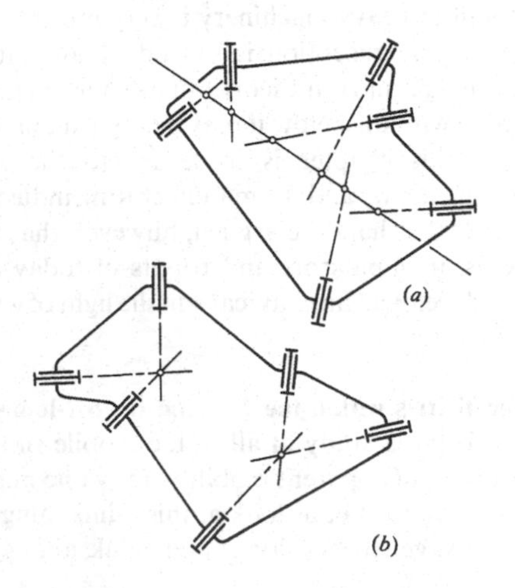

my drawing adapted from Altmann's photograph in [9] of his real apparatus, and that figure 20.07(*b*), also after Altmann, might well be taken here as a set of instructions for making a paper model of the kinematic chain. Altmann quotes Bricard (1927) and reports that, for one full revolution of the input crank at 2, the link 4 makes one full revolution about its own length. The reader might check this, either by eye if he can, or by means of his paper model. Other special spatial 4, 5 and 6-link loops mobile thoughout the full cycles of their motion had been found by other workers too, and a few of these were collected and reported by Altmann in his [9]. Altmann writes with a measure of wonder (or does he play the magician?) as he writes about such paradoxical linkages in his [9]. If however we study the theorem italicized at § 20.09, change the words *screw joint* to *revolute* throughout, and apply the resulting theorem here, we find that the Altmann linkage of the figures 20.07 – it does accord with the resulting theorem – is not such an impenetrable mystery. Bricard explains in volume 2 of his book (1927) that, before any 6*R*-loop can be mobile throughout a full cycle of its motion, its hinge axes (at every configuration of the loop) must reside as the members of some linear complex. All hinge axes of each of the famous 6*R*-loops of Bricard remain members of the same (though changing) linear complex of lines as the mechanisms move (§ 20.29); the hinge axes remain, sometimes, as members of the same special linear complex (§ 9.41); and the mechanisms are mobile throughout the full cycles of their motion. Refer in advance to § 20.29 *et seq.*, § 20.40 *et seq.* and § 20.45 *et seq.*, where these and a few of the other known mobile 6*R*-loops are discussed in greater detail. We find there, among other things, (*a*) that the Altmann linkage of [9] turns out to be a special case of the Bricard line-symmetric 6*R*-loop, and (*b*) that the group of linkages suggested by figure 20.06(*b*), the so-called double Hooke linkages, were not (despite their evident antiquity) a part of the repertoire of Bricard.

**Mobilities of the 6*R*-, 5*R*-, 4*R*-, and 3*R*-loops**

24. I would like to describe now one way of demonstrating that the lines of the six hinge axes of the 6*R*-loop must be members of the same linear complex of lines before the apparent mobility of the loop can be (at the instant) unity. Refer to figure 20.08. Break the sixth link 6 into 6*L* and 6*R* as shown and consider separately the possibilities for motion of the half-links 6*L* and 6*R*. The half-link 6*R* has only one possibility for movement; it can only rotate about the hinge at *F*; there is nothing else that it can do. The link 6 (namely 6*L* with 6*R* attached) can rotate as a whole about the hinge at *F* only if the portion 6*L* can also rotate about *F*. The half-link 6*L*

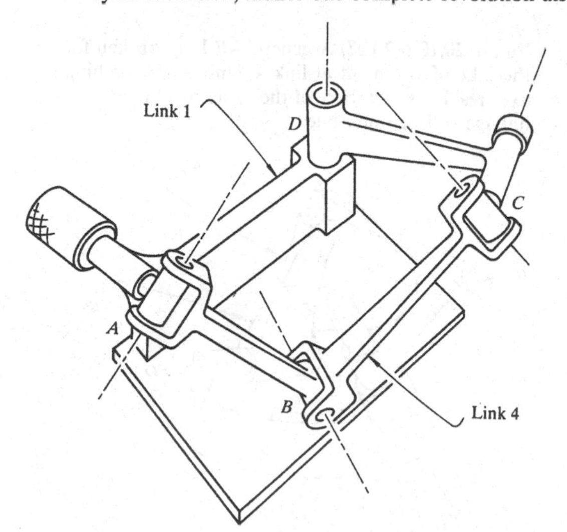


Figure 20.07(*a*) (§ 20.23) With minor modifications this figure has been taken by me directly from Altmann's photograph of his working model of what I have called here Altmann's linkage, overconstrained yet mobile throughout a cycle. Please notice that, although horizontal, the shaft 2 of the knurled knob at A is not parallel with the front edge of the rectilinear base plate. Become aquainted with the curious fact that, for each one revolution of this knob 2, the link 4, in making its corresponding cycle of motion, makes one complete revolution also.

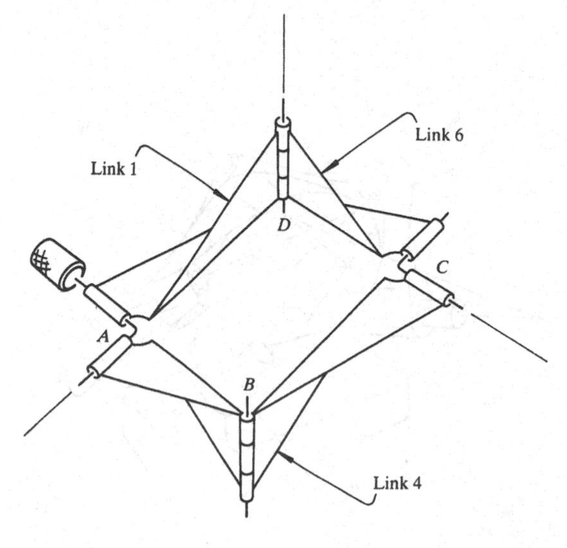


Figure 20.07(*b*) (§ 20.23) The geometrical essence of Altmann's linkage shown in figure 20.07(*a*). From this illustration a paper model of hinged-together tetrahedra may be made. Such a model will help to clarify the fact that, at a certain initial angle of say zero at the shaft at *A* (and again at 180° of that angle), the skew polygon *ABCD* becomes a flat rectangle.

can do this only if $F$ is a member of any one of the following congruences: $ABCD$ (in which case $E$ remains immobile), $ACDE$ (in which case $B$ remains immobile), $ABDE$ (in which case etc.), $ABCE$, $ABDE$ (...immobile). It can do this, in other words, if, and only if, the hinge axis at $F$ is a member of the complex defined by the five hinge axes at $A$, $B$, $C$, $D$, and $E$. That complex contains, of course, the five mentioned congruences. We have shown here that, *unless any sixth hinge axis of a 6R-loop is a member of the complex defined by the other five, the linkage will be immobile.* Refer to § 11.30.

Figure 20.08 (§ 20.24) A general 6*R* loop broken for the sake of argument at link 6. Unless all six hinge axes reside as members of the same linear complex the linkage will be immobile.

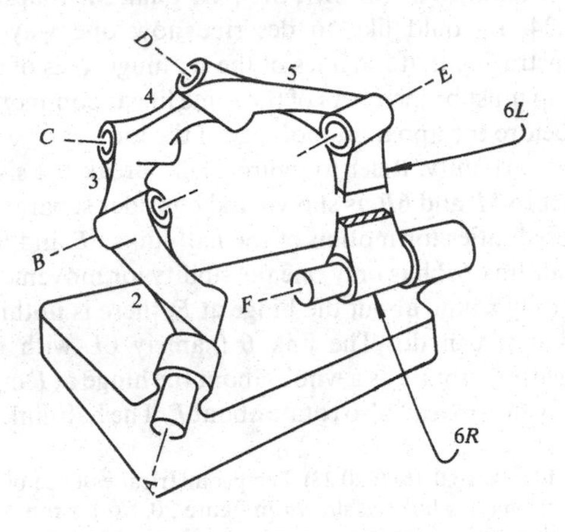


Figure 20.09 (§ 20.26) A general 5*R* loop broken for the sake of argument at link 5. Unless all five hinge axes reside as members of the same linear congruence the linkage will be immobile.

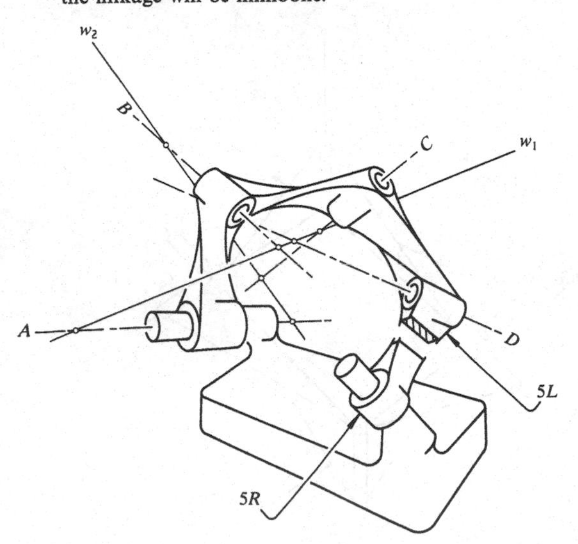


25. There are of course other ways of showing this. The recent literature is full of them. Read for example Hunt (1978). But the important thing to notice is that this argument – as with all the current others – relates to transient mobility only. There is no guarantee that, once established by means of a mechanical construction at a certain configuration, the conditions mentioned at § 22.24 will continue to prevail as the constructed mechanism begins to move. Necessary and sufficient conditions for the presence of *full-cycle* apparent mobility unity in the 6*R*-loop have not been discovered yet; they are, at the time of writing, unknown. Refer to further information about the mobile 6*R*-loop to passages from § 20.29 to § 20.48 inclusive.

26. I wish to take now the 5*R*-loop and examine the possibilities for its being mobile. Refer to figure 20.09. The four hinge axes of the open linkage connecting the half-link 5*L* to frame have only two transversals $w_1$ and $w_2$ (§ 11.20). These are the directrices of a hyperbolic congruence of *r*-lines, each one of which is an equivalent hinge about which the half-link 5*L* can rotate at the instant relative to frame. The four real hinge axes are of course themselves members of this congruence of *r*-lines, for they, indeed, define the congruence. The half-link 5*R* on the other hand has only one possibility for motion: it must rotate about the fixed hinge axis at *E*. Now if 5*L* and 5*R* were reconstructed into 5, the unbroken 5 would only be able to move if the hinge axis at *E* were a member of the mentioned congruence. Refer to Reynolds [10]. It follows that, *unless any fifth hinge axis of a 5R-loop is a member of the congruence defined by the other four, the linkage will be immobile.* Refer to § 20.49.

27. Take now the 4*R*-loop shown in figure 20.10. I have put this on a pedestal to emphasize our admiration

Figure 20.10 (§ 20.27) A general 4*R* loop broken for the sake of argument at link 4. Unless all four hinge axes reside as members of the same regulus the linkage will be immobile.

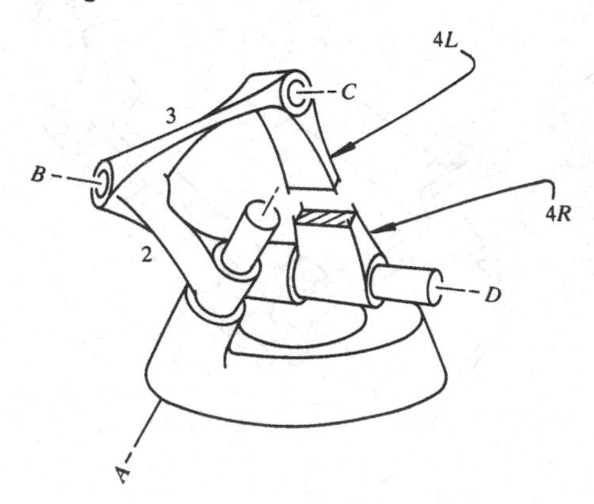

of it; but do we admire it for its simplicity and utility, or do we hold the simplistic view that it is, in a way, friendly, transparently easy to analyze? The hinge axes at $A$, $B$, and $C$ define a regulus of $r$-lines. If a reconstructed link 4 is to be able to move, the hinge axis at $D$ must belong to that same regulus (§ 10.57). It follows that, *unless any fourth hinge axis of a 4R-loop is a member of the regulus defined by the other three, the linkage will be immobile.* Refer to § 20.52.

28. I leave the 3$R$-loop as an exercise for the reader. There is no possibility with the 3$R$-loop (except for the occurrence in special cases of 'part chain' mobility) that full cycle apparent mobility unity might obtain. It should be possible to show however that, *unless all three revolute axes of the 3R-loop intersect at a common point or are parallel (or coaxial) and are coplanar, the linkage will be immobile.* Refer to § 20.54.

**The mobile 6$R$-loops of Bricard**

29. I shall mention here the five distinct kinds of mobile 6$R$-loops discovered and reported by Bricard (1927). The initial results were achieved by him during an earlier period beginning 1897. I refer to a recent algebraic analysis of these linkages made by Baker [11], and I use Baker's terms for the names of the five kinds. They follow: (1) the general line-symmetric loops; (2) the general plane-symmetric loops; (3) the trihedral loops; (4) the line-symmetric and the plane-symmetric octahedral loops, the former of which are a special case of (1); and (5) the doubly collapsible octahedral loops. Refer to figure 20.11. Baker numbers the joints, not the links, of his loops (§ 12.08, § 19.52). Common perpendiculars between the revolute axes, namely the fixed link lengths, he calls $a_{ij}$; the twist angles between the axes of the revolutes of the links, namely the fixed skew angles of the links, he calls $\alpha_{ij}$; the offsets of the links, not variable but fixed at each joint as shown because the joints are all revolutes (§ 19.52), he calls $R_i$; and the variable transmission angles, namely the variable joint-rotation angles between the link lengths, which occur of course at the joints, he calls $\theta_i$. All five kinds of the Bricard loop can be delineated by means of relationships between the fixed parameters $a_{ij}$, $\alpha_{ij}$, and $R_i$, all illustrated in figure 20.11. The reader is advised not simply to read the following descriptions of the linkages but somehow to make working models of them – refer to chapter 3 of Hartenberg and Denavit (1964) for models in general, to Baker [11] for metal and wooden models, and to Goldberg [17] and Baker [18] for wooden and paper ones. Whereas for the first three kinds of loop the link lengths $a_{ij}$ are in general finite non-zero, they are for the fourth and fifth kinds all necessarily zero (§ 20.34).

30. In the first kind (1), the general 6$R$ line-symmetric loop, the parameters are related as follows: $a_{12} = a_{45}, a_{23} = a_{56}, a_{34} = a_{61}; \alpha_{12} = \alpha_{45}, \alpha_{23} = \alpha_{56}, \alpha_{34} = \alpha_{61}; R_1 = R_4, R_2 = R_5, R_3 = R_6$. Refer to figure 20.12. This shows a general line-symmetric 6$R$ loop in one of its configurations and the line about which the linkage (in its present configuration) is symmetrical. Please be reminded here that the linkage of the Altmann device illustrated in figure 20.07 is a special case of this first kind. It is a special line-symmetric Bricard loop where, among other things, $a_{12} = a_{45} = 0$. But refer in advance to § 20.34 where all of the link lengths of the line-symmetric Bricard loop are zero. Refer also to my notes about the mounted double Hooke joint at § 20.55 and be aware that that loop is not a line-symmetric Bricard loop. It is, indeed, not any one of the Bricard loops; see figure 20.06($b$) and its discussion at § 20.23. The Sarrus loop, a special case of the double Hooke joint loop, is not one of the Bricard loops either; refer to § 20.48.

31. In the second kind (2), the general 6$R$ plane-symmetric loop, the parameters are related as follows: $a_{61} = a_{12}, a_{56} = a_{23}, a_{45} = a_{34}; \alpha_{61} + \alpha_{12} = \pi, \alpha_{56} + \alpha_{23} = \pi, \alpha_{45} + \alpha_{34} = \pi; R_1 = R_4 = 0, R_6 = R_2, R_5 = R_3$. Refer to figure 20.13. This shows a general plane-symmetric 6$R$ loop in one of its configurations and the plane about which the linkage is symmetrical. That the

Figure 20.11 (§ 20.29) These are the parameters employed by Baker for his algebraic analyses of overconstrained loops of revolutes. I have displayed here by means of examples not only the parameters themselves but also the sign conventions adopted. Notice that it is not the links but the joints that are consecutively numbered by Baker and that, because each link length is measured along the common perpendicular between its relevant pair of joint axes, offsets $R$ appear.

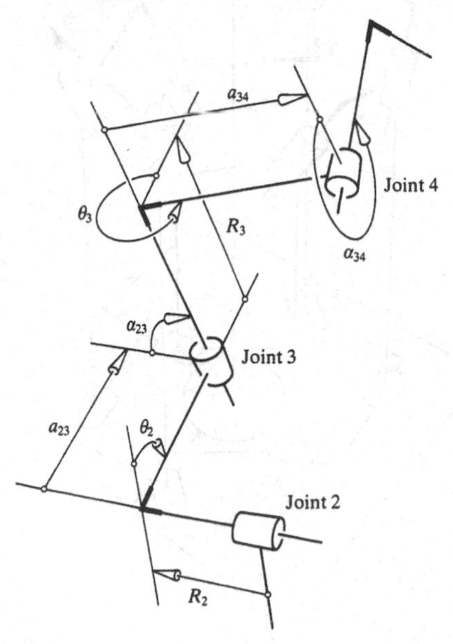

pairs of equal and opposite twist angles $\alpha$ add to $\pi$ is not an indication that the twist angles themselves are each $\pi/2$; it is an indication that the pairs of opposite angles are equal in magnitude but of opposite hand; please refer to the sign conventions illustrated in figure 20.11.

32. In the third kind (3), the trihedral 6*R* loop, the parameters are related as follows: $a_{12}^2 + a_{34}^2 + a_{56}^2 = a_{23}^2 + a_{45}^2 + a_{61}^2$; $\alpha_{12} = \alpha_{34} = \alpha_{56} = \pi/2$, $\alpha_{23} = \alpha_{45} = \alpha_{61} = 3\pi/2$; $R_1$, $R_2$, etc. are all zero. In this third kind we find that, throughout each whole cycle of movement, each of the two sets of three alternately taken joint axes is and remains a set of concurrent lines; refer to figure 20.14 and read the caption there. Please recall my remarks at § 20.23, refer again to figure 20.06(*b*), and be aware that the linkage there is not a trihedral loop (§ 20.55). I am unable to offer a direct example in engineering practice of the trihedral loop, but I can refer in advance to the top part of figure 20.20. There I have drawn a very special trihedral loop. In this special loop the links are of equal length, the twist angles are all 90° exactly, and the loop is shown in that particular configuration where, with all transmission angles 90°, the six joint axes are upon the edges of a cube.

33. Before I extract, again directly from Baker [11], the relations between the fixed parameters for the fourth and the fifth kinds of Bricard loop, the octahedral ones, I shall need to explain what an octahedron is in the present context. We are speaking here, not about the well known convex octahedra made up of eight outwardly facing triangles (like the octahedra of Euclid), but about the so-called *concave* octahedra. A description of a concave octahedron follows. Refer to figure 20.15. Set up as shown the two triangles *ABC* and *DEF* in space, perceiving them to be two rigid triangular plates. By means of hinges along the edges of the plates, attach four more triangular plates as follows: *EDA*, *ABD*, *BCF*, *FEC*. Thus we locate six of the eight faces. The remaining two faces of the concave octahedron are the pair of rigid triangles *ACE* and *DBF*. Provided we see

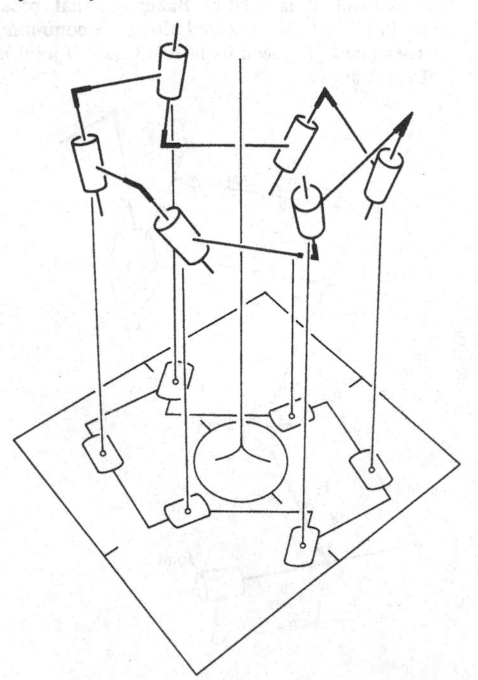

Figure 20.12 (§ 20.30) The general line-symmetric 6*R* loop of Bricard. Each point, joint or link of the loop is line-reflected in the line of symmetry. I have not only (*a*) drawn the linkage itself in space at the top of the figure, but also (*b*) projected it from there downwards upon a reference plane set normal to the line of symmetry. The seen shape of the projection should help to make clear the nature of line symmetry.

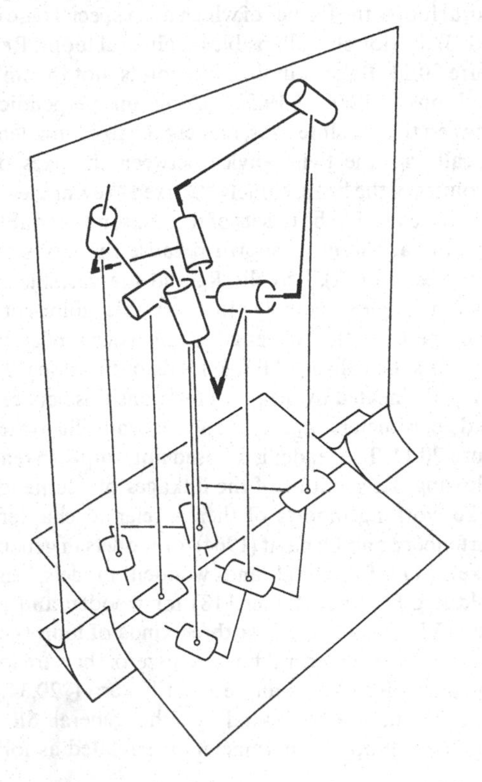

Figure 20.13 (§ 20.31) The general plane-symmetric 6*R* loop of Bricard. Each point, joint or link of the loop is plane-reflected (namely has its ordinary mirror-image) in the plane of symmetry. I have not only (*a*) drawn the linkage itself in open space at the top of the figure, but also (*b*) projected it from there downwards upon a reference plane set perpendicular to the plane of symmetry. The seen shape of the projection should help to make clear the nature of plane symmetry.

that all eight rigid triangular faces are interpenetrable (§ 1.06, § 1.07), we shall see that a concave octahedron may, under certain circumstances, be *deformable.* What this means is that all twelve hinges at the twelve edges of the concave octahedron may, under certain circumstances, be able to operate not only transitorily but throughout a whole cycle of movement of the octahedral linkage. It should be possible to see next that only six of the eight faces need to be actually occupied by the material of a real link for the possible motion (the deformation) of the octahedron to occur.

34. The fourth kind of the Bricard linkages (4) includes two different sub-kinds. They are *(a)* the line-symmetric, and *(b)* the plane-symmetric octahedral loops. Now it should be clear from § 20.34 and from the figure 20.15 (which figure is, indeed, a perspective view of a line-symmetric 6-plate octahedral loop, looking however otherwise than along the line of its symmetry) that all of the link lengths $a_{ij}$ must be zero in the octahedral loops. In his [11] Baker goes on to agree with Bricard that a special case, namely *(a)* above, of the general line-symmetric loop occurs when $a_{ij}$ are all zero, whereupon (and with a newly introduced sign convention) $R_1 + R_4 = R_2 + R_5 = R_3 + R_6 = 0$. With regard to *(b)* the plane-symmetric octahedral loops, Baker explains that, although these are derived from plane-symmetric octahedra, they (all sets of the six real links of the loops) are not in themselves plane-symmetric. The necessary relationships between the fixed parameters $\alpha_{ij}$ and $R_i$ are more complicated for these latter loops and I refer the reader for them directly to Baker [11].

35. In general the loops of § 20.34 do not collapse at any one of their configurations into a plane; but in the special cases when they do they do so only at one configuration in the cycle. The loops of the fifth kind (5), on the other hand, the doubly collapsible octahedral loops, always become coplanar in two separate configurations of their respective cycles. Refer to Baker [11].

**Mobile 6-link loops and the smoke ring**

36. Before mentioning some of the other known mobile 6*R*-loops, all of which are in general asymmetrical (§ 20.40 *et seq.*, § 20.45 *et seq.*), I wish to return now to the Altmann linkage. It is illustrated in figure 20.07 and already introduced at § 20.23. We perceive this linkage now to be a Bricard loop of the first kind (1), a line-symmetric loop, but with special features. Using the Baker-terminology and taking the joint at the input link as joint number 1, we see that the link lengths $a_{12}$ and $a_{45}$ are not simply finite and equal, but both zero. We see that the twist (or the skew) angles $\alpha_{ij}$ are all either $\pm \pi/2$, and that the offsets $R_i$ are all zero. The lines of the link lengths $a_{ij}$ intersect at the four points $A, B, C,$ and $D$

Figure 20.14 (§ 20.32) The general trihedral 6*R* loop of Bricard. This linkage may be seen to be drawn upon the edges of a hexahedron of plane quadrilateral faces, each quadrilateral changing of course but having two opposite angles that remain right angles. Alternately taken joint axes, taken three at a time, intersect as shown in the two points at the otherwise unoccupied apices of the said hexahedron. There is in general no line or plane of symmetry and the six link lengths are all different. See the special case at figure 20.20 where the hexahedron is a cube.

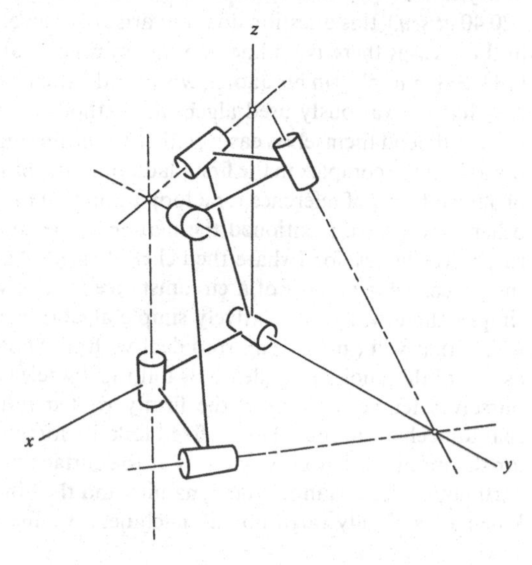


Figure 20.15 (§ 20.33) An octahedral 6-plate loop of Bricard. In this linkage all of the link lengths are zero and all of the offsets are non-zero. Although actual machinery is drawn here with real plates existing, the reader should be able to see that such plates should be able to interpenetrate. Refer to § 1.06, § 1.07 and § 20.35, and consider designing for full-cycle movement, namely for no obstructions due to clashing at a pair of stop surfaces (§ 1.29).

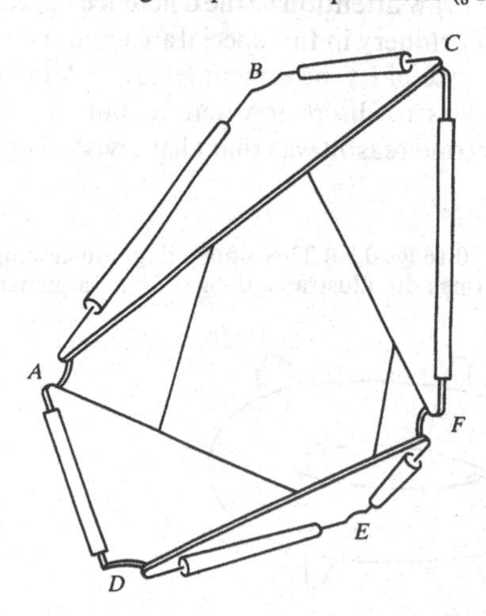

accordingly. These points, which are indicated in figure 20.07, are at the 'centres' of the relevant revolutes. Using the Baker terminology the joints 1 and 2 are at $A$, 3 is at $B$, 4 and 5 are at $C$, and 6 is at $D$. These particularities are illustrated in the Altmann communication [9] by means of a carefully drawn kinematic chain; this chain reappears as my illustration drawn for the sake of a paper model in figure 20.07(*b*). If the reader will make (from the mentioned figure) his own paper model he will see the following: firstly (*a*) that the model will need to be distorted once in a cycle to overcome a minor difficulty with the clashing of links (§ 1.07); and secondly (*b*) that, as the mechanism is driven at its input link, the deformable skew quadrilateral $ABCD$ remains a quadrilateral with its two pairs of opposite sides remaining equal, but twice in a cycle (one cycle occurring for each revolution at the input link) the quadrilateral collapses to a horizontal straight line segment along $AD$ and expands into a rectangle that lies in the horizontal plane of $AD$. Please erect along the fixed line segment $A$–$D$ your own origin-line for the measurement of the input angle $\theta$, and take it that that angle is set in figure 20.07(*a*) at about 30°. As the first collapsion from the rectangle to the line takes place (from $\theta = 0$ to $\theta = \pi/2$), the diagonal $A$–$C$ remains uppermost, while, as the second collapsion takes place (from $\theta = \pi$ to $\theta = 3\pi/2$), the diagonal $B$–$D$ remains uppermost. Let me point out here, as Altmann did, that the link 4, the 'coupler' link of this somewhat extraordinary 'crank-rocker' mechanism *rolls over upon its own axis one full revolution for each revolution of the input link*. As in many other mechanisms made up of revolutes, not all of the revolutes make a complete revolution. Some just oscillate (§ 2.42), the relative movement between the connected links being less than 360°.

37. I have carefully described the cycle of movement of this particular mechanism for two main reasons. First I wish to draw attention to the difference between a piece of real machinery in this special area and the mere geometric essence of its movement (§ 0.01) – Altmann's contribution was to discover where to put the metal (§ 1.07). My second reason was that I have wished by this

Figure 20.16 (§ 20.36) This smoke ring, functioning as smoke rings do, illustrates the gist of my argument at § 20.36.

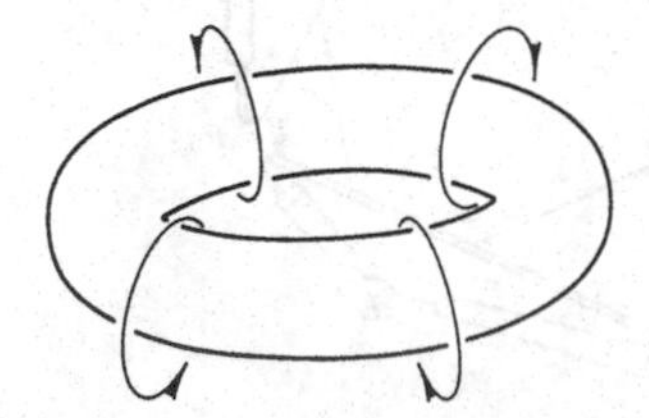

means to introduce the idea of the *smoke ring*. Please refer to figure 20.16.

38. I ask the reader now to take the Altmann linkage in his hands – not the actual machinery of figure 20.07, but the *linkage* (§ 1.10). Please hold link 1 in the left hand and link 4 in the right. Manipulating the loop as though it were a smoke ring now, and keeping the axis of the complex (wherever that may be) fixed in space, turn each of the two mentioned links through one half-revolution. Turn the loop only once, in other words, inside out. *Notice next that this single manoeuvre, executed with no link fixed (that is with none of the six possible inversions extant), has put the linkage through one whole cycle of its motion.*

39. I wish to discuss here, briefly, some possible consequences of this idea. I restrict myself in the following comments to the 6$R$-loops or to those 4-link and 5-link loops that may be expressed (by means of superimposed or intersecting revolutes) as 6$R$-loops. In view of the fact that the different inversions of a linkage are of little interest to those investigating the question of whether or not an overconstrained linkage is mobile [12], and in view of the fact that the algebraic closure equations written by the relevant workers do not need to be expressed explicitly in terms of input-output relations because the authors have no immediate interest in the various inversions of their loops [13], why not erect for analysis each line-symmetric loop about its fixed axis of the complex in this suggested way, and be aware that, as the loop is progressively turned like a smoke ring inside out, all revolute axes will at any one instant belong to the same linear complex? I am obliged to remark (in answer to this) that unfortunately the axis of the relevant complex is not collinear with the line of the line-symmetry [14]. In the plane-symmetric or the nonsymmetric 6$R$-loops (these latter are discussed at § 20.40 *et seq.*), this question does not arise of course, for in these cases there is no line of a line-symmetry at all [14]. Baker implies in his quoted work, and Yu shows in his, that the variously used algebraic methods do not appear to lend themselves easily, either to the finding of the axis of the complex in the first place, or to the basing of a fixed frame of reference for a loop upon that or any other axis. I have mentioned the matter of the smoke ring nevertheless, for I share the belief of many that, if the perceived geometry of a circumstance is relatively simple, there will be a relatively simple algebra somewhere that will (in the long run) deal with it. Another aspect of the smoke ring idea is as follows: by releasing ourselves temporarily from the firmly rooted notion that a mechanism must have a fixed link, by forgetting for the moment that gravity exists at the surface of the earth and is important in the dynamics and thus in the kinematics of only earth-bound machinery, we may be

able to make new gains in our understanding of mechanism. We may be able to pay a closer attention, for example, not to the gross motions of moving links in a fixed (terrestrial) space, *but to the relative movements between neighbouring links at joints.* Refer to recent works by Waldron and Baker on this important matter [15].

### The Goldberg mobile 6*R*-loops

40. Before beginning to understand the generally asymmetrical overconstrained yet mobile 6*R*-loops of Goldberg, one must be familiar with the most famous and mysterious overconstrained yet mobile loop of them all, the 4*R*-loop of Bennett. Otherwise known as the Bennett mechanism (or simply as the Bennett), this special loop was reported by its discoverer in 1903 [16]. First mentioned at the end of § 2.19 and illustrated pictorially in figure 2.01(*d*), the Bennett loop is discussed in detail at § 20.53. Please begin to read about the Bennett now.

41. Goldberg [17] presented his linkages in 1943. For his 6*R*-loops he took, both in his mind's eye and in the form of physical models, three selected Bennett loops. He first attached them in series 'back to back' and produced, in a manner to be shown, a mobile 6*R*-loop (§ 20.42). He 'syncopated' the result of this to produce another mobile 6*R*-loop (§ 20.43). He then took another three selected Bennett loops, and, attaching them together in his 'L-shaped' manner, produced another mobile 6*R*-loop (§ 20.44). This latter he syncopated also.

42. Refer to figure 20.17(*a*). This shows three Bennett loops attached in the Goldberg manner in series back to back. From § 20.53 we know that the first pair of opposite links $(a, \alpha)$ of a Bennett loop are identical, and we see in figure 20.17(*a*) that the three Bennett loops there all have their first pair of opposite links $(a, \alpha)$ equal. Each pair of identical links placed back to back is superimposed and *fused* (my word) into a single link as shown. The three Bennett loops, thus attached in series, are each (in the usual Bennett manner) mobile. It follows that either one of the two sets of three links mounted end to end, $(b, \beta)$, $(c, \gamma)$, and $(d, \delta)$, may be moved to become collinear; notice that such a configuration has been adopted in figure 20.17(*b*). In figure 20.17(*b*) we may now (*a*) make rigid or *freeze* (my word) the said straight line of three links into a new single link of length $b + c + d$ and twist angle $\beta + \gamma + \delta$, and (*b*) remove the two pairs of fused links $(a, \alpha)$ which are, at this stage of the argument, redundant. The two joints marked with the symbol O accordingly disappear. Thus we arrive in figure 20.17(*c*) at the first of the Goldberg mobile

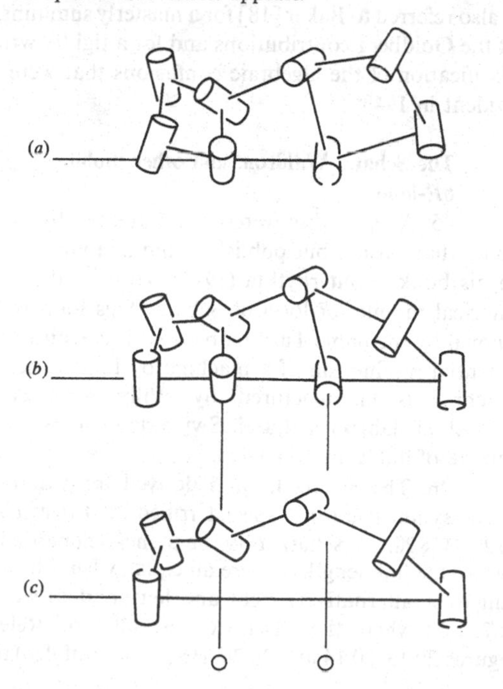


Figure 20.17 (§ 20.42) Formation of the Goldberg 6*R* series loop with zero offsets. At (*a*) three Bennett linkages are fused back to back in series, at (*b*) the three links put collinear as shown are ready for being frozen there, and at (*c*), after the two joints marked with the symbol O are omitted, the final mobile 6*R* loop with zero offsets appears.

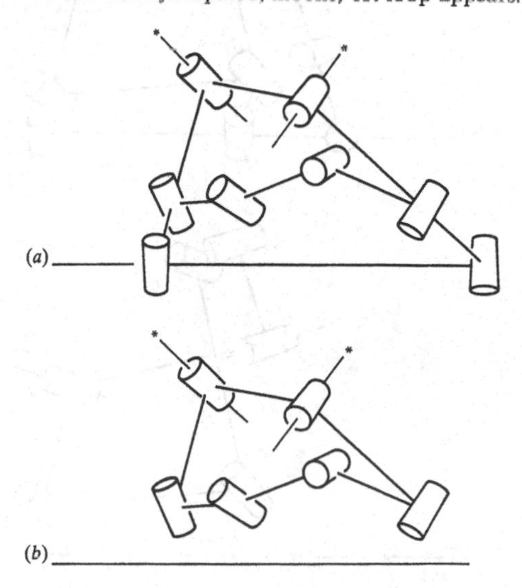


Figure 20.18 (§ 20.43) Formation of a Goldberg 6*R* series syncopated loop with zero offsets. At (*a*) the mobile 6*R* loop of figure 20.17(*c*) appears with its equal extended input and output links proportioned to complete the surrounding Bennett loop. The joints marked with the symbol * are the new joints introduced to complete this surrounding loop. At (*b*) the lower part of the linkage at (*a*) has been omitted and the said syncopated, mobile, 6*R* loop appears.

6*R*-loops. Notice that, in somewhat the same manner as the Altmann linkage did (§ 20.36), this Goldberg linkage passes through two different configurations in a cycle where all six link-lengths are collinear.

43. Refer to figure 20.18(*a*). This shows the mobile 6*R*-loop of the above paragraph with its single pair of identical links ($a,\alpha$) extended in length and altered in a manner consistent with the fact that was shown by Goldberg that the input-output relationship of this 6*R*-loop (with its long link fixed at the bottom of the figure) is the same as that of a Bennett mechanism with the same long link. He calculates, in other words, the necessary length and twist angle ($e,\varepsilon$) of the identical extended links, and introduces at the top of the figure, to join the tops of the extended links, a new link identical with the long link at the bottom. The two new joints marked with an asterisk accordingly appear. Thus the surrounding 4*R*-loop becomes another Bennett loop, and Goldberg argues next that the newly formed loop, shown separately in figure 20.18(*b*), is another mobile 6*R*-loop. He has thus, and in his words, syncopated the loop of § 20.42. Although De Jonge, in his long communication to [17], puts up an argument that the thus syncopated loop is not a 'new' loop, the possibility of the syncopation is itself a contribution to knowledge.

Figure 20.19 (§ 20.44) Formation of a Goldberg 6*R* L-shaped loop with zero offsets and another syncopated 6*R* loop. At (*a*) three Bennett linkages have been fused into an L-shaped arrangement and the two sets of links that are shown collinear have been drawn by me in the plane of the paper. First by freezing these latter and then by omitting the redundant joints, we can produce (as we did in figure 20.17) a Goldberg 6*R* L-shaped loop with zero offsets. Next, by inserting the two new joints marked with the symbol *, we can complete the surrounding Bennett loop in much the same way as we did in figure 20.18. At (*b*), by omitting a lower part of the linkage shown at (*a*), we are left with a syncopation, another mobile 6*R* loop with zero offsets.

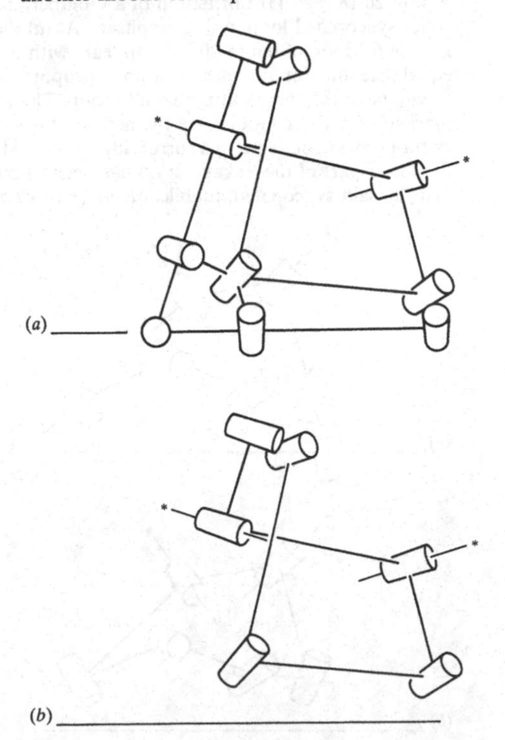


44. Refer to figure 20.19(*a*). Here I have fused together three Bennett loops into the above mentioned L-shaped arrangement. I have frozen in line each of the two pairs of links that form the back and the bottom of the mobile L and I have chosen for the sake of clarity to locate these links in the plane of the paper. Due to this latter the single revolute at the lower left appears as a circle. Next the surrounding Bennett loop may be introduced in the same way as it was in § 20.43, the new joints marked with the asterisk accordingly appear, and we see in figure 20.19(*a*) that the new surrounding linkage and thus the whole multi-looped conglomeration (with its rigid back and bottom) is overconstrained yet mobile. A next step that may be taken is illustrated in figure 20.19(*b*). It will be clear I hope that the loop remaining there is another mobile 6*R*-loop. These figures exemplify the addition and, as Goldberg also explains, the subtraction of Bennett loops by means of the L-shaped arrangement. In such ways the other mobile 6*R*-loops of Goldberg (as well as their several syncopations) are obtainable. Refer to [17]. The reader is also referred to Baker [18] for a masterly summing-up of the Goldberg contributions and for a tightly written clarification of the algebraic confusions that were still evident in 1943.

### The Schatz, Waldron, and other mobile 6*R*-loops

45. Already discovered and patented by Schatz somewhat earlier, but published and analysed by him in his book about rhythm (1975), we have the asymmetrical mobile 6*R*-loop of Schatz. This loop is also known by the name Turbula because it constitutes the essential mechanism of a machine by that name. The machine is manufactured by Willy A. Bachofen (Maschinenfabrik) in Basel, Switzerland. It is for the mixing of fluids and powders.

46. The Schatz loop is derived from a special plane-symmetric (or a special trihedral) Bricard loop (§ 20.31, § 20.32). Schatz took the plane-symmetric loop where the link-lengths $a$ were all equal, where the twist-angles $\alpha$ (alternatively right and left handed) were all $\pi/2$, and where the offsets $R$ were all zero. Refer to figures 20.13, 20.14 and 20.20. He saw it fruitful (*a*) to set

all $\theta$ at $\pi/2$, (*b*) to slice the said linkage at its plane of symmetry as shown, (*c*) to remove one half of the thus-cut linkage while keeping the other (namely the links 3, 4, and 5 as shown), (*d*) to insert a new link 1 of twist-angle zero and a new pair of parallel shafts 2 and 6 (two links of zero length) as shown, and thus (*e*) to arrive at a new asymmetrical mobile 6*R*-loop which is characterized by one dimension only. This linkage has been studied by Brat [19], by Baker and his students [20], by Wohlhart and his student Blaimschein [21], and by Yu [14]. Yu has shown how to find the axis of the moving and changing complex that contains at all times the six revolute axes of the Schatz linkage (§ 20.39). It is interesting to be aware that, if the Schatz linkage of figure 20.20 is driven at link 2 through a Hooke joint set at 45°, the overall cyclic motion is symmetrical, cyclic fluctuations at the output link 6 being then the same as at the input link 2. Refer to photograph 21. The Turbula machine itself however, for various practical reasons, is driven otherwise.

Figure 20.20 (§ 20.46) Formation of the Schatz linkage. Take the special plane-symmetric and/or trihedral Bricard 6*R* loop which is mentioned at the end of § 20.32. Keeping its plane of symmetry in the vertical plane, put the linkage into its mentioned cubical configuration and set the diagonal of the cube upon the horizontal line *a–a*. Next drop the front half of the linkage consisting of the links 3, 4 and 5 down to the new horizontal line *b–b* and connect the new links 1, 2 and 6 as shown. The centre lines of the shafts 2 and 6 are in the horizontal plane containing *b–b* and, given that the edges of the cube are of unit length, the distance along *b–b* from shaft 2 to shaft 6 is the square root of 3.

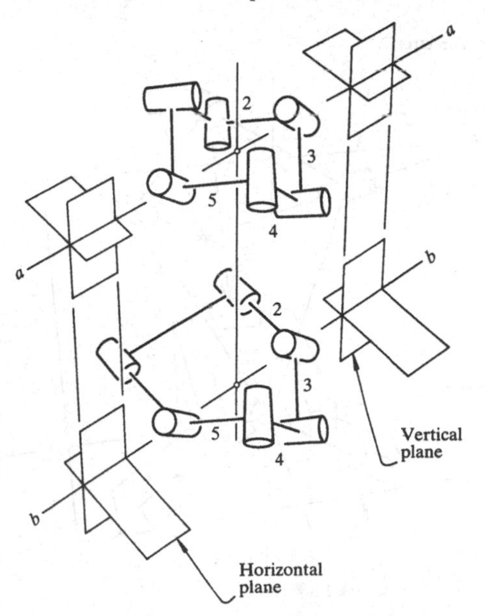


47. Another set of generally asymmetrical 6*R* loops is due to Waldron [22]. In [22] Waldron draws attention to a class of mobile 6-link loops employing only lower *f* 1 joints namely only helicals, prismatics and revolutes (§ 7.12). He calls the loops hybrid loops because he makes them up by combining in certain ways other 3, 4, and 5-link loops. Let me describe in detail here only one of these hybrid loops; it is a 6*R* and prominently displayed in [22] as an example. Refer to figure 20.21(*a*). Set up any two Bennetts in space in such a way that the axes of two of the revolutes, one axis from each of the Bennetts, are collinear. The points of intersection of the link lengths at the said revolutes need not coincide; in general an offset will obtain upon the mutually shared axis. This offset cannot be seen in my figure 20.21(*a*) because I have chosen there to put the shared axis normal to the plane of the paper. Next fuse pairs of links at the shared axis in the manner indicated at the numbers 1 and 4 in figure 20.21(*a*) and perceive that the resulting conglomeration is mobile with apparent mobility $M_a$ of only unity. Now reconstruct the fused links by replacing the old link lengths protruding from the shared axis by the common-perpendicular lengths drawn between the relevant pairs of revolutes. Next remove the old link lengths and the revolutes upon the shared axis as being redundant. The result, drawn according to the conventions of figure 20.11, is shown in figure 20.21(*b*). It is a hybrid mobile 6*R* loop with a

Figure 20.21 (§ 20.47) One of the Waldron hybrid 6*R* loops. Generally asymmetrical, it is formed by the addition in a certain way of two generally dissimilar Bennett loops. See text for an explanation.

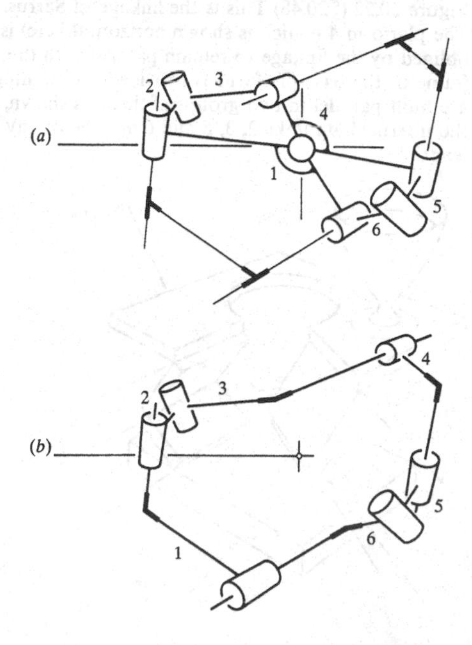

special characteristic: although the pair of new links (namely the fused and reconstructed ones) are not directly connected, they move with respect to one another as though they were pivoted.

48. Among others of the mobile 6*R*-loops there is the well known loop of Sarrus [23] – not Sarrut (Sarrut was a misprint in 1853). Refer to figure 20.22. The Sarrus loop permits relative pure translation to occur between the links 1 and 4 of the mechanism shown. The Sarrus loop may be seen as a limiting case of the double Hooke joint loop which is given a separate mention at § 20.55. The Sarrus loop may also be seen as a complex joint reciprocal to the simple prismatic joint or *P*-pair (§ 3.26, § 7.77). Any five of the six hinges which exist in parallel planes and connect links 1 and 4 define a first special 5-system of screws which is reciprocal to the first special 1-system of screws which characterizes the relative motion of 1 and 4 (§ 23.58, § 23.77). Remembering that there may exist kinds of 6*R*-loop that are not known, refer for a current list to Baker [24].

**Overconstrained yet mobile 5*R*-loops**

49. Myard [25] produced the first known of the overconstrained yet mobile 5*R* loops in 1931. As Waldron [26] explains, and as my figure 20.23(*a*) with its caption shows, this loop may now be seen as a special plane-symmetric form of one of the Goldberg 5*R* loops to be mentioned next (§ 20.24). Myard explains that the joint axes of the linkage all belong to the same hyperbolic linear congruence (§ 11.22, § 20.26) and, as figure 20.23(*b*) shows, the two directrices of this congruence may be located. One is normal to the plane of symmetry as Myard states, while the other is in that plane. Waldron [26] explains that, since the joints are all revolutes, the resultants of the paired axes and the unpaired axis all pass through the intersection with the plane of symmetry of that directrix which is normal to it; he points out that these three axes are, accordingly, members of a first special 2-system (§ 15.50); he draws from this and from symmetry the conclusion that the linkage is mobile. Please study § 20.50 now; return to look at the figures 20.23 later.

50. Goldberg [17] produced in 1943 not only the various mobile 6*R* loops already mentioned (§ 20.40), but a number of mobile 5*R* loops as well. Refer to figure 20.24. This shows at (*a*) the result of two Bennetts (not three) that he mounted back to back in the manner of figure 20.17(*a*), which result he arranged with the relevant adjacent links in line in the manner of

Figure 20.22 (§ 20.48) This is the linkage of Sarrus. The platform 4 (which is shown horizontal here) is obliged by the linkage to remain parallel with the frame or the base platform 1. Provided the six hinges are built parallel in two groups of three as shown, the intermediate links 2, 3, 5 and 6 may be of any length.

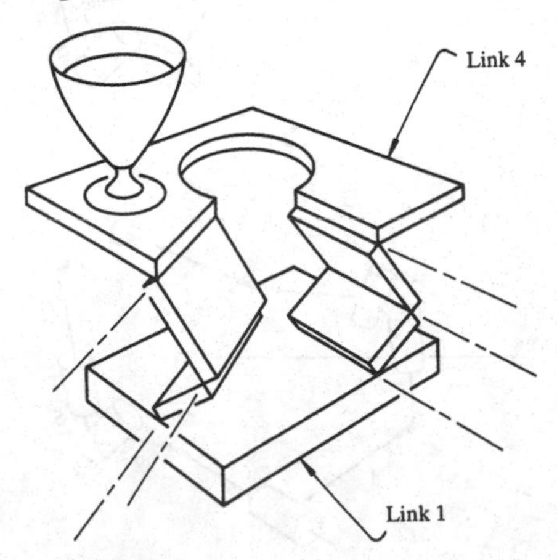


Figure 20.23(*a*) (§ 20.49) The symmetrical mobile 5*R* loop of Myard. In this picture the two contributing rectangular Bennetts may be seen. The longer of the two broken lines (the one in the plane of symmetry) shows where the Bennetts are fused together back to back, while the shorter of these two (the one along the axis for *y*) shows where the adjacent links of the Bennetts have been put, not merely in line with one another (§ 20.50), but coincident. Please note that link 1 (the rectangular block with its wedge-shaped extensions to carry two intersecting revolutes) is of zero length.

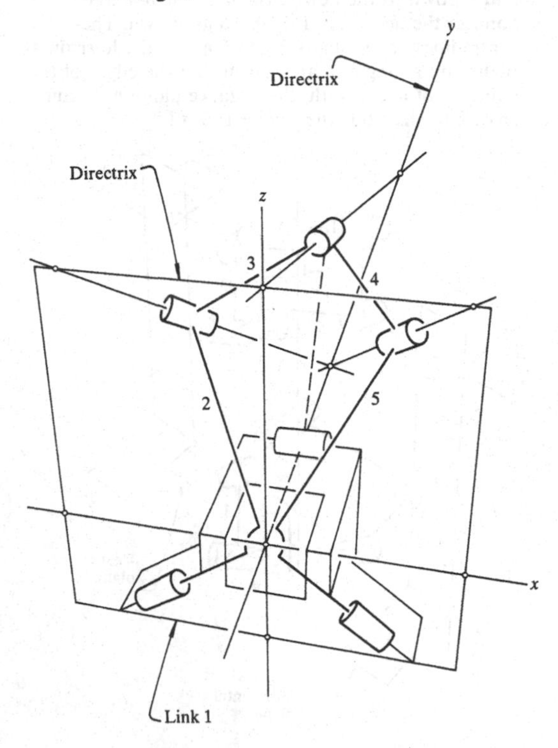

figure 20.17($b$), and which he reduced by omission of certain links and joints in the manner of figure 20.17($c$). At figure 20.24($b$) we see the introduction of the Goldberg surrounding loop with its two new joints (§ 20.43); and at ($c$) we see the obvious syncopation. Perceive now that if the two original Bennett loops were rectangular (§ 20.53), if they were also identical in their link lengths and twist angles but mirror images of one another, and if they were mounted back to back with their relevant adjacent links not merely in line but actually superimposed, we would arrive at the plane-symmetric Myard linkage already illustrated in figure 20.23($a$) and explained at the caption there. I should mention that, if the pair of Bennetts were not rectangular, the resulting loop would be, not the mobile 5*R*, but the mobile 6*R* plane-symmetric loop of Myard; there is no picture in this book of this loop [24].

51. Baker explains in the course of two papers [27] and another [18] why he thinks that the mobile 5-link loops in general (and thus the mobile 5*R* loops in particular) are not very thick on the ground; but nobody knows; these mentioned papers give a clear idea of our present limited knowledge of the mobile 5-link loops.

**Overconstrained yet mobile 4*R*-loops**

52. A number of just-constrained 4-link loops are presented as examples in chapter 1; refer to figures 1.03($g$), 1.11($a$), 1.11($b$). In figure 1.11($a$) an intermediate joint obtrudes; in the others, however, lower joints only are employed. Notice that, in these three examples and in all other 4-link loops of general mobility unity, the sum of the $f$-numbers at the joints $\Sigma f = 7$ (§ 1.46). There is an immense number of such loops. If a 4-link loop is mobile throughout a cycle of its motion while at the same time its $\Sigma f$ is less than 7, the loop is overconstrained yet mobile, and it falls into a special class. This class is also a wide one however, for there are multitudes of such overconstrained 4-link loops. During the first twenty years or so of this century Delassus [28] made a pioneering start upon the analysis of these linkages by producing a complete list of the overconstrained yet mobile 4-link loops which employ only the second three of the six lower joints, namely the lower $f1$ joints *HRP* in table 7.01 (§ 7.12). Having widened the field further we can now speak, in 1984, collectively and concisely of the overconstrained yet mobile 4-link loops which employ only the whole six of

Figure 20.23($b$) (§ 20.49) Three orthogonal views of the configuration of the Myard 5*R* loop in figure 20.23($a$). The technical features of this figure and some of the puzzling complexities of the deceptively simple Bennett linkage will become apparent when you ($a$) follow through the steps of the construction here, and ($b$) repeat them, using for the sake of exercise some value other than mine for the angle $\beta$. I have used $\beta = 35°$ for the construction here.

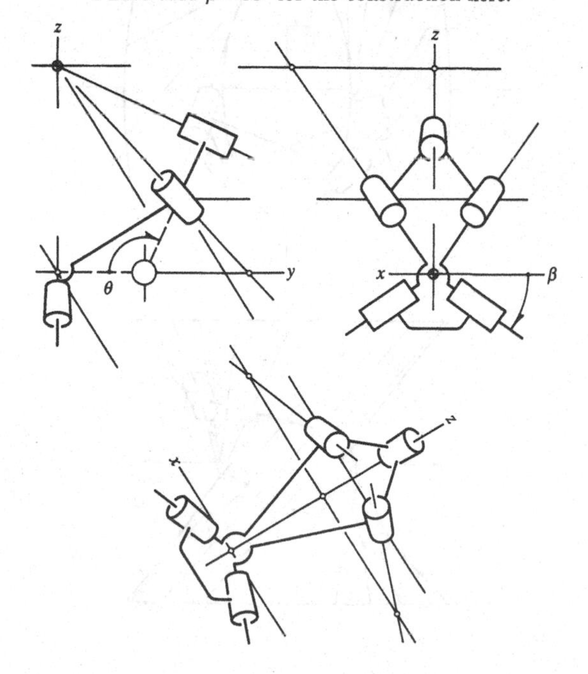


Figure 20.24 (§ 22.50) The Goldberg 5*R* loops. Although explained at § 22.50, the meanings of the three parts of this figure will become clearer if we think against the background of § 20.42 *et seq.* and the three figures 20.17, 20.18 and 20.19.

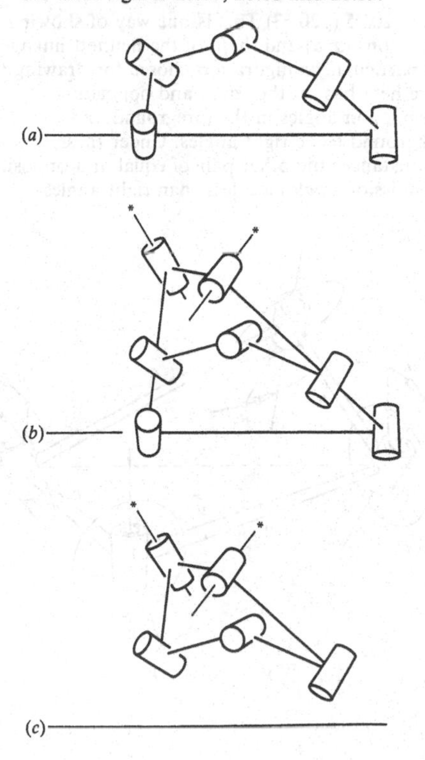

the lower joints. Refer to Table 1 of Waldron [29] for a near-definitive list of these loops; referring at the same time to Baker [29] for a few corrections. With certain mentioned exclusions made for the sake of brevity, Waldron's list goes to sixty-one. Not surprisingly we find among the listed loops in [29] the 4*R* planar loop (§ 2.36), the 4*R* spherical loop (§ 2.29), and the Bennett loop (§ 2.19). We find in Waldron's list no other 4*R* loops, for indeed there are none [29]. Now it is clearly possible to regard the planar and the spherical 4*R*-loops as special cases of the Bennett loop; we either put both of the two twist angles or both of the two link lengths to zero (§ 20.53), study the passages from § 19.40 to § 19.45 inclusive, and (at last for the sake of the present argument) come to that conclusion. Remember too that, whereas any three lines generally disposed define a regulus, any three parallel lines do not; we need four parallel lines to define a particular elliptical cylinder; thus we can get to the planar 4*R* loop. It is accordingly sufficient in the present context to describe all of the overconstrained yet mobile 4*R* loops by describing the Bennett loop. Refer to § 20.53.

53. Refer to figure 20.25 and see there the dimensions of the Bennett linkage. Pairs of opposite links are of equal lengths, $a$ and $b$. The links of length $a$ have twist angle $\alpha$ while the links of length $b$ have twist angles $\beta$. If the links $a$ are one-handedly twisted through $\alpha$ the links $b$ are other-handedly twisted through $\beta$. All offsets are zero. The following fundamental requirement, extracted from the first of [16], must obtain: $a \sin\beta = b \sin\alpha$. Given the requisite physical conditions for avoiding the clashing of links (such conditions may easily be achieved), each pair of opposite links can make complete revolutions relative to the other pair, all four links coming into collinearity once in a cycle. When the four links are collinear in this way, there is no change point, no uncertainty about the motion (§ 18.16); unless the angles $\alpha$ and $\beta$ are both zero (and neither of these can be zero without the other), the Bennett loop cannot escape at a change-point configuration from one mode of operation into another (§ 2.09). If the larger angle say $\alpha$ of the two angles $\alpha$ and $\beta$ is a right angle, the Bennett is said to

Figure 20.25 (§ 20.53) This is one way of showing the dimensions $(a, \alpha)$ and $(b, \beta)$ of the Bennett linkage. The particular configuration chosen for drawing the figure here has set the equal and opposite transmission angles in the foreground and background to be right angles. Under these circumstances the other pair of equal and opposite transmission angles are less than right angles.

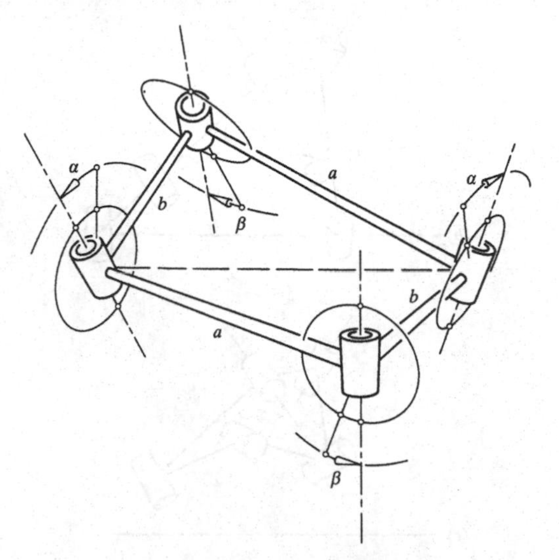


Figure 20.26 (§ 20.53) The Bennett linkage attaches to the regulus defined by any three of the central axes of its four revolute joints in the manner shown. The figure shows two orthogonal views of the regulus taken along two of its major axes. Contrary to expectation, the linkage does not encompass the regulus as a corset encompasses a waist; it clings, as it were, like a limpet at the side.

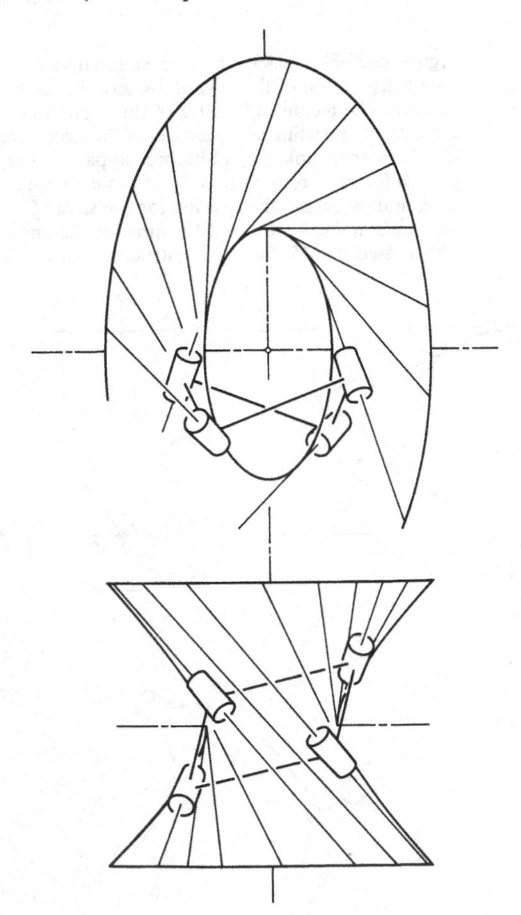

be *rectangular*. If the lengths $a$ and $b$ are equal, and if $\alpha$ and $\beta$ (which would need to be equal under the circumstances) were both right angles, the Bennett might, indeed, be said to be *square*; in this special case however it would be locked, immobile (§ 2.45, § 20.57). More recent studies of the Bennett linkage have been undertaken by Phillips and Hunt [30], Skreiner [31], Baker [18], and Yu [32]. These workers among others have revealed (*a*) that the four cylindroids fixedly mounted upon the four links in the manner described in § 12.33 intersect along the hinge axes and along the ISAs for the relative motions of opposite pairs of links and that the ISA for the relative motion of each pair of opposite links intersects the other pair of opposite links [30], (*b*) that the axodes for the relative motion of pairs of opposite links are not hyperboloidal [31], (*c*) that the closure equation for the Bennett linkage may be written unequivocally [18], and (*d*), referring to figure 20.26, that the Bennett linkage sits upon the regulus of its own hinge axes in an unexpected way [32], [33]. I made direct reference to figure 20.25 and 20.26 in § 10.58, and I now refer the reader back to those passages beginning at § 10.54. Persons interested in the Bennett mechanism must read the original Bennett papers [16], the hilarious discussion between De Jonge and Goldberg which was published at the end of [17] in 1943, and the rather more steady work of Baker [18] which failed to appear, as we now know, until 35 years later. Bennett, Delassus, Goldberg, Baker; these four and a few others have kept the Bennett alive since 1903. Its direct application in the ordinary business of practical machine design has been I suppose nil, but its continuing importance for our understanding of the overconstrained yet mobile linkages is immense.

**Overconstrained yet mobile 3-link loops**

54. Constrained and overconstrained yet mobile 3-link loops containing joints that are not all revolutes are often employed in ordinary engineering practice. Look first at the figures from 1.03(*a*) to 1.03(*j*) beginning at § 1.25, then cast your mind at the various simple gear sets – spur, bevel, hypoid etc.; see figure 22.01. See figure 2.01(*k*) and note that in the bevel set we have three pure angular velocities coplanar and intersecting at a point; these add, as we know they may under such circumstances, to rest (§ 10.15). In the spur gear set the relevant angular velocities are coplanar and not intersecting but parallel of course, and the addition of these three to rest (like the addition to equilibrium in the case of a set of three coplanar parallel forces) can be seen to obtain. In the hypoid set there are two angular velocities (those at the shafts with respect to frame) and a rate of twisting at the meshing teeth that add to rest; the latter displays itself at the $ISA_{23}$ (§ 22.18); examine the teeth at photograph 13 and refer to figure 13.06(*a*) (*b*) (*c*). The joint at the meshing of gear teeth is at least $f4$ and often $f5$ however, and this chapter has been restricted so far not only to lower joints but to the revolute joint in particular, so I must get on with the job of reporting the 3*R* loops. Before that however I would like to mention the two loops shown in figures 20.27(*a*) (*b*). The first should be compared with the loop I showed in figure 20.02 and be seen here as an extra part of the exercise mentioned in the caption there, while the second is the so-called differential screw [22]. This latter is a mobile 3-link loop of lower pairs; figure 20.27(*b*) is stylized, but a practical example is the well known turnbuckle used for tensioning cables of all kinds. Think also about the various ingenious kinds of cork extractor here. Figure 20.27(*c*) shows a 3*R* loop that, by virtue of the coplanarity of all three revolute axes and their intersection at a single point, is transitorily mobile (§ 10.15). Figure 20.27(*d*) shows a 3*R*-loop where the screw systems at the joints are superimposed in such a special way that we have a full-cycle apparent mobility of +2 (§ 2.38).

**More notes about the double Hooke joint**

55. Please find in figure 20.28 the layout of a double Hooke joint loop. The Hooke joints themselves are only schematically represented in the figure, one quadrant only of each of the cruciform pieces being sketched inside the vague spherical bulbs. The two Hooke joints are mounted in series and they connect a driver with a driven shaft. These latter rotate in bearings fixed to frame and their centre lines are skew. The shown

Figure 20.27 (§ 20.54) Various trivial 3-link loops.

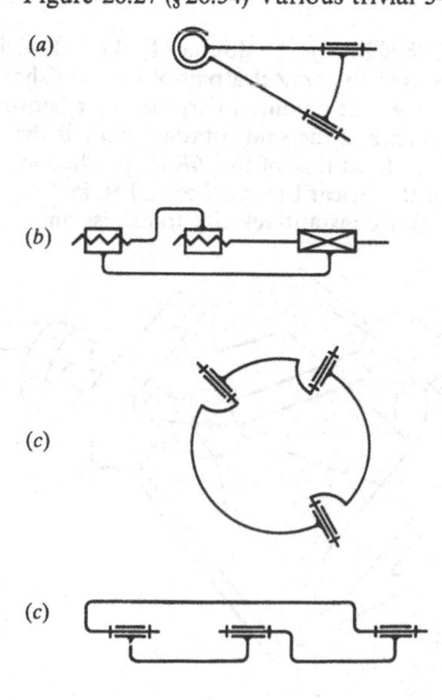

common perpendicular between the mentioned centre lines is the link-length of the frame link. Note that four of the six links are of zero length and that there are two non-zero offsets. Perceive that the arrangement is not a special case of either the line-symmetric or the trihedral loop of Bricard. Go back to figure 20.06(*b*) which shows the two sets of three successive intersecting axes which obtain in the present case but not the offsets which may occur. Did Bricard know about this unique arrangement and ignore it, or had he never noticed it? The double Hooke joint loop is a 6*R*-loop that is (at the time of writing) listed in a quite separate category of its own; refer to [24].

56. I wish to make some remarks about the conditions for constant-velocity transmission through the double Hooke joint drive. These conditions, worked out and stated by Yeo [34], are as follows: (*a*) that the two transmission angles shall be equal, and (*b*) *that the two Hooke joints shall be arranged in such a way that the countershaft is normal to the cruciform of each joint simultaneously*. Other authors have argued that the condition (*b*) should read as follows: that the yokes (or the clevis pieces) at either end of the countershaft shall be arranged to be coplanar. But this is correct only for the special case where all three of the shaft centre lines are coplanar. This latter (the coplanarity of the shafts) is not a necessary condition for constant-velocity transmission. I paraphrase also Tyson (1966): two shafts, says Tyson, which are neither parallel nor intersecting may be connected by a double universal joint and have uniform angular speed ratio providing (*a*) the connecting shaft makes equal angles with the driver and driven shafts, and (*b*) *the forks on the connecting shaft are so arranged that they lie simultaneously in the planes determined by the driver and the driven shafts with the connecting shaft*. In figure 20.28 I have constructed the layout in such a way that the centre lines of the two pins marked with an asterisk are set normal to the planes at the joints determined by the intersecting shaft axes. I have thus ensured that the other two pins of the cruciform pieces lie as Tyson (and Yeo by implication) say they should for constant-velocity transmission. Another way to state the two criteria is as follows: a double Hooke joint drive will transmit rotation with a constant angular velocity ratio provided (*a*) the transmission angles are equal, and (*b*) *the pins at each joint which are connected directly with the driver and the driven shafts respectively are set normal to the planes of the transmission angles there*. Will the reader notice that I have neglected in figure 20.28 to set the two transmission angles equal? I hope so, for I leave the necessary reconstruction of the drawing as an exercise: prove that the above criteria are sound and make a new clear drawing with no ambiguities; make it within the required space, and with despatch; allow say three hours. I have dealt with the matter here at some length because, as readers will know, it has been for a long time the subject of much debate.

Figure 20.28 (§ 20.55) The double Hooke joint drive. It may be used to connect a pair of rotating shafts that are skew. The common perpendicular between the central axes of the said rotating shafts is the link-length of the fixed link of this 6*R* loop. The loop is not one of the Bricard loops. See § 20.56 for conditions for constant-velocity transmission.

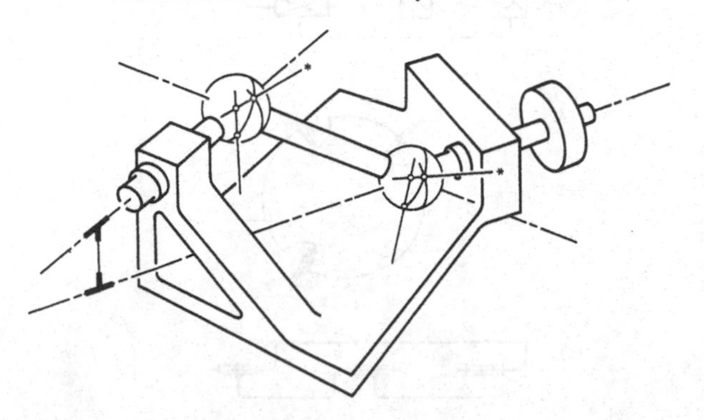

**Some doubly special symmetrical linkages**

57. My somewhat irrelevant but not unimportant remarks of the above paragraph, which relate not to questions of overconstraint but to questions of the fixed dimensions of that particular overconstrained yet mobile linkage there, lead in a new direction. So far I have tried to describe entirely generally the special linkages surveyed in the present chapter. In each special case I have paid only scant attention to the doubly special cases that might occur due to special choice of the fixed dimensions of the linkages. An example of such a doubly special linkage is mentioned in § 20.32 and § 20.46 where I draw attention to the special plane-symmetric (or the special trihedral) loop that, by virtue of its carefully chosen dimensions, can take up the mentioned cubical configuration displayed in the top part of figure 20.20. In this case the link lengths are all equal, the twist angles are all equal (all $\pi/2$), and the overconstrained linkage is reduced thereby to a highly symmetrical form with special geometrical properties. It is relatively easy to make a model of this particular linkage by folding paper into six connected tetrahedra; and such models are seen, from time to time, manufactured as toys. Refer to Schatz (1975) part I, section 9, *Der dreigeteilte Würfel* (The thrice-divided dice), and, being aware that the author links his (1975) with the well known anthroposophy of Rudolph Steiner, find that Schatz attibutes to linkages such as these mystical significance. Another such doubly special overconstrained linkage is mentioned at the

middle of § 20.53. There I spoke about the special Bennett linkage that I described as being square. I am inclined to argue that the Bennett in this particular case has its usual apparent mobility $M_a$ of unity, but that, due to the kind of circumstance explained at § 2.45, its range of movement is reduced to nil; although equipped with its mobility, it is immobile.

58. Two other examples of doubly special, overconstrained yet mobile loops might be taken for consideration now. I wish to take (*a*) the rectangular, and (*b*) the square, planar 4*R*-loop. See that, whereas in (*a*) there are transitory phenomena worthy of observation at the configuration where the revolute axes become coplanar, there is in (*b*) a quite spectacular series of different accidents which might occur at that configuration. Please notice that this chapter has ended where it began, with a consideration of the planar 4*R*-loop. The detailed mechanics of this loop, allegedly simple, is, unfortunately, beyond the scope of this book.

### Notes and references


[1] Waldron, K.J., The constraint analysis of mechanisms, *Journal of mechanisms*, Vol. **1**, p. 101–14, Pergamon, 1966. Refer § 20.01.

[2] Boden, H., Zum Zwanglauf gemischt räumlich ebene Getriebe, *Maschinenbautechnik* (*Getriebetechnik*), Vol. **11**, p. 612–15, 1962. Refer § 20.01.

[3] Baker, J. Eddie, On relative freedom between links in kinematic chains with cross-jointing, *Mechanism and machine theory*, Vol. **15**, No. 5, p. 397–413, Pergamon, 1980. On mobility and relative freedoms in multi loop linkages and structures, *ibid.*, Vol. **16**, No. 6, p. 583–97, 1981. Refer § 20.06, § 20.07.

[4] Davies, T.H., Mechanical networks: this is a series of three papers successively subtitled as follows: (1) Passivity and redundancy, (2) Formulae for the degrees of mobility and redundancy, (3) Wrenches on circuit screws: *Mechanism and machine theory*, Vol. **18**, No. 2, p. 95–112, Pergamon, 1983. Refer also Davies, T.H., Kirchhoff's circulation law applied to multi-loop kinematic chains, *ibid.*, Vol. **16**, No. 3, p. 171–83, 1981. Refer § 20.06, § 20.07.

[5] Waldron, K.J., A family of overconstrained linkages, *Journal of mechanisms*, Vol. **2**, p. 201–11, Pergamon, Oxford, 1967. See also, Symmetric overconstrained linkages, *Journal of engineering for industry*, Vol. **91**, (Trans ASME Series B), p. 158–62, 1969. Refer § 20.10.

[6] Hunt, K.H., Screw axes and mobility in spatial mechanisms via the linear complex, *Journal of mechanisms*, Vol. **2**, p. 307–27, Pergamon, 1967. This paper was soon followed in the same journal by another that included some corrections: Hunt, K.H., Note on complexes and mobility, *ibid.*, Vol. **3**, p. 199–202, Pergamon, 1968. Refer § 20.10.

[7] Baker, J.E. and Waldron, K.J., The C–H–C–H–linkage, *Mechanism and machine theory*, Vol. **9**, p. 285–97, Pergamon, 1974. It was in this paper that the authors wrote for the first time a closure equation for an overconstrained linkage containing helical joints without the simplification of parallelisms or intersections among the links. Refer § 21.10.

[8] Phillips, J.R. and Winter, H., Über die Frage des Gleitens in Kugel-Gleichganggelenken, *VDI-Zeitschrift*, Vol. **110**, No. 6, p. 228–33, February (III) 1968. Refer § 20.14.

[9] Altmann, F.G., His communication to a paper by Paul Grodzinski and Ewen M'Ewen, Link mechanisms in modern kinematics, *Proc. I. Mech. E.*, Vol. **168**, No. 37, p. 877–96, London, 1954. Refer § 20.23.

[10] Quentin Reynolds, an undergraduate at the University of Sydney, proposed this method of attack in 1974. Refer § 20.26.

[11] Baker, J.E., An analysis of the Bricard linkages, *Mechanism and machine theory*, Vol. **15**, p. 267–86, Pergamon, 1980. Refer § 20.29.

[12] I have in mind here the relevant works of Waldron, Waldron and Baker, Baker and Waldron, and Baker. A long list of papers has appeared in the journal literature beginning in 1967, and many of these have been quoted by me here in this chapter. Refer § 20.39.

[13] I have said this contrary to the works for example of Freudenstein, Hartenberg and Denavit, Chace, Uicker, Duffy, and others; these workers, who do indeed concern themselves with input-output relations, are quoted and discussed by me not here, but at various places in chapter 19. Refer § 20.39.

[14] Two papers here require to be mentioned. The first, in which some cogent remarks are made about the non-collinearity of the axes (*a*) of the line-symmetry, and (*b*) of the relevant complex in a special line-symmetric loop, is Baker, J. Eddie, Limiting positions of a Bricard linkage and their application to the cyclohexane molecule, *Mechanism and machine theory*, Vol. **21**, p. 253–60, Pergamon, 1986. The second, in which an algebraic calculation is made for the location of the axis of the complex in the Schatz mechanism (Turbula), is Yu Hon-Cheung, Geometrical investigation of a general octahedral linkage and the Turbula, *Mechanism and machine theory*, Vol. **15**, p. 463–78, Pergamon, 1968. Refer § 20.39, § 20.46.

[15] Baker, J.E., Screw system algebra applied to special linkage configurations, *Mechanism and machine theory*, Vol. **15**, p. 255–65, Pergamon, 1980. I should also mention here the important ASME preprint 74-DET-107: Baker, J.E. and Waldron, K.J., Limit positions of spatial linkages via screw system theory; this paper was prepared for the ASME-sponsored Design Engineering Technical Conference in New York, 1974. Refer § 20.39.

[16] Bennett, G.T., A new mechanism, *Engineering*, Vol. **76**, p. 777–8, London, 1903. Bennett, G.T., Deformable octahedra, *Proceedings of the London Mathematical Society*, 2nd series, Vol. **10**, p. 309–43, 1911. Bennett, G.T., The skew isogram mechanism, *ibid.*, 2nd series, Vol. **13**, p. 153–73, 1914. Refer § 20.40.

[17] Goldberg, Michael, New five-bar and six-bar linkages in three dimensions, *Trans. ASME*, Vol. **65**, p. 649–61, August 1943. Refer § 20.41, § 20.50.

[18] Baker, J. Eddie, The Bennett, Goldberg and Myard linkages – in perspective. *Mechanism and machine theory*, Vol. **14**, p. 239–53, Pergamon, 1978. Refer § 20.44.

[19] Brat, Vladimir, A six-link spatial mechanism, *Journal of mechanisms*, Vol. **4**, p. 325–36, 1969. Refer § 20.46.

[20] J. Eddie Baker, Tran Duclong, and P.S.H. Khoo, On attempting to reduce undesirable characteristics of the Schatz mechanism, *Journal of mechanical design*,

Vol. **104**, (Trans ASME) p. 192–205, Jan. 1982. Refer § 20.46.

[21] Wohlhart, Karl, On the synthesis of the Turbula, *Proceedings of the 5th World Congress IFToMM in Montreal 1979*, p. 747–50, ASME, 1979. Blaimschein, J., *Analyse des allgemeinen Turbula-Getreibes*, doctoral dissertation, Technische Universität Graz, 1980. Wohlhart, K., Dynamic analysis of the Turbula, *Transactions of the International Symposium on Gearing and Power Transmissions in Tokyo*, JSME, 1981, p. 425–30. Refer § 20.46.

[22] Waldron, K.J., Hybrid overconstrained linkages, *Journal of mechanisms*, Vol. **3**, p. 73–8, Pergamon, 1968. Refer § 20.47.

[23] Sarrut (but properly Sarrus), Note sur la transformation des mouvements rectilignes alternatifs, en mouvements circulaires; et reciproquement, *Academie des sciences, comptes rendus hebdomadaires des séances*, Vol. **36**, p. 1036–8, Paris, 1853. Refer § 20.48.

[24] In an Appendix somewhat oddly placed at the end of this particular paper, Baker gives a valuable list of the overconstrained yet mobile 6*R*-loops currently known to him: Baker, J.E., On 5-revolute linkages with parallel adjacent joint axes, *Mechanism and machine theory*, Vol. **19**, No. 6, p. 467–75, Pergamon, 1985. Refer § 20.48.

[25] During the period 1930–1 a series of papers appeared as follows: Myard, F.E., *Acadamie des sciences, comptes rendus hebdomadaires des séances*: Vol. **190**, p. 1491; Vol. **191**, p. 830; Vol. **192**, p. 1194, 1352, 1527. Appearing also, we find the following: Myard, F.E., Contribution à la géométrie des systèmes articulés. *Bulletin de la Societe Mathématique de France*, Vol. **59**, p. 183–210, 1931. Refer § 20.49.

[26] Waldron, K.J., *The mobility of linkages*, doctoral dissertation, Stanford University, 1969. Refer § 20.49.

[27] Baker, J. Eddie, Overconstrained 5-bars with parallel adjacent joint axes – I, *Mechanism and machine theory*, Vol. **13**, p. 213–18, 1978. Also by the same author. On 5-Revolute linkages with parallel adjacent joint axes, *ibid.*, Vol. **19**, p. 467–75, 1984. Refer § 20.51.

[28] Delassus, Et., Between the years 1900 and 1922 the following works appeared: Sur les systèmes articulés gauches, première partie, *Paris Ecole Normale Supérieure, annales scientifiques*, 3rd Series, Vol. **17**, p. 455–99, October, 1900; Sur les systèmes articulés gauches, deuxième partie, *ibid.*, 3rd Series, Vol. **19**, p. 119–52, March 1902; Les chaîns articulés fermées et déformables à quatre membres, *Bulletin des sciences mathématiques*, 2nd Series, Vol. **46**, p. 283–304, 1922. Refer § 20.52.

[29] Waldron, K.J., Overconstrained linkages, *Environment and planning B*, Vol. **6**, p. 393–402, 1979. Baker, J.E., A compendium of line-symmetric four-bars. *Journal of mechanical design*, Vol. **101**, (Trans ASME), p. 509–14, 1979. Refer § 20.52.

[30] Phillips, J.R. and Hunt, K.H., On the theorem of three axes in the spatial motion of three bodies, *Australian journal of applied science*, Vol. **15**, p. 267–87, CSIRO, Melbourne, 1964. Refer § 20.53.

[31] Skreiner, K. Michael, *Three dimensional kinematics and its application to linkages*, doctoral dissertation, University of Sydney, 1967; refer to chapter 5. Refer § 20.53.

[32] Yu Hon-Cheung, The Bennett linkage, its associated tetrahedron and the hyperboloid of its axes, *Mechanism and machine theory*, Vol. **16**, p. 105–14, 1981. Refer § 20.53.

[33] Eric Zapletal, an undergraduate at the University of Sydney, drew attention to the circumstance in 1977. Refer § 20.53.

[34] Yeo, Dallas S., *The variation of velocity ratio in a double Hooke's joint drive*, an undergraduate thesis in the Department of Mechanical Engineering, University of Western Australia, January, 1962. Refer § 20.56.

21. This working model of the Schatz linkage is mixing coloured beads inside the cylindrical drum. An overconstrained 6*R*-loop, the Schatz linkage comprises the essential mechanism of a machine that is called by its maker Turbula.

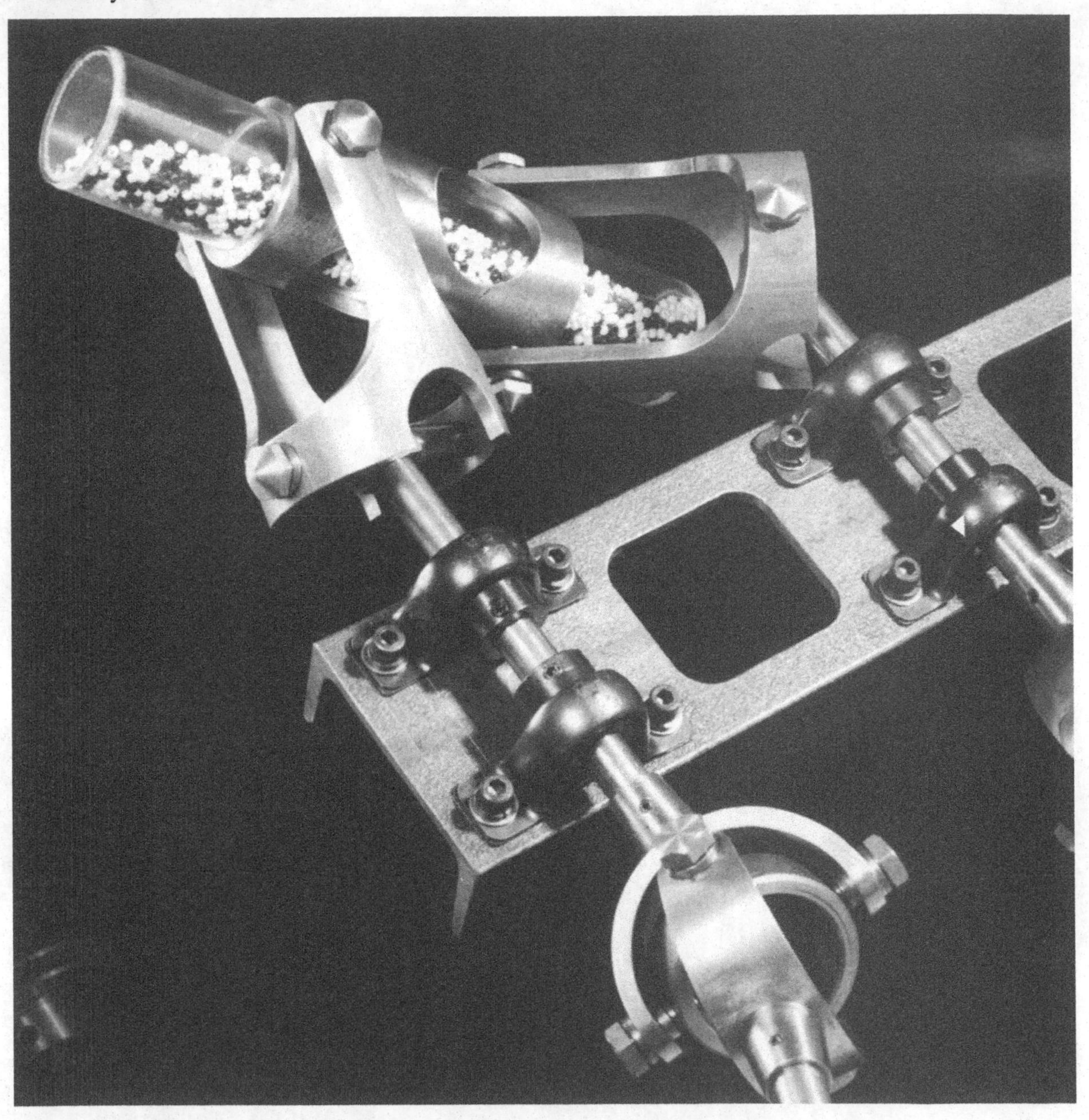

# 21 The helitangent lines in a moving body

## Introduction

01. The location of this chapter here among the chapters may seem to the reader odd. But the material is relevant, and relevant now. Nowhere yet have we explored among the lines in a body for a prevailing mechanical significance other than that that is possessed by the right lines. Refer to chapter 3 and recall here that the lines of the linear complex of right lines in a body constitute only one infinity-th of all of the lines that exist in the body (§4.10, §9.44), so, given the usual linear complex of right lines existing in a body, there is outside of that and in the body a wide-open space within which we may if we wish search for something else that may be interesting. As well as the so-called helitangent lines that figure largely in this chapter (these relate directly to the material of chapter 22), there will be found some other simple line-distributions of vector that relate to various special lines and instantaneous motions. Let me trust the forthcoming material (*a*) to be forthcoming easily, and (*b*) to be able to stand upon its own feet.

## The most general kind of line in a moving body

02. Refer to figure 5.01. The pair of points *A* and *B* are fixed in a rigid body that is moving. The figure shows an array of linear velocity vectors emerging in a characteristic way from its corresponding array of basepoints fixed in the straight line segment *A–B*. At §5.25 a summarizing remark is made about figure 5.01, and at §5.26 there is an early introduction to the material here. In figure 5.01 there are two skew lines. They are (*a*) the infinitely long line *AB* itself (§1.06), and (*b*) the infinitely long line of the tips or the endpoints of the vectors (§5.05). The actual distance between these lines depends upon the scale that is chosen for drawing the vectors. That distance, the common perpendicular distance, is not shown in figure 5.01.

03. It is however shown in figure 21.01. It appears there (in the lower, side-elevation view) as the broken line *d–d*. Now if we locate the basepoint Γ of that velocity vector whose tip is at *d*, we locate an important point in the line which may be defined and named as follows: *it is that unique point in every line in a moving body whose velocity (among the velocities of the points in that line) is least; the point* Γ *may be called the centre-point or the central point of the line.* Please note the angle δ at Γ; it is in general the smallest of the angles between the velocity vectors and the line. I should mention here the analogy which is fully discussed in chapter 10 between (*a*) the linear velocity at a point in a rigid body that is moving, and (*b*) the moment at a point in a rigid body that is subjected to a wrench. I invite the reader now to reread this and the above two paragraphs replacing on each occasion the words *linear velocity vector* with the words *moment vector* and thus to become aware that for the whole of this chapter one might if one wished write corresponding remarks about the special kinds of line in a body subjected to a wrench.

04. Refer to figure 8.05. See there that the point Γ in a line in a body is always at the foot of the common perpendicular *D*–Γ between the ISA for the body and the line. The same point Γ can be seen in figures 9.04 and 9.05, but the line *AB* in those particular cases is a *right*

Figure 21.01 (§21.03) Three orthogonal views of a general line in a moving body with its straight-line array of velocity vectors. Note the angle δ at the central point Γ of the line.

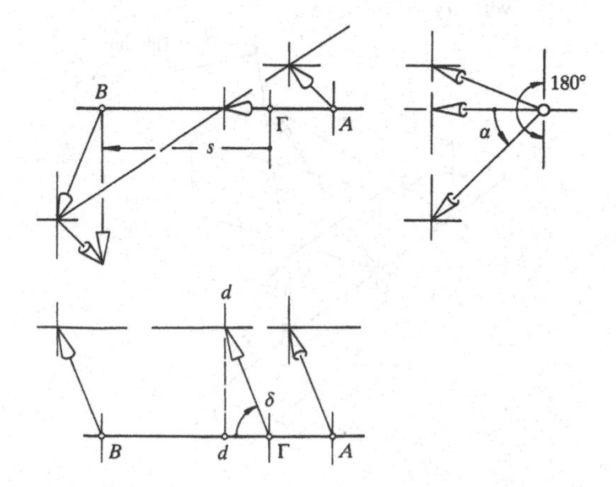

*line* in the moving body, the angle $\delta$ being a right angle (§ 3.01).

**Characterizing the various kinds of line**

05. Refer to figure 22.02 and see there the two angles $\delta$ and $\alpha$. Whereas $\delta$ (existing at $\Gamma$) is the least angle between a vector and the line and is unique in the line, the angle $\alpha$ is a variable in the line. The angle $\alpha$ depends upon $s$, the distance from $\Gamma$ to $Q$. The point $Q$ is a general point which might be imagined to be moving from right to left along the line $AB$ while the vector at $Q$ varies to sweep out the array. For all points $Q$ the component in the line $AB$ of the velocity vector at $Q$ remains the same (§ 5.03). The moving line segment $e$–$e$, which remains always perpendicular to $AB$, will coincide with the fixed line segment $d$–$d$ when $Q$ is at $\Gamma$. The infinitely long line $ee$ traces out, accordingly, an hyperbolic paraboloid, whose saddle point is at $d$ on $AB$, and whose main rectilinear axes, $dd$ and $AB$, intersect at $d$ on $AB$. It is important also to see that, as $Q$ moves from one end of the infinitely long line $AB$ to the other, the moving velocity vector at $Q$ never escapes that particular half of all space which is marked 180° in figure 21.01.

06. What parameters, then, characterize the line? It will be seen upon inspection that, independently of the scale, which might be measured in (mm sec$^{-1}$)/mm, and which determines the actual lengths of the drawn vectors, two convenient parameters which together are sufficient are (*a*) the angle $\delta$, and (*b*) the rate of change of $\alpha$ with $s$ namely $d\alpha/ds$ when $Q$ is at $\Gamma$. For any general line at an instant in the moving body, of which there will be an $\infty^4$ (§ 4.10), the angle $\delta$ will hold some value between zero and $\pi/2$ (rad) and the rate $d\alpha/ds$ will hold some value between positive and negative infinity (rad mm$^{-1}$), there being some convention adopted for distinguishing left handed from right handed arrays.

Figure 21.02(*a*) (§21.05) The fixed angle $\delta$ and the variable angle $\alpha$ among a straight-line array of velocity vectors.

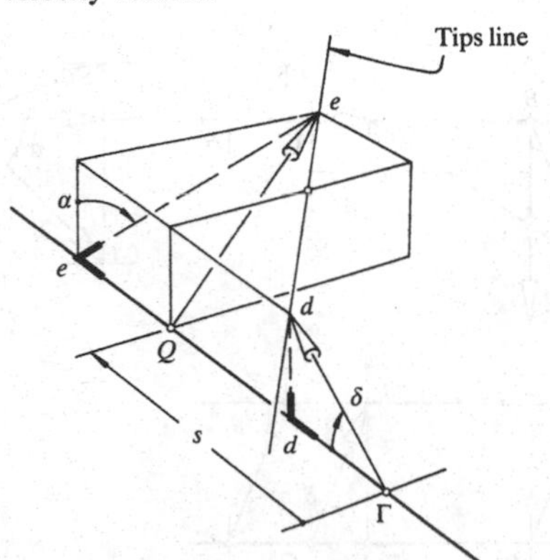


Table 21.01 (§ 21.07) *The discussion from § 21.02 to § 21.18 inclusive is summarized here. First read down for values of δ: F finite, R right angular, Z zero. Next read across for values of dα/ds: F finite, Z zero, I infinite. See ZF for the helitangent line.*

| FF | FZ | FI |
|---|---|---|
| *General line.* Discussed in chapter 5, there is an $\infty^4$ of them. They look like figure 5.01. | Parallel with the ISA, there is in general an $\infty^2$ of lines that look like figure 21.02(*b*). | These special lines occur when the pitch is zero. They all cut the ISA and they look like figure 21.02(*c*) |
| **RF** | **RZ** | **RI** |
| *Right line.* Discussed in chapter 3, there is an $\infty^3$ of them. They look like photograph 3. | These occur when the pitch is zero or infinite. They are parallel with or in planes normal to the ISA. See figure 21.02(*d*). | These special lines occur when the pitch is zero. They all cut the ISA and they look like figure 21.02(*c*). |
| **ZF** | **ZZ** | **ZI** |
| *Helitangent line.* Discussed in this chapter, there is an $\infty^3$ of them. They look like figure 21.03. | Along the ISA itself, or parallel with it when the pitch is infinite, there are lines that look like figure 21.02(*e*). | These special lines occur when the pitch is zero. They all cut the ISA and they look like figure 21.02(*c*). |

07. Among the $\infty^4$ of general lines, however, there will be special lines of one kind or another, and these will occur when $\delta$ is either zero or $\pi/2$, and/or when $d\alpha/ds$ is either zero or plus or minus infinity. A convenient way of exploring the whole range of possibilities is to tabulate the special values as shown at the table 21.01. The special values *F*, *Z*, *I*, and *R* are explained at the table, and I trust that, when I refer for example – and I do this only temporarily – to an *FZ* line, the symbolism *FZ* will be understood. I am sorry that, within the context here, I can think of no more descriptive, yet equally compact way to nominate the various kinds of special line.

**The right lines at an instant**

08. Since right lines are already mentioned here and already discussed at length in chapters 3 and 9 and elsewhere (§ 21.04), let us deal with these lines first. Please refer to that part of table 21.01 where the angle $\delta$ is listed *R*. We have there *RF*, *RZ* and *RI* lines, all of which are right lines. The first of these (the *RF*) are general, while the rest (the *RZ* and *RI*) are special kinds of right line.

09. As already mentioned the *RF* lines are the general right lines. There is at an instant an $\infty^3$ of these in any moving body (§ 9.16). They are always arranged in the form of a linear complex concentric with the relevant motion screw namely the relevant ISA in the body (§ 9.27). Among them however will sometimes occur *RZ* and *RI* lines, in both of which categories the vectors of the array are parallel and coplanar.

10. It would appear that *RZ* lines can occur only when the pitch of the rate of twisting (namely the pitch of the screwing) of the body is either zero or infinity. The array at an *RZ* line is arranged like the teeth of a comb; the vectors are equal and parallel and thus coplanar; see figure 21.02(*d*). When the pitch of the screwing is zero, *RZ* lines exist everywhere parallel with the motion screw, and there is then an $\infty^2$ of them. When the pitch is infinity, *RZ* lines exist everywhere in all planes normal to the direction of travel, and there is then an $\infty^3$ of them.

Figures 21.02(*b*), (*c*) (*d*) (*e*) (§ 21.12) Orthogonal views of various special, planar, straight-line arrays.

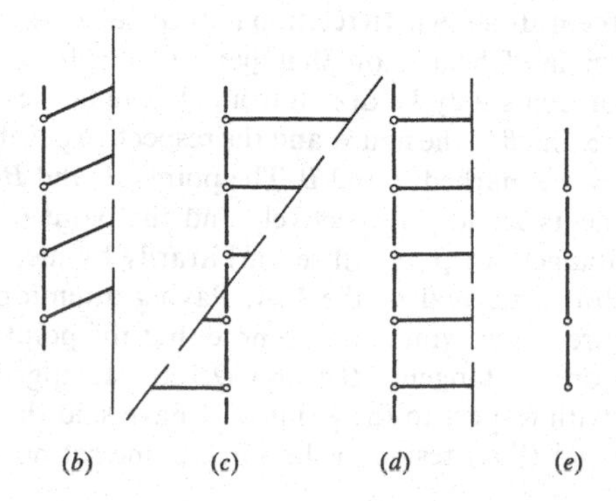


11. It would appear on the other hand that *RI* lines occur in a body only when the pitch of the screwing is zero. The array at an *RI* line is arranged as shown at figure 21.02(*c*); the vectors are coplanar. Whenever the pitch of a screwing is zero, the motion is purely rotational (§ 5.50). Any line in the rotating body which then cuts the axis of the motion screw at any angle is an *RI* line, and there is then an $\infty^3$ of them. They are arranged as the members of a special linear complex concentric with the motion screw whose pitch, as we know, like the pitch of the complex, will be zero (§ 9.41, § 11.33).

**Another special kind of line**

12. The *FZ* lines are interesting. In whatever way a body may be moving there are *FZ* lines parallel with the ISA. The vectors are parallel and equal and thus coplanar but in general the angle $\delta$ is not a right angle; see figure 21.02(*b*). There is in general an $\infty^2$ of such lines (§ 5.43).

13. The *ZZ* lines can be seen as a special case of the *FZ* which have just been mentioned. On a *ZZ* line all of the velocity vectors are equal and parallel and collinear with the line. See figure 21.02(*e*). There is in general only an $\infty^0$ of such lines, namely only one of them. It occurs collinear with the axis of the motion screw namely at the ISA itself (§ 5.43). When the pitch of the motion screw is infinite, however, there exists an $\infty^2$ of such lines, all parallel with the direction of travel. The *RZ* lines are another special case of the *FZ*, but they are also a special case of the right lines *RF*; they could be categorized either way.

**When $d\alpha/ds$ is infinite**

14. We see upon reflection that these three kinds of line – they are all special kinds of right line – are the same. Whenever it becomes necessary to envisage an angular velocity $\boldsymbol{\omega}$ infinity, it simply means in the context that the linear velocity of axial sliding $\tau$ must be zero (§ 5.49, § 8.23); it cannot be otherwise. Please read § 21.11 again, and see here why the sketch of the vector array in figure 21.01(*c*) applies in each of the three boxes that correspond with *RI*, *ZI*, and *FI* in table 22.01.

**The helitangent line *ZF***

15. We ought to speak here – and I ought to have spoken earlier – in terms of limits, when dealing in this way with zero and infinity. With $d\alpha/ds$ fixed at some finite value and $\delta$ small, it might be argued now that, as $\delta \to 0$, we arrive in the limit at that kind of line that could be called *helitangent line*. Refer to figure 21.03.

16. In figure 21.03 a coplanar array of velocity vectors is drawn. Both the line itself and the line of the tips are in the plane of the paper (§ 5.07). Please note that the two points Γ and $d$ ($d$ being the collapsion of the common perpendicular $d$–$d$ appearing in figure 21.02) are not coincident. This means that the least vector, occurring at Γ, is collinear with the line. Figure 23.03 is a picture of my definition of a helitangent line; and I wish to explain now why I call such a line by that name.

Figure 21.03 (§ 21.16) Illustrating a helitangent line. The array of velocity vectors is coplanar with the line, and the least velocity vector at Γ is collinear with the line.

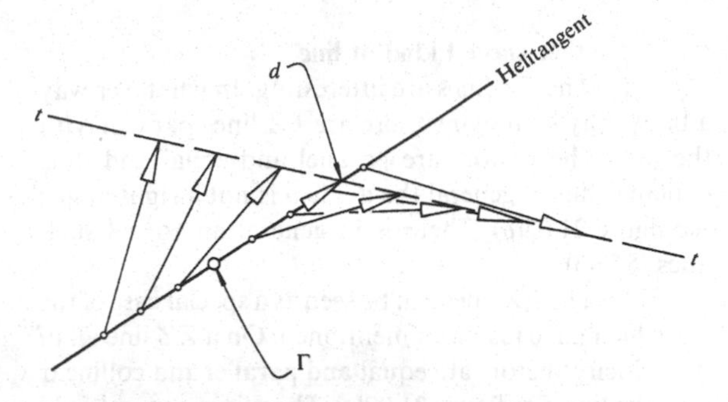


Figure 21.04 (§ 21.17) The osculating plane at $Q$, so called because it osculates with the surface of the helix at $Q$, contains both the helitangent which lies along the tangent to the helical curve at $Q$ and the helinormal which lies along the generator of the helix there. The helitangent, helinormal and helibinormal at $Q$ (which are also simply called by myself and others the tangent, normal and binormal there) are mutually perpendicular at $Q$.

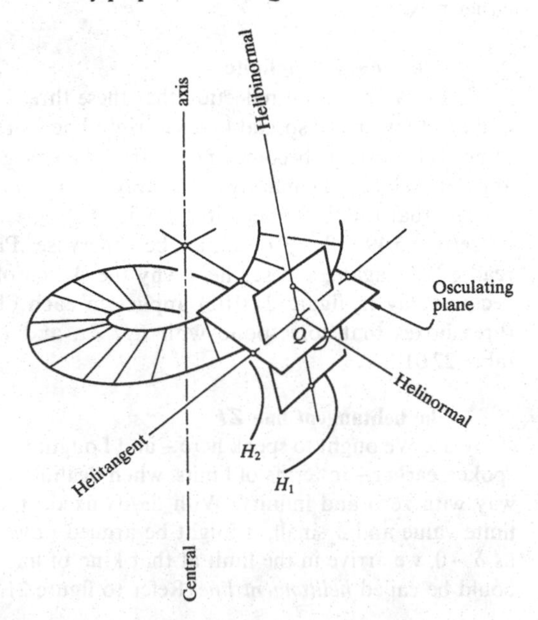


17. Refer to figure 21.04. It depicts a helix, a surface ruled of lines. The surface is drawn coaxial with a given motion screw (or ISA), and the pitch of the helix is the pitch of the rate of twisting (§ 10.47). The ruled helical surface, the helix, is truncated by two coaxial circular cylinders $H_1$ and $H_2$, generally chosen, that intersect it in the two helical curves that are shown. These curves – they are often called by the same name, *helix* – are designated similarly $H_1$ and $H_2$. A generally chosen point $Q$ on the helix $H_1$, which latter is itself a generally chosen helix from the single infinity of helices upon that cylinder (§ 9.05), is shown, and through $Q$ three lines are drawn. These lines, mutually perpendicular, are called in figure 21.04 by the following self explanatory names: *helinormal, helibinormal, and helitangent.* Please go to my notes about curvature beginning at § 5.52, consider not two but three points fixed in the moving body there, and consider whether my use throughout of the prefix *heli* here is fully justified.

18. Next we should note that, whereas the helinormal and the helibinormal at $Q$ and the whole planar pencil of lines defined by these two are right lines in the moving body (§ 9.09) and thus that another name for the right line in a body might be *helibinormal line* (§ 9.26), the helitangent at $Q$ is not a right line. It is not very difficult now to show, not only that the velocity vectors at all points along the helitangent at $Q$ are coplanar, but also that the plane in question is the osculating plane to the helix at $Q$. Figure 21.04 shows this plane; it contains the said helinormal and the said helitangent at $Q$ but is tangential to the helical surface at $Q$ only. Please note that, at $A$ in figure 5.09, there are two osculating planes to be careful of.

### Distribution of the helitangent lines

19. Figure 21.05 has been drawn to show that there is not only the unique helitangent at $Q$ (§ 21.18) but also a single infinity of other helitangents that may be drawn through $Q$. Please note that coaxial with the helix $H_2$ there is a single infinity of other helices of the same radius (§ 9.05). All of these could be inscribed upon the circular cylinder which appears in figure 21.05, but only two of them are shown. In relation to $Q$ these two are the unique pair of helices on that particular cylinder to which tangents may be drawn from $Q$. The helices are marked $\alpha$ and $\beta$ in the figure, and the respective points of tangency are marked $A$ and $B$. The points $A'$ and $B'$ on the tangents $\alpha$ and $\beta$ respectively and the point $Q'$ on the helitangent at $Q$ are all set (arbitrarily by me) in a plane drawn normal to the ISA. Paying attention to the nature of the symmetry, we note that the points $A'$ and $B'$ on the tangents $QA$ and $QB$ are arranged as shown with respect to the point $Q'$. I have said that $A'$ and $B'$ and $Q'$ all reside in the same plane set normal

to the ISA; but they also reside upon the periphery of an ellipse, the nature of which has yet to be explained (§ 21.20).

20. For a given $Q$ on a given helix $H_1$ there is a single infinity of cylinders $H_2$ whose radii are less than that of $H_1$. Accordingly there is a single infinity of pairs of tangents such as $QA$ and $QB$; and these, as can be shown, generate a second order (namely an elliptical) cone of helitangent lines at $Q$. That cone is illustrated in figure 21.05. Notice that, when the radius of the cylinder $H_2$ becomes zero, the cylinder becomes the ISA itself and the two tangents coalesce; they are tangent to the ISA at infinity and thus parallel with the ISA, as shown by the line $QC'$.

21. Now there are of course an $\infty^3$ of points such as $Q$ in space (§ 4.01), and at each of these there is a second order cone of helitangent lines, each of which contains an $\infty^1$ of tangents. Ensuring that each helitangent is not counted an infinity of times too often (§ 4.10), we can see that there is in all an infinity to the power $(3+1-1)$ namely an $\infty^3$ of helitangent lines.

Figure 21.05 (§ 21.19) This purports to clarify the layout of the quadratic complex of helitangent lines that exists in a moving body at an instant. The axis of the complex coincides with the motion screw (the ISA). The motion screw is drawn right handed here. Helitangent lines through a generally chosen point $Q$ form a cone with its vertex at $Q$. The particular helitangent lines through $Q$ which are ($a$) tangential to the helix of zero radius, and ($b$) tangential to the helix $\beta$ at $B$, are marked with asterisks, while the others of the helitangent lines through $Q$ relate to the others of the cylinders.

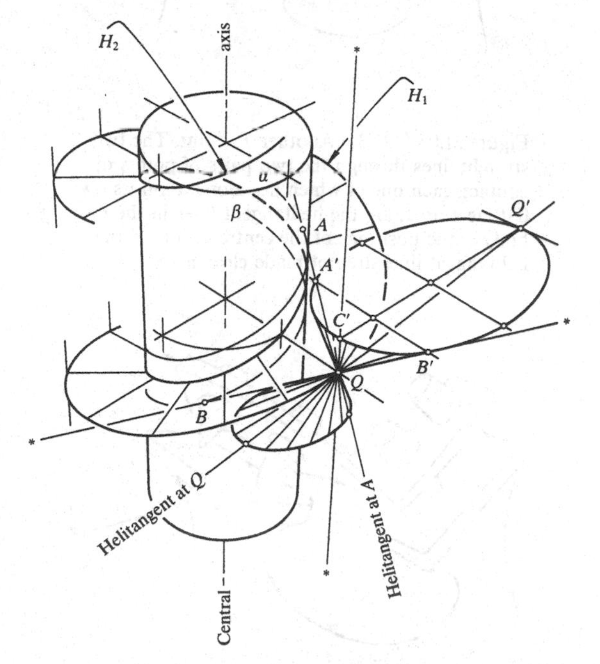


Please read for comparison and for a useful analogous argument § 9.18 and § 9.19, being aware, of course, that in figure 9.04 the two points $A$ and $B$ used there have no connection with the two points $A$ and $B$ used here.

**The quadratic complex of the helitangent lines**

22. Whereas the $\infty^3$ of right lines in a moving body are the generators of a *linear* complex coaxial with the ISA (§ 9.30, § 11.34), the $\infty^3$ of helitangents in a moving body are the generators of a *quadratic* complex coaxial with the ISA.

23. At § 11.64 I explained how the generic term quadratic complex applies to a range of possible arrangements of an $\infty^3$ of lines. Here I wish to describe the particular kind of quadratic complex that relates to helitangent lines. Its nature began to emerge in § 21.21.

24. Refer again to figure 21.05 and let the point $Q$ be free to move anywhere upon the generally chosen right circular cylinder $H_1$. At all points $Q$ a cone of helitangent lines will be extant, and each cone will be shaped and oriented as the cone at $Q$. It will be clear to see that the $\infty^3$ of lines which belong to these $\infty^2$ of cones constitute alone all of the lines of the complex. Of course we could also say that the complex comprises all of the lines on all of the cones on *all* of the cylinders but, if we counted thus, we would simply count the lines of the

Figure 21.06 (§ 21.25) An orthogonal view of some of the cones of helitangent lines whose vertices rest upon a radius of the complex. This view, which looks radially inwards towards the axis of the complex, shows the helix angle $\theta$ at $Q$.

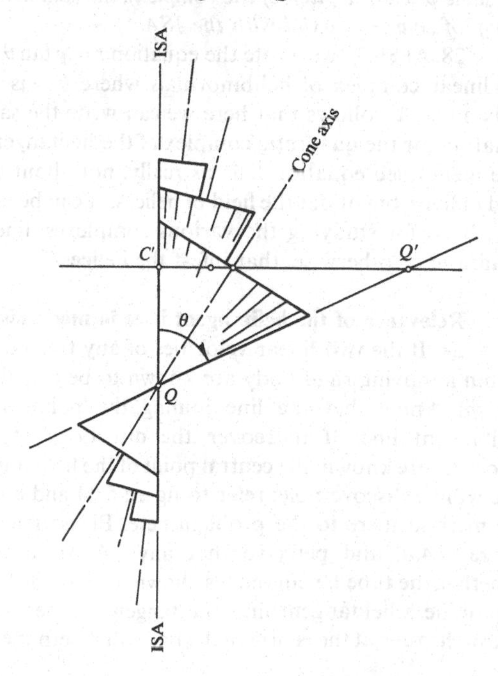

cones another infinity of times too often. This overall pattern of the helitangent lines – there is a second order cone of lines through every point in space (the shapes of the cones depending, as we shall see, upon the cylinder chosen) – continues unaltered as we go from one end to the other of the infinitely long ISA. Axi-symmetry prevails.

25. Figure 21.06 is an orthogonal view of figure 21.05 taken from $Q$ and against the ISA. Let $Q$ be free to move now, to and fro, along the helinormal, that is in and out of the paper at figure 21.06. As $Q$ approaches the ISA the projected angle $\theta$ of the cone diminishes to zero, while as it approaches infinitely far away from the ISA the same angle increases and becomes, in the limit, a right angle. This $\theta$ is also the helix angle.

26. Were we to look now at the cones upon a helinormal in the two other orthogonal directions, namely (*a*) projected against the plane containing the helinormal and the ISA and (*b*) projected against any plane normal to the ISA, we could make some further observations about the overall geometry. We would find that in general the cones are not circular cones, but we may find (given the pitch) that certain of the cones are circular.

27. I wish to recall the other quadratic complex mentioned in this book, namely the quadratic complex of the screws of the 4-systems (§ 6.29). It will of course be clear that the complex here (the quadratic complex of the helitangent lines) is a different quadratic complex, indeed a simpler one. This one here is, quite simply, *that* $\infty^3$ *of tangents drawn at the* $\infty^3$ *of points on the* $\infty^2$ *of helices of the same pitch* (*the pitch of the complex*), *the said helices being, of course, coaxial with the ISA.*

28. At § 9.31 we wrote the equation $r = p \tan\theta$ for the linear complex of helibinormals where $\theta$ was the helix angle. It follows that here we can write the same equation for the quadratic complex of the helitangents. The mentioned equation tells us really, not about any field of lines, but about the field of helices. It can be used as a basis for studying the various complexes, linear, quadratic or otherwise, that infest the helices.

### Relevance of the helitangent lines in mechanism

29. If the two linear velocities of any two points within a moving rigid body are known to be coplanar, we will know that the line joining the points is a helitangent line. If moreover the directions of the velocities are known, the central point of the helitangent line will be discoverable; refer to figure 5.04 and adapt the method there to the problem here. Please refer to figure 7.04(*a*) and perceive that any $f4$ ball-in-tube (whether the tube be curved, as shown, or straight) will determine a helitangent line: the tangent to the centre line of the tube at the centre of the ball will determine the helitangent line itself, while the centre of the ball will determine the central point of the helitangent line. Any $f4$ joint like the ones that are shown at figure 7.05(*a*) will determine a helitangent line: along the knife edge is the helitangent line itself, and across the plane surface of the joint is the plane of the coplanar velocity vectors. The reader will find other such examples.

30. Let a movable body 2 be connected to a fixed body 1 by means of two $f4$ ball-in-tube joints that are mounted not in series but in parallel. Refer to figure 21.07. The body 2 will have 2 degrees of freedom relative to the fixed body; the $f$-number of the combined single simple joint between the two bodies will be 2 in other words (§ 1.11). Refer to figure 5.01 and find there that the ratio of the magnitudes of the two velocities at

Figure 21.07 (§ 21.30) An $f2$ joint. The tangents to the centre-lines of the tubes at the centres of the balls are the helitangent lines in the moving body 2, and the centre-points of these helitangent lines are at the centres of the balls.

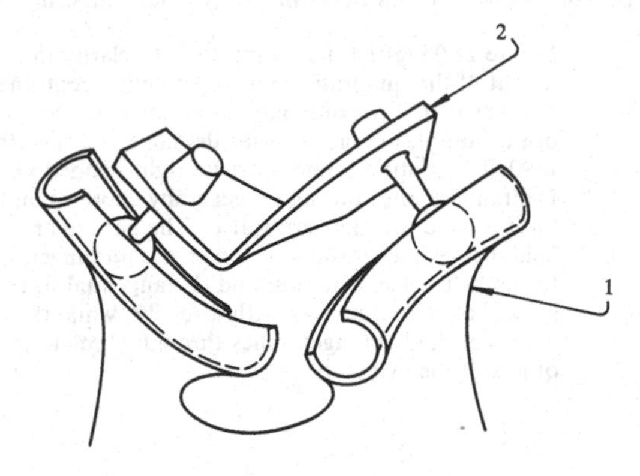


Figure 21.08 (§ 21.31) Another $f2$ joint. The two straight lines through the two pairs of points of contact, each one of which lies alone upon its own flat facet on 1, are the helitangent lines in the moving body 2; the positions of the centre-points of these helitangent lines are not made clear here.

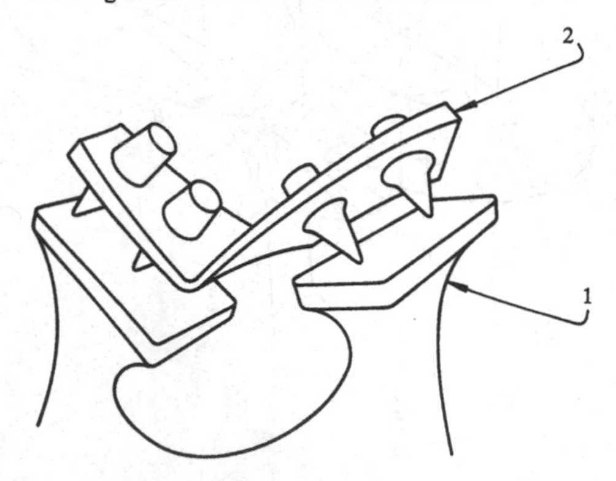

the centres of the balls will be determined by the configuration. Think first in terms of the two common-perpendicular lines that may be drawn with their feet at the centres of the balls; each one will run between its helitangent line and the single axis of a possible motion screw. Think then about these kinds of problem in general; but go next to figures 8.04, 8.05 and 16.01, and perceive there that the problem of locating the cylindroid of motion screws which must exist in figure 21.07 is fully explained in chapter 16.

31. Figure 21.08 shows a pair of another kind of $f4$ joints mounted in parallel between a pair of bodies. In this case we have two helitangent lines that are known at the instant with the planes of their vectors also known, but not their centre-points. Once again a cylindroid of motion screws must exist; but I leave the nature and the implications of this piece of apparatus as an open exercise-ground for the reader.

32. The idea of the helitangent line has great significance in gears (§ 22.17, § 22.41). The essential geometry in a skew gear set (namely a set wherein the axes of the two gear wheels are skew) is such that the line along which the characterizing hyperboloid of one wheel touches the ditto of the other may be called a *straight small tooth.* There are accordingly straight small teeth on both of the wheels, and the crux of the matter is that any operating straight small tooth on one of the wheels must, relative to the other, be a helitangent line. If the mentioned hyperboloids actually coincide with the axodes for the relative motion of the wheels, the straight small teeth in mesh coincide with the motion screw for the relative motion of the wheels and the helitangent line under these ideal circumstances becomes the motion screw itself; it becomes indeed the one and only $ZZ$ line that is available (§ 21.13). All of this is better explained in chapter 22.

22. The axodes of two skew gear sets. For both of them the centre distance $C$ is 40 mm, the gear transmission angle $\Sigma$ is 80° (or 100°), and the speed ratio is 3/5. Notice however that, for a given direction of rotation at the driver, the directions of rotations at the driven wheels are opposite.

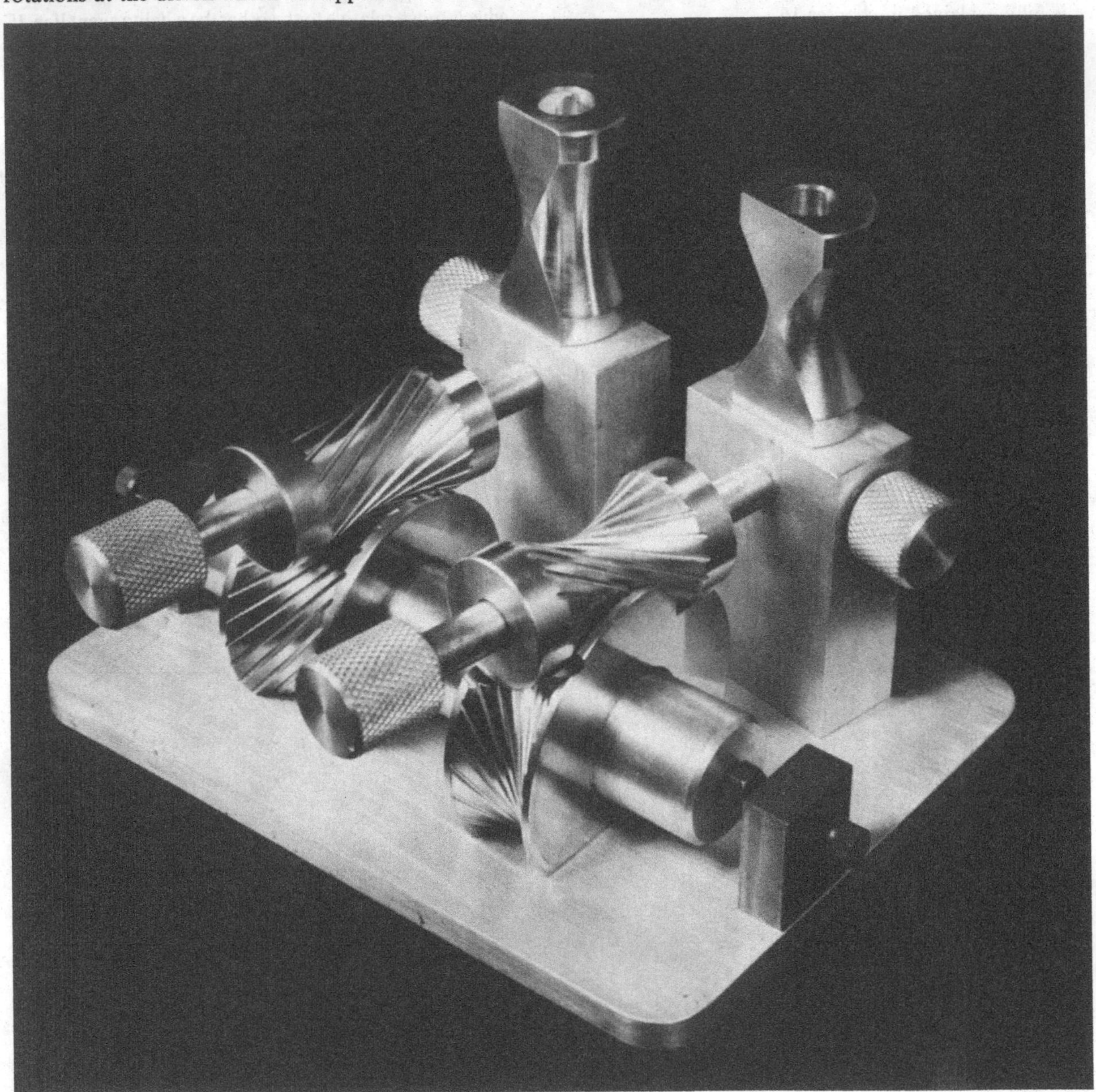

# 22 The cylindroid in gear technology

## Preliminary remarks

01. Let us call by the name *gear set* the combination of the two toothed wheels that mesh with one another and the frame that carries their respective shafts. There are thus in a gear set *three links*: the frame member 1 (or the frame), the driving wheel 2 (the driver), and the driven wheel 3 (the follower). These three form a 3-link loop; and the apparent mobility of the loop is always unity (§1.15, §2.07). Look at figure 2.01(*b*) where the 3-link loop there is mobile; it is however overconstrained because the teeth are held to be in line contact with one another (§2.05). Look next and in advance at the gear sets in figure 22.01; first inspection will reveal single points of contact between the sets of teeth; there is at each set an absence of overconstraint (§22.45). Notice that, whereas the shaft axes of the gear sets in figures 22.01(*b*)(*c*) and (*d*) are not intersecting, those in figures 2.01(*b*) and 22.01(*a*) are, and that the latter are special cases.

02. Most often in machinery, but not always, the function of a gear set is to transmit from one of the wheels to the other a continuous, steady, rotary motion in such a way (*a*) that sufficient power is transferred from wheel to wheel without excessive stress, (*b*) that there is some known *speed ratio* between the wheels, and (*c*) that there is no *pulsation*, no periodic fluctuation of the speed ratio as the successive teeth go into mesh with one another. *While the speed ratio of a gear set is determined wholly and exactly by the ratio of the numbers of teeth upon the wheels, the presence or absence of pulsation is determined by the actual shapes of the profiles of the teeth.*

03. Please notice that these remarks are excluding from consideration those sets we call for example cam, star-wheel, ratchet, and other 3-link sets used for producing the non-uniform, periodic, rotary, and other movements in machinery. Notice also that the connections by means of meshing teeth between some of the pairs of wheels in the various kinds of epicyclic trains have been excluded also. In such trains 4-link and even 5-link loops obtrude, and these complicate somewhat the question of the frames of reference that need to be chosen for the relative angular velocities of the wheels. Provided the wheels are circular wheels and evenly toothed in the accepted sense of the words however, the general principles outlined in this chapter are applicable to all those pairs of toothed wheels as well.

04. In the wide and very complicated area of gear set design we have the mechanics of the steady transmission of power; we have the periodically changing relative motions at the teeth, the changing forces at work at the changing patches of contact, the mechanical properties of the materials of the teeth, and the problems of achieving wedge-film lubrication. These are some of the fundamental matters that are at issue. Many of them are often outweighed for the designer, however, by his necessity for accepting whatever standard sets of tooth cutters are available, by the wholeness of whole numbers, and by the limited methods available for generating the shapes of the teeth in the cutting machines. It is nevertheless held that, in designing a gear set, one should aim attention at the three following matters in the listed order of priority: (1) the likelihood of pitting; this may occur at the tooth surfaces due to a too heavy concentration of the contact pressures across the limited Hertzian areas involved; (2) the strength of the teeth in bending; if this is inadequate the teeth will bend or break off; and (3) the occurrence or otherwise of scoring; scoring at the tooth surface is related with the relative sliding which is inevitable there and with the absence or otherwise of adequate lubrication.

## Common styles of the gear set

05. Whatever the respective shapes of the wheels and the relative arrangements of the cut teeth of a gear set, there are always the following two parameters which are together sufficient to describe the relative locations of the two shafts: (*a*) the common perpendicular distance between their central axes; and (*b*) the angle between those axes. The mentioned angle can be seen to be measured in one or other of the pair of parallel planes which can be drawn to contain the two axes. The pair of planes is of course unique, each plane

being normal to the common perpendicular distance.

06. In cases where the above-mentioned angle is neither zero nor a right angle it is common in gearing practice to measure that angle, often called Σ, according to an agreed convention. It is, after Merritt [1], 'the angle, viewed along the common perpendicular, about which the axis of the more distant shaft would have to be rotated in order that the shafts may be brought into parallelism whilst having opposite directions of rotation'; and it is, after Baxter [2], 'the angle between the axes which contains the relative angular velocity vector and lies between 0 and 180°'. This measure, distinguishing the acute from the obtuse angle, takes account of the relative directions of the two angular velocities of the shafts. The measure has the advantage that the angle Σ is, thereby, always the sum of the *helix angles* of the respective sets of teeth that are cut into the gear blanks (§ 22.29). This important yet somewhat confusing matter of the angle Σ of a gear set is mentioned again at § 22.62; but please refer just now, and in advance, to photograph 22. Be aware there that, whereas the angle $\Sigma_L$ for the gear set at the left of the picture is 80°, the corresponding angle $\Sigma_R$ for the set at the right is 100°. The pairs of shafts at photograph 22 are identically arranged, the speed ratio for the two sets is the same (3:5), but the relative directions of rotation of the shafts at the left and at the right are opposite.

07. Having seen the significance of this combination of (*a*) the common perpendicular distance and (*b*) the relevant angle between the shafts, namely the combination of the *centre distance C* and the *shaft angle* Σ, it becomes clear that each type or style of gear set can be characterized according to whether it can find its place in a group defined by a pair of special values of *C* and Σ. Some examples follow: (*a*) spur, internal-spur, helical, and herring-bone gears all belong in the category where *C* is non-zero while Σ is zero; (*b*) bevel gears, whether straight or spiral, belong where *C* is zero; (*c*) 'crossed helical' and other 'skew' gears belong where *C* and Σ are both non-zero; and (*d*) right-angular worm drives and the hypoid sets belong where *C* is non-zero while Σ is taking the special value 90°; for a rough impression of these latter refer to figure 22.01. All of the above kinds of gear set with their specially cut teeth of various kinds, many of which still reveal by their names their accidental origin or their origin by cunning in the history of technology, are well known in engineering practice. There is an enormous literature about the possibilities for manufacture, methods for kinematic and static analyses, means for lubrication, and methods and tables for tooth cutting and design. There was the book by Merritt (1942); older knowledge is outlined by Buckingham (1949); more recently Tuplin has written famously (1950); there is the well known handbook edited by Dudley (1962); and there is the recent book by Litvin (1968). I might also mention here Abbot [3], and Dyson (1969), both of whom made contributions to the theory, Abbot to the shapes of blanks, and Dyson to the shapes of teeth.

Figure 22.01 (§ 22.07) Some related gear sets of the kind where the centre distance *C* is finite non-zero and the angle Σ is a right angle. This chapter deals with the main parameters characterizing a gear set. It introduces various concepts from screw theory, it deals with the geometry of the meshing of the teeth, but it pays little attention to the actual shapes of the profiles of the teeth.

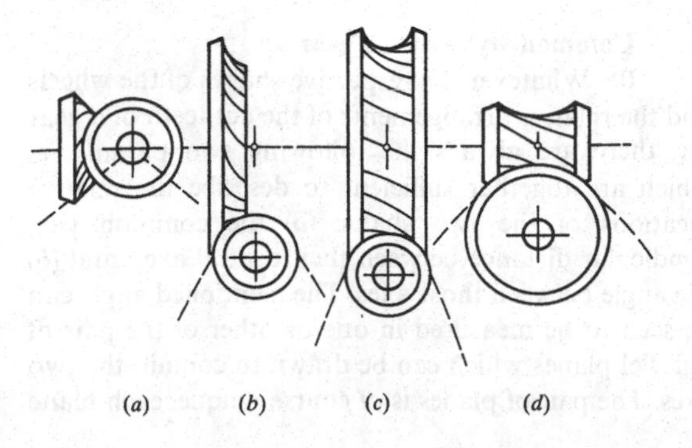


08. But in the literature there appears to be no thoroughgoing thread to bind the different kinds of gear set and the various tooth arrangements together from the kinematic point of view; and I say this despite for example the chapter 1 by Baxter in Dudley (1962). There seems to be no logical connection, for example, between the circumferentially moving tooth of a worm and the crosswise (namely the circumferential) sliding it suffers while in mesh, and the radially sliding tooth of an orthodox, involute spur wheel which suffers as it meshes no crosswise (namely axial in this case) sliding at all. There is, similarly, no easily understandable explanation for the fact that, while some teeth must be cut 'straight' (§ 22.11), other teeth may be cut 'curved' (§ 22.68). Is there a geometrical connection for example between the helix angle of a helical tooth at the pitch radius of a spur blank and the various helix angles of a worm at its various radii? Or are the two phenomena thoroughly unrelated? Upon what criteria might the helix angles be chosen for the two sets of teeth in mesh at a pair of skew gears? In view indeed of the curved nature of some cut teeth, must the helix angles for the teeth of a skew set remain constant across the cylindrical blanks, or may they vary? These are the kinds of question I would like to ask and then to attempt to answer in this chapter; and the best method that I can see for beginning to do that is to take a general numerical example (§ 22.18).

09. By way of introduction to the forthcoming numerical example, I would ask the reader to be aware

that a gear set of a certain speed ratio is required. It is needed to transmit rotation between a pair of skew shafts where the centre distance $C$ and the shaft angle $\Sigma$ are both already determined. For a combination of unlikely practical reasons that are irrelevant here, let us say that $C$ must be 40 mm, that the speed ratio $k$ must be $\pm 0.6$, and that the angle $\Sigma$ must be 80° (or 100° alternatively). Refer again to photograph 22 where the layout of the pairs of shafts and the cut wheels there conform with these dimensions. Please also see the wheels there as being equipped with straight teeth of infinitesimal depth and thickness, and be aware that I refer to such teeth as *straight small teeth*. The cut lines upon the wheels might be seen to be the *tooth lines* of representative ones of these straight small teeth (§ 22.10), and please see that those lines are so spaced upon the wheels that the pitches of the teeth can be seen to correspond at meshing.

10. The numerical example that follows – it begins at § 22.18 – will, and without any loss of generality or applicability I believe, ignore all questions of power transmission, distributions of pressure across the Hertzian areas of contact, tooth size, methods of cutting, profile geometry, and lubrication. I will assume in the first instance that the gear wheels are to be machine-cut by straight strokes of a pointed tool along successive generators of their ruled surfaces. In this way the mating teeth will be straight small teeth, and all of the usual practical problems concerning the finite region of mesh and the actual shapes of the profiles of the teeth will not obtrude. I will concentrate first upon the following two matters: (*a*) the layout of the lines of the straight small teeth, namely the layout of the tooth lines, and (*b*) the rate of crosswise sliding at the straight small teeth in mesh. The calculation will thus fall within the category (3) of § 22.04. It is hoped that the worked example will help to begin to show that there is a thoroughgoing thread of logic among the gear sets. I hope to show that this logic can, not only be made to be intelligible, but also be seen to be a skeletal base upon which the existing wide range of somewhat unrelated theory might be otherwise organized and thus be better understood.

**The mechanics of straight small teeth in mesh**

11. It is well known and accepted that for ordinary spur gears with involute teeth the $ISA_{23}$, which is a screw of zero pitch, is called the *pitch line*; refer for Kennedy's theorem to § 13.06. The tooth line of a straight small tooth coincides with this pitch line when the *contact line* (which is a straight, moving line, the line of contact between the faces of mating teeth) is passing through the *pitch point*. The pitch point is a mere geometrical thing. It has its meaning here in two dimensions only; the pitch point resides in the reference plane for the planar motion (§ 2.30); the pitch point is simply the trace upon the reference plane of the pitch line. Correspondingly the two *pitch circles*, which are tangential to one another at the pitch point, are simply the traces there of the two *pitch cylinders*. By revolving the pitch line about the two shaft axes in turn, we can generate the two pitch cylinders. These can then be seen as the two smooth wheels – they are the axodes for the relative motion – that roll on one another without slipping as the spur gear set engages in its action; *and the ratio of the radii of these axodes is the speed ratio of the spur gear set.*

12. In spur gears it is also well known (*a*) that the pitch line cannot be moved either closer towards or further away from either one of the shaft axes without increasing or decreasing the speed ratio, and conversely that, if the speed ratio is to be altered, the pitch line must be moved accordingly; (*b*) that such movements of the pitch line can only be permitted to occur in discrete steps; these steps must suit exactly each new vulgar fraction of the tooth ratio of the wheels; and (*c*) that under no circumstances would the pitch line be tilted away (in the plane of the two shaft axes) from its perpendicularity with the common perpendicular line.

13. If this latter, this (*c*), ever were imagined to occur and the tooth lines of the straight small teeth in mesh were tilted to try to suit that tilted line, the sets of teeth in mesh would simply destroy one another. The opposing conical shapes of the two wheels, in contact along their common tangent line, would not be compatible with one another. The relative motions of the mating wheels at the line of tangency would not be consistent with our idea of rolling; while circumferential slip was occurring in one direction at one end of a pair of mating wheels, an opposite circumferential slip would be occurring at the other end of the same pair, and that would clearly be inadmissible. Another way of making the same objection to the idea of a tilted pitch line would be to say that, under no circumstances, with straight small teeth in mesh in the plane of the two shaft axes, should tooth lines in mesh be other than parallel with the vector of the relative angular velocity $\boldsymbol{\omega}_{23}$. Otherwise the teeth would be *boring* at the meshing (§ 22.41).

14. Transporting these elementary ideas into the 3-dimensional arena where the two shaft axes are in general skew, we should see first of all that, before two straight small teeth can mate with one another satisfactorily, the two tooth lines of the teeth must coincide with one another at their intersections with the common perpendicular line in such a way that the two tooth lines themselves, and the line of tangency of the two wheels, coalesce. The wheels are not circular cylinders now but circular hyperboloids of revolution (§ 11.15), and, for the same reasons as those outlined above, at § 22.13, the line of tangency must cut the common perpendicular line at

no angle other than a right angle. If from nowhere else this follows from the material of chapters 5 and 13. There the general kinematics of the ISA in chapter 5 and the kinematical theorem of three axes in chapter 13 are discussed in sufficient detail to throw some early light onto this present matter.

15. Now for reasons yet to become apparent, and in the cases of all gear sets, *I wish to call the line of the $ISA_{23}$ the pitch line, and I wish to call the point of intersection of this line with the common perpendicular the pitch point P*. This is consistent with what I have said already about the special case of the spur gear set. It does imply also however that, although in the general case of the skew gear set the common tooth line of the teeth in mesh may be put in practice anywhere along the common perpendicular (§ 22.34), *the pitch line and the pitch point P will be unique.*

16. I refer now to the relevant theorem of three axes. Fully explained at chapter 13, it deals with the relative motion of three bodies. I say this: *if the common tooth line of the straight small teeth in mesh coincides with the $ISA_{23}$, that is if the tooth line of the teeth in mesh is seen by the designer not to be just anywhere along the common perpendicular line but exactly at the pitch point P, then (and I speak now from the purely kinematic point of view) the meshing of the straight small teeth will be entirely satisfactory*. Under these, the most ideal of circumstances, there will be no boring at the meshing (§ 22.13, § 22.41), and even very long, straight small teeth will be able to roll exactly, and thereby mesh with one another satisfactorily. The inevitable sliding at the straight small teeth, moreover, and this is the *crosswise sliding* I mentioned earlier (it occurs in the direction of the tooth line), is reduced thereby to its minimum. The ratio of the magnitude of the rate of crosswise sliding $\mathbf{v}_{2P3}$ (mm sec$^{-1}$) at the teeth and the magnitude of the relative angular velocity $\boldsymbol{\omega}_{23}$ (rad sec$^{-1}$) of the wheels will be, of course, the pitch $\mathrm{p}_{23}$ (mm rad$^{-1}$) of the screwing at the teeth; this comes directly from chapter 5; see § 5.47 to § 5.51 inclusive.

17. Chapter 21 also touches on practical matters (§ 21.32). I wish to make at this stage, and based on the material of chapter 21, the following as yet obscure, but considered statement: *while the tooth line of a straight small tooth in 3 need not necessarily pass through the pitch point as it passes through the centre-distance line, it must be a helitangent line in the body 2 as it does so; otherwise the meshing between that tooth and its mating tooth will be entirely unsatisfactory.* What I am intending by saying this now is (*a*) to make it quite

Figure 22.02 (§ 22.18) This is the data for the chosen problem in skew gear design. The variables $C$, $\Sigma$, and $k$ have been set at the values 40 mm, 80°, and $\pm 0.6$ respectively. A rather awkward trimetric projection is employed for the picture here, but please notice its exact correspondence with that at figure 22.03. Please peruse in advance the equations from (1) to (8) which are listed in table 22.01. The table is mounted next to figure 22.05.

Figure 22.03 (§ 22.20) The cylindroid for the gear set. Its size and location is wholly determined by the layout in space of the two shaft axes, namely by the centre distance $C$ and the shaft angle $\Sigma$. The speed ratio $k$ ($k$ in the worked example is $\pm 0.6$) determines which generator of the cylindroid becomes the pitch line. In the figure here straight small teeth are shown at the pitch line for each of the two different cases $L$ and $R$.

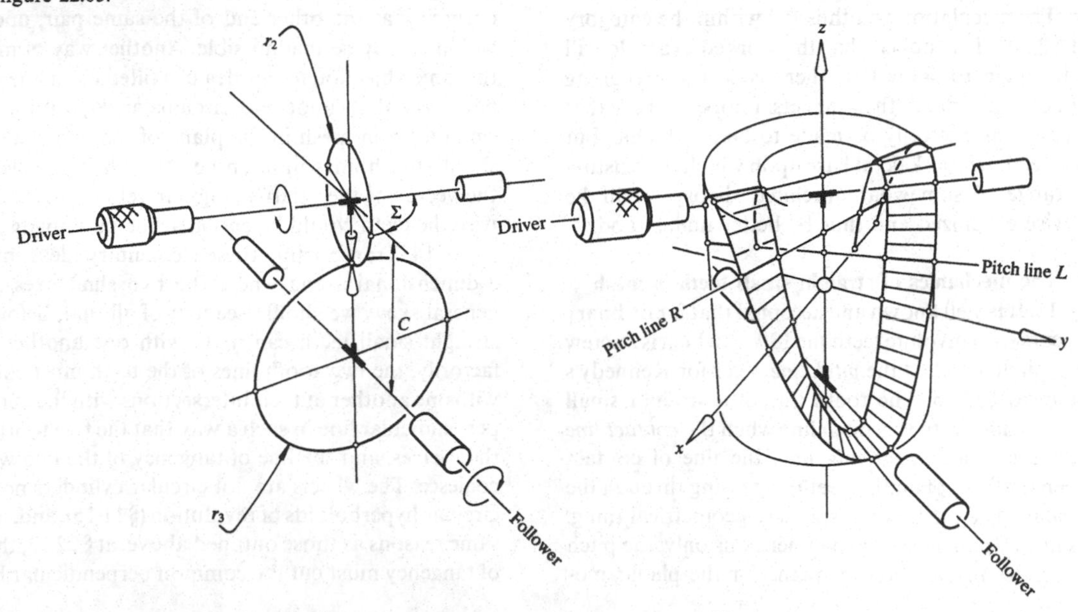

clear that the tooth line need not necessarily be at the pitch line, (*b*) to remove from our consideration the unwelcome possibility that *boring* at the meshing may occur (§ 22.13, § 22.41), (*c*) to let us see that proper *rolling* will occur about the tooth line under the mentioned circumstances, and (*d*) to imply that a measure of *rocking* at the meshing is acceptable (§ 22.42). I say this latter, this (*d*), in view of the facts that straight teeth are never infinitely long and that they can be, and usually are in practice, *straight-wound*. Refer for the concept of teeth that are straight-wound to § 22.35 and § 22.43. It might thus be seen in advance in gear-set practice that, *although the unique pitch line for a given set and a given velocity ratio is from many point of view the best location for the teeth to be in mesh, the teeth in mesh may be put, as they very often are, elsewhere along the centre-distance line.* This important question will be taken up again at § 22.34.

**Numerical example**

18. Refer to figure 22.02 where the data for the chosen problem in design has been displayed. The locations of the shaft centre-lines for the driver 2 and the follower 3 have been determined already, and the speed ratio is to be 3:5, $k = \pm 0.6$. Suitable angular velocities for the wheels have been predetermined also: they are to be $\boldsymbol{\omega}_{12} = 10$ rad sec$^{-1}$ for the driver 2 in the shown direction and $\boldsymbol{\omega}_{13} = 6$ rad sec$^{-1}$ for the follower 3. In the matter of the direction of rotation of the follower shaft, I shall do the forthcoming calculations according to both of the only possibilities, and these will be identified by the following symbols to be used subscriptively: L for the follower rotating as shown in figure 22.02; please examine the wheels at the left hand side of photograph 22; and R for the follower rotating in the opposite direction; please see the different set of wheels at the right hand side of the same photograph (§ 22.06).

19. I am aiming in the first instance at an entirely satisfactory meshing between straight small teeth (§ 22.16); so the first thing to do, in order to find the pitch line the $ISA_{23}$ in each case, is to determine the magnitudes and directions of the relevant angular velocities $\boldsymbol{\omega}_{23}$. I have done this in figure 22.04 by ordinary vector polygon (§ 12.12). The polygons are both mere triangles, the calculations are by ordinary trigonometry, and the results are shown: $(\boldsymbol{\omega}_{23})_L = 12.52$ rad sec$^{-1}$ inclined at $(\psi_3)_L = 51.85°$ to the follower shaft axes, and $(\boldsymbol{\omega}_{23})_R =$ 10.73 rad sec$^{-1}$ inclined at $(\psi_3)_R = 66.60°$ to the same shaft axis. The angles $(\psi_3)_L$ and $(\psi_3)_R$ are so named because they are automatically now the helix angles for the straight small teeth on the follower 3 at L and R respectively. Notice in passing that, while both sets of teeth are left handed sets at the gear set L, both sets are right handed sets at R. This is clear to see at photograph 22 where the actual wheels of the sets have been cut to coincide with the hyperboloidal axodes. By subtraction from $\Sigma_L$ and $\Sigma_R$ respectively, $(\psi_2)_L = 28.15°$, and $(\psi_2)_R$ $= 33.40°$; both of these are also shown in figure 22.04.

20. The next step, to discover the unique radii $(r_2)_L$ and $(r_2)_R$ of the hyperboloidal axodes, involves either the cylindroid formulae of chapter 15 employed directly, or an application of the theorem of three axes from chapter 13. The algebra is very simple in the case of the gear set, there being a particular 3-link loop (§ 22.01); the pitches of the rates of twisting at $ISA_{12}$ and $ISA_{31}$ are not only equal but both zero. I shall accordingly use the simple equations of chapter 15 directly. Please erect Cartesian axes $O\!-\!xyz$ according to figure 15.03 and take a view of the gear set looking downwards along the common perpendicular; see the centre part of the composite figure 22.04. There and in figure 22.02 it will be clear that, because the distance $b$ and the angle $\eta$ of figure 15.03 have been set at 20 mm and 40° respectively, the angle $\theta$ of figure 15.05 takes the values $\theta_L = 11.85°$ and $\theta_R = 73.40°$. Substituting these directly into the simple equation (8) which appears in the box at figure 15.06 we find that $z_\theta$, the displacement of the pitch line the $ISA_{23}$ upwards and away from the centre of the cylindroid, takes the values $(z_\theta)_L = 8.16$ mm and $(z_\theta)_R = 11.12$ nm. Refer now to figure 22.03. There the relevant cylindroid, which is wholly defined by the two motion screws of zero pitch at the two shaft centre lines, is shown superimposed upon those centre lines. The generators of the cylindroid at $(z_\theta)_L$ and $(z_\theta)_R$ are the two pitch lines for the L and the R respectively, and they are clearly shown in figure 22.03. For the sake of clarity, however, they are not dimensioned there. From the above-discovered values of $z_\theta$ it follows immediately by ordinary arithmetic that the unique radii $r_2$ and $r_3$ of the hyperboloidal axodes (the wheels in photograph 22) are as follows: $(r_2)_L = 11.84$ mm, $(r_2)_R = 8.88$ mm, $(r_3)_L$ $= 28.16$ mm, and $(r_3)_R = 31.12$ mm. These four radii are displayed in figure 22.04.

21. The total length $2B$ of the cylindroid, incidentally, which can be obtained from equation (3) in the box at figure 15.04, is only 40.62 mm; the length $2B$ is only slightly longer than the length $2b$ in other words; and the reason for this is that the angle $2\eta$, namely $\Sigma_L$ at the present instance, is only slightly less than a right angle (§ 15.49).

22. The striking fact that is now apparent is that, having calculated at § 22.20 the radii of the hyperboloidal axodes namely the pitch radii of the wheels, and knowing that straight small teeth on these axodes will give (*a*) an entirely satisfactory meshing, and (*b*) a minimum of crosswise sliding, the ratios of the radii bear no apparent or simple relationship with the speed ratio; the magnitude of the speed ratio for the set problem is

0.60, but the ratios of the pairs of the radii that we have found are 0.42 and 0.29 for L and R respectively. Refer also to the worked exercise (3) at figure 13.06 where the same phenomenon is apparent (§ 13.23).

23. The above most important statement throws into prominence, of course, the special case of the *spur* gear set which occurs when the angle Σ and thus the double angle $2\eta$ both become zero. Taking the same data otherwise, the cylindroid becomes of infinite length in those circumstances, the geometry degenerates, and it is not very easy to see, just yet, how the formulae of chapter 15 might give the well known, only answer in that case, namely that the ratio of the pitch circle radii must exactly equal the speed ratio. The matter of the special case of the spur gear set will come up again, at § 22.30.

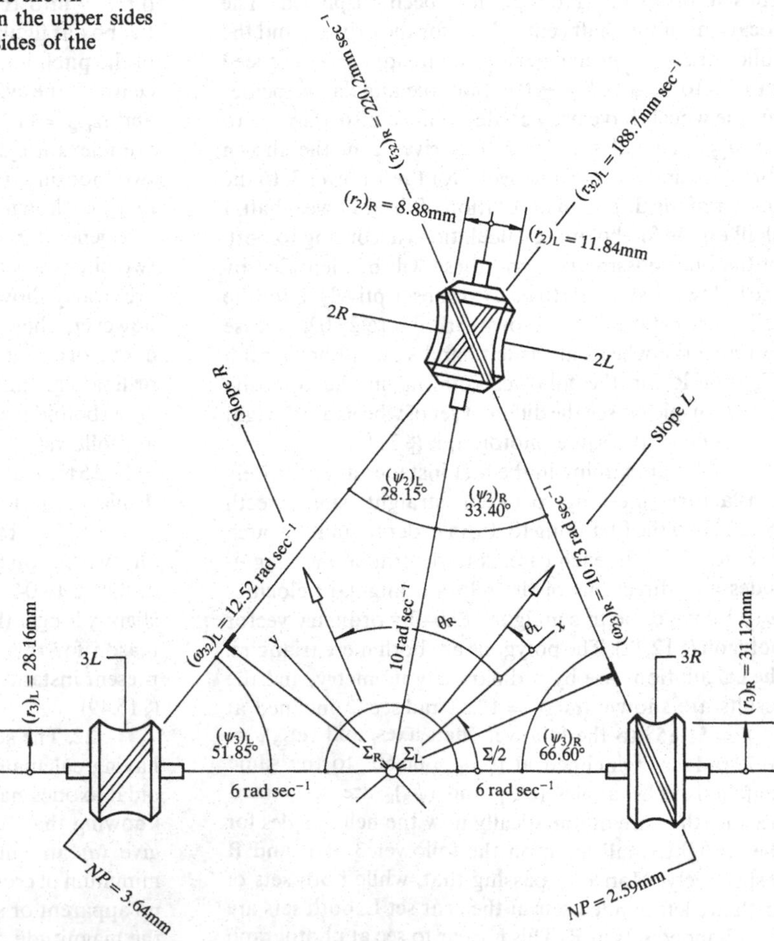


Figure 22.04 (§ 22.23) This omnibus diagram shows not only the two angular velocity polygons of the design problem which together determine almost everything else, but also the axes $O$–$x$ and $O$–$y$ and other results of the calculation. The normal pitches (NP) shown at the lower ends of the figure are calculated at the gorges of the wheels for numbers of teeth 18 and 30 at each gear set. These numbers accord with the given speed ratio 3:5 (±0.6) and with the straight small teeth that are actually cut in photograph 22. The few teeth that are shown diagrammatically in the figure are on the upper sides of the followers 3 but on the lower sides of the drivers 2.

**A first group of useful formulae**

24. The first three formulae appearing at table 22.01 are derived from arguments outlined in recent paragraphs. Equation (1) is an explicit expression for $\theta$ in terms of shaft angle $\Sigma$ and speed ratio $k$ where $k$ is the ratio less than unity; it is derived directly from the trigonometry of the angular velocity polygons in figure 22.04. Equation (2) is simply a rewrite in local terms of equation (8) in the box at figure 15.06; it gives the pitch radius $r_2$ in terms of centre distance $C$, shaft angle $\Sigma$ and the discovered angle $\theta$. Equation (3) is obvious. These three equations, derived as they are from cylindroid geometry, state the unique requirements for satisfactory meshing at straight small teeth. They fail to give expressions for $r_2$ and $r_3$ independently of $\theta$ however.

Table 22.01 (§22.24). *These sets of equations which relate to the figures 22.02 and 22.03 derive from various authors. Those of my own appear in the top box. Each of the authors has minimized the magnitude of crosswise sliding at the teeth, and all of the equations reconcile with one another.*

From the cylindroid at chapter 15:

$$(1)\quad \theta = \Sigma/2 - \arctan[(k \sin \Sigma)/(1 + k \cos \Sigma)]$$
$$(2)\quad r_2 = C/2[1 - \operatorname{cosec} \Sigma \sin 2\theta]$$
$$(3)\quad r_3 = C - r_2$$

Konstantinov:

$$(4)\quad r_2 = C\left[\frac{k^2 + k\cos\Sigma}{1 + 2k\cos\Sigma + k^2}\right]$$
$$(5)\quad r_3 = C\left[\frac{1 + k\cos\Sigma}{1 + 2k\cos\Sigma + k^2}\right]$$

Steeds:

$$(6)\quad \psi_2 + \psi_3 = \Sigma$$
$$(7)\quad \frac{\sin\psi_2}{\sin\psi_3} = k$$
$$(8)\quad \frac{r_2}{r_3} = \frac{\tan\psi_2}{\tan\psi_3}$$

They thus fail to show as clearly as they might the ratio of the pitch circle diameters.

25. By rewriting (1) in the form $\tan\theta = [(1-k)/(1+k)]\tan\Sigma/2$, by substituting this into (2) while noting at the same time that $\sin 2\theta$ may be expressed in terms of $\tan\theta$, we can arrive at equations (4) and (5). Each of these is self explanatory, and the pair of them together reveal the mentioned ratio. The ratio is neither $k$ nor $k^2$, incidentally, both of which simple expressions have been claimed by designers to be suitable for making a first choice in design. The ratio is, as can be seen, $(k^2 + k\cos\Sigma)/(1 + k\cos\Sigma)$. Unless $\Sigma$ is zero, in which case the skew set becomes the special spur gear set (§ 22.30), *the ratio of the pitch circle radii is not independent of* $\Sigma$.

26. Precisely these equations, namely (4) and (5),

Figure 22.05 (§ 22.27) The separate sinusoidal variations of $z$ and $p$ with $2\theta$, both extracted from chapter 15, are shown here. Superimposed upon the curves are the results of the numerical calculation (§ 22.32). What happens when, with a constant speed ratio $\pm k$, we depart by $\Delta r$, and consequently swing by $\Delta\psi$, is also shown by means of the tangent-function curves (§ 22.39).

$$z = (C/2)\operatorname{cosec}\Sigma\sin 2\theta$$
$$p = (C/2)\cot\Sigma - (C/2)\operatorname{cosec}\Sigma\cos 2\theta$$
$$\Delta r = p\tan\Delta\psi$$
$$(r_2)_L = 11.84\ \text{mm}$$
$$(r_2)_R = 8.88\ \text{mm}$$

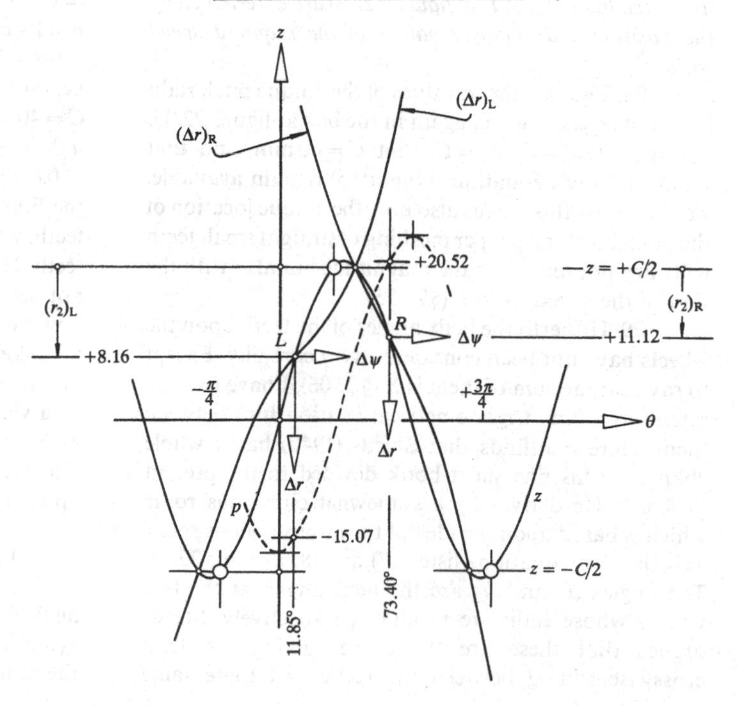

have been reported by Konstantinov (1980), and similar sets of equations appear in Litvin (1968). These authors arrived at their respective formulae by looking for that particular tooth line where the rubbing velocity at points along the tooth line might be in the direction of the tooth line itself, and in this respect their objective (and their result) has been the same as mine.

27. Refer to figure 22.05 where the distances $z$ measured axially from the centre of the cylindroid (and the pitches $p$ by means of the broken line) are plotted against the angle $\theta$. In the figure the shaft axes are shown to be set at their coordinates ($-40°$, $-20$ mm) and ($+40°$, $+20$ mm) as previously arranged (§ 22.18). The discovered pitch radii $(r_2)_L$ and $(r_2)_R$ appear at the points marked L and R. Note that the expression for $z$ (in the box at the figure) is an equation to the surface of the cylindroid; it is the same as equation (8) in the box at figure 15.06. The continuous sine curve drawn for $z$ is, indeed, an illustration of the surface of the cylindroid. If the magnitude of the required speed ratio $k$ were to be diminished from its given value 0.6, the points $L$ and $R$ would move unevenly but both upwards towards the crest of this sine curve; if increased they would move unevenly but both downwards along the same sine curve. It is important to see here that, *while this sine curve (indeed the whole of the cylindroid including the sinusoidal distribution of its pitches p) is solely and wholly determined by the relative locations of the two shaft axes and therefore does not change for a given pair of shaft axes, the positions of the points marked L and R upon it (the pitch lines L and R in figure 22.03) are determined by the positive and negative values of the required speed ratio.*

28. The calculated values of the unique pitch radii $(r_2)_L$ and $(r_2)_R$ are given again in the box at figure 22.05. Knowing that $r_2 + r_3 = C$, that $C = 40$ mm, and that Konstantinov's equations (4) and (5) remain available, we can see by this means also how the unique location of the pitch line for a proper meshing of straight small teeth will vary not only with the magnitude but also with the sign of the speed ratio $k$ (§ 22.25).

29. Hitherto the helix angles of the teeth upon the wheels have not been considered very carefully. Except to say that the sum of them is $\Sigma$ (§ 22.06), I have paid no attention to any trigonometrical relationships between them. Here one finds that Steeds (1940) has a whole chapter in his fine short book devoted to the present problem. He derives by a somewhat circuitous route which is based upon the idea of the instantaneous screw axis the two equations listed (7) and (8) at table 22.01. The angles $\psi_2$ and $\psi_3$ are the helix angles at the two wheels whose radii are $r_2$ and $r_3$ respectively. Steeds argues that these are the angles giving the least crosswise sliding between the teeth; and these same angles, I having employed my own terminology in describing Steeds, appear at the angular velocity polygons in figure 22.04. Steeds's equation (7) follows directly from the sine rule applied in these triangular polygons, while equation (8), as Steeds explains, follows directly from the fact that the normal circumferential pitch (NCP or NP) of the teeth upon the wheels (and this applies of course even for infinitesimally small teeth) must remain the same for proper meshing. Taking now equation (8) as a guide, drop perpendiculars from the centre point 1 of the polygons in figure 22.04 onto the sides marked $3L$-2 and $3R$-2, and perceive the following: *the feet of these perpendiculars divide the mentioned sides of the polygons into the proportions $r_2$ to $r_3$.*

### The special case of the spur gear set

30. Having collected the relevant formulae at table 22.01, and having constructed the curves for the numerical example in figure 22.05, it is possible now to fit the special spur gear set into the general scheme of things (§ 22.08, § 22.23). If the angle $\Sigma$ is taken all the way to zero the set becomes in the limit a spur gear set, the characteristic cylindroid becomes infinitely long (§ 15.54), and the equation (1) breaks down. The equations (2) and (3) also break down when $\Sigma = 0$, the radii $r_2$ and $r_3$ becoming indeterminate. Steeds's equations (6) (7) and (8) break down too. The alternative formulations of Konstantinov at (4) and (5), however, allow the radii $r_2$ and $r_3$ for a spur gear set to be determined: putting $\Sigma$ to zero reveals, from (4) that $r_2 = Ck/(k+1)$, and from (5) that $r_3 = C/(k+1)$. These results coincide with our experience: for a given speed ratio $\pm k$ and a given centre distance $C$ there is for a spur gear set a pair of pitch radii for each of the two cases. For $C = 40$ mm and $k = \pm 0.6$ we find (*a*) for $k = +0.6$ that $(r_2)_L = \pm 15$ mm and $(r_3)_L = +25$ mm, and (*b*) for $k = -0.6$ that $(r_2)_R = -60$ mm and $(r_3)_R = +100$ mm. In the first of these two cases both wheels have external teeth, while in the second the larger wheel has internal teeth. The set of the second case is a ring gear set. We are not yet at the end of this matter of the spur gear set however, for I have yet to explain why it is that, whereas for a skew gear set for a given speed ratio the actual radii may depart from the unique pitch radii $r_2$ and $r_3$ and still be a viable pair of radii from practical points of view, *such a departure from the pitch radii is not possible in the spur-gear set.* This somewhat puzzling matter is taken up again at § 22.40.

### Least crosswise sliding at the teeth

31. Given $C$, $\Sigma$, and the velocity ratio $k$, the above analysis has revealed a unique pair of pitch radii $r_2$ and $r_3$ and a unique pair of helix angles $\psi_2$ and $\psi_3$. These are the radii and the angles that must obtain if we wish to

reduce to its minimum the unavoidable crosswise sliding at the straight small teeth while achieving at the same time a proper meshing of those teeth without any *boring or rocking*. The terms boring and rocking are soon to be explained (§ 22.41). One basic premise of the above analysis was our stated idea that the two hyperboloidal axodes (the axodal surfaces or the *axodal sheets* as I now wish to call them) each carries upon its own infinity of straight-line generators its own infinity of the infinitesimally small, straight small teeth. These axodal sheets might be imagined now to be smooth in the directions of their generators but rough in all other directions so that they can easily engage in that combination of rolling and sliding which is, indeed, characteristic of their actual relative motion. To see the matters at issue here more clearly now, *invert* the 3-link loop of the gear set (§ 1.47). Holding the follower 3 at rest, rotate the frame; observe the driver 2 screwing with respect to the now stationary follower. Refer to § 5.43 and note the following: it is in the nature of the $ISA_{32}$ – or the $ISA_{23}$ which is the same (§ 5.58) – (*a*) that the points in body 2 which happen to be along it are precisely those points in the body 2 with least velocity, and (*b*) that the directions of the velocities of those points are all in the direction of the ISA. That is what I mean by the absence of boring and rocking: *no point anywhere along a small straight small tooth in mesh at the pitch line has its linear velocity relative to the mating wheel in a direction other than that of the line of the tooth itself.*

32. Before going on to discuss the departures from this ideal condition which abound in engineering practice, I should complete the calculations above and show how to obtain the velocity of the crosswise sliding that is the minimum and that occurs along the straight small teeth in mesh of the gear wheels at photograph 22. The four wheels in the photograph, incidentally, are manufactured to match exactly the results of calculation already assembled at figure 22.04. From equation (10) in the box at figure 15.07 we can write for the pitch $p$ of the relative screwing in our particular case equation (9) appearing in the box at figure 22.06. It refers to the pitch $p_{23}$ of the relative screwing described at § 22.30. Its numerical values at the relevant $\theta_L$ and $\theta_R$ turn out to be, on substitution, $-15.07$ mm rad$^{-1}$ and $+20.52$ mm rad$^{-1}$ respectively. These pitches occur in conjunction with the angular velocities namely the rates of rolling at the pitch lines $(\omega_{23})_L$ and $(\omega_{23})_R$; these are found from the polygons to be 12.52 rad sec$^{-1}$ and 10.73 rad sec$^{-1}$ respectively. It follows that the rates of crosswise sliding, from $\tau_{23} = p\omega_{23}$ (§ 5.50), are 188.7 mm sec$^{-1}$ and 220.2 mm sec$^{-1}$ respectively.

33. That these are relatively high is due to the fact that the angles $\Sigma$ are relatively high. I am not interested here to discuss any matters of friction, lubrication, or of power loss due to friction, along the straight lines of contact between the straight small teeth at the pitch line. Modification of such matters by the cutting of real teeth of finite size and other considerations and problems of design are not my immediate concern. I am however thinking here about the existence of worms and hypoid sets, and of other clear departures from the idealized devices presented hitherto, and I wish to discuss the gear-wheel geometry of such departed devices now.

**The concepts of departure $\Delta r$ and of swing $\Delta\psi$**

34. It is well known in engineering practice (*a*) that the actual teeth in skew gear sets are not cut straight, but often *helical*, and (*b*) that the ideal pitch circles at the radii $r_2$ and $r_3$ as calculated above are hardly ever calculated in the process of design, let alone adhered to in production. Reasons for (*a*) are many: first it is clear that straight teeth of finite size, unless they are spur gear teeth, must have an increasing normal pitch (NP) as we go outwards along any tooth from the throat of the hyperboloidal wheel; but second and most importantly it is a fact that unless the tooth line at the meshing is put at the pitch line the hyperboloids produced by its revolution in turn about the two wheel axes (which are, thereby, no longer the actual axodes for the relative motion, but what I wish to call the *departed* hyperboloidal sheets) will intersect at the tooth line and the small straight teeth will interfere with one another (§ 22.43). Reasons for (*b*) are also many: first it is often necessary with low values of $k$ (in worm drives for example) to depart upwards and away from $r_2$ in order to provide a hole big enough in a bored worm or a blank diameter big enough for strength; refer in advance to figure 22.07(*d*); and second there are geometrical considerations for finding a better approximation to 'line contact', for lubrication and for wear. Refer to Abbott [3].

35. I wish to leave aside just now all those other gear elements where the teeth are not helical but *spiral* in the sense that they are not straight-wound onto a circular cylinder (§ 22.43) but otherwise wound onto a cone or some similar non-cylindrical surface (§ 22.68). I wish to proceed here with discussing the necessary relationship that must exist in any skew set with cylindrical wheels between the actual radius $(r_2 + \Delta r)$ of either one of the wheels and the helix angle $(\psi_2 + \Delta\psi)$ of the infinitesimally small teeth of that wheel.

36. Looking at Steeds's equation (8) at table 22.01, we see that that is a necessary relationship whatever the radii of the wheels may happen to be. It derives from the necessity that the NP for *short* straight teeth, in the neighbourhood of the common perpendicular line, namely in the neighbourhood of the throats or at the 'gorges' of the wheels, must be equal with one another

for proper meshing. For given $C$, $\Sigma$ and $k$ we can see that, if we allow the actual radii to depart upwards and downwards from the pitch radii $r_2$ and $r_3$ by an amount $\Delta r$, we must adjust the helix angles $\psi_2$ and $\psi_3$ upwards and downwards by an amount $\Delta\psi$ in such a way that equation (10) in the box at figure 22.06 may be satisfied. Equation (9) in the box is mentioned at figure 22.05 and comes from chapter 15, equation (10) expresses the fact (for a given speed ratio) that the NP at the teeth must be the same for both wheels as the radii are varied, while equation (11) in the same box has yet to be derived.

37. Notice in equation (10) that, because $r_2 + r_3 = C$, a constant, and because $\psi_2 + \psi_3 = \Sigma$, another constant, both $\Delta r$ and $\Delta\psi$ require no subscripts. *Remember also that the p of equation (9), which is the pitch $p_{23}$ of the relative screwing at the pitch line (and therefore at the teeth themselves whatever the actual radii to the teeth may happen to be), is a constant for the gear set; this p, as already explained at § 22.07, is wholly determined by our choices for C, Σ and k.* Now for obvious reasons it would be convenient to have some names for the variables $\Delta r$ and $\Delta\psi$, and I propose to give them names as follows: let $\Delta r$ be called the *departure* from the pitch radii $r_2$ and $r_3$, and let $\Delta\psi$ be called the *swing* from the helixangles $\psi_2$ and $\psi_3$, that swing being necessary as a consequence of the departure. Refer to figure 22.06 and be aware that, as $\Delta r$ increases (or diminishes) from zero, $\Delta\psi$ is being made to vary in such a way (*a*) that the NP at the teeth upon the wheels remains the same at both wheels, and (*b*) that the speed ratio (and thus the ratio of the numbers of teeth upon the wheels) remains the same at all times.

Figure 22.06 (§ 22.37) Showing how the variation of swing $\Delta\psi$ with departure $\Delta r$ affects the layout of the teeth on one only of the two wheels. The variation on the other wheel. of course, is corresponding.

(9) $p = (C/2)\cot\Sigma - (C/2)\operatorname{cosec}\Sigma\cos 2\theta$

(10) $(r_2 + \Delta r)/(r_3 - \Delta r) = \tan(\psi_2 + \Delta\psi)/\tan(\psi_3 - \Delta\psi)$

(11) $\Delta r = p\tan\Delta\psi$

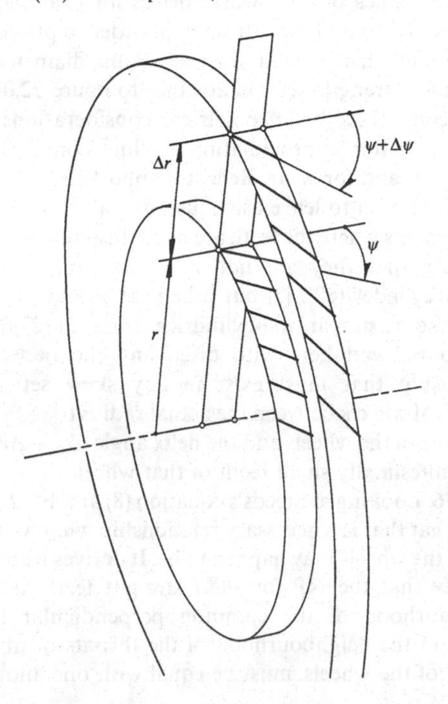


**On the special case of the spur gear set**

38. It will be clear, for a given $k$, that departure and swing must be related and that there will be, accordingly, some describable single infinity of pairs of helical wheels from which a single suitable choice may be taken. To find the mentioned relationship between departure and swing that must exist, one might first rewrite equation (10) in such a way that $\Delta r$ appears as an explicit function of $\Delta\psi$, and then, by substituting equations (3) (6) and (7), one might, especially with a foreknowledge of the result, easily arrive at the remarkably simple equation (11). This equation shows that, *unless we are to have a mismatch of the NP at the meshing of the teeth, tan Δψ must be directly proportional to Δr.* It shows moreover that the constant of proportionality is the fixed pitch $p_{23}$ of the screwing at the teeth.

39. The curves for equation (11) are plotted in figure 22.05. The curve for L is set at the origin marked *L*, and the curve for R is presented correspondingly. Note that, while $(p_{23})_L$ is negative due to the left handed screwing there (and this can be confirmed by contemplation of photograph 22), $(p_{23})_R$ is positive. The actual numerical values for $p$ were calculated at § 22.32, and they appear in figure 22.05 also: follow the broken line of the sine curve drawn for the pitch $p_{23}$. Note the positive directions of the axes drawn for $\Delta r$ and $\Delta\psi$: while $\Delta r$ was originally taken as an increase in $r_2$ (§ 22.34), $\Delta r$ is accordingly a decrease in $r_3$. Please note also that, whereas the full-line sine curve represents in three-dimensional reality the shape of the everpresent cylindroid of pitch lines illustrated in figure 22.03, the full-line tangent curves represent parabolic hyperboloids that sit with their centre-points at the two pitch points and with their main axes along the common perpendicular line and the pitch lines for the given $k$ respectively. Refer in advance to figure 22.08.

40. Looking at the tangent curves of equation (11) plotted for the numerical example in figure 22.05, and looking in advance at the parabolic hyperboloid sketched in figure 22.08 (this latter is also a graphic picture of the said equation), we can see now what happens in the special case of the spur gear set. In the case of the spur gear set the $p$ of equation (11) becomes

zero and the tangent curves at figure 22.05, and thus the parabolic hyperboloid at figure 22.08, become mere 'square' things where a sudden transition through all values from $\psi = -\pi/2$ to $\psi = +\pi/2$ occurs as the radius $(r_2 + \Delta r)$ passes across the pitch radius $r_2$. Refer to § 7.68, see the special regulus (when square) at figure 7.12(*d*), and refer to § 11.14; refer in advance to figure 22.08 and perceive that the lines of the special regulus (when square) at figure 7.12(*d*) reside upon their biplanar surface in precisely the same way as do the lines of the special parabolic hyperboloid we have here; perceive that the degenerate geometry in the plane at $r_2$ explains also the helical and herring bone sets of § 22.07. The sinusoidal curves of figure 22.05 become square too as the cylindroid becomes infinitely long, and this helps to explain of course the other set of solutions for the pitch radii for a given $k$ when $p = 0$, namely $r_2 = r_3 = \infty$. We need to envisage for this too the same special parabolic hyperboloid, a planar pencil of lines at $r_2$ intersected at its centre and perpendicularly by a plane of parallel lines, the lines in the plane being all perpendicular to the axes of the wheels.

**On the absence of boring**

41. Now what we have not checked in this discussion is the question of whether or not there might be *boring* at the meshing when the teeth are departed from the pitch line (§ 22.37), swung accordingly (§ 22.37), and wound helical (§ 22.43). We have not checked in other words that points in imagined straight small teeth remote from the common perpendicular line at meshing have no component of their relative linear velocities (relative to the mating wheel) in the direction that is perpendicular both to the tooth line and to the common perpendicular distance. Please refer to chapter 21 and to § 21.15 in particular where a special kind of line called a *helitangent line* is defined, and to § 21.28 where an equation is given which indicates all of the $\infty^3$ of helitangent lines which always exist in a moving body. It reads, $r = p \tan \theta$, where $r$ is the distance from the ISA, $p$ is the pitch of the screwing, and $\theta$ is the angle between the line in question and the ISA. It follows inevitably from all that and from what has been said here that any straight small tooth on 3 in mesh, which has been departed and swung according to equation (11), will be located along a helitangent line in 3 when 3 is regarded as the moving body screwing with respect to the stationary body 2. *Thus we can say that the tooth will not be boring at the meshing.* It will however be rocking at the meshing unless the tooth is a non-departed tooth (§ 22.43). Please hark back to § 22.17.

**Rocking and the need for straight winding**

42. In summary of the above discussion and for a fixed speed ratio we can say the following: (*a*) that it is impossible to build a working gear set following a departure and a corresponding swing where boring will be a result, but (*b*) that, *unless the teeth are set exactly at the pitch radii $r_2$ and $r_3$ of equations (4) and (5), rocking will be inevitable.*

43. This latterly mentioned phenomenon of *rocking* is the reason why departed teeth need to be *wound* somehow, onto a circular cylinder for example. In the most simple kind of winding, directly onto a circular cylinder, straight teeth become ordinary helical teeth with a fixed helix angle $\psi$ or $(\psi + \Delta\psi)$. Such wound teeth abound in engineering practice, and they may (for obvious reasons) be called *straight-wound* teeth. They abound for example on non-enveloping, cylindrical worms. In this and other ways the otherwise unhappy results of rocking are avoided by virtue of the curvatures. Refer to § 22.66 for other kinds of winding. Refer to figure 22.07(*a*) to see a straight-wound tooth on the wheel 2, and to figure 22.07(*c*) to take a first look at the kinematics of the rocking there. Refer to figure 22.07(*a*). I have already mentioned at § 22.34 that, as soon as the actual hyperboloidal wheels (or sheets) no longer coincide with the unique axodes (the axodal sheets) for the relative motion, that is, as soon as a departure $\Delta r$ from the pitch line at $P$ has occurred, the sheets will become what might be called *departed sheets* and intersect one another along the tooth line at $Q$. Although the surfaces of the departed sheets are tangential at the isolated point $Q$ they nevertheless intersect along the tooth line through $Q$ (§ 22.64). All of this is to say that, *unless the tooth line remains at the pitch line* (*and yet, most often, the designer will take it away from there*), *rocking will be inevitable.*

44. A much more effective argument to explain the occurrence of rocking might go as follows. Whenever a tooth line in mesh is put, not at the pitch line, that is, not upon the generator common to the axodal sheets (and thus, accordingly, not upon the unique hyperboloidal axodes for the motion), the direction of the relative angular velocity $\boldsymbol{\omega}_{23}$ will be inclined to the line of the tooth. Perpendicular to the tooth line and perpendicular to the common perpendicular distance, therefore, there will be a component of that relative angular velocity occurring. This is the component that constitutes the rocking; refer to figure 22.07(*a*); and this is the phenomenon related, incidentally, with the interference of straight small teeth with one another that was mentioned briefly at § 22.34. Because a tooth line is never inclined to the perpendicular distance at an angle other than a right angle, because, in other words, a tooth line is never *tilted* (§ 22.13), there is never a component of $\boldsymbol{\omega}_{23}$ in the direction of the common perpendicular distance. It is for this reason that I say that boring at the meshing is

always nil. Refer now to the figures 22.07 and be aware that, while angular velocities loom largely at (*a*) and are dealt with exclusively at the planar polygons at (*b*), linear velocities are in evidence also. Linear velocities $\mathbf{v}_{2A3}$ and $\mathbf{v}_{2B3}$ occur in the plane containing the tooth line and the common perpendicular distance, and a linear velocity $\mathbf{v}_{2Q3}$ (the velocity of crosswise sliding) occurs collinear with the tooth line; refer for these to the planar figure 22.07(*c*). That figure shows the straight small tooth on 3 and the common perpendicular distance by orthogonal projection, and it shows quite clearly the phenomenon of the rocking. The pitch point *P* and the point *Q* of the actual meshing may be seen at both of the figures 22.07(*a*) and 22.07(*c*).

45. Whenever a tooth is wound – and this is almost always done to avoid the ill effects of rocking (and rocking happens whenever a tooth has been departed) – the original, imagined, line contact along the straight small teeth in mesh at either *P* or *Q* is reduced of course to contact at a single point (§ 22.01). Refer again to figure 22.07(*a*); there, in a simplified way, the essences of rolling and boring and rocking and the winding of a departed tooth onto a cylindrical blank are together illustrated.

46. In practice of course such teeth as the straight tooth on the wheel 3 in figure 22.07(*a*) will need to be wound also. An actual practical gear set based on the dimensions illustrated in figure 22.07(*a*) might look like the set that is sketched in figure 22.07(*d*). Note that there I have drawn the teeth to be of finite size. This means that there is a finite region of the meshing and the question naturally arises: where exactly within this region might the point *Q* be? One could argue perhaps that the point *Q* might be taken to match the actual moving point of contact between the actual finite teeth as it crosses the common-perpendicular distance, but

Figure 22.07(*a*) (§ 22.43) Showing the absence of boring and the presence of rocking. The straight small tooth on the wheel 3, shown here with a wedge-shaped backing and designated *AQB* not here but in figure 22.07(*c*), has been departed and swung; it is neither collinear nor parallel with the pitch line; it occupies one of the helitangent lines of the system of such lines surrounding the pitch line (§ 22.40, § 21.28). The four points *A*, *Q*, *B* and *P* occupy the same vertical plane. The mating tooth on the wheel 2 has been straight-wound. The wheel 2 should be seen here as the fixed frame of reference for the linear velocities shown in figure 22.07(*c*).

Figure 22.07(*b*) (*c*) (§ 22.44) At (*b*) we see the planar angular velocity polygon which helps to explain the rolling and rocking shown at figure 22.07(*a*); the polygon lies in a horizontal plane. At (*c*) we see, existing wholly within the vertical plane *ABP*, an orthogonal view of the planar array of linear velocity vectors along the straight small tooth *AQB* – see figure 21.03. The frame of reference for these velocity vectors is not the frame 1 of the gear set but the wheel 2. It is this planar array which characterizes the phenomenon of rocking.

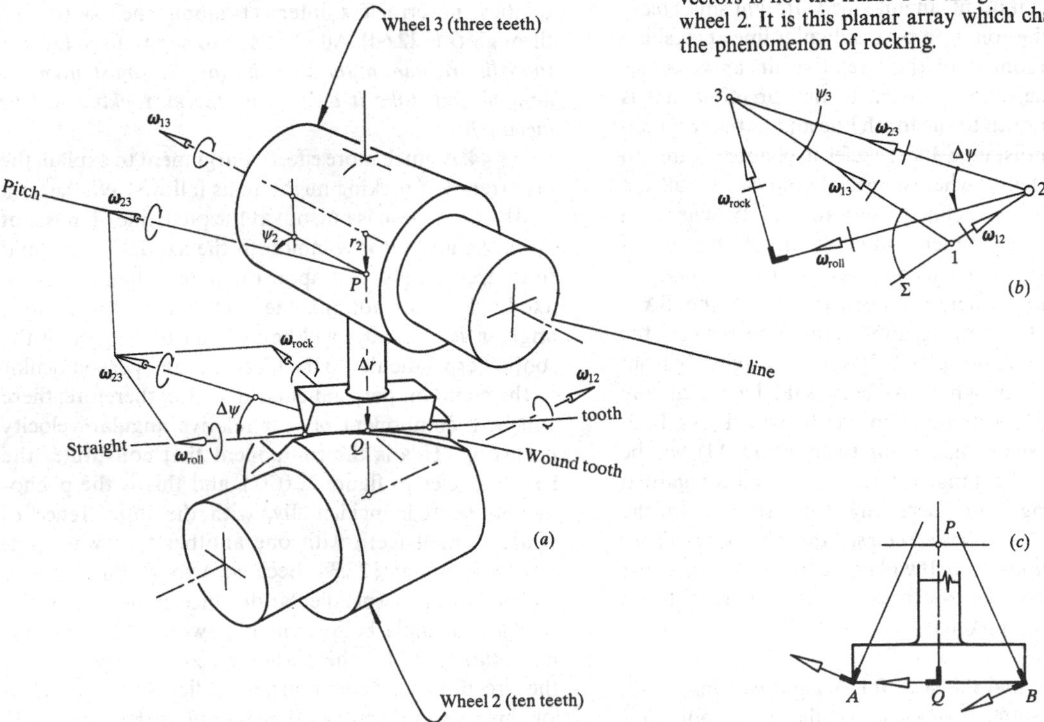

the geometry of real teeth in contact is very complicated, and the path of the actual moving point of contact might never cut that line. In any event such sets, or others more carefully designed to put, perhaps, $\Sigma$ to 90° or $\psi_2$ to zero, are commonly met in engineering practice. Having seen such sets and wondered, one might ask again the double question: why are the teeth of such sets departed and why, departed, are they so grossly departed? The answers will often lie in the matter of friction, and thus in the matter of non-reversibility, next to be examined.

**On the matter of reversibility**

47. Return to figure 22.07(*a*) and, instead of the gear set shown, imaging the gear set that would be built to employ as tooth line that unique screw for the continuous relative motion of the two wheels, the $ISA_{23}$ at the pitch point *P*. Imagine, in other words, that particular gear set wherein the actual hyperboloidal wheels have been chosen to coincide with the axodes for the motion. We might, for convenience, call this set the *ideal axode set*. The slower of the two rotating wheels (the wheel 2, the larger one with the ten teeth) could act as the driver then, and the faster of the two rotating wheels (the wheel 3, the smaller one with the three teeth) would follow. *Due to the presence of friction, however, the arrangement would not be reversible: owing to friction it would not be possible for the wheel 3 to drive the wheel 2.*

Figure 22.07(*d*) (§ 22.46) These projected pictures of a worm and its mating wheel should be read in conjunction with figure 22.07(*a*). They make clear the irreversibility of the drive from the worm 3 (which is the larger) to the wheel 2 (which is the smaller).

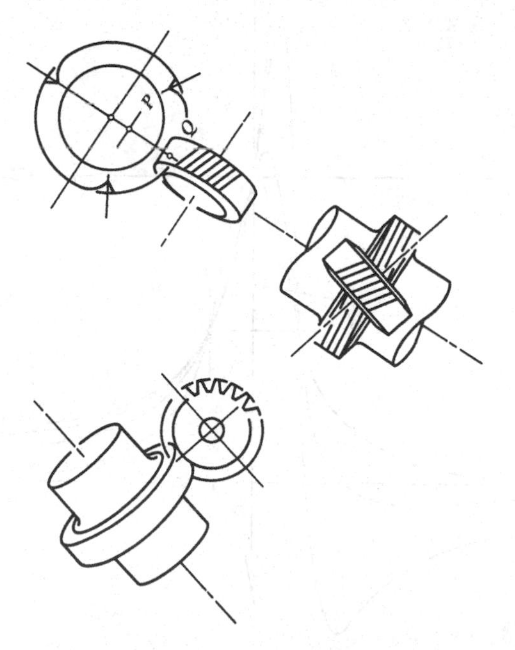


48. If we did require this, namely to drive in reverse from 3 to 2, we would need to depart the teeth – and (in the present example) to do so grossly, from *P* to *Q* say – to achieve it. Please note, in the workable gear set sketched in figure 22.07(*d*), (*a*) that the departed teeth at *Q* have been straight-wound at both wheels, and (*b*) that the slower rotating wheel (the wheel 2 with ten teeth) is in fact the smaller of the two wheels, while the faster rotating wheel (the wheel 3 with three teeth and now the 'worm') is in fact the larger of the two. Although it may be seen from this example that many gear sets in engineering practice are based upon the geometry of grossly departed teeth, it remains nevertheless a useful ploy to argue, as we have done in the present chapter, from the firm geometrical base of the ideal axode set.

49. In so far as this matter of friction and reversibility is concerned, it might be more logical to argue the matter of departure and swing the other way around: we might with a more convincing logic argue that, owing to the problem with friction caused by the awkward angle at which the applied force at the teeth will be acting in the case of the ideal axode set, we need firstly to swing the teeth; and then, due to the inescapability of the relationship $\Delta r = p \tan \Delta\psi$ (11), we need to depart the teeth to suit.

**The parabolic hyperboloid of the departed teeth**

50. Let us examine now the actual shape of the ruled surface that will be traced by a short straight small tooth as it becomes departed (in each direction) away from its ideal location. Represented by the point *P* at figure 22.07(*a*), its ideal location is the line of $ISA_{23}$. This line, the pitch line, is the common generator for, or the line of tangency of, the mating axodes for the relative motion of the wheels. By virtue of the inescapable relationship (11) we see that the single infinity of lines of the ruled surface is a special (rectilinear) regulus; it resides upon a parabolic hyperboloid. Refer to figure 22.08; refer to § 7.68, to figure 7.12(*g*), and, for further discussion, to § 11.14; refer also to item (3) in § 11.28. In what sense is the mentioned relationship (11) inescapable? If the straight small teeth are short, if we stay in the gorge of the hyperboloid namely upon a thin cylindrical blank, and if we insist (as we must) that the NP of the meshing teeth must match, then it is inescapable.

**Transitional cases of the skew gear set**

51. I turn now to phenomena associated with the changing of the speed ratio *k*. These are confusingly similar, although quite different, from those of departure and swing. With departure and swing, discussed above at § 22.34 *et seq.*, we kept constant both *C* and $\Sigma$; we kept constant the speed ratio *k*; we thus kept constant the ratio of the tooth numbers of the mating wheels; we

departed from the pitch radii $r_2$ and $r_3$ however; we added (and correspondingly subtracted) the extra radius $\Delta r$; then, in response to that, we swung the teeth by the consequent, necessary, extra angle $\Delta\psi$. The gist of the relationship between departure and swing is summarized in figure 22.08: *any single straight small tooth, in departure away from its pitch radius at the point P, must traverse the ruled surface of a parabolic hyperboloid.* In what follows next I wish to consider the confusingly similar although unrelated phenomena associated with the changing of the speed ratio $k$.

52. At § 22.25 I drew attention to the fact that the ratio of the pitch radii $r_2$ and $r_3$ is not, in general, equal to the speed ratio $k$. It is only in the special case of the spur gear set that that is so. It might be observed from the equations (4) and (5), or from the exact expression for the mentioned ratio given at § 22.25, that, for a given $k$, the ratio is double valued. To solve for either one of the ideal radii we solve an equation that is quadratic in $k$. This is a consequence of course of the nature of the cylindroid; at each displacement $z$ from the origin of coordinates there is not only one generator of the cylindroid but a pair of generators; see figure 22.03. It will moreover be seen from the curves in figure 22.09 that the radius $r_2$ for example may become either zero or equal to the centre distance $C$, while the speed ratio $k$ may not necessarily become zero (or infinite) on either or both of those occasions.

53. I hope that the curves in figure 22.09 are self

Figure 22.08 (§ 22.50) Showing the ruled surface of the parabolic hyperboloid generated by a short straight tooth as it becomes departed in both directions away from the pitch line and swung accordingly. The array of coaxial circles is coaxial with the common perpendicular line marked $c$–$c$, and at each of the circles a straight small tooth appears as a diameter. The ruled surface of the parabolic hyperboloid is accordingly truncated for the sake of clarity by a right circular cylinder coaxial with $c$–$c$. The line $x$–$x$ in the figure is drawn in such a way that the pitch line, the line $c$–$c$ and it are mutually perpendicular at the pitch point, so $x$–$x$ also lies at a diameter of the particular circle whose centre is at the pitch point. Each of the lines $c$–$c$, $x$–$x$, and the pitch line is an axis of symmetry of the ruled surface.

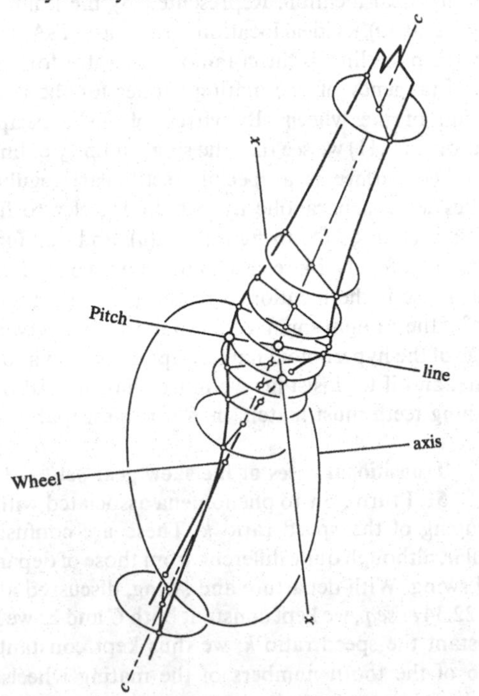


Figure 22.09 (§ 22.51) Curves showing the variation of $r_2$ with $k$, for various values of $\Sigma$. Note the special case $A$ of the spur gear set where $\Sigma$ is zero. Note also the special case $B$ of the 'square' set where $\Sigma$ is 90°; refer again to figure 22.01 and remember that whereas the sketches there depict a variety of square sets where the teeth are neither small nor straight the curves here are for minimum crosswise sliding at straight small teeth. Equations for the curves at $A$ and $B$ appear in the box below; these equations are directly derivable of course from equation (4) in the box at table 22.01.

At $\Sigma$ = zero (spur set), $r_2 = Ck/(k+1)$;
but at $\Sigma = 90$ (square), $r_2 = Ck^2/(k^2+1)$.

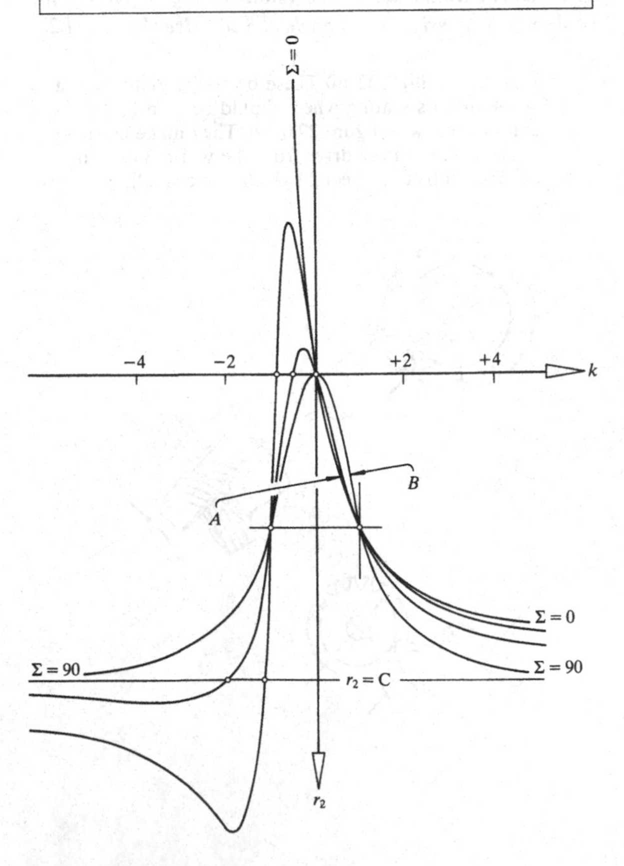

explanatory. The following comments may be helpful however. I have taken equation (4) for the value of $r_2$ in terms of $C$, $\Sigma$ and $k$. I might equally well have taken equation (5) for the value of $r_3$ however, for $r_2$ and $r_3$ add, as we know, to $C$. I have substituted into (4) a series of values for $\Sigma$, namely zero, 30, 60, and 90°, and plotted the corresponding curves. They may be seen. Please note that, when $\Sigma = 90°$, $r_2 = k^2/(k^2+1)$, the plotted curve is symmetrical about the axis for $r_2$, and the length $2B$ of the cylindroid is equal to $C$. Note that, whenever $0° < \Sigma < 90°$, the plotted curve is asymmetrical about the axis for $r_2$; note also that, for $r_2 = 0$ in these cases, the two solutions for $k$ are $k = 0$ and $k = -\cos\Sigma$; the length $2B$ of the cylindroid is greater than $C$ in all of these intermediate cases. When $\Sigma = 0$, $r_2 = k/(k+1)$, the plotted curve is no longer double valued in finite $k$ and, for each $k$, the ratio of $r_2$ and $r_3$ is equal to $k$; this special curve, asymptotic to the line $r_2 = C$ at one only of its two ends, refers of course to the spur gear set; the length $2B$ of the cylindroid in this particular case is infinite.

54. I should explain now that whenever the pitch radius $r_2$ becomes greater than $C$ or less than zero the nominated gear set changes from being an external set to being an internal one. Whenever the radius $r_2$ gets to be greater than $C$ or less than zero, the teeth on the wheel 2 become the internal teeth of a ring gear. Now the transition from internal set to external set is not very interesting at those points of transition that occur at the origin of coordinates; at those points, $r_2$ = zero, $\psi_2$ = zero, the wheel 2 is a mere line, and the speed ratio $k$ is zero too. At those points of transition that occur on the horizontal lines $r_2 = 0$ and $r_2 = C$ and remote from the axis for $r_2$, however, phenomena occur that bear a special mention.

55. Refer to figure 22.10. This illustrates one of the more interesting of the transition phenomena. We see there the axes of the two wheels at a centre distance $C$ apart and inclined at an angle $\Sigma$. The radius $r_2$ of the axode in 2 – this axode is conical – is zero, $ISA_{23}$ is exactly perpendicular to $ISA_{13}$, and the radius $r_3$ of the axode in 3 – this axode is flat – is $C$. It would appear that a truncated conical wheel 2 (a simple bevel wheel with its straight small teeth all intersecting at the vertex of the cone) could mate satisfactorily with a flat wheel 3 (with its straight small teeth all tangent to the circle of radius $C$ as shown). The set would be neither external nor internal, and its speed ratio $k$ would be $k = -\cos\Sigma$. I should mention that here the NP of the teeth are no longer constant along a tooth because we are no longer at the gorge of the hyperboloid (§ 22.50). The neighbourhood of the gorge has degenerated indeed to a single point (the vertex of the cone) and the short straight small teeth of § 22.50 have disappeared. At this juncture the reader might care to take as an exercise the transitional case of the simple bevel gear set (external or internal) which occurs when $C$ diminishes to zero in the presence of a finite $\Sigma$. There are, no doubt, other interesting transitional cases of the skew gear set.

Figure 22.10 (§ 22.55) Showing a transitional case of the skew gear set: when the pitch radius of the axode becomes zero in this special case, the axode becomes a cone, and we have the transition between external and internal gears.

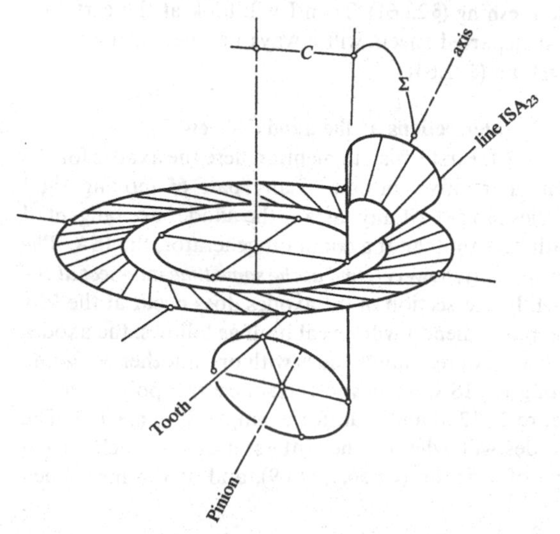


### A summary of the geometry

56. I would like to begin an interim summary here by saying this: whereas for a given speed ratio a tooth in departure from its pitch line must traverse the ruled surface of the parabolic hyperboloid sketched in figure 22.08 (§ 22.50), a tooth that might be shifted from its original pitch line to some new pitch line by virtue of required changes in speed ratio must traverse the ruled surface of the cylindroid sketched in figure 22.03 (§ 22.51).

57. We have at the straight small teeth in mesh at the pitch line the two axodes for the relative motion as well, so there are, accordingly, four different ruled surfaces intersecting along the line of a straight small tooth when that tooth is at the pitch line. The four surfaces are as follows: (*a*) the right circular hyperboloidal axode of radius $r_2$ coaxial with the axis of the wheel 2; (*b*) the tangentially arranged right circular hyperboloidal axode of radius $r_3$ coaxial with the axis of the wheel 3; (*c*) the characteristic cylindroid for the gear set whose central axis (its nodal line) is collinear with the common perpendicular distance $C$; its dimensions (its length $2B$) and the distribution of its pitches are wholly and solely determined by $C$ and $\Sigma$; and (*d*) the parabolic hyperboloid associated with the possibilities for departure; the central point of this special hyperboloid resides upon

the nodal line of the cylindroid and, while one of its principal axes is collinear with this nodal line, the other is collinear with the pitch line; its actual shape is determined by the pitch $p_{23}$ for the screwing at the teeth. Refer to figure 22.11 and see there the intersecting surfaces (*c*) and (*d*) namely the cylindroid and the parabolic hyperboloid. For the sake of clarity the surfaces (*a*) and (*b*), the axodes, are omitted from figure 22.11.

58. The surfaces (*c*) and (*d*) are not tangential at the pitch line. They intersect not only one another at the pitch line but also the axodes at the pitch line. Please note in this connection that the sine curve for $z$ passing through the point $L$ in figure 22.05 is not tangential with the tangent curve for $\Delta r$ that is passing through at the same point. These curves intersecting at $L$ in the planar figure 22.05 are representing in the spatial figure 22.11 the said characteristic cylindroid and the said parabolic hyperboloid respectively.

59. But notice from the data displayed in figure 22.05 that the surfaces (*c*) and (*d*) intersect again, at $z = \pm C/2$. Refer again to figure 22.11 and see that the cylindroid and the parabolic hyperboloid are shown to be intersecting again (and again along a common generator) at the levels $z = \pm C/2$. It is at these intersections – they are marked with an asterisk – that the kinds of transitional cases depicted in figure 22.10 occur.

Figure 22.11 (§ 22.57) The hyperboloidal axodes (*a*) and (*b*), coaxial respectively with the axes of the wheels 2 and 3, are tangential with one another along the pitch line, see figure 22.12(*b*); for the sake of clarity these axodes are not shown here. This figure shows only (*c*) the cylindroid of the straight small teeth at the pitch line when the pitch line moves in response to a variation in speed ratio, and (*d*) the parabolic hyperboloid of the possible straight small teeth that must be employed in the event of departure. Note the two secondary intersections at the radial lines marked with asterisks, and note the right angle shown which is, of course, a natural property of the cylindroid (§ 15.49).

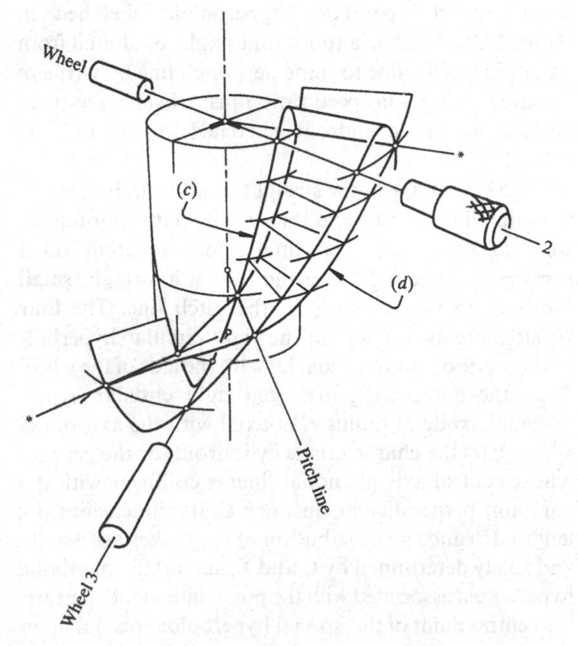


60. Having thus summarized the geometry in general I wish to reiterate now the fact that in the event of there being departed teeth there exist in the gear set not one but two relevant sets of rolling and sliding hyperboloidal sheets. There exists in any event the first set, the *axodal sheets.* These are, as I have already explained, the unique axodes for the motion; they are unique for each given velocity ratio; they are indeed the loci traced separately in the two wheels by the relevant ISA, the $ISA_{23}$. It is important to understand that independently of everything else these axodes or axodal sheets will always be there in a successful gear set, that they will continue to roll and slide upon one another in the well known comfortable manner of axodes in general, and that the presence or absence of actual metal in the space does not in any way inhibit this smooth rolling and sliding motion. If the teeth (the straight small teeth) are departed, however, the actual teeth themselves form another, a second set of hyperboloidal sheets; and these I have called the *departed sheets.* Unlike the axodal sheets these departed sheets are (*a*) backed on one side or the other with real metal, and (*b*) not tangential with one another. They *intersect* at the meshing. They cannot roll and slide upon one another in quite the same comfortable way as the axodes do; and there are, consequently, distinct extra mechanical difficulties with departed teeth. I wish to address these various matters now but under two separate headings. First I will examine the possibility that the axodal sheets, which are always tangential at the meshing, may at the same time be intersecting at the meshing (§ 22.61). Then I will look at the certainty that departed sheets will always be intersecting at the meshing (§ 22.64).

### Intersections at the axodal sheets

61. First I wish to mention here the axodes for the general relative motion of any pair of moving rigid bodies (§ 5.58). At any instant the axodes are tangential with one another at a common generator, the ISA. *The axodes may, however, and at the same time intersect at the ISA.* If intersection of the axodes does occur at the ISA the phenomenon will reveal itself as follows: the axodes will, as it were, 'change sides with one another' as, going along the ISA, we pass a certain central point; refer to figure 22.12(*a*) and read for example Skreiner [4]. The axodes will, whether they intersect or not, each have a line of striction (§ 7.56, § 11.09), and at the mentioned

central point upon the ISA the two lines of striction will intersect.

62. It is important to see in the case of gears that, independently of all of the phenomena previously discussed, and independently of tooth form and of tooth interference, there is the problem that the axodes will, under certain adverse circumstances, intersect. In the special case of the general gear set the axodes are not general in shape but special. They are each axisymmetric, consisting each of a circular regulus of lines upon a circular hyperboloid. Their lines of striction are circles (§ 5.58), and these circles intersect at the pitch point. The axodes (the axodal sheets) of a gear set will under all circumstances remain tangential at the common generator as I have said. But they may or may not be able to roll and slide upon one another at that generator in the same way as real wheels might. It is relatively easy to see that, *if the shaft angle Σ is greater than 90°*, the axodes will intersect at the mating generator, rolling and sliding like real wheels at the mating generator becomes impossible, and two intersections indeed take place, one at the said straight line (the pitch line, the common generator, the line of tangency), and one other at a curved line at another place. The curved line of the second intersection cuts the straight line of the first at the central point of the ISA; refer again to figure 22.12(*a*).

63. In gear sets the wheels drive one another not with their axodes but with their teeth of course. If the teeth are not upon the axodes (and this is the case most frequently), they are upon the departed sheets. The tooth ratio determines the velocity ratio; the velocity ratio determines the axodes; and the axodes, untroubled by the presence or absence of metal, roll and slide upon one another in their own characteristic way. The axodes, being mere surfaces (or sheets), are able to roll on one another about the ISA and slide on one another along the ISA even while, tangential at the ISA, they are intersecting there.

64. Please refer to photograph 22 where, at the right hand set, $\Sigma = 100°$. Refer next to figure 22.12(*b*) which is a drawing of the two axodes at the photograph, and please see there the stippled plane (which is a horizontal plane) and the overlapping triangles. The photograph 22 does not show the phenomenon, but parts of the lower (the bigger) wheel of the actual wheels at the right hand set have been removed to admit the whole of the upper wheel into a proper meshing at the straight small teeth. The lower wheel, in other words, partly hidden behind the upper wheel, is mutilated. More words are useless here. The reader should contemplate figure 22.12(*c*), refer to § 11.06, and perceive the truth of the matter: the axodes are tangential with one another at the pitch line but intersect one another both there and along another, a curved line; the curved line of intersection passes through and is symmetrical about the pitch point. Real hyperboloidal wheels such as the wheels at the right at the photograph 22 are unable to mesh in a proper manner unless the material bulk of the wheels interpenetrate, and that, of course, is not possible (§ 1.06, § 1.07). Real hyperboloidal wheels such as the wheels at the left, however, where Σ is less than 90°, suffer no such problem.

Figure 22.12(*a*) (§ 22.61) This sketch shows a pair of axodes for the general relative motion of a pair of rigid bodies. The axodes, which must be tangential along the ISA, may also intersect along the ISA, and they are shown here to be thus intersecting. The relative motion is such that the two lines of striction (which are not shown) always intersect at a point *C* somewhere along the ISA. The curved line of intersection marked *c–c* also passes through this *C*; it is the inevitable counterpart of the straight line of intersection along the ISA.

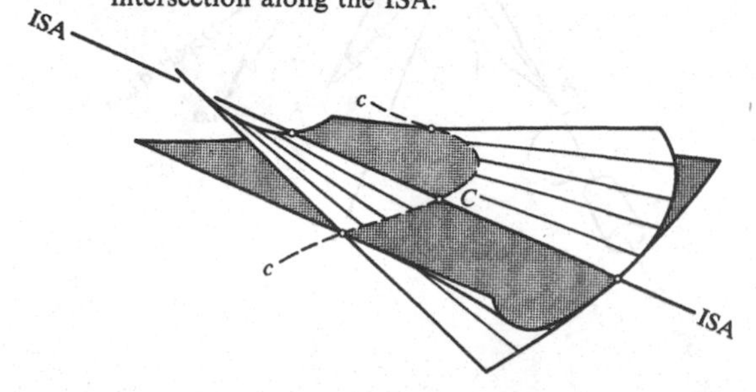


Figure 22.12(*b*) (§ 22.64) In this figure the shaft angle Σ is greater than 90°. The figure is showing that, although the axodes for the motion (the axodal sheets) are tangential at the pitch line, they intersect with one another there. This means that the actual hyperboloidal wheels (which are constructed here with their straight small teeth exactly upon the axodal sheets) interfere with one another; they cannot without a mutual mutilation come into mesh with one another.

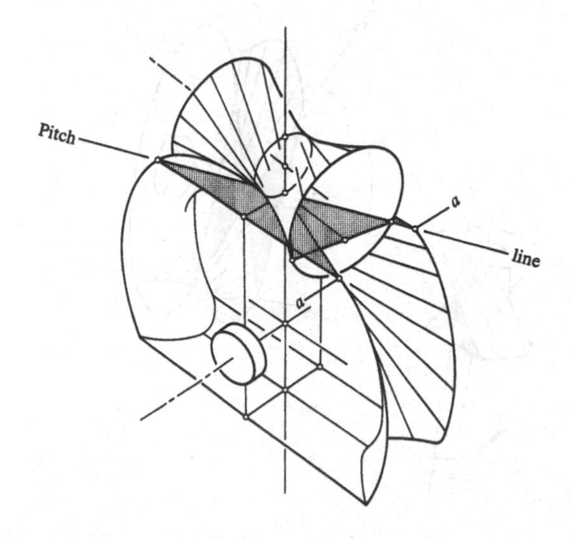

65. Practical considerations now lead to the obvious question of whether or not the phenomenon of axode-intersection is important. It is certainly important to understand that the phenomenon does occur in the mentioned circumstances. But it must be said – and here we go into the realm of losing line contact and this at the evident expense of strength – that, in the case of wheels where the actual straight small teeth are put to coincide with the pitch line and the axodes do intersect, a judicious winding of the teeth onto circular cylindrical blanks will (with all other things being equal) eliminate the difficulty.

**Intersections at the departed sheets**

66. Whereas axodes always exist and remain relevant whatever the circumstances, departed sheets exist only by virtue of departed teeth. Departed sheets are not axodes. There is, accordingly, no reason to expect that a pair of departed sheets will be tangential at the tooth line. There is, indeed, every reason to expect that they will merely intersect. Set up in the mind's eye two wheel axes 2 and 3 with their common perpendicular distance $C$ and their shaft angle $\Sigma$; refer to figure 22.13 and note that there, given the required directions of rotation, $\Sigma$ is less than 90°. Erect the unique cylindroid which is characteristic of that pair of wheel axes. Given next the gear ratio $k$, locate that particular generator of the cylindroid which is the pitch line. Note the pitch point $P$. Note that the figures 22.11 and 22.13 may be superimposed. Going now to the hyperbolic paraboloid (which is characteristic not merely of the layout of the wheel axes but of the velocity ratio as well) choose now some departed tooth line for the meshing. Let this cut the common perpendicular in $Q$. We know this $Q$; we have used it before. Choose now a general point $R$ along the departed tooth at $Q$. Erect at $R$ the plane normal to the tooth line and let it cut the axes 2 and 3 of the wheels at $A$ and $B$ respectively. Consider the linear velocities $\mathbf{v}_{1R2}$ and $\mathbf{v}_{1R3}$, both emanating at $R$. These are perpendicular to $B$–$R$ and the wheel axis 2, and $A$–$R$ and the wheel axis 3, respectively; they are not collinear. For correct tooth action the velocity difference $\mathbf{v}_{R2R3}$ must be parallel to the tooth line; or, better, the rubbing velocity at $R$ (which may be taken as either $\mathbf{v}_{2R3}$ or $\mathbf{v}_{3R2}$) must be collinear with the tooth line. For correct tooth action, in other words, the rubbing velocity $\mathbf{v}_{2R3}$ at $R$ must have no component perpendicular to the tooth line. But this can only happen if, at $R$, the bound vectors $\mathbf{v}_{1R2}$, $\mathbf{v}_{1R3}$ and the tooth line $RQ$ are coplanar. In general the line segments $A$–$R$ and $B$–$R$ are not collinear. Unless they are collinear, however, $\mathbf{v}_{1R2}$, $\mathbf{v}_{1R3}$ and the tooth line cannot be coplanar. Absence of interference at the tooth line requires, in other words, an absence of departed

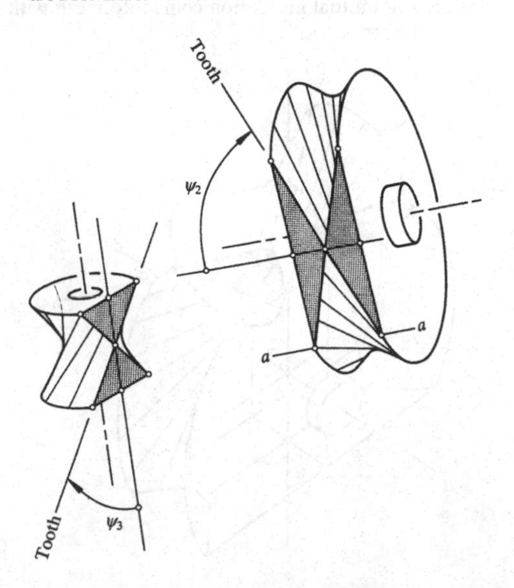


Figure 22.12($c$) (§ 22.64) Showing the matching faces of the wheels cut to correspond with those cut faces at figure 22.12($b$). That the helix angles $\psi_2$ and $\psi_3$ sum to a $\Sigma$ greater than 90° is the source of the trouble here.

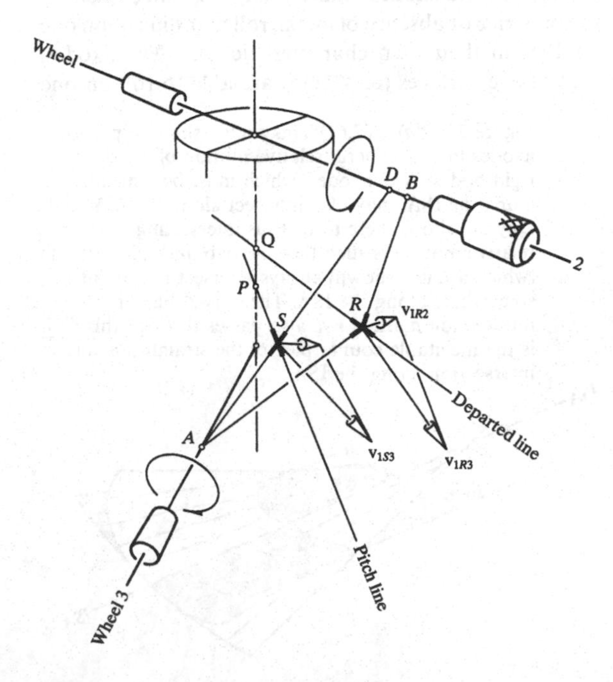


Figure 22.13 (§ 22.66) This figure, which may be superimposed upon figure 22.11, is illustrating the argument that it is only at contact points such as $S$ along a straight tooth through $P$ that entirely satisfactory meshing can occur. Note that the velocity-difference $\mathbf{v}_{S2S3}$ is parallel with the tooth line $SP$.

sheets which merely intersect; the departed sheets must also be tangential. Refer now to the two statements of Plücker's theorem concerning the straight lines that might be drawn perpendicular to a generally chosen generator of a cylindroid (§ 15.36, § 15.38). Our wheel axes here are a pair of conjugate generators of the characteristic cylindroid; their pitches are equal and both zero; we are trying to find a transversal of these two which will cut an available tooth line perpendicularly. By virtue of the given gear ratio the only available tooth lines are those upon the parabolic hyperboloid. But by virtue of Plücker's theorem the only available tooth line for correct meshing action is upon the cylindroid. It follows inescapably that, if we wish to avoid interference at straight small teeth at the meshing, we must choose $Q$ to be at $P$. We must put the departed tooth at the pitch line (where the departure is zero) in other words. Note now the shown transversal $AD$ which cuts the pitch line perpendicularly at $S$. At all points $S$ upon the pitch line, a plane normal to $PS$ will contain a transversal such as $AD$ and the rubbing velocity at $S$ will be in the direction of the tooth line.

67. The practical significance of this is that, whenever we choose to employ departed teeth, we must pay attention to the inevitable intersection of the departed sheets. Due to the intersecting sheets the straight small teeth will always *clash* at the meshing – they will *collide* or *interfere* with one another. This relates of course in retrospect to the matter of *rocking*; rocking was discussed at § 22.42 and illustrated in the figures 22.07. If we wish the wheels of a departed set actually to turn, we must adjust somehow the straightness of the straight small teeth; we must *wind* the teeth somehow. But this thoroughly undoes our stated plan to achieve line contact between the real, the actual finite teeth of the skew gear sets. And it is of course in the design and production of these real teeth that we find our many difficulties.

### Teeth that are otherwise shaped, curved, or modified

68. Except in the special case where the pitch radius was reduced to zero and the relevant axode became conical (§ 22.55), I have discussed so far only those gear sets that might be cut on the basis of the central portions (the gorges) of two general hyperboloidal wheels. I have moreover spoken exclusively about small straight teeth; the teeth have either been straight or what I have called straight-wound. I have been concerned so far to preserve a constant NP, either across the short gorges of the hyperboloidal wheels, or upon the circular cylindrical blanks onto which the straight teeth were straight-wound.

69. Now one may otherwise shape, curve, or modify the teeth of course, and I mean by the teeth here the straight small teeth. For a given velocity ratio the teeth on either wheel of a pair may be of any shape, provided the pattern is repeated tooth for tooth and the matching teeth on the other wheel are marked out to be conjugate. Please refer to Merritt (1954) [1] and read the whole of his chapter 4; this deals with his idea of *tooth spirals*; these are the curved lines upon the flanks of the actual teeth of helical spur and bevel wheels where the pitch surfaces and the flanks of the teeth intersect; the tooth spirals of Merritt might be seen in the context here as being the originally straight teeth upon the pitch surfaces which have been otherwise shaped, curved, or modified. But the NP may then vary irregularly and quite considerably along the curved length of a tooth, and one will begin to wonder about the practical means for cutting the teeth; and I mean by the teeth here the real teeth that are required for practical skew gears. Refer to figure 22.01 and note the various sketched items there: (*a*) is a 'spiral bevel' set, (*b*) is a 'hypoid' set, (*c*) is a 'worm' and its mating wheel, and (*d*) is an 'ordinary' pair of skew gears. Geometrically speaking they are, of course, substantially the same. They each have their sets of axodes or axodal sheets. Except for the bevel set where the wheel axes intersect and the designer had no choice in the matter, they each also have – and this is in so far as their straight small teeth are concerned – their departed sheets. Look also at photograph 13. This relates directly to the sketch at figure 22.01(*b*).

70. Except for the ordinary skew gear set at figure 22.01(*d*) where the teeth are straight, the teeth on all of the various wheels and pinions in figure 22.01 are *curved*. They are all curved, however, in such ways that practical advantages accrue. The alternate flanks of the teeth on the crown wheel of the hypoid set in photograph 13, for example, are cut with circularly moving fly cutters that traverse in a cunning way the smooth surface of an approximately hyperboloidal (namely a conical) blank; and by means of a different machining process, the curved teeth of the mating pinion are cut to match as well as possible the teeth of the wheel. An advantage of the curved teeth of the hypoid set is that, in the inevitable absence of line contact, the flanks of the actual finite teeth bed with one another in such a way that the curvatures (the curvatures of the curved teeth) encourage much larger areas of Hertzian contact at the flanks than would otherwise be possible. Similar remarks could be made about the curved teeth of the spiral bevel wheels at figure 22.01(*a*). With regard the worm and its wheel in figure 22.01(*c*), it is thoroughly likely that the teeth of the worm are departed. There is no doubt however that they are, also, otherwise shaped, namely curved. They are not straight teeth straight-

wound onto a cylindrical blank or otherwise wound onto an 'enveloping' blank. The whole shape of the wound wheel (the worm) has been changed by curving. The designer did this perhaps for a number of different reasons, but certainly one of the reasons was to avoid what otherwise would have been irreversibility due to friction (§ 22.47).

71. I write this paragraph with an even greater measure of caution then I did the last, for its contents are, indeed, quite beyond the intended scope of this chapter. I write it nevertheless, for I wish to complete the argument in some intelligible way. If a small curved tooth (or a tooth spiral) is curved at the meshing in such a way that the tangent to the tooth is *tilted* there (§ 22.13), there will be *boring* there as well (§ 22.41). This could be said more precisely perhaps (and here I offer a definition) as follows: if the tooth is tilted at $Q$, there will be a component of the relative angular velocity $\boldsymbol{\omega}_{23}$ not only about the tangent (which is the rolling) and about the normal (which is the rocking) but also about the binormal (which is the boring). I am aware that I have spoken here about one only of the mating tooth spirals at $Q$, of which of course there will be two. I can also see (*a*) that in the event of departure the two spirals that mate at $Q$ may not enjoy the same tangent, that is, that they may indeed either intersect or make no contact with one another at $Q$, and (*b*) that, even if they do enjoy a common tangent, there will be no guarantee that the pair of normals and thus the pair of binormals will coincide. With ordinary straight or straight-wound teeth this boring component is (*a*) more easily definable (§ 22.44), and (*b*) always zero. But with the otherwise shaped (the curved) teeth of the hypoid sets the component is not zero. Refer to the worked exercise at § 13.23 which deals with the hypoid set. This clearly demonstrates that, independently of the position of $Q$ and the locations and shapes of the flanks of the mating teeth and their mutual points of contact where the complicated rolling, rocking, boring and the consequent rubbing occurs, the ISA for the relative motion of the two wheels can be located unequivocably. Naturally and subsequently, this ISA can be used to advantage in those kinds of analyses that relate to the mentioned phenomena. I see no reason why not (§ 5.44). Such analyses would be accurate moreover, and after them we would know exactly the kinematics at least of what occurs at the actual (rigid) flanks of the teeth in these departed and otherwise-shaped arrangements. The reader may care to refer to Dyson (1969).

### Closure

72. Tuplin drops – and this at the very first sentence in the Preface to his book (1962) – the following pearl: 'Gear design is not an art or a science but just a routine.' He goes on to reflect, dourly it seems to me, on 'the earnest young man who collects an armful of text books and research papers and proceeds to handle his problem on strictly scientific lines'. It is needless to say that the older ordinary experienced gear designer of Tuplin's Preface, 'whose knowledge of gears is limited', wins the race and gets his gears designed by rules of thumb (and cut by existing standard tools and machines) quicker and much more effectively than the earnest young man might ever hope to do. It is not easy to disagree with the actual pith and core of these remarks; and such remarks abound, as we know, in the gear literature. But it is easy to argue (after Reuleaux) that our success with synthesis will depend upon our effort with analysis. It is certainly true to say, as Tuplin himself has, that the mechanics of gears (even the mere kinematics of gears) is poorly understood. I openly confess that the contents of this chapter will not help the busy machine designer today to design his gears before tomorrow, but I hope that it might in the long run shed some new light on the kinematics of skew gearing. I also hope that it begins to demonstrate the incisiveness that screw theory might be found to have in elucidating not only such matters as the axodes and the departed sheets of straight small teeth that were dealt with here but also the elusive intricacies of the points and lines of contact between the mating flanks of real gear teeth. There is no reason to believe incidentally that such matters are beyond our reach, nor any reason to believe that gears and gear-cutting tools and machines are so fixed in their form that further progress in the direction towards simplicity, harmony and effectiveness is impossible (§ 0.18).

### Notes and references


[1] Merritt, H.E., *Gears*, Sir Isaac Pitman and Sons, 3rd ed., London 1954. Refer § 22.06.

[2] Baxter, Meriwether L. Jr, Basic theory of gear-tooth action and generation. This work comprises Chapter 1 of the well known *Gear handbook*, edited by Darle W. Dudley, McGraw Hill, New York 1962. Refer § 22.06.

[3] Abbott, W., Worm gear contacts, *Proceedings of the Institution of Mechanical Engineers*, Vol. **133**, p. 349, London 1936. Refer § 22.07.

[4] Skreiner, K. Michael, A study of the geometry and the kinematics of instantaneous spatial motion, *Journal of mechanisms*, Vol. **1**, p. 115–43, Pergamon 1966. Refer § 22.61.

[5] Tuplin, W.A., *Gear design*, The Machinery Publishing Co., London 1962. Refer § 22.71.

23. This insect is thinking carefully and in real time where and in what order to put its feet. Computing problems were involved in the design of its brain, but much more severe mechanical problems were presented first: how to choose, proportion, build and articulate the various moving parts of its body. Photo by John Field, Australian Museum, Sydney.

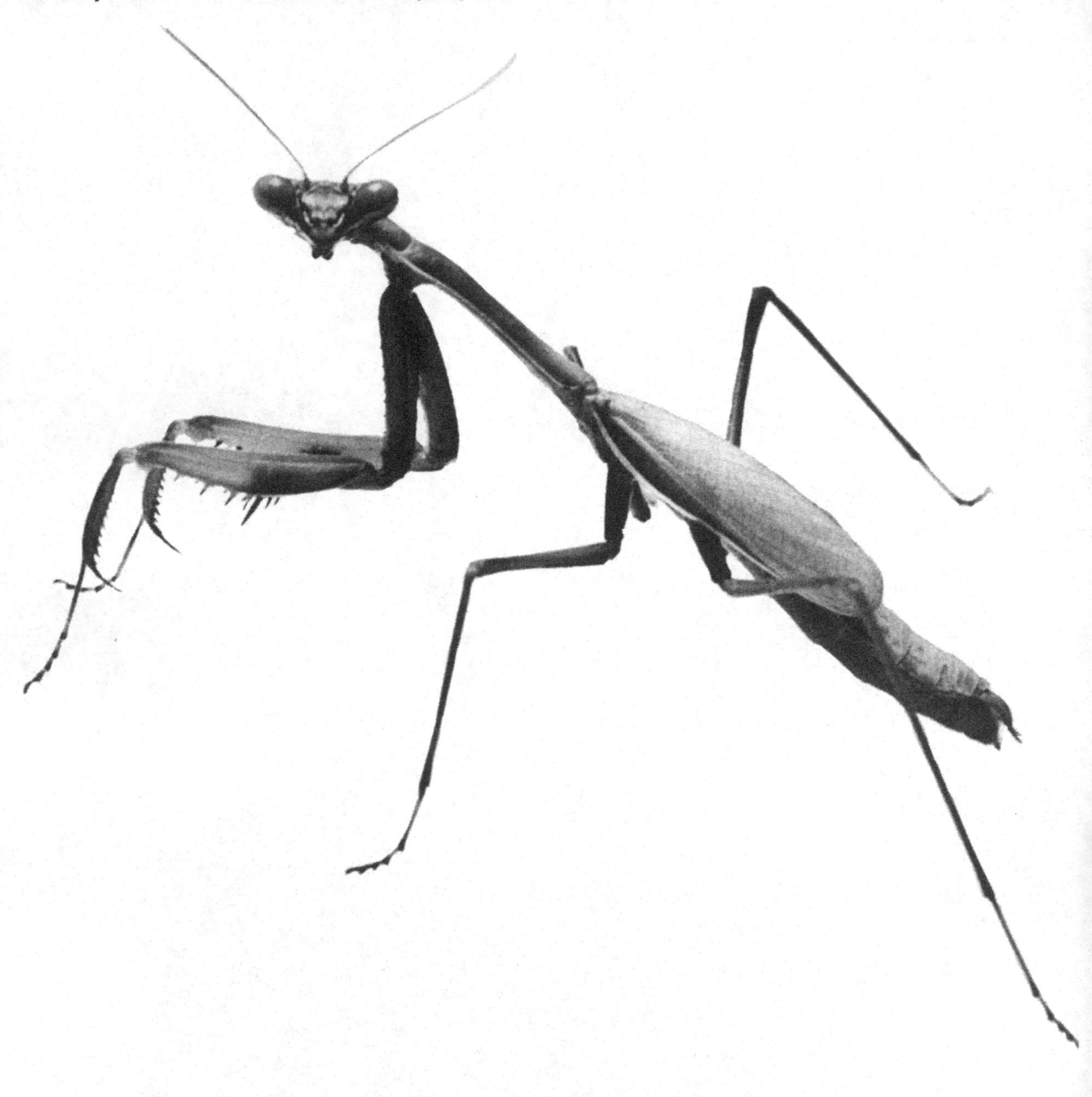

# 23 The general and the special screw systems

## Introduction

01. In this final chapter I try to present convincingly the generality and the elegance of the general and special screw systems. They were first glimpsed and then examined progressively by Plücker (1869), Ball (1900), Dimentberg (1965), and Hunt (1978). Whereas those writers have explored these systems with their various combinations of mechanical understanding, geometrical insight and powerful mathematical method, I propose to be, as it were, pushed along by them. I make no pretence at having independently arrived at all of the observations I have made in the following pages, but I do claim to have written a chapter that may be new in some respects, interesting, and even useful. I have paid careful attention to the reciprocities among the systems, introduced the terms action and motion, and accented the idea that moment is analogous with linear velocity. I have introduced a corresponding mechanics that helps the screw systems to be understood. I trust (*a*) that, by conducting an experiment in the imagination (which I do to begin), I have chosen a form of presentation that will not be unacceptable, (*b*) that the algebraic equations for the systems will not be overlooked (§ 23.47, § 23.52), and (*c*) that my verbal descriptions of the general and the special screw systems might be useful for mechanical analyses and applicable in design.

02. I have tried to give the systems substance in the sense that they may be visualized. I have avoided algebraic treatment to the extent that I merely explain the equations of others, making the differing algebraic devices of Descartes and Plücker clear. I do build upon the material of my chapter 6 however, claiming that the term quasi-general as I use it there was mine and that it has a useful meaning. This demands that I try to explain its meaning better now and to show how to step from the quasi-generality of chapter 6 and its appendices into the full arena of the general systems.

03. Generally in chapter 10 (and specifically in § 10.37 and § 10.50) I first suggested then later showed how to extend the idea of the linear dependence of sets of mere lines (screws of zero pitch) into the wider arena of sets of screws of the same pitch. One may always argue – one may always show, indeed, by physical experiment – that sets of screws of the same pitch may be handled geometrically in the same way as sets of screws of zero pitch. That screws may be linearly dependent is not only true for sets of screws of the same pitch however; the same is true for sets of screws of mixed pitch. I now propose to use those and other such aspects of natural philosophy to explain the natures of the general and the special screw systems and their reciprocals. Please refer for parallel reading to chapter 3 (and others) of Ball (1900), to chapter 6 of Dimentberg (1965), and to chapter 12 of Hunt (1978). In these places many of the same matters are presented and examined, but things are there done, of course, in the ways characteristic of those authors.

## An experiment

04. Let there be a rigid body suspended from the walls, the roof, and the floor of a room by a collection of rigid links and simple joints. Let the links be connected with the walls etc., with one another and with the body in such a complicated way that many complex joints are formed, and that the instantaneous capacity for motion of the body is not clear upon a first inspection. Please imagine in other words some complicated cross between the figures 1.10(*c*), 2.02, 3.08, and 19.10 to exist, and that blindfolded we are required to comment, first upon the body's degrees of freedom with respect to the room, and next upon all aspects of its ICRM (§ 3.34). To clarify the argument let there be no gravity and no friction, and let all of the joints be well made without appreciable clearances and be form-closed (§ 2.21, § 7.49). The reader may recall, at this juncture, Ball's public address of 1887, 'A dynamic parable'; it may be found at Appendix II of his book (1900). Conducted by him in the imagination, it was an experiment similar to this in some ways, but Ball had different methods and a different general objective. Whereas Ball was dealing with the dynamics of a rigid body constrained in various ways, subjected to impulsive wrenches, and initially at rest, I am dealing

here with the kinetostatics of a rigid body which is constrained in various ways but which remains in equilibrium. I consider no dynamics here.

05. Nominate first a reference location for the rigid body in the room, but check at the same time that none of the joints are immobilized in one direction or another by stop surfaces in contact (§ 1.29, § 7.36). This checking of the stop surfaces is not a necessary condition for the experiment but it will be convenient if all joints can move in all directions freely. Check also that no conditions exist such as that which is exemplified in figure 2.03, where the freedom to move of the link *BC* is limited by the geometry of the linkage. Check in other words that a tick within the body is safely within the laminated boundary zones of its finite zone of movement (§4.04, §10.66) and thus that, if its apparent connectivity with the room is greater than zero, it can indeed be moved in all of those respects (§ 2.39). A critical reader will be aware that I have requested a regime of checking here that is not easy to execute, or even easy to understand in all respects. It will nevertheless be clear, I hope, when I say that, except for full-cycle geometrical specialities (which are permitted here), no special conditions of the kind exemplified in figure 2.03 nor any transitory connectivities between the body and the wall should obtain accidentally (§ 2.08, § 2.09).

06. Having thus checked to remove from consideration various ambiguities that might otherwise occur, the degrees of freedom of the body may now be studied. The crux of the matter is this: *if the degrees of freedom of the body is f, there will be discoverable by experiment f separate ways of twisting or of screwing the body at the instant about f different motion screws of various pitches, each one of which is linearly independent of the others.* In the following paragraphs I deal first with the various eventualities in a preliminary way. I begin with zero degrees of freedom. I then take the other degrees of freedom in ascending numerical order.

**Zero degrees of freedom**

07. Grasp the rigid body to discover whether there are none, or only one, or an infinite number of different ways in which it is free to be screwed at the instant. If there are no ways, the body is fixed in space with respect to the room and the degrees of freedom is *zero*. The body cannot be moved. About every screw in space, of which there are an $\infty^5$, a wrench of the pitch of that screw may be applied to the body, and the body will not move. The screws of all of those wrenches constitute the 6-system of action screws. There is no motion screw, and this absence of motion screws constitutes the 0-system of motion screws which is reciprocal to all of the action screws.

**One degree of freedom**

08. If there is only one way in which the body can move, this will be detectable by discovering that every point in the moving body is obliged to travel a predetermined path (§ 1.18). One will find in the event of 1°F that, drawn through every point in the body, each point having its known position in the room when the body is at its reference location (§ 1.17), there will be a single twisted curve which is the predetermined path for that point. When each point in the body is obliged to travel thus, the degrees of freedom will be *one*.

09. Refer to figure 5.09 which shows the paths of two points *A* and *B* in a moving body 2 which, at its reference location, is screwing at the instant about the motion screw $ISA_{12}$. With our body here in this room, 2 is the body, and 1 is the room. In the event of 1°F, some particular one of an $\infty^2$ of coaxial helices, all of the same pitch as and all coaxial with the single available motion screw, will lie tangential at each point in the body with the path of that point (§ 9.06, § 10.47). The paths of the points will not themselves be helical; unless the motion screw remains at rest as the body moves, this being a special condition that may be achieved by design or by a substitution (§ 23.10), the paths of the points will simply lie tangential to the mentioned set of helices. The single motion screw in the event of 1°F is unique at the instant: *a body is free with 1°F to move at the instant about a single motion screw.*

10. Refer now to figure 3.05 and gain an impression there of an actual, equivalent, *screw joint* between the body with 1 °F and the room. Such an actual screw joint, collinear in location and equal in pitch with the single available motion screw, could (without removal as yet of the existing linkages) be physically inserted between the body and the room without destroying or altering either the 1 °F or the ICRM of the body. Inserted in this manner, the screw joint would render transitory the otherwise full-cycle connectivity of unity between the body and the room; but, because we are speaking instantaneously, these details of the circumstance hardly bear a motion (§ 2.09, § 2.46). Once inserted, however, the actual screw joint could next *replace* all of the other supporting links and joints of the apparatus, thus standing alone and becoming equivalent at the instant for all of them. This, in the general context of the present chapter, is what I mean by *substitution*. It should be mentioned in the present instance that, if such a substitution were to be made, all of the paths of all of the points in the movable body would be rendered helical; they would be coaxial with the axis of, and of the same pitch as, the actual screw joint.

11. If through every point in the body there is indeed a prescribed path for the point, there is intersecting

at every such point a single infinity of right lines; the right lines exist in the form of a planar pencil at the point, erected normal to the path. This matter was first mentioned at § 3.08, then discussed in greater detail beginning at § 9.08 and in conjunction with figure 9.02. Such right lines must be seen as motional right lines in the present context (§ 3.43); and this means that along any one, or any number of them, a pure force or pure forces may act without causing the body to move at its reference location. Systems of reaction forces mobilised at the constraints will always be called upon to resist any such pure force or group of pure forces and, as Ball says, destroy them. I explained in chapter 9 how all of these right lines – they were there, as they are here, motional right lines – together form a linear complex of lines, each line of which may also be called in the present context an action screw of zero pitch. This complex of action screws of zero pitch has itself a pitch (§ 11.35); it is the same as the pitch of the single motion screw. Alone, however, it is only a small part of the whole 5-system of action screws which is reciprocal to the 1-system, the single motion screw.

12. As well as the action screws of zero pitch, of which there are an $\infty^3$, there are action screws of all of the other pitches as well. There is indeed an $\infty^4$ of action screws belonging to the 5-system of action screws reciprocal to the 1-system of motion screws namely the single motion screw. The other action screws may be preliminarily located in the imagination by seeing, not only all of those pure forces which can act along the screws of zero pitch, but also all of those linearly dependent wrenches which can be produced by combinations of those pure forces (§ 23.03). It will be clear that any wrench produced by a combination of the pure forces will also be successfully resisted by reactions mobilized at the constraints, thus resulting in no movement of the body in response to an application of that wrench.

13. The beginnings of a clear explanation of where these other action screws of finite and infinite pitches exist may be seen at § 6.18 *et seq.*, at § 6A.01 *et seq.*, and in the figures 6.03 and 6A.01. At those places I am describing the $\infty^4$ of motion screws that are reciprocal to a single action screw of zero pitch. The discussion there is different – indeed it is the opposite, the reciprocal, discussion; but there are similarities there with what I must argue here, and the passages are thereby relevant. Notice in that discussion in chapter 6 that the pitch of the single screw of the 1-system was not some arbitrary finite number as it is here, but *zero*. Please become aware now that it was this zero which was responsible for what I called there the *quasi-generality* of the argument, and thus of the system of motion screws that was described.

14. Refer now to Ball's formula for reciprocity (§ 14.29) and find that, if the pitch of the single motion screw is $p_\alpha$ (*sic*), the equation for the layout of all of the action screws of pitch $p$, which are reciprocal, is as follows: $p = -p_\alpha + d \tan \theta$. This is the equation for a series of coaxial complexes, where the pitch $(p + p_\alpha)$ of each complex depends upon our free choice for $p$. All of the action screws of pitch zero for example reside upon a complex of pitch $+p_\alpha$, while all of the action screws of pitch $-p_\alpha$ reside upon a complex of pitch zero namely upon a special linear complex. This latter is consistent with our piece of common knowledge that, if two intersecting screws are reciprocal, they will be of equal but opposite pitch (table 14.01 at § 14.31). It will also be seen that action screws of pitch infinity – they carry pure couples – reside upon a complex of pitch infinity; this complex, coaxial with the others of course (but with no definable axis of its own), comprises all of the lines in all of the planes normal to the axis of the single motion screw (§ 9.43).

15. We can thus see that, while the single motion screw of the 1-system here has an axis with no centre-point and a pitch $p_\alpha$ (this latter single screw is called the principal screw of the 1-system), the 5-system which is reciprocal to it has the same axis with no centre-point and has its special linear complex composed of action screws of pitch $-p_\alpha$. The single so-called principal screw of the 5-system is a member of this special linear complex; it is at the axis of the axi-symmetry (§ 11.33), and its pitch is also $-p_\alpha$ (§ 7.34). While the 1-system is composed of only one namely an $\infty^0$ of screws, the 5-system reciprocal to it comprises an $\infty^4$ of screws arranged in an axi-symmetric pattern coaxial with it; and the pattern of this double ensemble of screws should now be quite familiar (§ 6.25). Refer to § 10.65 and reflect upon the apparatus there; each of the systems there is reciprocal to its corresponding system here.

**Equivalent supports in the event of 1°F**

16. I have claimed that a single screw joint can always be found to substitute for whatever motion with one degree of freedom may be found to exist at the instant (§ 23.10). This implies that I also claim that five points of contact between prongs on the body and facets fixed in the room could always be found to substitute for the screw joint at the instant and thus for the ICRM of the body (§ 7.21).

17. This further implies that five ball-ended rods joining the body to the room could always be found to substitute for the prongs and facets at the instant (§ 3.28). Please study figures 3.07(*a*) and 3.07(*b*) which are both without modification applicable here. Look also at figure 3.09(*a*) and see that that is dealing there with what I am calling here the quasi-general case where the pitch

of the motion screw has happened to be zero, and where the contact normals (the *n*-lines shown) are five linearly independent members of a special linear complex of action screws of zero pitch. In this special case in figure 3.09(*a*), $p_\alpha = 0$ and the pitch of the complex of action screws of zero pitch is also zero.

**Interlude**

18. Let me remark that, if there were gravity and the body were held to have mass and a mass centre, the five mentioned ball-ended rods, massless and rigid, would allow the mass-centre of the body to migrate to a local lowest point in its path. With the linkage in that configuration an equilibrium would prevail; refer to § 10.29. However I wish to speak often in this chapter about the reciprocal possibilities (*a*) for motion, and (*b*) for action; and on each occasion I will wish to do this at a generally chosen configuration of the linkage or of a substituted linkage. I therefore wish to pronounce the body and all of its actual or substituted supports massless (§ 23.04). In this way may line of argument, though no less general, can be far less complicated.

**Two degrees of freedom**

19. Please go back to § 23.07 and grasp the body again. If, on trying, there is found to be more than one way of moving the body, there will be at least a single infinity of different ways. If there is no more than that, namely no more than a single infinity of different motion screws available, the degrees of freedom will be *two*. This would be detectable by discovering that, except perhaps for a single infinity of special points in the body distributed along two straight lines, at which points the directions of small displacements would be predetermined (§ 3.50), each point in the body has the capacity at the instant to travel short paths confined to a plane; at any generally chosen point in the body, a planar pencil of possible small displacements would exist, all of which could be taken by the point. Or we could say this: *at a generally chosen point in a body with 2°F a certain, single, planar pencil of lines exists, and along each direction of that pencil the point may suffer* (*without resistance*) *a linear velocity*.

20. The two straight lines I mentioned (§ 23.19) – they may or may not be real – are of course the directrices of the linear congruence of action screws of zero pitch at the instant; the lines $n_A$ and $n_B$ in figure 6.04 are examples (in the quasi-special case) of such lines. The lines are also the conjugate generators of zero pitch of the cylindroid of motion screws that is extant. They are moreover the seats of the two equivalent hinges that could be inserted if they were real; see figures 3.10 and 3.11. In general, however, and we do have a general circumstance in our apparatus here, the two lines may be either coalesced or both unreal. If they are coalesced there is only one and if they are both unreal there are no equivalent hinges which could be substituted, for there are in these cases respectively only one and no motion screws of zero pitch available among the screws of the prevailing system; the pitches of the screws of the cylindroid are either all positive or all negative (§ 15.49). Refer in advance to figure 23.02 and see implicit there (and see this at the risk of some confusion) a cylindroid of action screws, the pitches of all of the screws of which are of the same sign.

21. In any event the reciprocal 4-system of action screws will contain screws of zero pitch; it will contain, indeed, screws of all pitches (§ 6.28). The screws of zero pitch will reside upon the lines of some linear congruence (§ 6B.05 to § 6B.08 incl.). In any general 4-system this congruence is either hyperbolic with two separated directrices (as it was in every case in chapter 6), parabolic with two coalesced directrices (§ 11.23), or elliptic with no real directrices (§ 11.24). Four lines chosen from any congruence will redefine the congruence. It follows that along any four of those lines of zero pitch – all $\infty^2$ of them are right lines, namely motional right lines, and all of them are determined by the constraints so they are all *n*-lines in the movable body (§ 3.05, § 3.43) – we can erect a prong, a platform and a contact normal. Four equivalent supports in the form of four prongs and platforms or four ball-ended rods could always be found to replace the existing constraints between the body and the room; and this applies however complicated the mechanical apparatus (the links and the joints connecting the body to the room) may happen to be.

22. It is fascinating to become aware, here, that the combination of the 2-system, the cylindroid of motion screws, and the superimposed 4-system of action screws, along with their common central point, central axis, and planes and lines of symmetry as discussed at § 6.29, springs into being at an instant as soon as the system of constraints (however complicated) is established. It should be remarked (or repeated) incidentally that the quasi-general systems of screws discussed at § 6.26 and § 6.52, namely the 4-system and the 2-system of chapter 6, are both indeed quite general systems. But the point is that, while the discussion beginning at § 6.26 threw up a 4-system whose $\infty^2$ of zero-pitch screws resided upon a linear congruence with real directrices, this is not necessarily a feature of the general 4-system. In any general presentation of the 4-system it should be made clear that the congruence of screws of zero pitch might be of any one of the three kinds. In the discussion also of the 2-system beginning at § 6.52 it was simply so that the cylindroid thrown up in the first instance was a cylindroid with a pair of screws of zero pitch. As duly explained at § 6.55 (and again a § 23.20), however, that

particular circumstance is not a required feature of the general 2-system.

**Three degrees of freedom**

23. In the event of there being an $\infty^2$ of motion screws available for the suspended body at the instant, the degrees of freedom will be *three*. As might be expected from the flow of argument hitherto, there will be in this case an $\infty^2$ (a star) of directions available for the unimpeded linear velocity of a generally chosen point in the suspended body; and this could be detected easily. There may however be a single hyperboloidal surface containing an $\infty^2$ of special points that do not have this freedom to move in all directions; the surface (if it exists) is the pitch quadric (§ 6.35); it may be real (in which case it will exist), or it may be not (§ 23.24). If it is real (it was for example at § 6.31), the $\infty^2$ of special points do exist and their possible short paths at the instant are restricted to being within a plane; this plane at each point is normal to the $n$-line there; see figure 7.08 and the discussion of it at § 7.53 *et seq.* for a quasi-special explanation of this phenomenon. *The mere existence of a star of possible velocities at each point in the body (the points being taken one at a time of course) will not be sufficient to indicate 3°F however, for such a star will occur with 4°F and 5°F also.* We will look again at this confusing aspect later, at § 23.26.

24. In chapter 6 and elsewhere hitherto (§ 6.42, § 6C.01 etc.) I have spoken about the three principal screws of the 3-system. Following the conventions of Hunt, I have always chosen $p_\alpha$ to be the greatest, $p_\beta$ to be the middling, and $p_\gamma$ to be the least. I have, moreover, always chosen hitherto $p_\alpha$ to be positive and $p_\gamma$ to be negative. *It was thus that I ensured for the sake of my argument in chapter 6 that screws of zero pitch did exist.* It was in doing this, in choosing to speak against the special background of points of contact at the simple joints, and thus to do what was necessary there for that discussion, that I am guilty of quasi-generality in chapter 6. But the fact of the matter is, and this is amply shown in the works of Ball and Hunt, that the three principal screws of the 3-system can take any values at all. We know that in the 3-system no screws exist whose pitches are greater than the greatest of those of the principal screws, and that none can exist whose pitches are less than the least; so it is quite possible, and this will occur in the event of the greatest and the least being either both positive or both negative, that the pitch quadric may be unreal.

25. Such a circumstance might very well happen with our apparatus supporting the body in the room, in which case (*a*) no set of three equivalent hinges would be found to substitute for the supporting apparatus at the instant, for no $r$-lines would be existing upon a real pitch quadric, (*b*) no set of three prongs and platforms would be found to substitute either, for no $n$-lines would be existing upon a pitch quadric, and (*c*) all the available motion screws would have pitches of the same sign, either all positive or all negative. In the absence of a pitch quadric there would be none of those mentioned special points in the body whereat the possible infinitesimal displacements are restricted to a plane (§ 23.24); at every point there would be a star of lines indicating the total freedom of that point to choose its own direction for a small displacement (or, that is, for a suffered velocity).

26. I promised at § 23.23 to clarify the puzzling question about the total freedom of the general point to suffer a velocity in any direction, and this in view of the confusing fact that the same phenomenon will be apparent with 4°F, 5°F, and 6°F. A necessary approach is to look at the reciprocal 3-system of action screws that is extant. It sits upon the same set of Cartesian axes, it is of the same 'shape', and it exhibits the same set of symmetries as does the 3-system of motion screws we have been studying (§ 6.51, § 10.69 *et seq.*). We can ask questions now about the various wrenches that can be applied upon the body and resisted at the constraints within the supporting apparatus and come to the following conclusion: *at a generally chosen point in a body with 3°F a star of lines exists, but along each direction of that star not only one but two things can happen, (a) the point may suffer without resistance a linear velocity, and (b) the body may suffer without rotation an applied couple.*

27. This latter aspect (*b*), in conjunction with its analogous aspect (*a*) of the matter, is the 'circle-completing' or complementary other part of the criteria for 3°F; please look in advance at figure 23.03. We can easily imagine some optical or other tripod-supported device to detect the presence and the direction of the possibility of an *unhindered velocity* at a point in the suspended body. It is not so easy however to imagine the nature of a corresponding device to detect the presence and direction of a *resisting moment* there (§ 10.41, § 10.43). But I offer this. Let there be fixed at the point in the body a ball surrounded by a movable socket and let the socket be attachable to the ball by a locking screw as shown at figure 23.01; let the ball be located in the room by an $\infty^2$ of radially arranged taut inextensible strings tried to the floor, the walls etc., and let the tensions in the strings be zero. Let there be hinged to the socket by means of two diametrically arranged pins a clevis as shown; and let a centrally arranged, peg - and - slotted cylindric joint connect to a tommy bar as shown. This is a crude version of what should correctly be a wrench applicator of infinite pitch – see figure 14.13 and adjust the slot for the roller

there to be parallel with the centre line of the instrument. The reader moreover might care to exchange the functions of the ball and the socket – this for the sake of a more believable design. With such an arrangement of strings to locate the ball (or the socket) and with such an instrument it will be possible, in the event of 3°F, to check that each of a whole star of pure couples applied at the ball will be able to mobilize such reactions at the constraints that a wrench with a resisting moment at the point in question (the centre of the ball) will appear. In each case the reaction wrench will destroy the applied couple, so that no rotation will occur. A tendency towards translation will occur in the direction of the total reaction force; but this will be negated (destroyed) by the necessary extra tension developing in only one of the taut inextensible strings. This matter of the *resisting moment at a point* I tackle again in the next paragraph, and I take it up again at 4°F (§ 23.30).

28. It may be instructive to ruminate as follows: (*a*) whenever a point in the body suffers an imposed, unhindered velocity in some direction, there will always be a rate of twisting (of the room relative to the body), a motion screw of some pitch, and some angular velocity of the whole room about that screw; and that angular velocity will be evident; (*b*) whenever a body suffers an applied couple at some point and in some direction which is resisted there, there will always be a wrench reaction from the constraints, an action screw of some pitch, and some force from the room upon the body along that screw; and that force will be evident. In each of these cases the relevant dual ($\hat{\mathbf{I}}$ or $\hat{\mathbf{W}}$), expressed with its basepoint at the point in question, will have its dual or its spatial part (its $\mathbf{v}_O$ or its $\mathbf{M}_O$ as the case may be) set equal and opposite to the interfering influence. The interfering person will detect thereon also the real or the global part of the dual (the angular velocity or the force), as the case may be. Whereas however he will on the one hand *see* the angular velocity, he will on the other *feel* the force (or destroy it, as I did above, with one of the taut inextensible strings). Grasp a light rod at one of its ends between your thumb and forefinger; holding the rod horizontal, put its other end under the edge of a table; keeping it horizontal (keeping, in other words, the 'ball' fixed in location at your thumb and forefinger), attempt to apply with those extremities a pure couple whose vector, perpendicular to the rod, it also horizontal; feel now not only the reaction couple, *but also the reaction force*; it acts, of course, vertically downwards upon your hand. Now orchestrate your ideas about this easily understood phenomenon against a more general spatial background and thus perceive the truth of what I have been saying here.

Figure 23.01 (§ 23.27) This simplified wrench applicator of infinite pitch has the capacity to apply a pure couple. We use the applied couple to detect the resisting moments if any at a generally chosen point in the suspended body. The radially arranged taut inelastic strings mentioned at § 23.27 are not shown here.

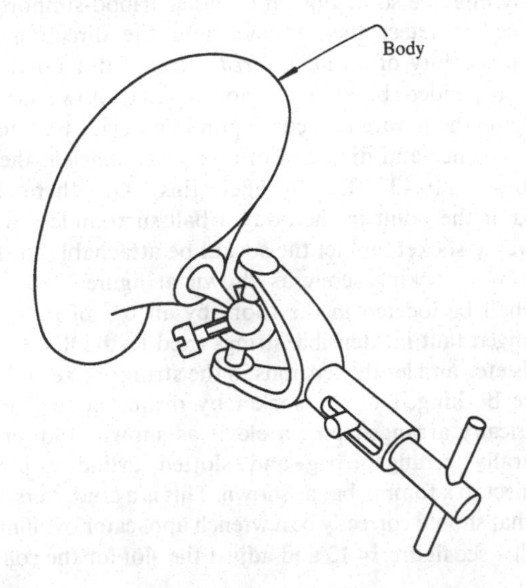


29. To complete this brief picture at 3°F we should mention the distribution of the resisting moments in the event of there being a real pitch quadric. In the event, there will be an $\infty^1$ of *r*-lines (motion screws of zero pitch) upon the pitch quadric in the body; and at each one of these an actual revolute joint, series connected to frame, might be substituted mechanically without causing an alteration in the effectiveness of the constraints (§ 3.63). At each of the $\infty^2$ of points upon the pitch quadric, not a star but only a planar pencil of possible resisting moments exists; the pencil at each point will be normal, not to the *n*-line there (as were the suffered velocities), but to the *r*-line there. Please go to § 6.35 for a description of the regulus of *n*-lines and the opposite regulus of *r*-lines upon the pitch quadric of the 3-system when we have the quasi-general case of 3°F, namely two bodies in contact at three points.

### Four degrees of freedom

30. If there were more than an $\infty^2$ of motion screws available, that is, more than the number that indicates 3°F (§ 23.23), it would be difficult to detect the degrees of freedom of the body merely by exploring the availability of the motion screws; it would be too confusing; it would be better to take some other tack. Whereas in § 23.19 we took a general point in the body and examined the possibilities for displacement there and thus explored the capacity for the whole body to suffer motion, it would be better here to examine, not that, *but the capacity of the constraints to constrain the*

*motion.* Please refer to § 10.35 where the moment at a point of a couple is discussed, to § 10.50 *et seq.*, to § 23.27 where devices for detecting a resisting moment are described, and to § 23.28. Now perceive the following: *at a generally chosen point in a body with 4 °F there exists, not only a star of lines along each line of which the point may suffer an unhindered velocity, but also a planar pencil of lines along each line of which the body may suffer (without rotation) an applied couple.*

31. This could be tested, at our experimental rig, by using at each point in the body the same complicated apparatus as that that I described at § 23.27 and considered again at § 23.28. Having thus tested, and thus having found the mentioned planar pencils, one would find that, normal to each pencil at its centre, there would exist a single line. This line is an actional right line, a motion screw of zero pitch (§ 3.43). As we know, there will be in a body with 4°F an $\infty^2$ of these lines, spread out in space as the lines of a linear congruence. If the congruence were a hyperbolic congruence with two real directrices, the two directrices would contain that $\infty^1$ of special points in the body where *only one* resisting moment will be detected. The two directrices would be two action screws of zero pitch, two *n*-lines indeed. At each of two points set anywhere along these two lines respectively we could mount a prong and a platform (§ 7.20); *and these two could substitute at the instant for the whole system of links and joints constraining the body in the room*; refer to figures 6.04, 14.10 and 15.01. If on the other hand the congruence were an elliptical one, there would be no real directrices of the congruence (§ 11.24), there would be no *n*-lines in the body, *and it would not be possible to substitute with prongs and platforms for the existing constraints.*

Figure 23.02 (§ 23.32) Four revolutes suspending a body with four degrees of freedom. Because the revolute axes are nearly parallel the congruence of motion screws of zero pitch is probably elliptic. Accordingly the reciprocal cylindroid of action screws will probably have no screws of zero pitch.

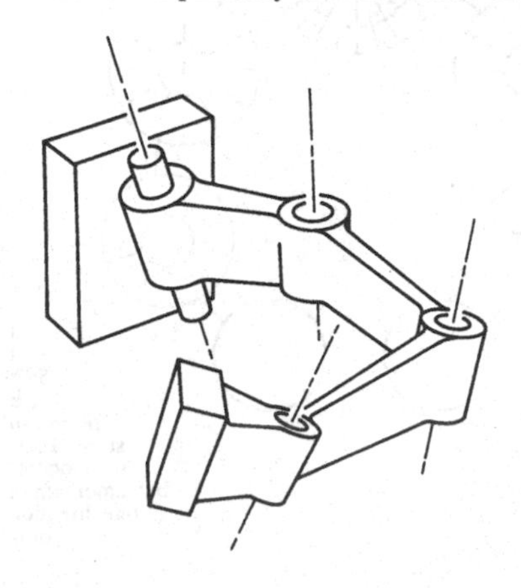

32. On the matter of equivalent hinges, however, any four lines of a congruence of any kind will redefine the congruence, so four hinges may be erected upon any four of the $\infty^2$ of motion screws of zero pitch in the body, and, connected together in series to frame, *they would always be able to substitute for the existing constraints.* Refer to figures 14.10 and 23.02.

33. I wish to mention here that figure 23.02 is drawn in such a way that the four hinge-axes shown are likely to define an elliptic congruence with no real directrices. Notice that any fourth hinge-axis of the four is almost (but not quite) a member of the regulus defined by the other three. *The figure is thus demonstrating a circumstance where there is no line of action existing along which a single pure force may act without causing the body 2 to move.* It is demonstrating a circumstance, in other words, where the reciprocal cylindroid of action screws has no generators of zero pitch, namely a circumstance where the pitches of the generators of the reciprocal cylindroid are either all positive or all negative, namely a circumstance where the possible reactions from the constraints are either all right handed or all left handed wrenches.

34. What are the chances, in this experiment and thus in general, that a cylindroid will be of this particular kind, namely one with no pair of generators of zero pitch, here or at § 23.20? Please refer to figure 15.07 and to my remarks at § 15.49. Refer to figure 3.08; remove one of the screw joints there to leave only two of them; then contemplate the thus determined cylindroid of motion screws which, as we know, might or might not have a pair of screws of zero pitch (§ 15.23). If the two remaining joints happen to be of opposite hand, the cylindroid will have a pair of screws of zero pitch because, obviously, it cannot be otherwise. If the two remaining joints are of the same hand, the cylindroid might or might not have screws of zero pitch, this depending, as we know, upon the pitches at the joints and other circumstances. The question might be non-question; I have asked it nevertheless; and I let the matter rest.

### Locating the axes of symmetry

35. Look again at figure 6.04 and become aware now of the marked quasi-generality of the argument beginning at § 6.26. Dealing with simple joints with two points of contact at § 6.26, I began there by simply asserting that two contact normals, two *n*-lines, two action screws of zero pitch, did exist; thus I asserted that a hyperbolic congruence of motion screws of zero pitch did exist in the relevant 4-system; and I was right. There was no mistake. The circumstance was quasi-general. It

might be observed however that § 6.29 still stands up fairly well. Except for its mention of the common perpendicular between the two contact normals, § 6.29 tells fully of the kinds of symmetries existing in the general 4-system and its embedded, reciprocal 2-system the cylindroid. But a question now arises: how can we tell, in this ongoing experiment in the room with 4°F and in the quite likely event of there being no hyperbolic congruence of motion screws of zero pitch, where the central point of the system is and where the principal axes are?

36. One answer is that we could look for any four of the motion screws of zero pitch. We could take, indeed, the axes of the four equivalent hinges presented in figure 23.02. We could then set out to find the central point and the principal axes of the elliptic congruence defined by them. This could be done as follows. Take the screws of zero pitch two by two, erecting cylindroids upon the common perpendiculars between them (§ 15.23). Select from the cylindroids a set of four screws of the same pitch, any pitch, which are intersected by two lines (§ 11.20). These lines are the directrices of the hyperbolic congruence of that pitch. They are also generators of the reciprocal cylindroid. This fact could be used to locate the centre of the cylindroid, its nodal line, and its principal axes. Thus we would find the central point and the principal axes of the 4-system of motion screws as well. Alternatively we could begin with any set of four motion screws of any pitches and proceed as explained above, erecting cylindroids upon these screws of generally unequal pitch taken two by two. Please refer to § 15.48, where a similar set of instructions is offered for the reciprocal situation.

37. In the event of 3°F being the case with no real pitch quadric existing, we would be faced with a similar problem (§ 23.24). We could however find by similar methods any three motion (or action) screws of the same pitch, then locate by the method of § 6.34 the central point and the principal axes of the relevant hyperboloid (§ 6.47). Thus the central point and the principal axes of the two 3-systems of the same shape, the one of motion screws and the other of action screws, would be found (§ 10.75, § 23.29).

#### Five degrees of freedom

38. In the event of five degrees of freedom there would be an $\infty^4$ of motion screws available (§ 6.16), and it would not be reasonable to try to distinguish this huge number from the earlier mentioned number for 4°F namely $\infty^3$. Please refer to figure 23.03. It is a stylistic chart which goes from zero to 6 and which is summarizing, clockwise from zero to 4, the results of this discussion so far. Figure 23.03 shows that the best tool for detecting 5°F will be the radiating strings and the couple applicator of § 23.27. *If at each point in the body it is found that the body can suffer without rotation a couple in one direction only, the degrees of freedom will be five.*

39. Refer now for comparison to § 23.08 through § 23.11, and perceive the following: normal to the direction of the resisting moment at every point within the body with 5°F there will be a planar pencil of special lines. About every one of these lines the body may infinitesimally rotate as though about a hinge, for the lines are, indeed, actional right lines or, as we know them also, motion screws of zero pitch. There is an $\infty^3$ of these lines, arranged in the way of a linear complex of some finite pitch. Completing the 5-system of motion screws, however, there is a complex of an $\infty^3$ of motion screws of each of the other pitches as well. There is indeed a single infinity of complexes of motion screws all coaxial (§ 6A.02).

40. The single axis which is common to all of the linear complexes of the 5-system of motion screws

Figure 23.03 (§ 23.38) This illustrates schematically the symmetry of the relationships of the other systems with the 3-system, the 3-system itself being characterised with what I have called its own knotty centrality. It also shows the equal importances of linear velocity and moment in the processes of detection that need to be used for determining the degrees of freedom of a body.

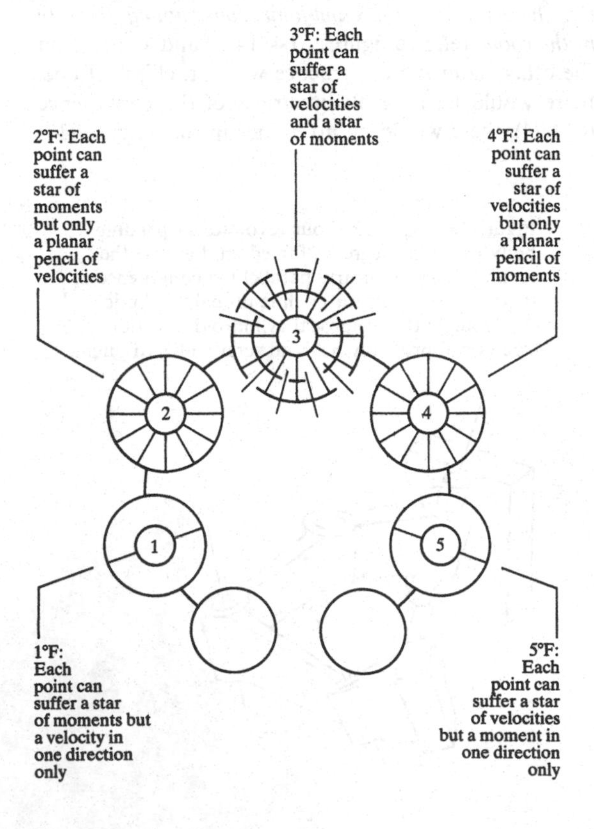
3°F: Each point can suffer a star of velocities and a star of moments
2°F: Each point can suffer a star of moments but only a planar pencil of velocities
4°F: Each point can suffer a star of velocities but only a planar pencil of moments
3
2
4
1
5
1°F: Each point can suffer a star of moments but a velocity in one direction only
5°F: Each point can suffer a star of velocities but a moment in one direction only

carries the single screw of the reciprocal system, the single action screw about which the constraints can offer a reaction wrench of varying magnitude. The reaction wrench remains at the same pitch of course, and this pitch is minus the pitch of the complex of motion screws of zero pitch (§ 6A.02).

41. If the reaction wrench happens to be of zero pitch we have the isolated, special case where the constraints can be substituted by a single prong and a platform (§ 6.18). In general however the reaction wrench will be of some finite non-zero pitch, either positive or negative, and the constraints could be substituted by a single unoperated wrench applicator or, as Hunt would call it, a single *wrench support*. It will be clear on the other hand that the constraints could always be substituted in this general case of 5°F with any five hinges from the linear complex of motion screws of zero pitch mounted and connected in series between the body and the room (§ 14.50). If the above mentioned reaction wrench were of infinite pitch (a pure couple) the five substituted hinges would need to occupy a set of five parallel planes all normal to the directed but non-located axis of the said wrench (§ 23.77).

42. On the question of quasi-generality I need only say here that, whereas in chapter 6 the linear complex of motion screws of zero pitch was itself of zero pitch, here, of course, its pitch might be of any finite value. Its pitch, indeed, will be minus the pitch of the reaction wrench (§ 6A.02, § 10.64, § 23.14).

**Summary of the simple mechanical substitutions**

43. In summary we can now say the following: the motion of any given body 2 with respect to any reference body 1 can be restricted in ordinary engineering practice by the provision of various mechanical connections between the bodies. These connections can only consist of links and joints or chains of links and joints. Unless the capacity for motion of the body 2 has been completely destroyed by the mentioned connections, in which case the body 2 is immobile with respect to its reference body 1 and has no degrees of freedom at all, the capacity for motion of the body 2 is characterized by some whole number of degrees of freedom and thus by some freedom number $f$ that will range from 1 to 6 inclusive. That number is also called the connectivity of the complex joint between the bodies incidentally.

44. If the body is thus constrained with respect to its reference body in such a way that the system of motion screws that is summarizing its capacity for motion at the instant is such that it contains no motion screws of zero pitch, there are no sets of lines in the body along which an appropriate number of hinges in series may be mounted to substitute for the mechanical connections between the bodies. When such a substitution is possible, namely when there are motion screws of zero pitch available, the appropriate number of hinges is, in fact, $f$. Note that we may write, if we wish, $f$°F. Please read § 3.37 again and become aware that that paragraph is dealing with a special, subtly different circumstance. At § 3.37 the mechanical connection between the bodies is decreed by definition to be in the form of multiple, direct points of contact between the bodies, there being no intermediate bodies, and that distorts the outlook, as we have seen, for 3°F. Refer in advance to table 23.01.

45. If on the other hand the body is thus constrained with respect to its reference body in such a way that the system of motion screws that is summarizing its capacity for motion at the instant is such that there are no screws of zero pitch in the reciprocal system of action screws, then there are no points within the body whose possible paths are restricted to a surface at the instant, and, thereby, no possibility of replacing the mechanical connection between the bodies by an appropriate number of direct points of contact between the bodies. When such a substitution is possible, that is when there are screws of zero pitch existing in the mentioned reciprocal system, the appropriate number of points of contact is, as we know, $c$, where $c$ is the degrees of constraint (§ 6.08). Note that we can write, if we wish, $c$°C. Remember also that a double ball-ended rod, suitably arranged with its centre line along the contact normal there, is equivalent at the instant to a direct point of contact between the bodies (§ 3.28).

46. The matters of certainty and uncertainty here can be summarized as follows. If the degrees of freedom is equal to or greater than 4, $f$ hinges in series can always be found to substitute for the mechanical connections; if it is less than 4, $f$ hinges may only perhaps be found for that purpose. If the degrees of constraint is equal to or greater than 4, $c$ direct points of contact can always be found to substitute for the mechanical connections; if it is less than 4, $c$ direct points of contact may only perhaps be found for that purpose. Refer to table 23.01. The contents of this table differ from the results of the discussion at § 3.37, accenting once again the 'knotty centrality' of the 3-system (§ 6.31).

**Cartesian equations to the general screw systems**

47. The two authors foremost in the business of describing the screw systems mathematically are Ball (1900) and Hunt (1978). Whereas Ball writes almost exclusively in Cartesian coordinates, however, and produces his most important equation for the 3-system, Hunt expands into the area of the Plücker, or line coordinates and extends the work of Ball to the 4-system. Naturally, of course, formulae for the 2-system, and corresponding formulae for the 1-system and the

5-system, follow, and Hunt presents Plücker formulae for all five of the systems in a systematic way (§ 23.51).

48. Let me summarize first the work of Ball. Ball explains at the early part of his chapter 14 that all screws of a given pitch within a 3-system lie upon the lines of a regulus upon a hyperboloid. He proves this easily and quickly at his § 172. He then addresses himself to the pitch quadric, the hyperboloid of the regulus of screws of zero pitch, and proves that its equation can be written,

$$p_\alpha x^2 + p_\beta y^2 + p_\gamma z^2 + p_\alpha p_\beta p_\gamma = 0. \tag{1}$$

There at his § 173, however, Ball omits to speak of the possibility that the pitch quadric may be unreal (§ 23.24).

49. At his § 174 Ball introduces his 'family of quadrics' which, as he proves, consists of an $\infty^2$ of hyperboloids concentric with the pitch quadric (§ 6C.04). He then shows, again quite easily and quickly, that the family of all of the hyperboloids (which is sketched by me in figure 6C.05) can be described by the single equation,

$$(p_\alpha - p)x^2 + (p_\beta - p)y^2 + (p_\gamma - p)z^2 + (p_\alpha - p)(p_\beta - p)(p_\gamma - p) = 0, \tag{2}$$

where $p$ takes all values between the minimum principal pitch $p_\gamma$ and the maximum principal pitch $p_\alpha$. Please note the similarity of this with equation (1) for the pitch quadric – put $p$ to zero for this, and see that equation (2) is, indeed, an equation describing the shape of the whole system. With the help of (2) I have been able to write the appendix C at chapter 6 more effectively than I otherwise could have done and, with the help of this same (2), Hunt was able to make his further progress.

50. At § 12.8 in Hunt (1978) one finds Hunt's explanation of his derivation of the following equation for the $\infty^2$ of hyperboloids for the reguli of the 4-system:

$$(p_\alpha - p)x^2 + (p_\beta - p)y^2 + gz^2 + (p_\alpha - p)(p_\beta - p)g = 0. \tag{3}$$

Hunt explains that the coefficient $g$ must be quoted at all values from minus to plus infinity to survey the entire system of the $\infty^2$ of hyperboloids, there being a single infinity of hyperboloids for each value of the pitch $p$. This for him is another way of saying that into the axis for $z$ he has introduced another, a fourth principal screw, whose pitch is of some value other than $p_\gamma$. Refer to my figure 6C.01 to see the actual mechanical effect of such an insertion. Imagine into the apparatus there a fourth screw joint of pitch $p_\delta$ to be introduced in such a way (*a*) that its axis is collinear with the existing screw joint in the axis for $z$ whose pitch is $p_\gamma$, and (*b*) that it is, mechanically speaking, in series with all three of the screw joints already existing. Provided $p_\delta \neq p_\gamma$, the body 2 may now screw relative to the body 1 with whatever pitch it pleases about the axis for $z$. This insertion of the fourth screw joint into the axis for $z$ has effectively destroyed the importance of the particular values of $p_\gamma$ and $p_\delta$; and the new system, the 4-system, is left with only two principal screws that are effective, $p_\alpha$ and $p_\beta$. Sometimes I will call these remaining screws of a system, there being in this case two of them, the *effective principal screws*. The process of similarly introducing yet another screw joint into the apparatus at figure 6C.01 – a fifth one, into the axis for $y$ – would alter the circumstance there to one of five degrees of freedom; there would be, then, *only one* effective principal screw – in the axis for $x$ – and its pitch would be $p_\alpha$ (§ 23.77). Equation (3) was naturally of help to me in the writing of my appendix B at chapter 6 because (as can be seen) the equation does describe, implicitly if not explicitly, all aspects of the shapes of all ruled surfaces involved in the 4-system (§ 6.29).

51. These Cartesian equations are written explicitly for the shapes of the hyperboloids only of the systems however. They are point-coordinate equations. They neither locate precisely the lines of the screws of the systems, nor do they help very much in certain other kinds of calculation. Please refer to my explanation of the Plücker line coordinates beginning at § 19.21, and then to my discussions in terms of these coordinates of the three linear line systems beginning at § 11.43. Go on to § 23.52.

Table 23.01 (§ 23.46). *This summarizes the possibilities for mechanical substitution (a) by means of revolutes in series only, or (b) by means of prongs and platforms in parallel only.*

| Degrees of freedom $f$ | 1 | 2 | 3 | 4 | 5 |
|---|---|---|---|---|---|
| Can $f$ hinges in series always be found to substitute for the mechanical connections? | No | No | No | Yes | Yes |
| Can $c$ direct points of contact always be found to substitute for the mechanical connections? | Yes | Yes | No | No | No |
| Degrees of constraint $c$ | 5 | 4 | 3 | 2 | 1 |

**Plücker equations to the general screw systems**

52. We see at § 11.44 that, if we write $P = kL$ where $k = -p$, we write the single Plücker equation to a linear complex whose pitch is $+p$. It is not a great step to see next that, if we write as Hunt implicitly does $P = (p_\alpha - p)L$, we will write the equation to the single infinity of coaxial complexes of the various pitches $p$ (with $p$ ranging from minus to plus infinity) that constitutes the general 5-system described by me from § 23.38 to § 23.42 inclusive. The single coaxial screw of pitch $-p_\alpha$ is the single principal screw of the reciprocal 1-system. This and the 5-system itself are coaxial with the axis for $x$.

53. In the 4-system on the other hand there is a single infinity not of complexes but of congruences of screws, each congruence containing screws of the same pitch. It follows that, if we write for the 4-system two equations in the form $P = (p_\alpha - p)L$ and $Q = (p_\beta - p)M$, we will successfully describe all of the congruences, one for each value of the pitch $p$ (§ 11.48). Values of the pitch may range, as we have seen, all the way from minus to plus infinity. Similarly for the 3-system, with its single infinity not of congruences but of reguli of screws of the same pitch, we can write the three equations $P = (p_\alpha - p)L$, $Q = (p_\beta - p)M$, and $R = (p_\gamma - p)N$. In this case, however, values of pitch outside the range from $p_\gamma$ to $p_\alpha$ will produce unreal reguli (§ 11.54); there is, as we have seen, a limited range for the single infinity of values for $p$; it extends from $p_\gamma$ to $p_\alpha$. Please refer in advance to table 23.02.

54. If we were wishing to write the four limiting conditions namely the four Plücker equations for a 2-system (a cylindroid) with its nodal line along the axis for $z$ (this is the same axis of symmetry that was chosen for the 4-system at § 23.53), we would see straight away that, because all screws of whatever pitch must be set parallel with the $xy$-plane and all screws of whatever pitch must be caused to intersect the axis for $z$, we must write first of all the following two conditions: $N = 0$ and $R = 0$. We could further see from the equations (1) and (2) in the box at figure 15.04 that $p = p_\beta - z \cot \theta$ and $p = p_\alpha - z \tan \theta$, where $p$ is the pitch of a pair of conjugate screws of the same pitch (say the screws marked 1 and 2 in figure 15.06), and where $z$ and $\theta$ have the meanings ascribed to them at chapter 15. But we can see by inspection that, for all pitches $p$, $\tan \theta = M/L$, $\cot \theta = L/M$, $P = -Mz$, and $Q = +Lz$; so we can write without any difficulty that $z \tan \theta = -P/L$ and that $z \cot \theta = +Q/M$. Accordingly we can write, also for all screws of the cylindroid, that $P = (p_\alpha - p)L$ and that $Q = (p_\beta - p)N$. These latter are the remaining two limiting conditions which, when applied to the screws of space, will determine the cylindroid whose two principal screws along the axes for $x$ and $y$ are of pitches $p_\alpha$ and $p_\beta$ respectively. The similarity of this result with the result for the 4-system produced at § 23.53 will now become apparent at table 23.02. The symbol * at the table means, not zero, but unspecified. The table is after Hunt (1978).

55. In the case of the 1-system, where we wish to exclude all of the screws of space except the one screw of pitch $p_\alpha$ that is collinear with the axis for $x$, we can write first of all $M = 0$ and $N = 0$; these exclude all screws of whatever pitch that are not parallel with the axis for $x$. Then we can write $Q = 0$ and $R = 0$; these exclude all screws of whatever pitch that do not reside in the axis for $x$. Finally we can write $P = (p_\alpha - p)L$; this ensures that there is only one available value for the pitch $p$ namely $p_\alpha$; for, as we can see, any unit vector in the line of the screw must have zero moment about the axis in which it resides, namely the axis for $x$ in this case. Please find these equations for the 1-system listed at table 23.02.

56. As I have said table 23.02 is after Hunt. He shows by this table the internal consistencies of his choices for axes, principal screws, and pitches to produce the orderly algebra of the screw systems. That this table

Table 23.02 (§ 23.54). *The Plücker equations (or the so-called limiting conditions) for defining the various screw systems. The symbol* * *means that the parameter there is unspecified. The scheme here and the layout is after Hurt. Refer also § 23.56.*

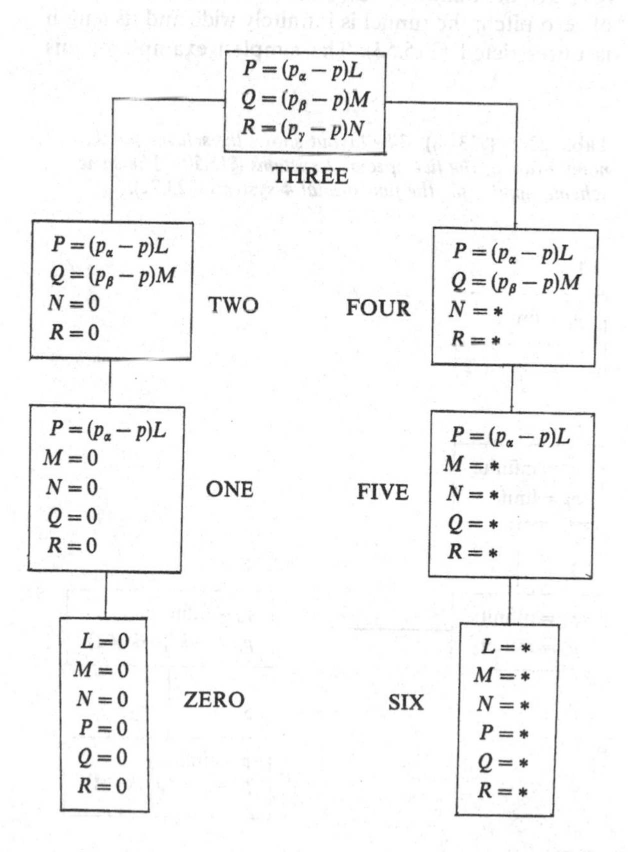

can be constructed in this way is one more striking piece of evidence highlighting the remarkably close relationship with nature of the six line (and the six screw) coordinates of Plücker (§ 10.39, § 11.43, § 19.28).

**The special screw systems**

57. I have mentioned hitherto the scheme for isolating and nominating the special cases of the general screw systems devised and adopted by Hunt (1978). In the following few paragraphs I give an account of this scheme as compactly as I can. I begin at § 23.58 with an account of the single special 1-system. I follow this at § 23.59 with (*a*) a summary of the five special 2-systems, referring the reader to the relevant parts of chapter 15, and (*b*) a summary of the ten special 3-systems, referring the reader for details to § 23.61 *et seq.* The five special 4-systems and the single special 5-system are dealt with in due course; my account of them begins at § 23.72.

58. *The first special 1-system*: The first (lone) special 1-system occurs when the pitch $p_\alpha$ of the single screw of the system becomes infinite. Under these circumstances the general 1-system (the single screw) degenerates into a parallel field of an $\infty^2$ of screws of infinite pitch all of which are parallel to the *x*-axis; the diameter of the bundle of these screws is unrestricted (§ 15.51); the bundle is surrounded by a tunnel of screws of zero pitch; the tunnel is infinitely wide and its length is unrestricted (§ 15.53). The simplest example of this lone first special 1-system might be found in the system of motion screws exhibited by the prismatic joint, the *P*-pair.

59. The first three of the five special 2-systems appear when (1) $p_\alpha = p_\beta$, (2) either $p_\alpha$ or $p_\beta$ is infinite, and (3) both $p_\alpha$ and $p_\beta$ are infinite; the fourth and the fifth special systems (4) and (5) both belong as subsets of the third; they exist in regions infinitely distant from the centre of a cylindroid that has degenerated. A geometrical account is given of these five special 2-systems in passages already written as parts of chapter 15; please read from § 15.50 to § 15.56 inclusive and refer to table 23.03. This table summarizes in a neat way, I hope, the origins of the five special 2-systems. Please remember the adopted convention that $p_\alpha$ is bigger than $p_\beta$ which is bigger (in the case of the 3-system) than $p_\gamma$ (§ 6.45, § 6C.01, § 15.49). Whatever the values may be of the pitches of the principal screws of a system, the convention can always be achieved by choosing the axes for *x* and *y*, or for *x*, *y* and *z*, accordingly. The first six of the ten special 3-systems occur when (1) $p_\beta = p_\alpha$, (2) $p_\gamma = p_\beta = p_\alpha$, (3) $p_\alpha = \infty$, $p_\beta$ and $p_\gamma$ remaining finite and different, (4) $p_\alpha = \infty$, $p_\gamma$ and $p_\beta$ remaining finite but equal, (5) $p_\beta = p_\alpha = \infty$, $p_\gamma$ remaining finite, (6) $p_\alpha$, $p_\beta$ and $p_\gamma$ are all infinite; four further special 3-systems (7) (8) (9) and (10) can be identified, two each as subsets of the fifth and the sixth; they exist in regions infinitely distant from the central point of a 3-system that has degenerated. There follows (after this next directory) a geometrical account of these special 3-systems. I take the systems one by one and in numerical order. Please refer in advance to table 23.04.

Table 23.03 (§23.58). *The layout shows the scheme for the nomination of the five special 2-systems (§15.50). The same scheme applies for the five special 4-systems (§23.72).*

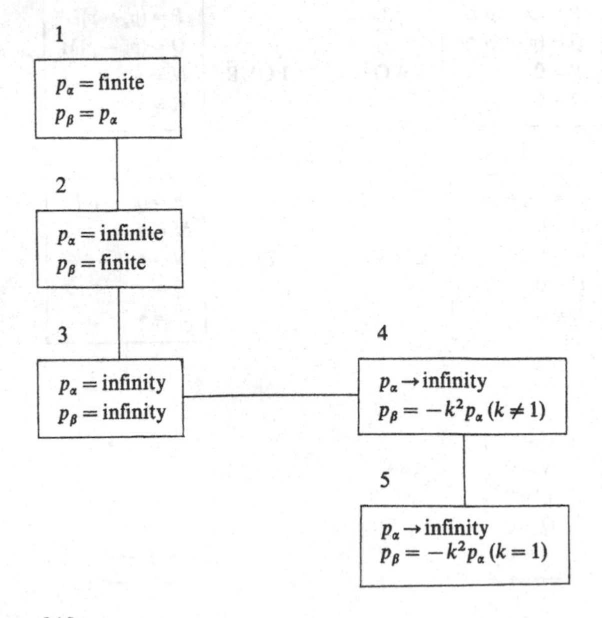


**Directory**

60. For summaries of the single special 1-system, the five special 2-systems, the ten special 3-systems, the five special 4-systems, and the single special 5-system, please go to § 23.58, § 23.59, and § 23.71. For expanded details of the special screw systems the reader should seek as follows: the one special 1-system is dealt with in § 23.58; the five special 2-systems begin at § 15.50, the ten special 3-systems begin at § 23.61, the five special 4-systems begin at § 23.72, and the single special 5-system is discussed at § 23.77. The necessary conditions for the middle three of these, the five special 2-systems, the ten special 3-systems, and the five special 4-systems, are summarized by means of the tables 23.03 and 23.04.

**The ten special 3-systems**

61. *The first special 3-system.* When $p_\beta = p_\alpha$, $p_\gamma$ remaining finite, all reguli of the system reside upon circular hyperboloids. The hyperboloids are of course concentric, but they are coaxial about a single axis as well; they are coaxial about the *z*-axis. The special

Table 23.04 (§23.60). *The layout shows the scheme for the nomination of the ten special 3-systems* (§ 23.61).

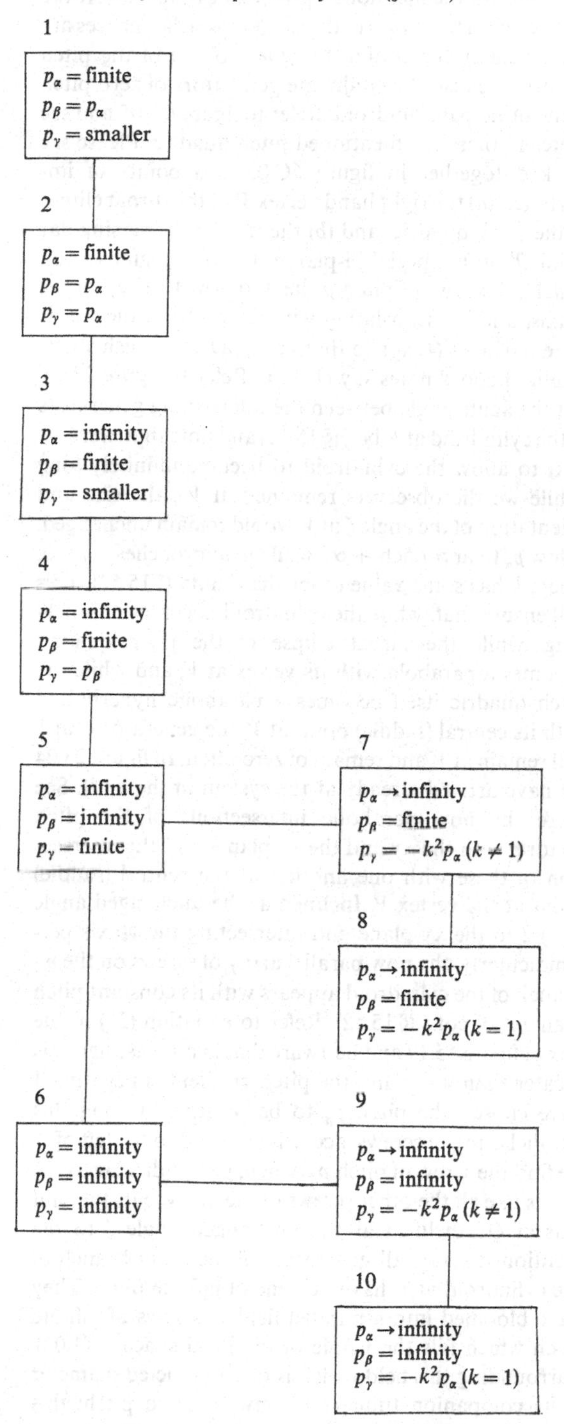


bi-planar regulus of screws of pitch $p_\beta$ (see figure 6.05(*d*)) has collapsed into a planar pencil in the *xy*-plane. The singular points $Q'$ and $Q''$ are coincident at the origin. The principal cylindroid whose nodal line is upon the *z*-axis has collapsed into the same planar pencile, and Hunt's sextic surface has collapsed into a mere circular area. Within this circular area in the *xy*-plane, intersections of two screws at a point occur.

62. *The second special 3-system.* When $p_\alpha$, $p_\beta$ and $p_\gamma$ are all equal and all finite, the 3-system has degenerated into a star of screws of the same pitch $p_\alpha$ at the origin. When the pitch is zero, that is when $p_\alpha$, $p_\beta$ and $p_\gamma$ are not only all equal but all zero, we have that particular second special 3-system which is mentioned for example at §7.61.

63. *The third special 3-system.* This system occurs when $p_\alpha = \infty$, $p_\gamma$ and $p_\beta$ remaining finite and unequal. There is a plane of symmetry at the *yz*-plane, and, cutting this plane normally, there is an $\infty^2$ of screws of infinite pitch all parallel with the *x*-axis. These screws occupy the whole of the local space. Surrounding the immense bundle of these screws, which is infinitely long and unrestricted in diameter, a tunnel of screws of zero pitch exists; the tunnel is infinite in diameter and unrestricted in length (§ 15.51, § 15.53). Contained within the pair of parallel planes $x = \pm\frac{1}{2}(p_\beta - p_\gamma)$, there is an $\infty^2$ of cylindroids whose nodal lines are collinear with the screws of infinite pitch, the length $2B$ of these identical cylindroids being $(p_\beta - p_\gamma)$. For each pitch $p$ between and including $p_\beta$ and $p_\gamma$ there is an $\infty^1$ of parallel screws existing in each of two parallel planes; these planes exist at those distances from the *yz*-plane which are the distances from the centres of the cylindroids of the two screws of pitch $p$ (§ 15.49). Except for the bundle of the $\infty^2$ of screws of infinite pitch and the accompanying tunnel of the $\infty^2$ of screws of zero pitch, no screws exist whose pitches are greater than $p_\beta$ or less than $p_\gamma$. The central point of the system is indeterminate; any point in the already mentioned plane of symmetry, the *yz*-plane, could be that point.

64. *The fourth special 3-system.* When $p_\alpha = \infty$, $p_\gamma$ and $p_\beta$ remaining finite but equal, the $\infty^2$ of parallel cylindroids of § 23.63 are of zero length. The two parallel planes at $x = \pm\frac{1}{2}(p_\beta - p_\gamma)$ have collapsed into the *yz*-plane, and the *yz*-plane is full with screws of pitch $p_\beta$, every line of that plane containing a screw of that pitch. The whole of the local space is occupied as before by the immense bundle of screws of infinite pitch which is parallel with the *x*-axis, and this is surrounded as before by the immense tunnel of screws of zero pitch (§ 23.63). Except for these screws, no screws exist other than those of the single finite pitch $p_\beta$. The central point of the system is again indeterminate; it can be anywhere upon the *yz*-plane. A fourth special 3-system with $p_\gamma = p_\beta =$

zero is exemplified at § 7.74 where the simple 3-ball polycentric joint in figure 7.14(*a*) is described; the system (an action system) occurs when the shown centre line of the balls is collinear with the axis 1 of the slots. Refer to § 7.76 and notice there the mentioned special configuration of the ball and slotted tube in figure 7.16(*a*). Go to figure 7.17 and find there (again with $p_\gamma = p_\beta = 0$) a fourth special 3-system of motion screws.

65. *The fifth special 3-system.* When $p_\alpha$ and $p_\beta$ both become infinite, $p_\gamma$ remaining finite, the system degenerates further. All line in all planes parallel with the *xy*-plane are occupied by screws of infinite pitch; there is an $\infty^3$ of these. In the same way as there was in the third special 2-system, tangent to the meridian circles of an infinitely large sphere whose polar axis is parallel to the *z*-axis, there exists an $\infty^3$ of screws of zero pitch (§ 15.54). There exists as well an $\infty^2$ of screws of pitch $p_\gamma$ all parallel to the *z*-axis and filling the whole of the local space. No other screws exist. When $p_\gamma$ is not simply finite but zero we find that the particular fifth special 3-system defined is the system of motion screws available at the well known ebene joint, the *E*-pair (§ 7.12).

66. *The sixth special 3-system.* When $p_\alpha$, $p_\beta$ and $p_\gamma$ are all infinite, and thereby equal, the system of screws has reduced itself to an $\infty^4$ of screws of pitch infinity. Every line in the local space, in whatever direction, is occupied by a screw of infinite pitch. To match these there is, touching the surface of a sphere of infinite radius at every point upon the surface and in every possible direction, an $\infty^4$ of screws of zero pitch; what this also means is that, if we were to go in any direction from the local space to the plane at infinity (§ 15.51), we would find that that plane (whose area is unrestricted) is filled with its $\infty^2$ of screws of zero pitch. No other screws exist. This system is the system of motion screws for a body which might be free to translate in any direction but not free to change its orientation with respect to some fixed frame. It is also the system of action screws for a body which might be free to suffer from its constraints a pure couple in any direction.

67. *The seventh special 3-system.* Among the special 3-systems, this seventh may be seen to derive from the fifth in somewhat the same way as, among the special 2-systems, the fourth derives from the third; compare the tables 23.03 and 23.04. Refer to § 15.52 and study there the nature of the pitch gradient $dp/dz$ in a cylindroid whose nodal line is upon the *z*-axis; study also the algebra that follows our definition of $k$ by writing $p_\beta = -k^2 p_\alpha$; and digest the implications of figure 15.17. Refer to figure 6.05(*c*) and see there marked $C_2$ one of the principal cylindroids of a general 3-system. Its nodal line is not upon the *z*-axis but upon the *y*-axis; its two principal screws of pitches $p_\alpha$ and $p_\gamma$ are upon the *x* and the *z*-axes respectively. See also in figure 6.05(*c*) that the major axis of the throat ellipse of the pitch quadric is shorter than the mentioned principal cylindroid. At the vertices $V$ on the *y*-axis of the said ellipse (the vertices are not labelled), the conjugate generators $m$ of the pitch quadric are also the conjugate generators of zero pitch of the principal cylindroid. Refer to figure 6C.05 and see sketched there the mentioned pitch quadric. Please see marked together in figure 6C.03 two points of importance, (*a*) the right hand vertex $V$ of the throat ellipse of the pitch quadric, and (*b*) the neighbouring singular point $Q'$ of the special bi-planar, bi-polar regulus upon which all screws of pitch $p_\beta$ lie. Go now to $V$ upon the *y*-axis, and, by translating without rotation the whole system of axes $O\text{–}xyz$ to that new place, establish a new origin of coordinates say $O'$ at $V$. Refer to figure 23.04. Let the acute angle between the intersecting generators of the cylindroid at $V$ be $\zeta$ (§ 15.52) and note that, were we next to allow the cylindroid to become infinitely long (while we the observers remained at $V$), the size and orientation of the angle $\zeta$ at $V$ would remain unchanged. Allow $p_\alpha$ to approach $+\infty$, while $p_\gamma$ approaches $-k^2\infty$, where $k$ has some value other than unity (§ 15.52). This will ensure that, while the cylindroid becomes infinitely long, while the throat ellipse of the pitch quadric becomes a parabola with its vertex at $V$, and while the pitch quadric itself becomes a parabolic hyperboloid with its central (saddle) point at $V$, the generator $m$ at $V$ will remain at $V$ and remain of zero pitch. In figure 23.04 we have arrived already at the system in the limit. See there the now parabolic intersections of the pitch quadric with the *xy* and the *yz*-planes and the intersection of these with one another at the central (saddle) point at the vertex $V$. Inclined at the unchanged angle $(\pi\text{-}\zeta)/2$ to the *xy*-plane and intersecting the *y*-axis perpendicularly, the now parallel array of screws on the *m*-branch of the cylindroid appears with its constant pitch gradient of $\cot\zeta$ (§ 15.52). Refer to equation (21) in the box at figure 15.17 and be aware that in our case here $k$ is greater than unity and the pitch gradient is negative. I have chosen the pitch $p_\beta$ to be positive. Among this parallel array of screws, accordingly, and to the left of $V$, we find the screw of pitch $p_\beta$ which, perpendicular to the *y*-axis like all the other screws of the array, cuts the said axis at $Q'$. Inclined at the unchanged angle $\zeta$ to the mentioned array, all generators of the other branch of the cylindroid at $V$ have become of infinite pitch. They have bloomed into a parallel field of screws of infinite pitch which fills the whole of the local space (§ 11.03). Surrounding this field which is of unrestricted diameter is its companion tunnel of screws of zero pitch; this tunnel is of infinite diameter and unrestricted length (§ 15.55). Let us rest here from the rigour of pure geometric imagination and have recourse to Ball's equation (2) at § 23.49. An algebraic examination of the

relevant special case of this equation will reveal the following extra features of this system. First there is the doubly special bi-planar regulus of screws of pitch $p_\beta$ which has a single polar point at $Q'$ and which is constructed in the manner of figure 7.12(*d*); the equations to its pair of planes are $x = \pm kz$; the screws in the plane $x = +kz$ intersect at $Q'$, while the screws in the other are parallel with $O\text{-}y$. Next for our attention is the single infinity of the remaining parabolic hyperboloids, each one of which carries its regulus of screws of the same pitch. Notice first that the central (saddle) points of these are distributed along the *y*-axis in exact correspondence with the already mentioned array of screws of finite pitch; their distances from $O'$ at $V$ vary directly with the relevant values of the pitch; $y = -(k^2 - 1)p/2k$. Notice second that the pairs of generators upon the surfaces that intersect at the saddle points all reside in the same pair of planes, namely the planes already mentioned, $x = \pm kz$. The surfaces of the hyperbolic paraboloids are moreover all tangent to the same pair of lines, $y = \pm kz$ at $x = 0$; I have illustrated this in the figure by drawing the traces of three representative surfaces as they intersect the *yz*-plane; the traces are, of course, parabolae. In each parabolic hyperboloid the screws of the relevant pitch reside upon the relevant regulus, and the generators of all of the relevant reguli reside in a single infinity of parallel planes (§ 11.06). These planes, parallel to the plane of the mentioned array of screws of finite pitch, occupy the whole of the local space. Symmetrically about a plane containing $Q'$ and parallel to the *zx*-plane, each regulus on the right has a counterpart (its mirror image) upon the left; see the generators drawn for the regulus of screws of pitch $2p_\beta$. Notice the overall symmetry, and notice that the screws of pitch $p_\beta$ of the special regulus correspond in their layout upon the special bi-planar hyperboloid with

Figure 23.04 (§ 23.67) The seventh special 3-system. One single infinity of screws of pitch $p_\beta$ forming a planar pencil intersect at $Q'$ upon the *y*-axis while another single infinity of screws of pitch $p_\beta$, coplanar and parallel, contain the *y*-axis. Together these two planes of screws make up the special regulus of screws of pitch $p_\beta$. The two planes $x = \pm kz$ enclose the angle $\pi - \zeta$, and the *xy*-plane bisects this angle. The origin is at $V$. In the plane $x = +kz$ containing the planar pencil there is also a linear array of screws of finite pitch with a fixed pitch gradient, all screws of which are parallel with the *zx*-plane. Screws of all finite pitches reside in planes parallel with the plane of this array, each single infinity of screws of the same pitch belonging to a regulus residing upon a parabolic hyperboloid whose central (saddle) point is upon the *y*-axis. Each parabolic hyperboloid cuts the *xy* and the *yz*-planes in parabolae and these curves cut one another perpendicularly at the saddle point; see the two parabolae of the pitch quadric intersecting at $V$; compare these with the intersecting curves at the vertex $V$ in figure 11.11(*a*) where the curves are there, for the general case of an hyperboloid, an ellipse (the throat ellipse) and an hyperbola. The single infinity of these parabolic hyperboloids are arranged plane-symmetrically, not about the plane $y = 0$, but about the plane $y = -(k^2 - 1)p_\beta/2k$, which is at $Q'$. In the *yz*-plane the hyperboloids are all tangential to the same pair of lines through $Q'$; the slopes $dy/dz$ of the lines are $\pm k$. Other aspects of this least easily visualized of the special 3-systems are explained at § 23.67.

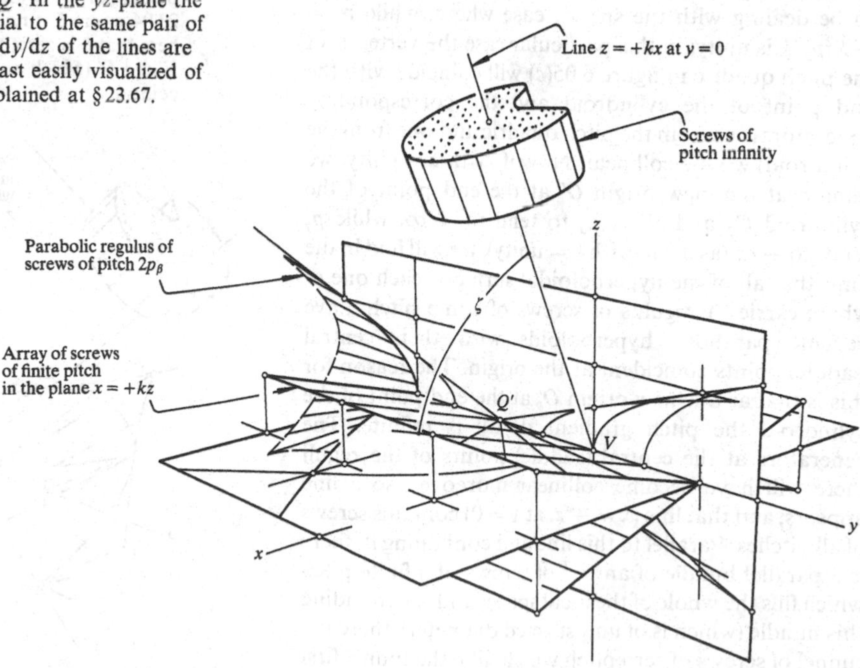

those of the others. It must now be said that the exact location in the immediate neighbourhood of $V$ of the shifted triad of axes $O'$–$xyz$ is not important (§ 15.55). It is also clear that there is a plane of symmetry and a central point, not at the vertex $V$ of the pitch quadric, but at $Q'$. Look for examples of this system at § 7.74 and § 7.76. As another example to show at least one aspect of the relevance not only of this special 3-system but of all other 3-systems as well (both general and special), the reader might (*a*) peruse the material of § 20.09 and § 20.10, (*b*) devise by superimposing upon figure 23.04 a 4-link loop containing four various lower $f1$ joints that will be transitorily mobile, and (*c*) examine that loop's other configurations to see whether or not its apparent mobility of unity might extend to full cycle.

68. *The eighth special 3-system.* This system (like the seventh) derives from the fifth special 3-system; it may also be seen as a special case of the seventh, occurring at a local region within the infinitely long cylindroid $C_2$ where the angle $\zeta$ is not merely acute but zero. It may, incidentally, also be seen as deriving from the fifth special 2-system; if a wholly new extra screw of finite pitch $p_\gamma$ is added at the origin of and perpendicular to the axis of the system explained at § 15.56, we can get this eighth special 3-system without difficulty (§ 23.50). We translate our triad $O$–$xyz$ this time, not to the screw of zero pitch at $V$ along the $y$-axis in figure 6.05(*c*) where the angle $\zeta$ is merely acute, but to a screw of zero pitch at the extreme generator of such a cylindroid as $C_2$ where the angle $\zeta$ is zero. In order to find a screw of zero pitch at the extreme generator of a cylindroid in this way we need to be dealing with the special case where, while $p_\gamma = -k^2 p_\alpha$, $k$ is unity; in this particular case the vertex $V$ of the pitch quadric in figure 6.05(c) will coincide with the end point of the cylindroid and the corresponding generators (one from the pitch quadric and one from the cylindroid) will be collinear. Now if, with $k$ = unity, we remain at our new origin $O'$ at the end point of the cylindroid $C_2$ and allow $p_\alpha$ to tend to $+\infty$, while $p_\gamma$ tends to $-\infty$ (as it will with $k$ = unity), we will find in the limit that all of the hyperboloidal surfaces, each one of which carries a regulus of screws of some pitch, have become parabolic hyperboloids with their central (saddle) points coincident at the origin. The reason for this is that at this new origin $O'$ at the end point of the cylindroid the pitch gradient $dp/dy$ is infinite. The generators at the central (saddle) points of the reguli there will have become collinear moreover; so a line appears; and that line ($x = +z$, at $y = 0$) contains screws of all pitches. Parallel to this line and containing it, there is a parallel bundle of an $\infty^2$ of screws of infinite pitch which fills the whole of the local space, and, surrounding this bundle (which is of unrestricted diameter), there is a tunnel of screws of zero pitch which, like the tunnel first described at § 15.53, is of infinite diameter and of unrestricted length. So far here, I have presented what amounts to being the contents of the fifth special 2-system. But as well as this there is the following. Perpendicular to the line $x = +z$ in the plane $y = 0$ (namely the line that contains the screws of all pitches) there is the line $x = -z$ in the same plane. One screw of every pitch passes perpendicularly through every point of this line. What this means is that all of the reguli are *rectilinear* reguli. I say this in the sense that the two intersecting generators of the saddle-shaped surface of each regulus at its central (saddle) point, namely the lines $x = \pm z$ at $y = 0$, are perpendicular there. This means

Figure 23.05 (§ 23.68) This unfinished drawing illustrates most aspects of the eighth special 3-system, which is that special case of the seventh which occurs when $\zeta$ is zero. Identify the rectangular box $ABCD$ whose front face $BCD$ is square and in the $zx$-plane. The rectilinear parabolic hyperboloids 1 and 2 (which are merely representative of a single infinity of such hyperboloids) contain the lines $x = \pm z$ at $y = 0$ namely the lines $VC$ and $VD$. Their vertices $V$ are coincident at $O$. Note that the bi-planar, unipolar, special parabolic hyperboloid which contains the screws of pitch $p_\beta$ also has its vertex (its polar point) at $O$ and contains these two lines. While the line $x = +z$ contains screws of all pitches, all screws of all pitches (with the exception of most of those whose pitches are infinity or zero) cut the line $x = -z$ perpendicularly. All screws of whatever pitch (with the exception of most of those of pitch zero) reside in planes parallel with the plane $x = +z$. See figure 23.04 where $\zeta$ is not zero and $k$ is not unity; then see here that $\zeta$ is zero, that $k$ is unity, and that the points $V$ and $Q'$ (which are separated at figure 23.04) coincide.

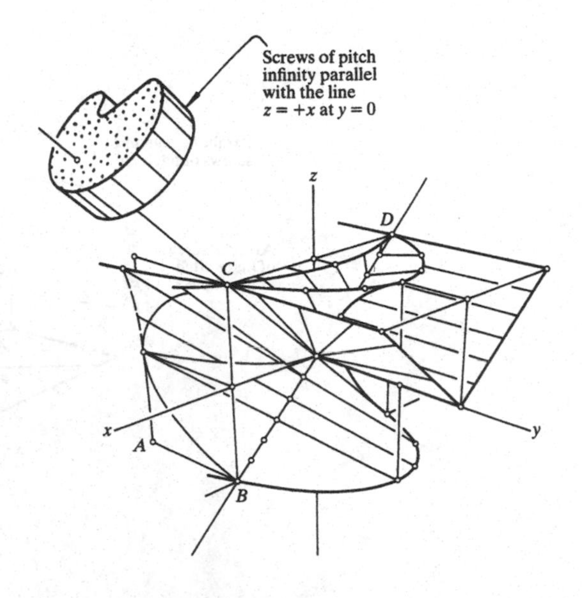

moreover that all of the surfaces of the reguli here have these two generators in common. Screws of pitch $p_\beta$ lie in the two intersecting planes $x = \pm z$; in the plane $x = +z$ the screws intersect at the origin, while in the plane $x = -z$ they are parallel to the $y$-axis. Notice that the two planes of this special bi-planar regulus for the screws of pitch $p_\beta$ are now mutually square with one another; refer to figure 7.12(*d*). Another way to picture this eighth special 3-system is to modify figure 23.04 by putting both $\zeta$ and the distance $O'Q'$ to zero; refer to figure 23.05. Refer to § 7.73 for a discussion of the eighth special 3-system exemplified in the two-ball cylindrical joint, and refer to § 7.76 to find mention there of the eighth special 3-system occurring in the joint of figure 7.16(*a*).

69. *The ninth special 3-system.* This occurs in the limit as $p_\alpha \to \infty$ while $p_\gamma = -k^2 p_\alpha$ where $k \neq 1$ and where $p_\beta$ is not finite but infinite. It may be seen accordingly as deriving from the sixth special 3-system as shown in table 23.04 or as a special case of the seventh where we replace a finite pitch $p_\beta$ with an infinite one. Taking the latter alternative for visualization here, refer to figure 23.04 and modify the scene there to suit the circumstance where $p_\beta$, the pitch of the principal screw along $O-y$, has been chosen to be infinity. The system in the limit is as follows: (*a*) every line parallel to the axis for $y$ contains a screw of infinite pitch; the origin has disappeared but the $zx$-plane continues to exist; (*b*) every line parallel to the direction $x = +kz$ in the $zx$-plane contains a screw whose pitch depends upon its distance from the $zx$-plane; the pitches are given by $p = -2ky/(k^2 - 1)$; this means that the whole of the local space is filled with a parallel field of screws of finite pitch but that the pitch gradient $dp/dy = -(k^2 - 1)/2k$ continues to exist; (*c*) all lines in all planes parallel to a plane $z = +kx$ contain screws of infinite pitch; these screws which number $\infty^3$ include the screws of item (*a*); note that the directions $x = +kz$ and $z = +kx$ are neither parallel nor perpendicular; they enclose the angle $\zeta$; (*d*) an $\infty^3$ of screws of zero pitch exist; these occupy all of the tangents at all of the points on all of the meridian circles of an infinitely large sphere whose polar axis is in the direction of a line $x = -kz$ in the $zx$-plane (§ 15.54); this polar axis is perpendicular to the screws of infinite pitch in the planes $z = +kx$ but not perpendicular to the screws of item (*a*).

70. *The tenth special 3-system.* This last of the special 3-systems occurs in the limit as $p_\alpha \to \infty$ while $p_\gamma = -k^2 p_\alpha$ where $k = 1$ and where $p_\beta$ is not finite but infinite. This relates to the ninth special 3-system in the same way as the eighth relates to the seventh. The angle $\zeta$ being zero, the pitch gradient at the new origin is now infinite and the system in the limit is as follows: (*a*) screws of all pitches occupy all lines parallel to a line $x = +z$ in the $zx$-plane; (*b*) screws of infinite pitch occupy all lines in all planes parallel to a plane $x = +z$; these number an $\infty^3$ and they include the screws of infinite pitch mentioned in item (*a*); note that the planes of the screws of infinite pitch are parallel to the lines of the screws of finite pitch; (*c*) an $\infty^3$ of screws of zero pitch exist; they occupy all of the tangents at all of the points on all of the meridian circles of an infinitely large sphere whose polar axis is in the direction of a line $x = -z$ in the $zx$-plane (§ 15.54); $x = -z$ is perpendicular to $x = +z$ so the polar axis of the said sphere is perpendicular to the screws of item (*a*).

**On a logical arrangement for the principal screws**

71. I need to mention next the five special 4-systems, which are symmetrical (via reciprocity) with the five special 2-systems, and the signale special 5-system which is similarly symmetrical with the single special 1-system. Before I do that, however, I wish to clarify here some overdue questions regarding the signs of the pitches $p_\alpha$, $p_\beta$ and $p_\gamma$ of the principal screws of the systems and of the reciprocal systems. The questions were left unanswered first at § 6B.14, then again at some incidental remarks I made at the beginning of § 7.34, and more recently again for example at § 23.40. I refer to the fact for example that, if we consider the reciprocal cylindroid of action screws which coexists with, and within, any 4-system of motion screws (§ 6B.14), the pairs of pitches of the two principal screws of the two systems, $p_\alpha$ and $p_\beta$, are of opposite sign. The first principal motion screw of pitch $p_\alpha$ of the 4-system and the first principal action screw of pitch $p_\alpha$ of the 2-sytem coexist in the axis for $x$, while the corresponding, second principal screws of pitch $p_\beta$ coexist in a similar way in the axis for $y$ (§ 6B.14). While the principal motion screw of pitch $p_\alpha$ might be, for example, right handed positive, the principal action screw of pitch $p_\alpha$ will be left handed negative, and the same applies for the two principal screws of pitch $p_\beta$. A similar discussion involving the pitches $p_\alpha$, $p_\beta$ and $p_\gamma$ of the three principal screws of a 3-system of motion screws and those for its coexisting 3-system of action screws might be reconstructed from the passages from § 6.31 to § 6.51 inclusive, and from § 10.69 to the end of chapter 10. Now Hunt (1978) after Ball (1900) deals with this question of the principal screws in a logical way by introducing the notation $h$ for the pitch of a screw and $h'$ for the equal but opposite pitch of a reciprocal screw which is collinear. I, however, have chosen not to do that, (*a*) for the sake of simplicity, and (*b*) because the peculiar logic of my approach has not required it. For a parallel reading in Hunt, go to the introductory parts of his § 12.11. He finds there, at that later stage of his argument, no further need for his double notation.

**The five special 4-systems**

72. *The first special 4-system.* This occurs when the pitches of the two principal screws of the reciprocal cylindroid ($-p_\alpha$ and $-p_\beta$), both finite, become equal. The reciprocal cylindroid is reduced to a mere pencil of screws in this circumstance (§ 15.50), and the 4-system itself appears accordingly. All of the reguli of § 6.29 reside upon an $\infty^2$ of circular hyperboloids coaxial about the $z$-axis. For all values of pitch other than $p_\alpha$, no one of the relevant single infinity of hyperboloids intersects another; this means that for all values of pitch other than $p_\alpha$ the relevant congruence is elliptic (§ 11.24). At $p = p_\alpha$, the congruence degenerates into all lines in the $xy$-plane and the whole star of lines though the origin; this is a special congruence of the first kind (§ 11.28). At $p = \pm\infty$, the congruence comprises all lines parallel to the $z$-axis; this is a special congruence of the fourth kind (§ 11.28). There is, surrounding this parallel field of screws of infinite pitch, an immense tunnel of screws of zero pitch (§ 15.53). Notice that the $z$-axis continues to contain screws of all pitches (§ 6B.03).

73. *The second special 4-system.* This occurs when the pitches of the two principal screws of the reciprocal cylindroid ($-p_\alpha$ and $-p_\beta$) are such that, with $p_\alpha$ finite, $p_\beta = \pm\infty$. The reciprocal cylindroid is infinitely long in this circumstance (§ 15.53), we remain at its centre (where the pitch gradient $dp/dz$ is zero), and the 4-system itself appears accordingly. The congruences of lines for the screws of each generally chosen pitch $p$ comprise sets of parallel lines in planes parallel with the $zx$-plane; these are special congruences of the third kind (§ 11.28). There is a characteristic angle for each pitch in each plane [1]. In plane $y = d$ screws of pitch $p$ make an angle $\phi$ with the $x$-direction, where $\tan\phi = (p_\alpha - p)/d$. This is a direct application of equation (2) in chapter 14; equation (2) relates to figure 14.04 and appears in the box at figure 14.05. At the special value $p = p_\alpha$, $\tan\phi = 0$, $\phi = 0$, and the congruence comprises all lines parallel to the $x$-axis and all lines in the $zx$-plane. This latter (seen here as a special case among its fellow congruences of the third kind) is a special congruence of the second kind (§ 11.28). Please remain aware while envisaging all of this that each screw of the 4-system we are constructing must be reciprocal to all screws of the chosen, reciprocal 2-system. Accordingly an $\infty^3$ of screws of infinite pitch may be seen to occupy all lines in all planes parallel with the $yz$-plane (see the bottom right-hand box of table 14.01 at § 14.31); the lines of these screws may be seen to comprise a linear complex of infinite pitch (§ 9.43) or an $\infty^2$ of special congruences of the sixth kind (§ 11.29). Surrounding these screws of infinite pitch, as tangents to the meridian circles of an infinitely large sphere whose polar axis is parallel to $O$–$x$, there is an $\infty^3$ of screws of zero pitch (§ 15.54). Notice that the $z$-axis continues to contain screws of all pitches (§ 6B.03).

74. *The third special 4-system.* This occurs when the pitches of the two principal screws of the reciprocal cylindroid ($-p_\alpha$ and $-p_\beta$) are such that, while $p_\alpha = \pm\infty$, $p_\beta = \pm\infty$. The reciprocal cylindroid is infinitely long in this circumstance (§ 15.54); and this is so whether the mentioned infinities are of the same or of opposite sign. We remain at the centre, and the 4-system itself appears accordingly. It consists of (*a*) screws of all pitches occupying all lines parallel to $O$–$z$, (*b*) an $\infty^4$ of screws of infinite pitch occupying all lines in the whole of space, and (*c*), surrounding these latter, occupying all of the tangents that may be drawn to touch an infinitely large sphere whose centre is within the local region at $O$, an $\infty^4$ of screws of zero pitch (§ 23.66). Notice that the $z$-axis continues to contain screws of all pitches (§ 6B.03).

75. *The fourth special 4-system.* This occurs when the pitches of the two principal screws of the reciprocal cylindroid ($-p_\alpha$ and $-p_\beta$) have gone to infinity in such a way that, as $p_\alpha \to \infty$, $p_\beta = -k^2 p_\alpha$ where $k \neq 1$; refer to § 15.52 and please digest there the definition and the implications of the constant $k$. We the observers must first have gone to some new origin $O'$ upon the $z$-axis within the reciprocal cylindroid where the angle $\zeta$ is neither a right angle nor zero. Each set of screws of a given pitch now lies upon the generators of a special congruence of the fifth kind (§ 11.29). All of the pencils of all of these special congruences are parallel with one another in planes parallel with the plane $y = +kx$, and the lines joining the centres of the pencils of screws of the same pitch are parallel also. These latter occupy positions along the $z$ axis in conformity with the pitch gradient $dp/dz = \pm 2k/(1-k^2)$, and they all lie in the same plane $y = -kx$. All lines in all planes parallel with the plane $x = +ky$ contain screws of infinite pitch. Surrounding these, as tangents to the meridian circles of a sphere of infinite radius whose polar axis is parallel to the direction $y = -kx$, screws of zero pitch exist. Notice that all lines parallel to $O$–$z$ in the plane $y = -kx$ contain screws of all pitches, and that thereby $O$–$z$ itself continues to contain screws of all pitches (§ 6B.03).

76. *The fifth special 4-system.* The most celebrated example of the fifth special 4-system will surely be the system of action screws displayed at the $f2$ cylindric joint, or the $C$-pair. Please refer to § 8.13 *et seq.*, to § 7.12, § 7.84, and § 7.94. Think first of the contact normals distributed across the cylindrical surface of the contact patch within this lower joint. There is an $\infty^2$ of these contact normals; they all cut the same line (the axis of the axi-symmetry at the joint) and they all cut that line perpendicularly. This system of contact normals, or of action screws of zero pitch, is a special congruence of the sixth kind (§ 11.29), whose one directrix is the axis of the axi-symmetry and whose other is perpendicular to it but

infinitely far away. This congruence is the central congruence of zero pitch of the special 4-system of action screws (a fifth special 4-system) that exists at the joint. A definitive description of the fifth special 4-system follows. It is that system reciprocal to the fifth special 2-system which exists, as we know (§ 15.56), at the extreme generator of a cylindroid that has become infinitely long. It occurs when the pitches of the two principal screws of the reciprocal cylindroid ($-p_\alpha$ and $-p_\beta$) have gone to infinity in such a way that, as $p_\alpha \to \infty$, $p_\beta = -k^2 p_\alpha$ where $k = 1$. The fifth special 4-system consists of a single infinity of parallel planar pencils of lines every line of which contains screws of every pitch. The planar pencils are arranged with their central points on a single straight line ($x = -y$ at $z = 0$), and this line is set normal to the planes of the pencils. In addition to the mentioned screws there is an $\infty^3$ of screws of infinite pitch residing in all lines in all of the planes of the pencils. Surrounding these, as tangents to the meridian circles of a sphere of infinite radius whose polar axis is the same single line ($x = -y$ at $z = 0$), an $\infty^3$ of screws of zero pitch exist. Notice that the $z$-axis, which now cuts the single line of line-symmetry of the system ($x = -y$ at $z = 0$) perpendicularly, continues to contain screws of all pitches (§ 6B.03). It will be noted that, in the example of the $f2$ cylindric joint that has been quoted, the joint is able to transmit without movement any wrench applied to cut the axis of the joint perpendicularly; this includes any couple applied with its vector anywhere perpendicular to the central axis of the joint; it evidently includes also any force in the plane of the axis of the joint that is infinitely far away. Refer to § 15.64.

**The lone (first) special 5-system**

77. *The first special 5-system.* This lone special 5-system occurs when the pitch $p_\alpha$ of the single effective principal screw of the 5-system becomes infinite (§ 23.50). It is the reciprocal of the first special 1-system (§ 23.58). Refer to § 6.18 *et seq.*, § 6A.02 *et seq.* and § 23.38 *et seq.* Contemplate the single infinity of coaxial complexes of screws within which each complex containing screws of pitch $p$ is itself of pitch $(p + p_\alpha)$ that makes up the general 5-system, and perceive the following: the first special 5-system consists of (*a*) an $\infty^4$ of screws of all pitches occupying all of the lines in all of the planes that are parallel with the $yz$-plane, that is, normal to the direction of $p_\alpha$, (*b*) an $\infty^4$ of screws of infinite pitch occupying all of the lines in the local space, and (*c*) an $\infty^4$ of screws of zero pitch occupying all of the tangents that may be drawn to a sphere of infinite radius that surrounds the local space (§ 15.51). A good example of a general 5-system is the screw of all of the wrenches that can be sustained by the $f1$ screw joint of $H$-pair. A good example of a general 5-system with $p_\alpha = 0$ is the screws of all of the wrenches that can be sustained by the $f1$ revolute joint or $R$-pair (§ 17.27). A good example of the first special 5-system is the screws of all of the wrenches that can be sustained by the $f1$ prismatic joint or $P$-pair. The six hinges (the six motion screws of zero pitch) of the Sarrus linkage of § 20.48 all belong to the same first special 5-system.

**Closing remarks**

78. It is fitting perhaps that the above short account of the special screw systems should bring this long book to a close. Its placement here accentuates what I have said before: it is not merely the generality of the several screw systems that we need to grasp for a proper understanding of machinery, but the details of their special forms. These do not present themselves for easy scrutiny however; unless we do a journey of discovery among the general wilderness of the screws, unless we grasp in this way the fundamental geometry of the generality, we can neither find nor effectively know the relevant special cases. This and what follows is what the main thrust of the work has been. It has been our careful explorations of the screw systems with a sufficient imagination to deepen our understanding of the details. It has been our studies in and the syntheses of joints and our formation of views about the fundamental role of reciprocity in joints and in machinery. It has been our incidental and limited examination of the over-constrained yet mobile linkages (including gears), our wrestles with the main other matters at issue in the realm of kinetostatics and, of course, our related studies in the dynamics of a link. The work has thrown up the division of kinetostatics into the finite and the infinitesimal (§ 19.36); this dichotomy can be repaired (§ 19.41). I hope that for the reader who finds himself in resonance with the aesthetics of machinery the work has been worthwhile.

**Notes and references**

[1] Please refer to note [5] of chapter 15 and be aware that, as was made clear in the mentioned correspondence during 1986, Professor Hunt was already aware of the error made by him in his description (1978) of the second special 4-system. The sentence noted [1] was, in fact, written by him. Refer § 23.73.

# Bibliography

[1968] Alexander, McNeil R., *Animal Mechanics*, Sidgwick and Jackson, London 1968. Refer § 10.64.

[1875] Ball, Robert Stawell, *The theory of screws; a study in the dynamics of a rigid body*, Hodges Foster, Dublin 1876.

[1900] Ball, Robert S., *A treatise on the theory of screws*, Cambridge University Press, 1900. This work is cited by me at many places.

[1953] Beyer, Rudolph, *Kinematische Getriebesynthese; Grundlagen eines quantativen Getriebelehre ebener Getriebe; u.s.w.*, Springer, Berlin and Heidelberg 1953. There is a translation into English (by H. Kuenzel) of this first work of Beyer, *The kinematic synthesis of mechanisms*, Chapman and Hall, London 1963.

[1956] Beyer, Rodolph, *Kinematisch-getriebeanalytisches Praktikum; Hand- und Übungsbuch zur Analyse ebene Getriebe; u.s.w.*, Springer, Berlin 1958.

[1963] Beyer, Rudolph, *Technische Raumkinematik; Lehr-, Hand-, und Übungsbuch zur Analyse räumliche Getriebe*, Springer, Berlin 1963. Refer § 1.55, § 12.07.

[1924] Blasche, W., *Vorlesungen über Differentialgeometrie*, Springer, Berlin 1923. Refer § 19.14.

[1964] Bogolubov, A.N., *Istoria mehaniki mashin*, (History of the mechanics of machines), Naykova Dymka, Kiev 1964. Refer note [1] of chapter 1, § 1.02.

[1979] Bottema, O. and Roth, Bernard, *Theoretical kinematics*, North Holland, Amsterdam 1979. Refer note [3] of chapter 0, § 11.53, § 19.12.

[1947] Brand, Louis, *Vector and tensor analysis*, Wiley, New York 1947. Refer § 19.07, § 19.14.

[1949] Buckingham, Earle, *Analytical mechanics of gears*, McGraw Hill, 1949. Refer § 22.07.

[1948] den Hartog, J.P., *Mechanics*, McGraw Hill 1948. See also a new edition, Dover, 1961. Refer § 10.18, § 12.06.

[1976] Dijksman, E.A., *Motion geometry of mechanisms*, Cambridge University Press, 1976. Refer § 19.60.

[1950] Dimentberg, F.M., *Opredelenie polozhenii prostranstvennikh mekhanizmov*, (Determination of location in spatial mechanism), Akad. Nauk., Moscow 1950.

[1965] Dimentberg, F.M., *Vintovoye ischisleniye i yego prilozheniya v mekhanike*, (The screw calculus and its application in mechanics), Nauka, Moscow 1965. There is a translation into English by the Foreign Technology Division of the US Air Force. It is Document no. FTD–HT–23–1632–67, dated April 1968. Refer note [6] at chapter 0, § 10.49, § 19.37.

[1971] Dimentberg, F.M., *Metod vintov v prikladnoi mekhanike*, (the method of screws in applied mechanics), Moscow 1971.

[1978] Dimentberg, F.M., *Teoriya vintov i ee prilozheniya*, (The theory of screws and its application), Nauka, Moscow 1978.

[1980] Duffy, Joseph, *Analysis of mechanisms and robot manipulators*, Edward Arnold, 1980. Refer note [9] of chapter 1, § 1.55.

[1969] Dyson, A., *Kinematics and geometry of gears in three dimensions*, Clarendon, Oxford 1969. Refer § 22.07.

[1765] Euler, Leonard, *Theoria motus coporum solidorum seu rigidorum . . .*, (Theory of motion of rigid bodies . . .), A.F. Röse, Rostock and Greifswald 1765. Refer note [10] at chapter 19, § 19.17.

[1928] Federhofer, Karl, *Graphische Kinematik und Kinetostatik des starren räumlichen Systems*. Springer, Wien 1928. Refer note [1] of chapter 1, § 1.02.

[1932] Federhofer, Karl, *Graphische Kinematik und Kinetostatik*. Springer, Berlin 1932. Refer note [1] of chapter 1, § 1.02.

[1917] Grübler, Martin, *Getriebelehre; Eine Theorie des Zwanglaufes und der ebenen Mechanismen*, Springer, Berlin 1917. Refer § 2.36, § 7.17.

[1921] Grübler, Martin, *Lehrbuch der Technischen Mechanik*. Published in 3 volumes, Springer, Berlin 1921–2. Refer § 1.34.

[1961] Hain, Kurt, *Angewandte Getriebelehre*, VDI-Verlag, Düsseldorf 1961. Translation into English by D.P. Adams and T.D. Goodman, *Applied Kinematics*, McGraw Hill, 1967. Refer § 7.17.

[1960] Hall, Alan S. Jr., *Kinematic synthesis of linkages*, Prentice Hall, Edgewood Cliffs 1960. Refer § 12.07.

[1889] Hamilton, W.R., *Elements of quaternions*, Cambridge University Press, 1899. Refer note [1] of chapter 0, § 0.10, § 19.13.

[1964] Hartenberg, R.S. and Denavit, J., *Kinematic synthesis of linkages*, McGraw Hill, New York 1964. Refer § 7.17, § 12.07, § 12.14

[1935] Hilbert, D. and Cohn-Vossen, S., *Anshauliche Geometrie*, Springer, 1935. There is a translation into English (by P. Nemenyi) of this work, *Geometry and the imagination*, Chelsea, 1956. Refer § 4.05.

[1962] Hirschhorn, Jeremy, *Kinematics and dynamics of plane mechanism*, McGraw Hill, New York 1962. Refer § 12.02, § 17.06.

[1967] Hirschhorn, Jeremy, *Dynamics of machinery*, Thomas Nelson, 1967. Refer § 12.60, § 17.06.

[1978] Hunt, Kenneth H., *Kinematic geometry of mechanisms*,

Clarendon, Oxford 1978. This work is cited by me in many places.

[1876] Kennedy, Alexander B.W., his translation of Reuleaux (1875), *Kinematics of machinery*, Macmillan 1876. See also the more recent edition with an important new introduction by Eugene S. Ferguson, Dover, 1963. Refer § 1.60.

[1886] Kennedy, Alexander B.W., *The mechanics of machinery*, Macmillan, 1886. Refer § 13.15, § 14.25.

[1908] Klein, Felix, *Elementar Mathematik vom höheren Standpunkt aus*, Springer, Berlin 1908. Translation by E.R. Hedrick and C.A. Noble (1939), *Elementary mathematics from an advanced standpoint, Volume 2, Geometry*, Dover, 1939. Refer § 9.49.

[1980] Konstantinov, Michail, *Teoria na mechanizmite i mashinite*, (Theory of mechanisms and machines), Technika, Sofia 1980. This book is in Bulgarian. Refer § 22.26.

[1954] Kraus, Robert, *Getriebelehre, Band 1, Einführung*, 2nd ed., VEB Verlag Technik, Berlin 1954. This quoted is the first volume only of a 3-volume work *Getriebelehre*, same publisher, 1952–6. Refer § 7.76.

[1788] Lagrange, Joseph Louis, *Mechanique analytique*, Veuve Desaint, Paris 1788. There have been many subsequent editions of this work. Refer § 0.13, note [1] of chapter 1, § 14.26.

[1912] Lamb, Horace, *Statics*, 1st ed., Cambridge University Press, 1912. Refer § 12.06.

[1966] Lebedev, P.A., *Kinematika prostransteneh mekanikov*, (Kinematics of spatial mechanism), Mashinostroenie, Moscow and Leningrad 1966. Refer § 19.12.

[1968] Litvin, F.L., *Teoria zybchateh zatseplenei*, (Theory of toothed links), Nauka, Moscow 1968. Refer § 22.07, § 22.26.

[1919] Mach, E., *The science of mechanics*, Count Publishing, Chicago 1919. This book is quoted by Synge and Griffith (1949), at a footnote on p. 16.

[1962] Maxwell, R.L., *Kinematics and dynamics of machinery*, Prentice Hall, 1962. Refer § 12.05.

[1837] Möbius, A.F., *Lehrbuch der Statik*, Goschen, Leipzig 1837. Refer § 0.08, § 10.19.

[1687] Newton, Isaac, *Philosophiae naturalis principia mathematica*, (The mathematical principles of natural philosophy), 3rd ed., London 1726. See Florian Cajori's revised version of the earlier translation by Andrew Motte (1729), University of California Press, 1934 and 1962. Refer § 0.09, § 19.17.

[1981] Paul, R.P., *Robot manipulators: mathematics, programming and control*, MIT Press, 1981. Refer § 19.54.

[1869] Plücker, Julius, *Neue Geometrie des Raumes gegründet auf die Betrachtung der Geraden Linie als Raumelement*, Teubner, Leipzig 1868–9. Refer § 0.08, note [3] of chapter 10, § 19.21, § 23.01.

[1848] Poinsot, Louis, *Elements of statics*, 9th ed., Bachelier, Paris 1848. First published in 1806, there were many subsequent editions of this book. Refer note [5] of chapter 10, § 10.51.

[1875] Reuleaux, Franz, *Theoetische Kinematik, Grundzüge einer Theorie des Maschinenwesens*. This is the title given a collected works published in conjunction with the technical journal *Berliner Verhandlungen* during the period 1874–5. That journal had previously published some of the chapters of the collected works serially. Refer § 1.47, § 1.60, § 14.25.

[1952] Rosenhauer, N. and Willis, A.H., *Kinematics of mechanisms*, Associated General Publications, Sydney 1953. See also a new edition, Dover, 1967. Refer § 5.56, § 7.17.

[1896] Routh, E.J., *A treatise on analytical statics, with numerous examples*, Analytical Statics, 2nd ed., in two volumes, Cambridge University Press, 1896.

[1915] Salmon, G.S., *A treatise on the analytical geometry of three dimensions*, 5th ed., Vol. 2, Longmans Green, 1915. There are also later editions of this book. Refer § 11.13.

[1975] Schatz, Paul, *Rhythmusforshung und Technik*, Freies Geistesleben, Stuttgart 1975. Refer § 20.45.

[1952] Semple, J.G. and Kneebone, G.T., *Analytical projective geometry*, Clarendon press, Oxford 1952. Refer § 10.08.

[1980] Shigley, Joseph and Uicker, John J., *Theory of machines and mechanisms*, McGraw Hill, 1980. Refer § 12.07.

[1934] Sommerville, D.M.Y., *Analytical geometry of three dimensions*, Cambridge University Press, 1934. Refer § 11.13.

[1940] Steeds, William, *Mechanism and the kinematics of machines*, Longmans Green, London and New York 1940, 2nd ed. 1947. Refer § 7.17, § 22.29.

[1948] Steeds, William, *Involute gears*, Longmans Green, London and New York 1948.

[1903] Study, Eduard, *Geometrie der Dynamen; Die zusammensetzung von Kräften und verwandte gegenstände der Geometrie*. Published in two parts. Teubner, Leipzig 1901–3. Refer § 0.10, § 10.49.

[1949] Synge, J.L. and Griffith, B.A., *Principles of mechanics*, 2nd ed., McGraw Hill, 1949. First edition McGraw Hill, 1942. Refer § 5.57.

[1968] Truesdell, Clifford A., *Essays in the history of mechanics*, Springer, Berlin and Heidelberg 1968. Refer note [10] of chapter 19, § 19.17.

[1950] Tuplin, William. Among this author's various books on involute geometry and gear design which appeared between 1950 and 1965 there are, for example, *Machinery's gear design handbook*, 2nd ed., Machinery Publishing Co., London 1950; *Involute gear geometry*, Chatto and Windus, London 1962; *Gear design*, Machinery Publishing Co., Brighton 1962. Refer § 22.07.

[1968] Volmer, Johannes and Collective, *Getriebetechnik (Lehrbuch)*, VEB Verlag Technik, Berlin 1968. Refer note [1] of chapter 1, § 1.02, § 2.42, § 5.33.

[1927] Whittaker, E.T., *A treatise on the analytical dynamics of particles and rigid bodies*, Analytical dynamics, 3rd ed., Cambridge University Press, 1927. The first edition of this book appeared in 1904. Refer note [1] of chapter 1 and note [6] of chapter 10.

[1923] Wittenbauer, Ferdinand, *Graphische Dynamik*, Springer, Berlin 1923. Refer § 1.03.

# Bibliography


[1968] Alexander, McNeil R., *Animal mechanics*, Sidgwick and Jackson, London 1968, Refer § 10.64.

[1963] Atanasiu, Mihail, *Mechanica tehnică statica*, Editura Technică, Bucharest 1963. Refer § 20.11.

[1875] Ball, Robert Stawell, *The theory of screws; a study in the dynamics of a rigid body*, Hodges Foster, Dublin 1876.

[1900] Ball, Robert S., *A treatise on the theory of screws*, Cambridge University Press, 1900. This work is cited by me at many places.

[1953] Beyer, Rudolph, *Kinematische Getriebesynthese; Grundlagen eines quantativen Getriebelhre ebener Getriebe; u.s.w.*, Springer, Berlin and Heidelberg 1953. There is a translation into English (by H. Kuenzel) of this first work of Beyer, *The kinematic synthesis of mechanisms*, Chapman and Hall, London 1963.

[1956] Beyer, Rudolph, *Kinematisch-getriebenalytisches Praktikum; Hand-und Übungsbuch zur Analyse ebene Getriebe; u.s.w.*, Springer, Berlin 1958.

[1963] Beyer, Rudolph, *Technische Raumkinematik; Lehr-, Hand-, und Übungsbuch zur Analyse räumliche Getriebe*, Springer, Berlin 1963. Refer § 1.55, § 12.07.

[1924] Blaschke, W., *Vorlesungen über Differentialgeometrie*, Springer, Berlin 1923. Refer § 19.14.

[1964] Bogolubov, A.N., *Istoria mehaniki mashin*, (History of the mechanics of machines), Naykova Dymka, Kiev 1964. Refer note [1] of chapter 1, § 1.02.

[1979] Bottema, O. and Roth, Bernard, *Theoretical kinematics*, North Holland, Amsterdam 1979. Refer note [3] of chapter 0, § 11.53, § 19.12.

[1947] Brand, Louis, *Vector and tensor analysis*, Wiley, New York 1947. Refer § 19.07, § 19.14.

[1927] Bricard, R., *Leçons de cinématique*, Vol. 2, Gauthier-Villars, Paris 1927. Refer § 20.23.

[1949] Buckingham, Earle, *Analytical mechanics of gears*, McGraw Hill, 1949. Refer § 22.07.

[1888] Burmester, L., *Lehrbuch der Kinematik*, A. Felix, Leipzig 1888. Refer § 19.41.

[1948] Den Hartog, J.P., *Mechanics*, McGraw Hill, 1948. See also a new edition, Dover, 1961. Refer § 10.18, § 12.06.

[1976] Dijksman, E.A., *Motion geometry of mechanisms*, Cambridge University Press, 1976. Refer § 19.60.

[1950] Dimentberg, F.M., *Opredelenie polozhenii prostranstvennikh mekhanizmov*, (Determination of location in spatial mechanisms), Akad. Nauk., Moscow 1950.

[1965] Dimentberg, F.M., *Vintovoye ischisleniye i yego prilozheniya v mekhanike*, (The screw calculus and its application in mechanics), Nauka, Moscow 1965. There is a translation into English by the Foreign Technology Division of the US Air Force. It is Document no. FTD-HT-23-1632-67, dated April 1968. Refer note [6] at chapter 0, § 10.49, § 19.37.

[1971] Dimentberg, F.M., *Metod vintov v prikladnoi mekhanike*, (The method of screws in applied mechanics), Moscow 1971.

[1978] Dimentberg, F.M., *Teoriya vintov i ee prilozheniya*, (The theory of screws and its application), Nauka, Moscow 1978.

[1980] Duffy, Joseph, *Analysis of mechanisms and robot manipulators*, Edward Arnold, 1980. Refer note [9] of chapter 1, § 1.55.

[1969] Dyson, A., *Kinematics and geometry of gears in three dimensions*, Clarendon, Oxford 1969. Refer § 22.07.

[1984] Erdman, Arthur G. and Sandor, George N. *Mechanism design, analysis and synthesis* (*Volume 1*) and Sandor, George N. and Erdman, Arthur G., *Advanced mechanism design, analysis and synthesis* (*Volume 2*), Prentice Hall, Edgewood Cliffs 1984. Refer § 12.7, § 12.87.

[1765] Euler, Leonard, *Theoria motus coporum solidorum seu rigidorum...*, (Theory of motion of rigid bodies...), A.F. Röse, Rostock and Greifswald 1765. Refer note [10] at chapter 19, § 19.17.

[1928] Federhofer, Karl, *Graphische Kinematik und Kinetostatik des starren räumlichen Systems*, Springer, Wien 1928. Refer note [1] of chapter 1, § 1.02.

[1932] Federhofer, Karl, *Graphische Kinematik und Kinetostatik*, Springer, Berlin 1932. Refer note [1] of chapter 1, § 1.02.

[1917] Grübler, Martin, *Getriebelehre; Eine Theorie des Zwanglaufes und der ebenen Mechanismen*, Springer, Berlin 1917. Refer § 2.36, § 7.17.

[1921] Grübler, Martin, *Lehrbuch der Technischen Mechanik*. Published in 3 volumes, Springer, Berlin 1921–2. Refer § 1.34.

[1961] Hain, Kurt, *Angewandte Getriebelehre*, VDI-Verlag, Düsseldorf 1961. Translation into English by D.P. Adams and T.D. Goodman, *Applied Kinematics*, McGraw Hill, 1967. Refer § 7.17.

[1961] Hall, Allen S. Jr., *Kinematics and linkage design*, Prentice Hall, Edgewood Cliffs 1961. Refer § 16.08.

# Bibliography


[1899] Hamilton, W.R., *Elements of quaternions*, Cambridge University Press, 1899. Refer note [1] of chapter 0, § 0.10, § 19.13.

[1964] Hartenberg, R.S. and Denavit, J., *Kinematic synthesis of linkages*, McGraw Hill, New York 1964. Refer § 7.17, § 12.07, § 12.14.

[1935] Hilbert, D. and Cohn-Vossen, S., *Anshauliche Geometrie*, Springer, 1935. There is a translation into English (by P. Nemenyi) of this work, *Geometry and the imagination*, Chelsea, 1956. Refer § 4.05.

[1962] Hirschhorn, Jeremy, *Kinematics and dynamics of plane mechanism*, McGraw Hill, New York 1962. Refer § 12.02, § 17.06.

[1967] Hirschhorn, Jeremy, *Dynamics of machinery*, Thomas Nelson, 1967. Refer § 12.60, § 17.06.

[1978] Hunt, Kenneth H., *Kinematic geometry of mechanisms*, Clarendon, Oxford 1978. This work is cited by me in many places.

[1903] Jessop, C.M., *A treatise on the line complex*, Cambridge University Press, 1903. Refer § 11.62, § 19.28.

[1876] Kennedy, Alexander B.W., his translation of Reuleaux (1875), *Kinematics of machinery*, Macmillan, 1876. See also the more recent edition with an important new introduction by Eugene S. Ferguson, Dover, 1963. Refer § 1.60.

[1886] Kennedy, Alexander B.W., *The mechanics of machinery*, Macmillan, 1886. Refer § 13.15, § 14.25.

[1908] Klein, Felix, *Elementar Mathematik vom höheren Standpunkt aus*, Springer, Berlin 1908. Translation by E.R. Hedrick and C.A. Noble (1939), *Elementary mathematics from an advanced standpoint, Volume 2, Geometry*, Dover, 1939. Refer § 9.49.

[1980] Konstantinov, Michail, *Teoria na mechanizmite i mashinite*, (Theory of mechanisms and machines), Technika, Sofia 1980. This book is in Bulgarian. Refer § 22.26.

[1954] Kraus, Robert, *Getriebelehre, Band 1, Einführung*, 2nd ed., VEB Verlag Technik, Berlin 1954. This quoted is the first volume only of a 3-volume work *Getriebelehre*, same publisher, 1952–6. Refer § 7.76.

[1788] Lagrange, Joseph Louis, *Mechanique analytique*, Veuve Desaint, Paris 1788. There have been many subsequent editions of this work. Refer § 0.13, note [1] of chapter 1, § 14.26.

[1912] Lamb, Horace, *Statics*, 1st ed., Cambridge University Press, 1912. Refer § 12.06.

[1966] Lebedev, P.A., *Kinematika prostransteneh mekanikov*, (Kinematics of spatial mechanism), Mashinostroenie, Moscow and Leningrad 1966. Refer § 19.12.

[1968] Litvin, F.L., *Teoria zybchateh zatseplenei*, (Theory of toothed links), Nauka, Moscow 1968. Refer § 22.07, § 22.26.

[1919] Mach, E., *The science of mechanics*, Count Publishing, Chicago 1919. This book is quoted by Synge and Griffith (1949), at a footnote on p. 16.

[1946] Maxwell, E.A., *The methods of plane projective geometry based on the use of general homogeneous coordinates*, Cambridge University Press, 1946. See also Maxwell, E.A., *General homogeneous coordinates in space of three dimensions*, Cambridge University Press, 1951.

[1962] Maxwell, R.L., *Kinematics and dynamics of machinery*, Prentice Hall, 1962. Refer § 12.05.

[1975] Meriam, J.L., *Engineering mechanics, statics and dynamics*, various editions the latest being 1980, John Wiley and Sons, New York 1980. Refer § 18.19.

[1985] Minkov-Petrov, Koljo, *Robotics, mechanics of manipulator systems*, in Bulgarian, Sofia University Kliment Ohridsky 1985. Refer § 19.54.

[1837] Möbius, A.F., *Lehrbuch der Statik*, Goschen, Leipzig 1837. Refer § 0.08, § 10.19.

[1687] Newton, Isaac, *Philosophiae naturalis principia mathematica*, (The mathematical principles of natural philosophy), 3rd ed., London 1726. See Florian Cajori's revised version of the earlier translation by Andrew Motte (1729), University of California Press, 1934 and 1962. Refer § 0.09, § 19.17.

[1981] Paul, R.P., *Robot manipulators: mathematics, programming and control*, MIT Press, 1981. Refer § 19.54.

[1869] Plücker, Julius, *Neue Geometrie des Raumes gegründet auf die Betrachtung der Geraden Linie als Raumelement*, Teubner, Leipzig 1868–9. Refer § 0.08, note [3] of chapter 10, § 19.21, § 23.01.

[1848] Poinsot, Louis, *Elements of statics*, 9th ed., Bachelier, Paris 1848. First published in 1806, there were many subsequent editions of this book. Refer note [5] of chapter 10, § 10.51.

[1875] Reuleaux, Franz, *Theoretische Kinematik, Grundzüge einer Theorie des Maschinenwesens.* This is the title given to a collected works published in conjunction with the technical journal *Berliner Verhandlungen* during the period 1874–5. That journal had previously published some of the chapters of the collected works serially. Refer § 1.47, § 1.60, § 14.25, § 19.60.

[1952] Rosenauer, N. and Willis, A.H., *Kinematics of mechanisms*, Associated General Publications, Sydney 1953. See also a new edition, Dover, 1967. Refer § 5.56, § 7.17, § 19.60.

[1896] Routh, E.J., *A treatise on analytical statics, with numerous examples*, Analytical Statics, 2nd ed., in two volumes, Cambridge University Press, 1896.

[1915] Salmon, G.S., *A treatise on the analytical geometry of three dimensions*, 5th ed., Vol. 2, Longmans Green, 1915. There are also later editions of this book. Refer § 11.13.

[1975] Schatz, Paul, *Rhythmusforshung und Technik*, Freies Geistesleben, Stuttgart 1975. Refer § 20.45.

[1952] Semple, J.G. and Kneebone, G.T., *Analytical projective geometry*, Clarendon Press, Oxford 1952. Refer § 10.08.

[1980] Shigley, Joseph and Uicker, John J., *Theory of machines and mechanisms*, McGraw Hill, 1980. Refer § 12.07.

[1934] Sommerville, D.M.Y., *Analytical geometry of three dimensions*, Cambridge University Press, 1934. Refer § 11.13.

[1940] Steeds, William, *Mechanism and the kinematics of machines*, Longmans Green, London and New York 1940, 2nd ed. 1947. Refer § 7.17, § 22.29.

[1948] Steeds, William, *Involute gears*, Longmans Green, London and New York 1948.

[1902] Study, Eduard, *Geometrie der Dynamen; Die zusammensetzung von Kräften und verwandte gegenstände der Geometrie.* Published in two parts. Teubner, Leipzig 1901–3. Refer § 0.10, § 10.49.

[1978] Suh, C.H. and Radcliffe, C.W., *Kinematics and mechanisms design*, John Wiley & Sons, 1978. Refer § 19.56.

[1949] Synge, J.L. and Griffith, B.A., *Principles of mechanics*, 2nd ed., McGraw Hill, 1949. First edition McGraw Hill, 1942. Refer § 5.57.

[1968] Truesdell, Clifford A., *Essays in the history of mechanics*, Springer, Berlin and Heidelberg 1968. Refer note [10] of chapter 19, § 19.17.

[1950] Tuplin, William. Among this author's various books on involute geometry and gear design which appeared between 1950 and 1965 there are, for example, *Machinery's gear design handbook*, 2nd ed., Machinery Publishing Co., London 1950; *Involute gear geometry*, Chatto and Windus, London 1962; *Gear design*, Machinery Publishing Co., Brighton 1962. Refer § 22.07.

[1966] Tyson, Howlett H., *Kinematics*, John Wiley & Sons, 1966. Refer § 20.56.

[1968] Volmer, Johannes and Collective, *Getriebetechnik (Lehrbuch)*, VEB Verlag Technik, Berlin 1968. Refer note [1] of chapter 1, § 1.02, § 2.42, § 5.33.

[1927] Whittaker, E.T., *A treatise on the analytical dynamics of particles and rigid bodies*, Analytical dynamics, 3rd ed., Cambridge University Press, 1927. The first edition of this book appeared in 1904. Refer note [1] of chapter 1 and note [6] of chapter 10.

[1923] Wittenbauer, Ferdinand, *Graphische Dynamik*, Springer, Berlin 1923. Refer § 1.03.

# Index of proper names

This index covers the whole work (the volumes 1 and 2) and lists the proper names of authors and other persons mentioned. The references – they do not always entirely exhaust the separate mentions made – are given in terms of chapter and paragraph. The overall scheme is explained more fully at the subject index following.

## Index of proper names

# Subject index

This index, covering both volumes of the completed work, is an improvement upon a similar index published in 1984 at the end of volume 1. This new index, like the earlier, now obsolete one, is itemized by means of a composite number that refers not to the relevant page but to the relevant chapter and paragraph. A double insertion such as 12.52–63 for example will refer the reader to the passages from §12.52 to §12.63 inclusive, while a simple insertion such as 10.24 will refer the reader to §10.24 exclusively.

# Subject index

## Subject index

# Subject index

## Subject index

# Subject index

# Subject index

## Subject index

# Subject index

# Subject index

# Subject index

# Subject index

# Subject index

Printed in the United States
By Bookmasters